Topics in Applied Physics

Volume 129

Topics in Applied Physics is a well-established series of review books, each of which presents a comprehensive survey of a selected topic within the broad area of applied physics. Edited and written by leading research scientists in the field concerned, each volume contains review contributions covering the various aspects of the topic. Together these provide an overview of the state of the art in the respective field, extending from an introduction to the subject right up to the frontiers of contemporary research.

Topics in Applied Physics is addressed to all scientists at universities and in industry who wish to obtain an overview and to keep abreast of advances in applied physics. The series also provides easy but comprehensive access to the fields for newcomers starting research.

Contributions are specially commissioned. The Managing Editors are open to any suggestions for topics coming from the community of applied physicists no matter what the field and encourage prospective editors to approach them with ideas.

Managing Editor

Dr. Claus E. Ascheron
Springer-Verlag GmbH
Tiergartenstr. 17
69121 Heidelberg
Germany
claus.ascheron@springer.com

Assistant Editor

Adelheid H. Duhm
Springer-Verlag GmbH
Tiergartenstr. 17
69121 Heidelberg
Germany
adelheid.duhm@springer.com

More information about this series at http://www.springer.com/series/560

Cheng-Chung Lee
Editor

The Current Trends of Optics and Photonics

Springer

Editor
Cheng-Chung Lee
Thin Film Technology Center/Department of Optics and Photonics
National Central University
Chung-Li
Taiwan

ISSN 0303-4216 ISSN 1437-0859 (electronic)
ISBN 978-94-017-9391-9 ISBN 978-94-017-9392-6 (eBook)
DOI 10.1007/978-94-017-9392-6

Library of Congress Control Number: 2014952783

Springer Dordrecht Heidelberg New York London

Printed on acid-free paper

Springer is part of Springer Science+Business Media (www.springer.com)

Preface

Life comes from light and so does human civilization. In our vocabulary, the terms optics and photonics were coined to describe light. Indeed, the evolution of human civilization goes hand in hand with optics and photonics; particularly, since the invention of the laser in the last century, optics and photonics have progressed extremely rapidly and influenced human civilization significantly. Today, our daily lives as we know them wouldn't be possible without optics and photonics, a fact that has moved the United Nations (UN) to declare 2015 the International Year of Light and Light-based Technologies (IYL 2015). Hopefully, this will help the light-based technologies to be promoted even further in solving the global problems concerning energy, education, agriculture, health, environment, etc. One of the most effective ways to achieve these goals is to provide updated and reliable content on the subject to the public, which is the aim of this book.

As opposed to covering all fields of optics and photonics, instead, the book will highlight a number of currently interesting topics that might be useful for further research and development. In Part I we present the latest developments concerning optical thin films and metamaterials. Optical thin films play a vital role in optics and photonics, and without them the two fields might well not be where they are today. We introduce the theory of and manufacturing technique for a gain layer to make coatings with reflectance greater than 100 %, then describe a method for producing large useful coating areas of filters for DWDM with record-breaking uniformity, better than +/−0.003 % over an area of 50 mm in diameter. We examine the design and production of antireflective coatings with hydrophobic and hydrophilic characteristics, residual color to decorate solar cells and blackness, then illustrate the advantages of coatings with a negative refractive index layer. Monitoring is the key technique to accomplishing new designs and is extremely important in the coating process; accordingly, five advanced monitoring methods are proposed.

Metamaterials are artificial structures designed for controlling electromagnetic and acoustic waves or fields, and exhibit exceptional properties that are unexpected on the basis of their chemical constituents. These properties lead to some fascinating phenomena such as negative refraction, sub-wavelength imaging, and field enhancement. A simple review on metamaterials research and discussion of the problems that still need to be resolved round out the second part of Part I (Chap. 2).

Since the invention of the laser in the 1960s of the last century, several kinds of lasers have been produced. A recently developed rare-earth-elements-doped fiber laser stands out for its special characteristics, such that of a movable focusing element for material process; its compact size, which makes it easy to install in a device; tunable spectrum and high output power; diffraction-limited and high-quality optical beam; low cost, etc. In this context, femtosecond Yb-doped fiber oscillators and amplifiers are the most important candidates, and their theory and development are examined in Part II (Chap. 3). Here, we demonstrate high-power femtosecond fiber lasers employing diode-pumping as well as novel mode-locking techniques for fiber-based oscillators. In fundamental mode, the ultimate fiber laser system generates a peak power of over 45 kW and burst pulse energy of 350 nJ. Its harmonics have peak powers of 6.7 kW (532 nm) and 0.6 kW (266 nm). The generation of a supercontinuum using noise-like pulses propagating through ~100 m of single-mode fiber in the normal dispersion region with a low threshold (43 nJ) and flat spectrum over 1050–1250 nm is also illustrated.

The optical free-space transmission scheme has recently been developed to provide high-speed and secure wireless connections, and visible light communication (VLC) systems in particular have proven to be a suitable candidate for delivering these wireless connections between fiber backbone networks and mobile devices. Therefore, a modern VLC communication system based on laser pointer lasers (LPL) for providing high transmission rates and long free-space links is introduced in Part III (Chap. 4). Different technical aspects of VLC, which include enhancing the transmission data rate, mitigating optical background noise, achieving a bi-directional transmission, and using AC-LED for VLC are demonstrated. In Chaps. 5–7, we subsequently propose a novel approach to nonlinear distortion compensation employing the subcarrier-to-subcarrier intermixing interference (SSII) cancellation technique in the context of an electroabsorption modulator (EAM)-based orthogonal frequency-division multiplexing and intensity-modulation-direct-detection (OFDM-IMDD) transmission system. In order to develop the universal transmitters for the next-generation wavelength division multiplexing passive optical network (WDM-PON), the colorless weak-resonant-cavity Fabry-Perot laser diode (WRC-FPLD), which is injection-locked with the help of a highly coherent master laser is used, reducing the relative intensity noise peak at 5 GHz by 18 dB and increasing its throughput frequency response by 5 dB. This enhances the signal-to-noise ratio from 10.5 dB to 18.9 dB and the on/off extinction ratio from 10.4 dB to 11.4 dB, enabling the error-free on-off-keying (OOK) transmission at 10 Gbit/s with a requested receiving power sensitivity of −15 dBm at a bit-error-rate (BER) of 10^{-9}.

Given their energy efficiency, longer service life, comparative environmental friendliness, and compact size, the light-emitting diodes (LEDs), are now utilized in many fields, such as lighting, displays, interferometry, biophotonics, biomedical applications, agriculture, transportation, art and decoration, communication, safety, and in the military, and will eventually replace, at least partially, the traditional fluorescent lamps and high-pressure halogen lamps, making them an important new light source. The key issues concerning LEDs for high lumen/watt efficiency are the internal quantum efficiency, light extraction, heat dissipation, and device packaging. Several techniques for addressing these issues, such as substrate pre-treatment, nano-pillar sapphire substrate, structure design with graded quantum wells, quantum dots, nano-particles and novel electron blocking layers, epitaxial technology, Joule heating minimization, transparent conductive layer patterning, and improved packaging and light extraction are described in Part IV (Chap. 8).

The depletion of fossil energy sources and dramatic change in our climate have sparked a new interest in renewable energies, with solar energy being one of the important among them. As solar cells are used to harvest solar energy and directly convert it to electricity, Part V (Chap. 9) discusses in great detail several types solar cell, including inorganic and organic variants. Recent advances are also briefly described.

Over the last few years, advances in display technology have greatly benefited people in their work and daily lives, and it would be fair to say that "displays" are among the most important and ubiquitous devices in today's society. One of the most essential elements for display is liquid crystal technology. In addition to displays, liquid crystals are also an important component applied in many optics and photonics context, such as holographic gratings, beam steering, q-plates for shaping laser beams, photo-alignment technology, optical attenuators, bio-sensors, lenses, polarization convertors, spatial filters, lasers, etc. The physics of liquid crystals and essential device skills, as well as the emerging new type of liquid crystal displays, are described in Part VI (Chaps. 10–12). Novel applications of cutting-edge liquid crystal technology in 3D displays, non-display applications and biomedical technology are also proposed (Chaps. 13–15).

Nanophotonics have opened up a new research field and allow us to manipulate light at the nanometer scale with characteristics that may not be limited by the physical and chemical composition of natural materials or the diffraction limit of electromagnetic waves. Now that highly interesting fundamental research works have been published, scientists and engineers have started searching for potential applications. In Part VII (Chaps. 16–25) several novel applications are illustrated such as plasmonic nano-lasers, nano-structure light-emitting diodes, one-dimensional photonic crystal nanowires with small footprints and ultrahigh Q-factors, nano-structured waveguides with slow light effect, thin films with a negative refractive index, anti-reflection with nano-structures for solar cells, high-sensitivity plasmonic biosensors, etc. We hope the illustrations in Part VII will inspire people to pursue further cutting-edge breakthroughs in nanophotonics.

The principles and applications of the interactions between photons and biological samples have recently been established as a field of study and are collectively referred to as biophotonics. Many novel photonic technologies and molecular probes have now been developed, making it possible to carry out multi-modality studies with unprecedented spatial and temporal resolutions that provide not only structural, but also functional information. In Part VIII (Chaps. 26–29), we present a brief historical/chronological overview to highlight the key technological developments in four specific biophotonics techniques based on laser microscopy, namely, Fluorescence Life-time Imaging/Foster Resonance Energy Transfer (FLIM/FRET) Microscopy; Optical Coherence Tomography (OCT); Super resolution Microscopy, and Harmonic Generation Microscopy, all of which have made tremendous strides in the last decade not only in terms of their technological development, but also in niche biological and biomedical applications.

Chung-Li, Taiwan, June 2014 Cheng-Chung Lee

Contents

Part IV Light Emitting Diodes

Part V Solar Cells

Contributors

Ching-Hung Chang Department of Electrical Engineering, National Chiayi University, Chiayi, Taiwan

Hung-Chih Chang Department of Electrical Engineering, Institute of Photonics and Optoelectronics, National Taiwan University, Taipei, Taiwan

Shoou-Jinn Chang Institute of Microelectronics and Department of Electrical Engineering, Advanced Optoelectronic Technology Center, Research Center for Energy Technology and Strategy, National Cheng Kung University, Tainan, Taiwan

Shu-Wei Chang Research Center for Applied Sciences, Academia Sinica, Nankang, Taipei, Taiwan; Department of Photonics, National Chiao-Tung University, Hsinchu, Taiwan

Wen-Hao Chang Department of Electrophysics, National Chiao-Tung University, Hsinchu, Taiwan

Chii-Chang Chen Department of Optics and Photonics, National Central University, Chung-Li, Taiwan

Hsing-Yu Chen Department of Photonics, National Chiao Tung University, Hsinchu, Taiwan

Hung-Ying Chen Department of Physics, National Tsing-Hua University, Hsinchu, Taiwan

Jyehong Chen Department of Photonics, National Chiao Tung University, Hsinchu, Taiwan

Kuo-Ju Chen Department of Photonics and Institute of Electro-Optical Engineering, National Chiao-Tung University, Hsinchu, Taiwan

Liang-Yi Chen Graduate Institute of Photonics and Optoelectronics, National Taiwan University, Taipei, Taiwan

Lih-Juann Chen Department of Materials Science and Engineering, National Tsing-Hua University, Hsinchu, Taiwan

Szu-Yu Chen Department of Optics and Photonics, National Central University, Zhongli, Taiwan

Ko-Ting Cheng Department of Optics and Photonics, National Central University, Zhongli, Taiwan

Yuh-Jen Cheng Research Center for Applied Sciences, Academia Sinica, Nankang, Taipei, Taiwan

Yu-Chieh Chi Graduate Institute of Photonics and Optoelectronics, Department of Electrical Engineering, National Taiwan University, Taipei, Taiwan

Chi-Wai Chow Department of Photonics and Institute of Electro-Optical Engineering, National Chiao Tung University, Hsinchu, Taiwan

Shi-Wei Chu Department of Physics, National Taiwan University, Taipei, Taiwan

Nima Dabidian Department of Physics, The University of Texas at Austin, Austin, TX, USA

Gitanjal Deka Institute of Biophotonics, National Yang-Ming University, Taipei, Taiwan

Andy Ying-Guey Fuh Department of Physics, National Cheng Kung University, Tainan, Taiwan

Shangjr Gwo Department of Physics, National Tsing-Hua University, Hsinchu, Taiwan

Hau-Vie Han Department of Photonics and Institute of Electro-Optical Engineering, National Chiao-Tung University, Hsinchu, Taiwan

Jr-Hau He Department of Electrical Engineering, Institute of Photonics and Optoelectronics, National Taiwan University, Taipei, Taiwan

Dar-Zu Hsu Department of Photonics, National Chiao Tung University, Hsinchu, Taiwan

Kung-Shu Hsu Department of Photonics, National Chiao Tung University, Hsinchu, Taiwan

Chi-Yen Huang Department of Physics and Graduate Institute of Photonics, National Changhua University of Education, Changhua, Taiwan

Jian-Jang Huang Graduate Institute of Photonics and Optoelectronics, National Taiwan University, Taipei, Taiwan

Yi-Pai Huang Department of Photonics, National Chiao Tung University, Hsinchu, Taiwan

Yi-Jun Jen Department of Electro-Optical Engineering, National Taipei University of Technology, Taipei, Taiwan

Fu-Jen Kao Institute of Biophotonics, National Yang-Ming University, Taipei, Taiwan

Jisun Kim Department of Physics, The University of Texas at Austin, Austin, TX, USA

Hao-Chung Kuo Department of Photonics and Institute of Electro-Optical Engineering, National Chiao-Tung University, Hsinchu, Taiwan

Wen-Chuan Kuo Institute of Biophotonics, National Yang-Ming University, Taipei, Taiwan

Wei-Chih Lai Department of Photonics, National Cheng Kung University, Tainan, Taiwan

Cheng-Chung Lee Thin Film Technology Center/Department of Optics and Photonics, National Central University, Chung-Li, Taiwan

Kuang-Li Lee Research Center for Applied Sciences, Academia Sinica, Taipei, Taiwan

Po-Tsung Lee Department of Photonics, National Chiao Tung University, Hsinchu, Taiwan

Bo-Hong Li Beijing National Laboratory for Condensed Matter Physics and Institute of Physics, Chinese Academy of Sciences, Beijing, China

Chien-Chung Lin Institute of Photonic System, National Chiao Tung University, Tainan, Taiwan

Ching-Fuh Lin Innovative Photonics Advanced Research Center, Graduate Institute of Photonics and Optoelectronics, Department of Electrical Engineering, National Taiwan University, Taipei, Taiwan

Chih-Hsuan Lin Department of Physics, Institute of Photonics Technologies, National Tsing Hua University, Hsinchu, Taiwan

Da-Wei Lin Department of Photonics and Institute of Electro-Optical Engineering, National Chiao-Tung University, Hsinchu, Taiwan

Gong-Ru Lin Graduate Institute of Photonics and Optoelectronics, Department of Electrical Engineering, National Taiwan University, Taipei, Taiwan

Tsung-Hsien Lin Department of Photonics, National Sun Yat-sen University, Kaohsiung, Taiwan

Yi-Hsin Lin Department of Photonics, National Chiao Tung University, Hsinchu, Taiwan

Hai-Han Lu Department of Electro-Optical Engineering, National Taipei University of Technology, Taipei, Taiwan

Ming-Yen Lu Department of Materials Science and Engineering, National Tsing-Hua University, Hsinchu, Taiwan

Tien-Chang Lu Department of Photonics and Institute of Electro-Optical Engineering, National Chiao-Tung University, Hsinchu, Taiwan

Tsan-Wen Lu Department of Photonics, National Chiao Tung University, Hsinchu, Taiwan

Yu-Jung Lu Department of Physics, National Tsing-Hua University, Hsinchu, Taiwan

Pi-Gang Luan Wave Physics and Wave Engineering Lab/Department of Optics and Photonics, National Central University, Chung-Li, Taiwan

Nirmal Mazumder Institute of Biophotonics, National Yang-Ming University, Taipei, Taiwan

Ci-Ling Pan Department of Physics, Institute of Photonics Technologies, National Tsing Hua University, Hsinchu, Taiwan

Xianggang Qiu Beijing National Laboratory for Condensed Matter Physics and Institute of Physics, Chinese Academy of Sciences, Beijing, China

Charlotte E. Sanders Department of Physics, The University of Texas at Austin, Austin, TX, USA

Chih-Kang Shih Department of Physics, The University of Texas at Austin, Austin, TX, USA

Min-Hsiung Shih Research Center for Applied Sciences, Academia Sinica, Taipei, Taiwan

Gennady Shvets Department of Physics, The University of Texas at Austin, Austin, TX, USA

Yu-Chuan Su Graduate Institute of Photonics and Optoelectronics, Department of Electrical Engineering, National Taiwan University, Taipei, Taiwan

Chi-Kuang Sun Department of Electrical Engineering, National Taiwan University, Taipei, Taiwan

Jin-Yi Tan Graduate Institute of Photonics and Optoelectronics, National Taiwan University, Taipei, Taiwan

Chun-Yuan Wang Department of Physics, National Tsing-Hua University, Hsinchu, Taiwan

Hsin-Ping Wang Department of Electrical Engineering, Institute of Photonics and Optoelectronics, National Taiwan University, Taipei, Taiwan

Chia-Chien Wei Department of Photonics, National Chiao Tung University, Hsinchu, Taiwan

Pei-Kuen Wei Research Center for Applied Sciences, Academia Sinica, Taipei, Taiwan; Institute of Biophotonics, National Yang-Ming University, Taipei, Taiwan

Chihhui Wu Department of Physics, The University of Texas at Austin, Austin, TX, USA

Shu-Han Wu Institute of Biophotonics, National Yang-Ming University, Taipei, Taiwan

Chien-Hung Yeh Information and Communications Research Laboratories, Industrial Technology Research Institute (ITRI), Hsinchu, Taiwan; Department of Photonics, Feng Chia University, Taichung, Taiwan

Yi-Jing You Department of Physics, Institute of Photonics Technologies, National Tsing Hua University, Hsinchu, Taiwan

Alexey Zaytsev Department of Physics, Institute of Photonics Technologies, National Tsing Hua University, Hsinchu, Taiwan

Part I
Optical Thin Films and Metamaterials

Chapter 1
Thin Film Optical Coatings

Cheng-Chung Lee

1.1 Introduction

Optical thin films are widely used and increasing their importance in nearly all technologies of optics and photonics. They not only improve the optical performance of optical devices, but also are vital parts in optical operations. The films can be composed of dielectric or metallic materials or a combination of the two. The optical thin films discussed in this chapter mainly have a thickness on the order of wavelengths so that optical interference take place, although some are in nano scale and come out from the interaction between electromagnetic waves or photons with electrons in the media and quantum effect has to be considered.

There are many thin films have been applied in optics and photonics, from daily consumer goods to defense and scientific research. The examples are thin films for cell phones, lattices—invisible touch panels, digital cameras, binocular telescopes, CD and DVD, projection TV, TFT LCD display, fax machines, eyeglasses, sunglasses, telecom, LED, Organic LED (OLED), solar cells, smart windows, green buildings, head-up displays, E-cars, security, anti-counterfeiting on documents, currency notes, decorations, telescopes, aircraft, satellites, interferometers, microscopes, fluorescence analysis, plasmonics and biosensors, lasers, nonlinear optics, ultra-short pulse laser mirrors, laser fusion, nanotechnology and metamaterial, etc., covering the spectral ranges from IR to EUV. In terminology those coatings are described as anti-reflection, anti-glare, transparent conduction, high reflection, omni-directional coating, beam splitting neutrally, color separation, UV-IR cut, polarizing separation, non-polarizing filter, long-wave pass, shortwave pass, band pass, band stop or minus or notch or rugate filters, gain flatten, chirped

C.-C. Lee (✉)
Thin Film Technology Center/Department of Optics and Photonics,
National Central University, 300 Chung-Da Road, Chung-Li 32001, Taiwan
e-mail: cclee@dop.ncu.edu.tw

C.-C. Lee (ed.), *The Current Trends of Optics and Photonics*,
Topics in Applied Physics 129, DOI 10.1007/978-94-017-9392-6_1

mirrors, high power laser, interferometry, lighting, display, optical communication, security, energy conversion, colorimetry, thermal management, biophotonics, structural and stratified films, etc.

Although powerful thin film software is very helpful to design an optical thin film assembly to meet the specification through modern numerical synthesis and refinement techniques, the basic theory of optical thin films is still the essential. Some theoretical aspects of optical interference coatings are developed essentially for scientific interest, others are required for progress in optics and photonics. However, the techniques to manufacture thin film interference coatings were developed to satisfy the needs of best performance in optics and photonics. Therefore the challenging requirement in theory and fabrication from optics and photonics are never stopped, particular when nanotechnology, quantum dots and meta-material are concerned.

In this chapter we are not intend to cover all topics, instead, some current technologies about theory, design and manufacture of optical thin films will be described.

1.2 Extra-High Reflection Coating with Negative Extinction Coefficient

A simple metallic film, such as Ag and Al can also act as a high-reflection coating. However, to have low absorption it is necessary to have an all-dielectric multilayer. One may find in nature such as the shell of abalone, scales that any periodic system of stratified layers with alternating high and low refractive indices will give high reflectance, but only the quarter-wave stack has the highest and broadest reflection zone. This is the basic structure for high reflection mirrors and laser cavity mirrors. The basic design of a quarter-wave stack is S/H(LH)^p/A, where S is the substrate, A is incident medium, air in general case, and H and L indicate the quarter-wave layers of high and low band filter for DWD. The equivalent refractive index of the quarter wave stack, n_E, can be expressed as in (1.1), where n_S, n_H, and n_L are the substrate index, high and low layers' refractive indices, respectively. It is easy to find that the higher the p, the higher the reflectance, R, from (1.2)

$$n_E = (n_H/n_L)^{2p} n_H^2/n_S. \tag{1.1}$$

$$R = [(1 - n_E)/(1 + n_E)]\,[(1 - n_E)/(1 + n_E)]^*. \tag{1.2}$$

Accordingly, the maximum of R is 0.99999.... but can't equal to 1, not even greater than 1. Recently, an extra-high reflector, eHR, by using multilayers with a negative k, commonly referred to as gain layers was demonstrated that R can be greater than 1 theoretically and experimentally [1].

In theory, the admittance locus of a perfectly dielectric single layer always traces clockwise and forms a circle while an absorbing layer shrinks the circle to

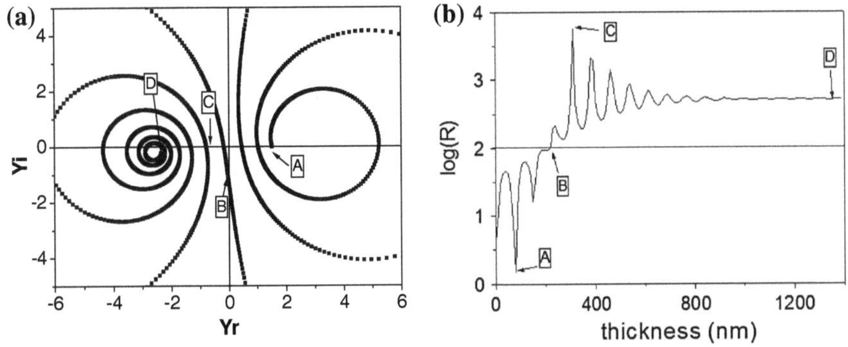

Fig. 1.1 **a** Admittance locus of a gain layers with refractive index n = 2.57 and negative k = −0.2; **b** Corresponding reflectance, *R*, is great than 1 after point B

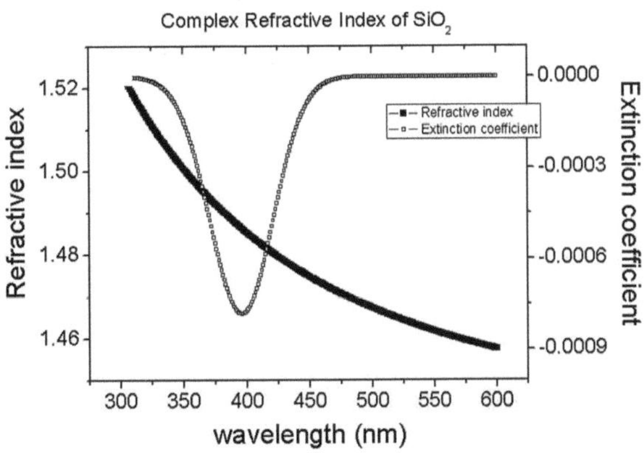

Fig. 1.2 Refractive index and negative extinction coefficient of SiO_2 layer with CdS quantum dots [1]

form a vortex [2]. Instead, the admittance locus of a gain layer expands the circle and spirals anticlockwise when the locus on the left side of the complex plane [3]. Figure 1.1 shows the admittance locus of a gain layers with refractive index n = 2.57, negative extinction coefficient $k = -0.2$ and optical thickness 30H deposed on a substrate with refractive index 1.515. The locus starts from point A, passes to negative side of real part of the admittance axis at point B where the reflectance is greater than 100 % and ends at point D.

The gain layer was fabricated by embedding CdS quantum dots in a dielectric, SiO_2, layer. The effective k value is around −0.0008, as shown in Fig. 1.2 [1]. Let L′ be the gain layer and deposited as the last second layer of a conventional quarter wave stack, $(HL)^8$ L′H. Admittance locus of L′ turns from a conventional clockwise direction into an anticlockwise one. The magnification of the square area of the sub-diagram demonstrates that this transition occurs in the gain layer after a conventional high reflector, HR, as shown in Fig. 1.3. R is greater than 1 at wavelength 393 nm as shown in Fig. 1.4 for HRs $(HL)^8$ L′. To push R even higher, a H layer is added as the last layer: $(HL)^8$ L′H.

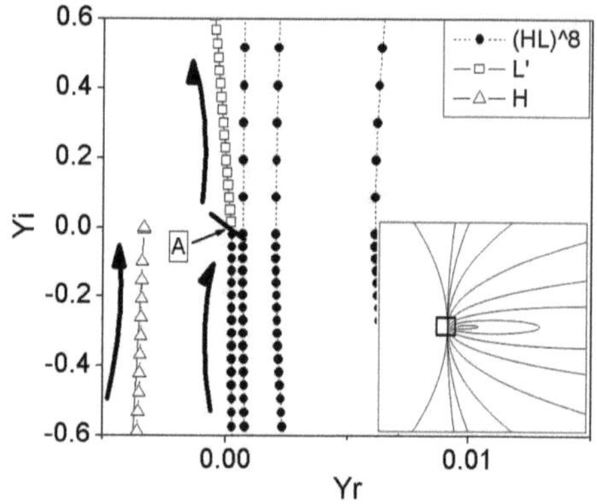

Fig. 1.3 Admittance locus of $(HL)^8$ L′H turns from a conventional clockwise direction into an anticlockwise one. The magnification of the square area of the sub-diagram demonstrates that this transition occurs in the gain layer, L′, after a conventional HR [1]

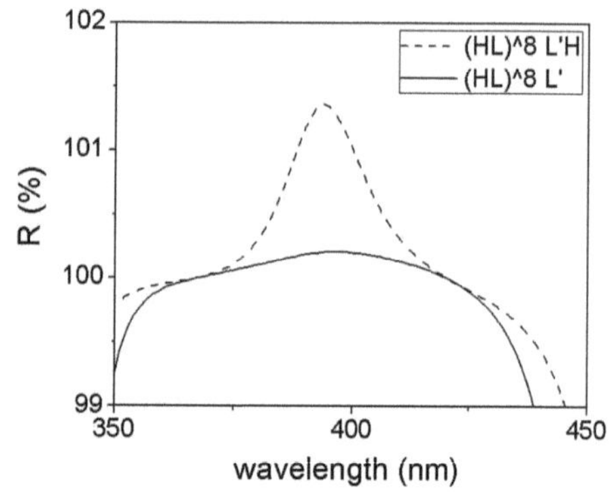

Fig. 1.4 Gain layer L′ improve the reflectance [1]

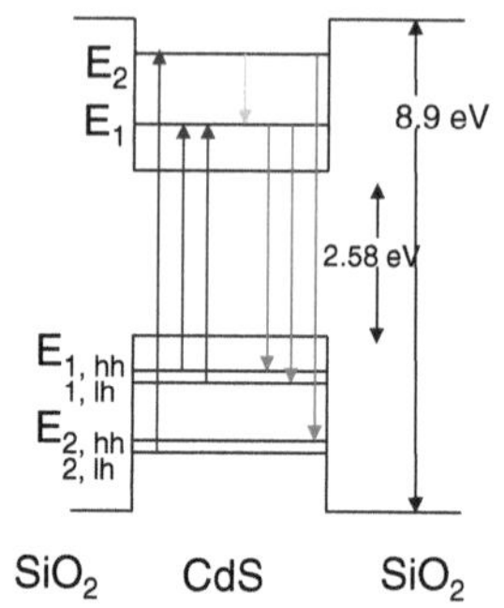

Fig. 1.5 Extra energy at emission wavelength is generated by the down transition owing to the discrete energy levels of quantum dots

A gain medium has a negative extinction coefficient at a specific wavelength. Such unusual behavior does not violate energy conservation. The extra energy at emission wavelength is generated by the down transition owing to the discrete energy levels of quantum dots, explaining why some of the energy associated with shorter wavelengths is transferred to longer wavelengths. Figure 1.5 indicates the energy transfer mechanism. This phenomenon implies that the media provided extra absorption and gain at some specific wavelengths. The application of gain layers is very forward-looking, such as efficiency increase in solar cell [4–7] and LED [8], LD [9], display [10], biology [11] and a loss compensation layer for metamaterials [12–14], etc.

1.3 Narrow Band Pass Filter for Optical Communication with the Half-Peak Bandwidth in Order of Nanometer

One of the techniques to increase the channels so that more optical carrier signals can be multiplexed into a light stream and transmitted on a single optical fiber is utilizing subnano-bandwidth narrow band pass filters, the so called dense wavelength division multiplexing/demultiplexing (DWDM) technology. The design of such filters is based on Fabry-Perot type filter **HR-Sp-HR**, where **HR** is a high reflector which is made of quarter stacks, $(HL)^qH$ or $(HL)^q$ and **Sp** is a multiple half-wave layer called a spacer, $m2L$ or $m2H$. The center wavelength λ_0 is determined by the optical thickness of the spacer nd and reflection phase shift of HRs ϕ_a and ϕ_b as indicated by (1.3), while half-peak bandwidth $\Delta\lambda_h$ is different for high and low refractive indices spacers and dependent on the order m and refractive indices of layers as expressed by (1.4) and (1.5), respectively [2].

$$\frac{1}{\lambda_0} = \frac{1}{2nd} \cdot \left(m + \frac{\phi_a + \phi_b}{2\pi} \right) \tag{1.3}$$

$$\frac{\Delta\lambda_h}{\lambda_0} = \frac{4n_L^{2x-1} n_S}{m\pi n_H^{2x}} \cdot \frac{(n_H - n_L)}{(n_H - n_L + n_L/m)} \quad \text{for a low - index spacer m2L;} \tag{1.4}$$

$$\frac{\Delta\lambda_h}{\lambda_0} = \frac{4n_L^{2x} n_S}{m\pi n_H^{2x+1}} \cdot \frac{(n_H - n_L)}{(n_H - n_L + n_L/m)} \quad \text{for a high - index spacer m2H.} \tag{1.5}$$

Taking an mth order spacer, we obtain a filter: $S/H(LH)^p m2L(LH)^p HL/A$, where the last layer L is an antireflection coating layer. It is called a single-cavity filter and is approximately triangular in shape. To have a square top and sharp edge, we need multiple cavities, as is the case for example in a three-cavity filter: |HR |Sp|HR|L|HR|Sp|HR|L|HR|Sp|HR|, where L is the coupling layer. Figure 1.6 depicts the spectra of the single (S), two (D) and three-cavity (T) narrow band-pass filters.

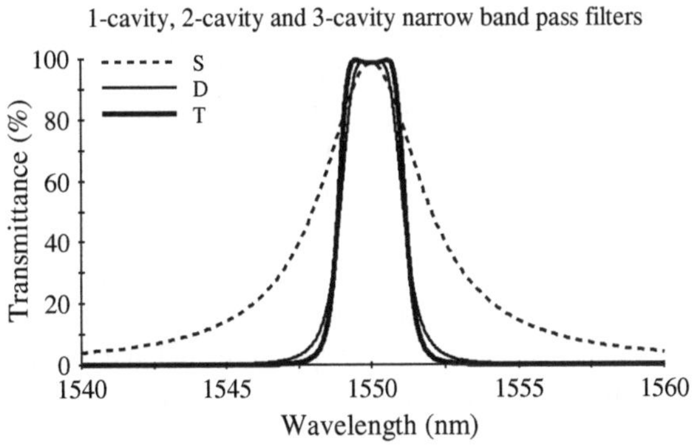

Fig. 1.6 Spectra of single (S), two (D) and three-cavity (T) narrow band-pass filters

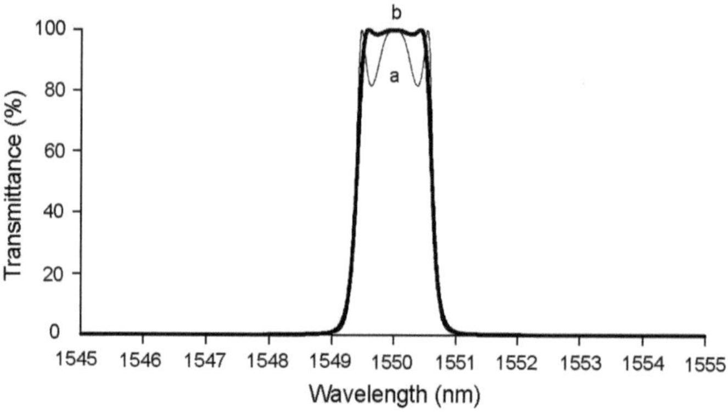

Fig. 1.7 Narrow band pass filter designed by using symmetric layers: **a** S | H L H L H L H L H L H L H L H L[(L H L H L H L H L H L H L H L H) L (H L H L H L H L H L H L H L)]^3 L H L H L H L H L H L H L H L H | S, $\lambda_0 = 1{,}500$ nm; **b** S | H L H L H L H L H L H L H L H[(H L H L H L H L H L H L H L H) L (H L H L H L H L H L H L H) (4H H L H L H L H L H L H L H) L (H L H L H L H L H L H L H 4H) (H L H L H L H L H L H L H) L (H L H L H L H L H L H L H)]H L H L H L H L H L H L H L H | S [16]

Another approach to design a narrow band pass filter is using the theory of symmetrical combinations of films and the Herpin index, derived by Epstein in 1952 [15]. The equivalent of a symmetric stack with long period such as [(L H L H L H L H L H L H L H L H) L (H L H L H L H L H L H L H L)]^3 changes very rapidly such that only a small range of wavelengths can have high transmission. Therefore, such a symmetric assembly potentially can be a narrow band pass filter. Adding matching layers H L H L H L H L H L H L H L H L on both sides of above symmetric assembly, creates a narrow band pass filter for dense wavelength-division multiplexing (DWDM): S/H L H L H L H L H L H L H L H L[(L H L H L H L H L H L H L H L H) L (H L H L H L H L H L H L H L)]^3 L H L H L H L H L H L H L H L H/S. The spectrum is shown in Fig. 1.7a. There are two dips in the pass band, yielding features like rabbit ears, which are due to the mismatch of refractive index between the symmetric assembly and its surrounding media. An absentee layer, a multiple of half-wave layer, sometimes called a flattening layer, such as 4H may solve this problem. By adjusting the symmetric assembly and adding 4H in the assembly, the design becomes: S/H L H L H L H L H L H L H L H[(H L H L H L H L H L H L H L H) L (H L H L H L H L H L H L H)(4H H L H L H L H L H L H L H) L (H L H L H L H L H L H L H 4H) (H L H L H L H L H L H L H) L (H L H L H L H L H L H L H)]H L H L H L H L H L H L H L H/S. The spectrum is shown in Fig. 1.17b. The rabbit ears are eliminated [16].

The fabrication of such narrow band pass filters for DWDM is very rigorous. Any tiny error may ruin the filter. A narrow band pass filter with layers thickness error 0.0001 will destroy the optical performance as show in Fig. 1.8. The solid line is the designed spectrum and the three dash lines are with standard deviation 0.0001. Figure 1.9 indicates the spectra of a 5-cavity narrow band pass

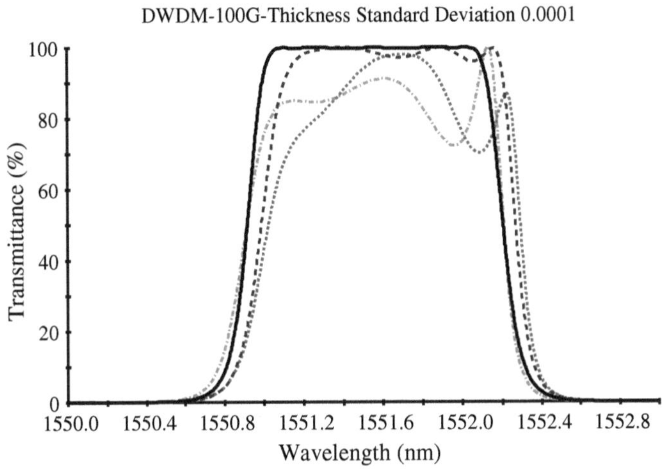

Fig. 1.8 Spectra of a narrow band pass filter with thickness error 0.0001 in each layers

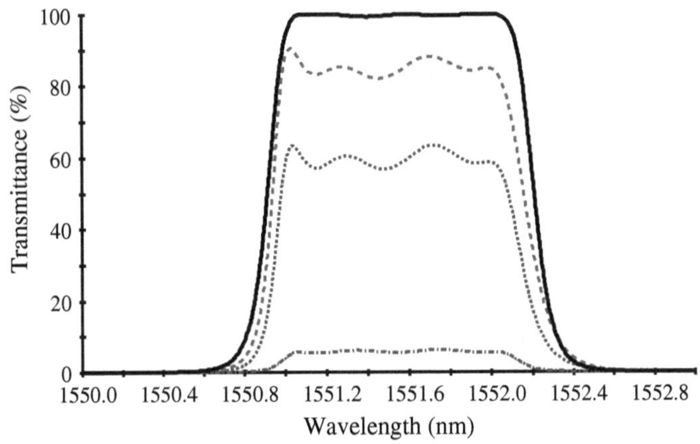

Fig. 1.9 Spectra of a 100 GHz DWD filter with thickness error 0.0005, 0.001 and 0.005 at the last spacer

filter with thickness error 0.0005, 0.001 and 0.005 at the last spacer. All of these explain the thickness error tolerance is very tight. Fortunately, by using error self-compensation effect through turning point optical monitoring, cutting layer with sensitive monitoring wavelength and in situ error compensation, [16–20] we are able to manufacture DWDM filter successfully. Figure 1.10 shows the comparison of designed and fabricated 5-caity narrow band filter for DWDM. The Half-maximum bandwidth $\Delta\lambda_h$ at 3 dB is less than 0.8 nm as required.

Since filters for is extremely sensitive to the optical thickness error in each composite layer. Therefore to have large useful coating area is extreme difficult because of the uniformity problem. For 100 GHz DWDM filters the uniformity

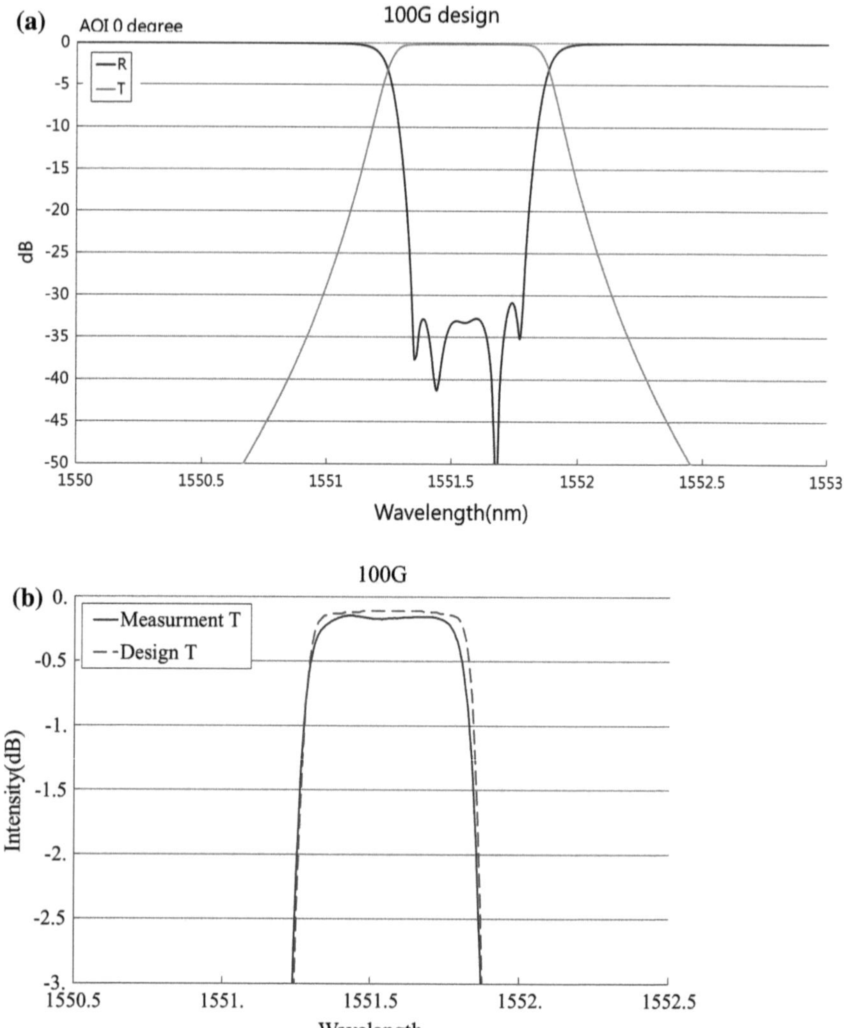

Fig. 1.10 Narrow band pass filter for DWDM, **a** design; **b** comparison of design and fabrication (courtesy of Apogee Optocom Co., Ltd., Taiwan)

must be better than ±0.005 %. The tooling factor, that is ratio of the optical thickness deposited outside the center to the optical thickness deposited at the center, of high and low layers at the different positions are different as indicated by Fig. 1.11 that makes coating a large useful area even difficult.

To enlarge the useful coating area it is necessary to improve the design and the fabrication of DWDM filters. By using an etching technique on the depositing layers with oxygen ion and a new technique, shaping tooling factor, STF, 100 GHz DWDM filters, $S/L(HL)^5\ H10LH(LH)^5\ [L(HL)^6\ H10LH(LH)^6]^3\ L(HL)^5$

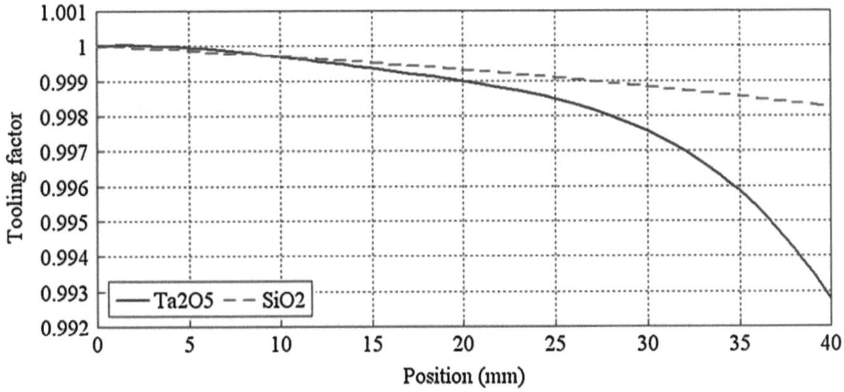

Fig. 1.11 Tooling factors of sources Ta_2O_5 (*solid line*) and SiO_2 (*dash line*)

H10LH(LH)^4 L′H′/Air have been successfully fabricated with the uniformity better than ±0.003 % over an area of 50 mm in diameter and better than ±0.0006 % over a 20 mm in diameter as illustrated by Fig. 1.12 [21].

Since the half-peak bandwidth $\Delta\lambda_h$ is determined by the refractive indices of layers as indicated by (1.4) and (1.5), $\Delta\lambda_h$ is can't be arbitrary. This sometimes will be a problem to make narrow band pass filters for DWDM, coarse WDM, CWDM and 4-skipt-0, etc. By adjusting the coating parameters such as the Ar flow of an ion source in an E-gun in an ion-assisted deposition, IAD, process, to vary the refractive index of the thin film material, one is able to fine tune the bandwidth of a narrow band pass filter to an arbitrary value [22]. The fabrication results are shown in Fig. 1.13 for 4-skipt-0 and CWDM narrow band pass filters.

1.4 Antireflection Coating (AR Coating)

Almost all optics components need anti-reflection coatings not only for increase transmission, but also to eliminate image deterioration due to ghost images and glare. Therefore there are millions of pieces of anti-reflection coatings are processed every day, not including other huge-volume anti-reflection coatings for non-imaging systems, such as solar cells for light absorption and LEDs for light extraction.

In theory a single-layer with a refractive index $n = \sqrt{n}$ can make reflectance $R = 0$. However, there is no robust material where n as small as $\sqrt{n}$. Therefore we need multilayers of alternating TiO_2 or Ta_2O_5 or Nb_2O_5 and SiO_2 to reduce the reflectance. Coated glasses and plastics are much clearer in transmission than an uncoated pair as shown in Fig. 1.14a, b [23]. However, there can still be a residual magenta or green color in reflection as shown in Fig. 1.15. Figure 1.14c shows a green residual color of an antireflection coating on plastics glasses. Coated optical

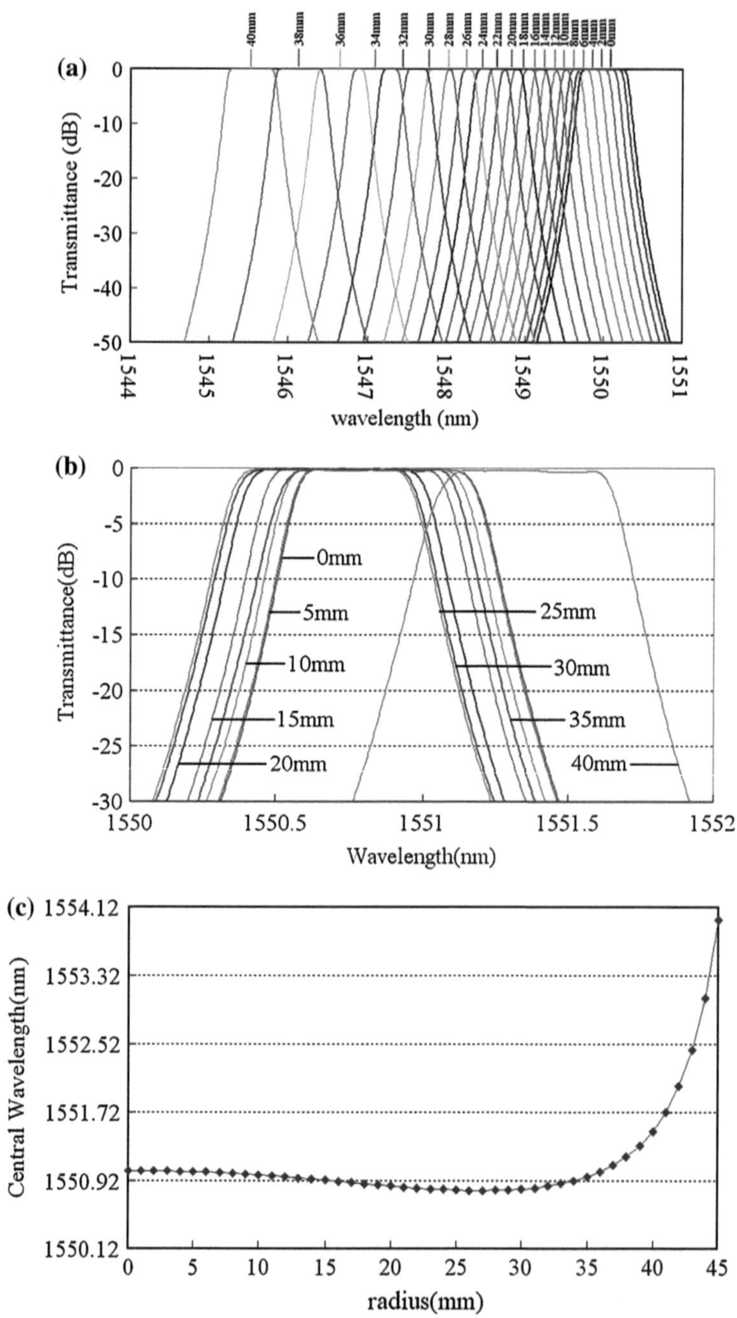

Fig. 1.12 Spectra of the 100 GHz filters measured every 5 mm from the center of the substrate: **a** fabricated by conventional method; **b** by modified method; **c** distribution of the center wavelength on a substrate [21]

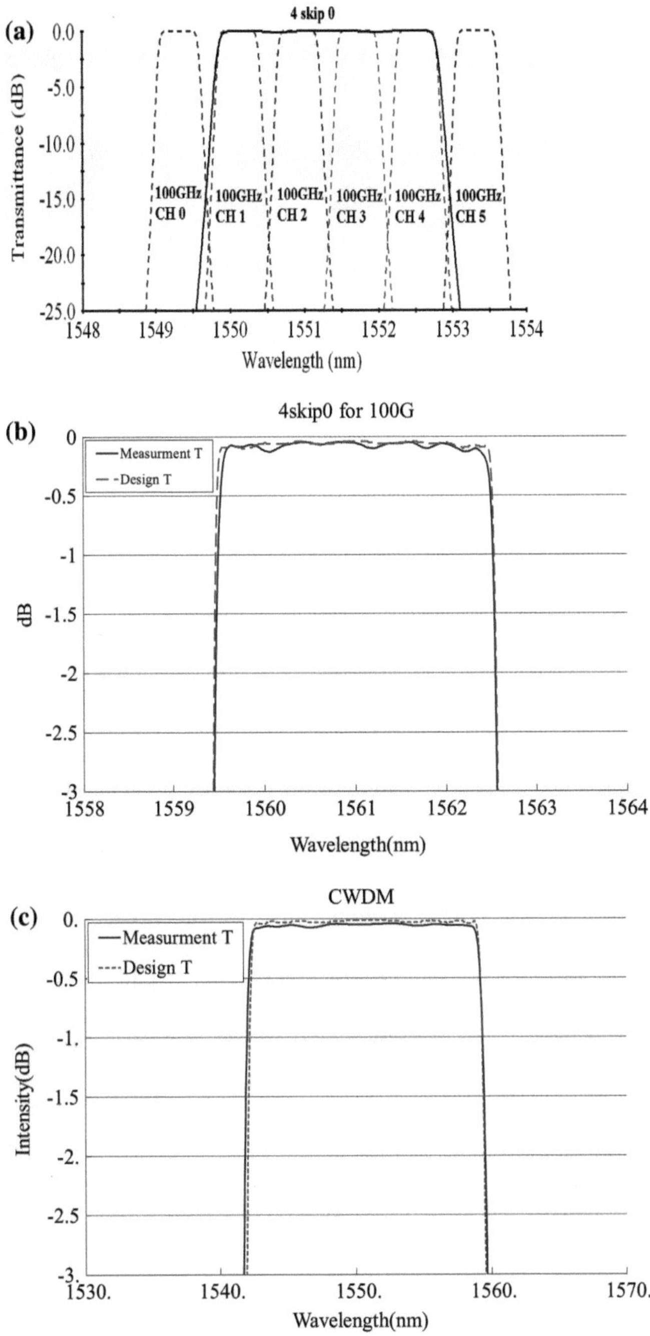

Fig. 1.13 Narrow band pass filter with arbitrary half-peak bandwidth: **a** Sketch of the 4-skip-0 [22]; **b** measured 4-skip-0 filter; **c** measured CWDM filter (courtesy of Apogee Optocom Co., Ltd., Taiwan)

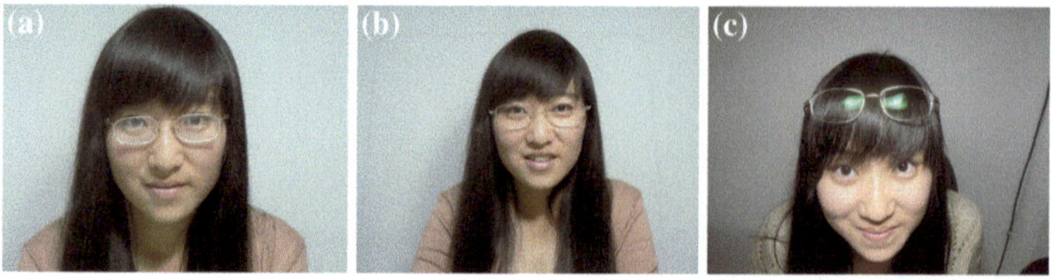

Fig. 1.14 Uncoated glasses **a** and coated glasses **b** and **c**, **c** has residual *green color* in the reflection [23]

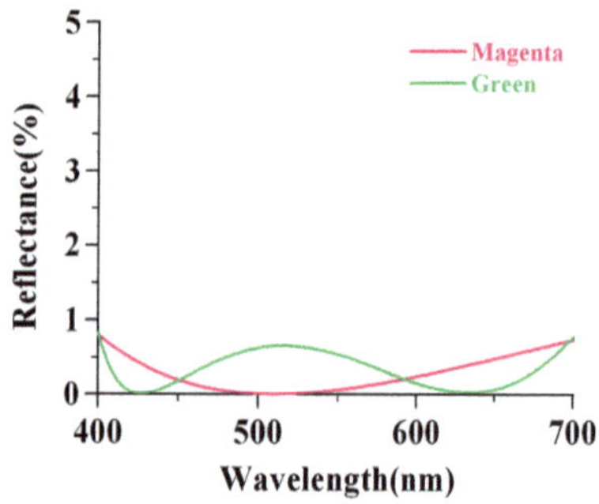

Fig. 1.15 Antireflection coating with residual *magenta* and *green colors*

surfaces in, for example, camera lenses, will frequently display some residual color and this is taken as an indication of the presence of an antireflection coating, no other instruments being necessary for its detection.

To remove the residual color usually implies more layers. Alternatively, an inhomogeneous or gradient index layer has also been proposed. As we understand that a single layer with gradient-refractive-index can represent a broadband wide-angle anti-reflection coating [16, 23, 24]. Such examples occur in nature, the moth eye for example. Researchers have fabricated such films by plasma etching [25, 26], nano-imprint lithography to produce a pyramid array [27], constructing an array of subwavelength-sized features in the form similar to a moth eye [28]. Alternatively, utilize negative refractive index layer may have a broader band and wider angle low reflection, if the negative refractive index layer could be realized [29]. The material with negative refractive index sometimes called as metamaterial which is not easy to fabricate on the conventional substrate with current technology unfortunately. Coatings designed with negative refractive index layer has been will be illustrated in Sect. 1.5. More description about material with negative refractive index will be addressed in Chap. 2.

Nevertheless, in some cases, we may take residual color for decoration. Figure 1.16 shows colored solar cells with different colors without too much loss of efficiency by adjusting the thickness of the layers for antireflection. That gives color appearance of the solar cells. Those solar cells are now in production and

Fig. 1.16 Colored solar cells with different colors installed on the roof of China Pavilion at the Shanghai World Expo 2010 (courtesy of LOF Solar Corp., Taiwan)

have been installed on the roof of the China Pavilion at the Shanghai World Expo 2010, 2012 Yeosu Expo, Italy Bari Airport, Ontario, Utah University, etc.

The anti-reflection and anti-glare coatings to reduce the annoying disturbance from external sources such as bright ambient lights, and to enhance the contrast of the images for the screens of flat panel displays need an absorbing layer combined

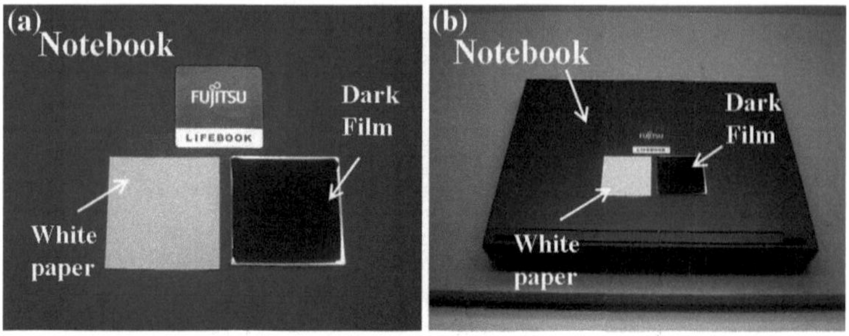

Fig. 1.17 Antireflection coating, *black dark* coating on aluminum made by thin film: **a** normal incidence; **b** oblique incidence [16, 23]

with dielectric layers such as Cr/CrO_x, Fe/FO_x, ITO/SiO_2, $TiNO_x/SiO_2$ [30], and TiN_xW_y/SiO_2 [31] To keep the screen free from contamination, as well as easy to clean with a soft cloth or tissue, a hydrophobic coating has been suggested. The hydrophobic coating should have low surface energy. Usually, the water contact angle should be greater than 110° [32]. Perfluoropolyether polymers modified with organofunctional silanes have been developed to present a hydrophobic surface [33, 34]. However, the durability of such material still needs to be improved. Alternatively, coating a film with nanoroughness has been proposed [35]. Contact angles as high as 152° have been obtained. Instead of hydrophobic coating, applying hydrophilic coating to increase the surface energy of the substrate may split the water droplets and reduce stray light and unwanted reflections. A hydrophilic coating can be realized by short wavelength, such as UV or near UV illumination of photo catalytic material, for example TiO_2 and Nb-doped TiO_2 (TNO) film, so that organic contamination, for example fingerprint, was decomposed through photo-catalysis.

An antireflection for metallic surface may be achieved by utilizing absorbing layers combined with dielectric layers. Figure 1.17 is an example antireflection coating on aluminum plate [16, 23], Al/Nb_2O_5 (36.78 nm)/Nb (12.67 nm)/Nb_2O_5 (28.70 nm)/SiO_2 (61 nm)/Air. Since there is almost no light reflected, it looks very dark and called as "black dark coating" or "dark film". It is darker than the black note book. It can be even darker if the metal surface is roughed before coating.

1.5 Coatings with Negative Refractive Index Layer

Normally, the phase thickness of optical thin film increases with the thickness of optical thin film. However, the phase thickness of negative refractive index material, NIM decreases as the thickness of the film increases. Therefore NIM might be possible be used for phase compensation in normal thin film design. The result could affect the spectra of the thin films have less sensitivity in wavelength and incident angle.

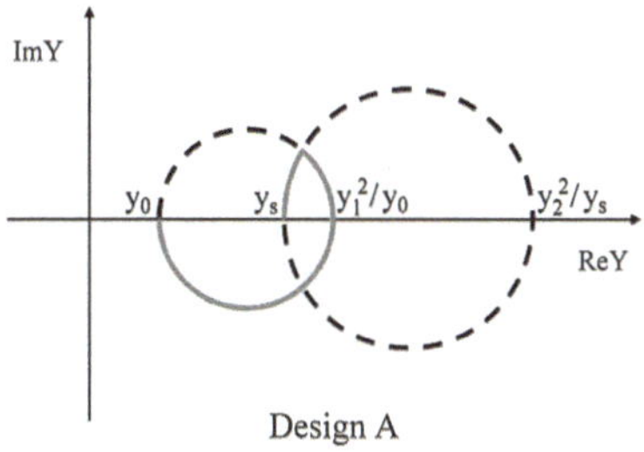

Fig. 1.18 Admittance loci of two-layer V-coat antireflection coating on a glass with refractive index 1.52. The first refractive indices of the first and second layers are 2.23 and 1.45 respectively

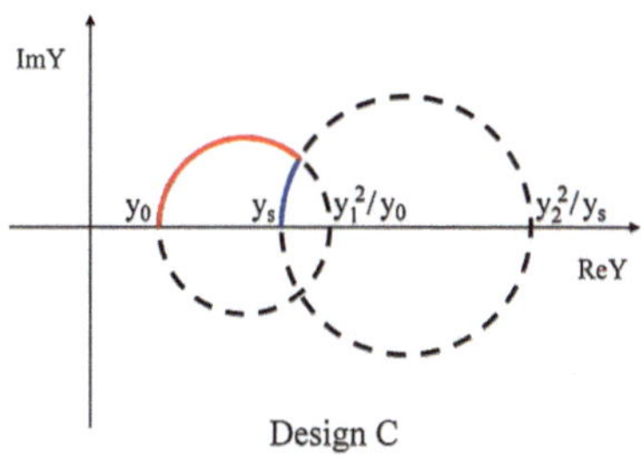

Fig. 1.19 Admittance loci of two-layer V-coat antireflection coating similar to Fig. 1.18, except the refractive index of the second layers is negative, −1.45 (*red color*) [29]

If the antireflection coatings are designed with NIMs, the low reflection range is broader than without NIM and the average reflection is lower. Another application for NIM is to replace the spacer layer using NIM in a narrow band pass filter. Thus the narrow band pass filter can be designed as a wavelength-non-shift filter with incident angle.

Although an antireflection using negative refractive index layer may be not achievable at this moment, it is still interesting to know the advantage of its application. Figure 1.18 shows admittance loci of a two-layer V-coat of antireflection on a glass with refractive index 1.52. The first refractive indices of the first and second layers are 2.23 and 1.45 respectively, let it be design A. All the loci trace in clockwise. Figure 1.19 shows admittance loci of a two-layer V-coat similar as Fig. 1.18 except the refractive index of the second layers is negative, −1.45, let it be design C. The locus of the second layer traces in counterclockwise so that it has phase compensation. It is obvious that the reflectance of design C is much lower over wider spectral range than that of design A as indicated by Fig. 1.20. Figure 1.21 shows that design C is less sensitive in the average reflection from 400 to 700 nm with incident angle than that of design C, particular for S-polarization light.

Usually a filter shows an annoying blue shift in spectrum at tilt angle and s-polarization and p-polarization shift differently. This is particular intolerable for a narrow band pass filter. By utilizing negative refractive index negative layer as

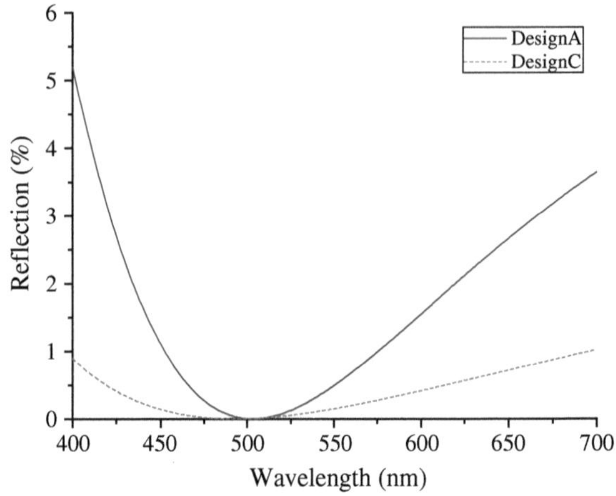

Fig. 1.20 Comparison of reflectance spectra of designs A and C [29]

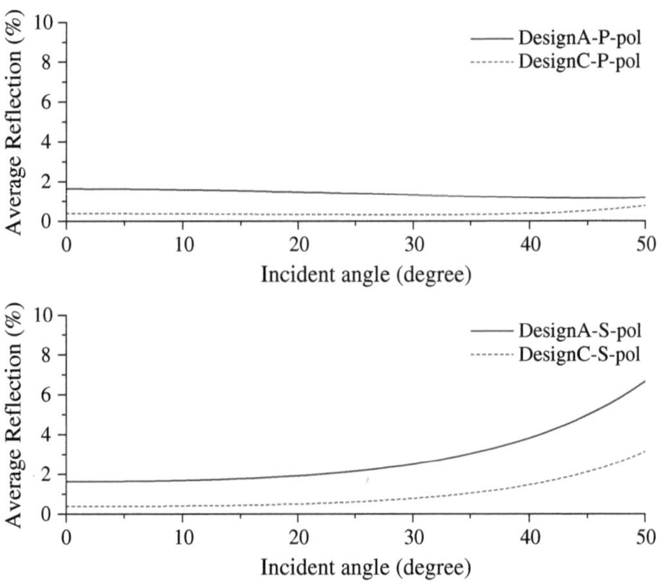

Fig. 1.21 Change of average reflection from 400 to 700 nm with angle for AR designs A and C [29]

one of the spacer can solve this problem. Figure 1.22 is a narrowband pass filter design with NIM layer in the spacer: Sub|(HL)^7 H 3NL 6H 3NL H (LH)^7|Air. The spacer 3NL 6H 3NL, inserted two three-quarter-wave layers 3NL with negative refractive index -1.47. The shift in peak wavelength is less than 0.1 nm for incident angle 10°, and less than 0.4 nm for s-polarize light and almost no shift for p-polarize light for incident angle up to 18°.

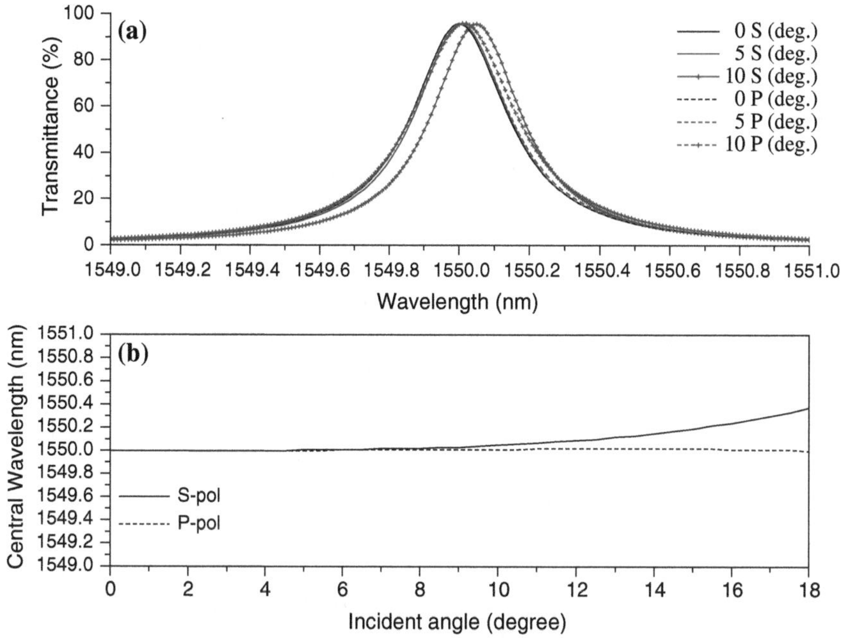

Fig. 1.22 Narrowband pass filter: Sub|(HL)^7 H 3NL 6H 3NL H (LH)^7|Air. H and L are quarter wave layers with refractive indices 2.3 and 1.47 respectively. 3NL is a three-quarter wave layer with negative refractive index -1.47 [29]

1.6 Optical Monitoring

A reliable monitoring for thin film deposition is very important to achieve a theoretical multilayer design. Various monitoring methods, such as time counting, color recognition, quartz monitoring and optical monitoring, have been developed to control thickness during thin-film deposition [17, 18, 36]. Monitoring by time counting is simpler than other monitoring methods, but it needs a very stable deposition rate. Quartz monitoring is also simple but may not be suitable for a coating in plasma environment or a coating with many layers. Besides, in time counting and quartz monitors, the errors in the layer thickness cannot be compensated for and will accumulate with each successively deposited layer. Monitoring by color appearance only works for a simple design. And most of times the operator may loss his/her color cognitive sensitivity after a long looking at the colors. Optical monitoring is based on changes of reflectance (R)/transmittance (T) as the film thickness increases and it has several advantages that to realize a sophisticate design. For example, the turning point method, TPM, is the most commonly used, and has the advantage of error compensation for a coating which composed of quarter-wave layers and multiple of quarter-wave layers [17]. The turning point monitoring is the layer deposition terminated when the signal of R/T reaches its extreme value. For nonquarterwave stacks, the wavelength of the monitoring system must be changed to an appropriate wavelength for monitoring each individual layer. In this case, the level monitoring method, LM, is more suitable [36]. The so called level monitoring is the deposition of a layer terminated when the signal of reflectance/transmittance reaches its theoretical predicted level. General speaking, the monitoring sensitivity of TPM is low, while in LM, the monitoring wavelength can be selected to make the signals sensitive to changes in thickness near the termination point of the deposition, but the error compensation ability is lower. If *R*/*T* are measured on the coating to be fabricated, we call it is a direct monitoring. It is usually applied on coatings which composed of quarter-wave layers and multiple of quarter-wave layers because of a powerful self-compensation effect. For nonquarterwave stacks, to avoid a rapid error acumination, layer thicknesses are monitored on one or several separate witness chips, and called as indirect monitoring, is applied. Several numerical methods and the optimum trigger point method have been developed to improve the performance, sensitivity and compensation associated with the TPM and the LM methods [18, 37–41].

1.6.1 In Situ Sensitive Optical Monitoring with Error Compensation (ISMEC)

The conventional monitoring methods use runsheet diagrams to control the thickness of the films, based on their transmittance or reflectance, and no phase information can be provided. The fact is that the refractive index *n*, the extinction coefficient *k*, and the thickness *d* are not independently determined in real-time, which make phase extraction difficult. In a growing thin-film stack, the refractive index of the materials may change with time due to variations in the coating parameters or environmental fluctuations in the chamber, meaning we cannot terminate the current layer at the point specified in the original design. Therefore real-time determination of *n, k* and *d* is necessary for precision coatings. Ellipsometric and phase extraction methods have been proposed to solve this problem. Ellipsometric monitoring [42, 43] requires algebraic computation to obtain the optical constants that fit the measurements, but require so many parameters that they are difficult to solve analytically. In the case of multilayer coatings, this could lead to the accumulation of errors, which would cause the final optical performance to differ from that of the original design. To avoid this type of error accumulation, in situ sensitivity monitoring with proper error compensation, ISMEC, has been proposed to find the correct refractive index and physical thickness of each deposited layer [20]. An operator can choose the sensitivity monitoring wavelength that offers better error compensation for the reference wavelength. Figure 1.23 shows the experimental results of the monitoring of a narrow band-pass filter with a center wavelength at 800 nm by the ISMEC, TPM and LM methods [20]. The better performance of ISMEC method is clear. The Half-maximum bandwidth $\Delta\lambda_h$ and the position of the maximum

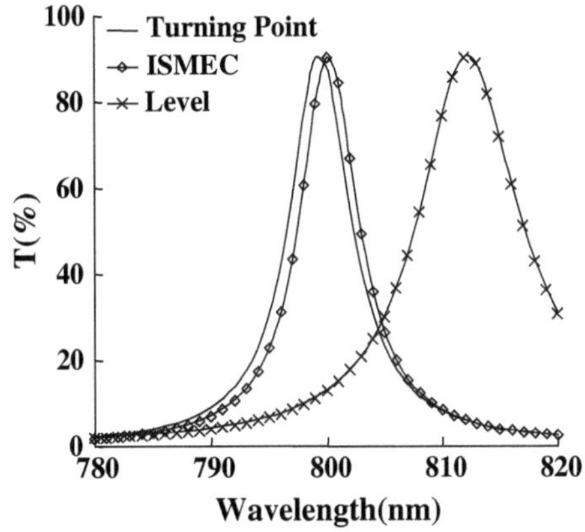

Fig. 1.23 Spectra of narrow band-pass filters monitored by the three methods [20]

transmittance are closer to the designed values than are those obtained using the other two methods. This is because TPM offers error compensation only for previous layers, but lacks sensitivity at termination points, which results in a broadening of the $\Delta\lambda_h$. The level monitoring method has higher sensitivity, but no error compensation, which results in a shift of the output spectrum. ISMEC method on the other hand, combines the advantages of both methods, the sensitivity at termination points and the error compensation for previous layers.

1.6.2 Optical Monitoring and Real Time Admittance Loci Calculation Through Dynamic Interferometer

To get rid of the influence of the variations in the coating parameters caused by environmental fluctuations in the chamber, a polarization Fizeau interferometer-based monitoring system has been developed [44]. This is a vibration-insensitive in situ system, whereby one can instantly obtain the real-time phase and magnitude of the reflection coefficient and/or the transmission coefficient. The layout of a reflection monitoring system with a short coherence light source is shown in Fig. 1.24. The test surface is located on the side of thin film growth, while the other side of the substrate is the reference surface, as shown in Figs. 1.25 and 1.26 for transmission and reflection monitoring system, respectively. The tunable retarder is composed of two wedge x-cut uniaxial birefringent crystals. The crystal-induced phase difference between two orthogonally polarized beams has to match the difference in the optical phase induced by the reference surface and the test surface. The interference will therefore occur only between either the S-polarized reference beam and the P-polarized test beam or the P-polarized reference beam and the S-polarized test beam. Other reflections from the interfaces will be suppressed, because of the short coherence length of the light

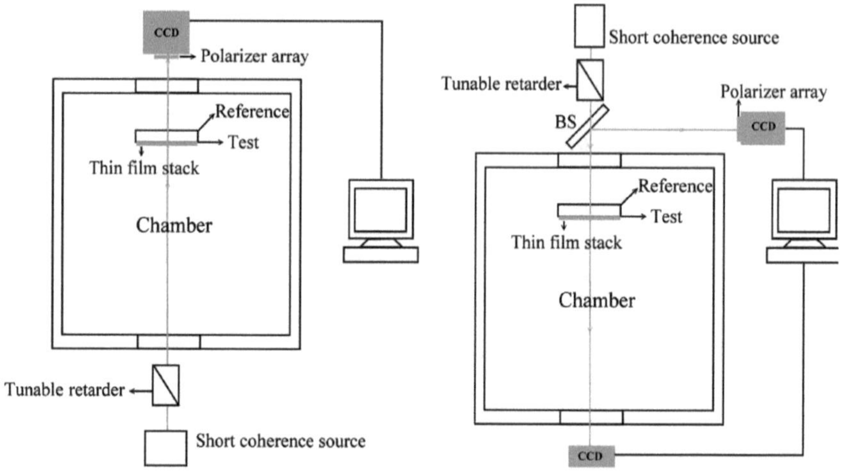

Fig. 1.24 Schematic representation of the transmission and reflection optical monitoring system through dynamic interferometer [44]

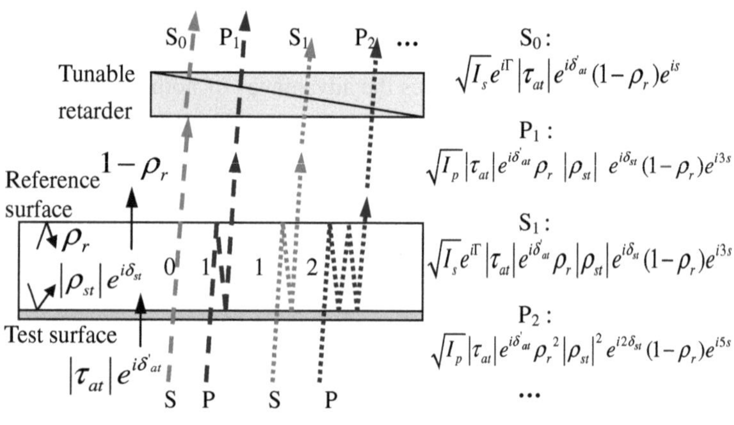

Fig. 1.25 Multiple reflections in the substrate for transmission monitoring system [44]

source [45]. Interference will occur when the transmission axes of the polarizers are oriented 45° and −45° to the fast axis of the birefringent crystal. These two polarizers, as well as the other two (whose transmission axes are oriented 0° and 90° to the fast axis of the birefringent crystal) are combined, and detected by the compact polarizer detector array, as shown in Fig. 1.27. After mathematical analysis using (1.6–1.7), the magnitude of the reflection coefficient and the phase can be acquired from the phase-shifted interferograms of two orthogonal polarization beams on the detector array, where we assumed the transmission coefficient of the film stack (from air to films) is $\tau_{at} = |\tau_{at}|\exp(i\delta_{at})$, the

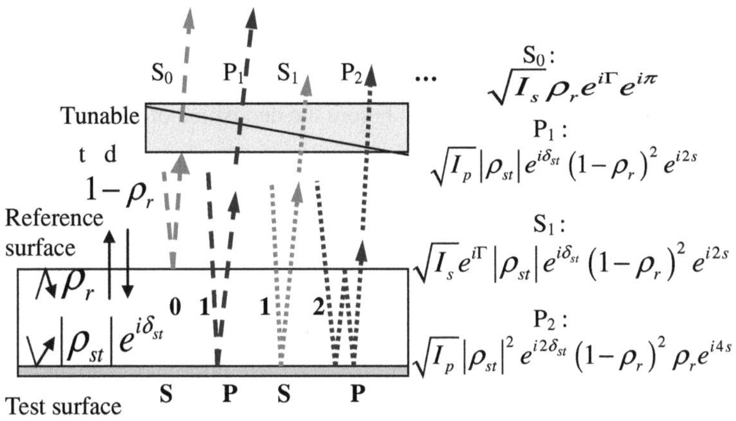

Fig. 1.26 Multiple reflections in the substrate for reflection monitoring system [44]

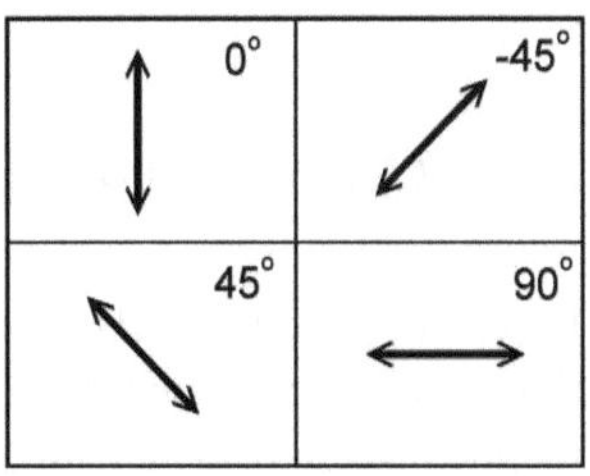

Fig. 1.27 Layout of the polarizer detector array

reflection coefficient of the film stack (from substrate to the deposited film stack) is $\rho_{st} = |\rho_{st}| \exp(i\delta_{st})$, and the reflection magnitude of reference surface (from the substrate to air) is equal to ρ_r. Only the paired beams that are path matched and drawn in the same type of lines will interfere with each other, since the light has short coherence length. The reflection magnitude of the reference surface is assumed equal to ρ_r, and the optical phase difference between two orthogonal polarizations induced by one pass through the substrate and the retarder are s and Γ, respectively. The complete information about the growing film stack thus can be obtained, and the thickness of the whole stack can be monitored accurately.

$$I_{0^\circ} = I_s|\rho_r|^2 + \frac{I_s(1-|\rho_r|)^4|\rho_{st}|^2}{1-|\rho_{st}|^2|\rho_r|^2}, \quad I_{90^\circ} = I_p|\rho_r|^2 + \frac{I_p(1-|\rho_r|)^4|\rho_{st}|^2}{1-|\rho_{st}|^2|\rho_r|^2},$$
$$I_{+45^\circ} = \frac{1}{2}(I_{0^\circ}+I_{90^\circ}) + \sqrt{I_{0^\circ}I_{90^\circ}}|\rho_r||\rho_{st}|\left(1-\frac{|\rho_r|^2+(1-|\rho_r|)^2}{\sqrt{I_{0^\circ}I_{90^\circ}}}\right)\cos\theta,$$
$$I_{-45^\circ} = \frac{1}{2}(I_{0^\circ}+I_{90^\circ}) - \sqrt{I_{0^\circ}I_{90^\circ}}|\rho_r||\rho_{st}|\left(1-\frac{|\rho_r|^2+(1-|\rho_r|)^2}{\sqrt{I_{0^\circ}I_{90^\circ}}}\right)\cos\theta, \tag{1.6}$$

where I_j is the reflectance of $j = 0°, -45°, +45°$ and 90° on the detector array; and $\theta = \phi_{st} + \phi_0$. $\phi_0 = 2s - \Gamma$, is the phase shift caused by the optical path difference between two orthogonal polarizations as they pass through the substrate, s, and the retarder, Γ. ϕ_0 can be acquired before the deposition of the thin film.

$$\phi_{st} = \{\arccos[\frac{\left(I_{+45^\circ} - I_{-45^\circ}\right)}{2\sqrt{I_{0^\circ}I_{90^\circ}}|\rho_r||\rho_{st}|\left(1 - \frac{|\rho_r|^2+(1-|\rho_r|)^2}{\sqrt{I_{0^\circ}I_{90^\circ}}}\right)}]\} - \phi_0. \tag{1.7}$$

Thus by applying an interferometer the optical phase information for a growing film stack can be obtained. Both the magnitude and phase of the reflection coefficient can be found and give us access to the complete information about reflection coefficient, optical admittance, refractive index and actual thickness of the growing film stack at every moment. The optical admittance during the coating process can then be extracted. This system is simple, compact, and will be useful in the fabrication of precision optical coatings.

1.6.3 Optical Monitoring Using Admittance Diagram

Better monitoring performance can be achieved from an admittance diagram than a runsheet diagram, because of the higher sensitivity and larger amount of physical and visual information. It has therefore been proposed that they be used in an optical monitoring method [44, 46–48]. The diagram is obtained by plotting the locus of the effective admittance during the deposition process. Changes in the refractive index during the coating process can be easily observed and immediately compensated for from layer to layer as shown in Fig. 1.28. An error occurred at the first layer, b is the predicted termination point and B is

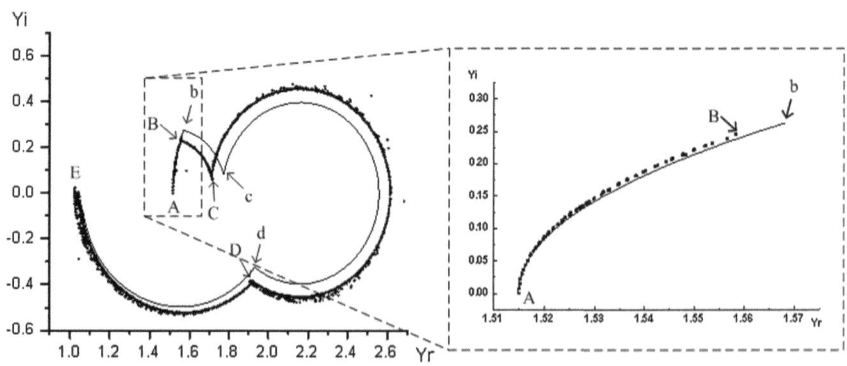

Fig. 1.28 An error occurred at the first layer, *b* is the predicted termination point and *B* is the deposited point. *Thin-solid line* is the simulated loci. In experiment, thickness error was compensated resulting in a new locus, *bold-dot* [48]

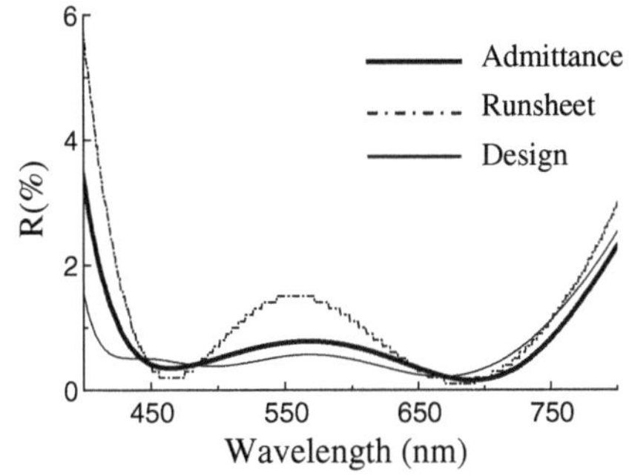

Fig. 1.29 Optical performance of an antireflection coating monitored by the admittance diagram and the conventional runsheet method [48]

the deposited point. The refractive index of the deposited film is higher than the designed film, since the locus is outward expansion. Thin-solid line is the simulated loci. In experiment, thickness error was compensated resulting in a new locus, bold-dot. Figure 1.29 shows plots of the optical performance of an antireflection coating monitored using a real-time admittance diagram and the conventional runsheet method.

1.6.4 Broadband Monitoring Through Equivalent Optical Admittance Loci Observation

The precision of single wavelength monitoring is lower than that of the broadband monitoring, BM, because only one wavelength signal is provided. Not only is it not as comprehensive as broadband monitoring, but the signal errors for the monitoring wavelength dominate the monitoring results. Therefore it is sometimes difficult to obtain the equivalent optical admittance precisely from one single wavelength measurement. Thanks to the highly developed processing technology such as charge-coupled device (CCD) and data extraction and analysis software, the real-time BM is easy to install in a conventional coating machine [49, 50]. In BM, T and/or R values are evaluated simultaneously in different wavelengths. The proper termination time is typically thought to be when the real-time spectrum is the closest to the design spectrum of the termination point. Therefore, a novel method to obtain the refractive index and thickness from the broadband spectrum measurements instead of from a single wavelength measurement and convert this information into the equivalent optical admittance loci of a single wavelength to process the monitor, called as equivalent optical admittance monitoring, EOAM, can be realized [51, 52]. The admittance values can be acquired with higher precision and less noise. This method combines the advantages of both broadband and single wavelength monitors, to let operators terminate the deposition process at one predicted point, with good error compensation. A 4-layer antireflection coating with physical thicknesses from the first to forth layer are 14 nm (Ta_2O_5), 33 nm (SiO_2), 130 nm (Ta_2O_5), and 85 nm (SiO_2) has been fabricated. In the third and fourth layers, a phase of π was added to the reflection phase, to improve the monitoring sensitivity as shown by Fig. 1.30. The experimental results in Fig. 1.31 clearly show that EOAM performs better than the other methods. A 20 % thickness excess in the 2nd layer has been compensated by the successively deposited layers (Fig. 1.30b). EOAM combines the advantages of the broadband and single-wave monitoring methods without their drawbacks. A phase-adding technique increases the monitoring sensitivity. This should greatly improve the fabrication precision and decrease the defection rate of optical filters, especially for the fabrication of multilayer filters where good error compensation and higher monitoring precision are needed.

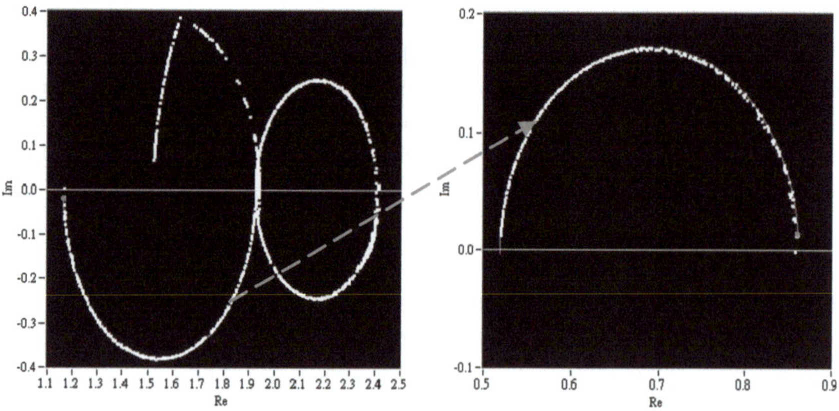

Fig. 1.30 EOAM monitoring loci and the forth layer monitoring locus with phase-shifting technique

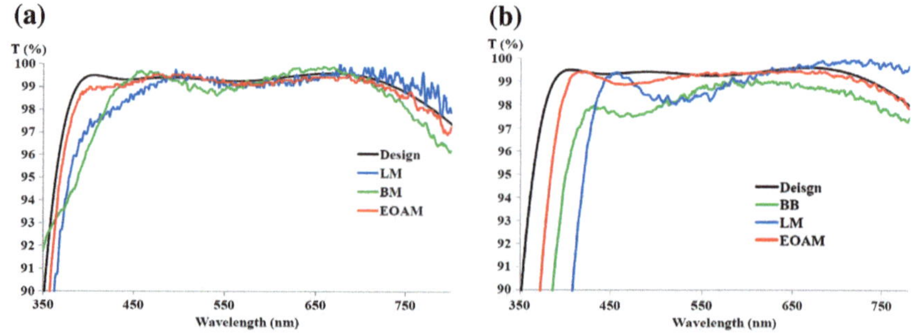

Fig. 1.31 Optical performance of an antireflection coating monitored by various methods: **a** physical thicknesses from the first to forth layer were 14, 33, 130, and 85 nm; **b** 20 % thickness excess in the 2nd layer [51, 52]

1.6.5 Reflection Coefficient Monitoring Through Broadband Spectrum

In this section the reflection coefficient locus will be introduced. Comparing with Turning-Point Monitoring, TPM, Level Monitoring, LM, and Broadband Monitoring, BM, the reflection coefficient monitoring method, RCM, of one single wavelength was acquired through the real time broadband spectrum and it then was used for providing better error compensation and easier termination judgment during the multilayer deposition [53]. BM monitors the real-time broad-band spectrum and terminates the deposition when the real-time spectrum coincides with the design spectrum at the termination point. Therefore, there may be many local minimums of the difference between the real-time spectrum and the design spectrum during the deposition process. In RCM, one wavelength is selected for the reflection coefficient loci monitoring during the coating process. The reflection coefficient, r, is a function of refractive indices n, thicknesses, d, and the monitoring wavelengths. As the thickness grows, the value of r is changed. The monitoring signal changes with respect to the change in thickness is called monitoring sensitivity. If there is noise interference, the sensitivity determines whether the deposition termination point is easy to be correctly judged. Figure 1.32 shows a comparison of the simulation results for the signal change per 5 nm thickness for different optical monitoring methods. The loci are plotted for one layer of Ta_2O_5 film deposited on a glass. The figure shows that the transmittance locus and optical admittance locus monitoring will have low monitoring sensitivity at certain areas. On the other hand, the reflection coefficient has more uniform monitoring sensitivity than the other monitoring methods. For multilayer stacks, the difference in the uniformity of the monitoring sensitivity between the monitoring methods will be enormous in the later deposited layers. The operator can choose a monitoring wavelength that has a larger reflection coefficient change near the deposition termination point for easier judgment of termination. Since only one point can be determined as the termination point, the best choice for termination will be if the deposition is ended when the reflection coefficient at the monitoring wavelength is the closest to its designed value. Similarly, for some coatings which are sensitive to phase variations, the reflection phase provided in this method can be used to precisely terminate the locus at the designed phase.

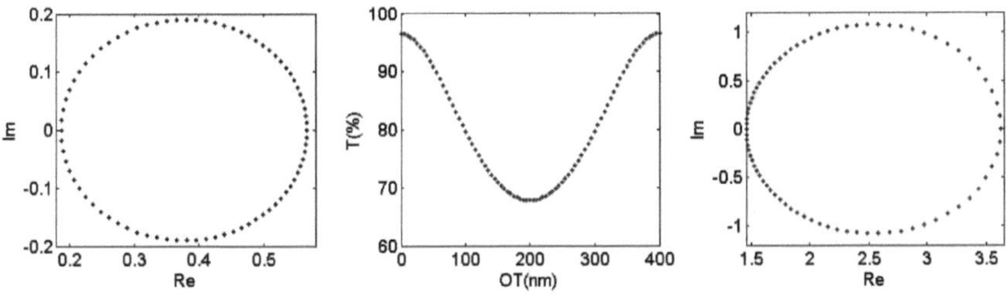

Fig. 1.32 Reflection coefficient, transmittance, and optical admittance loci (from *left* to *right*) [53]

In the monitoring, one wavelength is selected for the reflection coefficient loci monitoring during the coating process. However, for higher precision calculation, the broadband transmittance or reflectance spectrum is used to calculate the real time refractive indices *n*, and thicknesses, *d*, in order to acquire the corresponding reflection coefficient value *r* of the selected wavelength. The deposition of each layer should be terminated when the reflection coefficient value is the closest to the design value. The selected wavelength could be the reference wavelength of the design.

The refractive index, *n*, and thickness, *d*, then can be obtained from the real time broadband spectrum, and derived the reflection coefficient, both of the reflection phase and amplitude, from the obtained refractive index and thickness. The more wavelengths that are used in the calculations, the more accurate will be the *n* and *d* obtained through the numerical algorithm. Then, re-optimization of the design is carried out after the termination of the deposition of each layer. The selected wavelength for the reflection coefficient monitoring may not be one of the wavelengths for T or R measurements. Unlike the general BM using multi-wavelength signals to evaluate the differences in the design, only the reflection coefficient of the monitoring wavelength is used for the monitor. In this way we can avoid ambiguity in finding the minimum of the merit function.

Long-wave pass filters with a total of 14 layers have been fabricated by various monitoring methods with only one monitoring chip through the whole coating process, i.e., direct monitoring, for comparison. An ion-beam sputtering deposition system with an ion beam source 6-cm in diameter was employed to prepare the filters. Ta_2O_5 and SiO_2 were used as the high and low refractive index materials, respectively. The substrate was B270 glass. The intensity of the monitoring light passing through the monitoring chip was measured by a CCD spectrometer. The edge of the long-wave pass filter was designed at 550 nm. Figure 1.33 shows the results of comparisons between different monitoring methods. We can see that the reflection coefficient monitoring, RCM, has the

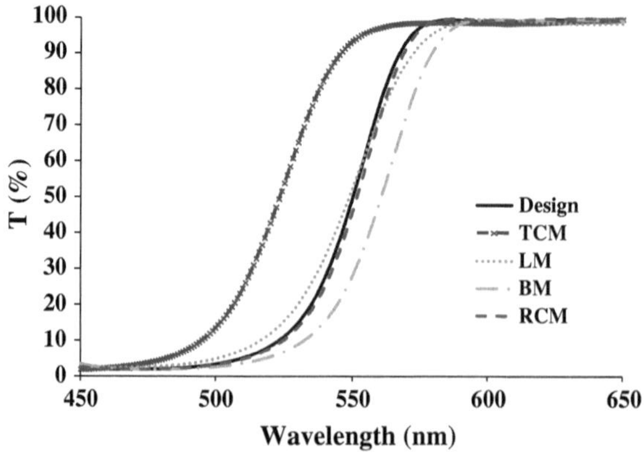

Fig. 1.33 Long-wave pass filters made by various monitoring methods [53]

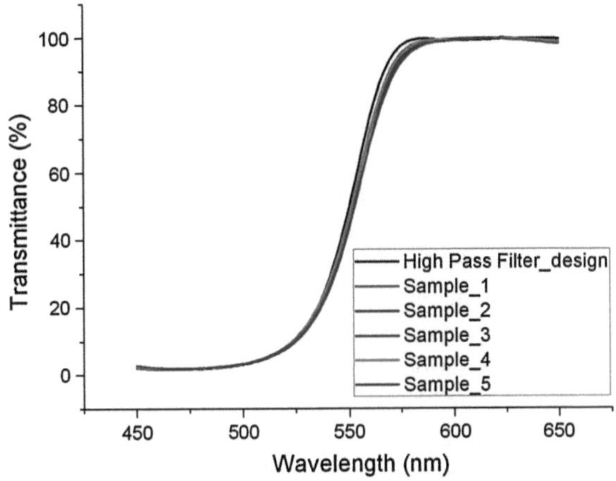

Fig. 1.34 Long-wave pass filters made with direct monitoring for five continuous runs by RCM

best performance. Time Counting Monitoring, TCM, in a stable ion-beam sputtering system has a similar efficiency to quartz monitoring. The accumulated errors introduced a large shift from the design. In Level Monitoring, LM, the position of the spectrum edge was close to the design, but there was obvious distortion in the shape of the spectrum possibly arising from the low precision of thickness control. Broadband monitoring, BM, gives the correct spectrum shape, but the spectrum shift coming from the excess deposition shows the defects. The spectrum of the coating made by RCM was 82 % closer to the design than that made by BM.

When the experiments were done in five continuous runs, the results show that the reflection coefficient monitoring has good error compensation ability, as shown in Fig. 1.34. The shift of the edge from 550 nm (design) was lower than 0.5 %. The reproducibility is also remarkable. In five continuous runs, the shift among each run was below 0.3 %.

In conclusion, the correct choice of monitoring method is extreme important to realize multilayer thin film optical coatings. If thickness error and refractive index change during the deposition are unavoidable, the strategies about optical monitoring for thin film coatings always be the issues. More examples about the strategies of optical monitoring may be found in several articles [54].

1.7 Summary

Optical thin films not only improve the optical performance of optical devices, but also are vital parts in optical operations. The progress of the theory, design and manufacture techniques of optical thin films is never suspended to promote the growth of optics and photonics. Some current technologies about theory, design and manufacture of optical thin films have been described.

The reflectance can be transcended over the limited of quarter wave stacks and photonic crystals, by using gain layer with negative extinction coefficient. Coatings with reflectance more than 100 % have been realized. The gain layer can be made by embedding quantum dots, QDs, in a dielectric film. More information about thin film embedded with QDs can consult with Flory et al. [55–57]. The extra energy at emission wavelength is generated by the down transition owing to the discrete energy levels of quantum dots, so that such unusual behavior does not violate energy conservation. The energy associated with shorter wavelengths is transferred to longer wavelengths. The application of gain layers is very promising, such as efficiency increase in solar cell and LED, LD, display, biology and loss compensation for metamaterials.

The design theory and fabrication techniques of subnano-bandwidth narrow band pass filters are very important and rigorous for dense wavelength division multiplexing/demultiplexing, DWDM, technology. The detail of theory has been introduced and the techniques of coatings has been explained and demonstrated. A useful coating area has been enlarged with the uniformity better than ±0.003 % over an area of 50 mm in diameter and better than ±0.0006 % over a 20 mm in diameter, by using an etching technique on the depositing layers with oxygen ion and a new technique, shaping tooling factor, STF.

Anti-reflection coatings are indispensable for optics elements, since not only transmission can be increased but also image deterioration due to ghost images and glare can be eliminated. The design principle has been explained and some practical antireflection coatings with hydrophobic and hydrophilic characteristics have been investigated. The utilization of the residual color of anti-reflection coatings as a decoration is interesting and has been applied to the solar cells so that the appearance of solar cells is bright and lively but not so dim and gloomy. A black film based on antireflection coating has many applications such for better contrast in display has been designed and fabricated and shown that the black film is darker than a conventional black notebook since almost no light is reflected.

To obtain a theoretical multilayer design needs a reliable monitoring during the thin film deposition. The thickness error of deposited layer and change of refractive during the deposition during the deposition are inevitable and unavoidable. The monitoring technology to achieve the coating as the design one is very crucial. Various advanced monitoring methods, such as in situ sensitive optical monitoring with error compensation, monitoring with real time admittance loci calculation through dynamic interferometer, optical monitoring using admittance diagram, broadband monitoring through equivalent optical admittance loci observation and reflection coefficient monitoring through broadband spectrum have been introduced. Several excellent coatings have been successfully fabricated by utilized the proposed monitoring methods.

Nevertheless, the requests of new technologies about theory, design and manufacture of optical thin films will be tougher and more challenging with the rapid progress in the optics and photonics.

References

1. Y.-J. Chen, C.-C. Lee, S.-H. Chen, F. Flory, Extra-high reflection coating with negative extinction coefficient. Opt. Lett. **38**, 3377–3379 (2013)
2. H.A. Macleod, *Thin-Film optical Filter, 4th ed* (CRC Press, New York, 2010)
3. H.A. Macleod, Gain optical coatings: part 1. Bull. Soc. Vac. Coaters **Issue Fall**, 22–27 (2011)
4. A.J. Nozik, Quantum dot solar cells. Phys. E **14**, 115–120 (2002)
5. Prashant.V. Kamat, Quantum dot solar cells. Semiconductor nanocrystals as light harvesters. J. Phys. Chem. C **112**(48), 18737–18753 (2008)
6. H.-J. Lin, S. Vedraine, J. Le-Rouzo, S.-H. Chen, F. Flory, C.-C. Lee, Optical properties of quantum dots layers: application to photovoltaic solar cells. Sol. Energy Mater. Sol. Cells **117**, 625–656 (2013)
7. J. Wu, Z.M. Wang (eds.), *Quantum Dot Solar Cells* (Springer, New York, 2014)
8. J. Zhao, J.A. Bardecker, A.M. Munro, M.S. Liu, Y. Niu, I.-K. Ding, J. Luo, B. Chen, A.K.-Y. Jen, D.S. Ginger, Efficient CdSe/CdS quantum dot light-emitting diodes using a thermally polymerized hole transport layer. Nano Lett. **6**(3), 463–467 (2006)
9. H. Liu, T. Wang, Q. Jiang, R. Hogg, F. Tutu, F. Pozzi, A. Seeds, Long-wavelength InAs/GaAs quantum-dot laser diode monolithically grown on Ge substrate. Nat. Photonics **5**, 416–419 (2011)
10. T.-H. Kim, K.-S. Cho, E.K. Lee, S.J. Lee, J. Chae, J.W. Kim, D.H. Kim, J.-Y. Kwon, G. Amaratunga, S.Y. Lee, B.L. Choi, Y. Kuk, J.M. Kim, K. Kim, Full-colour quantum dot displays fabricated by transfer printing. Nat. Photonics **5**, 176–182 (2011)
11. S. Jin, Y. Hu, Z. Gu, L. Liu, H.-C. Wu, Application of quantum dots in biological imaging. J. Nanomaterials **2011**(834139) (2011). doi:10.1155/2011/834139
12. A. Fang, T. Koschny, M. Wegener, C.M. Soukoulis, Self-consistent calculation of metamaterials with gain. Phys. Rev. B **79**, 241104(R) (2009)
13. S.M. Xiao, V.P. Drachev, A.V. Kildishev, X.J. Ni, U.K. Chettiar, H.K. Yuan, V.M. Shalaev, Loss-free and active optical negative-index metamaterials. Nature **466**, 735–740 (2010)
14. M. Decker, I. Staude, I.I. Shishkin, K.B. Samusev, P. Parkinson, V.K.A. Sreenivasan, A. Minovich, A.E. Miroshnichenko, A. Zvyagin, C. Jagadish, D.N. Neshev, Y.S. Kivshar, Dual-channel spontaneous emission of quantum dots in magnetic metamaterials. Nat. Commun. **4**(2949) (2013). doi:10.1038/ncomms3949
15. L. Ivan Epstein, The design of optical filter. J. Opt. Soc. Am. **42**, 806–810 (1952)
16. C.-C. Lee, *Thin Film Optics and Coating Technology*, 7th edn. (Yi Hsien, Taipei, 2012)
17. H. Angus Macleod, Turning value monitoring of narrow-band all-dielectric thin-film optical filters. Optica Acta **19**, 1–28 (1972)
18. H.A. Macleod, Monitoring of optical coatings. Appl. Opt. **20**, 82–89 (1981)
19. C.-C. Lee, W. Kai, C.-C. Kuo, S.-H. Chen, Improvement of the optical coating process by cutting layers with sensitive monitor wavelengths. Opt. Express **13**, 4854–4861 (2005)
20. C.-C. Lee, W. Kai, In situ sensitive optical monitoring with error compensation. Opt. Lett. **32**(15), 2118–2120 (2007)
21. C.-C. Lee, S.-H. Chen, C.-C. Kuo, Fabrication of DWDM filters with large useful area. SPIE Optics and Photonics, paper #6286-15, San Diego, USA, 13–17 Aug 2006
22. C.-C. Lee, S.-H. Chen, C.-C. Kuo, C.-Y. Wei, Achievement of an arbitrary bandwidth for a narrow bandpass filter. Opt. Express **15**, 15228–15233 (2007)
23. C.-C. Lee, Optical interference coatings for optics and photonics (Invited). Appl. Opt. **52**, 73–81 (2013)
24. S.R. Kennedy, M.J. Brett, Porous broadband antireflection coating by glancing angle deposition. Appl. Opt. **42**, 4573–4579 (2003)
25. Y.F. Huang, S. Chattopadhyay, Y.J. Jen, C.Y. Peng, T.A. Liu, Y.K. Hsu, C.L. Pan, H.C. Lo, C.H. Hsu, Y.H. Chang, C.S. Lee, K.H. Chen, L.-C. Chen, Improved broadband and quasi-omnidirectional anti-reflection properties with biomimetic silicon nanostructures. Nat. Nanotechnol. **2**, 770–774 (2007)

26. U. Schulz, C. Präfke, C. Gödeker, N. Kaiser, A. Tünnermann, Plasma-etched organic layers for antireflection purposes. Appl. Opt. **50**, C31–C35 (2011)
27. B. Päivänranta, T. Saastamoinen, M. Kuittinen, A wide-angle antireflection surface for the visible spectrum. Nanotechnology **20**, 375301 (2009)
28. J.-Q. Xi, M.F. Schubert, J.K. Kim, E.F. Schubert, M. Chen, S.-Y. Lin, W. Liu, J.A. Smart, Optical thin-film materials with low refractive index for broadband elimination of Fresnel reflection. Nat. Photonics **1**, 176–179 (2007)
29. J.-L. Tsai, Applications of negative refraction index materials for antireflection and narrow band pass filters. MS Thesis, Department of Optics and Photonics, National Central University, Taiwan, 2009
30. H. Ishikawa, B. Lippey, in *Two layer broad band AR coating*. Proceedings of 10th International Conference on Vacuum Web Coating (Bakish Materials Corporation, 1996), pp. 221–233
31. G. McHale, N.J. Shirtcliffe, M.I. Newton, Contact-angle hysteresis on super-hydrophobic surfaces. Langmuir **20**, 10146–10149 (2004)
32. L. Mascia, T. Tang, Polyperfluoroether-silica hybrids. Polymer **39**, 3045–3057 (1998)
33. K.C. Camargo, A.F. Michels, F.S. Rodembusch, M.F. Kuhn, F. Horowitz, Visibly transparent and near infrared, wideangle, anti-reflection coatings with simultaneous selfcleaning on glass. Opt. Mater. Express **2**, 969–977 (2012)
34. M. Flemming, A. Duparre, Design and characterization of nanostructured ultrahydrophobic coatings. Appl. Opt. **45**, 1397–1401 (2006)
35. K. Zhang, F. Zhu, C.H.A. Huan, A.T.S. Wee, Indium tin oxide films prepared by radio frequency magnetron sputtering method at a low processing temperature. Thin Solid Films **376**, 255–263 (2000)
36. B. Vidal, A. Fornier, E. Pelletier, Wideband optical monitoring of nonquarter wave multilayer filter. Appl. Opt. **18**, 3851–3856 (1979)
37. F. Zhao, Monitoring of periodic multilayer by the level method. Appl. Opt. **24**, 3339–3343 (1985)
38. C.J. van der Laan, Optical monitoring of nonquarterwave stacks. Appl. Opt. **25**, 753–760 (1986)
39. B. Bobbs, J.E. Rudisill, Optical monitoring of nonquarterwave film thickness using a turning point method. Appl. Opt. **26**, 3136–3139 (1987)
40. C. Zang, Y. Wang, W. Lu, A single-wavelength monitoring method for optical thin-film coatings. Opt. Eng. **43**, 1439–1443 (2004)
41. A.V. Tikhonravov, M.K. Trubetskov, Eliminating of cumulative effect of thickness errors in monochromatic monitoring of optical coating production: theory. Appl. Opt. **46**, 2084–2090 (2007)
42. J. Lee, R.W. Collins, Real-time characterization of film growth on transparent substrates by rotating-compensator multichannel ellipsometry. Appl. Opt. **37**, 4230–4238 (1998)
43. S. Dligatch, R. Netterfield, B. Martin, Application of in-situ ellipsometry to the fabrication of multi-layered coatings with sub-nanometre accuracy. Thin Solid Films **455–456**, 376–379 (2004)
44. C.C. Lee, K. Wu, S.H. Chen, S.J. Ma, Optical monitoring and real time admittance loci calculation through polarization interferometer. Opt. Exp. **15**, 17536–17541 (2007)
45. B. Kimbrough, J. Millerd, J. Wyant, J. Hayes, Low coherence vibration insensitive Fizeau interferometer. Proc. SPIE **6292**, 62920F (2006)
46. Y.R. Chen, Monitoring of film growth by admittance diagram. Master Thesis, National Central University, Taiwan, 2004
47. B.J. Chun, C.K. Hwangbo, J.S. Kim, Optical monitoring of nonquarterwave layers of dielectric multilayer filters using optical admittance. Opt. Express **14**, 2473–2480 (2006)
48. C.-C. Lee, Y.J. Chen, Multilayer coatings monitoring using admittance diagram. Opt. Express **16**(9), 6119–6124 (2008)
49. S. Wilbrandt, N. Kaiser, O. Stenzel, In-situ broadband monitoring of heterogeneous optical coatings. Thin Solid Films **502**, 153–157 (2005)

50. B. Badoil, F. Lemarchand, M. Cathelinaud, M. Lequime, Interest of broadband optical monitoring for thin-film filter manufacturing. Appl. Opt. **46**, 4294–4303 (2007)
51. W. Kai, C.-C. Lee, T.-L. Ni, Advanced broadband monitoring for thin film deposition through equivalent optical admittance loci observation. Opt. Express **20**, 3883–3889 (2012)
52. C.C. Lee, K. Wu, T.L. Ni, *Optical Admittance Loci Monitoring for Thin Film Deposition* (Lambert Academic, Saarbrücken, 2012) (ISBN 978-3-659-00198-7)
53. C.-C. Lee, W. Kai, M.-Y. Ho, Reflection coefficient monitoring for optical interference coating depositions. Opt. Lett. **38**, 1325–1327 (2013)
54. A.V. Tikhonravov, T.V. Amotchkina, in *Optical Thin Film and Coatings*, ed. by A. Piegariand, F. Flory. Optical monitoring strategies for optical coating manufacturing, Chap. 3 (Woodhead, Cambridge, 2013). (ISBN 978-0-85709-594-7) (2013)
55. F. Flory, Y.J. Chen, C.C. Lee, L. Escoubas, J.J. Simon, P. Torchio, J. Le Rouzo, Optical properties of dielectric thin films including quantum dots. Appl. Opt. **50**, C129–C134 (2011)
56. F. Flory, L. Escoubas, G. Berginc, optical properties of nanostructured materials a review. Nanophotonics **5**(1), 052502 (2011). doi:10.1117/1.3609266
57. A. Sytchkova, in *Optical Thin Film and Coatings*, ed. by A. Piegariand, F. Flory. Complex materials with plasmonic effects for optical thin film application, Chap. 5 and F. Flory, Y.J. Chen, H.L. Lin, Optical thin films containing quantum dots, Chap. 12 (Woodhead, Cambridge, 2013). (ISBN 978-0-85709-594-7)

Chapter 2
Metamaterials and Transformation Optics

Pi-Gang Luan

2.1 Introduction

Metamaterial is not a well-defined terminology. In fact, this terminology does not mean any specific material, but instead a new way of thinking. Usually this term means carefully engineered material structures composed of carefully designed inclusions that can exhibit unusual electromagnetic properties not inherent in the individual constituent components [1]. These properties include, for example, artificial magnetism [2], negative permeability [3], negative index of refraction [4], and hyperbolic dispersion [5, 6]. These properties lead to many fascinating phenomena such as negative refraction of light [7], sub-wavelength imaging [8], field enhancement [9], and evanescent-to-propagating wave mode conversion [5, 6], etc. Novel devices such as superlens [8–10], hyperlens [5, 6], invisibility cloak [11, 12], and plasmonic waveguide [13] based on these ideas have been designed, fabricated, and tested in the last decade.

The original purpose of initiating metamaterials research might be to construct artificial structures that can respond to electromagnetic waves intensively at any frequency we desire [2]. After this goal had been achieved, researchers then made a metallic periodic structure which behaves like an effective medium for EM waves having both negative permittivity [14] and negative permeability [3], and hence has a negative refractive index [4], realizing the 'bending light to the wrong way' phenomena predicted by V. G. Veselago more than 4 decades ago [7]. Furthermore, Pendry pointed out that a slab of negative refraction medium having appropriate negative permittivity and permeability not only cancels the phase accumulation of propagating waves but also amplifies the evanescent

P.-G. Luan (✉)
Wave Physics and Wave Engineering Lab/Department of Optics and Photonics,
National Central University, 300 Chung-Da Rd, Chung-Li 32001, Taiwan
e-mail: pgluan@dop.ncu.edu.tw

C.-C. Lee (ed.), *The Current Trends of Optics and Photonics*,
Topics in Applied Physics 129, DOI 10.1007/978-94-017-9392-6_2

waves, thus can focus the light from a tiny source into a spot narrower than one wavelength, overcoming the diffraction limit [8, 9]. This 'supelensing effect' stems from the coupling between the source's near fields (evanescent waves) and the surface plasmon-polariton (SPP) waves propagating along the slab boundaries. Pendry also argued that even if the relative permeability is nonnegative (close to 1), subwavelength imaging can still occur if the 'quasi-static condition' is satisfied [10].

After these concepts and predictions had been studied at the early stage, a tremendous amount of theoretical/numerical works have been done and a lot of experimental tests have been implemented. These developments finally helped to confirm the reality of negative refraction and subwavelength imaging phenomena [7–9] in the microwave frequency regimes. Recent studies further convinced that these notions still hold for optical waves [15]. However, some subtleties of metamaterials unnoticed before have been revealed [16–18]. For example, although a superlens can focus the light of a tiny source into a subwavelength spot; the image is located at the near filed zone and cannot be further processed by conventional optical devices. The efforts to resolve this problem then led to the development of hyperlens [5, 6].

In addition to the above mentioned developments, there are many related research works, which include: negative refraction and subwavelength imaging of EM and acoustic waves by photonic crystal slab [19–28] and sonic crystal slab [29–32], negative refraction and subwavelength imaging of water waves [33, 34], acoustic metamaterials [35–37] and acoustic cloaking devices [38, 39], matter wave cloaks [40, 41], and plasmonic devices [42].

In this chapter we introduce some important topics in the metamaterials research and explain the essential physics related to them. However, we are not able to discuss all topics of this research area because it covers phenomena of too wide range and is evolving too fast. We provide very detailed discussions about wire array and split-ring resonator array structures in this chapter and derive the effective permittivity and permeability formulas for them, which might be helpful to a beginner.

2.2 Negative Refraction, Flat Lens, and Perfect Lens

A. Negative refraction and left handed media

Now we discuss the refraction behavior of a beam of light incident from an empty space to a left-handed medium (LHM) or double-negative medium (DNM). By DNM we mean the medium's permittivity ε as well as its permeability μ are negative. In Fig. 2.1, the medium above the interface (which is represented by a black horizontal line) is an LHM, and the medium below the interface is an empty space. The red arrows, white cross circles, and the orange colored arrows indicate the directions of the electric field $\mathbf{E}$, magnetic field $\mathbf{H}$, and Poynting vector $\mathbf{S} = \mathbf{E} \times \mathbf{H}$. For simplicity

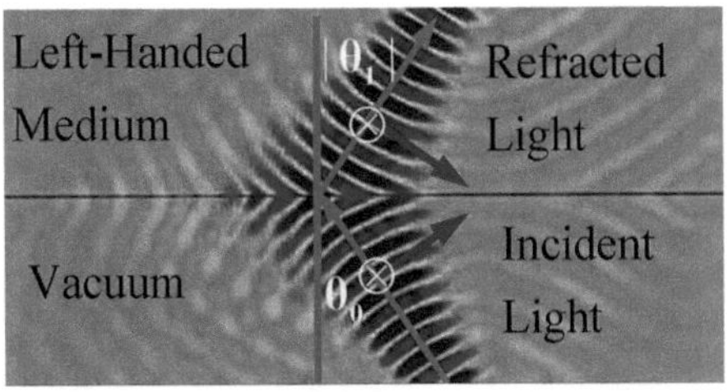

Fig. 2.1 Negative refraction. Above the *black line* is the left-handed medium with $\varepsilon < 0$ and $\mu < 0$

we assume that $\varepsilon \approx -\varepsilon_0$ and $\mu \approx -\mu_0$ so that the impedance of the LHM is matched to that of the empty space, and the reflected waves can thus be neglected. According to Maxwell's equations, four boundary conditions (not all of them independent) should be satisfied at the interface, they are: the continuity of the normal components of the electric displacement field (D_n) and magnetic induction (B_n); and the continuity of the tangential components of the electric field (E_t) and magnetic field (H_t).

Set the orientations of the x, y, z axes to be along the rightward, upward, and outward directions of the page respectively, and choose the polarizations of the **E** and **H** fields as that shown in the empty space region of Fig. 2.1 (TM polarization), then the incident light has $E_x^{inc} > 0$, $D_y^{inc} = \varepsilon_0 E_y^{inc} > 0$, $H_z^{inc} < 0$, and $S_x^{inc} < 0$, $S_y^{inc} > 0$. Using the boundary conditions for D_n, B_n, E_t and H_t, we conclude that the refracted light has $E_x^{ref} = E_x^{inc} > 0$, $D_y^{ref} = D_y^{inc} > 0$, and $H_z^{ref} = H_z^{inc} < 0$. However, since the LHM has $\varepsilon < 0$, so $D_y^{ref} = D_y^{inc} > 0$ implies $E_y^{ref} = D_y^{ref}/\varepsilon < 0$, and thus $S_x^{ref} > 0$, $S_y^{ref} > 0$, as indicating in the LHM region of Fig. 2.1. Now, if we apply the Snell's law to the incident and refracted beams, a negative refractive index ($n < 0$) must be assigned to the LHM because the refracted beam bends to the 'wrong way'. This argument explains why a DNM is a negative index medium (NIM), and similar argument can be applied to the TE polarization case. Furthermore, since the wave phase at the interface must be continuously connected, the wave front (the phase front) in the LHM region must propagate towards the interface, that is, the wave vector **k** is antiparallel to the Poynting vector **S**. Remember that in a usual medium with positive permittivity and positive permeability the **E**, **H**, and **k** triplet forms a right-handed coordinate system. However, the same triplet in a DNM forms a left-handed coordinate system. This explains why a DNM is called a LHM [7].

B. LHM and metamaterials

Experimentally the LHM can be realized by fabricating artificial structures called metamaterials (see Fig. 2.2). In general metamaterils are not naturally exsiting materials or their simple mixtures or chemical products, instead, they are carefully designed structures such like thin metallic wires array, split-ring resonators (SRR) array, metal-dielectric multilayers [5, 6], fishnet structures [43], and helical metallic resonators array [44] etc. We will discuss in this chapter about how to realize negative permittivity and permeability by using wires and SRR arrays, and how to realize hyperbolic metamaterial [45] or indefinite medium [46] using metal-dielectric multilayers.

Fig. 2.2 Two typical metamaterials consisting of metallic structures, both have negative refractive index in a certain frequency range. The metallic wires or rods provide the negative permittivity and the split-ring resonators provide the negative permeability. (Images: Physics Today, May 2000, 17; Science Vol. 292: 77, 2001)

In general metamaterials are dispersive media, which means the response of a metamaterial medium to the applied fields is frequency dependent. Usually they are also lossy or absorptive, which will degrade the NIM properties seriously. In fact, in the early stage of metamaterial research some researchers argued that due to the dispersive and absorptive properties, the negative refraction phenomenon in a metamaterial violates causality because it implies the possibility of superluminal propagation of signals. Besides, they even claimed that although the continuous waves of single frequency can deflect negatively, wave packet or beats always propagate positively [47]. However, after more careful analysis finally these arguments were found to be incorrect and thus these claims have been denied [48–50].

A simulation about beam propagation through a slab of LHM is shown in Fig. 2.3. The field strengths at 6 different times are revealed in the subplot (a) to (f), respectively. Comparing (b) with (c) or (e) with (f), one can find that the beam

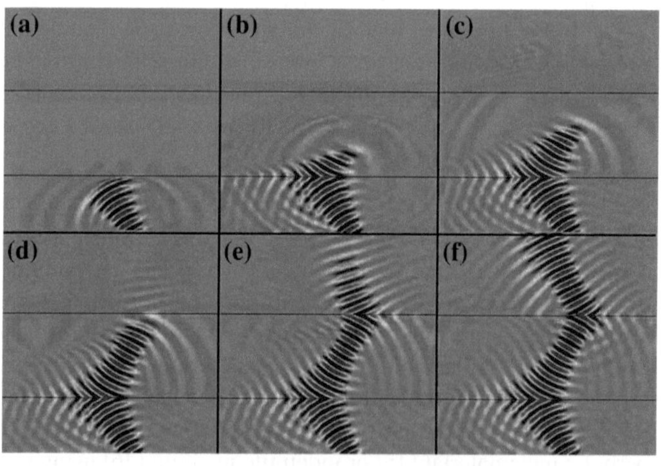

Fig. 2.3 A Gaussian beam penertrates through an LHM slab that has a refractive index close to -1 ($n \approx -1$). **a–f** are the results of 6 successive times

propagating direction changes a little during the time interval between two successive instants. This fact implies that the LHM is a dispersive medium so the beam cannot keep its propagation direction unchanged.

C. Electromagnetic energy density in metamaterials

There is another interesting issue concerning the LHM, that is, the electromagnetic energy density inside the metamaterials. We learned in the electrodynamics course that the electromagnetic energy density of time varying electromagnetic fields in a nondispersive medium is written as [51]

$$W = \frac{1}{2}\varepsilon E^2 + \frac{1}{2}\mu H^2 \tag{2.1}$$

Here E^2 and H^2 are the square of the instantaneouse **E** and **H** fields. Now if the permittivity and permeability become negative, what will be the expression of energy density? The above formula seems to imply the energy density becomes negative in such a medium, however, this is incorrect. In fact, for a dispersive medium without loss, Brillouin [52] and Landau [53] have already provided the following formula for the time average of the energy density (here **E** and **H** are the complex vector representation of the electric and magnetic field, respectively)

$$\langle W \rangle = \frac{1}{4}\varepsilon_0 \frac{\partial(\omega\varepsilon_r)}{\partial\omega}|\mathbf{E}|^2 + \frac{1}{4}\mu_0 \frac{\partial(\omega\mu_r)}{\partial\omega}|\mathbf{H}|^2 \tag{2.2}$$

which, if substituted the effective (relative) permittivity and permeability of the wire-SRR medium (referring to (2.12) and (2.26)), assuming the loss has been turned off, we get a positive result. For the cases with finite loss and the specific expressions for the electric and magnetic dipoles inside the medium are given, instantaneous energy density can also be derived, which is always positive and include three parts: the pure EM energy, the electric dipole energy, and the magnetic dipole energy [54]. In the latter two parts, the dipole energies may contain both the electric energy stored in the capacitors as well as magnetic energy stored in the inductances of the resonators that consist the medium.

D. Flat lens and perfect lens

A very interesting consequence of negative refraction is that a slab of NIM is in fact a flat lens (see Fig. 2.4), which can focus the light emanating from a point source to two images, one inside and one outside of the slab, if the slab is thick enough [7]. This phenomenon can be understood in two ways, based on the languages of geometrical optics or wave optics. For simplicity we assume the refractive indices for the lens and the medium outside are -1 and 1, respectively. According to geometrical optics, light rays refract negatively across the interface between the positive and negative index materials. Since the flat lens has two interfaces, negative refraction happens twice, and the intersection points of different rays yield the two images. On the other hand, a focal point according to wave optics is a point having stationary wave phase. Since the regions inside and outside the slab lens have refractive indices -1 and 1, the two image points are just the two points having the same phase as that of

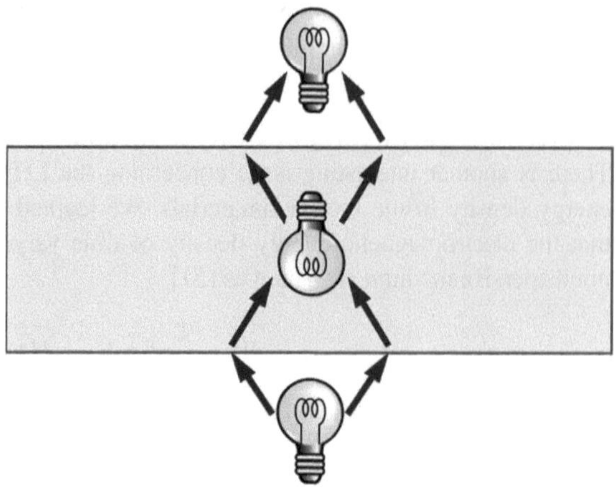

Fig. 2.4 Flat lens made of a slab of LHM. It can focus the light from a point source

the source point. Light propagating outside the slab accumulates positive phase increment along the propagation direction, while negative phase difference is accumulated inside. The two images locate at the two points where the positive and negative phase accumulations cancel with each other completely.

Although the possibility of making a flat lens from a NIM had already been found by Veselogo in 1967, people did not know before Pendry claimed that such a flat lens is a 'perfect lens', and it can be used to break the diffraction limit [8]. According to the traditional concepts of optics, light can be focued into a small spot, but the spot size (spot width) cannot be made much smaller than the wavelength. Such a restriction comes from the wave nature of light, and is sometimes called 'diffraction limit'. In 2000, Pendry studied the EM properties of a slab made of a LHM having $\varepsilon_r = \mu_r = -1$, and found that such a slab has $n = -1$ and it not only can cancel the phases of propagating waves, but also can amplify the amplitudes of evanescent waves. Here $\varepsilon_r = \varepsilon/\varepsilon_0$ and $\mu_r = \mu/\mu_0$ are the relative permittivity and relative permeability. This finding is astonishing because it implies that it is possible to achieve perfect imaging using such a flat lens. According to Pendry's analysis, diffraction limit is mainly due to the fact that the information encoded in the evanescent waves of the light source is lost in the imaging process because of its exponential decay characteristic. However, the evanescent waves from the source can be amplified by the slab so they can contribute to the image. If the propagating and evanescent waves can be added properly at the image plane without lossing any information, which Pendry claimed an $n = \varepsilon_r = \mu_r = -1$ slab in an empty space can do, then perfect imaging will happen. Notice that the choice of $\varepsilon_r = \mu_r$ is for the purpose of impedance matching so there would be no reflection and no information can be lost, whereas $n = -1$ is for perfect phase cancellation of the propagating waves and perfect amplitude compensation of the evanescent waves.

The amplification of evanescent waves is relying on the mechanism of exciting the surface plasmon polaritons (SPPs). The condition $\varepsilon_r \approx \mu_r \approx -1$ in fact implies that the effective surface plasmon-polaritons (SPP) corresponding to both the electric polarization resonance and the magnetization resonance have been excited efficiently by the evanescent fields from the source. Pendry further argued that even if the permeability is nonnegative, if $\varepsilon_r \approx -1$ is satisfied, subwavelength imaging can still occur if the slab thickness is much smaller than the wavelength (quasi-static condition). This kind of flat lens using only negative permittivity has already been fabricated and tested [10]. A flat lens having the ability of focusing light into a subwavelength spot no matter it uses perfect LHM or not is now called a 'superlens' [55].

The physics of surface plasmon can be understood as follows. In a metallic material of high conductivity such as silver or gold the conduction electrons or free electrons inside the material can move a long distance without being scattered. If we apply sinusoidally varying (harmonic) electromagnetic fields to the material, the free electrons would oscillate with the applied fields and move freely before being scattered by the phonons, disloacations, and defects, etc. However, if these electrons were moving close to the boundary of the metal, they would not be able to move outside far from the boundary and escape from the metal if they have no enough energy to overcome the 'work function'. Note that inside the metal these electrons can move freely because the attractive forces due to the positive ions form all directions cancel with each other. However, once an electron move to the boundary, the positive ions appear only on one side (the interior side) so a net 'restoring force' acts on it, forbiding the escaping of the electron from the metal. Note that the restoring forces acting on these electrons work just like many springs connected with them, which provide a resonance mechanism for forming SPPs. In the case of LHM slab the metal is replaced by a metamaterial having negative permittivity and negative permeability, and the true plasmon waves are replaced by the effective plasmon waves caused by the resonances of electric polarization and magnetization resonance of the medium.

No matter how great a perfect lens sounds like, any realization of such an ideal lens suffers from a number of restrictions due to the materials used or the structures being chosen. The first limitation comes from loss. A real material must absorb a part of the light energy, this effect limits the amplification of the evanescent waves, degrades the coherence of the light, and distortes the field distribution. All of them restrict the possibility of further reducing the spot size of the image. Another restriction is from the space period or lattice constant of the metamaterial. Such a characteristic length plays the role of cutoff length, and which implies that it is impossible to make an image narrower than the lattice constant [17]. In fact, in the early stage of perfect lens study, some researchers argued that the idealest perfect lens operating at the condition of $\varepsilon = \mu = -1$ is physically impossible because the energy stored in such a lens would be infinite, or some boundary conditions such as the continuity condition of wave phase would not be possible to satisfy [16]. This kind of controversies have never stopped but all of them are not directly related to the practical issues concerning applications [18]. Now we know that in practice the not-so-perfect negative refaction and subwavelength imaging are indeed realizable when properly designed metamaterials are used in making the lens (see Fig. 2.5).

Besides the above mentioned restrictions, people also found that the original designs of metamaterials which work well in the microwave frequency regime

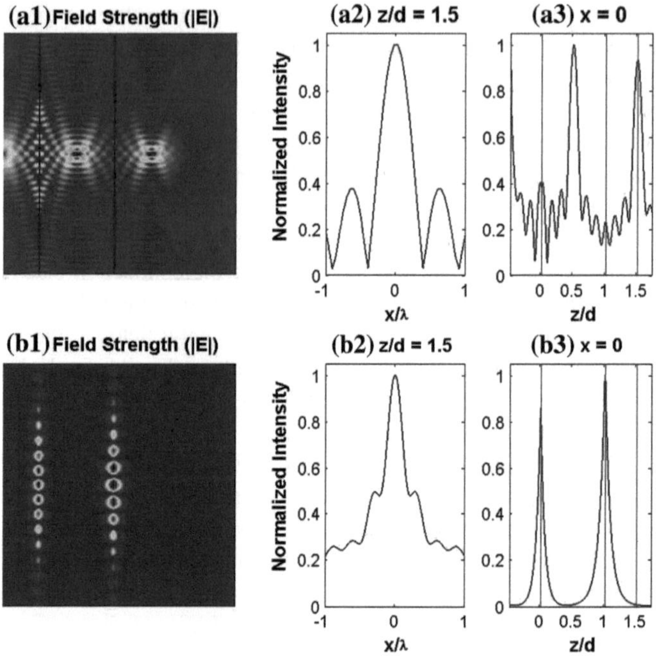

Fig. 2.5 Field strengths (|E|) and images made by a LHM flat lens. **a1**–**a3** show the case that the image size is within the diffraction limit, whereas **b1**–**b3** show the case of subwavelength imaging. The two boundaries of the lens are parallel to the x (vertical) axis, and the horizontal axis is the z axis. **a2** and **b2** are the (normalized) field strengths at the image plane. **a3** and **b3** are the (normalized) field strengths at the YZ plane

cannot do as well in the much higher frequency regime because the original circuit models for deriving the effective permittivity and permeability should be modified. For example, the resonance effect of SRRs becomes weaker at higher frequency due to electron's nonzero inertial mass, so that the desired negative permeability property becomes more difficult to achieve [56]. Thus for realizing negative refraction at optical or visible frequency regime some different structures must be used [57].

2.3 Photonic Crystals and Subwavelength Imaging

Negative refraction and subwavelength imaging can also be observed in photonic crystals (PhCs) made of nondispersive dielectric materials (see Fig. 2.6). The beam propagation direction inside a PhC is determined from the equal-frequency-surface (EFS) or constant frequency curve of the PhC [19–28]. The group velocity of the beam is along the direction normal to the constant frequency curve.

There are two kinds of slightly different mechanisms being used to achieve the unusual refraction/deflection and subwavelegth imaging phenomena. The first

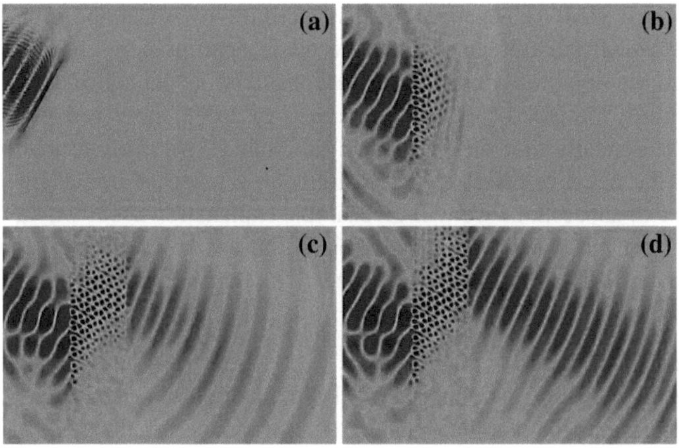

Fig. 2.6 A Gaussian beam penertrates through an slab of photonic crystal which has a negative effective refractive index. **a–d** are the results of 4 successive times

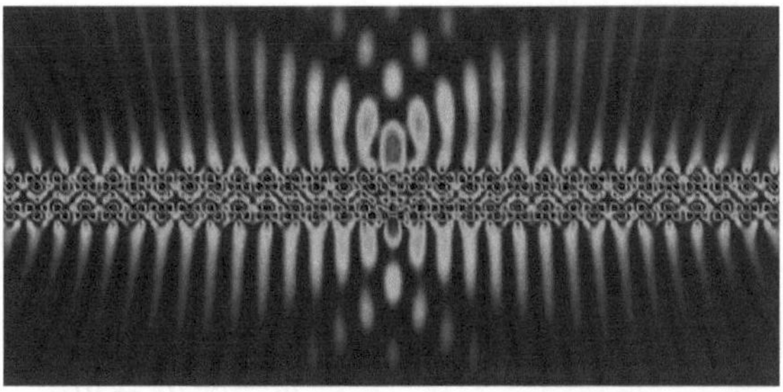

Fig. 2.7 Subwavelength imaging using photonic crystal slab

is the negative group index usually happening in the PhC with triangular lattice structures, the other is called all-angle negative refraction or canalization effect commonly encountered in the PhC with square lattice structures [20]. Deep subwavelength imaging is more easily to be achieved by using the canalization effect but the image always appears at the near field zone close to the PhC surface (see Fig. 2.7). However, negative refraction phenomenon is more easily to be observed using the negative group index mechanism [21]. Interested readers can learn the detailed knowledge from the textbook by Joannopoulos et al. [58].

2.4 Resonance, Constraints, and Metamaterials

The first experimental realization of DNM is a metallic structure consisted of periodically arranged metallic wires and split-ring resonators (SRRs) [3, 4]. In 1996, Pendry et al. [14] found that an array of thin wires behaves like an artificial plasma, which yields a negative effective permittivity under the influence of an incident wave, if the frequency of the wave is appropriately chosen. This finding opened the doors of constructing these artificial media called 'metamaterials' using carefully designed metallic structures. In 1999, these researchers further showed theoretically that the periodically arranged Swiss roll structures or SRR array provide negative effective permeability in a range of frequency [2]. In the following two subsections we will give simple explanation about why a wire-array is an effective plasma medium, and then in the next subsection we will further argue why an SRR array can provide negative effective permeability in a certain range of frequency.

A. Realization of negative permittivity

We first review briefly the concept of plasma medium. For simplicity we define plasma medium as the medium containing a lot number of charges that can move long distances without scattering. Many metals (for example, gold and silver) can be treated as plasma media at appropriate frequencies. The simplest model for describing the electromagnetic behaviors of plasma is the Drude model. According to this model, the free charges (electrons) are accelerated by the applied external (electric) field and move freely until being scattered by the defects, dislocations, and phonons etc. inside the medium (the metal). The scattering causes the relaxation of the kinetic energy of these charges, and the generation of heat. In Drude model, every free charge is assumed to be moving independently, and the energy relaxation process is described effectively by including a phenomenological damping force in the equation of motion.

For simplicity, we assume the applied electric field is along the x axis, and each charge is q. The equation of motion is written as

$$m\frac{dv}{dt} + bv = qE \tag{2.3}$$

Here $v = \frac{dx}{dt}$ is the velocity of the charge, $-bv$ is the damping force, x is the displacement of the charge. Under the condition of harmonic (single frequency) applied field, we adopt the complex representation (phasor representation) of the physical quantities and assume the time factor being $e^{i\omega t}$, that is, $x = x_0 e^{i\omega t}$ and $E = E_0 e^{i\omega t}$, we can find the solution

$$x = -\frac{qE}{m}\left(\frac{1}{\omega^2 - i\Gamma\omega}\right) \tag{2.4}$$

where $\Gamma = b/m$ is a dissipation coefficient, corresponding to the damping force.

We now derive the relative permittivity $\varepsilon(\omega)$. The constitutive relation between the E and D fields is

$$D = \varepsilon_0 E + P = \varepsilon_0 \varepsilon(\omega) E \tag{2.5}$$

were P is the polarization field. For the sake of simplicity, here we have dropped the subscript r of the relative permittivity notation. Assume that the concentration of the free charge is N, we get $P = Nqx$, here qx is the electric dipole of a single charge. From (2.4) and (2.5) we find

$$\varepsilon(\omega) = 1 + \frac{P}{\varepsilon_0 E} = 1 - \frac{\omega_p^2}{\omega(\omega - i\Gamma)} \tag{2.6}$$

where ω_p is the plasma frequency, defined by

$$\omega_p = \sqrt{\frac{Nq^2}{m\varepsilon_0}}. \tag{2.7}$$

When Γ is negligible, we have $\varepsilon(\omega) < 0$ if $\omega < \omega_p$. That is, for electromagnetic wave with $\omega < \omega_p$, the plasma medium has negative permittivity.

Now we turn to the discussion of wires array medium. For simplicity we consider only the two dimensional case. We assume the wires are located at the lattice points of a square lattice, and the applied field is parallel to the wires (along the direction of the z axis). Denote the lattice constant as a, the radius of the wires as r, and assume the charge concentration inside the wires is N. The array is in fact a metallic photonic crystal (MPC), thus the electromagnetic behaviors of this structure can be understood from the results of its photonic band structure (PBS). However, in 1996, Pendry et al. found that if the wavelength of the applied field is much longer than the lattice constant, then an effective medium theory for this MPC can be constructed without referring to the PBS. Here we show how to derive the relative permittivity of this effective medium. Before we derive this result, first note the following differences between the 'true plasma medium' and the wire array structure: 1. The charges are confined inside the wires, and the current caused by the charge motion can only flow along the z direction, 2. The cross section area πr^2 of each wire is much smaller than the unit cell area a^2 of the lattice, thus the effective concentration $N_{eff} = \pi r^2 N/a^2$ is also much smaller than the true concentration N inside the wires, 3. Every wire has a non-negligible self-inductance because there is an azimuthal directed magnetic field around the wire, and this magnetic field is not appearing in the original Drude model. Pendry argued in [14] that the vector potential of this magnetic field provides a modification to the charge's canonical momentum, and if we treat the canonical momentum as a kinetic momentum without vector potential, than the effective mass of the electron should be redefined as a function of a and r. Here we provide an alternative derivation of Pendry's result as follows.

Consider a segment of a wire, which has length l, self-inductance L, and resistance R. Apply the harmonic E field to the wire, than the potential drop along this segment is $V = El$. Suppose the current in the wire is I, than we have the circuit equation

$$L\frac{dI}{dt} + RI = El. \tag{2.8}$$

Simple observation indicates that the (2.8) can be cast into the same form as (2.3)

$$m_{eff}\frac{dv}{dt} + b_{eff}v = qE. \tag{2.9}$$

provided that we define the effective mass m_{eff} and damping coefficient b_{eff} as

$$m_{eff} = \frac{qLI}{lv}, \quad b_{eff} = \frac{qRI}{lv}. \tag{2.10}$$

The drift velocity v of the charges is related to the current I through $I = \pi r^2(Nqv)$, so

$$m_{eff} = \frac{N\pi r^2 L}{l}q^2, \; b_{eff} = \frac{N\pi r^2 R}{l}q^2. \tag{2.11}$$

From (2.9) we find the formula of relative permittivity

$$\varepsilon_{eff}(\omega) = 1 - \frac{\omega_p^2}{\omega(\omega - i\Gamma)} \tag{2.12}$$

where the plasma frequency ω_p and dissipation coefficient Γ are given by

$$\omega_p = \sqrt{\frac{N_{eff}q^2}{m_{eff}\varepsilon_0}} = \frac{1}{a}\sqrt{\frac{l}{L\varepsilon_0}}, \quad \Gamma = \frac{R}{L}. \tag{2.13}$$

The permittivity formula in (2.12) is indeed the same form as that of the plasma media in (2.6). We can further calculate approximately the self-inductance $L = \Phi/I$ of the wire and rewrite the effective mass m_{eff} and plasma frequency ω_p using geometric parameters. To estimate the magnetic flux Φ around a wire, we ignore the contribution from neighboring wires and that inside the wire, and calculate the magnetic flux through the rectangle defined by the conditions (using cylindrical coordinates): $z = 0$ to $z = l$ and $R = r$ to $R = a$

$$L = \frac{\Phi}{I} = \frac{\mu_0}{I}\int_r^a HldR = \frac{\mu_0 l}{I}\int_r^a \frac{I}{2\pi R}dR = \frac{\mu_0 l}{2\pi}\ln\left(\frac{a}{r}\right). \tag{2.14}$$

According to (2.11), this leads to

$$m_{eff} = \frac{\mu_0 N r^2 q^2}{2}\ln\left(\frac{a}{r}\right), \quad \omega_p = \sqrt{\frac{2\pi c^2}{a^2\ln\left(\frac{a}{r}\right)}}. \tag{2.15}$$

These results are the same as those obtained in [14].

In fact, from (2.8) and (2.9) the following energy relations can be derived by multiplying them with I and v, respectively

$$\frac{d}{dt}\left(\frac{LI^2}{2}\right) + RI^2 = IV, \quad \frac{d}{dt}\left(\frac{m_{eff}v^2}{2}\right) + bv^2 = qEv \tag{2.16}$$

Note that the magnetic energy $LI^2/2$ is stored in the magnetic field around a wire, and its value is the same as $N\pi r^2 l m_{eff} v^2/2$, which is the effective kinetic energy of the charges inside the volume of a unit cell la^2:

$$N\pi r^2 l \frac{m_{eff} v^2}{2} = \frac{LI^2}{2} \qquad (2.17)$$

This equality implies that in the effective theory of wire-array medium the magnetic energy around the wire is identified with the kinetic energy carried by the massive charged particles in the corresponding Drude model (see (2.9)). This identification transforms the wire array structure to the effective plasma medium described by the equations of motion in the Drude model (see (2.3) and (2.9)).

B. Realization of negative permeability

We have shown in the previous subsection that the free moving charges inside the metal provide a means to realize the negative permittivity of the medium effectively. However, such a mechanism is not possible for realizing negative permeability because there has no magnetic monopole been found yet. One thus should use different method to make the effective permeability negative. From the constitutive relation

$$\mathbf{B} = \mu_0(\mathbf{H} + \mathbf{M}) = \mu_0 \mu(\omega)\mathbf{H} \qquad (2.18)$$

which relates the **B** field (magnetic induction), **H** field (magnetic field) and **M** field (magnetization), one can imagine that if resonance effect can be used to make $\mu_0 M > B$, then the directions of **B** and **H** become opposite to each other, and the desired $\mu(\omega) < 0$ result can be realized. Since magnetization **M** is defined as the magnetic dipole moment per unit volume, and a magnetic dipole is physically a current loop, thus negative permeability can be realized in principle by an array of resonant current loops. The array of split-ring resonators (SRR array) is the most studied structure for realizing the negative permeability experimentally [3, 4]. If a harmonic **H** field perpendicular to the ring plane is applied to the SRR array, alternating current can be induced in each ring if the frequency is properly chosen. For each ring the split gaps of the ring stop the currents and help to accumulate the positive and negative charges on the two sides of each gap, thus they provide the capacitance C, whereas the currents circulating in the ring generate magnetic field and hence provide the inductance L [2]. Therefore, each SRR is an LC resonator having resonance frequency $\omega_0 = \frac{1}{\sqrt{LC}}$. When the frequency ω of the applied **H** field approaches the resonance frequency ω_0, the resonance of currents occurs, and the condition $\mu_0 M > B$ can be satisfied. This leads to the negative effective permeability.

We now derive the effective relative permeability $\mu(\omega)$[2]. The details may refer to [59]. Consider the SRR structures shown in Fig. 2.8. A number of SRRs are piled up in the y direction to form an SRR stack, which can be viewed as a solenoid. These SRR stacks are periodically arranged in the xz plane at a square lattice of lattice constant a. The y spacing between two successive SRRs in one stack is l. This array structure can be viewed as solenoids array.

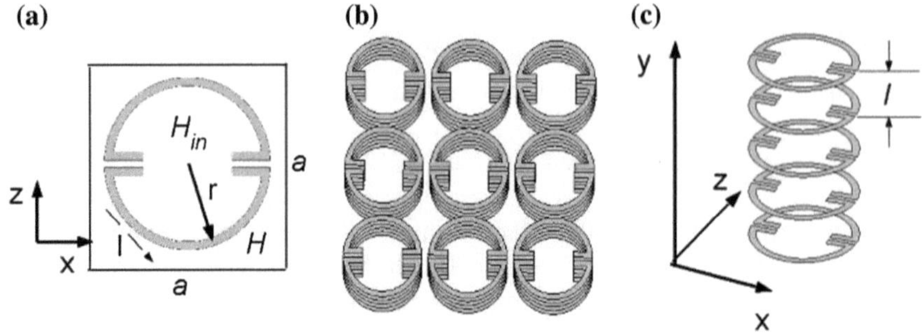

Fig. 2.8 Array of split-ring resonators (SRRs). **a** A single SRR in a unit cell. **b** Array of SRR stacks. **c** An SRR stack formed by several SRRs arranged along the y direction

Now we apply an external harmonic H field of y direction, strength H_0 to this SRR-stacks array. According to Faraday's Law, a current I is induced in each ring, thus a ring acquires a magnetic dipole moment $m = I\pi r^2$, here r is the radius of the ring. In a real structure, the number of rings in a stack is finite. Thus the field lines of the induced magnetic field which relate to the induced currents of the these rings 'spill out' from the terminals of the SRR stacks, contributing to the 'depolarization field'. If the wavelength of the incident H_0 field is much longer than the lattice constant a and the y-spacing l, we can approximate the magnetic fields inside and outside the solenoid as two homogeneous fields H_{in} and H_{out}, respectively. According to Ampere's Law, the difference of them is given by

$$H_{in} - H_{out} = \frac{I}{l}. \tag{2.19}$$

Approximately assume that the depolarization flux is homogeneously distributed in the entire xz plane, and remember that the total flux caused by the induced currents is zero (because each of these magnetic field lines is closed), we get

$$FH_{in} + (1 - F)H_{out} = H_0, \tag{2.20}$$

where $F = \pi r^2/a^2$ is the area filling fraction of a stack in a unit cell of the xz plane.

From (2.19) and (2.20) we find

$$H_{in} = H_0 + (1 - F)\frac{I}{l}, \quad H_{out} = H_0 - F\frac{I}{l}. \tag{2.21}$$

The magnetic induction B is defined as magnetic flux per unit area, therefore $B = \mu_0 H_0$. Besides, from (2.18) and the meaning of magnetization we have

$$H_0 = H + M, \quad M = \frac{m}{la^2} = F\frac{I}{l} \tag{2.22}$$

where M is the magnetization and la^2 is the volume occupied by a ring. Comparing (2.21) with (2.22), one can make the identification: $H = H_{out}$. That is, the magnetic field H in the effective medium is just the average magnetic field H_{out} in the 'connected region' outside the SRR stacks [2].

Now we calculate the magnetization M and derive the explicit expression of $\mu(\omega)$. The self-inductance $L = \mu_0 F a^2/l$ of an SRR in a chosen SRR-stack can be derived from the formula $L = \Phi_s/I$, where $\Phi_s = \pi r^2 \mu_0 I/l$ is the flux through the SRR. In addition to L, the SRR also has a resistance R, a capacitance C, and a mutual inductance $\mathcal{M} = -FL$ due to the depolarization fields spilled out from all the other SRR stacks in the array [2, 54, 59]. Under the influence of the H_0 field, the current I in the SRR satisfies (note that the charges stored in the capacitance of the SRR is $q = \int I\, dt$)

$$(1-F)L\frac{dI}{dt} + RI + \frac{q}{C} = -\frac{d}{dt}\left(Fa^2\mu_0 H_0\right), \tag{2.23}$$

Using the relation $LI = \mu_0 M a^2$ and (2.22), (2.23) becomes

$$\frac{dM}{dt} + \gamma M + \omega_0^2 \int M dt = -F\frac{dH}{dt}, \tag{2.24}$$

where the dissipation coefficient γ and the resonance frequency ω_0 are defined as

$$\gamma = \frac{R}{L},\ \omega_0 = \frac{1}{\sqrt{LC}}. \tag{2.25}$$

Since the applied field is a harmonic field with frequency ω, every dynamical quantity acquires the same time factor $e^{i\omega t}$. Thus from (2.18) and (2.24) we get the relative permeability:

$$\mu(\omega) = 1 - \frac{F\omega^2}{\omega^2 - \omega_0^2 - i\gamma\omega}. \tag{2.26}$$

This result is the relative permeability of the SRR medium. If γ is ignorable, $\mu(\omega)$ becomes negative in the frequency range $\omega_0 < \omega < \frac{\omega_0}{\sqrt{1-F}}$. Note that in this derivation we did not use the specific expression of the capacitance C [2, 59].

2.5 Indefinite Media/Hyperbolic Metamaterials and Hyperlens

We have shown before that negative permittivity and negative permeability can be realized by using periodically arranged wires and SRRs, respectively. Simple observation tells us that these structures do not provide isotropic permittivity and permeability automatically if we do not arrange the orientations of these inclusions properly. For example, if in the 2D wire array all the wires are parallel to the z axis, than the standard effective plasma behavior will appear only when the electric field is also parallel to the z axis. In fact, if the wires are really thin enough, and we apply an **E** field perpendicular to them, then the structure responds to the incident field just like the background medium (i.e., the empty space without these wires) does. Based on this consideration, it is not difficult to predict that the effective permittivity tensor for the array consisting of merely z-oriented wires has two different eigenvalues: $\varepsilon_z = \varepsilon_\parallel$ and $\varepsilon_x = \varepsilon_y = \varepsilon_\perp$, so the z-oriented wires array is an uniaxial anisotropic medium. In this medium $\varepsilon_\perp$ is in general positive but $\varepsilon_\parallel$ takes negative value if the frequency is lower than the effective plasma frequency (see (2.12)). When the permittivity tensor of a uniaxial medium gives negative determinant, this medium is called 'indefinite medium' or 'hyperbolic medium'. Metamaterials behave like hyperbolic media are called hyperbolic metamaterials. A multilayer structure made of one-dimensional photonic-crystal which contains one dielectric layer and one metal layer in one unit cell (one space period) also belongs to this category if the operating frequency is correctly chosen. In this chapter we will derive the effective permittivity of this structure and introduce the most important application of hyperbolic metamaterials: the hyperlens.

We have learned in the previous sections that a LHM has negative permittivity ($\varepsilon < 0$) and negative permeability ($\mu < 0$), leading to the negative refraction phenomena. Here we show that indefinite media or hyperbolic metamaterials with $\varepsilon_\perp < 0$ can also bend the light beam to the 'wrong way' and gives an apparent negative refractive index referring to Snell's law.

We refer to Fig. 2.1 again and consider TM polarized light, but now we replace the LHM region by an indefinite medium with $\varepsilon_\parallel > 0$ and $\varepsilon_\perp < 0$. Here $\varepsilon_\parallel$ and $\varepsilon_\perp$ are the permittivity along the directions parallel and perpendicular to the interface, respectively. When this TM beam penetrates through the indefinite medium, the horizontal components of the **E** and **H** fields do not change. The vertical component of the **D** field does not change too. However, since we have $\varepsilon_\perp < 0$, so the vertical component of the **E** field changes sign just like that in the LHM case. We thus conclude that the Poynting vector **S** (the light beam) bends negatively. Notice that in this argument the key point is the sign change of the vertical component of the **E** field, and this happens only in the TM wave case ($E_z = H_x = H_y = 0$) and does not apply to the TE wave case ($H_z = E_x = E_y = 0$). Furthermore, we stress here that the phase propagation direction (i.e., the direction of the **k** vector) in general is not along the direction of the beam direction (i.e., the direction of the Poynting vector **S**), so indefinite media are not LHM.

Now we derive the effective permittivity of the multilayer structure. Suppose the period of the structure is along the x direction, so the interface between the empty space and the structure is the yz plane. Consider a TM wave (P wave) incident from the empty space, and we choose the plane of incidence to be the xy plane, so the **H** field is parallel to the z axis. Let the permittivities of the metal and dielectric layers to be ε_m and ε_d, and their thicknesses are a_m and a_d, respectively, so the period or lattice constant is $a = a_m + a_d$. To satisfy the continuity condition of the wave phase, k_y must be the same in each layer. We denote the k_x in the two layers as k_m and k_d, and use K to represent the Bloch wave number. When the wavelength of the incident wave is much longer than the lattice constant, $\omega a/c$,

$k_m a_m$, $k_d a_d$ and Ka take very small values. Under this condition, starting from the dispersion relation of one-dimensional photonic crystals [60]

$$\cos Ka = \cos k_m a_m \cos k_d a_d - \frac{1}{2}\left(\frac{k_m \varepsilon_d}{k_d \varepsilon_m} + \frac{k_d \varepsilon_m}{k_m \varepsilon_d}\right) \sin k_m a_m \sin k_d a_d \quad (2.27)$$

and expand the cosine and sine terms up to second order, we get the effective dispersion relation

$$\frac{K^2}{\langle \varepsilon \rangle} + \left\langle \frac{1}{\varepsilon} \right\rangle k_y^2 = \frac{\omega^2}{c^2} \quad (2.28)$$

where

$$\langle \varepsilon \rangle = \frac{a_m \varepsilon_m + a_d \varepsilon_d}{a}, \left\langle \frac{1}{\varepsilon} \right\rangle = \frac{a_m/\varepsilon_m + a_d/\varepsilon_d}{a} \quad (2.29)$$

Now define $k_\parallel = k_y$, $k_\perp = K$, $\varepsilon_\parallel = \langle \varepsilon \rangle$, and $1/\varepsilon_\perp = \langle 1/\varepsilon \rangle$, we get the dispersion relation for this anisotropic effective medium

$$\frac{k_\perp^2}{\varepsilon_\parallel} + \frac{k_\parallel^2}{\varepsilon_\perp} = \frac{\omega^2}{c^2} \quad (2.30)$$

Choosing the filling fraction of the metal layer $f = a_m/a$ and frequency (remember that $\varepsilon_m = \varepsilon_m(\omega)$) properly, we can make $\varepsilon_\parallel > 0$, $\varepsilon_\perp = -|\varepsilon_\perp| < 0$, and thus the dispersion relation in (2.30) becomes the hyperbolic form

$$\frac{k_\perp^2}{\varepsilon_\parallel} - \frac{k_\parallel^2}{|\varepsilon_\perp|} = \frac{\omega^2}{c^2} \quad (2.31)$$

This dispersion relation implies that both the propagating waves with $|k_\parallel| \le \omega/c$ and the evanescent waves with $|k_\parallel| > \omega/c$ in an empty space can excite the propagating waves inside this indefinite medium, because for a finite ω/c, $k_\parallel$ and $k_\perp$ can take arbitrary large values.

We have explained in the previous sections that a slab of negative permittivity material such as a silver thin film behaves like a superlens, which means it can make a subwavelength image of a tiny light source. However, since the imaging mechanism of a slab lens relies on the evanescent waves, the image can appear only in the near field zone. Besides, the subwavelength image cannot be processed by conventional optical devices easily, thus a device that can make amplified image at the far field zone is desired. Hyperlens is indeed this kind of device to fulfill this requirement. A hyperlens is in fact a cylindrical device formed by curling up the multilayered structure of the dielectric-metal 1D photonic crystal (see Fig. 2.9). In a hyperlens, the $\varepsilon_\parallel$, $\varepsilon_\perp$, $k_\parallel$, and $k_\perp$ are replaced by ε_θ, ε_r, k_θ, and k_r, so we have $\varepsilon_\theta > 0$, $\varepsilon_r < 0$, and the dispersion relation (2.31) becomes $\frac{k_r^2}{\varepsilon_\theta} - \frac{k_\theta^2}{|\varepsilon_r|} = \frac{\omega^2}{c^2}$.

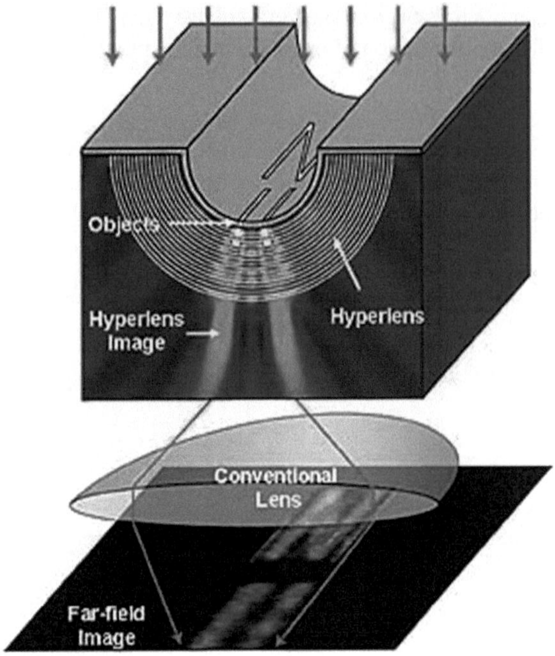

Fig. 2.9 Hyperlens and imaging through it. (Images courtesy of Zhang group, UC Berkeley)

A small light source close to the inner surface of the hyperlens if emitting TM polarized light, then its evanescent waves can be coupled into the hyperlens and transformed into propagating waves inside. However, unlike in the flat lens case, the propagating modes inside this cylindrical structure are the products of Bessel function $J_m(kr)$ and Neuman function $N_m(kr)$ with the phase factor $e^{im\theta}$, where m is the order of the Bessel/Neuman function and $k = \frac{\omega}{c}n$ is the wavenumber. The azimuthal component of the wave vector $k_\theta = \frac{2\pi}{\lambda_\theta} = \frac{2\pi}{(2\pi r/m)} \sim m/r$ reduces as the radius increases. If the outer radius is large enough so that $k_\theta(R) = m/R < \omega/c$, then the image would be made of propagating waves and can appear in the far field zone [5, 6].

2.6 Invisibility Cloak and Transformation Optics

Now we introduce the concept of invisibility cloak and transformation optics [11, 12]. An invisibility cloak by definition is a shell made of carefully designed materials, and an object hidden inside as well as the shell itself would not be observed from outside. Light incident upon the cloak would not be scattered, and it can only propagate along the shell and goes back to the free space without penetrating through the cavity region enclosed by the shell. After leaving the cloak the light would propagate along the same direction as the original incident light, and no shadow will be formed (see Fig. 2.10). Since light does not penetrate into the

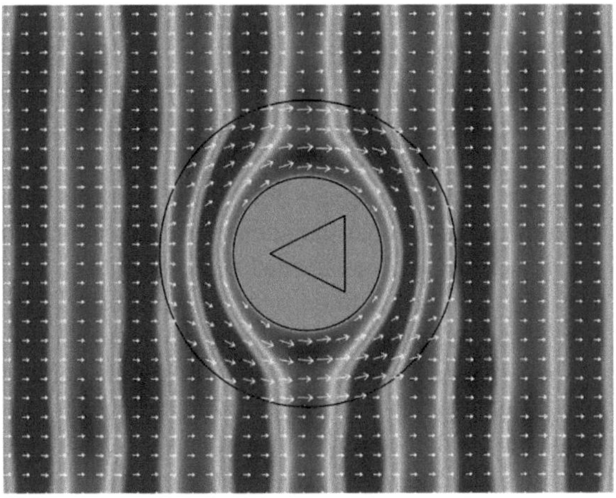

Fig. 2.10 Electromagnetic waves propagating around an invisibility cloak. The *white arrows* indicate the direction and magnitude of the Poynting vectors. The *triangle* represents the object being hidden inside the cavity region

cavity region, the objects hidden inside would not have any electromagnetic interaction with the outside world.

In optics we know that the refractive indice of media can influence the propagating direction of light. If the refractive index of a medium can be made a smooth function of position, then in principle a carefully arranged distribution of refractive index can guide light around a finite cavity region just like the desired invisibility cloak can do. However, this guiding would never be perfect and light scattering caused by internal reflection would make the cloak visible.

Is this fate unavoidable? To answer this question, remember that according to the electromagnetic theory of light, the refraction and reflaction behaviors of light are in fact influenced by the relative permittivity $\varepsilon_r = \varepsilon/\varepsilon_0$ and relative permeability $\mu_r = \mu/\mu_0$, or equivalently by the refractive index $n = \sqrt{\varepsilon_r \mu_r}$ and relative impedance $Z_r = \sqrt{\frac{\mu_r}{\varepsilon_r}}$. For usual optical media the magnetic response is weak so we can assume $\mu_r = 1$ and get $n = \sqrt{\varepsilon_r} = \frac{1}{Z_r}$. This reduces the independent material parameters from two to one and restricts the optical behaviors of the medium. In fact, the interface reflection of EM wave is caused by the impedance change, and not by the refractive index change. Thus if both the permittivity and permeability can be used, then it would be possible to guide light around an object without internal reflection.

To realize an invisibility cloak, John Pendry et al. proposed in 2006 a physically plausible scheme based on the idea of coordinate transformation [11]. In fact, it is known that the form of Maxwell's equations is invariant under a continuous and smooth coordinate transformation, although the expressions of the permittivity and permeability tensors after this transformation become anisotropic and inhomogeneous. That is, they are not simple constants or scalars but tensor densities [61]. After this early development, the same idea has been generalized to design other novel devices for controlling the light flows. Such kinds of researches are now categorized as 'transformation optics'(TO) [62, 63].

Now we explain why coordinate transformation can tell us how to design an invisibility cloak or other TO devices. We first choose a 'target virtual space' and define a coordinate transformation between the original space and this virtual one. The electromagnetic fields, light trajectories, and the media parameters (permittivity and permeability) would get their new expressions, directions, and values in this virtual space after the coordinate transformation. We then can treat the transformed media parameters as the true medium parameters in the original space and thus the transformed trajectories become the true trajectories of light in this original space if filled with the transformed medium. If the original space is empty, and we transform the space points inside a sphere of radius b to the shell region between a and b ($a < b$) without changing the points outside the sphere, then a light ray originally goes straight through the sphere would become curved inside the shell and around the cavity $r < a$, and goes back to its original direction after leaving the shell. Thus the shell filled with the transformed medium is nothing but the desired invisibility cloak (see Fig. 2.11).

Now we derive the material parameters for the spherical cloak. First note that in the inner region $r < b$, the empty space has material parameters $\varepsilon = \mu = 1$. Here ε and μ represent the relative permittivity and relative permeability, respectively. Now transform every point inside the sphere from (r, θ, φ) to (r', θ', φ') using the transformation:

$$r' = \left(\frac{b-a}{b}\right) r + a, \quad \theta' = \theta, \quad \varphi' = \varphi \tag{2.32}$$

A cavity of radius a is thus created, which corresponds the central point of the sphere. The shell region $a < r' < b$ is from the whole interior region inside the sphere, except the central point. The new material parameters can be calculated according to the formula [64]

$$\varepsilon' = \frac{J\varepsilon J^T}{\det J}, \quad \mu' = \frac{J\mu J^T}{\det J} \tag{2.33}$$

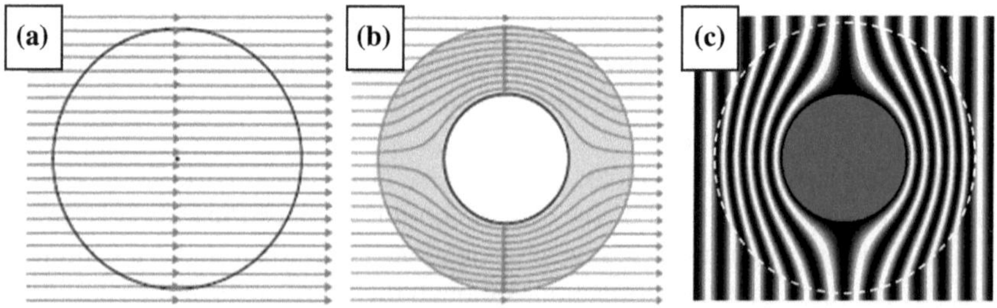

Fig. 2.11 Invisibility cloak, the ray trajectories, and the phase of the electromagnetic waves. **a** A spherical region of empty space. **b** and **c** represent the effect of a invisibility cloak designed according to coordinate transformation. The curves with arrow heads are the ray trajectories, and the *white* and *black colors* in **c** represent the wave crests and wave troughs. (Images **a**, **b**: Science 313, 1399 (2006))

which yields

$$\varepsilon_{r'} = \mu_{r'} = \frac{b}{b-a}\left(\frac{r'-a}{r'}\right)^2 \tag{2.34}$$

$$\varepsilon_{\theta'} = \mu_{\theta'} = \frac{b}{b-a} = \varepsilon_{\varphi'} = \mu_{\varphi'} \tag{2.35}$$

These arbitray material parameters in principle can be realized by using metamaterials. We have learned in the previous sections that in the long wavelength limit, periodic structures consisted of arrays of small electric and magnetic resonantors such like wires and SRRs can work like a homogeneous medium having certain unusual permittivity and permeability, no matter it takes positive, negative, close to zero, or very large values. We can replace the smooth TO devices by an artificial structure consisted of a large amount of 'metamaterial elements' and approximate the material parameters as continuous functions if the changes of these parameters from an element to its neighboring element are really small enough.

Experimental realization of a 2D cylindrical cloak was first demonstrated in 2006 by David R. Smith's group at Duke University [12]. They designed the 2D cloak for cloaking microwaves about 10 GHz, and their experiments confirmed that the cloak, though not perfect, can reduce the scattered wave dramatically. However, the parameters they used were not referring to the theoretical values directly from the coordinate transformation. The theoretical parameters were replaced by a set of reduced parameters, in which only the permeability tersor is varied along the radial direction, and the permittivity is a constant. A light ray propagating inside the shell region (now replaced by a ring region) would have the same trajectory as in the perfect cloak. However, reduced parameters do not satisfy the impedance matching conditions at the outer surface of the ring, so the scattered waves cannot be eliminated completely and thus the cloak is not perfect. In fact, designing transformation optics devices using metallic metamaterials with resonance property has some disadventages. First, resonance implies the loss. As we have mentioned before, the energy loss of the electromagnetic wave causes the modification of the strength and phase of the wave, and they lead to the degration of the device function such as the invisibility. Besides, resonance implies narrowband, so an invisibility cloak based on resonance mechanism can only work in a very narrow bandwidth, which is undesireable. Finally, it is difficult to fabricate metamaterials from nanostructures of resonant type, and as mentioned before, their resonance property are usually not good enough.

In order to avoid the above mentioned disadvantages, new designs of cloaking devices often use the idea of gradient index and they are built with common dielectric materials. One of them is the carpet cloak [65–67]. When covers this kind of cloak on an object lying on a table, the cloak can cancel the scattered light from the object completely, changing the table optically equivalent to a flat surface.

In addition to the various cloaking devices for controlling the electromagnetic waves, recently the similar ideas have been utilized and generalized for designing devices for cloaking other kinds of waves. To build an acoustic cloak, we have to fabricate the acoustic metamaterials having the required effective mass density and bulk modulus or Lame's constants. Such kind of cloaks have already been built and tested [38, 39]. People have also designed a cylindrical shaped water wave cloak for cloaking water waves by varying the water depth gradually along the radial direction around the cavity region [68]. This kind of cloaks may someday be used for protecting coastlines from tsunamis by making the land invisible to the incoming waves. Quantum wave cloaks or matter wave cloaks have also been considered, which guides the propagation of matter waves around a cloaked region by designing the effective mass and potential functions in the Schrodinger equation [40, 41]. A very interesting novel device called time cloak has also been proposed recently [69]. The fundamental concept of the previously mentioned cloaking devices is this: creating a hole in the space, and guide the waves around the hole by using properly designed artificial materials. The basic idea of time cloak is instead creating a gap in the time, and cheat the prob wave, making it unable to detect the event happening in the time gap. This description also explains why such a device is also called 'history editor' [70]. Reader may refer to the original papers and review articles for understanding these new developments [71].

2.7 Summary

In this chapter we have reviewed some topics in metamaterials research, which include: negative refraction, left-handed media, perfect lens or superlens and their relation with subwavelength imaging, the energy density problem, wire array and negative permittivity, SRR array and negative permeability, indefinite media or hyperbolic metamaterials and hyperlens, invisibility cloak and other transformation optics devices.

Metamaterials is in fact not materials but is a new way of thinking. With this new way of thinking people try every possible method to design structures for application and use them as materials. The most used mechanisms in designing metamaterials include resonance (SRR array), constraints for current flows (wire array), and high anisotropies (hyperbolic metamaterials). The effective media theories are useful but only accurate enough at the long wavelength limit. If the wavelength is not long enough we must treat metamaterials as what they really are, for example, photonic crystals consisting of periodic arranged metallic wires and split-rings. In that situation we must know the band structures or dispersion relations before we can make any predictions about the wave propagation properties inside these structures. Another very important issue is how to design broadband metamaterial devices which can work in a wide rage of frequency. We are not able to discuss all the topics in this research area in only one section because it covers phenomena of too wide range and is evolving too fast. However, readers may find the materials provided in this chapter are essential and useful which can help them to pass through the main obstacles of understanding metamaterials.

References

1. W.J. Padilla, D.N. Basov, D.R. Smith, Mater. Today **9**, 28 (2006)
2. J.B. Pendry, A.J. Holden, D.J. Robbins, W.J. Stewart, IEEE Trans. Microw. Theory Tech. **47**, 2075–2084 (1999)
3. D.R. Smith, W.J. Padilla, D.C. Vier, S.C. Nemat-Nasser, S. Schultz, Phys. Rev. Lett. **84**, 4184–4187 (2000)
4. R.A. Shelby, D.R. Smith, S. Schultz, Science **292**, 77–79 (2001)
5. Z. Jacob, L.V. Alekseyev, E. Narimanov, Opt. Express **14**, 8247 (2006)
6. Z. Liu, H. Lee, Y. Xiong, C. Sun, X. Zhang, Science **315**, 1686 (2007)
7. V.G. Veselago, Sov. Phys. Usp. **10**, 509 (1968)
8. J.B. Pendry, Phys. Rev. Lett. **85**, 3966–3969 (2000)
9. R. Merlin, Appl. Phys. Lett. **84**, 1290–1292 (2004)
10. N. Fang, H. Lee, C. Sun, X. Zhang, Science **308**, 534 (2005)
11. J.B. Pendry, D. Schurig, D.R. Smith, Science **312**, 1780 (2006)
12. D. Schurig, J.J. Mock, B.J. Justice, S.A. Cummer, J.B. Pendry, A.F. Starr, D.R. Smith, Science **314**, 977 (2006)
13. W.L. Barnes, A. Dereux, T.W. Ebbesen, Nature **424**, 824 (2003)
14. J.B. Pendry, A.J. Holden, W.J. Stewart, I. Youngs, Phys. Rev. Lett. **76**, 4773 (1996)
15. V.M. Shalaev, Nat. Photonics **1**, 41 (2007)
16. N. Garcia et al., Phys. Rev. Lett. **88**, 207403 (2002)
17. D. Maystre, S. Enoch, J. Opt. Soc. Am. A: **21**, 122 (2004)
18. P.G. Luan, H.D. Chien, C.C. Chen, C.S. Tang, WSEAS Trans. Electron. **1**, 236 (2004)
19. M. Notomi, Phys. Rev. B **62**, 10696 (2000)
20. C. Luo, S.G.Johnson, D.J. Joannopoulos, J.B. Pendry, Phys. Rev. B **65**, 201104 (2002)
21. S. Foteinopoulou, C.M. Soukoulis, Phys. Rev. B **67**, 235107 (2003)
22. E. Cubukcu, K. Aydin, E. Ozbay, S. Foteinopoulou, C.M. Soukoulis, Nature **423**, 604 (2003)
23. E. Cubukcu, K. Aydin, E. Ozbay, S. Foteinopoulou, C.M. Soukoulis, Phys. Rev. Lett. **91**, 207401 (2003)
24. S.S. Xiao, M. Qiu, Z.C. Ruan, S. He, Appl. Phys. Lett. **85**, 4269 (2004)
25. K. Guven, K. Aydin, K.B. Alici, C.M. Soukoulis, E. Ozbay, Phys. Rev. B **70**, 205125 (2004)
26. A. Martnez, H. Mguez, A. Griol, J. Mart, Phys. Rev. B **69**, 165119 (2004)
27. P. Vodo, P.V. Parimi, W.T. Lu, S. Sridhar, Appl. Phys. Lett. **86**, 201108 (2005)
28. P.-G. Luan, K.D. Chang, J. Nanophotonics **1**, 013518 (2007)
29. A. Sukhovich et al., Phys. Rev. Lett. **102**, 154301 (2009)
30. E. Di Gennaro et al., Phys. Rev. B **77**, 193104 (2008)
31. L.-S. Chen, C.-H. Kuo, Z. Ye, Appl. Phys. Lett. **85**, 1072 (2004)
32. P.-G. Luan, C.Y. Chiang, H.Y. Yeh, J. Phys.: Condens. Matter **23**, 035301 (2011)
33. X. Hu, Y. Shen, X. Liu, R. Fu, Z. Jian, Phys. Rev. E **69**, 030201 (2004)
34. X. Hu, C.T. Chan, Phys. Rev. Lett. **95**, 154501 (2005)
35. Z. Liu, X. Zhang, Y. Mao, Y.Y. Zhu, Z. Yang, C.T. Chan, P. Sheng, Science **289**, 1734 (2000)
36. J. Li, C.T. Chan, Phys. Rev. E **70**, 055602 (2004)
37. S. Zhang, L. Yin, N. Fang, Phys. Rev. Lett. **102**, 194301 (2009)
38. S. Zhang, C. Xia, N. Fang, Phys. Rev. Lett. **106**, 024301 (2011)
39. L. Zigoneanu, B.I. Popa, S.A. Cummer, Nat. Mater. **13**, 352 (2014)
40. S. Zhang, D.A. Genov, C. Sun, X. Zhang, Phys. Rev. Lett. **100**, 123002 (2008)
41. D.-H. Lin, P.G. Luan, Phys. Rev. A **79**, 051605 (2009)
42. Nature Photonics 6, p. 707–p. 794 (2012)
43. J. Valentine, S. Zhang, T. Zentgraf, E. Ulin-Avila, D.A. Genov, G. Bartal, X. Zhang, Nature **455**, 376 (2008)
44. Z. Li, M. Mutlu, E. Ozbay, J. Opt. **15**, 023001 (2013)
45. A. Poddubny, I. Iorsh, P. Belov, Y. Kivshar, Nat. Photonics **7**, 948 (2013)
46. D.R. Smith, P. Kolinko, D. Schurig, J. Opt. Soc. Am. B **21**, 1032 (2004)

47. P.M. Valanju, R.M. Walser, A.P. Valanju, Phys. Rev. Lett. **88**, 187401 (2002)
48. J.B. Pendry, D.R. Smith, Phys. Rev. Lett. **90**, 029703 (2003)
49. D.R. Smith, D. Schurig and J. B. Pendry **81**, 2713 (2002)
50. S. Foteinopoulou, E.N. Economou, C.M. Soukoulis, Phys. Rev. Lett. **90**, 107402 (2003)
51. J.D. Jackson, Electrodynamics, 3rd edn. (Wiley, New York, 1998)
52. L. Brillouin, Wave propagation and group velocity (Academic Press, New York, 1960)
53. L.D. Landau, E.M. Lifshitz, *Electrodynamics of continuous media*, 2nd edn. (Pergamon Press, New York, 1984)
54. P.-G. Luan, Phys. Rev. E **80**, 046601 (2009)
55. X. Zhang, Z. Liu, Nat. Mater. **7**, 435–441 (2008)
56. J. Zhou, T. Koschny, M. Kafesaki, E.N. Economou, J.B. Pendry, C.M. Soukoulis, Phys. Rev. Lett. **95**, 223–902 (2005)
57. C.M. Soukoulis, M. Wegener, Nat. Photonics **5**, 523 (2011)
58. J.D. Joannopoulos, S.G. Johnson, J.N. Winn, R.D. Meade, *Photonic Crystals: Molding the Flow of Light*, 2nd Edn. (Princeton University Press, Princeton, 2008)
59. H. Chen, L. Ran, J. Huangfu, T.M. Grzegorczyk, J.A. Kong, J. Appl. Phys. **100**, 024915 (2006)
60. P. Yeh, Optical waves in layered media (Wiley-Interscience, New York, 2nd edn, 2005)
61. Ulf Leonhardt, T.G. Philbin, New J. Phys. **8**, 247 (2006)
62. P. Sheng, Science **313**, 1399 (2006)
63. H. Chen, C.T. Chan, P. Sheng, Nat. Mater. **9**, 387 (2010)
64. S.G. Johnson, http://www-math.mit.edu/~stevenj/18.369/coordinate-transform.pdf (notes for the course 18.369 at MIT)
65. J. Li, J.B. Pendry, Phys. Rev. Lett. **101**, 203901 (2008)
66. T. Ergin, N. Stenger, P. Brenner, J.B. Pendry, M. Wegener, Science **328**, 337 (2010)
67. M. Gharghi, C. Gladden, T. Zentgraf, Y. Liu, X. Yin, J. Valentine, X. Zhang, Nano Lett. **11**, 2825 (2011)
68. M. Farhat, S. Enoch, S. Guenneau, A.B. Movchan, Phys. Rev. Lett. **101**, 134501 (2008)
69. J.M. Lukens, D.E. Leaird, A.M. Weiner, Nature **498**, 205 (2013)
70. M.W. McCall, A. Favaro, P. Kinsler, A. Boardman, J. Opt. **13**, 024003 (2011)
71. Y. Liu, X. Zhang, Nanoscale **4**, 5277 (2012)

Part II
Progress in Short-Pulse Yb-doped Fiber Oscillators and Amplifiers

Chapter 3
Progress in Short-Pulse Yb-Doped Fiber Oscillators and Amplifiers

Ci-Ling Pan, Alexey Zaytsev, Chih-Hsuan Lin and Yi-Jing You

3.1 Introduction

Recently, high-speed and high-power fiber lasers have emerged not only as a workhorse for information technology but also as a viable alternative to current generation of main-frame lasers for industrial and scientific applications [1, 2]. Laser machining and processing of materials, e.g., welding, drilling and precision cutting, are among the most important conventional applications of the laser. In the past two decades, femtosecond laser ablation [3] was shown to exhibit unique capabilities for micro- and nano-machining in a variety of materials, including transparent materials [4]. Machining platforms based on ultrafast fiber lasers was reported recently [5].

Ytterbium (Yb)-doped fiber-based laser systems [6–8] are currently the laser system of choice for the applications mentioned above. This is primarily due to their efficiency, broad gain bandwidth, and operational wavelength near 1,060 nm (that of Nd: YAG lasers, one of primary workhorses of the industry) [8]. Figure 3.1 shows the energy level diagram of Yb-doped silica fiber and the corresponding absorption and emission spectra. The laser transitions occur between the sublevels of $^4F_{5/2}$ and $^4F_{7/2}$ states in Yb-doped silica. YDF has a very broad absorption and emission band and can be pumped efficiently by 915 nm or 975 nm laser diode. Absorption at 975 nm is much stronger than that at 915 nm though its bandwidth is comparatively narrow. As a result, relatively complicated temperature stabilization schemes are required for the pump laser diodes emitting at the wavelength of 975 nm versus that at 915 nm. The fluorescence lifetime of the upper laser level in YDF, $\tau_f = 0.8$ ms.

C.-L. Pan (✉) · A. Zaytsev · C.-H. Lin · Y.-J. You
Department of Physics, Institute of Photonics Technologies, National Tsing Hua University, Hsinchu, Taiwan
e-mail: clpan@phys.nthu.edu.tw

C.-C. Lee (ed.), *The Current Trends of Optics and Photonics*,
Topics in Applied Physics 129, DOI 10.1007/978-94-017-9392-6_3

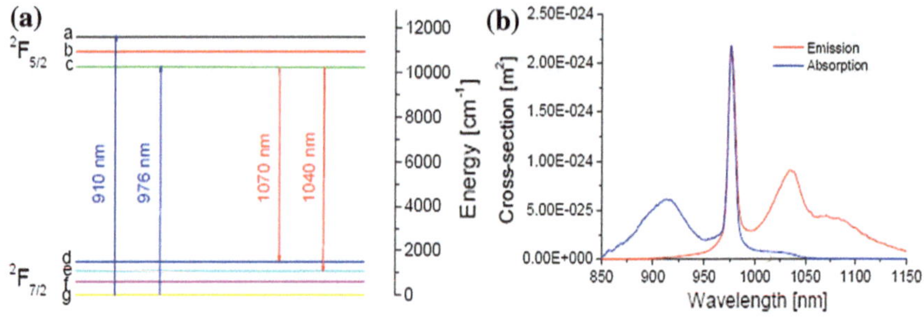

Fig. 3.1 Energy level diagram (**a**) and Absorption (*solid*) and emission cross-sections (*dashed*) of Yb in silica fiber (**b**) as a function of wavelength [6]

Although the long length and tight beam confinement make active fibers suitable for high-power amplification, output of fiber laser systems soon reach peak powers approaching or surpassing thresholds for undesirable nonlinear effects. Therefore achieving high peak power associated with high-power ultrashort pulses with good beam quality constitutes a major challenge. Techniques such as chirped pulse amplification [9, 10] were introduced to reduce peak power in the fiber amplifier and consequently minimize nonlinear effects. This method is well suited for achieving high-energy pulses but typically involves a complex optical arrangement. In practice, direct amplification appears more practical for scaling average power because of its simplicity. This implies that adequate management of nonlinearities in the YDF laser systems tailored to specific applications, e.g., nonlinear frequency conversion, is essential. Nevertheless nonlinearities such as self-phase modulation (SPM) can also be very attractive for the generation of high-power pulses and their subsequent compression.

In this chapter, we describe our recent work on picosecond and femtosecond Yb-doped fiber oscillators and amplifiers. Beginning with a brief description of the master-oscillator power amplifier technology and its capabilities, Sect. 3.2 reviews some of the basic theories useful to understand amplification in fiber amplifiers and describes the various nonlinearities encountered by short pulses propagating in an optical fiber. Such analyses are useful for optimum design of the YDF laser systems. Experimentally, we demonstrated direct fiber amplification of a picosecond passively mode-locked diode-pumped solid-state (DPSS) seeder through a chain of amplifiers using low-cost non-polarization-maintaining (non-PM) YDF. The YDF laser system generated ~11 ps-wide pulses at 250 MHz and an average power as high as 60 W with 73 W pumping. Moreover, the output beam quality was excellent, with M^2 ~ 1.6. Detrimental effects such as stimulated Raman Scattering (SRS) and amplified spontaneous emission (ASE) were minimized.

In Sect. 3.3, we showed that a ring-type YDF laser cavity with nonlinear polarization evolution (NPE) port can be configured to generate either continuous-wave mode-locked pulses or noise-like pulses (NLP) with a super-broadband spectrum. Simulation results showing the buildup dynamics for the two regimes were also presented. Self-starting noise-like operation of the laser can be maintained over a relatively large range of pumping powers (4–13 W). The maximum 3 dB spectral bandwidth of the noise-like pulses was ~48 nm which the output energy can be as high as 47 nJ, limited by optical damage of the fiber components. Generation of supercontinuum (SC) by NLPs propagating through ~100 m of single-mode fiber in the normal dispersion region was also demonstrated. The SC exhibits low threshold (43 nJ) and a flat spectrum over 1,050–1,250 nm.

Finally, we consider amplification of picosecond bursts in Yb-doped fiber amplifiers its impact on frequency conversion applications. A theoretical and experimental comparison between regular pulse trains and pulse bursts shows significant enhancement in terms of conversion efficiencies.

3.2 The MOPA Approach to High-Power Short-Pulse Yb-Doped Fiber Lasers

High-power fiber lasers have been increasingly used in scientific applications [11, 12] as well as material processing [5, 13]. These systems are typically based on master oscillator power amplifier (MOPA) designs. For wavelengths around 1 μm, ytterbium-doped fiber is widely adopted for fiber-based high-power lasers and amplifiers [6]. Yb-doped double-cladding (DC) fibers can be used to build laser systems with output power up to several kilowatts. Such laser systems also exhibit desirable characteristics such as high gain, good efficiency and excellent beam quality [14–19]. For example, C. Zheng et al. reported an all-fiber master oscillator power amplifier (MOPA) system which can generate ~10 ns pulses with energies of 1.2 mJ/pulse [19]. The output beam quality $M^2 \approx 1.4$.

The broad gain bandwidth of Yb-doped fiber is also suitable for the amplification of ultrashort pulses [17, 18]. High-power short-pulse fiber lasers, however, must meet challenges from the onset of unwanted nonlinear effects, like self-phase modulation or stimulated Raman scattering, which limit the maximum output power and degrade output pulse shape and duration. Recently developed photonic crystal fiber (PCF) is able to overcome many of the difficulties mentioned above. PCFs, however, are still very expensive. If a linearly-polarized high-power short-pulse laser is required, it can be based on a master oscillator fiber amplifier (MOFA) design where polarization-maintaining (PM) fibers are used. Further, the PM fibers are well suited to amplify and deliver linearly polarized optical beams to targets with excellent extinction ratio. The PM fibers, however, are much more expensive than conventional non-PM fibers. Further, the thresholds of nonlinear effects in non-PM fibers are higher [20, 21].

A stable seed laser is required for the MOFA. Currently, several research groups adopted fiber mode-locked oscillators as seed lasers [22, 23]. For the moment, however, long-term reliability of modern fiber oscillators is still the main challenge which may affect on the overall MOFA performance. Vertical external-cavity surface-emitting semiconductor lasers (VECSELs) producing nearly transform-limited ps pulses and quasi-diffraction limited beams are suitable for MOFAs [24]. The output power of VECSEL, however, is well below 1 mW. As a result, multiple amplification stages are required. On the other hand, diode-pumped solid-state (DPSS) lasers can provide transform-limited pulses at moderate power levels. In this section, we presented progress of relatively high-power picosecond MOFAs using non-PM Yb-doped fiber amplifiers seeded by a compact and robust passively mode-locked DPSS Nd:$GdVO_4$ laser which can provide 12 ps duration, at ~200 MHz repetition rate with average power of ~400 mW [25]. An average power of ~28 W at a repetition frequency of 250 MHz was achieved with a single stage of 10-μm-core Yb-doped fiber amplifier in the MOFA. A pulse duration of ~1.4 ps was obtained after recompression with a grating pair [26]. With two stages of non-PM Yb-doped fiber amplifiers, Stimulated Raman scattering (SRS) and amplified spontaneous emission (ASE) can be suppressed by optimizing the active fiber lengths and pumping powers of the amplification stages. We achieved an average output power as high as 60 W, with an optical conversion efficiency 82 % [26]. The SRS peak at 1,120 nm is suppressed by 30 dB. No distortion or other unwanted effects were found at the system output. The output beam quality is measured to be $M^2 \sim 1{:}6$.

3.2.1 Theoretical Analysis and System Design

For theoretical analysis and optimum design of the MOFA, we employed the analytical model for rare-earth-doped fiber amplifiers and lasers first proposed by Saleh et al. [27]. Certain simplifying assumptions have to be used in these models: (1) The power extracted by ASE must be negligible. This is valid for a fiber amplifier with input seed power significantly higher than equivalent ASE noise in the channel. (2) The field and dopant distributions must be homogeneous. This is typically the case for single mode doped fiber cores. (3) Excited state absorption must be negligible, which is true for Ytterbium-doped fibers. Further, we neglected the influence of optical feedback on the signal. This is reasonable because angle-cleaved fiber ends were employed in our experimental setup. Following Barnard et al. [28], we can write down the steady-state rate equations for the amplifier as follows:

$$N_1 + N_2 = N_t \tag{3.1}$$

$$\frac{\partial P_p}{\partial z} = \Gamma_p \left[\sigma_p^e N_2 - \sigma_p^a N_1 \right] P_p \tag{3.2}$$

$$\frac{\partial P_s}{\partial z} = \Gamma_s \left[\sigma_s^e N_2 - \sigma_s^a N_1 \right] P_s \tag{3.3}$$

$$\frac{N_2}{N} = \frac{(\sigma_s^e + \sigma_s^e)\sigma_p^a P_p P_s^{sat} + \left(\sigma_p^a + \sigma_p^e\right)\sigma_s^a P_s P_p^{sat}}{\left(\sigma_s^a + \sigma_s^e\right)\left(\sigma_p^a + \sigma_p^e\right)\left(P_p P_s^{sat} + P_p^{sat} P_s^{sat} + P_p P_s^{sat}\right)} \tag{3.4}$$

In (3.1–3.4), N_t is the doping density of Yb^{3+} ions in the active fiber. N_1 and N_2 are the population densities of lower and upper lasing levels, respectively. P_p is the pump power, z is the spatial coordinate along the fiber axis, P_s is the signal power. $\sigma_s{}^e$ and $\sigma_p{}^e$ are the emission cross-sections of the signal and the pump light, respectively. Similarly, $\sigma_s{}^a$ and $\sigma_p{}^a$ are the absorption cross-sections of the signal and the pump light, in that order. Other parameters used include the lifetime of the lasing transition, τ; the overlap integral of pump and signal, Γ_s and Γ_p; the effective area of the fiber core, A; Planck constant, h; the velocity of light, c. Equations (3.2–3.4) can be solved analytically and expressed as follows [29]:

$$\frac{P_p^{IN}}{h\nu_p}\left(1-\left(\frac{G}{G\mathrm{max}}\right)^{\delta}\right)=\frac{P_s^{sat}}{h\nu_s}\left(\Gamma_s N\sigma_s^a L+\ln G\right)+\frac{P_s^{IN}}{h\nu_s}(G-1), \qquad (3.5)$$

where $\delta=\frac{P_s^{sat}/h\nu_s}{P_p^{sat}/h\nu_p}$ is the ratio of saturation powers, G is a signal gain, $G\mathrm{max}=\exp\left[\left(\frac{\Gamma_p N\sigma_p^a}{\delta}-\Gamma_s N\sigma_s^a\right)L\right]$ is the maximum gain. P_p^{in}, P_s^{in} are total power of pump and signal lights at the input of the amplifier, respectively.

Using the formula summarized above and the physical parameters of Yb-doped double-cladding fibers provided by fiber manufacture (Liekki), we calculated the output power as the function of active fiber length (see Fig. 3.2). The calculations were made assuming fiber's cross-sections from [6], 100 mW of average seed power and pumping power at 20 W. These are summarized in Table 3.1.

Simulations indicated that the optimal fiber length for a 10-μm-core fiber amplifier stage would be ~10–12 m. Similarly, the optimal fiber length of 30-μm-core diameter fiber is ~3–5 m. These are plotted in Fig. 3.2.

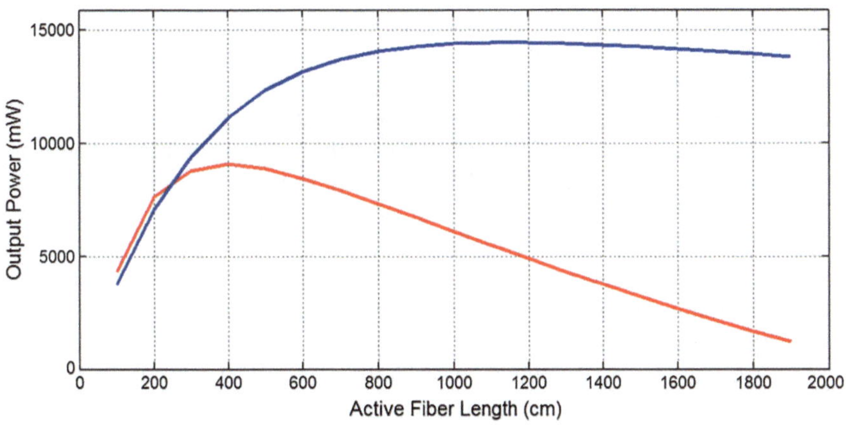

Fig. 3.2 Simulation for the dependence of output power as a function of fiber length of two different Yb-doped fibers (YB1200-10/125DC, Liekki with 10 μm core diameter—*blue curve*, and YB1200-30/250DC, Liekki with 30 μm core diameter—*red curve*) for single stage amplification. The input power and pump power are assumed to be 100 mW and 20 W, respectively, for both fibers

Table 3.1 Main parameters of the 10-μm-core Yb-doped double-clad fiber

Parameter	Value	Parameter	Value
N	9.5×10^{25} m^{-3}	Γ_s	0.82
τ	0.84 ms	Γ_p	0.01
σ_s^a	5×10^{-27} m^2	λ_p	915 nm
σ_s^e	3.4×10^{-25} m^2	λ_s	1,064 nm
σ_p^a	7.8×10^{-25} m^2		
σ_p^e	3.1×10^{-26} m^2		

In comparison, the 10-μm-core fiber amplifier stage with optimal fiber length appears to exhibit higher power than the optimized 30-μm-core fiber amplifier stage (see Fig. 3.2). The above analysis, however, does not consider the nonlinear effects such as the stimulated Raman scattering (SRS) and self-phase modulation (SPM). These nonlinearities are non-negligible at high signal powers and result in pulse distortion and other undesirable consequences. A rough estimate shows that the Raman threshold for our input condition and the optimal fiber length is below a peak power of 2.3 kW or an average signal power of 7 W. Therefore, the output power of a single 10-μm-core fiber amplifier stage is limited in order to avoid significant pulse distortion and efficiency degradation. On the other hand, the output power of a 30-μm-core fiber amplifier stage is relatively small (~0.4 W, see Fig. 3.2) with the same input signal and pump power. Due to the large core size, a seed signal power higher than 1 W is required to achieve a good pump to signal conversion efficiency. It is reasonable, therefore, to consider the design of a dual-stage amplifier system, using a length of 10-μm-core fiber as the preamplifier and another length of 30-μm-core fiber for the main amplifier.

For optimization of such a dual-stage fiber MOPA system, we first used the formalism above to optimize each stage separately in order to achieve the highest gain. It is well known that larger core size and shorter fiber length will increase the nonlinear thresholds. Therefore, effects such as SRS and SPM are expected to be more likely to happen in the first or the preamplifier stage, which employs fibers of smaller core size. The SPM will broaden the spectrum, which is distributed about the signal wavelength. The SRS will create new wavelengths which are far from the signal. The signal power limitation of SRS is more serious than the SPM. Using the formula in [30] and assuming a Raman gain coefficient $g_R \sim 3.2 \times 10^{13}$ mW^{-1} [31], we estimated that the SRS threshold is ~2.3 kW for a 10-μm-core fiber with optimized length of 7 m. For a 30-μm-core-fiber and ~5 m in length, the SRS threshold increases by a factor of more than ten to ~27 kW.

In our design, we set the output power of the preamplifier stage at ~3 W, which is below the SRS threshold for the 10-μm-core fiber preamplifier. Assuming 2 W of power was injected into the 2nd or main-amplifier stage. In Fig. 3.3, we have calculated and plotted the normalized gain of the first stage (10-μm-core fiber) and second stage (30-μm-core fiber) of MOPA as a function of their respective

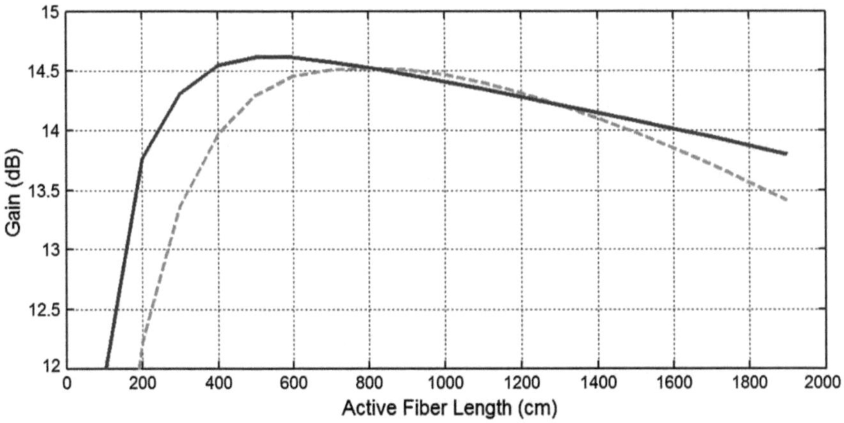

Fig. 3.3 Simulated gain as a function of the fiber length of the 1st (*dashed red curve*) and 2nd (*solid blue curve*) stage amplifiers in a dual-stage Yb-doped fiber amplifier. The core diameters are 10 and 30 μm, respectively. The input signal power is 100 mW in the 1st stage and 2 W in the 2nd stage, respectively. The pump power is 5 W in the 1st stage and 80 W in the 2nd stage

length. The optimal fiber length of the main amplifier stage is ~5 m (see solid curve of Fig. 3.3). Actually, the optimal fiber length of the 10-μm-core fiber can be reduced in the dual-stage amplifier system. In order to achieve an output power of ~3 W in the preamplifier stage, the needed fiber length is only ~5 m (see dashed curve of Fig. 3.3), further reducing the possible detrimental nonlinear effects.

3.2.2 Picosecond Laser Seeder

For the master laser, we designed and constructed a compact and robust passively mode-locked DPSS $Nd:GdVO_4$ laser which can provide 12 ps duration, at ~250 MHz repetition rate with average power of ~400 mW [25].

The laser employed a linear cavity, as shown in Fig. 3.4. A Brewster-cut $Nd:GdVO_4$ laser crystal with a dopant concentration of 1.5 % Nd and the

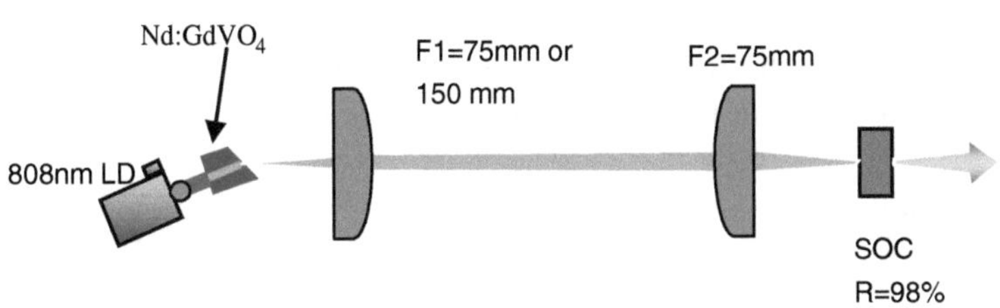

Fig. 3.4 Schematic of the DPSS laser. *SOC* Saturable Output Coupler

dimensions of 2 × 2 × 3 mm^3 was used. One of its light-passing facets was coated for high reflection (HR) at the lasing wavelength of 1,064 nm and high transmittance (HT) at the pump wavelength of 808 nm, the other side (Brewster-cut) was coated for HT at 1,064 nm. The laser was pumped by a C-mounted 808 nm laser diode (LD) with integrated fast axis collimator and maximum output power of ~2 W. For simplicity and cost reduction, we did not use any optics for re-imaging the pumping spot into the laser crystal. Instead, we placed laser crystal at a short (~0.2 mm) distance from the LD chip surface. The LD and $Nd:GdVO_4$ crystal, wrapped with indium film, were embedded in a copper heat sink that is part of a compact module [25]. The cavity contains a saturable output coupler (SOC, made by BATOP Gmbh, Weimar, Germany) that serves as the saturable absorber as well as the output coupler. In our experiments, we used a SOC of 2 % transmission not to achieve the lowest possible threshold but to find the laser condition optimally matched with the laser diode used for pumping.

The HR-coated crystal's facet, two HT-coated convex lenses (F1 and F2), and the SOC form the two-mirror cavity of the laser. We adjusted the cavity length from 0.6 to 1 m and experimented with F1 lens of which the focal lengths, f_1, were 75 and 150 mm, respectively. The focal length of F2 was fixed at 75 mm. The calculated and experimentally measured critical average powers P_{out} for several variations of the cavity design are summarized in Table 3.2. Here, P_{out} is the minimal output average power corresponded to the CW mode-locked state [25]. An example of measured laser spectrum in the mode-locked state is shown in Fig. 3.6a.

We found that measured P_{out} is slightly higher than theoretically predicted values. This is tentatively attributed to additional intracavity loss mechanisms, e.g., reflection losses, scattering losses, mode areas mismatch. From Table 3.2, it is clear that increasing the cavity size and using F1's with shorter focal length result in increasing intracavity pulse energy, reduced mode-locking threshold, and suppression of Q-switching instabilities. On the other hand, longer cavity length is undesirable for the design of a compact system and also reduces the repetition rate.

In Fig. 3.5, we have plotted laser output powers versus pumping power for different cavity configurations. Note that the cavity with f_1 = 150 mm exhibited higher efficiency than ones with f_1 = 75 mm. We attribute this to enlarging the mode area and better mode matching with the pumping beam in the former. The improvement in efficiency, however, is accompanied by higher mode-locking thresholds.

The envelope of the interferometric autocorrelation trace with the Gaussian fitting curve is shown in Fig. 3.6b. From fitting parameters, we can estimate the

Table 3.2 Calculated and measured critical average powers

Cavity length	1 m	60 cm	1 m	60 cm
f_1 (mm)	75	75	150	150
Calculated P_{out} (mW)	170	250	340	510
Measured P_{out} (mW)	185	260	350	NA

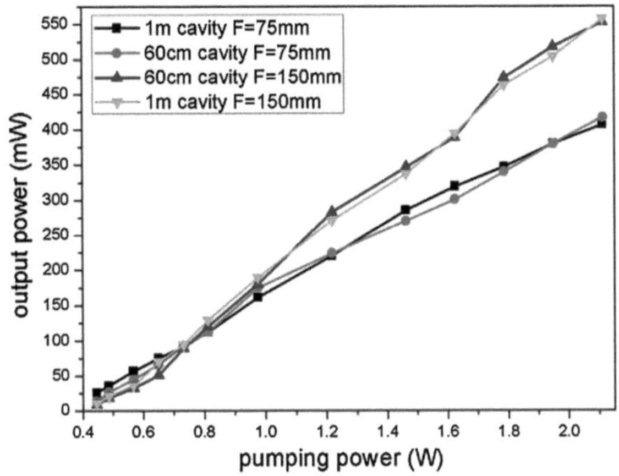

Fig. 3.5 Output power versus the pump power curves for cavity configurations listed in Table 3.1

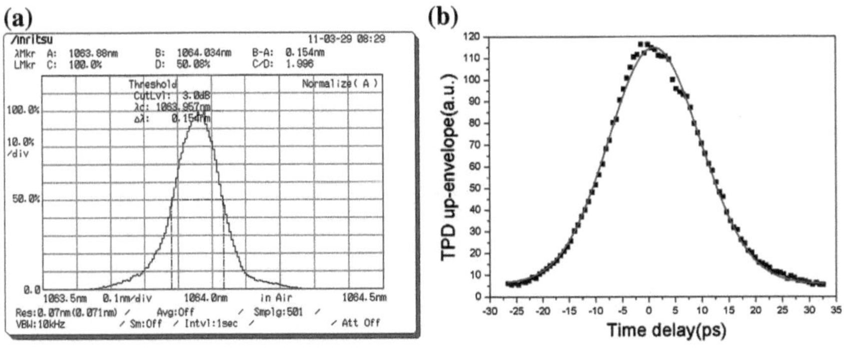

Fig. 3.6 **a** Measured optical spectrum of mode-locked Nd:$GdVO_4$ laser; **b** The measured envelope of an interferometric autocorrelation trace and its *Gaussian fitting curve*

pulse duration at ~12 ps. The calculated time-bandwidth product 0.49 obtained by using the measured optical spectrum (see Fig. 3.6a) shows that DPSS laser irradiates almost transform-limited pulses.

In the case of optimal condition (60 cm cavity and $F_1 = 75$ mm) the laser generates ~12 ps pulse trains at ~250 MHz (see Fig. 3.7) with an average output power P_{out} of about 260 mW when the pumping power is 1.3 W.

For the condition that the output power of the laser was ~270 mW which was slightly higher than that of the CW mode-locking threshold, the fluctuation of average output power for 11 h operation was recorded (see Fig. 3.8). For comparison, we also presented the fluctuation data if active cooling was used to stabilize the temperature the basement of LD and laser crystal mount at 20 °C (red dot).

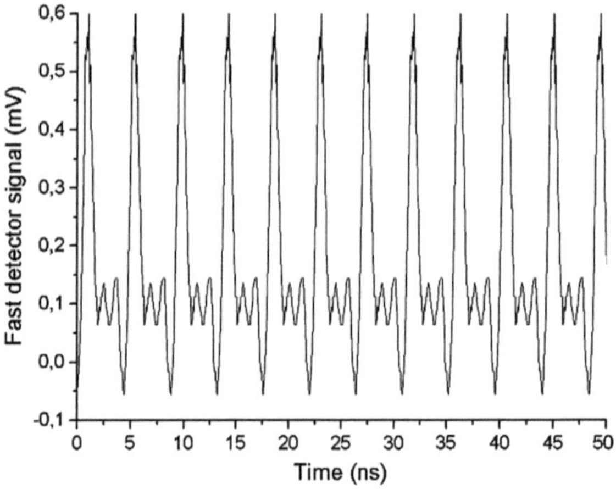

Fig. 3.7 The pulse train at 60 cm cavity

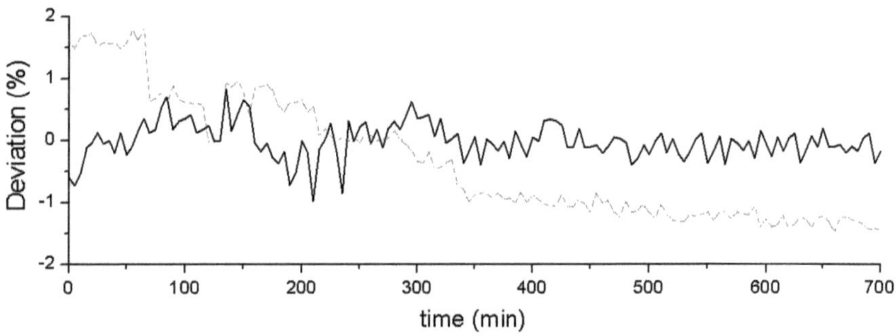

Fig. 3.8 Long-term fluctuations of laser output with active cooling (*solid*) and without of active cooling (*dot*)

We can observe from Fig. 3.8 that the average power fluctuations in long-term operation do not reach more than ±2 % which was slightly higher than in the case when active cooling was used (±1 % fluctuations). We attribute this superior performance to several factors. First of all, we achieved good thermal stability through reduction of excessive absorbed heat and good thermal sinking of the pumping module consisting of the laser crystal and the pumping LD [25]. Secondly, the cavity configuration is simple and thus mechanically stable. The last feature also suggests easer alignment of laser cavity.

3.2.3 Multi-Stage Fiber Laser Amplifiers

The experimental setup of the proposed high power picosecond laser system is shown schematically in Fig. 3.9 [32]. It is constructed in a dual-stage MOFA configuration. The mode-locked DPSS seed laser has been reported previously (see Fig. 3.4).

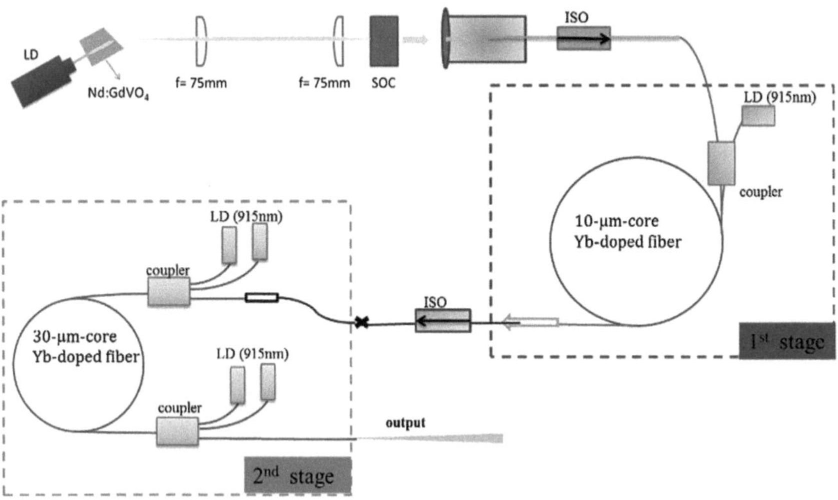

Fig. 3.9 Schematic drawing of the MOFA system. *LD* laser diode; *SOC* saturable output coupler, *ISO* isolator

In the two fiber amplifier stages, all fibers are non-PM fibers. We used the theoretical model in Sect. 3.2.1 to determine the optimum fiber length with a given pump power for maximum power output while minimizing undesired nonlinear effects. Thus, a 7-m-long, 10-μm-core double-cladding Yb-doped fiber (YB1200-10/125DC, Liekki) was used in the preamplifier stage. The absorption per unit length of this fiber 1.8 dB/m at 915 nm. Another 5-m-long 30-μm-core double-cladding Yb-doped fiber (YB1200-30/250DC, Liekki) was used in the main-amplifier stage. The absorption per unit length for either fibers was 3.6 dB/m at 915 nm. A $(2 + 1) \times 1$ optical combiner was used to couple the signal and pumping light together before being sent into the active fiber. The signal fiber core size of the combiner was matched with that of the active fiber core. Due to the large wavelength tolerance of absorption near 915 nm, pump laser diode at this wavelength were used in order to provide stable absorption in the active or gain fibers [33]. Between the two stages, we used a polarization-independent optical isolator. This is essential to prevent the reflection from the output end and avoid self-oscillation in the amplifier stages. At the pigtail of fiber amplifiers, we also used angle-cleaved ends to prevent any feedback. A home-made second-harmonic intensity autocorrelator and an optical spectrum analyzer (MS9780A, ANRITSU) were used for pulse characterization. The beam quality was determined by a home-made M square measurement setup.

In the main or second amplifier stage, either forward or backward pumping is possible. The pumping efficiencies for the two cases are shown in Fig. 3.10. Clearly, backward pumping is superior. This is consistent with reports by previous workers [34, 35].

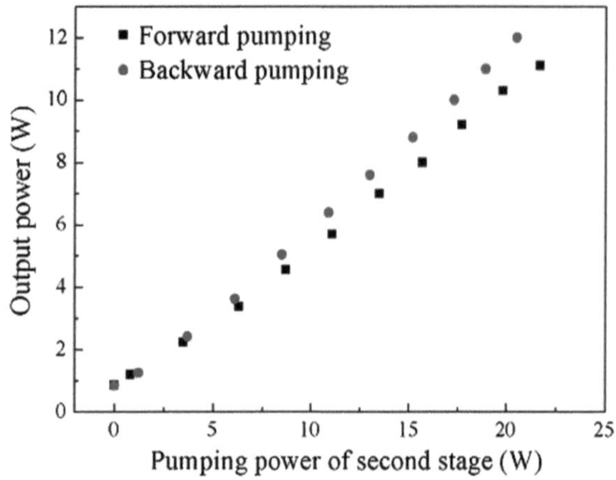

Fig. 3.10 Experimentally measured output power of the main-amplifier as a function of the pumping power for forward (*black squares*) and backward (*red circles*) pumping schemes

3.2.4 MOPA Performance

The output power of the first stage fiber amplifier or the preamplifier is plotted as a function of pumping power in Fig. 3.11a. The solid curve is the theoretical prediction according to (3.5). A maximum average output power of 28 W was obtained by employing a 7-m-long 10-μm-core double-cladding Yb-doped fiber pumped at 50 W [26]. The corresponding optical conversion efficiency is ~56 %. At low pumping powers (below 12 W), the experimental results are in good agreement with the theoretical curve. At higher pumping powers, the experimentally measured output powers are lower than the simulation results. This is attributed to the SRS effect. Simple estimation shows, when the pumping power is up to 13 W, the peak power from the preamplifier is $\leq$2.2 kW, which is close to the estimated SRS threshold of 2.3 kW (see Sect. 3.2).

Figure 3.11b shows the output spectrum of the preamplifier for several values of the pump power. The peak at 1,120 nm wavelength, corresponding to the first Raman Stokes line [21], became visible when pumping power reaches 13 W. Larger pump power will cause more serious Raman amplification in the spectral region around 1,120 nm. As a result, signal depletion at 1,064 nm became evident. This is not desirable.

Moreover, at high pump powers, self-phase modulation (SPM) significantly broadens the output spectrum. On the one hand, SPM is unwanted phenomena because it may cause pulse distortion. On the other hand, it opens the way to recompress the pulse by employing an anomalous dispersion line. In [26] we demonstrated such a possibility. A gratings pair in Littrow configuration was used to compensate the positive phase shift in fiber. We found that the minimal full-width half-maxima (FWHM) of autocorrelation trace of the pulses was ~2 ps. This was observed at a grating separation of ~30 cm. A comparison between autocorrelation traces at 10 W average output power measured on the amplifier output before

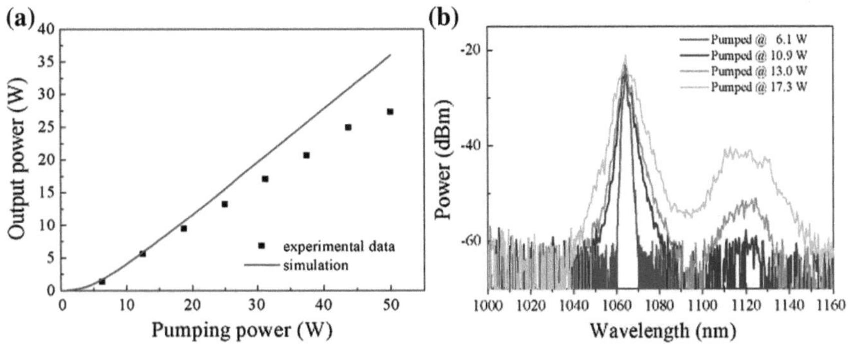

Fig. 3.11 Performance of the preamplifier: **a** Output power versus (*forward*) pumping power. The *solid curve* is the theoretical prediction according to (3.5); **b** optical spectrum of the output of the preamplifier at different pumping power levels

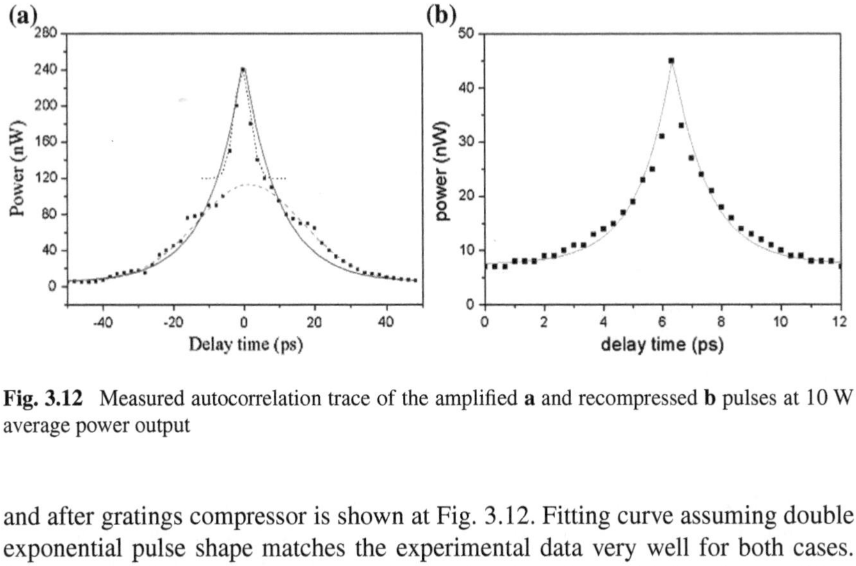

Fig. 3.12 Measured autocorrelation trace of the amplified **a** and recompressed **b** pulses at 10 W average power output

and after gratings compressor is shown at Fig. 3.12. Fitting curve assuming double exponential pulse shape matches the experimental data very well for both cases. From extracted parameters we can estimate that the pulse width duration at ~12 ps before and ~1.6 ps after compression, respectively. Further increasing of output power of the pre-amplifier beyond 10 W results in pulse broadening that cannot be compensated by adjusting the gratings compressor. Obviously, at the higher power levels the spectrum shape at the amplifier output is beyond typical SPM and in order to get shorter pulses the high-order nonlinearity should be taken into account and more comprehensive study should be done.

To obtain higher output power and suppress the SRS effect, the 30-μm-core fiber was chosen for the gain fiber of the second amplification stage. About 2 W of average power was measured at the input of the main or second-stage amplifier (after the second fiber-type optical isolator). This is significantly below the SRS

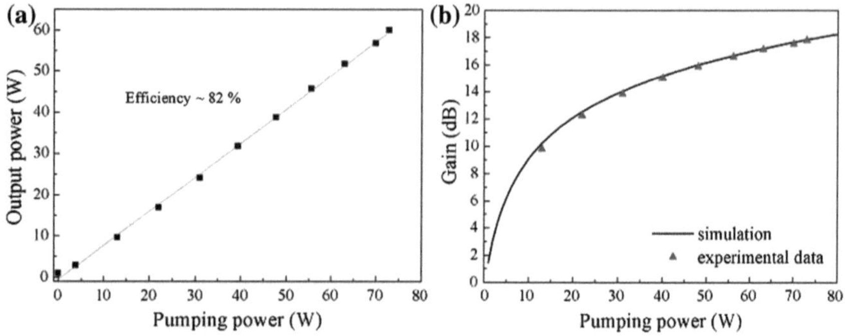

Fig. 3.13 Performance of the 2nd-stage or main amplifier with 30-μm-core fiber: **a** experimentally measured output power versus pumping power; The *solid red line* is the fitting curve, corresponding to an efficiency of ~82 %. **b** the comparison of simulated and experimentally measured gain

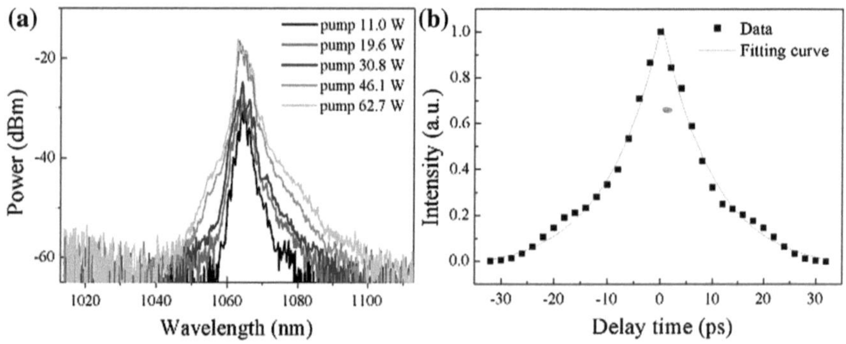

Fig. 3.14 Spectral and temporal characteristics of the output of the 2nd-stage or the main amplifier: **a** optical spectra as a function of pumping powers; **b** the second-harmonic intensity autocorrelation trace of the pulse train

threshold. The pump-power scaling of the output of the main amplifier stage is shown in Fig. 3.13a. With 73 W of pumping power, an average output power of 60 W or a peak power ~21.8 kW was measured at the output of the main amplifier. The corresponding optical conversion efficiency was 82 %. The simulation results, according to (3.5), are plotted as the solid curve in Fig. 3.13b. Clearly, the experimental results are in good agreement with the simulation curve.

Figure 3.14a above shows the spectrum of the MOFA output as a function of the pumping power. The spectrum broadens from 1.6 to 6.4 nm when the pumping power was 62.7 W. This is in good agreement with theoretical estimation of the broadened spectral width, which is 8 nm. The spectral broadening is attributed primarily to self-phase modulation in the active fiber.

The peak power of the MOFA output was lower than that for the SRS threshold, estimated to be 27 kW. The SRS was therefore negligible in our MOFA system. This is consistent with the observation that SRS peak is at least 30 dB smaller than the main peak at the highest pumping power level used (see Fig. 3.14a). Therefore, most of the output power concentrates at the seed laser wavelength. The second-harmonic intensity autocorrelation trace of the MOFA output is shown in Fig. 3.14b. It can best be fit by a double-exponential function. The corresponding FWHM of the electric field waveform of the pulse was ~11 ps (see Fig. 3.14b). This is about the same as that of the seed laser.

The beam quality or propagation factor (M^2) of the MOFA, as measured by a home-made setup, was ~1.6 (See Fig. 3.15).

For now, the output power of our MOFA system was limited by the pump power. That is, the main amplifier had not yet reached saturation. Making sure that we operate below the SRS threshold of the main amplifier stage, we estimate that the maximum output power of the present MOFA can reach 75 W, if we provide enough pump power up to ~92 W.

In summary, an average power as high as 60 W with 73 W of pumping was achieved with an optimized pico-second ytterbium-doped non-PM fiber-based dual-stage amplifier (MOFA) system seeded by a compact modelocked solid-state laser. The corresponding optical conversion efficiency is 80 %. The laser system generates a steady pulse train at the repetition rate of 250 MHz with pulse width of 11 ps or a peak power of 21.8 kW. The output beam quality $M^2 \approx 1.6$. The length and pumping power for the Yb-doped fibers were optimized to suppress stimulated Raman scattering (SRS) and amplified spontaneous emission (ASE) while maintaining desirable output characteristics.

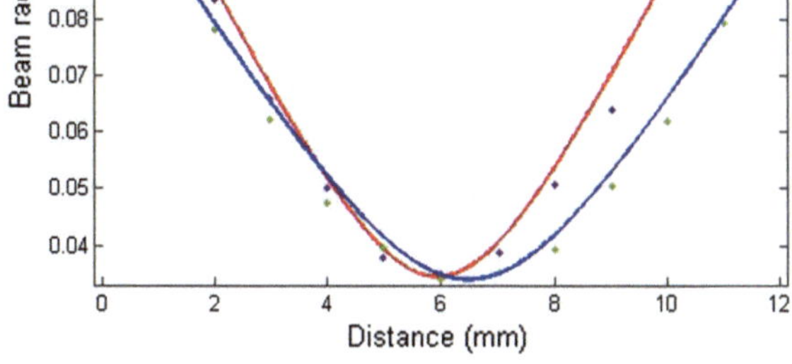

Fig. 3.15 Beam profile measurement of the MOFA system. W_x and W_y are the beam radii in horizontal directions, respectively

3.2.5 Mode-Locked Yb-Doped Fiber Laser

Mode-locking technique is commonly employed to generate ultrashort pulse train from a laser oscillator. One popular approach, passive mode-locking is achieved by using a fast saturable absorber and a gain medium with broad gain bandwidth. In the fiber laser, fast saturable absorber (SA) action can be realized using nonlinear polarization evolution (NPE) [36] due to the optical Kerr effect in the fiber. It can be easily implemented with the help of a polarizing beam splitter (PBS) and a few wave plates as shown in Fig. 3.16. Fast saturable absorption action based on NPE can be explained as follows. The linear polarization after the PBS is made elliptical using a quarter wave plate. As the pulse propagates through the fiber, the axes of the polarization ellipse rotate continuously from the peak to the valley of the pulse due to Kerr nonlinearity. At the fiber exit, the polarization directions are adjusted using a quarter wave and a half wave plates so that the peak of the pulse sees high transmission and the wings are attenuated.

In this way, the NPE acts like a fast saturable absorber and leads to the intensity discrimination and narrowing of the pulse. Usually, a unidirectional ring cavity is required for NPE to be effective and for self-starting operation of the laser. There are, however, several important characteristics that distinguish NPE from the more traditional SESAM mode-locking mechanism. First of all, it differs from a saturable absorber material since high intensities can "overdrive" the transmittance, causing the negative slope regions (see Fig. 3.17). Secondly, the NPE transmission function becomes periodic at high energies. This can be explained by the dependence of NPE on the angle of rotation of the elliptically polarized light. This over-driving of the mode-locking creates limitations in the attainable pulse energy. Further, because it is a polarization-based mechanism, NPE is extremely sensitive to environmental changes. Any disturbances (mechanical, temperature) may cause unstable operation.

According to the fluctuation mechanism of mode-locked pulse generation using a saturable absorber, there are two stages of pulse formation. In the linear stage, a large number of axial modes interfere to create a strong burst of intensity

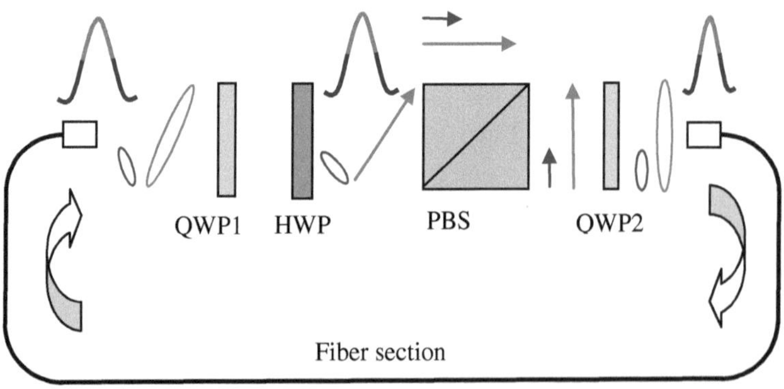

Fig. 3.16 Fast saturable absorption action based on nonlinear polarization rotation

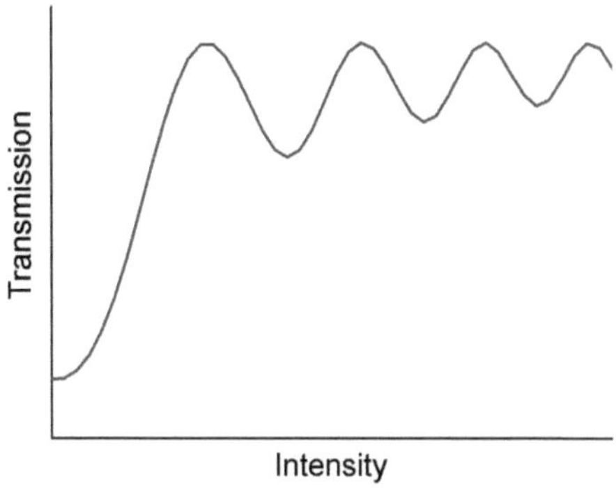

Fig. 3.17 Transmittance of an artificial saturable absorber such as NPE port versus intensity

fluctuation. In the nonlinear stage, when the saturable absorber is bleached, the most intense fluctuation peak is compressed in time and amplified further. Pulse narrowing by the SA is balanced by the gain filtering effect of the active medium leading to the formation of bandwidth-limited mode-locked pulse [21]. In a fiber laser, however, the stable pulses will be generated only when the interplay of nonlinearity, gain and dispersion will be such that the chirping in the pulse can be managed with the help of a dispersion compensating elements, like a grating pair, so that the pulse shape repeats itself after each round trip. Hence, a prior simulation is required for designing the mode-locked fiber oscillator to find the operating regime at which the stable mode-locked pulses can be evolved from the initial intensity noise fluctuations. The propagation of noise fluctuation in each segment of the oscillator can be modeled by the generalized coupled nonlinear Schrodinger equations (NLSE) given by [37]

$$\left\{\begin{array}{l} \frac{\partial A_x}{\partial z} = i\gamma\left\{|A_x|^2 A_x + \frac{2}{3}\left|A_y\right|^2 A_x + \frac{1}{3}A_y^2 A_x^*\right\} + g\left(E_{pulse}\right)A_x - \frac{i}{2}\beta_2\frac{\partial^2 A_x}{\partial t} \\ \frac{\partial A_y}{\partial z} = i\gamma\left\{\left|A_y\right|^2 A_y + \frac{2}{3}|A_x|^2 A_y + \frac{1}{3}A_x^2 A_y^*\right\} + g\left(E_{pulse}\right)A_y - \frac{i}{2}\beta_2\frac{\partial^2 A_y}{\partial t} \end{array}\right. . \tag{3.6}$$

Equation (3.6) also includes the effects of gain and gain filtering due to the finite gain bandwidth. If the parameters given for the oscillator are favorable for mode-locking, then a stable pulse shape will evolve from the initial intensity noise fluctuation after many round trips.

Various types of ultrashort pulse shapes can be generated by mode-locking Yb-doped fiber oscillator by adjusting the interplay of dispersion, nonlinearity and gain. YDF oscillator can be operated mainly in three different mode-locking regimes, namely stretched pulse, similariton and all-normal dispersion depending on the net value of GVD in the cavity as discussed below:

Stretched pulse mode-locked YDF laser: The stretched pulse YDF laser was first demonstrated by Lim et al. [38]. It is so named because the mode-locked pulse undergoes periodic stretching and compression in each round trip. Under the condition of negative GVD, the natural solution of (3.6) is a hyperbolic secant pulse. Such pulses are referred to as soliton pulses as the negative GVD and the

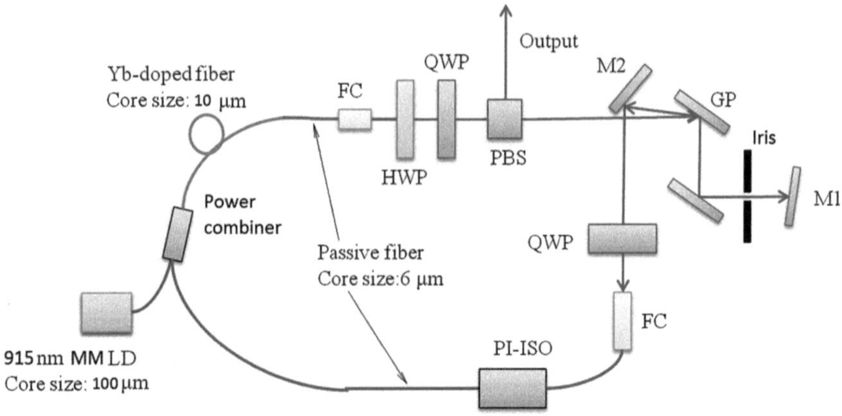

Fig. 3.18 Schematic of dispersion managed soliton fiber laser: *FC* fiber coupler; *HWP* half-wave plate; *QWP* quarter-wave plate, *GP* grating pair; *PI-ISO* isolator; M1, M2, mirrors

nonlinearity cancel each other leading to the pulse shape whose temporal and spectral profiles do not change with the propagation distance. Under normal dispersion (for wavelength ~1 μm in silica fibers), conventional soliton pulse cannot be generated. Instead, soliton-like pulses known as the dispersion managed (DM) soliton or stretched pulses can be formed by employing a negative dispersing element like a grating pair in the cavity to cancel the phase introduced by the positive GVD and SPM in the fiber [39].

The schematic of a typical stretched pulse YDF laser is shown in Fig. 3.18 [40]. The laser set-up comprises of a segment of single-mode silica fiber followed by a gain segment. NPE is implemented by a PBS and wave plates. A grating pair in the near-Littrow configuration is placed after the PBS to provide negative dispersion in the cavity. The grating is therefore also known as the dispersive delay line (DDL). The net GVD of the cavity is kept slightly positive. An optical isolator is placed in the air space to ensure unidirectional operation of the laser so that the mode-locking process is self-starting. By adjusting the wave plates, stable mode-locked pulses are readily observed. As the pulse propagates through the SMF and the gain segment, the temporal and spectral widths increase with a positive chirp due to the interplay of GVD and nonlinearity. The pulse becomes negatively chirped after passing through the grating pair. Afterwards, propagation in the SMF adds positive dispersion. This leads to spectral and temporal compression of the pulse and bringing it back to its original shape. The output can be taken directly from the NPE ejection port (Fig. 3.19).

In Fig. 3.20, we show the simulation results of the temporal and spectral buildup dynamics of the stretched pulse mode-locked YDF laser. It illustrates the following process: An initial burst of broadband thermal noise photons serves as the seed starting the laser oscillation. The nonlinearity begins to affect propagation of these noise photon burst only after its flux reaches some value (after few

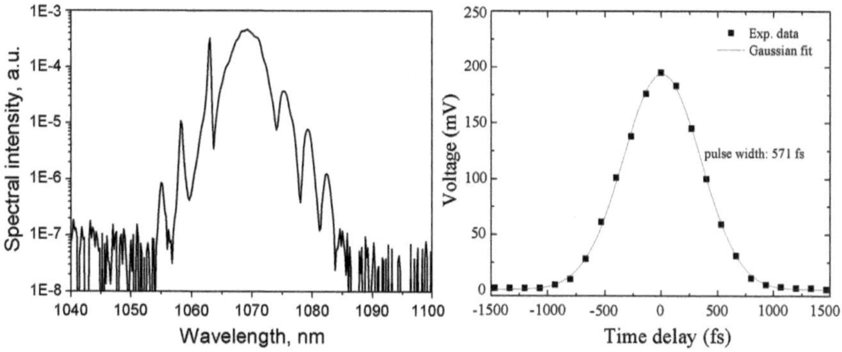

Fig. 3.19 Spectrum (*left*) and autocorrelation trace (*right*) of the dispersion-managed soliton fiber laser like on Fig. 3.18

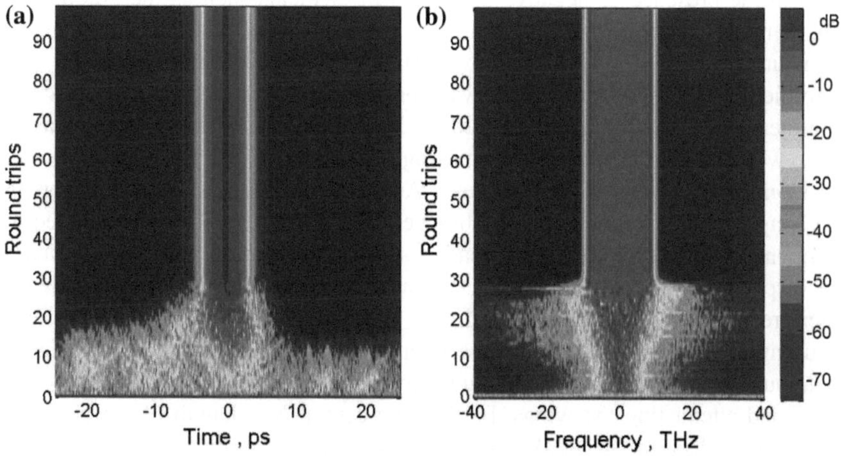

Fig. 3.20 Simulated build-up dynamics in time **a** and spectral **b** domains of the fiber oscillator based on NPE SA

roundtrips). Then, NPE mechanism gets involved in the pulse formation process. Finally, regular pulses start to circulate in the laser cavity after several tens of round trips.

The pulses are chirped inside the gain segment in a stretched pulse mode-locked YDF laser. As a result, their energies can be scaled up significantly as compared to that of a soliton pulse. Pulse energy of 5 nJ with a compressed pulse duration of 50 fs has been demonstrated from a stretched pulse YDF laser [41]. The shortest pulse duration of 33 fs from a YDF oscillator was also achieved using the stretched pulse configuration by employing simultaneous third-order-dispersion (TOD) compensation [42]. Further energy scaling is unfortunately limited by the wave breaking of the pulses and over-driving of the NPE due to strong nonlinear effects in the fiber.

Similariton laser: Similariton pulses are the asymptotic solution of the NLSE under normal dispersion. That is, if certain conditions are satisfied, any input pulse shape would evolve as a parabolic one after propagating through the fiber [43]. Mathematically, a similarton pulse has a parabolic temporal shape as well as quadratic phase profile. This would naturally lead to linear frequency chirping. A similariton pulse propagates through the fiber in a self-similar manner. That is, the spectral and temporal profiles are always a scaled version of itself even in the presence of high gain or loss. The configuration of this class of lasers is similar to that of the dispersion managed soliton laser (See Fig. 3.18) but has a large amount of net positive GVD obtained by increasing the undoped single-mode fiber length. The purpose of the grating pair is to compress the pulse to its initial width. For the self-similar pulse to evolve to the steady state, its pulse energy should be high enough (requirement of high gain) and should pass through long distance in the undoped fiber. It should be noted that the pulse always has a linear positive chirp even in the presence of negative dispersion of elements such as the grating pair. The temporal shape of a similariton pulse has a quadratic phase profile, a unique signature of such a pulse. Therefore, similariton pulses are best-suited for fiber amplifiers as they are resistant to wave-breaking even in the presence of strong nonlinearity. Due to the linear chirp, there is no limitation on energy scaling as the pulse would not suffer from wave-breaking.

All-normal dispersion (ANDi) laser: An all-normal dispersion laser is perhaps the simplest kind of mode-locked Yb-doped fiber oscillators. There is no need for components for dispersion compensation such as the grating pair [44]. Typically, the dispersion in the cavity of an ANDi laser is managed by a narrow-bandwidth interference filter. The wings of a highly chirped pulse will be cut by the filter, resulting in a shorter pulse. Thus, the filter leads not only to pulse width management but also to a strong self-amplitude modulation by transmitting the peak of the pulse and attenuating the wings. Further, the central wavelength of the pulse can be tuned by changing the peak transmission wavelength of the filter. As there is no negative dispersing element, the net GVD in the cavity is very high and positive. As the pulse propagates through the SMF, the pulse spectrum broadens significantly and develops a structured profile with steep edges due to strong SPM. The narrow band filter cuts down the edges of the spectrum and hence the pulse wings to restore it to its initial form. Even with highly structured spectral profile, the temporal shape is smooth with a linear chirp near the center of the pulse. The breathing of spectrum and the temporal shape are not as large as the other regimes of the mode-locking.

3.2.6 Noise-Like Pulse Generation from a YDF Laser

There has been increasing interests recently in the study of a special regime of passively mode-locked fiber lasers, that of the so-called noise-like pulses [45–49]. These pulses are relatively long (sub-ns) wave packets, which exhibit a fine inner structure of sub-ps pulses with randomly varying amplitude and duration [50, 51].

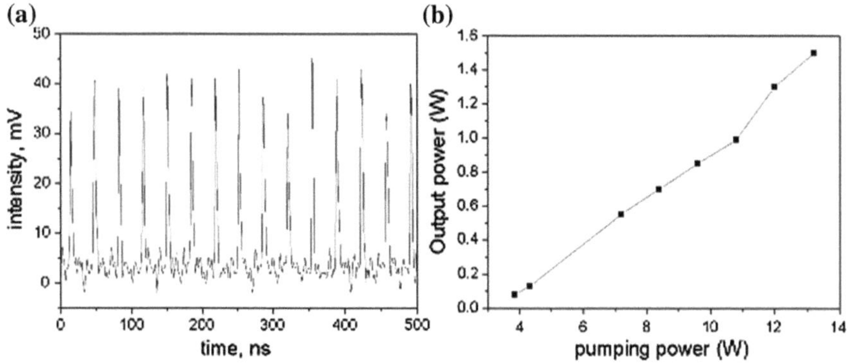

Fig. 3.21 Oscilloscope trace of the train of noise-likes pulse (**a**); Output power of the noise-like pulse laser versus pumping power (**b**)

The properties of such kind of pulses include: (1) a very large optical bandwidth (usually of several tens of nm); (2) a double-scaled autocorrelation trace with a sub-ps peak riding a wide sub-ns pedestal; (3) low temporal coherence. Moreover, it was found [50] that narrow peak in the autocorrelation trace of noise-like pulses or NLPs could be maintained even after the pulses propagated through a long dispersive medium. This unique property of noise-like pulses contrasts drastically with the propagation characteristics of transform-limited pulses of similar bandwidth. Light sources with such narrow autocorrelation trace and broadband spectrum which can propagate for a long distance are very useful in such areas as optical metrology (for example, in optical coherent tomography or lidar), where short coherence lengths are essential and ultimately required [52]. It was also demonstrated [53, 54] that amplified noise-like pulses could be successfully used for SC (supercontinuum) generation.

Recently, it was found [45–47, 49, 55] that different laser structures can operate in noise-like regime. The current challenges are focused on new wavelength bands [47], higher pulse energy generation [48, 55] and tuning the temporal and spectral properties of the noise-like pulses [45, 51]. To the best of our knowledge, 28 nJ as the highest pulse energy was achieved for noise-like pulses [48]. This was achieved using an ultra-long laser cavity (>100 m).

In our studies, we used a modified ring fiber laser cavity, similar to those used for the stretched pulse mode-locked YDF laser (see Fig. 3.18). The same components, e.g., an Yb-doped fiber, NPE-port, dispersive delay line, and an isolator, were employed. Details can be found in [40].

After the single-pulse operation was obtained in this laser, we tuned one of the wave plates while holding all the other experimental conditions unchanged. It is possible to shift the laser operation from the conventional single-pulse operation into the noise-like pulse emission regime. This behavior was also reported by previous workers [46, 48]. Figure 3.21a shows the oscilloscope trace of the laser output as measured by a fast InGaAs detector. It is found to be typical for noise-like

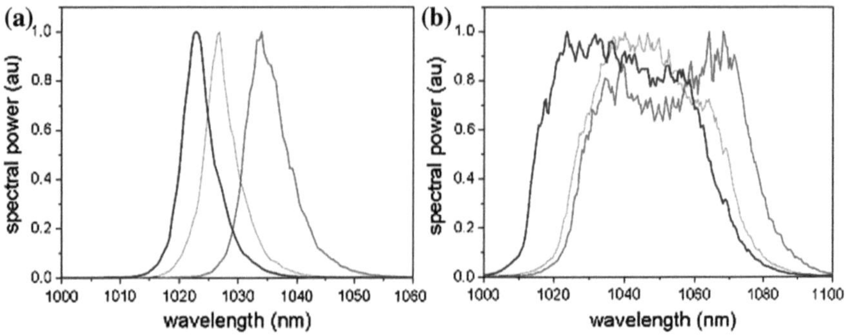

Fig. 3.22 Optical spectra of the measured pulses for different positions of the iris with 1 mm (**a**) and 3 mm (**b**) widths

pulse emission state [41–43, 46]. We observed, however, self-starting noise-like laser operation for a relatively large range of pumping powers. The output powers of noise-like pulses can be varied from <0.1 to >1.6 W (see Fig. 3.21b).

We attribute this improved stability to the laser design where nonlinear polarization evolution is accompanied by self-amplitude modulation induced by spectral filter. The role of negative group velocity dispersion (GVD) induced by gratings pair is also important. In our experiments, this term was set to -0.11 ps^2. According to our estimation, that value is smaller than the total net positive GVD due to the fiber (~0.15 ps^2). Finally, the observed maximal pulse energy at ~13 W pumping power reaches 45 nJ. After increasing pumping power higher than 14 W, one of the fiber couplers was damaged.

In the experiments, we found that by translating the iris transversely across the laser axis, it is possible to modulate the central wavelength of generated pulses. Figure 3.22 demonstrates such results for both narrowband (left) and broadband (right) noise-like pulses. The tuning range for both cases reaches 12 nm. That could be important for some applications. For example, in the case of further amplification of generated pulses by fiber amplifiers, it is necessary to match the bandwidth of oscillator with the gain spectrum of amplifier.

We also found that the diameter of the iris determining filter bandwidth also affects (see Fig. 3.23) the output spectral bandwidth (left) and duration (right) of generated noise-like pulses. In order to understand NLP formation, we simulate the buildup dynamics of our laser cavity by recognizing it as consisting of several connected fiber components. Pulse propagation in each fiber section was described by the corresponding nonlinear Schrodinger coupled-mode equations [see (3.6)]. Dispersive delay line was modeled by introducing a negative GVD of ~0.02 ps^2. High-order dispersion terms were neglected. Wave plates and the polarization beam splitter (PBS) in the cavity were represented by their equivalent Jones matrices in the calculation. For modeling of propagation through each fiber section, we employed the split-step Fourier method [20]. Assuming a thermal Gaussian noise as a source of oscillations, we found the conditions where repeated noise-like

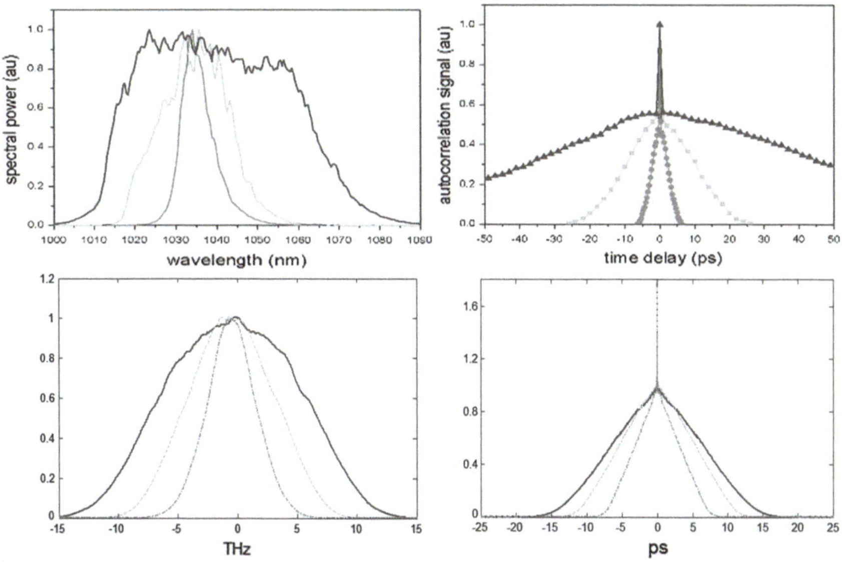

Fig. 3.23 Measured (*top*) and simulated (*bottom*) optical spectra (*left*) and autocorrelation traces (*right*) of output pulses for different (1, 2, and 3 mm) widths of the iris

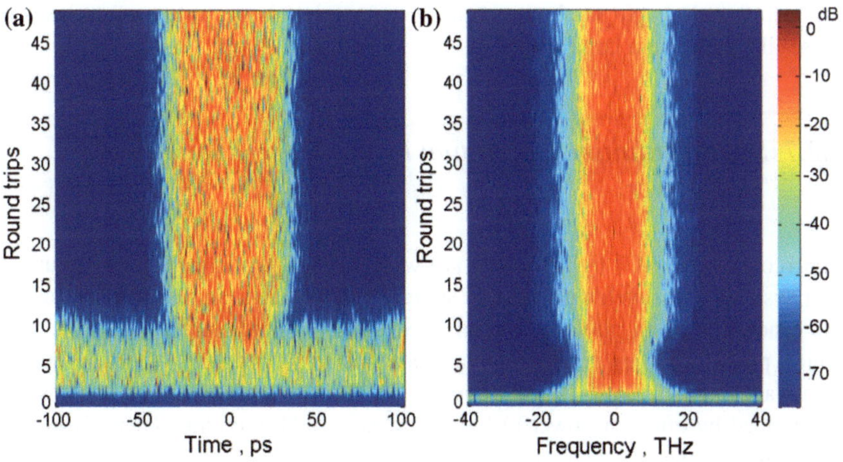

Fig. 3.24 Building up dynamics in time (**a**) and spectral (**b**) domains of a fiber ring oscillator generating NLPs

pulses begin to circulate in the laser cavity after just a few round-trips (see Fig. 3.24). It is significant that all nonlinear terms (self-phase modulation, cross-phase modulation, four-wave mixing, etc.) in (3.6) contribute to NLP creation.

The simulation results for the steady state are in good agreement with experimental observations. The estimated iris filter bandwidth, corresponded iris diameter, output bandwidths and bunch duration are listed in Table 3.3.

Table 3.3 The summary of measured and simulated results for NLPs

Iris diameter (mm)	Iris bandwidth (THz)	Measured output bandwidth (THz)	Calculated output bandwidth (THz)	Measured autocorrelation trace half-width (ps)	Calculated autocorrelation trace half-width (ps)
1	5	2.1	5	5	7
2	10	4.2	9	20	12
3	20	13.4	15	90	20

We noted that larger output bandwidth corresponds to smaller duration of waveforms. This is in agreement with observation by previous workers [45]. The durations of wave packets in our case are, however, an order of magnitude shorter than those reported in [45]. Tuning the 3-dB wavelength bandwidth of the NLP laser from 7.6 to 48.2 nm, we were able to vary the duration of NLPs from 90 to 5 ps.

In summary, by using cladding-pumped Yb-doped active fibers and high-power multimode laser diodes, we demonstrated that a passively mode-locked fiber laser can be tuned to generate noise-like pulses with relatively large energies. To the best of our knowledge, the pulse energy of 45 nJ obtained in this work is the highest for noise-like pulses generated in fiber oscillators reported to the date. The achieved energy is only limited by damage threshold of available fiber couplers. We also found simple method to control temporal and spectral characteristics of noise-like pulses in a large range.

3.2.7 Supercontinum Generation by the Noise-Like Pulsed YDF Laser

Supercontinuum (SC) light sources are widely employed in optical communication systems, as wavelength tunable sources, gas sensing, and optical metrology. SC generation using photonic crystal fibers (PCF) or microstructured fibers [56], which provide higher nonlinearity and allow one to blue-shift the zero-dispersion wavelength (ZDW), is quite popular. PCF-based supercontinum sources can be pumped by efficient powerful light sources at ~800 nm and ~1 μm. However, PCFs are still expensive and not widely available.

Standard single-mode fibers (SMF) for optical communication, on the other hand, are inexpensive and easy to integrate. Compared to PCFs for generation of SC, which normally feature a non-circular symmetry, light propagated through SMF usually exhibits much better spatial beam profile. On the other hand, SC generation in SMFs operating in the normal dispersion regime is not so efficient. It requires much more pulse energy to excite SC compared to the anomalous regime. Also, the SC spectrum so generated exhibits very strong oscillations [57].

Recently, SC generation was reported in a piece of standard fiber (SMF-28) using as the pump a train of NLPs at the central wavelength of 1.5 μm [54]. To excite a flat broadband SC, the energy threshold was as low as ~12 nJ.

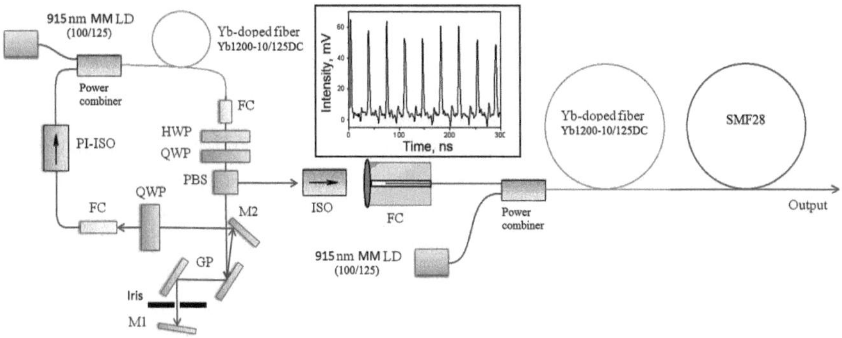

Fig. 3.25 Schematic of the experimental setup: *FC* fiber coupler; *HWP* half-wave plate; *QWP* quarter-wave plate, *GP* grating pair; *PI-ISO* polarization-insensitive isolator; *ISO* Faraday isolator; *M*1 and *M*2 mirrors; *MM LD* multi-mode laser diodes

In the previous section, we demonstrated that a dispersion-mapped Yb-doped fiber laser based on the ring cavity design can generate noise-like pulses with energies as high as ~45 nJ and controllable characteristics at ~1 μm wavelength. The use of negative dispersion delay line and the spatial spectral filter were found to be important for such high-power noise-like operation (see Fig. 3.25). In the NLP regime and pumped at ~10 W, the oscillator irradiates typically a pulse train (see inset at Fig. 3.25) with repetition rate of ~31.5 MHz, average power of ~800 mW and average noise-like bunch duration of ~35 ps.

The iris in the cavity was used to tune the center wavelength from ~1,030 to ~1,070 nm with a bandwidth of ~11 nm (FWHM) [37]. These noise-like pulses were then boosted up to ~3 W in a single 2.3 m-length Yb-doped amplifier stage and used as a pump to excite the SC. A spool of 100 m length of standard SMF (SMF28, POFC, Taiwan) was directly spliced to the output fiber end of the amplifier.

We studied theoretically SC generation in a standard SMF by pumping with either NLP or mode-locked Gaussian pulses of similar pulse energy. The results are calculated by solving the general nonlinear Schrodinger equation in the frequency domain [58]

$$\frac{\partial A'(z,\omega)}{\partial z} = i\frac{\gamma\omega}{\omega_0}\exp\{-L(\omega)z\}F\left\{A(z,T)\int_{-\infty}^{\infty} R(T')\left|A(z,T-T')\right|^2 dT'\right\}, \tag{3.7}$$

where A and A' represent the electric field envelopes [58], F{} denotes the Fourier transform, R(t) is the Raman response function [20]

$$R(t) = (1-f_R)\delta(t) + f_R\frac{\tau_1^2+\tau_2^2}{\tau_1\tau_2^2}\exp\left(-\frac{t}{\tau_2}\right)\sin\left(-\frac{t}{\tau_1}\right)\Theta(t), \tag{3.8}$$

where $f_R = 0.18$ is the fractional contribution of the delayed Raman response; $\tau_1 = 12.2$ fs, and $\tau_2 = 32$ fs. $\Theta(t)$ is the Heaviside step function and $\delta(t)$ is the Dirac delta function. In (3.7), $\boldsymbol{L}(\omega)$ is the linear operator representing dispersion and absorption behaviors of the SMF, given by $\boldsymbol{L}(\omega) = i(\beta(\omega) - \beta(\omega_0) - \beta_1(\omega_0)(\omega - \omega_0)) - \alpha(\omega)/2$. GVD and third-order dispersion (TOD) were assumed to be 0.02 ps^2/m and 25×10^{-6} ps^3/m, typical for SMF28. The non-linear coefficient γ was assumed to be 0.0025 $(W\ m)^{-1}$. Absorption in fiber was neglected. This approach allows us to treat (3.7) as an ordinary differential equation (ODE) and to use powerful numerical methods, e.g., the 4th order Runge-Kutta method, to solve it. The simulation results for three different types of pulses with pulse energy of 200 nJ each and propagating through 50 m of SMF are presented in Fig. 3.26.

Examining Fig. 3.26, we find that SC evolution in the case of NLP pumping is different from those of mode-locked picosecond or femtosecond Gaussian pulses. It is interesting to note that NLPs, in comparison to mode-locked Gaussian pulses of similar waveform duration (~40 ps), exhibit similar broadening in the time domain after propagation [see Fig. 3.26b and d]. On the other hand, NLPs can generate very broad SC covering the spectral range from 1,030 to 1,700 nm [see Fig. 3.26a and c]. Note that picosecond Gaussian pulses (~40 ps) can only generate three clear Raman peaks in its spectrum (Fig. 3.26c). As for the shorter mode-locked Gaussian pump pulse (200 fs), it can also generate SC from 830 to 1,500 nm (Fig. 3.26e) but the peak pump power needed is 2 orders of magnitude higher than the corresponding NLP of similar pulse energy (~200 nJ). Also, 200 fs-wide Gaussian pump pulses excite SC in first tens of centimeters of the SMF (see inset in Fig. 3.26). Then, its spectrum does not change (see Fig. 3.26e). The evolution of NLP-excited SC takes longer distance in the fiber. Moreover, the spectral broadening occurs mainly in the longer-wavelength (relative to the pump wavelength) region. Recently, it was theoretically predicted that Raman-induced spectral shift, which may happen even in the normal dispersion region is responsible for such an asymmetry [59]. Because of the relative narrow band of the NLPs, this could not be the most effective process for SC generation. The distinctive Raman orders in Fig. 3.26c for mode-locked pulses suggested that multiple Raman processes, i.e., cascaded Raman scattering could have happened for NLP pumped SMF. Therefore, we have tentatively attributed our calculated behavior of the broadband SC emission to Raman amplification of noise in the SMF [56]. The smooth SC spectrum in the case of NLP pumping can be explained by broadband pumping [60], which is a feature of NLP [50].

The spectra of the SC output for different values of average power at the fiber input are shown in Fig. 3.27a (dotted lines). The corresponding simulated SC spectra are plotted as solid lines in Fig. 3.27a. The characteristic asymmetrical spectra hinted that the broadband emission is due to Raman amplification of NLPs [50]. The pulse energy threshold (~43 nJ) to achieve significant spectral broadening, which corresponds to ~1 W of average pump power, is ~4 times larger than that found in the case of work in the anomalous dispersion regime [54]. On the other hand, compared to other reports of SC generation pumped

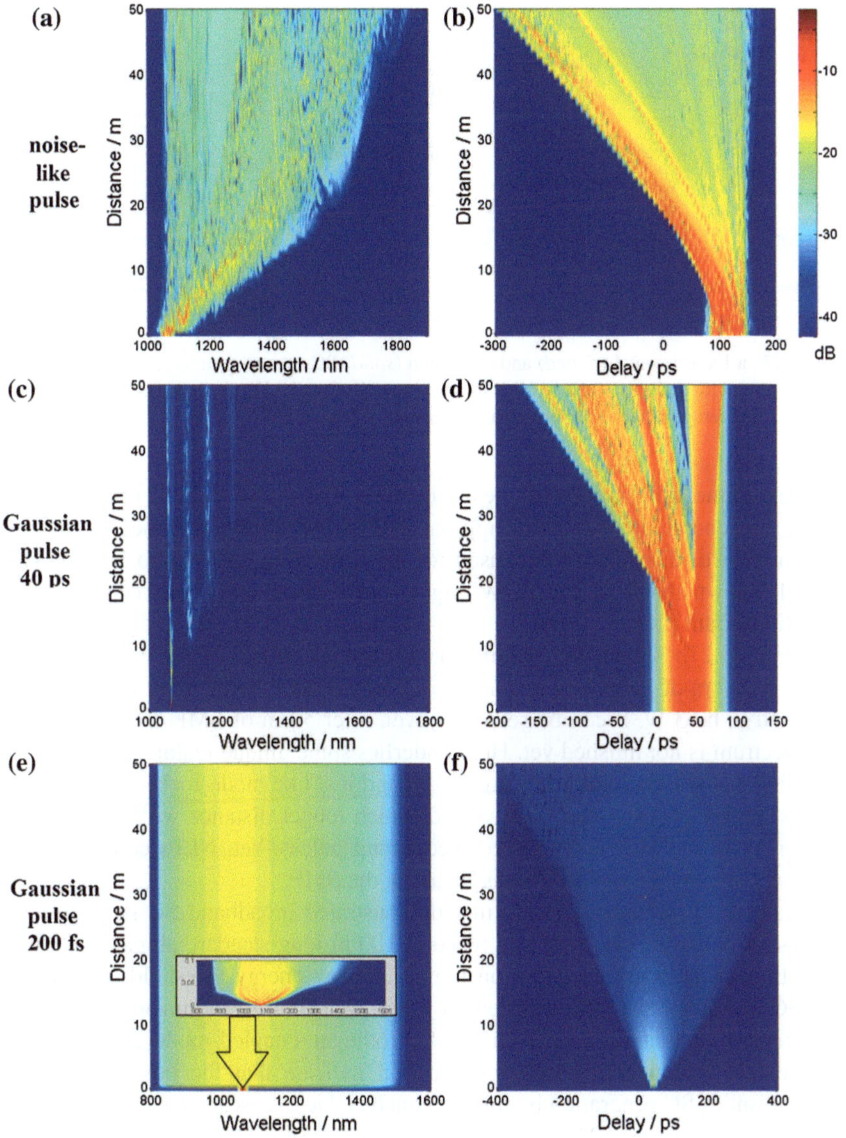

Fig. 3.26 Calculated SC evolution in spectral (**a, c, e**) and time (**b, d, f**) domains for different pump pulses: noise-like pulse (**a, b**), 40 ps-wide mode-locked Gaussian pulses (**c, d**), and 200 fs-wide mode-locked Gaussian pulses (**e, f**)

in the normal-dispersion regime [57, 60–62], the measured threshold is from 1 to 3 orders of magnitudes lower. The measured spectra generated by cascaded Raman scattering is quite uniform, in particular for the 1,050–1,250 nm region. Also, as it can be seen from Fig. 3.27a that the simulated SC spectra matched

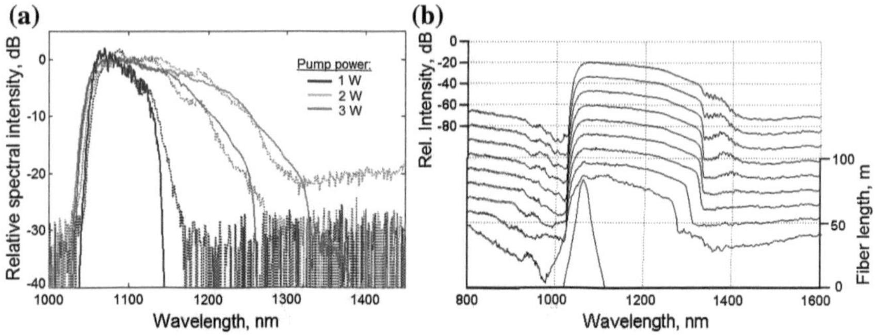

Fig. 3.27 **a** Experimental (*dotted*) and simulated (*solid*) SC spectra generated in 100 m of SMF by pumping with different average NLP input powers (1, 2 and 3 W); **b** simulated spectral evolution of 3 W average power NLP propagating through 100 m of SMF

with experimentally measured ones quite well. The discrepancy is only observed when the generated components approach ZDW. We think that is because of limited number of dispersion terms used for the simulation (only up to 3rd order). It is well known that near the ZDW, higher-order dispersion terms become much more important [20].

Figure 3.27b shows the simulated spectral evolution of NLPs propagating through 100 m of SMF. The average power of the NLP at the input of the SMF is assumed to be 3 W. It can be seen that even after 50 m of SMF, the evolution of SC spectrum is not finished yet. Here underlies some unique features of SC generation by NLPs. First of all, they are broadband just like mode-locked femtosecond pulses. Secondly, they can propagate for much longer distance without distortion and breakup just like mode-locked picosecond pulses. Yet, NLPs contribute new spectral components as they propagate along the SMF.

In summary, we have successfully demonstrated broadband SC generation by noise-like pulses propagating in a piece of 100 m-long standard single-mode fiber operating in the normal dispersion regime. A low energy threshold (43 nJ) and flat SC spectrum over the wavelength range of 1,050–1,250 nm were achieved. Theoretical simulations based on the Schrodinger coupled-mode equations and the general nonlinear Schrodinger equation indicate that the possible physical mechanism of SC generation by this approach is due to cascaded Raman scattering and Kerr effect in the SMF. The Raman effect is responsible for significant spectral broadening the longer wavelength regions whereas the Kerr effect results in smoothing of SC generated spectrum pumped by NLPs.

We believe the achieved low energy threshold and flat SC spectrum are caused by special properties of NLP (broadband spectral range and ability to propagate over a long distance). This new SC light source exhibit attractive characteristics that are potentially competitive to those of the currently used technologies including PCF-based approaches.

3.2.8 Nonlinear Conversion of Picosecond Bursts from Yb-Doped Fiber Laser Amplifiers

Nowadays, UV lasers with the central wavelength at 266 nm are widely used in the fields of medical surgery, chemistry, photolithography, and material sciences. For example, UV light source provide higher resolution and better alignment control in photolithography. Also, the semiconductor industry requires optical inspection tools using lasers with shorter and shorter wavelengths. Because of the continuous reduction in feature size, such sources are essential in order to reach the necessary imaging resolution for wafer inspection [63].

UV laser sources are also popular for scribing and cutting of sapphire and GaN [63]. LED manufacturers increasingly employed either the Q-switched (Nd:YAG or Yb-doped fiber) lasers or cutting tools with diamond blades for wafer scribing. By using the Q-switched laser, the precision of scribing is smaller than 100 μm. This is much better than that achieved by using the diamond blade for scribing. The peak power of the high-repetition-rate Q-switched Nd:YAG or Yb-doped fiber laser, however, is typically not high enough to reach the ablation threshold of many materials. To get better performance in material processing, high pulse energy and shorter wavelength laser sources are desirable.

Recently, laser ablation of metals using bursts of picosecond pulses has been investigated [64–67]. Compared with nanosecond single pulses, there are some benefits for material micromachining in terms of material removal efficiency and the surface quality by using bursts of picosecond pulses. Most of ultrafast laser systems have pulse trains with pulses that are equally spaced in time. There are much interests in lasers that generated a burst of limited number of short pulses. One of the reasons why burst pulses are advantageous for material ablation is that they can keep thermal effect minimized. Thermal effect always depends on the average power irradiated but ablation depends on the peak power. By controlling the repetition rate of the burst, one can easily reduce the unwanted thermal effect. On the other hand, highly repetitive pulses with high pulse energy within the burst not only can enhance the efficiency of material ablation but also improve the efficiency of frequency conversions.

To date, burst-pulse lasers are almost always based on solid-state laser systems. In this work, we constructed Yb-doped fiber based laser amplifier systems that can generate bursts of high-repetition rate picosecond pulses. The energy stored in the gain fiber can be extracted efficiently. We also found the improvement of frequency doubling and frequency quadrupling efficiencies by using LBO and BBO nonlinear crystals respectively using such burst pulses.

Our theoretical analysis is based on our previous work [32] and summarized in Sect. 3.2.1. The small-signal gain G_0 and the saturation energy E_{sat} for the active optical fiber can be determined accordingly. Extending similar works

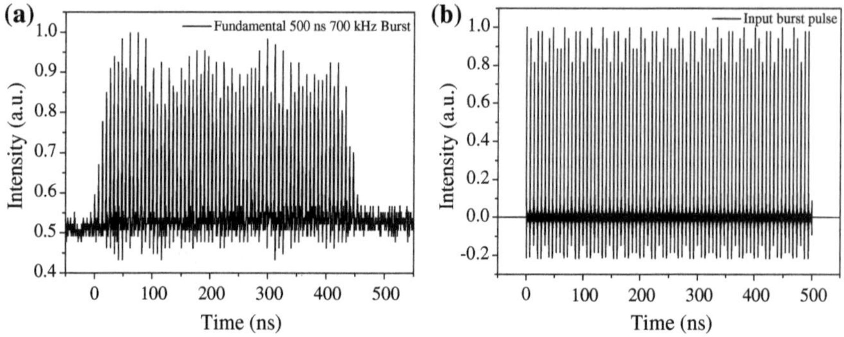

Fig. 3.28 Experimental measurement (**a**) and simulated data (**b**) of input burst pulse with burst duration 500 ns and 700 kHz repetition rate within 10 ps pulse duration and 150 MHz repetition rate

on nanosecond burst pulses, [68–70] the output power P_0 due to amplification of the picosecond burst pulses with energies of $E_i(t)$ can be calculated using the following:

$$P_0(t) = \frac{P_i(t)}{1-(1-1/G_0)e^{(-E_i(t)/E_{sat})}} \tag{3.9}$$

where

$$E_i(t) = \int_{-\infty}^{t} P_i(t')dt' \tag{3.10}$$

The energy $E_i(t)$ of the input picosecond burst can be calculated by integration of input burst in time from 0 to the current time t. From our previous experiment [28], we estimated that the saturation energy of the YDF, E_{sat} ~ 36.7 μJ. The small signal gain is about 30 dB. Assuming rectangular shape for the input burst, the The experimentally measured and simulated input burst pulses are shown in Fig. 3.28a and b, respectively. In the simulation, we assume 500 ns-wide burst at a repetition rate 700 kHz. Within each burst, there are 10 ps-wide Gaussian pulse at a repetition rate of 150 MHz. These are in accordance with our experimental conditions.

Figure 3.29a and b, on the other hand, show the experimentally measured and theoretically simulated amplified bursts. They indicate a dynamically saturated behavior of fiber amplifier. Compared Figs. 3.28 and 3.29 it can be seen that initially quasi-rectangular envelope input bursts are transferred to exponentially-modulated ones on the amplifier output. That behavior is similar to the case of ns-pulse amplification [68].

Assuming the fundamental beam from the amplified YDF output is not depleted, the intensity of the second-harmonic signal generated in the plane wave approximation is given by

$$I_{2\omega}(z) = 2Z\kappa^2 I_{\omega}^2(0)z^2 \sin c^2(\Delta kz/2), \tag{3.11}$$

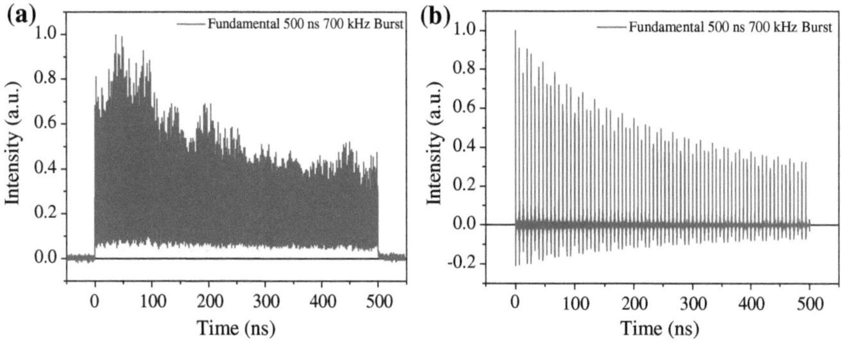

Fig. 3.29 Experimental measurement (**a**) and simulated data (**b**) of output burst pulse train after amplification

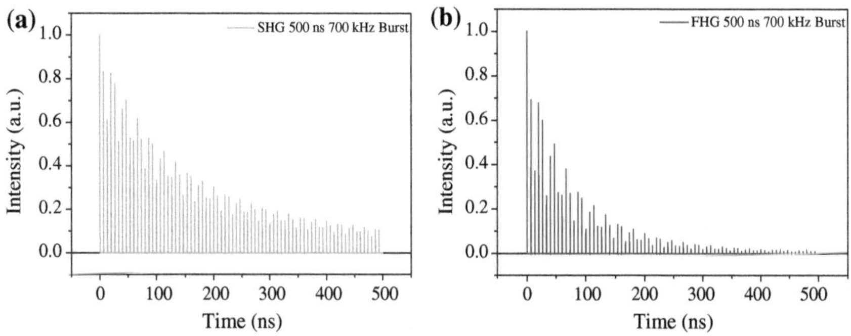

Fig. 3.30 Theoretically simulated results for frequency-doubled (**a**) and quadrupled (**b**) pulse bursts from the MOFA

where $Z = (\mu_0/\varepsilon_0)^{1/2}/n$ is the impedance of the crystal, $\kappa = d_{eff}\ \omega/c_0\ n$ is the coupling coefficient, Δk is the wave vector mismatch and z is the crystal length. The intensity of fourth-harmonic signal can be represented in a similar way. Figure 3.30 shows the simulated pulse trains after frequency doubling (a) and quadrupling (b) of the fundamental pulse burst (see Fig. 2.29b). It can be seen that the shapes of doubled and quadrupled pulse burst become more and more asymmetrical.

We then estimate the nonlinear conversion efficiency for SHG by the pulse burst by integrating the converted SHG within the burst:

$$\eta = \frac{\int_0^{\tau} P_{2\omega}(z,t)dt}{\int_0^{\tau} P_{\omega}(z,t)dt} = \frac{\int_0^{\tau} \frac{2Z\kappa^2 P_i^2(0,t)z^2\ \sin c^2(\Delta kz/2)}{A}dt}{\int_0^{\tau} \frac{P_i(t)}{1-(1-1/G_0)e^{(-E_i(t)/E_{sat})}}dt}, \quad (3.12)$$

where τ is the period of the burst and A is the effective area of the beam on the nonlinear crystal. Equation (3.12) predicts a conversion efficiency η of about

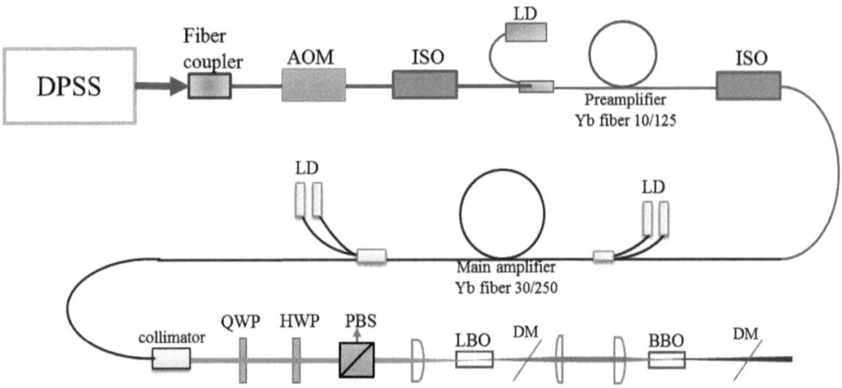

Fig. 3.31 Schematic drawing of the MOFA system and frequency conversion. *AOM* acousto-optic modulation. *ISO* isolator; *LD* laser diode; *QWP* quarter-wave plate. *HWP* half-wave plate. *PBS* polarization beam splitter. *DM* dichroic mirror

17.6 % for the pulse burst versus 8 % for the CW pulse train at 150 MHz. Frequency quadrupled output of the pulse burst can similarly be simulated. The experimental setup is shown schematically in Fig. 3.31.

The MOFA has been described in the previous section. LBO and BBO crystals were used for frequency doubling and quadrupling of the MOFA output. Since non-PM fibers were used, we realized the desired polarization states with appropriate wave plates and polarizing beam splitters. An acousto-optic modulator (AOM) was used to create the desired pulse bursts and control their repetition rate.

To enhance the frequency conversion efficiency, higher peak power from the amplifier output is desirable. For the system described in [25], the output peak power is limited by the pumping power. To enhance the peak power without further pumping, we take advantage of dynamically saturated gain of the fiber amplifier. That is, the front part of pulse burst would experience much larger gain than the tail part of the burst.

We plotted the average output power of the MOFA as a function of pumping power in Fig. 3.32a for the case of regular pulse train (without AOM) and for the case of pulse burst (with AOM). It should be mentioned that the only the spectral output near 1,064 nm was taken into account. From Fig. 3.32a, it is apparent that the output signal power (wavelength at 1,064-nm) become saturated when the pulse burst was amplified at relatively high pumping power (larger than 50 W). Figure 3.32b indicates that the SPM and SRS became significant at these high pumping powers. Thus saturation of the amplifier output is attributed to the presence of these nonlinear effects.

In order to optimize the amplifier performance, we need to prevent the amplified pulse burst from reaching thresholds of these nonlinear effects. One possible way is to reduce the fiber length [20]. The fiber lengths for the preamplifier and the main-amplifier were reduced to 3 and 3.5-m long, respectively. Figure 3.33a

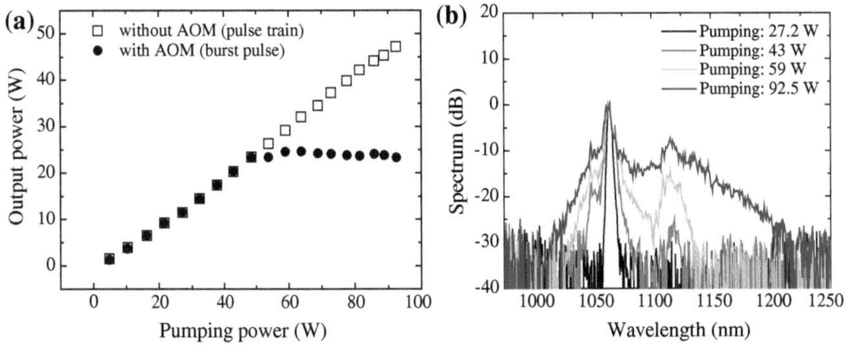

Fig. 3.32 **a** Output power (in a *narrow band* around 1,064 nm) as a function of pumping power with pulse train and burst pulse; **b** Optical spectrum of the burst pulse output

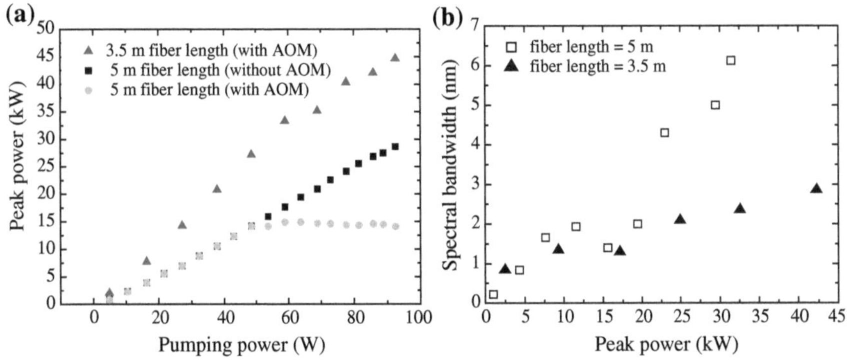

Fig. 3.33 **a** Output peak power as a function of pumping power; **b** the spectral bandwidth as a function of peak power of the 1,064 nm pulse

illustrates that the output peak power of the 3.5-m long fiber length in main-amplifier with using AOM is significantly higher than in the other cases. This is advantageous for applications such as frequency conversion.

In Fig. 3.33b, we show the spectral bandwidths of the burst-mode MOFA as a function of the peak powers for fiber lengths of 5 and 3.5 m in the main amplifier. The bandwidth for the MOFA with shortened fiber can be controlled to ~3 nm at the highest pumping power studied. This is suitable for frequency doubling. The pulse width of the amplified burst was measured by a second-harmonic intensity autocorrelator. The typical second-harmonic intensity autocorrelation trace and corresponding optical spectrum of the burst pulse are shown in Fig. 3.34. The autocorrelation trace is best fit by a double-exponential with FWHM of the pulse intensity profile ~18 ps.

For SHG, we focused the burst-mode MOFA output to a waist size of ~60 μm in the center of a critically phase-matched lithium triborate (LBO) nonlinear crystal. The LBO crystal was 10-mm-long ($\theta = 90°$, $\phi = 11.2°$) and operated at room

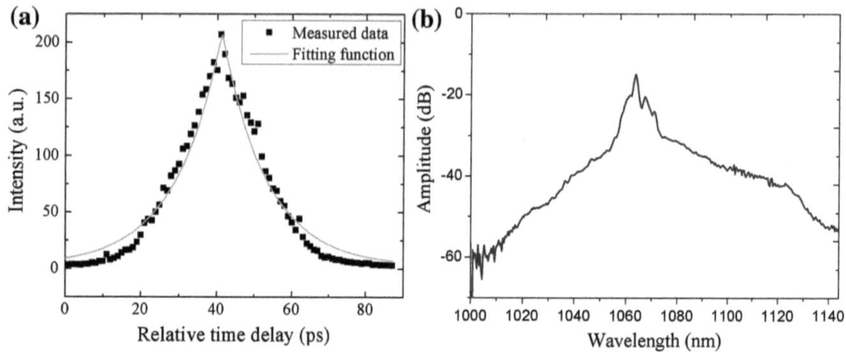

Fig. 3.34 The autocorrelation trace (**a**) and corresponded optical spectrum (**b**) of burst pulse on the MOFA output at ~80 W pump power

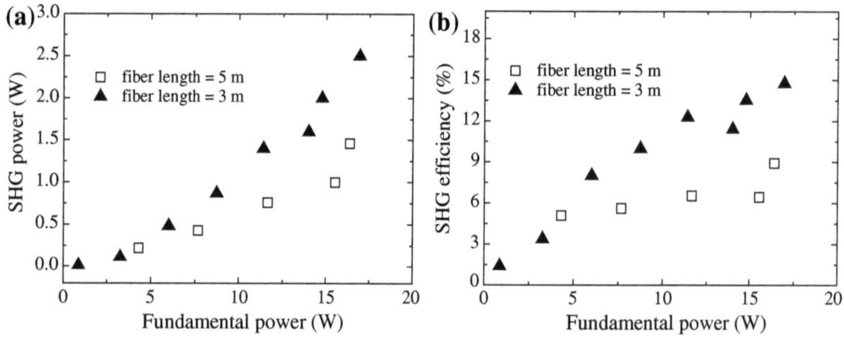

Fig. 3.35 The dependence on fundamental power of **a** average second-harmonic power and **b** efficiency generated by the burst-mode MOFA with two different fiber length for the main amplifier

temperature. After the LBO crystal, a dichroic mirror (DM) is employed to remove the unconverted fundamental light. Lenses were installed to reimage the second-harmonic beam frequency quadrupling.

Figure 3.35 shows the second-harmonic signal generated by the burst-mode MOFA with two different fiber length for the main amplifier. The results show that the output power for MOFA with 3.5-m long Yb-doped fiber is larger than that by the 5-m long Yb-doped fiber in the main-amplifier stage. This is due to the higher fundamental peak power (Fig. 3.33a) and the smaller spectral width (Fig. 3.32b) output generated by the shorter active fiber. Both of these characteristics contribute to the higher conversion efficiency of SHG process (Fig. 3.34b). The 10-mm LBO crystal is expected to have spectral bandwidth acceptance for phase matching of around 3 nm. This matched the spectral bandwidth of the fundamental output from the MOFA system with the shorter active fiber, as can be seen from Fig. 3.32b.

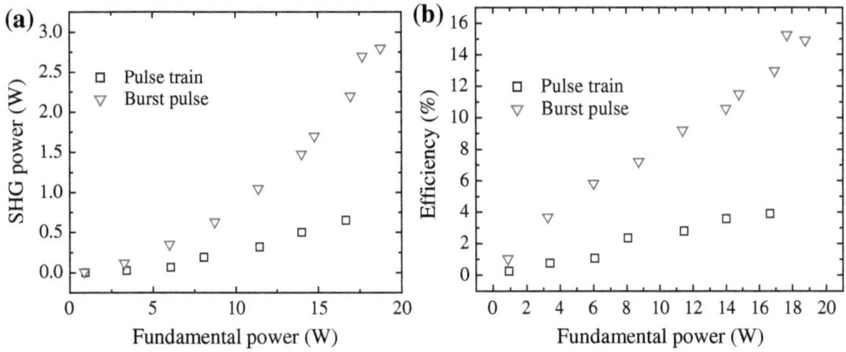

Fig. 3.36 SHG Experimentally measured **a** SHG output power and **b** conversion efficiency of the MOFA as a function of fundamental power. Data for the regular pulse train and burst pulse cases, respectively

We have compared performance of SHG by the regular pulse train output and burst-mode MOFA using the 3.5-m long Yb-doped fiber in the main-amplifier stage. The results are presented in Fig. 3.36. It is obviously to see that the burst pulse have better results than the regular pulse train. The main reason is not only the burst pulse provides the higher peak power, but also the lower thermal influence on crystal when beam passing through LBO. That is caused by lower average fundamental power applied for nonlinear crystal. So, the temperature stabilization is more stable than in the case of regular pulses. The results show that we can obtain ~3 W of SHG power in the burst pulse condition.

We also report fourth harmonic generation (FHG) of our YDF MOFA system. The nonlinear crystal for generation of 266-nm UV light is a beta-barium borate (BBO) crystal, 5-mm-long and cut for type I critical phase matching ($\theta = 47°$, $\phi = 0°$) and operated at room temperature. The second DM installed after the BBO crystal serves to remove the unconverted green light.

Because the efficiency of FHG in the case of regular pulse train is quite low (~1–2 % at maximum), we only present results for the burst-mode of the MOFA. At an average second harmonic input power of 0.5–2.5 W, the typical power of fourth harmonic beam is 20–200 mW. The efficiency of frequency doubling from 532 nm beam is 2–8 % (see Fig. 3.37a). Figure 3.37b summarized the pulse energy and efficiency for generated second and fourth harmonic as a function the fundamental pulse energy. From the figures, we can see that peak powers in excess of 6.7 kW for the second harmonic (532-nm) and 0.6 kW for the fourth harmonic (266-nm) of the MOFA in burst-mode were generated.

In summary, we present an experimental and theoretical study of the generation of picosecond bursts by an Yb-doped fiber MOFA system, which was used to generate 3 W of green and 200 mW of UV light with the conversion efficiencies are 16 and 8 %, respectively. The enhancement of conversion efficiency by the

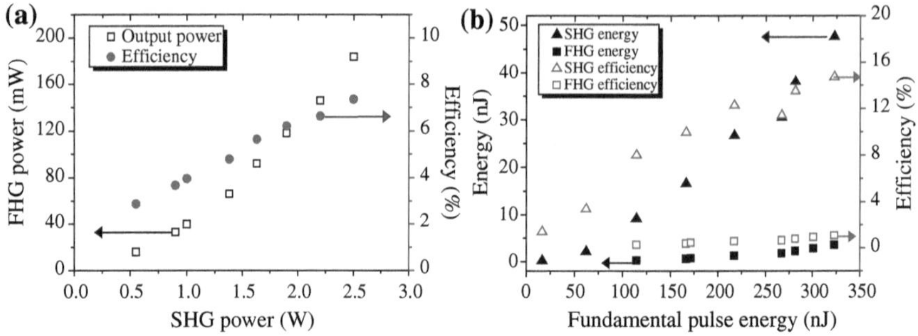

Fig. 3.37 **a** Performance of fourth harmonic generation; **b** pulse energies and efficiencies of SHG, FHG of the output of the dual-stage MOFA

pulse burst compared with regular pulses was analyzed and attributed to dynamically saturated gain of the pump-power-limited fiber amplifier. The length and output power of Yb-doped fiber were optimized to suppress stimulated Raman scattering (SRS) and self-phase modulation (SPM). The ultimate fiber laser system generates a peak power of fundamental higher than 45 kW and burst pulse energy of 350 nJ. Its harmonics have the peak powers of 6.7 kW (532 nm) and 0.6 kW (266 nm).

3.3 Conclusions

In this chapter, we first considered practical aspects in design and construction of relatively high-average-power picosecond Yb-doped fiber laser systems. Employing a highly stable diode-pumped solid-state laser as the seed source together with proper design of the fiber amplifiers, we were able to achieve an average output power of ~60 W with 73 W pumping using just 2 amplifier stages based on regular non-PM Yb-doped fiber. Applying modulation technique to generate pulse bursts at 700 kHz allowed us to optimize dynamically saturated amplifier and extract higher energies from the MOFA. This was used to improve the nonlinear conversion efficiency in the cases of second (16 % vs. 4 %) and fourth (8 % vs. 2 %) harmonic generation compared to regular pulse trains at ~250 MHz.

We also described mode-locking techniques of fiber-based oscillators based on ring type cavities with NPE port. Our simulation results based on coupled nonlinear Schrödinger equations showed the possibility to generate either regular mode-locked pulse trains or noise-like pulses in such oscillators. Novel scheme of supercontinuum generation by noise-like pulses in normally dispersive single-mode fibers was demonstrated. The SC exhibits low threshold (43 nJ) and flat spectrum over the wavelength range of 1,050–1,250 nm.

References

1. F. Röser, D. Schimpf, O. Schmidt, B. Ortaç, K. Rademaker, J. Limpert, A. Tünnermann, 90 W average power 100 μJ energy femtosecond fiber chirped-pulse amplification system. Opt. Lett. **32**(15), 2230–2232 (2007). doi:10.1364/OL.32.002230
2. O. Schmidt, J. Rothhardt, F. Röser, S. Linke, T. Schreiber, K. Rademaker, J. Limpert, S. Ermeneux, P. Yvernault, F. Salin, A. Tünnermann, Millijoule pulse energy Q-switched short-length fiber laser. Opt. Lett. **32**(11), 1551–1553 (2007). doi:10.1364/OL.32.001551
3. G. Smith, K. Kalli, K. Sugden. Advances in femtosecond micromachining and inscription of micro and nano photonic devices, in *Frontiers in Guided Wave Optics and Optoelectronics*, ed. by B. Pal. InTech. doi:10.5772/39542 (2010)
4. R.R. Gattass, E. Mazur, Femtosecond laser micromachining in transparent materials. Nature Photon. **2**(4), 219–225 (2008)
5. M. Mielke, D. Gaudiosi, K. Kim, M. Greenberg, X. Gu, R. Cline, X. Peng, M. Slovick, N. Allen, M. Manning, M. Ferrel, N. Prachayaamorn, S. Sapers, Ultrafast fiber laser platform for advanced materials processing. J. Laser Micro/Nanoeng. **5**(1), 53–58 (2010). doi:10.2961 /jlmn.2010.01.0012
6. R. Paschotta, J. Nilsson, A.C. Tropper, D.C. Hanna, Ytterbium-doped fiber amplifiers. IEEE J. Quantum Electron. **33**(7), 1049–1056 (1997). doi:10.1109/3.594865
7. H.M. Pask, R.J. Carman, D.C. Hanna, A.C. Tropper, C.J. Mackechnie, P.R. Barber, J.M. Dawes, Ytterbium-doped silica fiber lasers: versatile sources for the 1–1.2 μm region. Sel. Top. Quantum Electron. **1**(1), 2–13 (1995). doi:10.1109/2944.468377
8. D.J. Richardson, J. Nilsson, W.A. Clarkson, High power fiber lasers: current status and future perspectives. JOSA B **27**(11), B63–B92 (2010). doi:10.1364/JOSAB.27.000B63
9. T. Eidam, S. Hanf, E. Seise, T.V. Andersen, T. Gabler, C. Wirth, T. Schreiber, J. Limpert, A. Tünnermann, Femtosecond fiber CPA system emitting 830 W average output power. Opt. Lett. **35**(2), 94–96 (2010). doi:10.1364/OL.35.000094
10. F. Röser, J. Rothhard, B. Ortac, A. Liem, O. Schmidt, T. Schreiber, J. Limpert, A. Tünnermann, 131 W 220 fs fiber laser system. Opt. Lett. **30**(20), 2754–2756 (2005). doi:10.1 364/OL.-30.002754
11. J. Thieme, Fiber laser—new challenges for the materials processing. Laser Tech. J. **4**(3), 58–60 (2007). doi:10.1002/latj.200790168
12. M.E. Fermann, I. Hartl, Fiber laser based hyperspectral sources. Laser Phys. Lett. **6**(1), 11–21 (2009). doi:10.1002/lapl.200810090
13. A. Tuennermann, S. Nolte, J. Limpert, Femtosecond vs. picosecond laser material processing. Laser Tech. J. **7**(1), 34–38 (2010). doi:10.1002/latj.201090006
14. L. Goldberg, J.P. Koplow, R.P. Moeller, D.A.V. Kliner, High-power superfluorescent source with a side-pumped Yb-doped double-cladding fiber. Opt. Lett. **23**(13), 1037–1039 (1998). doi:10.1364/OL.23.001037
15. J. Limpert, T. Clausnitzer, A. Liem, T. Schreiber, H.J. Fuchs, H. Zellmer, E.B. Kley, A. Tünnermann, High-average-power femtosecond fiber chirped-pulse amplification system. Opt. Lett. **28**(20), 1984–1986 (2003). doi:10.1364/OL.28.001984
16. J. Yoonchan, J. Nilsson, J.K. Sahu, D.N. Payne, R. Horley, L.M.B. Hickey, P.W. Turner, Power scaling of single-frequency ytterbium-doped fiber master-oscillator power-amplifier sources up to 500 W. Sel. Topics Quantum Electron. **13**(3), 546–551 (2007). doi:10.1109/JS TQE.2007.896639
17. X. Wang, P. Li, H. Yang, T. Jiang, Y. Ma, Z. Fan, G. Niu, J. Yu, A. Wang, Z. Zhang, Microjoule level femtosecond optical pulses with double-cladding fiber-based nonlinear chirped-pulse amplification. Laser Phys. **21**(11), 1941–1944 (2011). doi:10.1134/ S1054660X11190315
18. A. Fernández, L. Zhu, A.J. Verhoef, D. Sidorov-Biryukov, A. Pugzlys, A. Galvanauskas, F.Ö. Ilday, A. Baltuška, Pulse fidelity control in a 20-μJ sub-200-fs monolithic Yb-fiber amplifier. Laser Phys. **21**(7), 1329–1335 (2011). doi:10.1134/S1054660X11130111

19. C. Zheng, H.T. Zhang, W.Y. Cheng, M. Liu, P. Yan, M.L. Gong, All-fiber millijoule energy and nanoseconds pulse operation of a high beam quality multi-stage pulse-pumped Yb-doped amplifier cascade. Laser Phys. **22**(3), 605–608 (2012). doi:10.1134/S1054660X12030309
20. G.P. Agrawal, *Nonlinear Fiber Optics* (Academic, San Diego, 2007)
21. R.H. Stolen, Polarization effects in fiber Raman and Brillouin lasers. IEEE J. Quantum Electron. **15**(10), 1157–1160 (1979). doi:10.1109/JQE.1979.1069913
22. F. Krausz, M. Ivanov, Attosecond physics. Rev. Mod. Phys. **81**(1), 163–234 (2009). doi:10.11 03/RevModPhys.81.163
23. X. Liu, D. Du, G. Mourou, Laser ablation and micromachining with ultrashort laser pulses. IEEE J. Quantum Electron. **33**(10), 1706–1716 (1997). doi:10.1109/3.631270
24. K. Sugioka, M. Meunier, A. Pique, *Laser Precision Microfabrication* (Springer, Berlin, 2010)
25. A.K. Zaytsev, C.L. Wang, C.H. Lin, C.L. Pan, Robust diode-end-pumped Nd:GdVO4 laser passively mode-locked with saturable output coupler. Laser Phys. **21**(12), 2029–2035 (2011). doi:10.1134/S1054660X11210316
26. A.K. Zaytsev, C.L. Wang, C.H. Lin, Y.J. You, F.H. Tsai, C.L. Pan, Effective pulse recompression after nonlinear spectral broadening in picosecond Yb-doped fiber amplifier. Laser Phys. **22**(2), 447–450 (2012). doi:10.1134/S1054660X12020259
27. A.A.M. Saleh, R.M. Jopson, J.D. Evankow, J. Aspell, Modeling of gain in erbium-doped fiber amplifiers. IEEE Photon. Tech. Lett. **2**(10), 714–717 (1990). doi:10.1109/68.60769
28. C. Barnard, P. Myslinski, J. Chrostowski, M. Kavehrad, Analytical model for rare-earth-doped fiber amplifiers and lasers. IEEE J. Quantum Electron. **30**(8), 1817–1830 (1994). doi:10.1109/3.301646
29. T. Pfeiffer, H. Bulow, Analytical gain equation for erbium-doped fiber amplifiers including mode field profiles and dopant distribution. IEEE Photon. Tech. Lett. **4**(5), 449–451 (1992). doi:10.1109/68.136482
30. J.W. Dawson, M.J. Messerly, R.J. Beach, M.Y. Shverdin, E.A. Stappaerts, A.K. Sridharan, P.H. Pax, J.E. Heebner, C.W. Siders, C.P.J. Barty, Analysis of the scalability of diffraction-limited fiber lasers and amplifiers to high average power. Opt. Express **16**(17), 13240–13266 (2008). doi:10.1364/OE.16.013240
31. M.J.F. Digonnet, *Rare-Earth-Doped Fiber Lasers and Amplifiers* (Marcel Dekker, NY, 2001)
32. Y.-J. You, C.-H. Lin, A. Zaytsev, F.-H. Tsai, C.-L. Wang, C.-L. Pan, Optimal design of a high-power picosecond laser system using a dual-stage ytterbium-doped fibre amplifier. Laser Phys. **23**(7), 075114 (2013). doi:10.1088/1054-660X/23/7/075114
33. M.J. Hekmat, M.M. Dashtabi, S.R. Manavi, E. Hassanpour, R. Massudi, Selection of suitable pump diode laser parameters and their effects on efficiency and optimum length of Yb-doped double clad fiber lasers. Laser Phys. **22**(10), 1581–1585 (2012). doi:10.1134/ S1054660-X12100088
34. R.I. Laming, J.E. Townsend, D.N. Payne, F. Meli, G. Grasso, E.J. Tarbox, High-power erbium-doped-fiber amplifiers operating in the saturated regime. IEEE Photon. Tech. Lett. **3**(3), 253–255 (1991). doi:10.1109/68.79772
35. E. Desurvire, Analysis of gain difference between forward- and backward-pumped erbium-doped fiber amplifiers in the saturation regime. IEEE Photon. Tech. Lett. **4**(7), 711–714 (1992). doi:10.1109/68.145247
36. M. Hofer, M.H. Ober, F. Haberl, M.E. Fermann, Characterization of ultrashort pulse formation in passively mode-locked fiber lasers. IEEE J. Quantum Electron. **28**(3), 720–728 (1992). doi:10.1109/3.124997
37. A. Zaytsev, C.-H. Lin, Y.-J. You, C.-C. Chung, C.-L. Wang, C.-L. Pan, Supercontinuum generation by noise-like pulses transmitted through normally dispersive standard single-mode fibers. Opt. Express **21**(13), 16056–16062 (2013). doi:10.1364/OE.21.016056
38. H. Lim, F.Ö. Ilday, F.W. Wise, Generation of 2-nJ pulses from a femtosecond ytterbium fiber laser. Opt. Lett. **28**(8), 660–662 (2003). doi:10.1364/OL.28.000660
39. K. Tamura, E.P. Ippen, H.A. Haus, L.E. Nelson, 77-fs pulse generation from a stretched-pulse mode-locked all-fiberring laser. Opt. Lett. **18**(13), 1080–1082 (1993). doi:10.1364 /OL.18.00-1080

40. A.K. Zaytsev, C.H. Lin, Y.J. You, F.H. Tsai, C.L. Wang, C.L. Pan, A controllable noise-like operation regime in a Yb-doped dispersion-mapped fiber ring laser. Laser Phys. Lett. **10**(4), 045104 (2013). doi:10.1364/OE.21.016056
41. F. Öllday, J.R. Buckley, H. Lim, F.W. Wise, W.G. Clark, Generation of 50-fs, 5-nJ pulses at 1.03?m from a wave-breaking-free fiber laser. Opt. Lett. **28**(15):1365–1367. (2003) doi:10.13 64/OL.28.001365
42. J.R. Buckley, S.W. Clark, F.W. Wise, Generation of ten-cycle pulses from an ytterbium fiber laser with cubic phase compensation. Opt. Lett. **31**(9), 1340–1342 (2006). doi:10.1364 /OL.31.001340
43. M.E. Fermann, V.I. Kruglov, B.C. Thomsen, J.M. Dudley, J.D. Harvey, Self-similar propagation and amplification of parabolic pulses in optical fibers. Phys. Rev. Lett. **84**(26), 6010–6013 (2000)
44. A. Chong, J. Buckley, W. Renninger, F. Wise, All-normal-dispersion femtosecond fiber laser. Opt. Express **14**(21), 10095–10100 (2006). doi:10.1364/OE.14.010095
45. O. Pottiez, R. Grajales-Coutiño, B. Ibarra-Escamilla, E.A. Kuzin, J.C. Hernández-García, Adjustable noiselike pulses from a figure-eight fiber laser. Appl. Opt. **50**(25), E24–E31 (2011). doi:10.1364/AO.50.000E24
46. S. Kobtsev, S. Kukarin, S. Smirnov, S. Turitsyn, A. Latkin, Generation of double-scale femto/pico-secondoptical lumps in mode-locked fiber lasers. Opt. Exp. **17**(23), 20707–20713 (2009). doi:10.1364/OE.17.020707
47. Y. An, D. Shen, W. Zhao, J. Long, Characteristics of pulse evolution in mode-locked thulium-doped fiber laser. Opt. Comm. **285**(7), 1949–1953 (2012). doi:10.1016/j.optcom.2011.12.001
48. S.M. Kobtsev, S.V. Smirnov, Fiber lasers mode-locked due to nonlinear polarization evolution: Golden mean of cavity length. Laser Phys. **21**(2), 272–276 (2011). doi:10.1134/S1054660-X11040050
49. L.M. Zhao, D.Y. Tang, J. Wu, X.Q. Fu, S.C. Wen, Noise-like pulse in a gain-guided soliton fiber laser. Opt. Exp. **15**(5), 2145–2150 (2007). doi:10.1364/OE.15.002145
50. M. Horowitz, Y. Barad, Y. Silberberg, Noiselike pulses with a broadband spectrum generated from an erbium-doped fiber laser. Opt. Lett. **22**(11), 799–801 (1997). doi:10.1364/OL.22.000-799
51. M. Horowitz, Y. Silberberg, Control of noiselike pulse generation in erbium-doped fiber lasers. IEEE Photon. Tech. Lett. **10**(10), 1389–1391 (1998). doi:10.1109/68.720270
52. Y. Li, A. Hoskins, F. Schlottau, K.H. Wagner, C. Embry, W.R. Babbitt, Ultrawideband coherent noise lidar range-Doppler imaging and signal processing by use of spatial-spectral holography in inhomogeneously broadened absorbers. Appl. Opt. **45**(25), 6409–6420 (2006). doi:1 0.1364/AO.45.006409
53. J.C. Hernandez-Garcia, O. Pottiez, J.M. Estudillo-Ayala, Supercontinuum generation in a standard fiber pumped by noise-like pulses from a figure-eight fiber laser. Laser Phys. **22**(1), 221–226 (2012). doi:10.1134/S1054660X1123006X
54. J.C. Hernandez-Garcia, O. Pottiez, J.M. Estudillo-Ayala, R. Rojas-Laguna, Numerical analysis of a broadband spectrum generated in a standard fiber by noise-like pulses from a passively mode-locked fiber laser. Opt. Comm. **285**(7), 1915–1919 (2012). doi:10.1016/j.optcom.2011.12.069
55. L.M. Zhao, D.Y. Tang, Generation of 15-nJ bunched noise-like pulses with 93-nm bandwidth in an erbium-doped fiber ring laser. Appl. Phys. B **83**(4), 553–557 (2006). doi:10.1007/s00340-006-2179-0
56. J.M. Dudley, G. Genty, S. Coen, Supercontinuum generation in photonic crystal fiber. Rev. Mod. Phys. **78**(4), 1135–1184 (2006). doi:10.1103/RevModPhys.78.1135
57. C. Lin, V.T. Nguyen, W.G. French, Wideband near-I.R. continuum (0.7–2.1 μm) generated in low-loss optical fibres. Electron. Lett. **14**(25), 822–823 (1978). doi:10.1049/el:19780556
58. J.M. Dudley, J.R. Taylor, *Supercontinuum generation in optical fibers* (Cambridge University Press, Cambridge, 2010)
59. J. Santhanam, G.P. Agrawal, Raman-induced spectral shifts in optical fibers: general theory based on the moment method. Opt. Comm. **222**(1–6), 413–420 (2003). doi:10.1016/S0030-4018(03)01561-X

60. I. Ilev, H. Kumagai, K. Toyoda, I. Koprinkov, Highly efficient widebandcontinuum generation in a single-modeoptical fiber by powerful broadband laser pumping. Appl. Opt. **35**(15), 2548–2553 (1996). doi:10.1364/AO.35.002548
61. R.S. Watt, C.F. Kaminski, J. Hult, Generation of supercontinuum radiation in conventional single-mode fibre and its application to broadband absorption spectroscopy. Appl. Phys. B **90**(1), 47–53 (2008). doi:10.1007/s00340-007-2812-6
62. H. Chen, Y. Lei, S. Chen, J. Hou, Q. Lu, Experimentally investigate the nonlinear amplifying process of high power picoseconds fiber amplifier. Opt. Laser Tech. **47**, 278–282 (2013). doi:10.1016/j.optlastec.2012.09.010
63. D. Karnakis, E.K. Illy, M.R.H. Knowles, E. Gu, M.D. Dawson, High-throughput scribing for the manufacture of LED components. SPIE 5366 (2004) doi:10.1117/12.531685
64. P. Deladurantaye, A. Cournoyer, Drolet M, Desbiens L, Lemieux D, Briand M, Taillon Y (2011) Material micromachining using bursts of high repetition rate picosecond pulses from a fiber laser source. Proc. SPIE 7914. doi:10.1117/12.875265
65. H. Kalaycioglu, K. Eken, F.Ö. Ilday, Fiber amplification of pulse bursts up to 20 μJ pulse energy at 1 kHz repetition rate. Opt. Lett. **36**(17), 3383–3385 (2011). doi:10.1364/OL.36.003383
66. H. Kalaycioglu, Y.B. Eldeniz, Ö. Akçaalan, S. Yava, K. Gürel, M. Efe, F.Ö. Ilday, 1 mJ pulse bursts from a Yb-doped fiber amplifier. Opt. Lett. **37**(13), 2586–2588 (2012). doi:10.1364/OL.-37.002586
67. Y. Ren, C.W. Cheng, J.K. Chen, Y. Zhang, D.Y. Tzou, Thermal ablation of metal films by femtosecond laser bursts. Int. J. Therm. Sci. **70**, 32–40 (2013). doi:10.1016/j.ijthermalsci.2013.03.003
68. D.N. Schimpf, C. Ruchert, D. Nodop, J. Limpert, A. Tünnermann, F. Salin, Compensation of pulse-distortion in saturated laser amplifiers. Opt. Exp. **16**(22), 17637–17646 (2008). doi:10.1364/OE.16.017637
69. A. Malinowski, K.T. Vu, K.K. Chen, J. Nilsson, Y. Jeong, S. Alam, D. Lin, D.J. Richardson, High power pulsed fiber MOPA system incorporating electro-optic modulator based adaptive pulse shaping. Opt. Express **17**(23), 20927–20937 (2009). doi:10.1364/OE.17.020927
70. A. Agnesi, L. Carrà, P. Dallocchio, F. Pirzio, G. Reali, S. Lodo, G. Piccinno, 50-mJ macropulses at 1,064 nm from a diode-pumped picosecond laser system. Opt. Exp. **19**(21), 20316–20321 (2011). doi:10.1364/OE.19.020316

Part III
Optical Communications

Gong-Ru Lin and Yin-Chieh Lai

Introduction

The earliest history of optical communication technologies can be dated back to ancient times when smoke and beacon fires were used for signaling attacks from the enemy. One historical example is the ancient defensive system of the Great Wall in China, where beacon towers are built around every 5 km to observe enemies and to send/transmit messages. It is commonly believed that the oldest beacon towers actually appeared earlier than the Great Wall and can be dated back to the West Zhou Dynasty (Eleventh Century BC–771 BC). When enemy attacks were observed, alarm messages were sent by releasing smoke in daytime or by lighting fires at night. The amount of smoke or the number of fires indicated the size of the enemy. In modern terminology, this ancient system is an optical wireless communication system which utilized an incoherent light source (sun or firelights) with the "intensity" modulation scheme aiming for propagation distance around 5 km. The equivalent data rate was of course very low. Other visual techniques like hydraulic telegraphs, ship flags, and semaphore lines were also the earliest forms of optical communication. The modern break-through in optical communication occurred when A.G. Bell invented and demonstrated the first photophone in 1880. It was still an optical wireless system which utilized an incoherent light source (initially sunlight and later arc lamps). The modulation scheme is by using a flexible mirror attached to a speaking tube. The mirror oscillates between convex and concave when there are voices, encoding the audio vibration onto the optical beam divergence. The receiver was a parabolic mirror with a photoconductor (the selenium cells) placed at its focal point. The optical beam divergence modulation signal is converted into the electrical amplitude modulation signal due to the altering amount of light incident on the photoconductor through the optical setup. The equivalent data rate was now the audio rate and the achieved propagation distance is about 200 m initially and eventually up to 14 km by 1935. With the concept of low loss optical fibers proposed in 1966 by C.K. Kao and practically demonstrated in 1970 by Corning, the early commercial optical fiber telephone systems were successfully developed around 1977. The system was now an optical wire system which utilized a coherent light source (GaAs semiconductor laser around the wavelength of 850nm). The graded-index multimode fiber is used as the transmission medium and the employed modulation scheme should be the fundamental digital intensity modulation scheme (on-off keying). The data rate was of the order of 45 Mb/s and the optical propagation distance (or the repeater spacing) was up to 10 km. In the 1980s, the light source for fiber communication was switched to InGaAsP semiconductor lasers around the wavelength of 1300 nm to reduce fiber loss and single-mode fiber was used to eliminate modal dispersion for achieving longer propagation distance. The bit rates were now of the order of Gb/s with repeater spacing up to 50 km. In the 1990s, with the availability of Er-doped fiber amplifiers and wavelength-division multiplexing (WDM) technologies, the optical wavelength for long-distance fiber transmission was shifted to wavelength around 1550 nm. The dispersion-shifted fiber was developed to reduce fiber dispersion at this wavelength. By properly engineering the dispersion map of the fiber link for reducing the impacts of fiber nonlinearity, the maximum optical propagation distance was extended long enough to across the biggest ocean with use of Er-doped fiber amplifiers to compensate for fiber loss at every 10 km. The maximum system capacity was multiplied using the WDM and other multiplexing technologies up to the Tb/s level with reasonable long propagation distance. Continual development in the 2000s has further increased the possible system capacity to the tens Tb/s level with the use of advanced optical modulation formats like differential phase shift keying (DPSK), quadrature amplitude modulation (QAM), and orthogonal frequency division multiplexing (OFDM) for better performance and higher spectral efficiency. In particular, the coherent optical OFDM modulation scheme allows the direct/adaptive compensation of various channel linear/nonlinear responses through digital signal processing techniques to achieve greater flexibility and better performance. In 2010, with the recent space division multiplexing (SDM) techniques that utilize multi-core or multimode fibers for another multiplexing degree of freedom, net transmission rates in the hundreds Tb/s level per fiber have been made feasible. These heroic optical transmission achievements clearly indicate that we have already developed systematic methods and efficient technological solutions to utilize the wide bandwidth offered by optical fibers.

Besides the goal of achieving the highest data rates at longer propagation distance, which mainly aims at core network applications, optical communication technologies are also driven by another important goal of achieving cost-effective high data rates at shorter propagation distance, which mainly aims toward access network applications. The passive optical network (PON) structure has been well accepted as the solution for fiber access networking systems. To achieve higher data rates, coarse wavelength-division multiplexing (CWDM) or even dense wavelength-division multiplexing (DWDM) technologies will also be needed in access networks. Since the cost issues become more dominant when the technologies are brought closer to end users, cost-effective optical communication solutions have become the R&D focus for access network applications. Among the various challenges that can be investigated, cost-effective light sources for CWDM/DWDM PON systems are certainly much desired. Distributed feedback (DFB) semiconductor lasers have been the mature light sources for optical communication. The possibility of using other lower cost light sources for fiber communication systems has been continually investigated in many research studies. Some notable examples include the various proposals of utilizing light emitting diodes (LED), amplified spontaneous emission (ASE) broadband sources, semiconductor optical amplifiers (SOA), as well as injection-locked Fabry-Perot (FP) semiconductor lasers for transmitting optical data with the propagation distance of the 1–100 km order. On the other hand, optical communication technologies also become more important for applications that require high data rates at even shorter propagation distances. Optical fiber cables for ethernet and high speed USB links have been commercially available. Parallel optical fiber transmitter/receiver modules for computer board-to-to-board connection have also been practically used. For these applications, the propagation distance is of the order of meters and the data rate is at the level of 1–10 Gb/s per fiber. Vertical cavity surface emitting lases (VCSEL) around 800nm are typically employed with the use of multi-mode fibers for reducing the packaging cost. In these applications, the cost-effective VCSEL and other surface emitting type laser sources also play a central role for related technology development.

In recent years, optical wireless communication for in-building applications emerges as another R&D focus. The propagation distance is of the 1–10 m order with the free space as transmission medium. This development trend is mainly motivated by the success of radio wireless communication for networking applications. Among the possible approaches, visual light communication (VLC) has attracted a lot of research interest mainly because the possibility of using lighting LEDs for transmitting data potentially offers a new cost-effective light source for new optical communication applications. The light source is now an incoherent one and the channel bandwidth is severely limited by the LED modulation bandwith. The challenges are thus how to cost-effectively achieve high enough data rates under severe modulation bandwith constraints. Coding and signal processing techniques are the main tools that can be explored to boost the performance, besides the improved design of the LED modules. It is also possible to explore the use of cheap visible lasers for transmitting optical wireless data. The modulation bandwith constraint can be greatly relaxed in this way, although the possible combination with lighting applications has been sacrificed. These proposals may eventually lead to whole new applications of optical communication in our daily life.

In view of the above technology trends, in this chapter we present some of our recent research results in the following three focus area:

(1) Visual light communication based on lighting LEDs
(2) Visual light communication based on laser pointer lasers
(3) Colorless laser diode sources for DWDM PON applications

The emphases are on the development of innovative approaches including new optical sources and new transmission schemes for solving emergent optical communication problems. The main contents are summarized below:

(1) Visual light communication based on lighting LEDs

LED-based VLC offers many transmission advantages, making it a promising technology for future short-range communication applications. A common advantage offered by VLC is that the visible electromagnetic spectrum is not regulated (license-free), which makes it relatively easy to develop new products for VLC. Most importantly, LED-based VLC can also be further integrated with lighting/signaling systems to develop innovative optical communication solutions. We first discuss the VLC R&D activities that are happening in the world. Then we discuss important technological aspects of VLC, including the main challenges and possible solutions. Topics to be covered include enhancing the transmission data rate of VLC, mitigation of optical background noise, achieving bidirectional transmission, and finally, using AC-LED for VLC.

(2) Visual light communication based on laser pointer lasers

We provide an overview of modern VLC systems based on laser pointer lasers (LPL) that can offer higher transmission rate and longer free-space link than those based on high-brightness LED (HB-LED), red-green-blue (RGB) LED, and phosphor-based LED. The methodologies used to improve the LPL free-space transmission are theoretically discussed. With the simultaneous assistance of the preamplifier and adaptive filter, the amplitude and phase errors can be nicely compensated. The signal-to-noise ratio (SNR) and bit error rate (BER) of the systems can also be further improved. In addition to the discussion of VLC communication systems, the feasibility of integrating various bidirectional passive optical fiber networks with the LPL free-space VLC transmission schemes are also explained. These integrated transport systems are shown to be distinguished not only because of the simplicity in hybrid integration of PON and VLC applications, but also because of the convenience for installation.

(3) Colorless laser diode sources for DWDM PON applications

The weak-resonant-cavity Fabry-Perot laser diode (WRC-FPLD) injection-locked by three master sources with different degrees of coherence, and directly modulated by either the on-off-keying (OOK) at up to 10 Gbit/s or the 16 quadrature-amplitude-modulation (QAM) orthogonal frequency division multiplexing (OFDM) at 12.5 Gbit/s are demonstrated. The injection master coherence shows impacts on noise, bandwidth, and bit-error-rate (BER) performances of the slave WRC-FPLD transmitter. By using the highly coherent master, the injection-locked WRC-FPLD reduces its relative intensity noise peak at 5 GHz by 18 dB and increases its throughput frequency response by 5 dB. This enhances the signal-to-noise ratio from 10.5 dB to 18.9 dB and the on/off extinction ratio from 10.4 dB to 11.4 dB, enabling the error-free OOK transmission at 10 Gbit/s with a requested receiving power sensitivity of -15 dBm at BER of 10^{-9}. Simulated eye diagrams from modified rate equations are also performed to show such improvements. With highly coherent injection-locking, the back-to-back transmitted BER of the WRC-FPLD carried 16-QAM OFDM data with a modulation bandwidth of 3.1 GHz can be minimized to 1.6×10^{-13}.

In comparison, the WRC-FPLD injection-locked WRC-FPLD pair with partial coherence can also provide an 8-Gbit/s OOK with a receiving power sensitivity of -19 dBm at BER of 10^{-9}. Such a cost-effective colorless source also supports a 12.5-Gbit/s 16-QAM OFDM transmission with a BER of 5.6×10^{-12} to serve multi-channel passive optical network in next generation.

Chapter 4
Visible Light Communication

Chi-Wai Chow and Chien-Hung Yeh

4.1 Worldwide VLC Activities

Visible light communication(VLC) is a kind of optical communication using visible wavelength of the electromagnetic spectrum from 380 to 780 nm. VLC based onlight emitting diode (LED) lamps has attracted much attenuation recently for the future secure and high speed data transmissions. The continuous improvement in power efficiency and cost reduction have made LED lamp gradually replacing the traditional lamps. According to the IMS Research, by 2013, the LED lamp market is the largest in the lighting market, and has 40 % market share of the total [1]. The IMS Research has forecast there will be a significant increase in LED lamp sales in coming years due to the "banning" of incandescent lamps in many countries [1].

LEDbased VLC provides many transmission advantages, making it a promising technology for future short-range communication. One advantage offered by VLC is that it is using the visible electromagnetic spectrum, which is not regulated, or license-free. Hence, it is relatively easy to develop new products for VLC. Besides, VLC can be used to broadcast real-time traffic information using traffic lights to the vehicles; hence, drivers can be updated continuously for the real-time traffic conditions [2]. Moreover, it can be a secure channel between different

C.-W. Chow (✉)
Department of Photonics and Institute of Electro-Optical Engineering,
National Chiao Tung University, Hsinchu 30010, Taiwan
e-mail: cwchow@faculty.nctu.edu.tw

C.-H. Yeh
Information and Communications Research Laboratories,
Industrial Technology Research Institute (ITRI), Hsinchu 31040, Taiwan
e-mail: yehch@fcu.edu.tw

C.-H. Yeh
Department of Photonics, Feng Chia University, Taichung 40724, Taiwan

C.-C. Lee (ed.), *The Current Trends of Optics and Photonics*,
Topics in Applied Physics 129, DOI 10.1007/978-94-017-9392-6_4

mobile devices, since the light beam is visible and users can securely limit the coverage of illumination for communication. In addition, VLC does not penetrate through walls; hence, there is no interference among users in different rooms [3]. VLC is also desirable to provide optical wireless communication (OWC) in some areas where radio frequency (RF) communication is prohibited, such as in hospitals or aircrafts. In addition, radio waves do not propagate for a long distance under water; hence VLC is particularly suitable for under water communication [4]. Apart from communication purposes, VLC can be used in a three dimensional position measuring system with high accuracy [5, 6].

As mentioned before, VLC offers many transmission advantages and it has attracted much attenuation recently. VLC is being extensively studied by a number of universities and companies worldwide. In November 2003, the Visible Light Communication Consortium (VLCC) was established in Japan to explore different applications of VLC. It invited companies to contribute in VLC development [7]. In 2007, the VLCC proposed two standards: Visible Light Communication System Standard and Visible Light ID System Standard to the Japan Electronics and Information Technology Industries Association (JEITA), which then accepted these standards as JEITA CP-1221 and JEITA CP-1222 [8]. In 2008, the OMEGA Project [9] funded by the European Commission under the Seventh Research Framework Program (FP7) started to develop a home area network capable of delivering high-bandwidth services and content at a transmission speed of 1 Gb/s. The OMEGA Project consisted of 20 European partners from industries and academia. In 2010, the Center for Ubiquitous Communication by Light (UC-Light) was established in the United States [10]. Its mission was to enable wireless communications by embedding signals into the light emitted by next-generation LED systems. Besides, the Smart Lighting Engineering Research Center was established to promote researches leading to smart lighting systems with adaptive and controllable properties [11]. Communication using light to send information and enable lighting information processing was within the scope of the Smart Lighting Engineering Research Center. In 2011, an international platform, called Li-Fi Consortium was formed [12]. Its goal was to foster the development and distribution of optical wireless technologies such as communication, navigation, natural user interfaces and others. An IEEE standard (IEEE 802.15.7—Short-Range Wireless Optical Communication Using Visible Light) was finalized in 2011 [13], enhancing the prospects for commercializing VLC. The standard covers both the physical (PHY) layer interface and the medium-access control (MAC) layer. However, several aspects, such as flicker mitigation, dimming control, implementation of advanced modulations, background noise mitigation, and effective bi-directional transmission have not been specified [14]; and further research and development for the VLC are still going on.

In this chapter, we discuss different technical aspects (the challenges and possible solutions) of VLC, which include enhancing the transmission data rate of VLC, mitigation of the optical background noises, achieving a bi-directional transmission, and using AC-LED for VLC. Finally, a conclusion was presented to summary the section.

4.2 Different Technical Aspects of VLC

4.2.1 Enhancing Transmission Data Rate

When the VLC is incorporated into the lighting system, white-light LED is need. There are two kinds of white LED used in general lighting. One type consists of a blue LED chip with a phosphor layer on top of it. The emitted blue light is partly absorbed by the phosphor layer to produce yellow light. Hence, the blue and yellow lights result in white light. Another type of white LED is fabricated using three primary colored chips: red, green and blue (RGB). The combination of the RGB produces white light. The phosphor-based LED is low cost, but the limited response time of phosphor makes the direct modulation speed of this phosphor-based LED only a few MHz. On the other hand, the RGB white LED is higher cost; however the direct modulation speed of each chip can be as high as several tens of MHz.

When using the phosphor-based LED for VLC, a blue optical filter at the receiver (Rx) is used to remove the slow yellow component. However, this increases the attenuation of the VLC transmission link and decreases the transmission distance and performance. Electrical pre-equalization or post-equalization for the LED driving and Rx circuits can be used [15–22] respectively to enhance the transmission data rate of the VLC. Yeh et al. [23] demonstrated using quaternary-amplitude-shift-keying (4-ASK) modulation and digital filtering to increase the transmission data rate of a white-light LED VLC system by 20 times. A white-light LED commercially available for lighting application with a direct modulation speed of 1 MHz was used. Figure 4.1 shows the normalized frequency response of the white phosphor-based LED used in the experiment.

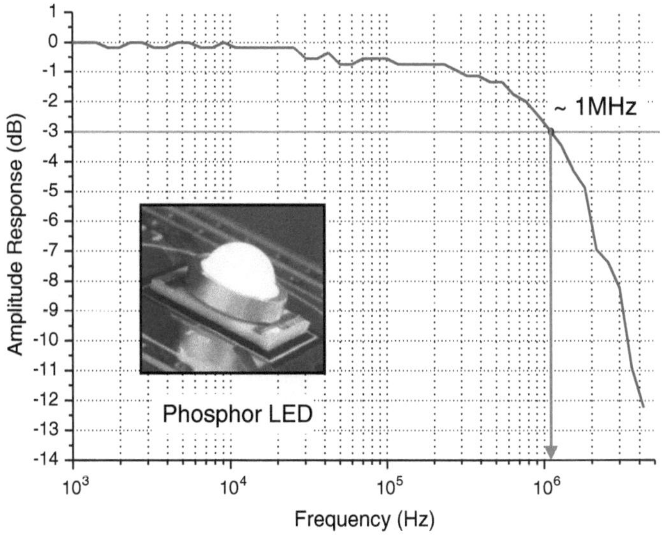

Fig. 4.1 Normalized frequency response of the white phosphor-based LED used in [23]

Figure 4.2 shows the diagrams for generation of the 4-ASK modulation format. The binary data logic sequence was mapped to the 4-ASK symbol having different amplitude levels. The up-sampling process in-creased the sampling rate by inserting zeros between the original sample points. Then, the digital finite impulse response (FIR) filter provided a frequency domain compensation for the system channel response. Besides, matched square-root raised cosine (SRRC) filters were used in the transmitter (Tx) and Rx, respectively.

Figure 4.3 shows the experimental bit-error-rate (BER) measurement by using the proposed scheme. No optical blue filter was used in this experiment. A clear and wide open 4-ASK eye diagram can be achieved, as shown in Fig. 4.3a, in which the total transmission data rate was 20 Mb/s and the transmission length was 1 m. BER of $<10^{-10}$ can be achieved. Figure 4.3b shows the eye diagram without using the proposed scheme. The measured eye-diagram was completely

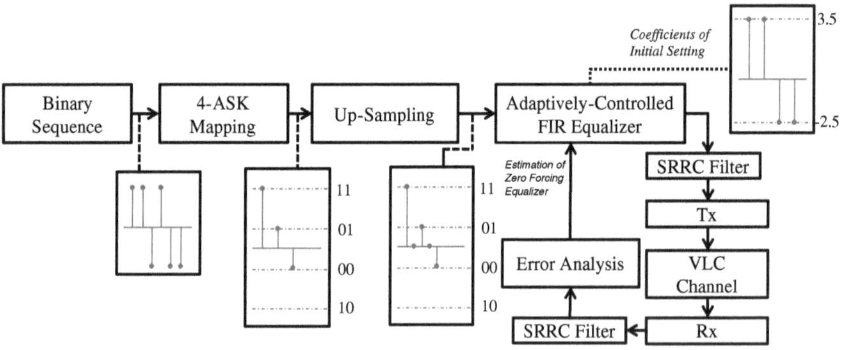

Fig. 4.2 Generation of the 4-ASK modulation format

Fig. 4.3 BER performance against different peak-to-peak driving voltages. Insets: Measured eye-diagrams of **a** with and **b** without the FIR digital filtering

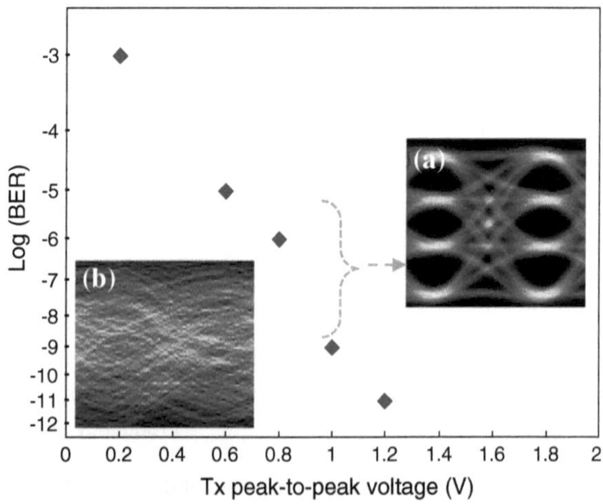

closed. This is because the data rate of the applied signal was much beyond the direct modulation bandwidth of the white-light phosphor-based LED. No BER can be measured in this case.

Advanced modulation, such as orthogonal frequency division multiplexing (OFDM) provides high spectral efficiency; hence allowing high data rate (>1 Gb/s) transmission in the bandwidth limited LED communication. Figure 4.4 shows the logic flow diagram of the OFDM LED VLC described in [24]. At the Tx, the pseudo-random binary sequence (PRBS) data was first mapped to quadrature phase shift keying (QPSK) or quadrature amplitude modulation (QAM). Then the data was serial-to-parallel converted, render to N − 1 symbol streams and modulate N − 1 OFDM subcarriers (note that the DC subcarrier was un-modulated). To pro-duce real-value signal for the inverse fast Fourier transform (IFFT), conjugate symmetry was needed. By performing the conjugate symmetry conversion, the $X_0 = X_N = 0$; $X^*_n = X_{2N-n}$, where X and X* were the input symbol and its conjugate symmetry respectively. Hence at the output, 2 N OFDM symbols were obtained. After the IFFT, a section of the 2 N symbols were inserted for the cyclic prefix (CP). After the CP insertion, the OFDM symbols were re-sampled two times. Then the OFDM data was applied to the LED via a digital-to-analog converter (DAC), which was an arbitrary waveform generator (AWG). The received VLC signal was captured by a real-time oscilloscope (RTO), which was an analog-to-digital converter (ADC). The digital signal was then down-sampled, CP removed and FFT. The channel estimation was implemented at the Rx side to perform the one-tap equalization for enhancing the channel capacity. After the channel estimation and equalization, the QAM de-mapping and decoding were performed. Then the average signal-to-noise ratio (SNR) was calculated by averaging the SNRs of all the subcarriers, and the BER was calculated based on the average SNR [24].

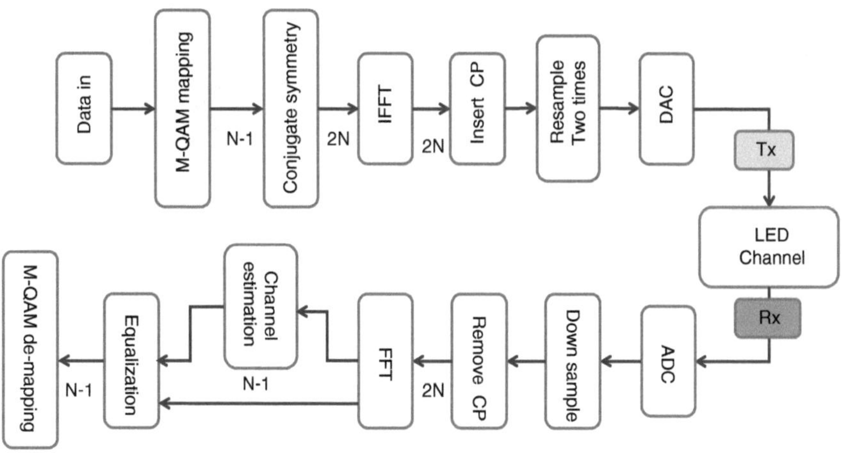

Fig. 4.4 Flow diagram of the LED VLC using OFDM

The OFDM system needs precise synchronization for the IFFT/FFT operations. Besides, channel estimation is also needed to provide parameters for signal equalization enhancing the transmission performance. One method to achieve synchronization and equalization is to use training symbol. The idea of training symbol is to use a known data to estimate the channel response. However, one or more OFDM symbol is used to carry the training symbol; hence, the effective capacity of the system may be reduced. Another scheme to achieve equalization is using pilot-tone. Since the pilot-tone is embedded into some or all OFDM symbols, the particular OFDM symbol carrying the training sequence can be used for carrying the payload data. As the channel response (both amplitude and phase) of the VLC system is relatively flat and continuous, channel estimate using pilot-tone could be effective. In the OFDM system, CP is copied from the end of the OFDM symbol to the front for reducing the inter-symbol interference (ISI). It can be assumed that the front part of OFDM symbol is still similar to the end after the channel. Then, auto-correction can be used to create a period relationship for synchronization.

Figure 4.5 shows the experimental results for using adaptive bit-loading for different OFDM subcarriers in a RGB LED. In adaptive bit-loading, the number of bits that can be transmitted in each subcarrier is determined by the SNR of that subcarrier. As SNR varies from user to user on the same subcarrier and also changes over time, channel utilization and data rate can be improved by using adaptive bit-loading. In the demonstration, RGB LED with data rate ~1 GHz can be achieved.

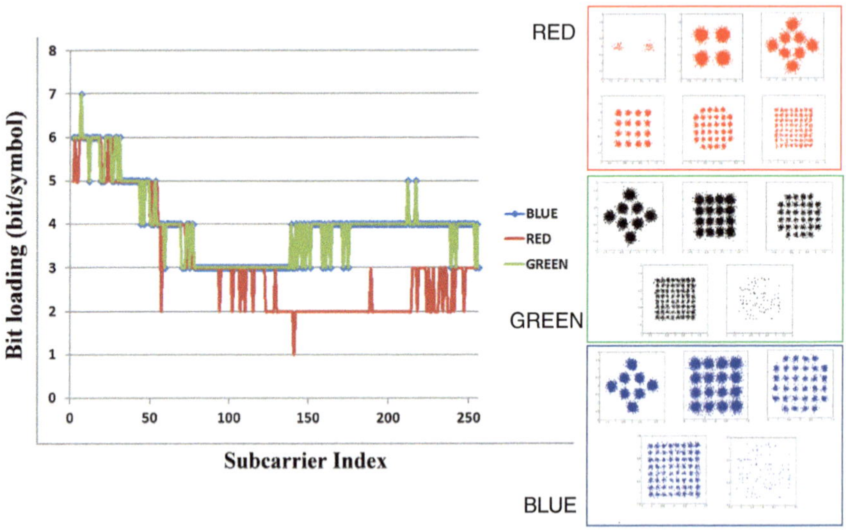

Fig. 4.5 Adaptive bit-loading for different OFDM subcarriers in a RGB LED. *Insets* Constellation diagrams for the *red*, *green* and *blue* LEDs

4.2.2 Mitigation of Optical Background Noises

Background optical noises can affect the performance of the VLC significantly. Scenarios may happen when the LED lamps and conventional fluorescent lamps or AC-LED lamps coexist in the same place. Chow et al. [24] reported the experimental results of the background noise mitigation by using white-light LED OFDM VLC. 64 OFDM subcarriers were used (each subcarrier was in 4-QAM). The transmission data rate was 12 Mb/s, and the bandwidth was 6.25 MHz. The OFDM carrier spacing was 97.66 kHz. A fluorescent lamp was used to produce different optical interference noise powers. The spectral characteristics of the fluorescent lamp noise were shown in Fig. 4.6. The gas discharge lamps, like the fluorescent lamps, needed a ballast to operate. The ballast converted the main supply 60 Hz frequency to higher frequencies for efficient lighting. The fluorescent lamp had a dominant frequency tone at 90 kHz, and harmonic tones at 180, 270 kHz.

Figure 4.7a, b show the SNR of each OFDM subcarrier at the data rate of 12 Mb/s. Different fluorescent noise powers (5 and 15 dB) were introduced in the VLC link. These noise powers were defined as the ratio of fluorescent power to the Rx thermal noise. The Rx thermal noise power was measured when all light sources (LED and fluorescent light) were switched off. When the fluorescent noises were added, the SNR at lower frequency OFDM subcarriers were affected, since the dominate tone of the fluorescent noise was at 90 kHz. It corrupted the

Fig. 4.6 RF spectra of the fluorescent light at noise powers of **a** 5 dB, **b** 15 dB

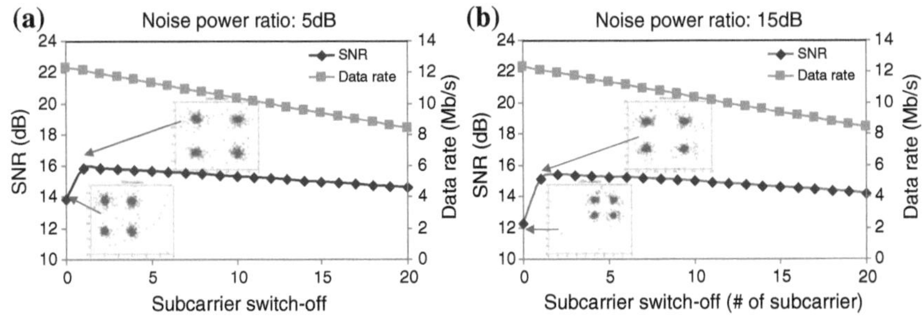

Fig. 4.7 SNR when different OFDM subcarriers were switched-off when the fluorescent noise was **a** 5 dB, and **b** 15 dB

lower frequency OFDM subcarriers. The fluorescent noise can be mitigated by dynamically control the number of OFDM subcarriers. This can be easily achieved in the Tx by not applying the data onto these subcarriers, or neglecting these subcarriers at the Rx. Since the first OFDM subcarrier was at 97.66 kHz, by switching off the first subcarrier, the VLC link can be restored. However, the total capacity was decreased from 12.3 to 12.1 Mb/s.

Besides using OFDM, modulation format such as Manchester coding can also mitigate the optical background noise. Chow et al. [25] demonstrated using Manchester coding to mitigate the optical noise. Another advantage of the Manchester coding is that it can provide signal synchronization and enhance the clock recovery. Figure 4.8 illustrates the encoding and decoding of the Manchester signal. In Manchester coded signal as shown in Fig. 4.8b, the transition from low-to-high represents logic "1", and the transition from high-to-low represents logic "0". The Manchester signal was generated by using exclusive-or (XOR) operation of the original non-return-to-zero (NRZ) data and the clock. At the Rx, the Manchester signal was power divided (Fig. 4.8b), one part was half-bit delayed as shown in Fig. 4.8c. Then, one part subtracted the other part using different amplifier for decoding. Finally logic decision can be made at the time interval as shown in Fig. 4.8d. By comparing the original NRZ signal shown in Fig. 4.8a, it illustrates that the received signal can be correctly decoded.

Conventional NRZ coding VLC was compared with the Manchester coding VLC, and the signals were experimentally evaluated using Q-factors under different "fluorescent noise power/thermal noise power" ratios in [25]. It is observed from Fig. 4.9 that Manchester coding can significantly mitigate the noise. The Manchester coded signal showed error-free optical wireless communication (Q > 6) in all the measurements, while the transmission of the NRZ coded signal produced high error even at low noise level. The insets show the corresponding eye-diagrams. The eye-shape of the received Manchester signal had a half-bit

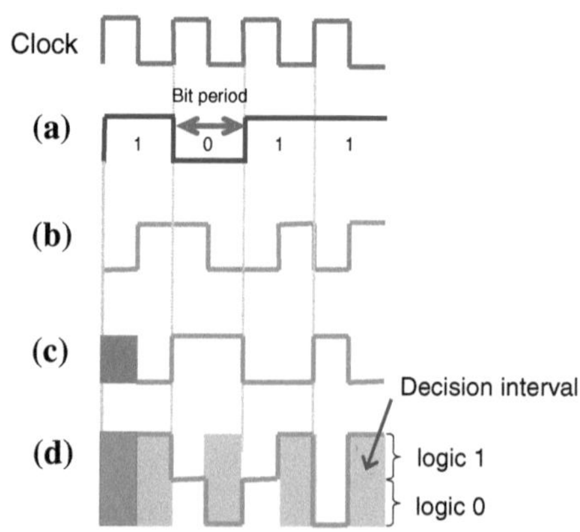

Fig. 4.8 Schematic bit-pattern: **a** NRZ signal, **b** Manchester signal, and **c** half-bit delayed Manchester signal and **d** decoded Manchester signal

period of the NRZ. This was due to the received Manchester signal subtracted its half-bit delayed signal in the demodulation.

The explanation of the noise mitigation principle of the Manchester signal was as follow. Figure 4.10a, b show the simulated power spectra of the NRZ coding and Manchester coding, both with the optical noise at frequency of 0.08/τ. Due to the high spectral overlap of the optical noise with the power spectrum of the NRZ signal, high signal distortion was observed in NRZ case. In the Manchester

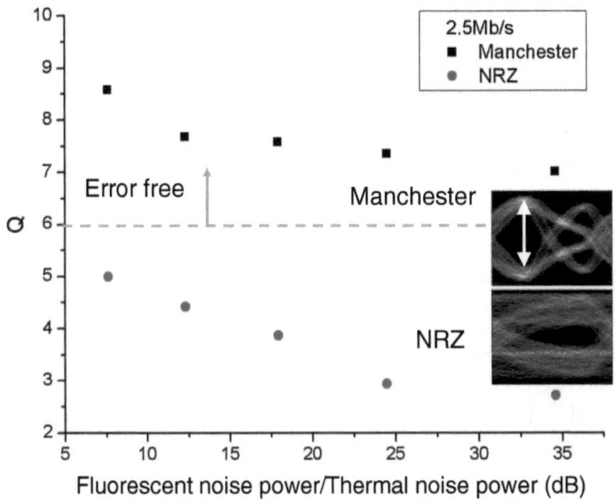

Fig. 4.9 BER of NRZ and Manchester coded signal

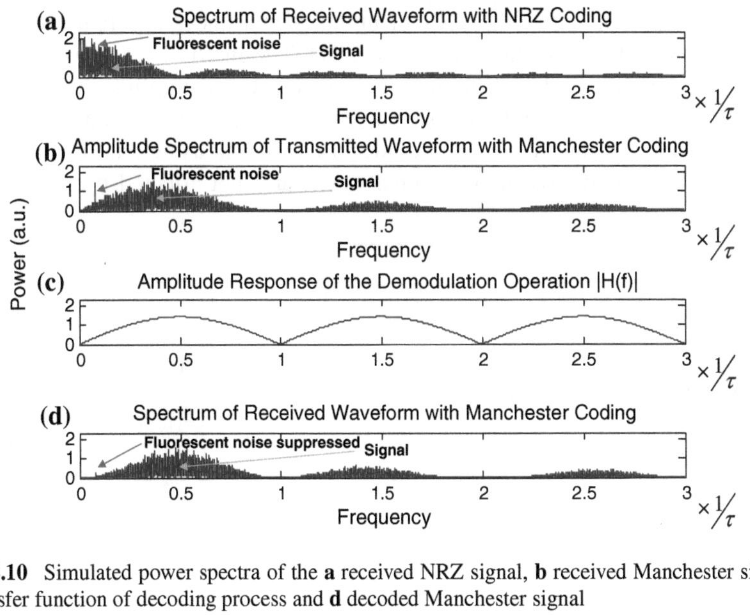

Fig. 4.10 Simulated power spectra of the **a** received NRZ signal, **b** received Manchester signal, **c** transfer function of decoding process and **d** decoded Manchester signal

decoding process, the received signal went through the decoding, which was equivalent to passing through a periodic band-pass filter as shown in Fig. 4.10c. As a result, the optical noise can be effectively suppressed as shown in Fig. 4.10d.

4.2.3 Bi-Directional Transmission

Since VLC is basically a broadcast transmission, providing an uplink path in VLC is challenging. Komine et al. [26] described an uplink VLC path by using modulated retro-reflecting link; however the available modulators are generally low speed and costly. Besides, the location of the Rx needs further study. Hou and O'Brien [27] described using radio-frequency (RF) to provide the uplink path; however it cannot be used in some RF restricted areas. Liu et al. [28] demonstrated a bi-directional transmission link using white-light LED VLC in both downlink and uplink paths. Time-division-duplex (TDD) was implemented to significantly eliminate the reflection interference in VLC.

Figure 4.11 shows the experimental setup of TDD VLC system reported in [28]. The downlink and uplink data were generated by function generator (FG1 and FG2) respectively. The uplink signal from FG2 was applied to a single white-light LED, while the downlink signal from FG1 was applied to a 5 × 8 LED array with a DC bias of 10 V (the 40-LED array had a maximum driving voltage of 12 V). Both the downlink and uplink signals were modulated using NRZ for-mat. To study the reflection interference, two mirrors were placed at about 25 cm away from the LEDs as shown in Fig. 4.11.

Figure 4.12 shows the uplink path analysis in the bi-directional TDD VLC experiment [28]. It shows the SNR of uplink signal under different interference power ratios with and without using the TDD. The interference power ratio was defined as the power ratio of the interference signal to the uplink signal. By tuning the mirror angles, the interference powers were adjusted accordingly. The SNR degraded when the interference power increased without using the TDD mode. By using the TDD mode, the uplink signal was nearly not affected by the reflected

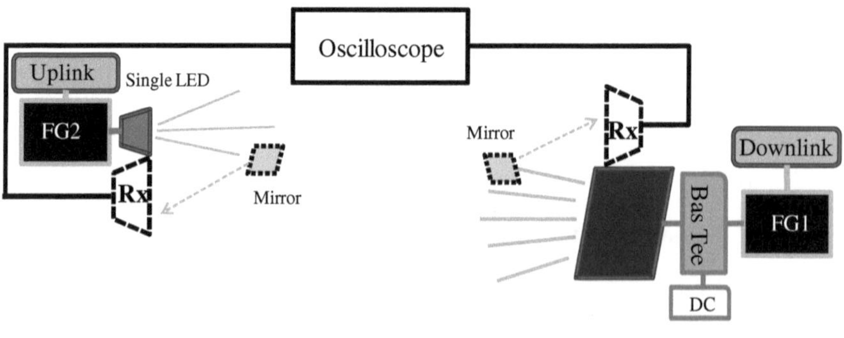

Fig. 4.11 Experimental setup for the TDD VLC system

interference signal, showing the proposed scheme can significantly eliminate the reflection interference even at high reflection power.

Besides using visible light to provide the uplink path in the bi-directionalcommunication, infrared (IR) can also be used to provide the uplink path. Figure 4.13 shows the bidirectional VLC system together with its corresponding module design demonstrated by Industrial Technology Research Institute (ITRI) and

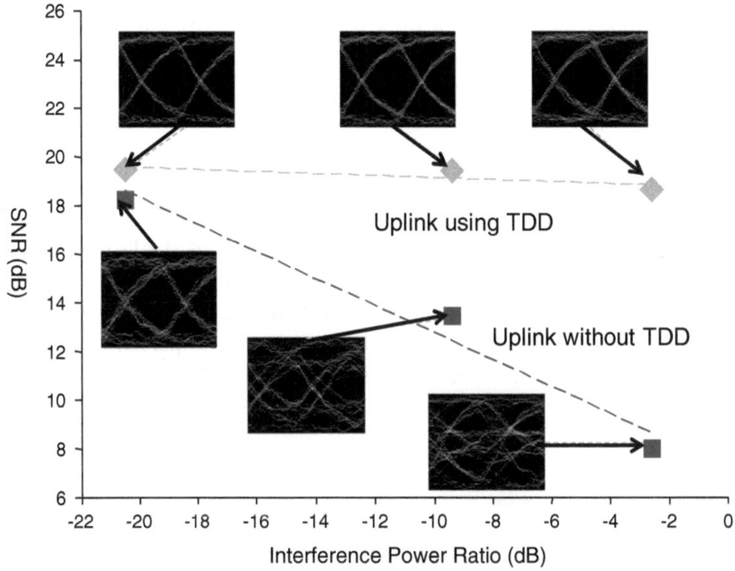

Fig. 4.12 SNR against different interference power ratios with and without using TDD

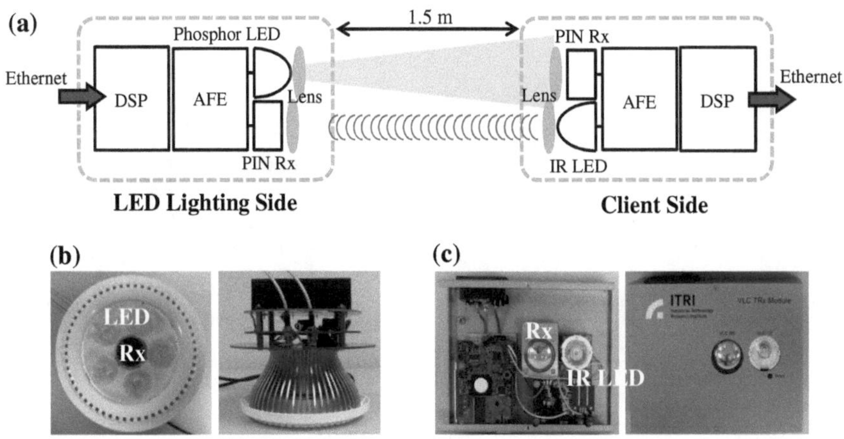

Fig. 4.13 **a** Bi-directional LED VLC system using IR uplink, **b** LED lighting and Tx side and **c** Rx client side

National Chiao Tung University (NCTU). In LED lighting side, phosphor based white-light LEDs served as the Tx for lighting and VLC simultaneously. In client side, an 850 nm IR LED was used as Tx for uplink traffic. Two corresponding optical PIN Rxs with 50 MHz bandwidth were placed in lighting and client sides. There were two main designs for the LED lighting and client sides. One was the analog front end (AFE), and the other was digital signal processing (DSP), as illustrated in Fig. 4.13. In AFE part, it combined the DC power and modulation data and received the optical signal. Pre-equalization technology was performed in the AFE part. In the DSP part, real-time OFDM-based DSP chip was used. A real-time video demonstration was also performed, as shown in Fig. 4.14.

4.2.4 Using AC-LED for VLC

It is more energy-efficient to use AC-LED instead of DC-LED for lighting, and many AC-LED technologies have been developed. It is highly desirable using AC-LED for the VLC. One main challenge for the AC-LED VLC is the synchronization issue for modulating the data during the turn-on period of the AC-LED. A microcontroller was used in previous research of using AC-LED for VLC [29, 30]. Liu et al. [31] demonstrated an AC-LED based system, which was designed using

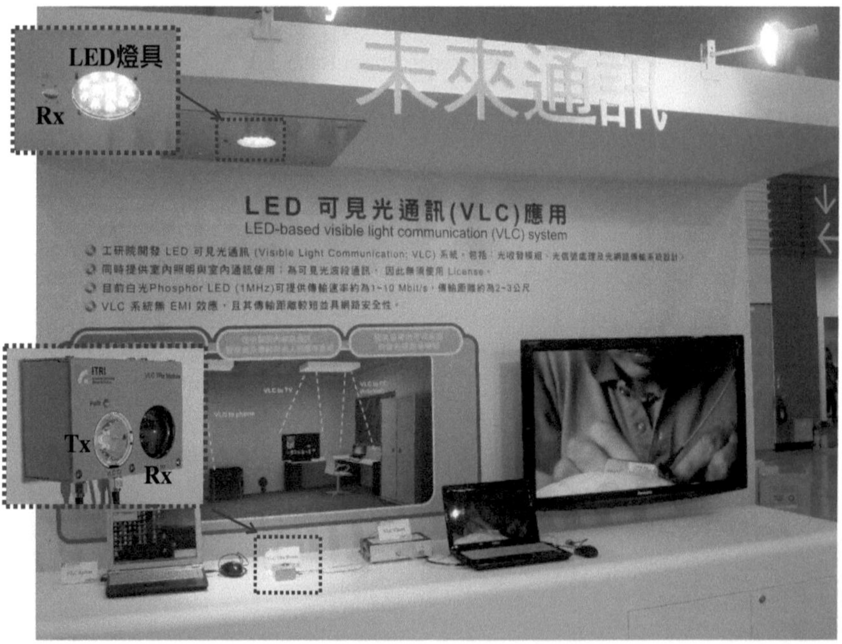

Fig. 4.14 Real-time bi-directional LED VLC system using IR uplink

a synchronized signal modulation with clock recovery and bias-tee circuits. It was verified to support 5 ms time slot without distortion caused by the threshold voltage limitation of LED.

Figure 4.15 shows the experimental setup for AC-LED based system in [31]. The 110 Vrms with a frequency of 60 Hz was provided from power outlet and converted to 9 Vrms by transformer (optional). A 60 Hz square wave synchronized to the AC power bias signal was generated by the clock recovery circuit, with duty cycle of ~50 %. Then, the signal was used to synchronize the AWG. The sinusoidal wave was rectified to have positive voltage only at the output of the full-wave rectifier. Finally, the bias-tee circuit combined the signals from the AWG and the full-wave rectifier for driving the LEDs.

The received waveform of the AC-LED VLC system is shown in Fig. 4.16a using QPSK formats. The rectified AC power bias signal was modulated by the

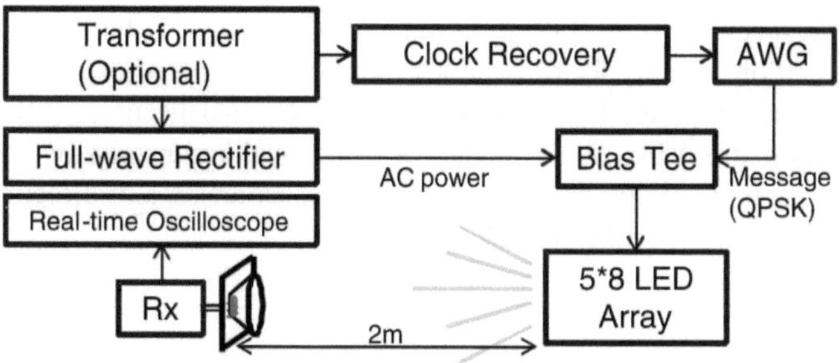

Fig. 4.15 Experimental setup of the AC-LED biased communication system

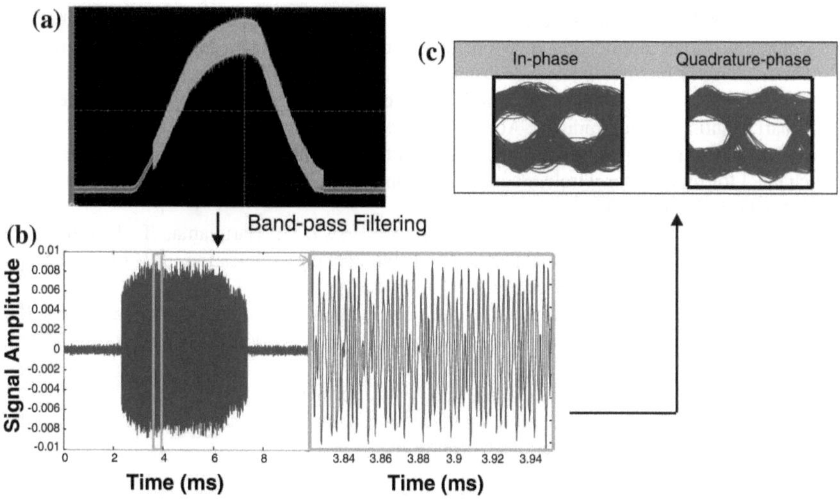

Fig. 4.16 **a** Received waveform of the AC-LED VLC system, **b** filtered waveform and **c** the corresponding eye-diagrams

QPSK data signal. Then, by using a band-pass filter, the received signals can further be processed to remove the 60 Hz AC power signal. The filtered waveform and the eye-diagram were shown in Fig. 4.16a, b, respectively.

Recently, power line communication (PLC) has emerged as a promising wired communication for in-home communication using already installed mains power cables. It is convenient over traditional Ethernet since no additional cables are needed. Researchers are planning to integrate the PLC and the VLC [32]. Hence, the further integration of PLC and AC-LED could increase the network flexibility and scalability.

4.3 Summary

VLC is a kind of optical communication using visible wavelength of the electromagnetic spectrum from 380 to 780 nm. VLC using LED lamps has attracted much attenuation recently for providing secure and high speed data transmissions. We have discussed the worldwide VLC activities. We also discussed different technical aspects of the VLC. These include enhancing the transmission data rate of VLC, mitigation of the optical background noises, achieving a bi-directional transmission, and using AC-LED for VLC.

References

1. http://www.imsresearch.com/press-release/Succeeding_in_the_Global_Retail_LED_Lamp_Market
2. A. Cailean, B. Cagneau, L. Chassagne S. Topsu, Y. Alayli, J.-M. Blosseville, Visible light communications: application to cooperation between vehicles and road infrastructures, in *Proceedings of Intelligent Vehicles Symposium*, 2012 pp. 1055–1059
3. T. Little, Exploding interest in visible light communications: an applications viewpoint, Smart Lighting Annual Industry-Academia Days, 2012
4. Y. Ito, S. Haruyama, M. Nakagawa, Rate-adaptive transmission on a wavelength dependent channel for underwater wireless communication using visible light LEDs. IEICE Tech. Rep., SIP **105**, 127–132 (2006)
5. H. Uchiyama, M. Yoshino, H. Saito, M. Nakagawa, S. Haruyama, T. Kakehashi, N. Nagamoto, Photogrammetric system using visible light communication, in *IEEE 34th Annual Conference of Industrial Electronics* (IECON), 2008 pp. 1771–1776
6. T. Tanaka, S. Haruyama, New position detection method using image sensor and visible light LEDs, in *IEEE Second International Conference on Machine Vision* (ICMV 2009), 2009 pp. 150–153
7. Visible Light Communications Consortium, www.vlcc.net
8. S. Haruyama, Japan's visible light communications consortium and its standardization activities, Visible Light Communications Consortium
9. OMEGA project, http://www.ict-omega.eu
10. Center for Ubiquitous Communication by Light, http://www.uclight.ucr.edu/
11. Smart Lighting ERC, www.smartlighting.rpi.edu
12. The Li-Fi Consortium website http://www.lificonsortium.org

13. IEEE Standard for Local and Metropolitan Area Networks–Part 15.7: Short-range wireless optical communication Using Visible Light, IEEE Std 802.15.7-2011, 2011 pp. 301–309
14. S. Rajagopal, R.D. Roberts, S.-K. Lim, IEEE 802.15.7 visible light communication: modulation schemes and dimming support. IEEE Comm. Mag. **50**, 72–82 (2012)
15. H. Le-Minh, D. O'Brien, G. Faulkner, L. Zeng, K. Lee, D. Jung, Y. Oh, High-speed visible light communications using multiple-resonant equalization. IEEE Photon. Technol. Lett. **20**, 1243–1245 (2008)
16. H. Le-Minh, D. O'Brien, G. Faulkner, L. Zeng, K. Lee, D. Jung, and Y. Oh, 80 Mbit/s visible light communications using pre-equalized white LED, in *Proceedings ECOC*, Paper P.6.09, 2008
17. H. Le-Minh, D. O'Brien, G. Faulkner, L. Zeng, K. Lee, D. Jung, Y. Oh, 100-Mb/s NRZ visible light communications using a post-equalized white LED. IEEE Photon. Technol. Lett. **21**, 1063–1065 (2009)
18. J. Vučić, C. Kottke, S. Nerreter, K. Habel, A. Buttner, K.-D. Langer, and J.W. Walewski, 125 Mbit/s over 5 m wireless distance by use of OOK-modulated phosphorescent white LEDs, in *Proceedings ECOC*, Paper 9.6.4, 2009
19. J. Vučić, C. Kottke, S. Nerreter, K. Habel, A. Buttner, K.-D. Langer, and J.W. Walewski, 230 Mbit/s via a wireless visible-light link based on OOK modulation of phosphorescent white LEDs, in *Proceedings OFC*, Paper OThH3, 2010
20. C.W. Chow, C.H. Yeh, Y. Liu, Y.F. Liu, Digital signal processing for light emitting diode based visible light communication, (Invited paper) in *IEEE Photonics Society Newsletter*, vol. 26, (2012) pp. 9–13
21. Y.F. Liu, Y.C. Chang, C.W. Chow, C.H. Yeh, Equalization and pre-distorted schemes for increasing data rate in in-door visible light communication system, in *Proceedings of OFC*, Paper JWA083, 2011
22. C.W. Chow, C.H. Yeh, Y.F. Liu, Y. Liu, Improved modulation speed of LED visible light communication system integrated to main electricity network. Electron. Lett. **47**, 867–868 (2011)
23. C.H. Yeh, Y.F. Liu, C.W. Chow, Y. Liu, P.Y. Huang, H.K. Tsang, Investigation of 4-ASK modulation with digital filtering to increase 20 times of direct modulation speed of white-light LED visible light communication system. Opt. Express **20**, 16218–16223 (2012)
24. C.W. Chow, C.H. Yeh, Y.F. Liu, P.Y. Huang, Background optical noises circumvention in LED optical wireless systems using OFDM. IEEE Photon. J. **5**, 7900709 (2013)
25. C.W. Chow, C.H. Yeh, Y.F. Liu, P.Y. Huang, Mitigation of optical back-ground noise in light-emitting diode (LED) optical wireless communication systems. IEEE Photon. J. **5**, 7900307 (2013)
26. T. Komine, S. Haruyama, M. Nakagawa, Bidirectional visible-light communication using corner cube modulator. IEIC Tech. Rep. **102**, 41–46 (2003)
27. J. Hou, D. O'Brien, Vertical handover-decision-making algorithm using fuzzy logic for the integrated radio-and-OW system. IEEE Trans. Wirel. Comm. **5**, 176–185 (2006)
28. Y.F. Liu, C.H. Yeh, C.W. Chow, Y. Liu, Y.L. Liu, H.K. Tsang, Demonstration of bi-directional LED visible light communication using TDD traffic with mitigation of reflection interference. Opt. Express **20**, 23019–23024 (2012)
29. F.L. Jenq, Y.C. Pu, Z.C. Weng, An AC LED smart lighting system with white light FSO communication, in *Proceedings of Computer Comm. Control and Auto.*, 2010 pp. 488–491
30. F.L. Jenq, T.J. Liu, F.Y. Leu, An AC LED smart lighting system with visible light time-division multiplexing free space optical communication, in *Proceedings of Computer Comm. Control and Auto.*, 2010 pp. 589–593
31. Y.F. Liu, C.H. Yeh, C.W. Chow, Alternating-signal biased system design and demonstration for visible light communication. IEEE Photon. J. **5**, 7901806 (2013)
32. T. Komine, M. Nakagawa, Integrated system of white LED visible-light communication and power-line communication. IEEE Trans. on Consum. Electron. **49**, 71–79 (2003)

Chapter 5
Fiber-Wireless Communication

Hai-Han Lu and Ching-Hung Chang

Optical free-space transmission scheme is recently developed by engineers and researchers to provide high-speed and secure wireless connections, and visible line communication (VLC) systems are particularly proofed to be an appropriate candidate to deliver such wireless connections among fiber backbone networks and mobile devices which are connected to VLC in-building networks. This chapter is therefore providing an overview of modern VLC communication systems based on laser pointer lasers (LPL) that can provide higher transmission rate and longer free-space link than that in high-brightness LED (HB-LED), red-green-blue (RGB) LED, and phosphor-based LED. The methodologies utilized to improve the LPL free-space transmission are theoretically discussed in the chapter. With the assistance of preamplifier and adaptive filter simultaneous, the pars of amplitude and phase errors can be compensated as well as the systems' signal-to-noise ratio (SNR) and bit error rate (BER) can be further improved. In addition to discuss the VLC communication systems, the feasibility of integrating various bidirectional passive optical networks(PONs) with the LPL free-space VLC transmissions schemes is also clarified. These integration transport systems are shown to be a distinguished one not only to present its simplicity in PON integration with VLC application but also to reveal its convenience to be installed. This chapter will help readers get close to the development of modern VLC communication systems.

H.-H. Lu (✉)
Department of Electro-Optical Engineering, National Taipei University of Technology, Taipei, Taiwan
e-mail: hhlu@ntut.edu.tw

C.-H. Chang
Department of Electrical Engineering, National Chiayi University, Chiayi, Taiwan
e-mail: chchang@mail.ncyu.edu.tw

C.-C. Lee (ed.), *The Current Trends of Optics and Photonics*,
Topics in Applied Physics 129, DOI 10.1007/978-94-017-9392-6_5

5.1 Development Progress on Fiber-Wireless Communications

During this decade, optical communication systems have been deployed extensively with high expectations to provide broadband integrated services. Optical fiber with its low attenuation, high capacity, and electromagnetic noise interference (EMI) free characteristics provides an proper pathway to distribute broadband signals to their destinations [1–3]. With the rapid development of various communication technologies, the increasing requirements in bandwidth raise the needs for high-speed and secure communication links, not only for the single-mode fiber (SMF)-based backbone networks, but also for the free-space visible light communication (VLC)-based in-house network. SMF is widely employed in fiber optical communication systems, and has already established an undisputable position to distribute high quality signals over long-haul transmissions. As a widespread medium, SMF provides good performances in terms of attenuation and security et al. However, when the SMF is deployed toward in-house networks, the installation convenience and cost are beyond disputed issues needed to be solved. To overcome the challenge, a new kind of in-building network medium is required. Recently, wireless communication network is a promising technology to provide broadband services in the consumers' premises. Optical free-space transmission scheme is particularly developed recently by researchers and engineers to create high-speed and long-haul free-space link. Such optical free-space transmissions can provide many benefits, like composing wireless communication link in specific areas in which radio frequency (RF) communication signals are prohibited, such as in the aircraft or hospital [4–7]. However, due to their high attenuation characteristics in the air, employing wireless communication to bridge the access point (AP) and the head-end will place serious limitations on the allowed repeaterless distance. Instead, it is much more possible to link the AP and the head-end by long-haul fiber link. By integrating optical fiber with optical free-space transmission systems, the broadband signal is converted into the optical signal format and delivered to the remote APs by fiber link, in which providing broad bandwidth and low attenuation characteristics [8, 9].

The schematic diagram of bidirectional passive optical network (PON) integration with free-space VLC is illustrated in Fig. 5.1. This brilliant system is the potential candidate to solve the problem of in-building connection and is an ideal scheme to integrate in-building networks with fiber backbone ones. For the application at the premises, VLC systems use modulated light wavelengths emitted and received by a variety of suitably adapted standard sources. Generally, the VLC light source can be classified into two categories: the diffused system and the line-of-sight (LOS) one. The former utilizes diffused beam to cover a wide service area and to provide the mobile service to the end-users, primarily through the use oflight emitting diode (LED). In the published studies, high-brightness LED (HB-LED), red-green-blue (RGB) LED, and phosphor-based LED are employed as the light sources for VLC systems [6, 7, 10]. However,

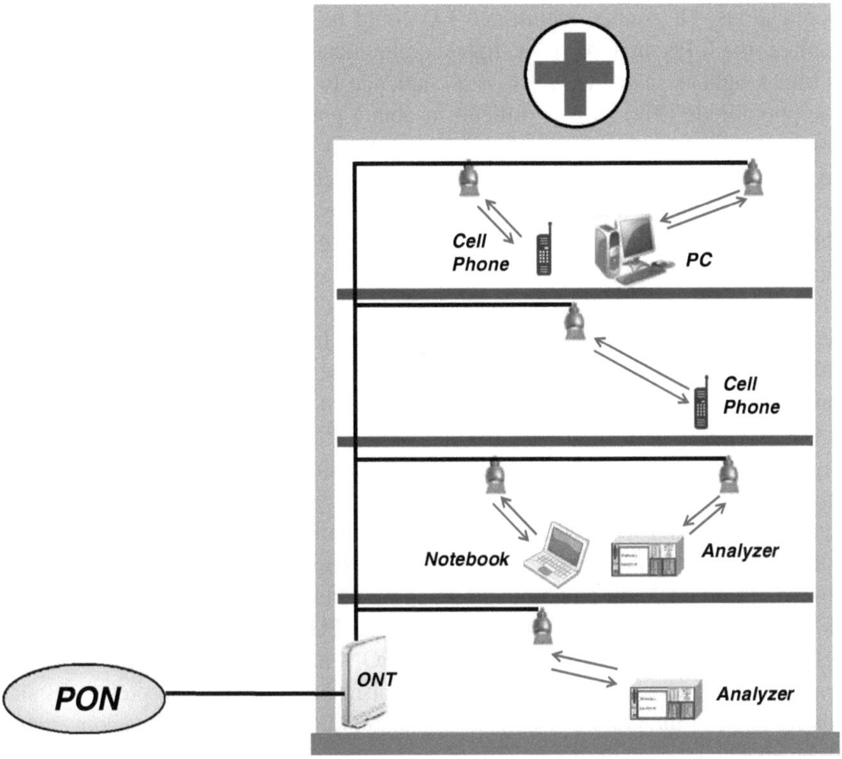

Fig. 5.1 The schematic diagram of bidirectional PON integration with free-space VLC systems [1]

those VLC communication systems are difficult to obtain long free-space link and high-speed transmission rate due to low optical power per unit area. Alternatively, the line-of-sight (LOS) one employs a convergence beam, such as laser pointer lasers (LPL), to establish a point-to-point long free-space link. Nevertheless, no mobility is provided in the LOS system even though it can provide high-speed transmission rate. Because, rapid performance degradation may happen in the LOS system as blocking occurs. However, with the rapid progress of wireless broadband services, the increasing bandwidth requirements raise the needs for high-speed transmission rate and long free-space link. The LOS system employing LPL light source is the potential candidate to meet the demands. As to the mobility (non-line-of-sight) problem, optical signal auto-tracking scheme and optical coupling system [11–15] in which consisting of multiple lenses and a fiber collimator, can be added at the receiving site to overcome it. As a result, the LPL free-space transmission techniques, with high optical power and light beam convergence characteristics, are shown to be a prominent one to present their advantages in wavelength-division-multiplexing (WDM) VLC applications. Of course, theinfrared LD could be engaged as the light source to replace the LPL in free-space light communication systems. Nevertheless, the infrared light is invisible, yet it is a challenge to aim the invisible laser light at the photodiode (PD) so it is difficult to obtain good free-space transmission performance due to infrared laser light misalignment between the transmitter and the receiver. As laser light misalignment problem occurs, rapid performance degradation will happen in the connection links. Thereby, the LPL is more suitable than the infrared LD to be employed as the light source in free-space light communication systems.

To guarantee successful design of a full-duplex lightwave transport system, system designer will have to optimize the overall architecture to obtain the best transmission performances. This section illustrates the progress of modern researches on the LPL-based VLC communication systems in the following sequence. Firstly, a WDM VLC system employing red and green LPLs with directly modulating data signals is illustrated and discussed. In this discussion, the methodology of employing a preamplifier and an adaptive filterat the receiving sites to improve the free-space transmission performance is illustrated and their operation principles and functionalities are theoretically discussed. Consequently, to provide suitable end-to-end communication environment among a central office (CO) and consumers' devices, the feasibility of various bidirectional PON integrations with free-space VLC are illustrated and discussed. Such PON integration with VLC lightwave transport systems are shown to be a distinguished one not only to present its simplicity in PON integration with VLC application but also to reveal its convenience to be installed.

5.2 WDM Visible Light Communication Systems

LED VLC systems are recognized as creating possible valuable portfolios for future generations of technology, which have the potential to use light for communication at data rate larger than that in the current wireless communication systems. In the published schemes, HB-LED is employed not only as the lighting devices but also as the light sources for LED VLC systems [16, 17]. The dual functions of HB-LED, for lighting and communication, emerges many new and interesting applications. Nevertheless, HB-LED array and convex lens are required for longer free-space link in LED VLC systems [4, 10]. To overcome the limitations, a basic LPL-based VLC system is developed as shown in Fig. 5.2 [14]. A LPL is a small portable device with a power source embedded (usually a battery). A LPL can emit a visible light with very narrow coherent light beam to highlight something of interest by illuminating it with a small bright spot of colored light. LPLs, with high optical power and light beam convergence characteristics, are expected to further improve the performance of the LED-based VLC systems with longer transmission distance and higher data rate. Furthermore, with the assistance of preamplifier and adaptive filter at the receiving sites to eliminate parts of nonlinear distortions and to compensate transmission errors, low bit error rate (BER) at 10 m length 500 Mbps

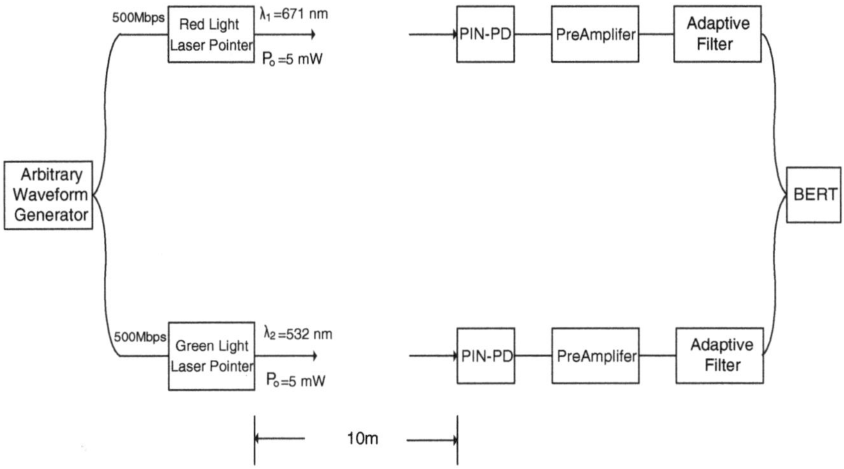

Fig. 5.2 Experimental configuration of a basic LPL-based WDM VLC systems employing *red* and *green* LPLs with directly modulating data signals over a 10-m free space link [14]

data rate operation is experimentally achievable for each wavelength [13, 14]. Such LPL features create a new category of good performance with high-speed data rate, long transmission length and easy handling and installation.

In the basic LPL-based VLC system, the schematic diagram of the preamplifier (push-pull amplifier), is illustrated in Fig. 5.3. Since the even-order harmonic distortions of the systems can be eliminated by the push-pull amplifier, the preamplifier output can be given by [14]

$$v_o = a_1 \cdot v_i + a_3 \cdot v_i^3 + a_5 \cdot v_i^5 \tag{5.1}$$

where v_o is the preamplifier output voltage, v_i is the preamplifier input voltage, and a_1, a_3, a_5 are the amplitude coefficients (a_3 and a_5 are coefficients characterize nonlinearities). Generally, a VLC system with high order nonlinear distortions can be described as

$$q = b_1 \cdot n + b_3 \cdot n^3 + b_5 \cdot n^5 \tag{5.2}$$

Fig. 5.3 A schematic diagram of the preamplifier (push-pull amplifier) [14]

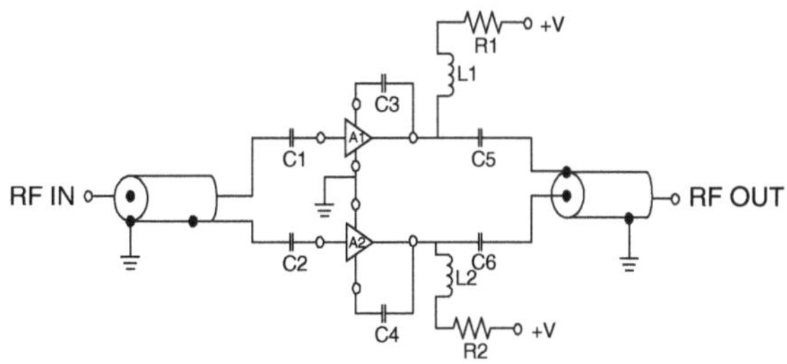

where q is system's output voltage detected from PIN-PD, n is system's input voltage, and b_1, b_3, b_5 are the amplitude coefficients (b_3 and b_5 are coefficients characterize nonlinearities). It is clear that q is equal to v_1. If the (5.2) is substituted into the (5.1) and neglecting higher order nonlinear terms, then yields

$$v_o = (a_1 \cdot b_1) \cdot m + (a_1 \cdot b_3 + a_3 \cdot b_1^3) \cdot m^3 + (a_1 \cdot b_5 + a_5 \cdot b_1^5) \cdot m^5 \tag{5.3}$$

While achieving linearity means cancelling out the nonlinear terms, a RF amplifier pre-distorter would have to cancel out the third-order nonlinear term by setting the appropriate nonlinear coefficient to

$$a_1 \cdot b_3 = -a_3 \cdot b_1^3 \tag{5.4}$$

Then (5.3) can be changed as

$$v_o = (a_1 \cdot b_1) \cdot m + (a_1 \cdot b_5 + a_5 \cdot b_1^5) \cdot m^5 \tag{5.5}$$

It is clear that, from (5.5), the third-order nonlinear distortion can be removed by properly adjusting the nonlinear coefficient. Furthermore, the amplitude of the harmonic distortion decreases with the increases of the harmonic order. Thus, the amplitude of the 5th harmonic distortion will very small, and it will not induce any serious distortion in the VLC systems.

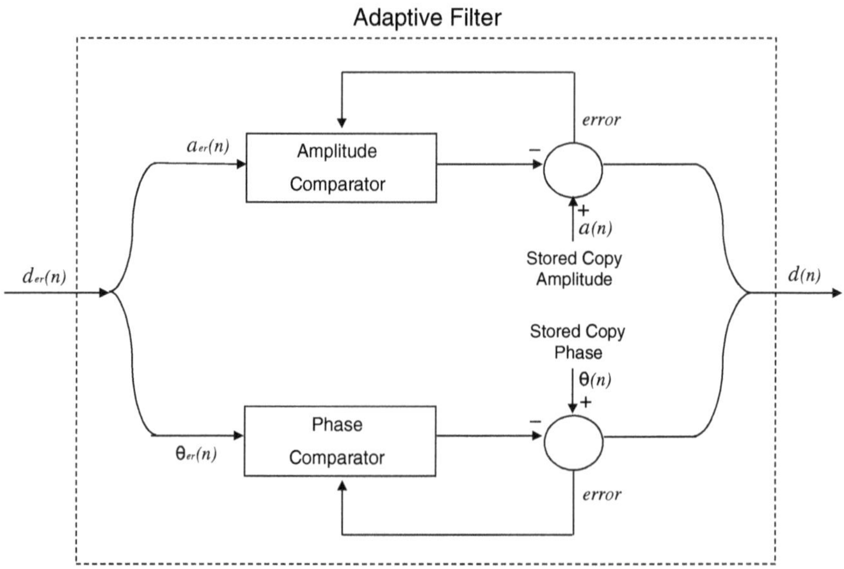

Fig. 5.4 A functional block of the adaptive filter, in which including an amplitude/phase comparator [14]

In parallel with the application of push-pull amplifier, the functional block of the adaptive filter is illustrated in Fig. 5.4, in which an amplitude/phase comparator is included to make error corrections. In implementing the adaptive filter, the transmitter has to firstly send out an arbitrary data pattern as a protocol, and at the receiver site, the adaptive filter must have a pre-stored copy of data signal in the adaptive filter before starting communication. Let the transmitted signal, $d(n)$, has an amplitude $a(n)$ and phase $\theta(n)$, then the signal can be displayed as [13, 14]

$$d(n) = a(n)e^{j\theta(n)} \tag{5.6}$$

After transmission through free-space links, the received signal $d_{er}(n)$ may have a distorted amplitude $a_{er}(n)$ and phase $\theta_{er}(n)$, and can be illustrated as

$$d_{er}(n) = a_{er}(n)e^{j\theta_{er}(n)} \tag{5.7}$$

If the power of a transmitted symbol is $P(n)$, and a received symbol is $P_{er}(n)$:

$$P(n) = a^2(n)/2 \tag{5.8}$$

$$P_{er}(n) = a_r^2(n)/2 \tag{5.9}$$

To make error corrections, the adaptive filter has to estimate the $d(n)$ from the $d_{er}(n)$. The errors between the pre-stored copy of the arbitrary data pattern and the output of the comparators are fed into one input of the comparators, and the received data signals are applied to another input. For amplitude comparison, the output of the amplitude comparator is compared with the pre-stored copy of the $a(n)$, the amplitude comparator has to estimate the $a(n)$ from the $a_{er}(n)$. For phase comparison, the output of the phase comparator is compared with a pre-stored copy of the $\theta(n)$, the phase comparator has to estimate the $\theta(n)$ from the $\theta_{er}(n)$. During communication, an adaptive algorithm will update the amplitude and phase errors every time so that the errors can be minimized. Amplitude and phase errors compensation are crucial for ensuring maximum nonlinear distortion suppression, the use of adaptive filter offers significant amplitude and phase errors compensation.

The functionalities of the preamplifier and adaptive filter can be further ensured by the BER performance. At a free-space transmission distance of 10 m and data rate of 500 Mbps; without the assistance of the preamplifier and the adaptive filter, the BER is around 10^{-5}; with preamplifier scheme alone and adaptive filter alone, the BER are all about 10^{-7}, and with the assistance of both the preamplifier and the adaptive filter simultaneously, the BER can reached to 10^{-9} [13, 14]. It is clear that as the preamplifier and the adaptive filter are employed in the system, impressive BER performance improvement (10^4 orders) can be achieved. This means that the VLC systems with the assistance of the preamplifier scheme alone or with adaptive filter scheme alone, their BER performance improvement is limited. The results indicate that the preamplifier and the adaptive filter play important roles for error correction functions, and they can further improve systems' signal-to-noise ratio and BER performance for 10^4 orders to accomplish high performance VLC systems.

5.3 Integrating FTTH and Free-Space VLC Transport Systems

As discussed above, the free-space VLC system can provide many benefits, like providing communication links in specific areas in which RF communication is prohibited. Nevertheless, the VLC transport system is a short distance communication technique. To provide suitable end-to-end communication environment among a central office (CO) and clients' devices, integrating fiber to the home (FTTH) and free-space VLC transport systems together would be a suitable solution to accomplish the expectation by directly delivering broadband multimedia signals among the CO and the clients' devices. This section is therefore sequentially introducing various modem methodologies in integrating FTTH and free-space VLC transport systems.

5.3.1 Integrating OFDM FTTH and Free-Space VLC Transmission

The schematic diagram of a bidirectional passive optical network (PON) integration with free-space LPL VLC is illustrated in Fig. 5.5 and its experimental configuration is displayed in Fig. 5.6, where a 2.5 Gbps/2.5 GHz 16-QAM OFDM signal is transmitted over a 20 km SMF link, detected by a PIN-photodiode (PIN-PD), and supplied to the 15 m VLC systems. In this structure, broadband services originated from a CO can be communicated over a span of SMF to consumers' premises and then wirelessly indicated into mobile devices via the LPL VLC link. The upstream signal can be sent back by reversed processes using different LPL and optical fiber to avoid the crosstalk of the downstream signal. In this configuration, laser light will send out the data signal wirelessly, so the maximum light intensity should be limited to accomplish eye safety. The European authority has put an additional prerequisite for the maximum light intensity, 25 W/m^2. If an eye pupil diameter is 7 mm, the maximum light intensity for each eye is roughly 0.96 mW. In this condition, directly facing to the class IIIa lasers lights could be dangerous. However, the risk of injury is very small as the LPL conformed to the FDA Class IIIa limit is used. Besides, the natural motion of a person who might be exposed makes a person difficult to expose his eyes for a long period of time. People also have a natural aversion to bright lights and are likely to close their eyes or turn their heads away if exposed. It can be seen from Fig. 5.5 that the bidirectional optical paths in each room are transmitted among the ceiling and the desks, so people will not look straight at the LPL when stay in their office. Therefore, it is safe for the eyes to use an LPL conformed to the FDA Class IIIa limit as the VLC light source under such conditions. Experimental results shown in the [13, 14] proof that such integrated 20 km FTTH and 15 m free-space VLC in-building networks can provide good BER performances and constellations maps, as the data rate is 2.5 Gbps.

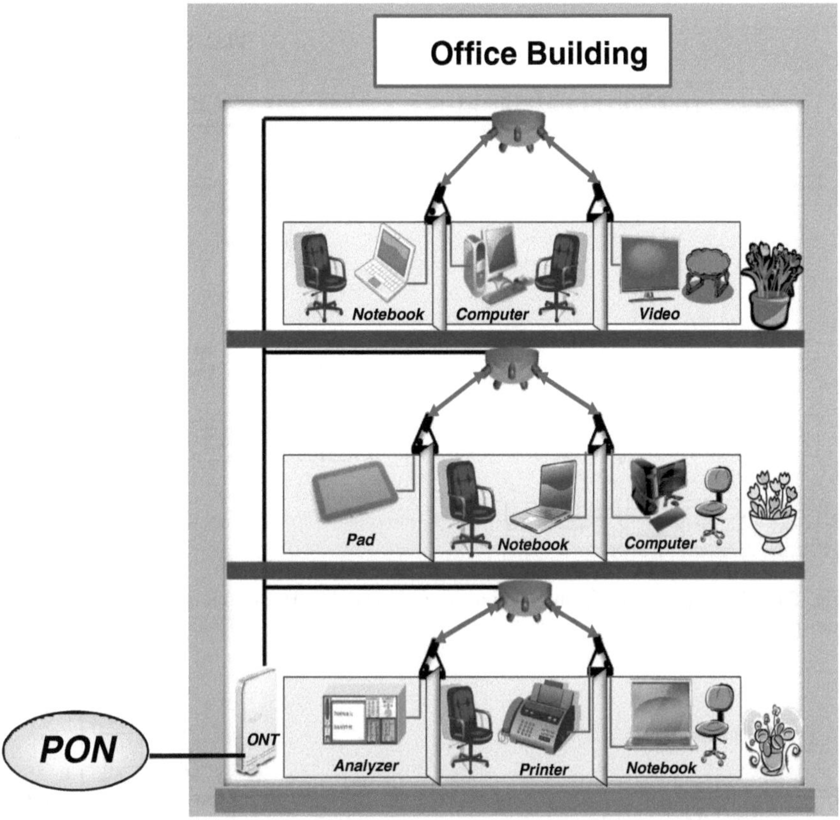

Fig. 5.5 The schematic diagram of a bidirectional PON integration with free-space VLC [3]

5.3.2 *Optimizing FTTH and Free-Space VLC Integration System*

In fiber optical communication systems, phase modulation (PM) schemes are proofed to be an efficient method to reduce noise and distortion induced by communication pathway, so the performance of the previously discussed FTTH and free-space VLC integrated transport system could be further promoted by replacing the intensity modulator by a phase modulator. Nevertheless, the phase-modulated signal needs a delay interferometer (DI) to transfer it into the intensity-modulated one before been received by an optical receiver. Even the overall transmission performances are greatly promoted, the sophisticated and expensive DI will be a serious limitation in promoting such systems. To simplify the network structure by cutting down the complexity of the required DI, an injection-locked distributed feedback laser diode (DFB LD) could be employed as a duplex transceiver to receive phase-modulated downstream signal

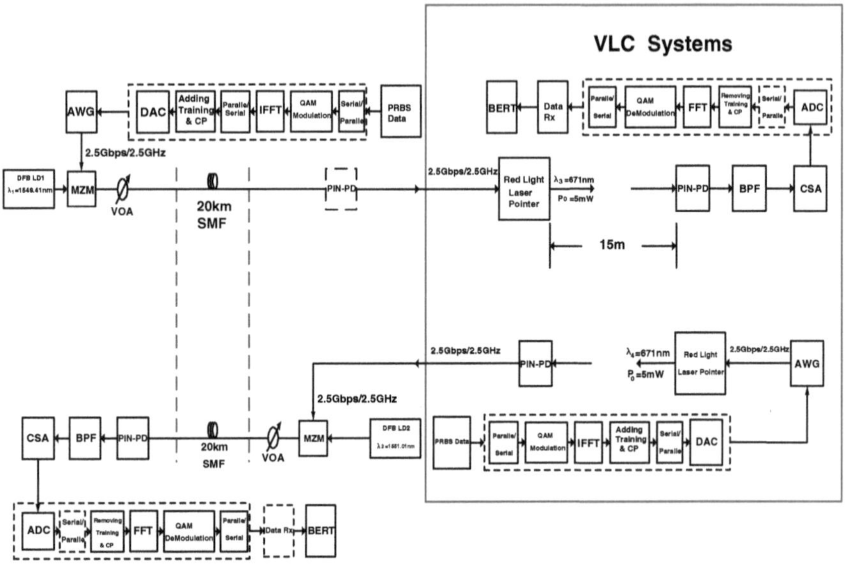

Fig. 5.6 The experimental configuration of an OFDM FTTH and free-space VLC integrating transport system [3]

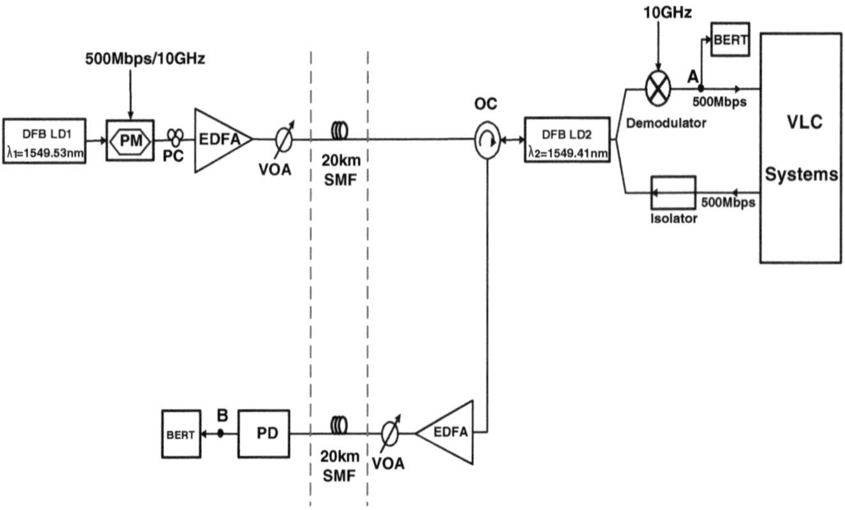

Fig. 5.7 The experimental configuration of the optimized FTTH and free-space VLC integration system [3]

and to transmit upstream data in the system, as displayed in Fig. 5.7 [1]. With the operation characteristic of the PM scheme, the noise and distortion induced amplitude fluctuation effect can be reduced dramatically resulting in better transmission performances. It should be noted that the transmitted RF signal

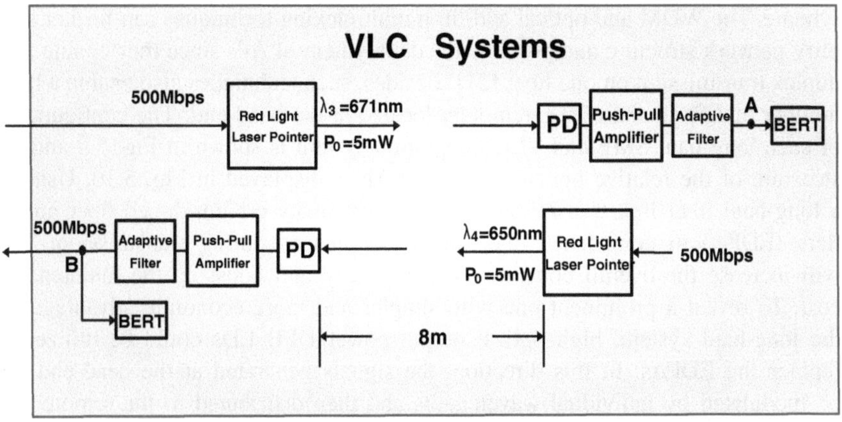

Fig. 5.8 The schematic diagram of a bidirectional VLC system employing two red LPLs with directly modulating data signals over an 8-m free-space transmission [1]

will be detected only when the DFB LD is properly injection-locked to the phase-modulated signal. An optimum injection locking can be achieved when the wavelength of the master laser (DFB LD1) is roughly 0.12 nm longer than that of the slave laser (DFB LD2) [18–20]. As a slave laser is injection-locked, its output optical spectrum will slightly shift to longer wavelength direction, matching to that of the master laser.

Similar with the previous discussed VLC systems, thePM-based FTTH and free-space VLC integration system also employ red and green LPLs in the VLC systems, as illustrated in Fig. 5.8. Generally, the visible light LD (LPL) in the free-space VLC systems could be replaced byinfrared LD. Nevertheless, when an infrared LD is employed to communicate signal, the VLC system is difficult to obtain proper transmission performance since the infrared light is invisible, yet it is a challenge to aim the invisible laser light at the PD. As laser light misalignment problem occurs, rapid performance degradation will happen in the systems. Thereby, the LPL is much more suitable than theinfrared LD to be employed as the light sources in free-space VLC systems.

5.3.3 Long-Haul SMF and Optical Free-Space Transmissions

To deliver optical signal in urban area, the transmitted optical fiber length in most of the published FTTH transport system is generally set to 25 km only. Nevertheless, when the communication area is extended from urban area to rural area, a longer transmission length could be set based on WDM and optical add-drop multiplexing (OADM) techniques, as well as optical free-space transmission scheme. The WDM and optical add-drop multiplexing techniques can further simplify network structure and promote the deployment of APs since they enable full-duplex transmission on one fiber [21]. Besides, such techniques also enable a large number of APs to share LDs remotely located at the head-end. The configuration of such long-haul SMF and VLC integration system is shown in Fig. 5.9 and the structure of the relative optical free-space AP is displayed in Fig. 5.10. Usually, a long-haul fiber link transmission will employ many erbium-doped fiber amplifiers (EDFA) to compensate the optical attenuation during transmission. This will increase the overall construction complexity and boost up the maintenance cost. To reveal a prominent one with simpler and more economic advantages in the long-haul system, high optical output power DFB LDs could be utilized to replace the EDFAs. In this direction, the signals generated at the head-end can be modulated by individual wavelengths and then distributed to the remote APs by OADMs. Similarly, the upstream wavelength can also be added into the fiber backbone through the same OADM. When the optical carrier is dropped by the OADM, the downstream wavelength can be directly transmitted to the air by a fiber transmitter, and detected by a broadband PD. As shown in the Fig. 5.10, the fiber transmitter can be composed by a fiber end, a lens, and a coupling subsystem. The lens is utilized to enlarge the divergence of the emitted optical beam to cover a wider area, and the coupling subsystem is employed to focus the diverged optical beam into a receiver point. Similar with the discussed LPL VLC systems, the quality of the transmitted signal can be amplified by a push-pull amplifier and passed through an adaptive filter for error correction.

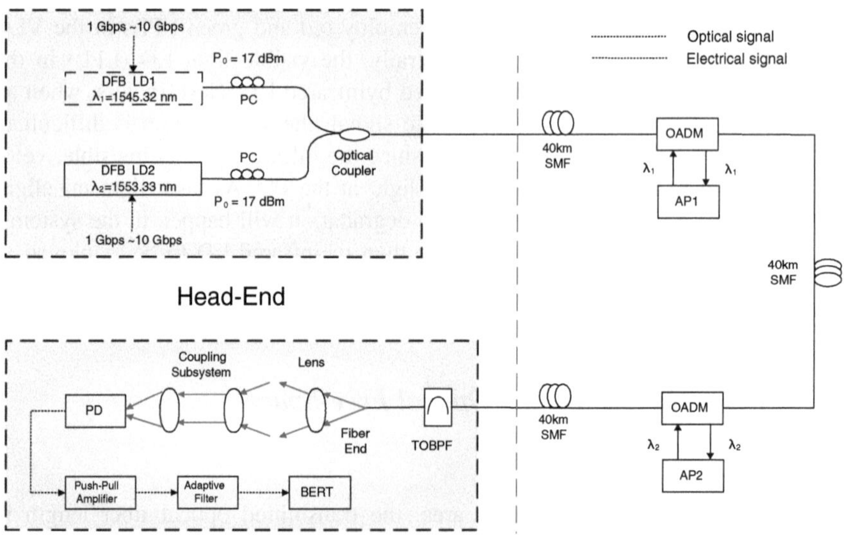

Fig. 5.9 The schematic diagram of a full-duplex lightwave transport systems employing WDM and optical add-drop multiplexing techniques, and optical free-space transmission scheme

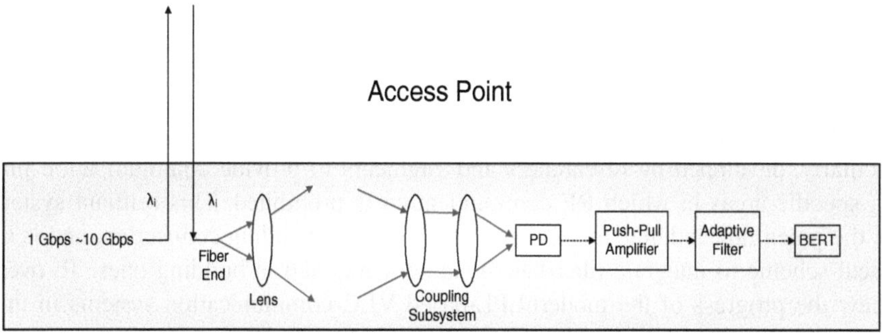

Fig. 5.10 The schematic diagram of the relative AP in a long-haul SMF and VLC integration system

In this kind of long-haul transmission systems, crosstalk (XT) between the downstream and upstream optical signals will limit the overall BER performance. Crosstalk that arises from the imperfect isolation of the OADM add/drop channels can be expressed as [22, 23]:

$$XT = 10 \cdot \log \left[\frac{K_{ad} \frac{P_d}{P_a}}{1 + K_{ad} \frac{P_d}{P_a}} \right]^2 \tag{5.10}$$

where P_a is the optical power of add channel, and P_d is the optical power of drop channel. The effective isolation factor, K_{ad}, is the ratio of the power transmission of drop channel to the add channel. It is expected that the lower crosstalk level from the adjacent channel the better BER performance will present, so system designers must ensure a proper add/drop channel isolation property of the OADM to prevent crosstalk from the add/drop channel.

In addition to provide suitable isolation property of the OADM, utilizing push-pull amplifier and adaptive filteralso play import roles for error correction. The relationship between SNR and BER is given by [24]:

$$BER = \frac{1}{2} erfc \left(\sqrt{\frac{SNR}{2}} \right) \tag{5.11}$$

Since the push-pull amplifier and the adaptive filterschemes can eliminate parts of the unwanted noise form the system, an enlarged signal-to-noise ratio (SNR) value will cause system with better BER performance and leading to an improvement of free-space transmission distance. Furthermore, to obtain a suit transmission performance a trade-off between the data stream and the beam radius as well as a trade-off between the data stream and the distance from beam center are required.

5.4 Summary

Recently, wireless communication techniques are developed to provide broadband service to the premises, and optical free-space VLC transmission scheme is, in particularly, developed by researchers and engineers to provide communication link in specific areas in which RF communication is prohibited. This brilliant system is the potential candidate to solve the problem of in-building connection, and is an ideal scheme to integrate fiber backbone networks and in-building ones. To overview the progress of the modernLPL-based VLC communication systems in this chapter, a WDM VLC system employing red and green LPLs with directly modulating data signals is firstly illustrated and discussed. Its methodologies of employing a preamplifier and an adaptive filter to eliminate parts of unwanted noises at the free-space transmission links are theoretically discussed. Consequently, the feasibility of integrating bidirectional PON with free-space VLC is clarified by transmitting 2.5 Gbps/2.5 GHz 16-QAM OFDM signal over 20 km SMF and 15 m free-space VLC links. This integration system not only presents a simplicity methodology in integrating PON with VLC application but also reveals its convenience to be installed. Following this, a bidirectional lightwave transport system employingPM scheme and light injection-locked DFB LD as a duplex transceiver for PON; as well as employingLPLs with directly modulating data signals for WDM VLC is discussed. Such bidirectional system is proofed to providing better transmission performance than IM transport scheme since the amplitude fluctuation caused by downstream noise and distortion can be reduced dramatically. Finally, a long-haul full-duplex lightwave transport system employing WDM and OADM techniques, as well as optical free-space transmission scheme is discovered. Such integration architecture can not only present its advancement in lightwave application, but also reveal its simplicity and convenience for the real implementation.

References

1. C.Y. Chen, P.Y. Wu, H.H. Lu, Y.P. Lin, C.H. Chang, H.C. Lin, A bidirectional lightwave transport system based on PON integration with WDM VLC. Opt. Fiber Technol. **19**, 405–409 (2013)
2. C.Y. Chen, P.Y. Wu, H.H. Lu, Y.P. Lin, M.C. Gao, J.Y. Wen et al., Bidirectional phased-modulated hybrid cable television/radio-over-fiber lightwave transport systems. Opt. Lett. **38**, 404–406 (2013)
3. C.Y. Chen, P.Y. Wu, H.H. Lu, Y.P. Lin, J.Y. Wen, F.C. Hu, Bidirectional 16-QAM OFDM in-building network over SMF and free-space VLC transport. Opt. Lett. **38**, 2345–2347 (2013)
4. Y.F. Liu, Y.C. Chang, C.W. Chow, C.H. Yeh, Equalization and pre-distorted schemes for increasing data rate in-door visible light communication system, in *Conference on optical fiber communication (OFC)*, pp. JWA83, 2011
5. C.W. Chow, Y.H. Lin, Convergent optical wired and wireless long-reach access network using high spectral-efficient modulation. Opt. Express **20**, 9243–9248 (2012)
6. C.H. Yeh, Y.F. Liu, C.W. Chow, Y. Liu, P.Y. Huang, H.K. Tsang, Investigation of 4-ASK modulation with digital filtering to increase 20 times of direct modulation speed of white-light LED visible light communication system. Opt. Express **20**, 16218–16223 (2012)

7. Y. Wang, Y. Wang, N. Chi, J. Yu, H. Shang, Demonstration of 575-Mb/s downlink and 225-Mb/s uplink bi-directional SCM-WDM visible light communication using RGB LED and phosphor-based LED. Opt. Express **21**, 1203–1208 (2013)
8. C.Y. Li, H.S. Su, C.Y. Chen, H.H. Lu, H.W. Chen, C.H. Chang et al., Full-duplex lightwave transport systems employing phase-modulated RoF and intensity-remodulated CATV signals. Opt. Express **19**, 14000–14007 (2011)
9. C.Y. Li, H.S. Su, C.H. Chang, H.H. Lu, P.Y. Wu, C.Y. Chen et al., Generation and transmission of BB/MW/MMW signals by cascading PM and MZM. IEEE/OSA J. Lightw. Technol. **30**, 298–303 (2012)
10. J. Vucic, C. Kottke, S. Nerreter, K.D. Langer, J.W. Walewski, 513 Mbit/s visible light communications link based on DMT-modulation of a white LED. IEEE/OSA J. Lightw. Technol. **28**, 3512–3518 (2010)
11. C.W. Chow, C.H. Yeh, C.H. Wang, F.Y. Shih, C.L. Pan, S. Chi, WDM extended reach passive optical networks using OFDM-QAM. Opt. Express **16**, 12096–12101 (2008)
12. K. Wang, A. Nirmalathas, C. Lim, E. Skafidas, "High-speed optical wireless communication system for indoor applications. IEEE Photon. Lett. **23**, 519–521 (2011)
13. W.Y. Lin, C.H. Chang, H.H. Lu, P.C. Peng, Y.P. Lin, C.Y. Chen et al., A hybrid CATV/OFDM long-reach passive optical network architecture. Opt. Express **20**, 4219–4224 (2012)
14. W.Y. Lin, C.Y. Chen, H.H. Lu, C.H. Chang, Y.P. Lin, H.C. Lin et al., 10 m/500Mbps WDM visible light communication systems. Opt. Express **20**, 9919–9924 (2012)
15. Y.F. Liu, C.H. Yeh, C.W. Chow, Y. Liu, Y.L. Liu, H.K. Tsang, Demonstration of bi-directional LED visible light communication using TDD traffic with mitigation of reflection interference. Opt. Express **20**, 23019–23024 (2012)
16. H.L. Minh, D.O. Brien, G. Faulkner, L. Zeng, K. Lee, D. Jung et al., 100-Mb/s NRZ visible light communications using a postequalized white LED. IEEE Photon. Lett. **21**, 1063–1065 (2009)
17. J. Vucic, C. Kottke, S. Nerreter, A. Buttner, K.D. Langer, J.W. Walewski, White light wireless transmission at 200 + Mb/s net data rate by use of discrete-multitone modulation. IEEE Photon. Lett. **21**, 1511–1513 (2009)
18. G. Yabre, Effect of relatively strong light injection on the chirp-to-power ratio and the 3 dB bandwidth of directly modulated semiconductor lasers. IEEE/OSA J. Lightw. Technol. **14**, 2367–2373 (1996)
19. W.Y. Lin, H.H. Lu, S.J. Tzeng, K.H. Chang, Y.C. Hsiao, Employing mutually injection-locked FP LDs scheme over full-duplex radio-on-fiber transport systems. Opt. Commun. **282**, 550–553 (2009)
20. H.C. Peng, H.H. Lu, C.Y. Li, H.S. Su, C.T. Hsu, Integration of FTTH and GI-POF in-house networks based on injection locking and direct-detection techniques. Opt. Express **19**, 6749–6755 (2011)
21. D.C. O'Brien, Visible light communications: challenges and potential, in *IEEE Photon.* pp. 365–366, 2011
22. M.R. Phillips, D.M. Ott, Crosstalk due to optical fiber nonliearities in WDM CATV lightwave systems. IEEE/OSA J. Lightw. Technol. **17**, 1782–1792 (1999)
23. M.R. Phillips, D.M. Ott, Crosstalk caused by nonideal output filters in WDM lightwave systems. IEEE Photon. Lett. **12**, 1094–1096 (2000)
24. H.H. Lu, S.J. Tzeng, Y.L. Liu, Intermodulation distortion suppression in a full-duplex radio-on-fiber ring network. IEEE Photon. Lett. **16**, 602–604 (2004)

Chapter 6
Colorless Laser Diodes for DWDM-PON Transmission

Gong-Ru Lin, Yu-Chuan Su and Yu-Chieh Chi

6.1 Historical Review and Challenges on Injection-Locked Laser Diode Transmitters for DWDM-PON Transmission

6.1.1 Historical Review on the Roadmap of Injection-Locked Transmitters for DWDM-PON

Since the requirement on data transmission capacity of the broadband internet access network is persistently increasing, the remarkable efforts have been paid on both optical wavelength and electrical frequency usages of the data transmitters emerged for developing the next-generation passive optical network (NGPON). The dense-wavelength-division-multiplexed passive-optical-network (DWDM-PON) is regarded as one of the promising candidates for the fiber-to-the-home network owing to its simple architecture with extremely large transmission capacity at low cost of power budget [1, 2]. In consideration of practical optical distribution within a short-distant metropolitan network ($\leq$20 km), the wavelength injection-locked colorless laser diode is usually considered as the universal transmitter at user end due to its broadband gain-spectral range that ensures single-mode operation with enhanced modulation bandwidth at multiple DWDM channels [3, 4]. Versatile master light sources were employed to approach the injection-locking of different colorless transmitters to meet the demand of the DWDM-PON system. The historical progress and existed problem of different injection-locked transmitter with various master light sources are shown in Fig. 6.1. In 2000, Kim et al. [5] proposed an amplified spontaneous emission

G.-R. Lin (✉) · Y.-C. Su · Y.-C. Chi
Graduate Institute of Photonics and Optoelectronics, Department of Electrical Engineering, National Taiwan University, Taipei, Taiwan
e-mail: grlin@ntu.edu.tw

C.-C. Lee (ed.), *The Current Trends of Optics and Photonics*,
Topics in Applied Physics 129, DOI 10.1007/978-94-017-9392-6_6

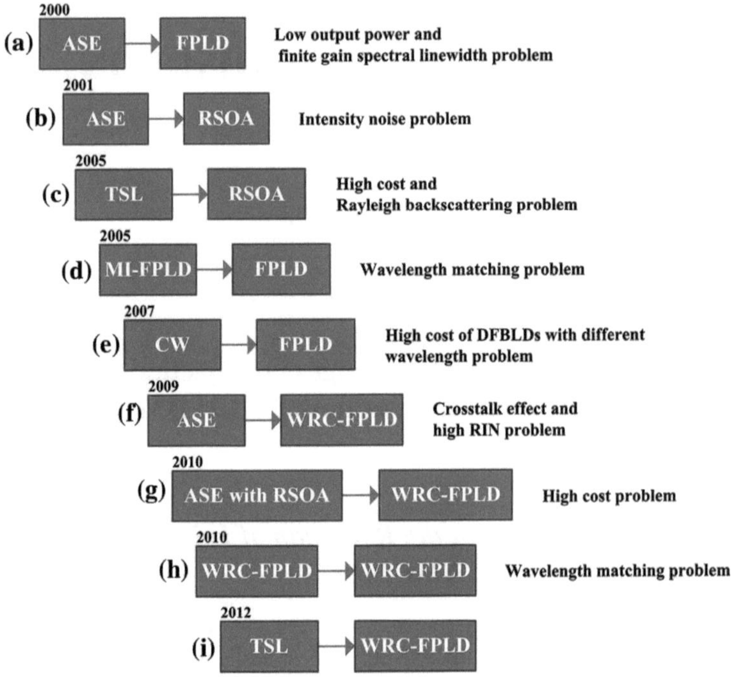

Fig. 6.1 The roadmap of DWDM-PON transmitters with external injection light source

(ASE) injection-locked Fabry-Perot laser diode (FPLD) with single-mode output as a low-cost DWDM transmitter. The side-mode suppression ratio (SMSR) and the extinction ratio (ER) of the wavelength-locked FPLD were larger than 29 and 13 dB, respectively. Owing to the high injection budget, lower output power and thermally unstable behavior, the FPLD with a finite gain spectral linewidth is hard to be externally injection-locked and thus not an ideal transmitter for broadband spectral sliced DWDM-PON system (Fig. 6.1a). Later on, the ASE injection-locked reflective semiconductor optical amplifier (RSOA) based DWDM-PON transmitter [6] was proposed to successfully provide eight DWDM channels with 1 nm wavelength spacing. With each channel modulated at a data rate of 1.25 Gbit/s, the non-return-to-zero (NRZ) on-off-keying (OOK) data transmission is shown to propagate over 25-km. However, the transmission performance is limited by the bad relative intensity noise (RIN) performance of the ASE injection source (Fig. 6.1b).

In order to reduce the ASE noise and release the limited injection-locked wavelength range, the single-mode laser (SML) injection-locked RSOA has been demonstrated. Since 2005, Lee et al. have proposed a bidirectional DWDM-PON by using a SML injection locked RSOA at gain saturation mode [7], which achieved 1.25 Gbit/s for upstream and 2.5 Gbit/s for down-stream data transmission over 20 km. Although the upstream data carried by the SML injection-locked RSOA can be generated by remodulating the data-erased [8–10] down-stream carrier, such a down- and up-stream wavelength preserved solution is not practical for DWDM-PON due to its high cost with the need of an additional data eraser and its increasing noise caused by Rayleigh backscattering (Fig. 6.1c). In the meantime, the mutually injection-locked [11, 12] AR-coated FPLDs pair has preliminarily emerged as an alternative broadband light source (BLS) for controlling the wavelength of the slave AR-coated FPLD at user-end optical network unit (ONU) [13]. The mutually injection-locked AR-coated FPLD master exhibits a narrow channel spacing of 0.2 nm associated with an extremely low RIN, however, which requires two identical FPLDs with exactly same end-facet reflectance and cavity length to guarantee the flattened master gain spectrum with equivalent mode spacing (Fig. 6.1d). Later on, the continuous-wave (CW) injection-locked FPLD has demonstrated in 2007 [14] to solve the wavelength matching problem and provide up- and down-stream transmission up to 10 Gbit/s, due to the highly-coherent injection (Fig. 6.1e). However, the injection-locking of commercial FPLDs with high end-facet reflectance (R = 30 %) usually requires a relatively large injection level, which inevitably causes the high power budget and high equipment cost from the highly coherent single-mode master lasers. This makes the CW injection-locked FPLD less comparable with an ideal cost-effective transmitter for DWDM-PON system.

6.1.2 Development of a Promising Universal Transmitter for Colorless Operation in DWDM-PON

Recently, the AR-coated colorless FPLD with long weak-resonant-cavity (hereafter referred as WRC-FPLD) has been considered as a more potential candidate to provide dense and weak longitudinal modes for next-generation DWDM-PON system than previous works because of its broad gain-spectrum, which provides more DWDM transmission channels than common FPLD and better throughput coherence than the RSOA. The illustrated WRC-FPLD spectrum and the WRC-FPLD and traditional FPLD channelized by the 200-GHz arrayed waveguide grating (AWG) are shown in Fig. 6.2a. With the 200-GHz channel spacing, the WRC-FPLD possesses more modes in one channel than the traditional FPLD due to its smaller mode spacing. As more longitudinal modes can be modulated in one channel, the ASE injection-locked WRC-FPLD transmitter provides a better transmission performance [15]. On the other hand, the WRC-FPLD can be channelized by the AWG based multiplexer/de-multiplexer with narrower channel spacing to providing more DWDM channels in its finite gain spectrum. Figure 6.2b shows the illustrated spectrum of WRC-FPLD with ASE, WRC-FPLD and tunable single-mode laser (TSL) injection-locking. Hence, the ASE injection-locked WRC-FPLD used for DWDM-PON was proposed in previous work [15, 16], providing a bit-error-rate (BER) of 10^{-13} at receiving power of −31 dBm with a bit rate of 622 Mbit/s at least.

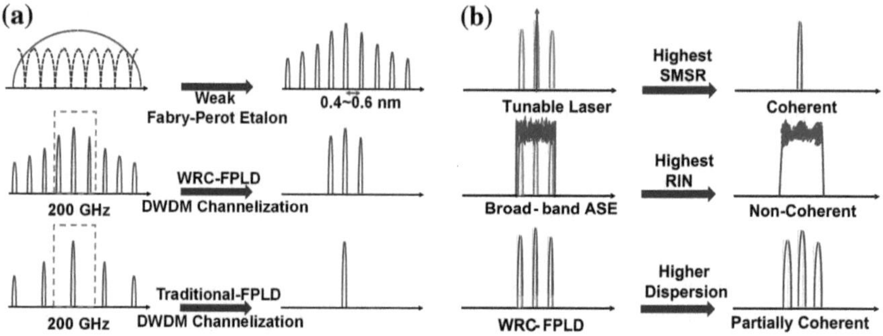

Fig. 6.2 **a** The illustrated WRC-FPLD spectrum and WRC-FPLD or traditional FPLD channelized by 200-GHz AWG. **b** Schematic of slave WRC-FPLD with TSL, ASE and master WRC-FPLD injection-locking

Owing to the fluctuant noise of the ASE source, the AWG channelized ASE injection-locked WRC-FPLD presents the highest noise, which was then considered to be reduced by changing the ASE source with the partial-coherent master WRC-FPLD. However, it is inevitable to inject-lock two or three mode within one channel when the channel spacing of AWG is not narrow enough, which may result in the unexpected dispersion during long-distant transmission. This dispersion effect can be solved by using the dispersion compensation fiber (DCF) [16] but is not practical for the existing long-haul fiber network. Another solution is the use of the TSL to injection-lock WRC-FPLD, which shows a perfectly single-mode output with the highest SNR, narrowest linewidth and lowest dispersion to provide a better transmission quality. When comparing the conventional FPLD (front facet reflectance of 30 %) with the WRC-FPLD (front facet reflectance of 1 %) [17], the WRC-FPLD possess the higher SMSR at the same injection power level and the wider detuning of injection-locked wavelength [18] than the traditional FPLD [19] owing to its weak-resonant-cavity feature.

To effectively increase the network capacity of the DWDM-PON system with the injection-locked WRC-FPLD transmitter, both the 200-GHz and the 50-GHz AWG were considered to slice the ASE injection source with their chirp, noise and transmission performances compared each other [20]. Although the DWDM-PON system with 50-GHz AWG can provide more channels with lower negative frequency chirp than those with the 200-GHz AWG, the 200-GHz AWG channelized ASE injection minimizes the intensity noise of WRC-FPLD transmitter to perform better transmission performance. With the additional AWG filtering, the RIN induced by ASE source and the intraband crosstalk [21] between the upstream transmitted data and the reflected ASE signal are two important issues of the DWDM-PON system with the ASE injection-locked WRC-FPLD transmitter. A solution has been proposed in previous work by using the SOA based noise reduction [22] to squeeze the ASE noise at local ONU part to solve these problems, and the RIN and the crosstalk effect can be reduced by 4.5 and 6.3 dB, respectively [23]. Although the noise of the ASE injection source could be minimized by the saturable RSOA, the extra RSOA in each ONU channel inevitably raises the cost for the DWDM-PON system. After considering the noise and cost of the ASE injection-locked WRC-FPLD applied to the DWDM-PON system, the mutually injection-locked FPLD [24] was proposed to meet these demands. Afterwards, the partially coherent WRC-FPLD injection-locked WRC-FPLD has also emerged [25], which shows an SMSR of up to 40 dB and a Q factor of 9.5 dB to provide a receiving sensitivity of −24.4 dBm at a BER of 10^{-12} in the 2.5-Gbit/s DWDM-PON transmission. More recently, the highly coherent tunable CW fiber laser was employed to solve the wavelength matching problem between the master and slave lasers [26]. Particularly, the transmission performance of the injection-locked WRC-FPLD transmitters with different front-facet reflectances were also investigated [27]. In contrast to the WRC-FPLD injection-locked WRC-FPLD, the TSL injection-locked WRC-FPLD has also been proposed, as shown in Fig. 6.1h, which offers the highest SNR to obtain the ultimate transmission performance when comparing with the ASE and the WRC-FPLD injection-locking cases.

6.1.3 Using Long-Cavity Colorless Laser Diodes for OOK/OFDM Transmission in DWDM-PON

When considering the back-to-back and 25-km single-mode fiber (SMF) transmissions, the measured eye-diagrams of the free-running and ASE injection-locked WRC-FPLDs under direct on-off keying (OOK) modulation are compared in Fig. 6.3. The back-to-back transmitted eye-diagram of WRC-FPLD without injection-locking is clearer than that of the ASE injection-locked WRC-FPLD, since all longitudinal modes of the free-running WRC-FPLD under direct modulation are received without spectral slicing. However, the eye-diagram of the free-running WRC-FPLD is seriously distorted after 25-km SMF transmission because these longitudinal modes covering a broadband gain spectrum results in strong modal dispersion after long-distant transmission. In contrast, the single-mode injection-locked WRC-FPLD exhibits a narrower linewidth to minimize the dispersion effect and provide a clearer eye-diagram after the 25-km transmission. For increasing the transmission data rate, it is necessary to enhance the modulation bandwidth of such a broadband WRC-FPLD transmitter by enlarging its bias current or injection-locking power. An alternative approach of carrying the high capacity data in finite bandwidth can be achieved by using the orthogonal frequency-division multiplexing (OFDM) modulation format [28, 29]. Transmission of the OFDM data were successively demonstrated with the distributed feedback laser diodes (DFBLDs) [30, 31], the vertical-cavity surface emitting laser diodes (VCSELDs) [32, 33], the TSL injection-locked WRC-FPLD [34] and the down-stream reusable RSOA [35, 36]. The OFDM

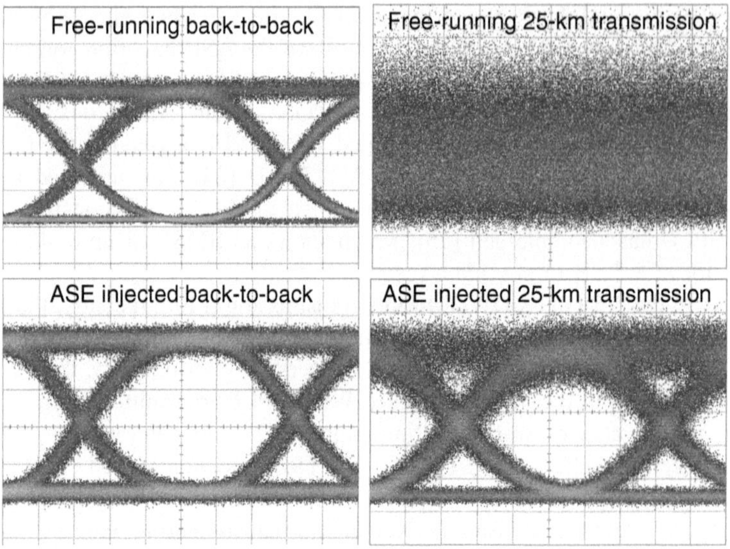

Fig. 6.3 The measured back-to-back and 25-km transmitted eye-diagrams of the WRC-FPLDs without and with spectrum-sliced ASE injection

modulated WRC-FPLD or RSOA transmitter provides the high spectral usage to achieve the data rate of 4 or 7.5–10 Gbit/s by efficiently using the modulation bandwidth of only 1 GHz. Not long ago, the TSL injection-locked WRC-FPLD with the data transmission rate up to 20 Gbit/s by pre-leveling technique was reported [37]. It is known that the OFDM modulation format consisted of intensity and phase signals is sensitive to the coherence of transmitters; however, the comparisons on the OFDM performance of the WRC-FPLD transmitters injection-locked by master sources with different degrees of coherence have never been discussed.

Although the laser throughput properties and transmission performance can be improved by enhancing the coherence of injection light source, higher coherent injection light also inevitably raises the equipment cost. In this work, the effect of injection coherence on the noise, bandwidth and error rate performances of the long-cavity colorless WRC-FPLD with an end-facet reflectance of 0.5 % is demonstrated for OOK and OFDM transmission. By using three kinds of master sources with different degrees of coherence including an AWG channelized ASE, free-running WRC-FPLD and TSL, several key parameters of the wavelength injection-locked slave WRC-FPLD transmitter such as the RIN, the frequency response, the signals to noise ratio (SNR) and the bit error rate (BER) of OOK and 16-QAM OFDM transmitted data are characterized. In addition to the throughput response and transmission performance, the simulated eye diagrams from a set of modified rate equations are compared with the experimental results.

6.2 OOK or OFDM Data Transmission Performances of Directly Modulated Slave WRC-FPLD Injection-Locked by Master Sources with Different Degrees of Coherence

6.2.1 Methods for Building up the DWDM-PON with Directly Modulated Slave WRC-FPLD Injection-Locked by Master Sources with Different Degrees of Coherence

The Fig. 6.4 illustrates a DWDM-PON system based on the up-stream WRC-FPLD transmitter with a front-facet reflectance of 0.5 %. The slave WRC-FPLD is injection-locked by the channelized ASE light source, master WRC-FPLD or tunable laser (Agilent, 8164A) in the central office, respectively. The external injection-locking was performed by seeding the non-coherent, partial coherent or high coherent light sources with fixed output power of −6 dBm into the WRC-FPLD through a polarization controller, and passing through a 100-GHz AWG multiplexer.

In the transmission part of the ONU, the DC bias current and the modulation signal with OOK or OFDM modulation format was coupled into a bias-tee (Mini-circuits, ZX85-12G-S+) for driving the WRC-FPLD. The bias current of WRC-FPLD directly modulated by a 1.25-Gbit/s OOK data-stream with a pattern length of $2^{23}-1$ was set nearby 34 mA corresponding to 2 I_{th} to achieve the optimized ER for transmission diagnosis. The peak-to-peak amplitude of the electrical PRBS digital data was set at 500 mV. In the receiver part of the ONU, the eye-diagram and BER of the slave WRC-FPLD based up-stream transmitted data was analyzed by a digital sampling oscilloscope (Agilent, 86100A + 86109B) and commercial error detector (Agilent, 70843A). The OFDM signal was sent to an arbitrary waveform generator (Tektronix 7102B) with a central frequency at 1.5625 GHz, a carrier number of 122, a sampling rate of 12.5 GHz, a total bit rate at 12.5 Gbit/s and

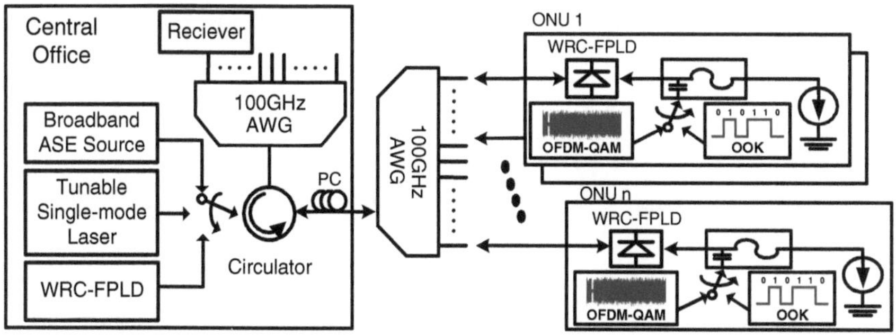

Fig. 6.4 A DWDM-PON system with a coherent laser, WRC-FPLD or ASE injection-locking source located at central office

a modulation bandwidth of 3.125 GHz. The transmitted OFDM signal delivered by the slave WRC-FPLD injection-locked with different master sources were analyzed by a real-time digitized oscilloscope analyzer (Tektronix CSA7404B) with a sampling rate of 100 GS/s.

6.2.2 The Effect of the Injection Coherence on Mode, Noise Characteristics and Frequency Response

The full-band spectra of the WRC-FPLD at free-running condition or under injection-locked by using non-coherent ASE, partially coherent WRC-FPLD and highly coherent TSL are compared in the upper row of Fig. 6.5. The free-running WRC-FPLD exhibits an extremely wide gain spectrum with numerous weak-resonant longitudinal modes covering the wavelength from 1,530 to 1,585 nm. Such a broad lasing spectrum of the WRC-FPLD is comparable with the gain spectrum of a free-running SOA, indicating that the extremely weak Fabry-Perot etalon effect is introduced by the relatively small end-facet reflectance of 0.5 %. With same injection level of −6 dBm, all the injection-locked spectra reveal single mode operation. However, the zoom-in lasing spectra (see middle row of Fig. 6.5) of the single-mode WRC-FPLD injection-locked by three master sources with different degrees of coherence can still be distinguished from one another. After injection-locking with the broad-band ASE source channelized by using an AWG

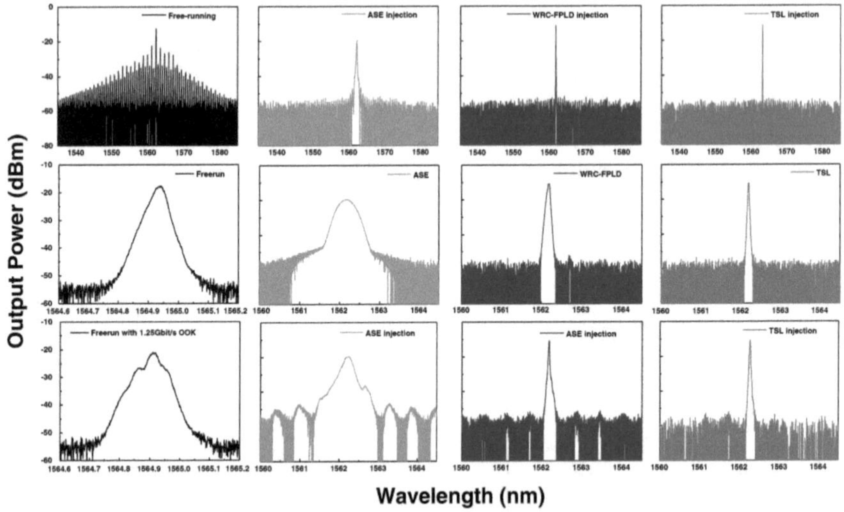

Fig. 6.5 The continuous-wave operated (*upper* for full-band and *middle* for zoom-in spectra) and the OOK modulated (*lower*) optical spectra of the slave WRC-FPLDs under free-running (*first column*) and injection-locked with non-coherent ASE (*second column*), partially coherent WRC-FPLD (*third column*) and highly coherent TSL (*forth column*)

demultiplexer with a channel spacing of 100 GHz (0.8 nm), the slave WRC-FPLD presents a much broader mode linewidth of up to 0.724 nm. In contrast, the WRC-FPLD injection-locked WRC-FPLD pair results in a mode linewidth as narrow as 0.146 nm. In comparison with those master sources with lower coherence, the highly coherent TSL consists of more coherent photons at a cavity resonant wavelength to shrink the slave WRC-FPLD linewidth to 0.086 nm and suppress its relative intensity noise. Owing to the higher coherence and lower ASE noise of TSL, the TSL injection-locked WRC-FPLD presents a narrowest mode spectrum.

After 1.25 Gbit/s OOK modulation, the OOK modulated modal linewidths of the WRC-FPLD injection-locked by 100-GHz AWG channelized ASE, master WRC-FPLD and TSL are 0.8, 0.21 and 0.09 nm respectively (see the lower row of Fig. 6.5). Because the 100-GHz AWG channel spacing (0.8 nm) is much broader than the mode spacing (0.5 nm) of the WRC-FPLD, the adjacent side-modes near by the major injection-locked mode of the WRC-FPLD are inevitably injected by channelized ASE slightly. The inevitably injected side-modes and broader linewidth of laser transmitter are not beneficial for long-distance transmission due to the mode-beating and dispersion effects. Technically, the adjacent side-modes of the slave WRC-FPLD injection-locked by the ASE and the master WRC-FPLD can be filtered out by using an AWG demultiplexer with channel spacing of 50 GHz in DWDM-PON. However, the AWG channelized ASE injection-locked WRCFPLD also degrades its transmission performance when narrowing down the AWG bandwidth as which inevitably increases the intensity noise [20]. In comparison with the ASE injection-locking case, the master WRC-FPLD and TSL not only suppress the side-modes of the slave WRC-FPLD but also lead to a narrower linewidth. With the same injection power, the side-mode suppression ratios (SMSRs) of WRC-FPLD injection-locked by ASE, master WRC-FPLD and TSL are 27, 40.5 and 45.3 dB, respectively.

Besides, the wavelength tunability of the master WRC-FPLD and TSL enables the exact wavelength coincidence with the slave WRC-FPLD. The injection master with higher coherence ensures more stimulated emitting photons created in the injection-locked mode, which provides the slave WRC-FPLD transmitter better coherent throughout and higher SMSR. Undoubtedly, the non-coherent ASE master consists of large intensity and phase noise, and the phase noise also cross-correlates with the intensity noise through the rate equations [38]. Improving the master coherence effectively reduces both phase and intensity fluctuations due to the enhanced stimulated emission. As evidence, the RIN spectra of master ASE, WRC-FPLD and TSL at constant power of −6 dBm after passing through the 100-GHz channelized AWG are shown in Fig. 6.6. The non-coherent ASE shows the RIN floor as high as −95 dBc/Hz, which is 5–12 dB worse than the WRC-FPLD and TSL masters. The WRC-FPLD master exhibits comparable RIN with the TSL at frequency higher 5 GHz, whereas its low-frequency RIN level is degraded by 5 dB with frequency decreasing from 5 to 1 GHz. After injection-locking by masters with different degrees of coherence, the RIN spectra of slave WRC-FPLDs under injection-locking powers of −12 and −6 dBm are compared in Fig. 6.7.

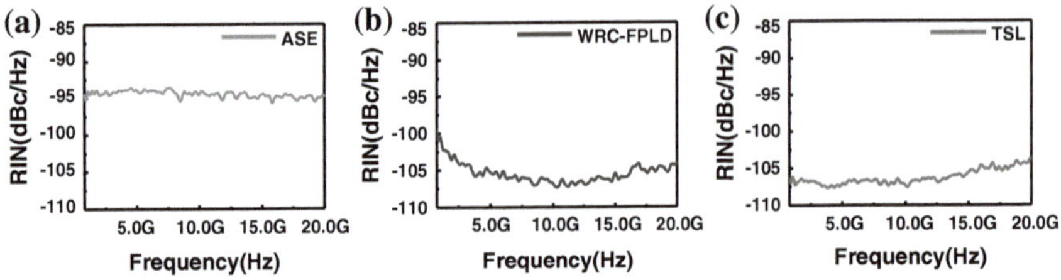

Fig. 6.6 The RIN spectra of master ASE, WRC-FPLD and TSL after the 100-GHz AWG

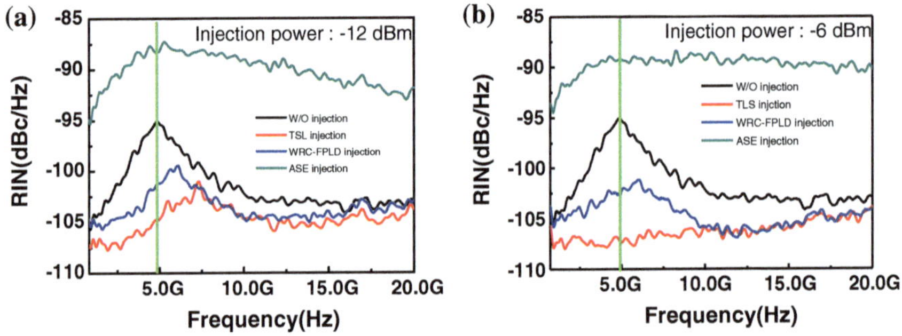

Fig. 6.7 The RIN of slave WRC-FPLD injection-locked by TSL, WRC-FPLD and ASE at injection powers of **a** −12 dBm and **b** −6 dBm

The free-running WRC-FPLD reveals a relaxation oscillation peak with intensity of −95 dBc/Hz at 5 GHz. Improving the coherence of master not only suppresses the RIN floor but also shift the relaxation oscillation peak of slave WRC-FPLD under injection-locking. The ASE injection-locking fails to improve the RIN of the slave WRC-FPLD. At −12 dBm injection, the slave WRC-FPLD shows a RIN peak that is 7 dB larger than that of the original ASE master. The non-coherent injection greatly degrades the RIN by raising its floor up to 10 dB, which cannot be improved simply by enlarging the injection-locking power, as shown in Fig. 6.7b. In contrast, the injection of a partially coherent WRC-FPLD master slightly up-shift the RIN peak to 6 GHz and reduces its intensity by 10–15 dB as compared to the ASE injection-locking case. The 6-dB increment on the injection level could further reduce the entire noise background linearly. Both the WRC-FPLD and TSL maters improve the RIN of the slave WRC-FPLD; however, the TSL shows a superior capability on suppressing the RIN background at smaller frequencies. Furthermore, the relaxation oscillation peak up-shifts and even disappears with enlarging injection level. Such a lowest RIN floor explains why the highly coherent TSL inject-locked WRC-FPLD exhibits the highest SMSR.

Subsequently, the analog modulation response of the slave WRC-FPLD injection-locked by different masters with injection powers of −12 and −6 dBm are compared in Fig. 6.8a, b. With injection-locking, the modulation response in

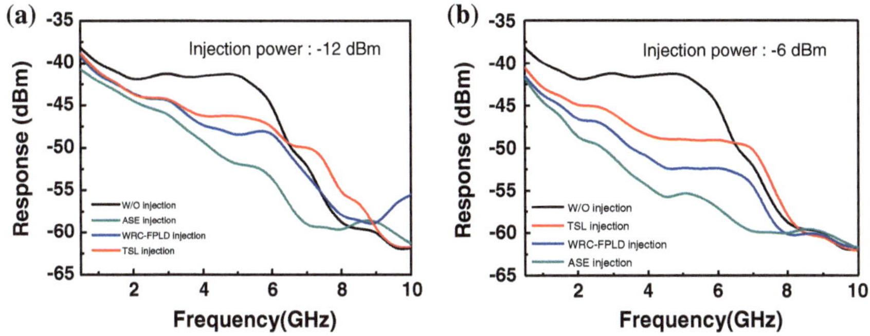

Fig. 6.8 The frequency response of slave WRC-FPLD injection-locked by TSL, WRC-FPLD and ASE at injection powers of **a** −12 dBm and **b** −6 dBm

high-frequency region can be enhanced at a cost of sacrificing the throughput power in low-frequency region. The enhanced coherence of master simultaneously suppresses the RIN floor and improves the modulation bandwidth with up-shifted relaxation oscillation frequency. The ASE injection-locked WRC-FPLD shows lowest modulation throughput with its power-to-frequency slope increasing from −2.2 to −3.4 dB/GHz by enlarging the injection from −12 to −6 dBm. The injection-locked master-slave WRC-FPLD pair performs an increment on the relaxation oscillation frequency from 5 GHz (free-running case) to 6–6.5 GHz; however, the enlarged injection power from −12 to −6 dBm pays more the throughput power attenuation than the frequency bandwidth extension. The highly coherent TSL injection-locks the slave WRC-FPLD and further up-shifts its relaxation oscillation frequency from 5 to 7 GHz. In comparison, the slave WRC-FPLD exhibits a larger throughput at lower injection level, and a trade-off is set between injection level and modulation bandwidth.

6.2.3 The Effect of Injection Coherence on OOK Transmission Performances of the Slave WRC-FPLD Laser Transmitter with Different-Locking Master Sources

The measured and simulated eye-diagrams of the directly modulated and AWG channelized WRC-FPLD injection-locked by versatile light sources with different degrees of coherence are compared in Fig. 6.9a, b. When comparing with other two cases, the measured eye-diagrams of the ASE injection-locked WRC-FPLD shows an ER of 8.02 dB and a relatively low SNR of 6.07 dB owing to its intense noise at on-level. That is, the ASE process continues in the slave WRC-FPLD to cause more non-coherent photons with randomized phases. Increasing the degree of coherence by changing the master from ASE to WRC-FPLD could effectively

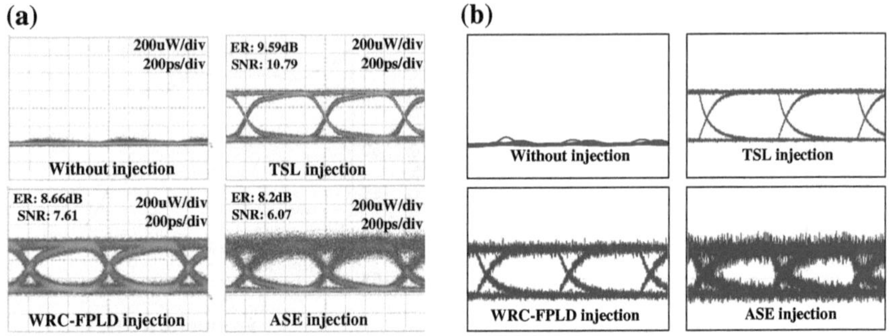

Fig. 6.9 **a** The measured and **b** the simulated eye-diagrams of WRC-FPLDs without and with TSL, WRC-FPLD and ASE injection

improve the SNR of the slave WRC-FPLD transmitted data from 6.07 to 7.61 dB at same power level. Obviously, the TSL injection-locked WRC-FPLD shows greatly optimized ER of 9.6 dB and SNR of 10.8 dB with a injection-locking level of −3 dBm. Note that the increment of ER is attributed to the noise suppression of off-level data, and the enhancement on SNR is owing to the reduction of the RIN at on-level. Increasing the coherence of injection-locking master effectively contributes to a significant suppression on the on- and off-level noises such that the better enhancement of both the SNR and the ER can be achieved.

In simulation, the following rate equations adopting by the theoretical model of a single-wavelength FPLD [38] under external injection-locking condition are shown below,

$$\frac{dN(t)}{dt} = \frac{\eta_i I}{qV} - \frac{N(t)}{\tau_s} - \frac{v_g a}{V}[N(t) - N_{tr}]S(t), \tag{6.1}$$

$$\frac{d\phi(t)}{dt} = \frac{\alpha}{2}\{\frac{\Gamma v_g a}{V}[N(t) - N_{tr}] - \frac{1}{\tau_p}\} - \kappa\sqrt{\frac{S_{inj} + \Delta S \cdot Rand(t)}{S(t)}} \sin\phi(t) - \Delta w_{inj}, \tag{6.2}$$

$$\frac{dS(t)}{dt} = \frac{1}{2}\{\frac{\Gamma v_g a}{V}[N(t) - N_{tr}] - \frac{1}{\tau_p}\}S(t) + \kappa\sqrt{[S_{inj} + \Delta S \cdot Rand(t)]S(t)} \cos(\phi(t)), \tag{6.3}$$

where $N(t)$, $\Phi(t)$ and $S(t)$ are the time dependent functions of carrier number, phase and photon number of the slave WRC-FPLD, respectively. In these equations, I denotes the biased current, η_i the internal quantum efficiency, a the differential gain, Γ the optical confinement factor, v_g the velocity, τ_p the photon lifetime, τ_s the spontaneous carrier lifetime, κ the coupling efficiency, α the linewidth enhancement factor, S_{inj} the injection photon and $\Delta\omega_{inj}$ the detuning frequency.

To present all simulated eye-diagrams of the WRC-FPLD injection-locked by master sources of different degrees of coherence, the related parameters are

Table 6.1 The parameters of the WRC-FPLD injection-locked by master sources of different degrees of coherence

Master sources	ASE	WRC-FPLD	TSL
The percentage of non-coherent photons (%)	60	40	<10
Active layer thickness (μm)	0.081		
Cavity length (μm)	681		
Active layer width (μm)	1.5		
Rear facet reflectance	93 %		
Front facet reflectance	0.5 %		
Carrier number at transparency	8.27E7		
Carrier number at threshold	1.67E8		
Carrier lifetime (ns)	1.00		
Photon lifetime (ps)	1.39		
Optical confinement factor	0.23		
Mirror loss (cm^{-1})	78.75		
Threshold gain (cm^{-1})	364.14		
Group velocity (cm/s)	8.57E9		

summarized in Table 6.1. The degrees of injection coherence indicate the phase consistency of photons in WRC-FPLD, which directly induces the noise fluctuation through the phase-intensity conversion in the WRC-FPLD or after passing through dispersion material such as optical fiber. In approximation, the degree of coherence can be approached by changing the weighting factor of randomized intensity term from master source. During simulation, the injection light with different coherences is approached by substituting the S_{inj} into the $S_{inj} + \Delta S \cdot Rand(t)$, where $Rand(t)$ is a random function presenting the noise fluctuation, ΔS the percentage of the non-coherent photon number. As expected, the decreased coherence of injection also increases the fluctuation of stimulated emitting photon number as well as the intensity noise in slave WRC-FPLD cavity, as shown in Fig. 6.9(b). The eye-pattern of the free-running WRC-FPLD is hardly recognized due to the spreading modulation power over all longitudinal modes of the WRC-FPLD. Nevertheless, the eye-diagrams can be significantly improved by implementing the injection-locking with enhanced coherence.

With the definition of $Q = [(ER\text{-}1)/(ER + 1)](M \bullet SNR)^{0.5}$ (which decides the lowest BER in a DWDM-PON with an optical receiver gain of M). The least Q factor to perform error-free transmission at BER $< 10^{-9}$ is 6, and the ASE injection-locked WRC-FPLD with degraded ER and SNR shows a smallest Q of 6.2. The slave WRC-FPLD can enhance its Q factor from 6.95 to 8.5 by changing the master from partially coherent WRC-FPLD to highly coherent TSL. Under back-to-back transmission at bit rates of 1.25 and 2.5 Gbit/s, the BER responses of the slave WRC-FPLD injection-locked with different masters are compared in Fig. 6.10. Non-coherent ASE injection-locked WRC-FPLD causes a large amount of the transmission errors to limit its minimal BER of 10^{-9} at a receiving power of 21.8 dBm. The partially coherent WRC-FPLD injection achieves better ER and

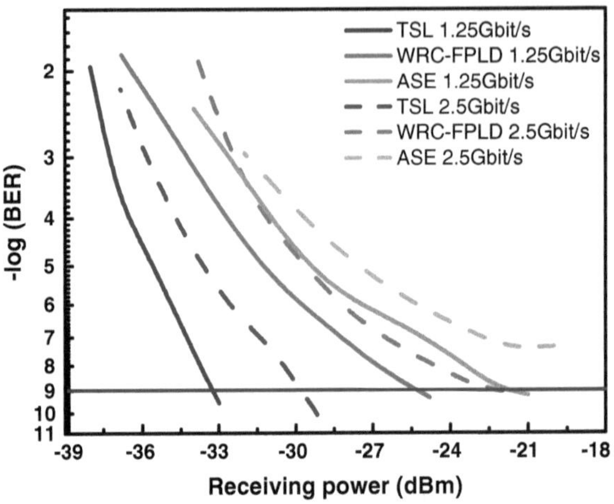

Fig. 6.10 BER of ASE, WRC-FPLD and TSL injection-locked WRC-FPLDs

SNR, thus providing a receiving power as low as −25.5 dBm for BER of 10^{-9}. The WRC-FPLD injection-locked by a highly coherent TSL greatly promotes its receiving sensitivity to −33 dBm at a BER of 10^{-9}. When comparing the 1.25- and 2.5-Gbit/s transmissions carried by the slave WRC-FPLD injection-locked with masters of different coherences, the receiving power penalty of 2–3 dB is observed by upgrading the bit rates from 1.25 to 2.5 Gbit/s. The ASE injection-locked WRC-FPLD transmitter at a bit rate of 2.5 Gbit/s shows a BER floor at around 3×10^{-7} even with the receiving power enlarging up to −20.5 dBm.

On the other hand, the performances of 16-QAM OFDM carried by using the slave WRC-FPLD injection-locked with different masters are compared with their transmitted data-stream in time and frequency domains shown in Figs. 6.11 and 6.12. In contrast to the electrical 16-QAM OFDM shown in Fig. 6.11a with a sampling rate of 12.5 GS/s, the ASE injection-locked WRC-FPLD delivers a distorted and noisy OFDM waveform with its cyclic prefix and training symbol lost during transmission (see Fig. 6.10b). The OFDM signal carried by the master WRC-FPLD injection-locked WRC-FPLD shows smaller peak-to-peak amplitude with clearer cyclic-prefix and training symbol (Fig. 6.11c), which is correlated with its broadened modal linewidth with larger intensity noise at lower frequency region. When injection-locked by partially coherent WRC-FPLD or non-coherent ASE, a nonlinearly distorted OFDM trace in time domain may by induced by modulation distortion and frequency chirp of the slave WRC-FPLD with a wider modal linewidth [16]. The completely non-distorted OFDM waveform carried by directly modulated WRC-FPLD is received when injection-locking by the highly coherent TSL master.

Since OFDM utilizes multiple subcarriers to encode the digital data-stream in parallel, the SNR feature at all subcarriers plays an important role on the overall OFDM transmission performance. From the OFDM spectrum in frequency

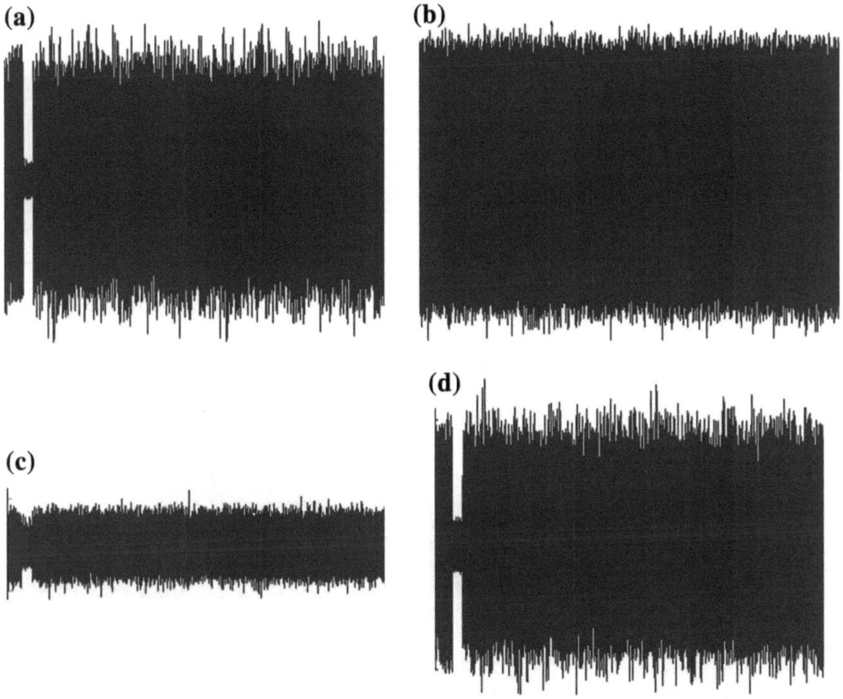

Fig. 6.11 Time-domain waveforms of 16-QAM OFDM data-streams carried by lave WRC-FPLD injection-locked by different masters. **a** Electrical waveform. **b** ASE. **c** WRC-FPLD. **d** TSL

domain shown in Fig. 6.12, the OFDM data-stream carried by ASE injection-locked WRC-FPLD reveals a totally reshaped and didtorted spectrum dominated by RIN. The whole data carried by both low- and high-OFDM subcarriers are covered with RIN and the ASE carrier is too noisy to provide high-SNR OFDM transmission, as shown in Fig. 6.12b. Owing to its lowest frequency response as well as the highest RIN, the received OFDM spectrum from the ASE injection-locked WRC-FPLD is similar to the RIN performance of the ASE injection source shown in Fig. 6.7a. In contrast, the master WRC-FPLD injection-locked WRC-FPLD scales down its noise floor with the greatly suppressed RIN (see Fig. 6.12c); however, the lower OFDM spectral power is observed due to its low modulation throughout. With the modulation response improved by highly coherent TSL injection-locking, the 16-QAM OFDM data-stream can be correctly analyzed with lowest noise floor as well as distinguished SNR at high subcarriers, as shown in Fig. 6.12d.

The constellation plots of the decoded 16-QAM OFDM data also present different degrees of amplitude and phase errors by injection-locking the WRC-FPLD with masters of different coherences, as shown in Fig. 6.13. The phase noise is mainly attributed to the frequency chirp affected by the modal linewidth of the

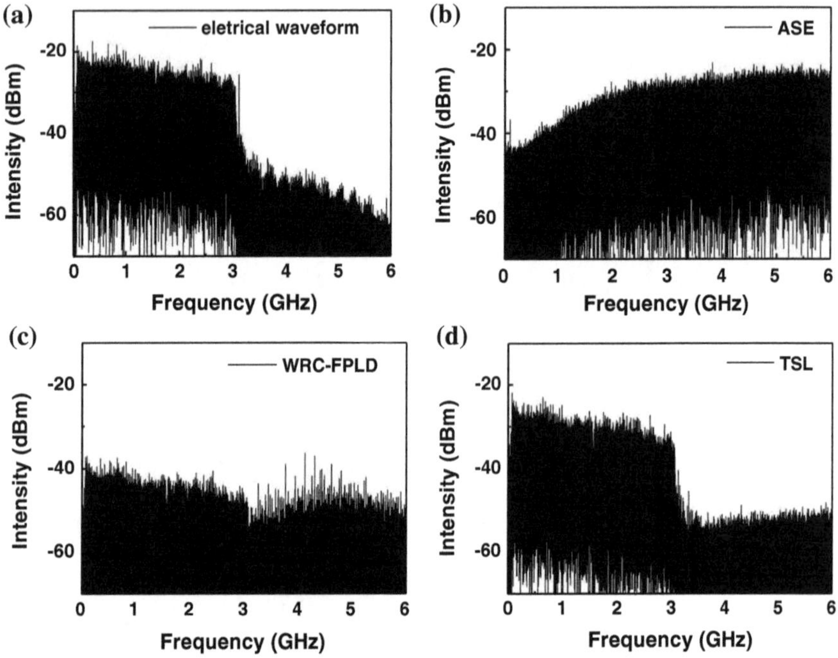

Fig. 6.12 RF spectra of 16-QAM-OFDM data-streams carried by lave WRC-FPLD injection-locked by different masters. **a** Electrical waveform. **b** ASE. **c** WRC-FPLD. **d** TSL

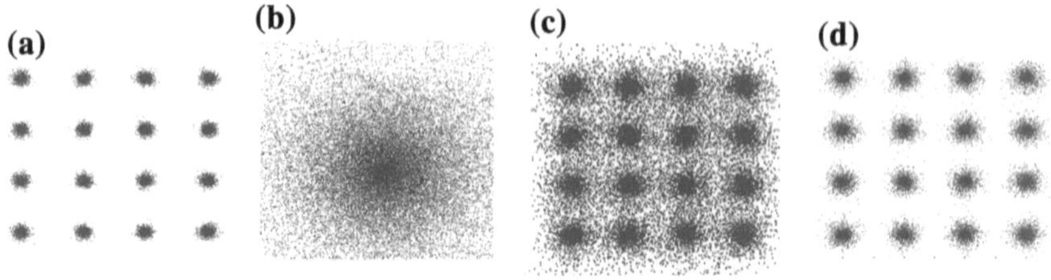

Fig. 6.13 The constellation plots of up-stream 16-QAM OFDM data-streams carried by the slave WRC-FPLD injection-locked with ASE, WRC-FPLD and TSL masters. **a** Electrical waveform. **b** ASE. **c** WRC-FPLD. **d** TSL

injection-locked slave WRC-FPLD, whereas the amplitude error results from the clamped OFDM output with limited modulation response of the WRC-FPLD. In Fig. 6.12b, the ASE injection-locked WRC-FPLD with largest RIN and lowest bandwidth fails to demonstrate a clear constellation plot, leading to an extremely large EVM as compared to its original value of 2 % (see Fig. 6.12a). The master WRC-FPLD injection-locked WRC-FPLD performs a resolvable constellation plot with an EVM of 6.9 %, as depicted in Fig. 6.12c. The clearest constellation plot of the 16-QAM OFDM data carried by the highly coherent TSL injection-locked

Table 6.2 BER and SNR of 16-QAM OFDM signals transmission with TSL, WRC-FPLD and ASE injection

Source	TSL	WRC-FPLD	ASE
Coherence	Complete	Partial	None
EVM (%)	4.5	6.9	≫20
SNR (dB)	22.8	14.7	10.2
BER	4.38E-09	5.90E-03	1.50E-01

WRC-FPLD with a minimized EVM of 4.5 % is obtained and shown in Fig. 6.12d. In summary, the EVM, SNR and BER of the decoded OFDM data carried by the slave WRC-FPLD injection-locked with masters of different coherences are listed in Table 6.2.

Obviously, the ASE injection-locked WRC-FPLD shows the worst BER response for both OOK and OFDM transmissions because the disordered OFDM waveform experiences finite modulation response with enormous noise. The master WRC-FPLD injection-locked slave WRC-FPLD can transmit the 16-QAM OFDM data-stream with a acceptable BER of 5.9×10^{-3}. The best OFDM transmission is obtained with the TSL injection-locked WRC-FPLD, which presents the narrowest mode with lowest noise to carry the OFDM data with a receiving BER as low as 4.38×10^{-9}. Note that the WRC-FPLD injection-locked WRC-FPLD pair with partial coherence could also provide an OOK receiving power sensitivity of −25.5 dB at a BER of 10^{-9} and a 16-QAM OFDM transmission BER of 5.9×10^{-3}, which is considered as the most cost-effective master source to concurrently support multi-channel wavelength injection-locking when comparing with other candidates.

6.3 Summary

The effect of master coherence on the OOK and 12.5-Gbit/s OFDM signal transmission of the injection-locked slave WRC-FPLD in a 100-GHz channelized DWDM-PON system has been investigated. Increasing the coherence of the master source improves the frequency response and suppresses the RIN of injection-locked WRC-FPLD. By changing the injection master from ASE to TSL, the RIN of the injection-locked WRC-FPLD can be decreased by at least 15 dB. The SNR of the OOK signal carried by the slave WRC-FPLD increases from 6.07 to 10.8 by enhancing the injection coherence from non-coherence to high coherence. Increasing the master coherence suppresses the on-level noise of the transmitted OOK data such that the SNR is enhanced more than ER. When comparing the 1.25- and 2.5-Gbit/s transmissions carried by the slave WRC-FPLD injection-locked with masters of different coherences, the receiving power penalty of 3 dB is observed by upgrading the bit rates from 1.25 to 2.5 Gbit/s. On the other hand, the BER of the 16-QAM OFDM data is significantly enhanced from 1.5×10^{-1} to

4.38×10^{-9} by increasing the master coherence. Although the use of QAM-OFDM data could enhance the spectral efficiency, the degradation of the transmission performance with decreasing the injection coherence is more significant than that using the OOK data format. The QAM-OFDM data reveals a stronger correlation with the enhanced master coherence than the OOK for the injection-locked WRC-FPLD. Therefore, the coherence of injection light source is more important for the OFDM than OOK signals carried by injection-locked WRC-FPLD. Although the highest coherence leads to the best transmission performance, the TSL injection-locked WRC-FPLD is inappropriate for a cost-effective DWDM-PON, whereas the partially coherent WRC-FPLD injection-locked WRC-FPLD pair with a favorable cost could be considered as a more practical transmitter for the DWDM-PON system.

References

1. S.-J. Park, C.-H. Lee, K.-T. Jeong, H.-J. Park, J.-G. Ahn, K.-H. Song, Fiber-to-the-home services based on wavelength-division-multiplexing passive optical network. J. Lightwave Technol. **22**(11), 2582–2591 (2004)
2. D. Gutierrez, W.-T. Shaw, F.-T. An, Y.-L. Hsueh, M. Rogge, G. Wong, L.G. Kazovsky, Next-generation optical access networks. J. Lightwave Technol. **25**(11), 3428–3442 (2007)
3. J. Wang, M.K. Haldar, L. Li, F.V.C. Mendis, Enhancement of modulation bandwidth of laser diodes by injection locking. IEEE Photon. Technol. Lett. **8**(1), 34–36 (1996)
4. A. Murakami, K. Kawashima, K. Atsuki, Cavity resonance shift and bandwidth enhancement in semiconductor lasers with strong light injection. IEEE J. Quantum Electron. **39**(10), 1196–1204 (2003)
5. H.-D. Kim, S.-G. Kang, C.-H. Lee, A low-cost WDM source with an ASE injected Fabry-Perot semiconductor laser. IEEE Photon. Technol. Lett. **12**(8), 1067–1069 (2000)
6. P. Healey, P. Townsend, C. Ford, L. Johnston, P. Townley, I. Lealman, L. Rivers, S. Perrin, R. Moore, Spectral slicing WDM-PON using wavelength-seeded reflective SOAs. Electron. Lett. **37**(19), 1181–1182 (2001)
7. W. Lee, M.-Y. Park, S.-H. Cho, J. Lee, C. Kim, G. Jeong, B.-W. Kim, Bidirectional WDM-PON based on gain-saturated reflective semiconductor optical amplifiers. IEEE Photon. Technol. Lett. **17**(11), 2460–2462 (2005)
8. Y.-H. Lin, C.-J. Lin, G.-C. Lin, G.-R. Lin, Saturated signal-to-noise ratio of up-stream WRC-FPLD transmitter injection-locked by down-stream data-erased ASE carrier. Opt. Express **19**(5), 4067–4075 (2011)
9. H. Takesue, T. Sugie, Wavelength channel data rewrite using saturated SOA modulator for WDM networks with centralized light sources. J. Lightw. Technol. **21**(11), 2546–2556 (2003)
10. G.-R. Lin, Y.-H. Lin, C.-J. Lin, Y.-C. Chi, G.-C. Lin, Reusing a data-erased ASE carrier in a weak-resonant-cavity laser diode for noise-suppressed error-free transmission. IEEE J. Quantum Electron. **47**, 676–685 (2011)
11. G.-R. Lin, Y.-H. Lin, C.-J. Lin, Y.-C. Chi, G.-C. Lin, Reusing a data-erased ASE carrier in a weak-resonant-cavity laser diode for noise-suppressed error-free transmission. IEEE J. Quantum Electron. **47**(5), 676–685 (2011)
12. G.-R. Lin, Y.-H. Lin, Y.-C. Chang, Theory and Experiments of a Mode-Beating Noise-Suppressed and Mutually Injection-Locked Fabry-Perot Laser Diode and Erbium-Doped Fiber Amplifier Link. IEEE J. Quantum Electron. **40**(8), 1014–1022 (2004)
13. K.-M. Choi, J.-S. Baik, C.-H. Lee, Broad-band light source using mutually injected Fabry-Pérot laser diodes for WDM-PON. IEEE Photon. Technol. Lett. **17**(12), 2529–2531 (2005)

14. Z. Xu, Y.-J. Wen, W.-D. Zhong, C.-J. Chae, X.-F. Cheng, Y. Wang, C. Lu, J. Shankar, High-speed WDM-PON using CW injection-locked Fabry-Pérot laser diodes. Opt. Express **15**(6):2953–2962 (2007)
15. G.-R. Lin, T.-K. Chen, Y.-H. Lin, G.-C. Lin, H.-L. Wang, A weak-resonant-cavity Fabry–Perot laser diode with injection-locking mode number-dependent transmission and noise performances. J. Lightw. Technol. **28**(9), 1349–1355 (2010)
16. T.A. Birks, D. Mogilevtsev, J.C. Knight, P.S.J. Russell, Dispersion compensation using single-material fibers. IEEE Photon. Technol. Lett. **11**(6), 674–676 (1999)
17. Y.-S. Liao, H.-C. Kuo, Y.-J. Chen, G.-R. Lin, Side-mode transmission diagnosis of a multichannel selectable injection-locked Fabry-Perot laser diode with anti-reflection coated front facet. Opt. Express **17**(6), 4859–4867 (2009)
18. X. Jin, S.-L. Chuang, Bandwidth enhancement of Fabry-Perot quantum-well lasers by injection-locking. Solid-State Electron. **50**(6), 1141–1149 (2006)
19. G.-R. Lin, H.-L. Wang, G.-C. Lin, Y.-H. Huang, Y.-H. Lin, T.-K. Cheng, Comparison on injection-locked fabry-perot laser diode with front-facet reflectivity of 1 % and 30 % for optical data transmission in WDM-PON system. J. Lightwave Technol. **27**(14), 2779–2785 (2009)
20. G.-R. Lin, T.-K. Chen, Y.-C. Chi, G.-C. Lin, H.-L. Wang, Y.-H. Lin, 200-GHz and 50-GHz AWG channelized linewidth dependent transmission of weak-resonant-cavity FPLD injection-locked by spectrally sliced ASE. Opt. Express **17**(20), 17739–17746 (2009)
21. H. Kim, H.-C. Ji, C.-H. Kim, Effects of intraband crosstalk on incoherent light using SOA-based noise suppression technique. IEEE Photon. Technol. Lett. **18**(14), 1542–1544 (2006)
22. A.D. McCoy, B.C. Thomsen, M. Ibsen, D.J. Richardson, Filtering effects in a spectrum-sliced WDM system using SOA-based noise reduction. IEEE Photon. Technol. Lett. **16**(2), 680–682 (2004)
23. Y.-H. Lin, G.-C. Lin, H.-L. Wang, Y.-C. Chi, G.-R. Lin, Compromised extinction and signal-to-noise ratios of weak-resonant-cavity laser diode transmitter injected by channelized and amplitude squeezed spontaneous-emission. Opt. Express **18**(5), 4457–4468 (2010)
24. H.-C. Ji, I. Yamashita, K.I. Kitayama, Cost-effective WDM-PON delivering up/down-stream data and broadcast services on a single wavelength using mutually injected FPLDs, in *Proceedings of optical fiber communication conference*, Paper OTuH2, Feb 2008
25. G.-R. Lin, Y.-S. Liao, Y.-C. Chi, H.-C. Kuo, G.-C. Lin, H.-L. Wang, Y.-J. Chen, Long-cavity Fabry-Perot laser amplifier transmitter with enhanced injection-locking bandwidth for WDM-PON application. J. Lightw. Technol. **28**(20), 2925–2932 (2010)
26. C.-L. Tseng, C.-K. Liu, J.-J. Jou, W.-Y. Lin, C.-W. Shih, S.-C. Lin, S.-L. Lee, G. Keiser, Bidirectional transmission using tunable fiber lasers and injection-locked Fabry–Pérot laser diodes for WDM access networks. IEEE Photon. Technol. Lett. **20**(20):794–796 (2008)
27. S.-Y. Lin, Y.-C. Chi, Z.-W. Liao, H.-L. Wang, G.-C. Lin, G.-R. Lin, WDM-PON transmission using WRC-FLDs with AR coating reflectance of 0.5 % and 1.2 %, in *Proceedings of OECC*, pp. 83–84, 2012
28. J. Armstrong, OFDM for optical communications. J. Lightw. Technol. **27**(3), 189–204 (2009)
29. A.J. Lowery, Fiber nonlinearity pre- and post-compensation for long-haul optical links using OFDM. Opt. Express **15**(20), 12965–12970 (2007)
30. J.-M. Tang, P.-M. Lane, K.-A. Shore, High-speed transmission of adaptively modulated optical OFDM signals over multimode fibers using directly modulated DFBs. J. Lightw. Technol. **24**(1), 429–441 (2006)
31. X.Q. Jin, R.P. Giddings, E. Hugues-Salas, J.M. Tang, Real-time demonstration of 128-QAM-encoded optical OFDM transmission with a 5.25 bit/s/Hz spectral efficiency in simple IMDD systems uti-lizing directly modulated DFB lasers. Opt. Express **17**(22), 20484–20493 (2009)
32. E. Hugues-Salas, R.P. Giddings, X.-Q. Jin, J.-L. Wei, X. -Zheng, Y. Hong, C. Shu, M.-J. Tang, Real-time experimental demonstration of low-cost VCSEL intensity-modulated 11.25 Gb/s optical OFDM signal transmission over 25 km PON systems. Opt. Express **19**(4), 2979–2988 (2011)
33. A. Schimpf, D. Bucci, B. Cabon, Optimum biasing of VCSEL diodes for all-optical up-conversion of OFDM signals. J. Lightw. Technol. **27**(16), 3484–3489 (2009)

34. Y.-C. Chi, Y.-C. Li, H.-Y. Wang, P.-C. Peng, H.-H. Lu, G.-R. Lin, Optical 16-QAM-52-OFDM transmission at 4 Gbit/s by directly modulating a coherently injection-locked colorless laser diode. Opt. Express **20**(18), 20071–20077 (2012)
35. R.P. Giddings, E. Hugues-Salas, X.-Q. Jin, J.-L. Wei, J.-M. Tang, Experimental demonstration of real-time optical OFDM transmission at 7.5 Gb/s Over 25-km SSMF using a 1-GHz RSOA. IEEE Photon. Technol. Lett. **22**(11), 745–747 (2010)
36. T. Duong, N. Genay, P. Chanclou, B. Charbonnier, A. Pizzinat, R. Brenot, Experimental demonstration of 10 Gbit/s upstream transmission by remote modulation of 1 GHz RSOA using adaptively modulated optical OFDM for WDM-PON single fiber architecture, in *Proceedings of European conference on optical communication*, Paper Th.3.F.1, Sep 2008
37. G.-R. Lin, Y.-C. Chi, Y.-C. Li, J. Chen, Using a L-band weak-resonant-cavity FPLD for subcarrier amplitude pre-leveled 16-QAM-OFDM transmission at 20 Gbit/s. J. Lightwave Technol. **31**(7), 1079–1087 (2013)
38. K.-Y. Park, S.-G. Mun, K.-M. Choi, C.-H. Lee, A theoretical model of a wavelength-locked Fabry-Pérot laser diode to the externally injected narrow-band ASE. IEEE Photon. Technol. Lett. **17**(9), 1797–1799 (2005)

Chapter 7
Cost-Effective OFDM Transmission Technologies for Long-Reach PONs

Dar-Zu Hsu, Chia-Chien Wei, Hsing-Yu Chen and Jyehong Chen

7.1 Introduction to the OFDM for Long-Reach PONs

As the development of broadband services, such as IPTV, HDTV, blue ray disk, telepresence, 3D HDTV, and super HDTV, increases, so is people's demand for high bandwidth and a high quality entertainment experience. So far, we have many access network solutions to support the broadband services, such as Fiber to the x (FTTx), WiMAX, LTE, and so on. Among these, FTTx is the most expensive, due to costly fiber deployment. However, optical fiber can provide huge bandwidth, ultra-low power loss of transmission, and is not necessary to be upgraded for many years, thus it is a long-term solution. Therefore, many operators have invested in a lot of budget to deploy the FTTx infrastructures.

Figure 7.1 shows the typical architecture of passive optical network (PON) [1], which is composed of optical line terminal (OLT) in central office side, optical distribution network (ODN), and optical network units (ONUs) in user side. PON is the most popular and cost-effective FTTx solution [1–3], since ODN is passive, and operators don't need to maintain ODN and provide the electricity. Besides, the network topology of PON is point-to-multi-point (P2MP), and the cost of OLT and ODN can be shared by many ONUs, thus the cost can be further reduced.

Currently, 1.25-Gbps Ethernet PON (EPON) and 2.5-Gbps Gigabit PON (GPON) are being rapidly deployed and 10-Gbps 10GEPON and XGPON have been ready for field trails [2–4]. However, with the exponentially increasing of customer needs for broadband services, the international PON standard organization Full Service Access Network (FSAN) has been developing the standard for next-generation PON (NG-PON) to provide higher bandwidth, said at least 40-Gbps aggregated capacity and 1-Gbps data rate for each ONU, to meet the bandwidth requirements in the future [5–7]. Besides, recently, there have been

D.-Z. Hsu · C.-C. Wei · H.-Y. Chen · J. Chen (✉)
Department of Photonics, National Chiao Tung University, 1001, Ta Hsueh Rd., Hsinchu 30051, Taiwan
e-mail: jchen@faculty.nctu.edu.tw

C.-C. Lee (ed.), *The Current Trends of Optics and Photonics*,
Topics in Applied Physics 129, DOI 10.1007/978-94-017-9392-6_7

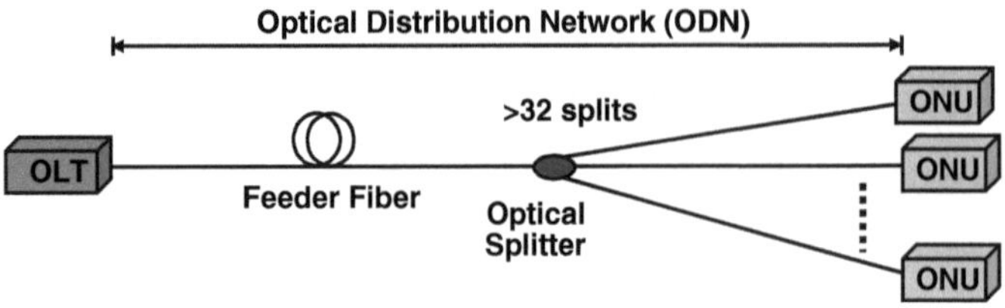

Fig. 7.1 Passive optical network (PON) [1]

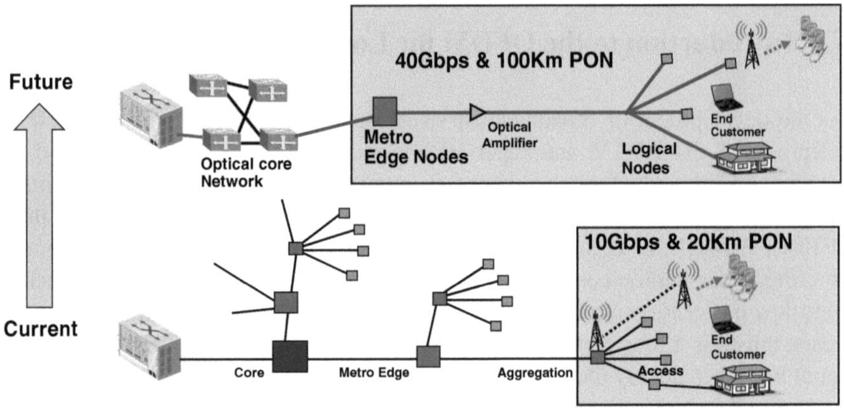

Fig. 7.2 Development tends of optical access networks

growing interests in the new type of optically amplified large-split long-reach PON (LR-PON) [4, 8–12]. Thus, as shown in Fig. 7.2, integrated the access network with the metro network within the 100-km target range, LR-PON claims to considerably reduce the capital and operational expenditures by increasing the coverage of the central office and supporting large number of ONU (≥128), and consolidating the O/E/O conversion interfaces inside the existing networks. Besides, LR-PONs can simplify the network hierarchies and thus reduce the network latency, which is very important for the real-time broadband services and cloud computing. Thus, the trend of next generation PON networks will be high bandwidth LR-PONs, which can support at least 40-Gbps capacity and 100-km standard single mode fiber (SSMF) transmission [5, 6].

Current EPON/GPON and 10GEPON/XGPON use non-return-to-zero (NRZ) on-off keying (OOK) modulation format [2]. If the 40-Gbps LR-PON still uses NRZ-OOK modulation format, the bandwidth requirement of optical transceiver will be about 30–40 GHz. However, the opto-electronic performance and dispersion tolerance degrade quickly with bandwidth over 10 GHz. As shown in Fig. 7.3, for example, the equivalent input noise density of 40-Gbps PIN-TIA is 3 times bigger than 10-Gbps PIN-TIA, and the responsivity of the photodiode is also

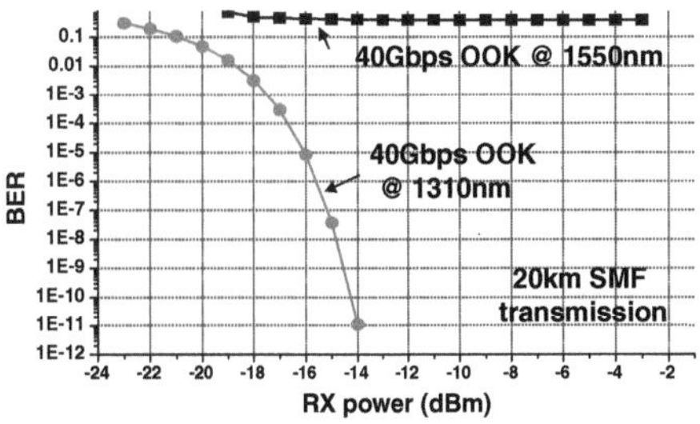

PIN-TIA NRZ data rate (Gbps)	2.5	10	40
Equivalent input noise density	8	10	30
PD Responsivity	0.9	0.9	0.75

NRZ data rate (Gbps)	10	40
Chromatic dispersion tolerance (2dB penalty)	±800ps/nm	±50ps/nm

Fig. 7.3 The challenges for 40-Gbps NRZ-OOK modulation

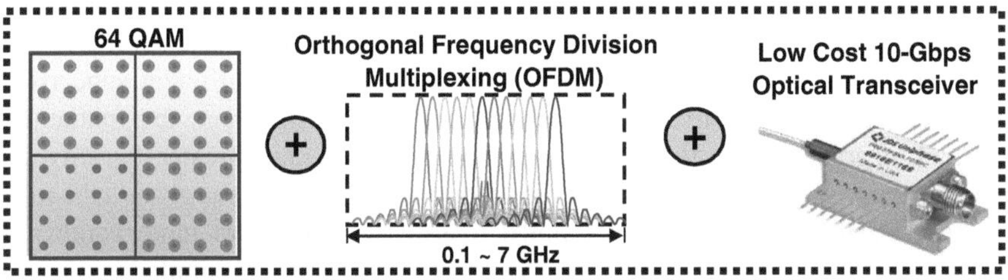

Fig. 7.4 40-Gbps optical OFDM with QAM modulation and low-cost 10-Gbps optical transceiver

degraded from 0.9 to 0.75 A/W. These two issues will seriously reduce the signal to noise ratio (SNR) of received signal. Besides, fiber chromatic dispersion tolerance is inverse proportional to the data rate and the occupied bandwidth, thus dispersion tolerance of 40 Gbps is only 1/16 one of 10 Gbps. In Fig. 7.3, we can see that the chromatic dispersion tolerance of 40 Gbps is 50 ps/nm, which only supports less than 10-km SSMF transmission. From the simulation results as shown in Figs. 7.1, 7.2 and 7.3, we can also see the performance of 40-Gbps OOK optical signal at wavelength of 1,550 nm is seriously limited by the chromatic dispersion. If we change the wavelength to 1,310 nm, which has zero chromatic dispersion, the performance is also degraded quickly because of poor SNR. Therefore, a higher spectral efficiency modulation format is necessary. If the spectral efficiency can be improved, we can get better component and transmission performance, like receiver thermal noise, linearity, dispersion, and also the cost-effective optical components.

Against this backdrop, as shown in Fig. 7.4, thanks to advances in digital signal processing technology, one such prominent modulation candidate for next-generation PONs is optical orthogonal frequency division multiplexing (OFDM) [13–17]. Optical OFDM can achieve high data rate by adopting higher order quadrature-amplitude modulation (QAM) for each sub-carrier. OFDM is known for its high spectral efficiency, which brings a benefit of a lower optical-component bandwidth requirement [13–17]. For example, based on 64-QAM, one can achieve a 40-Gbps data rate by using a low-bandwidth 7-GHz optical transceiver as shown in Fig. 7.4, or based on 16-QAM and 10-GHz optical transceiver. Due to the lower bandwidth usage, as a result, the optical OFDM system not only increases the transmission data rate in a cost-effective manner, but it is also more immune to both chromatic dispersion (CD) [15, 16] and polarization mode dispersion (PMD) [17], and the receiver thermal noise, linearity, and dispersion, can be also reduced.

To realize next-generation high-bandwidth LR-PON systems, several multiplexing schemes are viable candidates to support more ONUs with higher data rate, such as time-division multiplexing (TDM), OFDM, and wavelength-division multiplexing (WDM). In particular, OFDM LR-PON, which boasts to offer high spectral efficiency and flexible bandwidth allocation, has attracted a lot of attention recently [18–32]. In an optical OFDM LR-PON, in the downstream, all the OFDM subcarriers are broadcasted to the all ONUs due to using P2MP topology of PON systems based on an optical splitter, and ONUs will choose their own subcarriers. Thus, multi-user access is enabled by assigning different subcarriers to different users. For frequency-domain bandwidth assignment, the fundamental unit of granularity is the OFDM per-subcarrier data rate. Greater capacity can be achieved by adopting a higher-order modulation format on a given subcarrier or several subcarriers can be combined to satisfy ONUs with more bandwidth demand. In the upstream, multiple access can be supported by similar frequency-domain subcarrier assignment, or by dedicated timeslot for each active ONU, just the same as traditional time-division multiplexing access (TDMA)-PON, or such that an ONU group can flexibly share a group of subcarriers in combined OFDMA and TDMA mode. If TDMA is used for upstream multiple access, the burst mode optical receiver is necessary in the OLT [22].

Both TDM and OFDM LR-PONs can easily support numerous ONUs by using an optical splitter with large splitting ratio and can broadcast the aggregated data to all ONUs on a single wavelength, thus have the benefit of simple wavelength management [11]. However, the aggregated data rate on a single wavelength will be very high, and each ONU has to process very high-speed aggregated data in order to receive and transmit a small portion of data (e.g. <1/128) [19]. In [21] and [23], a 108-Gbps OFDM PON over single wavelength is demonstrated employing two Mach-Zehnder modulators, polarization multiplexing and two receivers. For cost-sensitive ONUs, the proposed scheme will be too complicated and expensive to be a practical solution in the near future.

An alternative solution is WDM LR-PON [33, 34], where each ONU can be assigned a wavelength, thus each ONU doesn't have to process the aggregated data and the speed requirement of each transceiver can be considerably lower. Besides, instead of using a large-splitting-ratio optical splitter, WDM LR-PON can use a wavelength multiplexer/demultiplexer (waveguide grating router, WGR), and the optical power loss of WGR is much lower than the optical splitter and insensitive to wavelength number. Unfortunately, even though the speed requirement of each transceiver is lower in a WDM system, numerous color optical transmitters are costly and undesired. To preserve the colorless upstream architecture, an RSOA (reflective semiconductor optical amplifier) is employed for upstream service [33–35]. Currently, employing RSOAs are still facing some technical

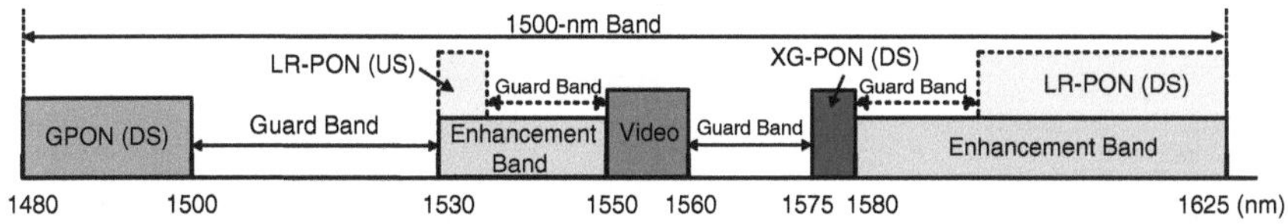

Fig. 7.5 Wavelength allocations of current G-PON/XG-PON and proposed LR-PON. (*DS* downstream; *US* upstream)

challenges such as limited bandwidth, ASE noise, and Rayleigh backscattering, needed to be overcome. Besides, the systems fail to allocate bandwidth flexibly among ONUs of different wavelengths. In particular, since a LR-PON is needed to be coexisted with the current PONs, such as GPON and XGPON. As shown in Fig. 7.5 [36], excluding guard bands, the available enhancement bands provide very limited bandwidth at C-band for future LR-PONs.

For the reasons stated above, a more practical solution will be a hybrid WDM/OFDM LR-PON, which optimizes the trade-off between the number of wavelength channels and the cost of transceivers. For instance, if a cost-effective transceiver can provide 40-Gbps capacity, only four wavelength channels are required for each direction (downstream and upstream) to support a LR-PON with 128 ONUs and 1-Gbps data rate per ONU.

Based on the trade-off between wavelength channel number and transceiver cost and the requirements of NG-PON discussed in [5–7], some criteria have been contemplated for a feasible LR-PON. Though some of the proposed numbers are open to discussion and may change with the development of technologies, these are the important issues that need to be addressed prudently: (a) Sustainable symmetric data rate of each ONU is 1 Gbps for supporting future broadband multimedia services [5–7]; (b) Power splitting ratio (N) is 32 for a reasonable optical loss budget and the aggregated data rate on a single wavelength is at least 40 Gbps; (c) Supporting ONU number is $\geq$128 for sharing the capital and operational expenditures; (d) Using WDM technology to stack four (W = 4) 40-Gbps PON, thus at least 128 (N $\times$ W) ONUs can be achieved by four wavelengths for each direction, and (e) Cost effective transceiver design is achieved by a 10-GHz intensity-modulation-direct-detection (IMDD) scheme. Notably, four wavelengths for each direction (downstream and upstream) are recommended due to very limited wavelength resources, as shown in Fig. 7.6. Thus, the proposed architecture of a LR-PON and the corresponding parameters of the above criteria are illustrated in Fig. 7.6, and the proposed LR-PON can meet the requirements of NG-PON [5–7] by using a cost-effect and feasible architecture. Wavelength management of four channels is not technique challenging, but the most crucial part turns out to be achieving the expected data rate of $\geq$40 Gbps via a cost-effective transceiver.

To reach the target transmission distance of up to 100 km, nonetheless, will require better performance and more expensive components. To further reduce costs, optical IMDD scheme is expected. Additionally, it is desirable to generate high-speed signals by low-bandwidth cost-effective transceivers, such as

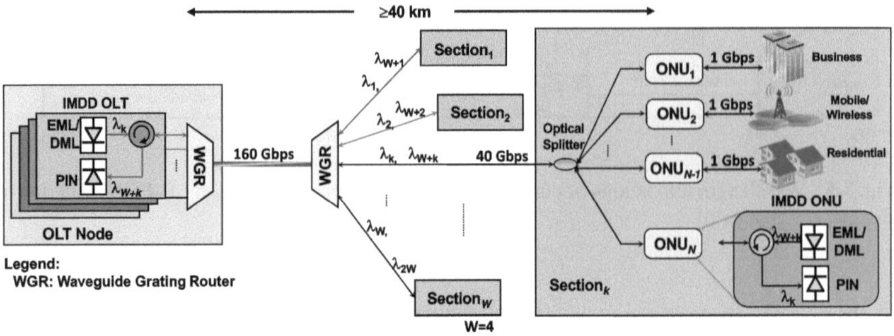

Fig. 7.6 Proposed feasible architecture for a cost-effective WDM/OFDM LR-PON

commercially matured 10-GHz directly modulated DFB lasers (DMLs) or electro-absorption modulated lasers (EMLs) and 10-GHz PIN detectors, assisted by spectrally efficient modulation format, such as QAM. Nonetheless, one critical drawback of DMLs or EMLs is the generated optical signal will be double-side band (DSB), namely, an electrical OFDM subcarrier will be translated into two conjugated optical subcarriers on both sides of an optical carrier. After transmission in fiber link, however, the residual dispersion destroys their conjugated property and results in a well-known power fading [144]. Moreover, since an optical signal modulated by a DML or an EML is chirped, the additional phase modulation also weakens the conjugated property between subcarriers. Accordingly, both frequency chirp and fiber dispersion will affect the received power [24–28]. As a result, after 100-km transmission over SSMF, the bandwidth of the transmission system is limited to a few GHz [24–26, 29]. Thanks to the advance in DSP technology, both power levels and modulation levels of OFDM subcarriers can be adaptively allocated to efficiently utilize very limited transmission bandwidth at baseband, and make the best of bandwidth efficiency to overcome the limitations of bandwidth by detrimental dispersion- and chirp-related power fading.

High spectral efficiency orthogonal frequency-division multiplexing (OFDM) with quadrature amplitude modulation (QAM) format has become one of the viable schemes in LR-PONs, because it can provide high capacity with low-bandwidth requirement of components. However, intensity modulation, employing an EAM, generates optical double-sideband and frequency-chirped OFDM signals. The signals suffer series damages, such as dispersion-induced power fading and nonlinear transmission distortion. Since the power fading is periodic, the available bandwidth of a fading channel can be efficiently used by employing the dynamic multi-band optical OFDM scheme. In this scheme, the discrete passbands can be dynamically allocated by OFDM subcarriers with adaptive modulation formats to make the best bandwidth utilization. Thus, the power fading may not be the major bottleneck in the IMDD systems. Besides, OFDM signals will also be degraded by the dispersion- and chirp-related nonlinear transmission distortion, which has been theoretically and experimentally investigated as subcarrier-to-subcarrier intermixing interference (SSII) [32, 36]. Based on the SSII model, an SSII cancellation scheme has been proposed to eliminate the dispersion-induced distortion and to improve the transmission performance. Alternatively, nonlinear Volterra filtering has also been proposed to compensate the dispersion-induced distortion [37]. Since an OFDM signal has inherently high peak to average power ratio (PAPR) and the transfer curve of an EAM has rather narrow linear region, the optical modulation index (OMI) is usually kept relatively small to avoid severe nonlinear distortion. To combat with the modulator nonlinearity, several pre-distortion schemes have been proposed to linearize optical transmitters [38, 39]. Nonetheless, no previous works have presented the nonlinear compensation of an OFDM-IMDD system with considering the combination of the modulator nonlinearity and the dispersion-induced distortion.

Based on the 2nd-order approximation, this work proposes a novel small-signal model to analyze the combined nonlinear distortion of an EAM-based OFDM-IMDD system, consisted of the modulator nonlinearity and the dispersion-induced distortion. The combined nonlinear distortion is regarded as SSII in the proposed model, and therefore, it can be estimated by calculating the theoretical SSII. Furthermore, an SSII cancellation technique is also employed to compensate the combined distortion and experimentally demonstrated in an EAM-based OFDM transmission system for the first time. Thus, the EAM nonlinearity and the dispersion-induced nonlinear distortion can be eliminated at the receiver simultaneously. Accordingly, employing the proposed SSII cancellation technique, the OMI of the system can be increased to reach better performance, and the experimental results reveal the performance improvement in both configurations of optical back-to-back (BtB) and fiber transmission. To accommodate possible longer distance of current metro networks and to accomplish the data rate of 1 Gb/s for each of 32 ONUs, this work also demonstrates a 32-Gb/s OFDM system over 150-km uncompensated SSMF. By incorporating the proposed SSII cancellation scheme into the system, the maximum of 5-dB signal-to-noise ratio (SNR) improvement and a 3.3-dB sensitivity improvement at the FEC limit are experimentally demonstrated.

7.2 Theory of SSII and SSII Cancellation Technique

To simplify the analysis of the dispersion-induced distortion of a chirped OFDM signal, a 2nd-order small-signal model has been used to approximate the distortion as the SSII. In the experimental demonstration, the model in has been modified to more precisely estimate the chirp-related nonlinear distortion. Another underlying assumption of the model is its linear intensity modulation, i.e. constant modulation slope, but a practical EAM shows deceasing modulation slope as increasing reverse bias (see for instance [40]) to result in nonlinear E/O conversion. Hence, the EAM nonlinearity cannot be neglected, especially when the OMI is increased. This section will extend the 2nd-order small-signal model to include the EAM nonlinearity. At the optical transmitter, the modulation voltage of OFDM signals can be expressed as $V = V_b + \Sigma_{n=1}^{N} \Re\{v_n e^{jn\omega t}\}$, where V_b is the bias voltage, N is the subcarrier number, $\omega/(2\pi)$ is the subcarrier spacing, v_n is the encoded complex information of the nth subcarrier, and $\Re\{\cdot\}$ represents the real part. If an EAM-based transmitter is a linear E/O system, the power of each optical subcarrier is linearly proportional to $\Re\{v_n \cdot H_T(n)\}$, where $H_T(n)$ is the whole frequency response of the transmitter. To consider only the 2nd-order nonlinearity for simplicity, the normalized optical power could be approximated as $P \cong 1 + X + p_2 X^2$, where $X = \Sigma_{n=1}^{N} \Re\{x_n e^{jn\omega t}\}$ is the normalized AC power envelop, $x_n = (v_n/V_b)\cdot[H_T(n)/H_T(0)]$, and p_2 is the weighting of the 2nd-order nonlinear distortion induced by an EAM. Moreover, since the measured chirp parameter of the EAM [11] is approximately a linear function of the modulation voltage, it could be represented as $\alpha \cong \alpha_0 + \alpha_1 X$. As a result, the normalized envelop of the chirped optical field, $E = \sqrt{P}\exp(j\alpha \ln(P/2))$, can be approximated as

$$E \cong 1 + \frac{(1 - j\alpha_0)}{2} X - \frac{(1 + \alpha_0^2 - 4p_2) + j2(\alpha_1 + 2\alpha_0 p_2)}{8} X^2, \tag{7.1}$$

where X^2 is the 2nd order term, and the higher order terms of x_n are neglected. Considering only chromatic dispersion, the response of fiber transmission is $H_{CD}(n) = \exp(jn^2\omega^2\beta_2 L/2)$, where L and β_2 are the fiber length and the dispersion parameter, respectively. Thus, if $\Theta\{\cdot\}$ represents the effect of dispersive transmission, it follows, for instance, $\Theta\{X\} = X \cdot H_{CD}(n) = \sum_{n=1}^{N} \Re\{x_n e^{jn\omega t}\} H_{CD}(n)$ and $\Theta\{X^2\} = X^2 \cdot H_{CD}(n)$. After square-law photo-detection and the 2nd-order approximation, the normalized received signal is

$$R = |\Theta\{E\}|^2 \cong 1 + \Re\{(1 - j\alpha_0)\Theta\{X\}\} \\ \underbrace{+ \frac{(1+\alpha_0^2)}{4} \times \left[|\Theta\{X\}|^2 - |\sec\theta_p| \cdot \Re\left\{ \Theta\left\{X^2\right\} e^{j\theta_p} \right\} \right] + p_2 |\sec\theta_p| \cdot \Re\left\{ \Theta\left\{X^2\right\} e^{j\theta_p} \right\}}_{SSII}, \tag{7.2}$$

where $\theta_p \equiv \tan^{-1}(4(\alpha_1 + \alpha_0 p_2)/(1 + \alpha_0^2 - 4p_2))$. The 2nd term at the RHS of R corresponds to the desired OFDM signal; the 3rd and 4th terms indicate the intermixing terms among subcarriers, and the combination of the 2nd order terms is the modified SSII, including the nonlinear distortion induced by an EAM. In addition, at optical BtB ($L = 0$), the SSII will turn into $p_2|X|^2$, which will be 0 without considering the EAM nonlinearity. From the new SSII model, after the related parameters are obtained, the combined distortion of an EAM-based OFDM transmission can be estimated.

To mitigate the influence of nonlinear distortion, we developed an SSII cancellation technique at the receiver and the block diagram is shown in Fig. 7.7 [27]. The received OFDM data with the nonlinear distortion are demodulated and detected as the normal OFDM demodulation process. Then the detected data are used as X to calculate the SSII based on the SSII model and (7.2), and the calculated result is further fed back to carry out the SSII cancellation. Then the

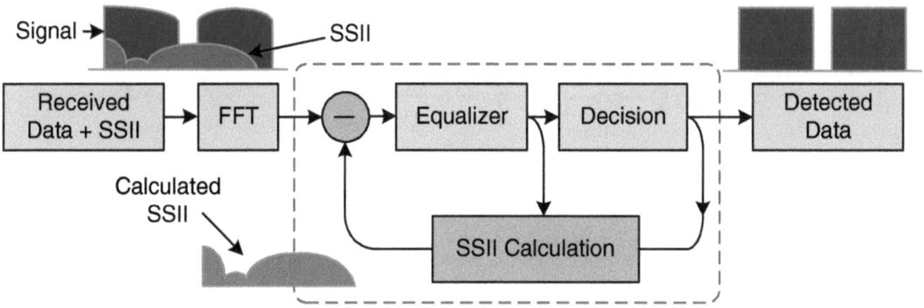

Fig. 7.7 The block diagram of the SSII cancellation technique

OFDM data after the SSII cancellation are demodulated again to get the more correct detected data. When the SSII dominates the error, the correlation between calculated SSII and error will be high, and the required parameters can be decided. Based on (7.2), α_0, α_1 and p_2 are combined into a single parameter θ_p, thus we can decide a θ_p to achieve maximum value of correlation. Hence, the nonlinear distortion induced by the EAM nonlinearity can be compensated optimally by adjusting θ_p as applying different OMI. Besides, through measuring the responses of optical BtB and optical fiber transmission by the training symbols, the responses of H_T and H_{CD} can be estimated.

7.3 Experimental Setup and Results

The experimental setup is shown in Fig. 7.8. The baseband electrical OFDM signal is generated by an arbitrary waveform generator (AWG, Tektronix® AWG7122) with 12-GS/s sampling rate and 8-bit resolution using Matlab® programs. As shown in inset (a) of Fig. 7.1, the channel-1 OFDM signal is located in the 1st passband, which consists of 23.4375-MSym/s 16-QAM and symbols encoded at subcarriers 2nd–182th. The channel-2 OFDM signal is modulated at the same symbol rate and same order of QAM and is encoded at subcarriers 2nd–61th, as shown in inset (b) of Fig. 7.8, and then, it is up-converted to 247th–306th and 310th–369th located in the 2nd passband via a local oscillator with the frequency of 7.22 GHz (308th subcarrier), as shown in inset (c) of Fig. 7.8. Consequently, as shown in inset (d) of Fig. 7.1, two OFDM bands are then combined to modulate an EAM (CIP 10G-LR-EAM-1550), and there are two tunable attenuator to adjust the power of each band. The EAM has small signal 3-dB bandwidth of 14-GHz, and dynamic extinction ratio under NRZ modulation of 9 dB. Then, a variable attenuator is used to control the amplitude of the electrical driving voltage and the OMI of the optical OFDM signal. After EAM-based modulation, 150-km SSMF transmission and direct-detection, the received electrical signal, as shown in inset (e) of Fig. 7.1, is recorded by a digital oscilloscope (Tektronix® DPO 71254) with a 50-GS/s sampling rate and a 3-dB bandwidth of 12.5 GHz and

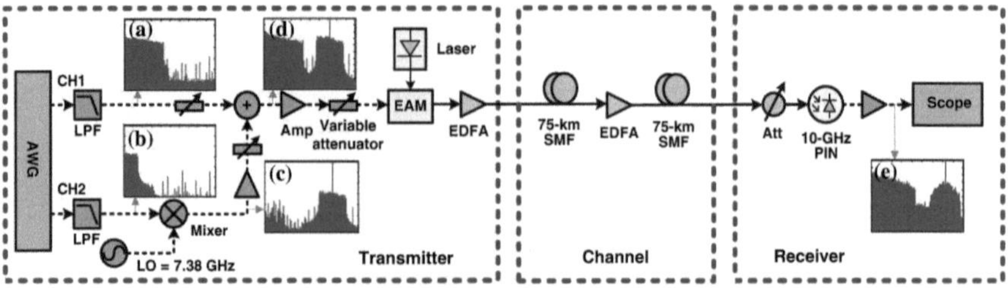

Fig. 7.8 Experimental setup with spectral illustrations. (*a*) channel 1, (*b*) channel 2, (*c*) channel 2 after up-conversion, (*d*) combination of channels 1 and 2, and (*e*) after photo detector

demodulated by an off-line Matlab® DSP program. Then, the SSII, including the nonlinear distortion induced by the EAM, is calculated and cancelled based on the SSII cancellation technique shown in Fig. 7.7. Lastly, we measure the SNR and count the number of errors to calculate the bit error rate (BER).

Figure 7.9 shows the electrical spectra of the OFDM signals and the estimated SSII at optical BtB or after 150-km transmission. Four different modulation conditions, presented in Table 7.1, are adopted to investigate the effect of nonlinear distortion induced by the EAM, yet the detected optical powers are fixed for all conditions in Fig. 7.9. Lower attenuation indicates that the OFDM signal has higher modulated power and higher OMI, which would induce more nonlinear distortion. As shown in Fig. 7.9a, higher signal power is achieved with higher OMI at the price of more nonlinear distortion of the EAM nonlinearity at optical BtB. Since the dispersion is zero at optical BtB, the SSII will be $p_2|X|^2$ as mentioned in Sect. 7.2, and the SSII mostly locates at the baseband and degrades

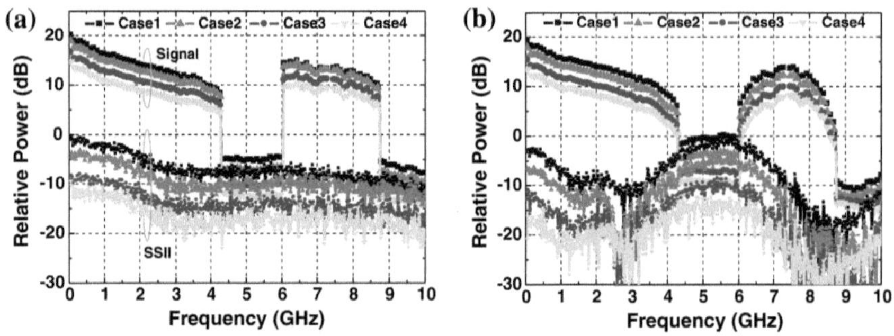

Fig. 7.9 Electrical spectra of the two-band OFDM signals and the calculated SSII **a** at optical BtB and **b** after 150-km SMF transmission

Table 7.1 Driving conditions of the EAM

	Case 1	Case 2	Case 3	Case 4
Attenuation	0 dB	2 dB	4 dB	6 dB
Driving voltage (RMS)	190 mV	150 mV	115 mV	90 mV

the performance of the 1st passband. After 150-km transmission, as shown in Fig. 7.9b, the estimated SSII turns out to be the combined nonlinear distortion, and in addition to the SSII at the baseband mainly contributed by the EAM nonlinearity, the dispersion-induced SSII is centered at ~6 GHz [25]. Since the SSII is related to the square of the modulated signal, the SSII power grows more quickly than the signal power. Thus, higher OMI by increasing the amplitude of the driving signal would result in worse signal-to-interference ratio in both configurations of optical BtB and fiber transmission. Ceaselessly decreasing the amplitude of the driving signal, however, does not imply better performance, because the signal of the lower OMI and a fixed optical power would suffer from more noise relatively. Figures 7.10a, b exhibit the demodulated SNR with −8-dBm received power at optical BtB and after 150-km SSMF transmission, respectively. When the amplitude is decreased from Case 1 to Case 3, evident SNR improvement can be observed in both configurations of Figs. 7.10a, b. Nevertheless, from Case 3 to Case 4, in which additive noise turns to dominate the performance, very limited SNR improvement is shown in Fig. 7.10a, and some subcarriers suffer from SNR penalty in Fig. 7.10b. Figure 7.11 shows the SNR improvement by SSII

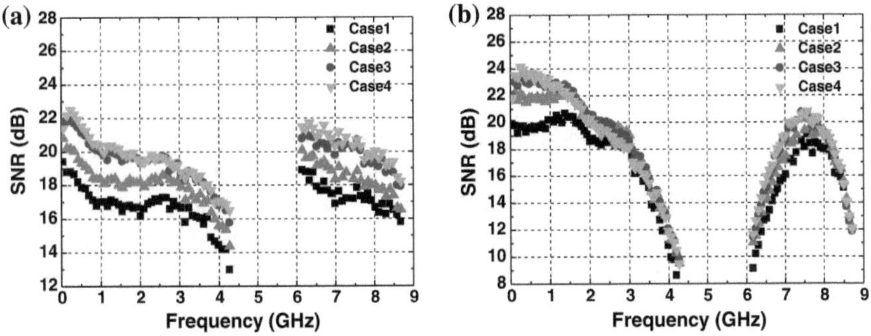

Fig. 7.10 SNR of each subcarrier **a** at optical BtB and **b** after 150-km SSMF transmission without SSII cancellation

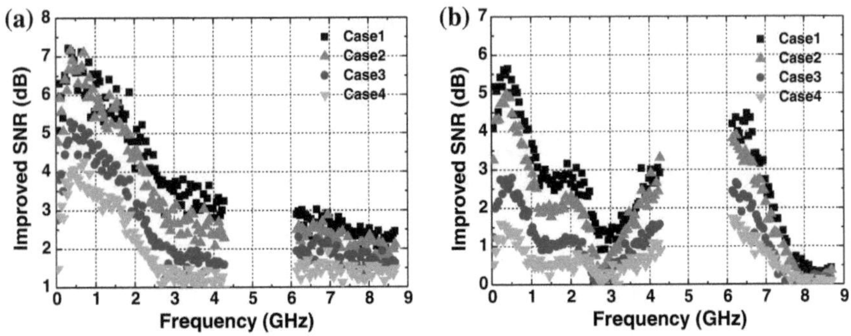

Fig. 7.11 SNR improvement by SSII cancellation **a** at optical BtB and **b** after 150-km SSMF transmission

cancellation. Notably, the cancellation is implemented with the full knowledge of the transmitted data, since the fixed 16-QAM is used and the SNRs of some subcarriers are relatively low to result in too many decision errors. Thus, the iteration is not required in this work. For a practical application [27], the required knowledge of the transmitted data would be replaced by the detected data, when the bit-loading technique is applied to reach the equal BER of all subcarriers. For the training process, the iteration would be needed with high BER, thus increasing the complexity. Since the block of emulated SSII calculation involves the discrete convolution of the frequency components of the corresponding 1st-order terms, it dominates the overall extra computational complexity. In (7.2), $|\Theta\{X\}|^2$ and $\Theta\{X^2\}$ include the discrete convolution, and the discrete convolution can be realized by multiplying their inverse Fourier transforms in the time domain followed by applying Fourier transform. Each transform and inverse transform requires $O((2N) \times \log_2(2N))$ operations, and the multiplication requires $O(2N)$ operations. Thus, the extra computational complexity of $O((2N) \times \log_2(2N))$ is required for each iteration. The number of iteration depends on the required performance of received signal. As expected, the improvement is increased from Case 4 to Case 1 owing to more distortion by increasing the amplitude of the driving signal. Compared to Fig. 7.11a, in which the improvement is mainly at the baseband, Fig. 7.11b exhibits extra SNR improvement at higher frequencies of ~6 GHz, corresponding to the dispersion-induced distortion. Moreover, the SNR after SSII cancellation is plotted in Fig. 7.11. Compared to the SNR in Fig. 7.10, in addition to better SNR, the best conditions are different thanks to the less nonlinear distortion. When the nonlinear distortion is lessened by the SSII cancellation, higher OMI would be preferred. At optical BtB, the best cases are Case 3 and Case 4 in Figs. 7.12a and 7.3a, respectively. After 150-km uncompensated SSMF transmission, the best case is Case 2 in Fig. 7.12b, while the best one is Case 3 or Case 4 in Fig. 7.10b.

Lastly, adopting 2-dB attenuation of Case 2 and applying the bit-loading technique to set the data rate of 32-Gb/s, excluding 1.5 % CP, the SNR and the

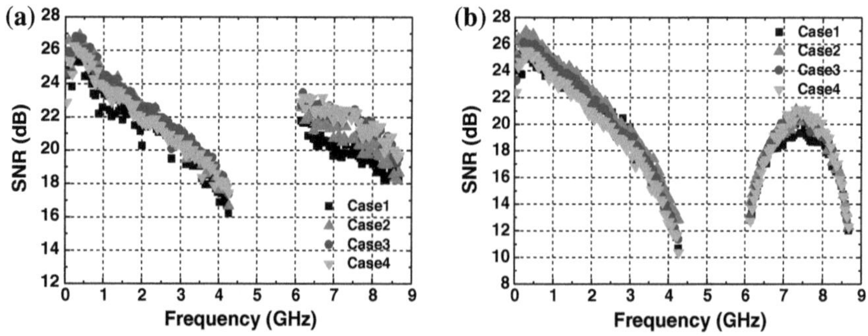

Fig. 7.12 SNR of each subcarrier **a** at optical BtB and **b** after 150-km SSMF transmission with SSII cancellation

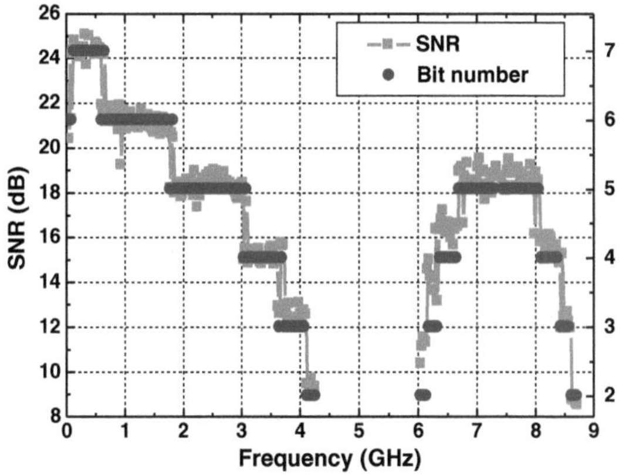

Fig. 7.13 The SNR and the corresponding number of bit of a 32-Gb/s OFDM signal with SSII cancellation at receiver power equal to −13-dBm after 150-km SSMF transmission

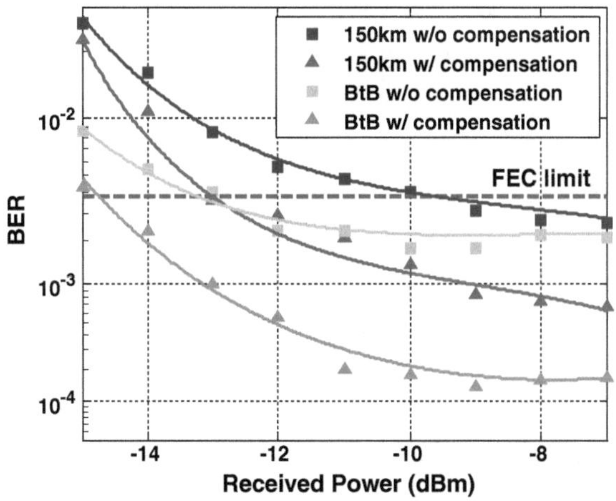

Fig. 7.14 The BER curves of a 32-Gb/s OFDM signal with and without SSII cancellation at optical BtB and after 150-km SSMF transmission

corresponding number of bit as shown in the Fig. 7.13. The measured BER curves are plotted in Fig. 7.10, where the cancellation is implemented without the knowledge of the transmitted data. The total number of OFDM blocks of 1,000, which including 1.388×10^6 bits, are used to calculate the BER. At the FEC limit (the BER of 3.8×10^{-3}), the sensitivities of the OFDM signals with and without SSII cancellation are −14.8 and −13.1 dBm at optical BtB, respectively. The sensitivities after 150-km transmission with and without SSII cancellation are −13 and −9.7 dBm, respectively. As a result, the 1.7 and 3.3-dB sensitivity improvements are successfully achieved at optical BtB and over 150-km SSMF transmission, respectively. Furthermore, Fig. 7.15 exhibits the constellations of 64- and 128-QAM of the 32-Gb/s OFDM signal over 150-km SSMF transmission at the received power of −8 dBm. The clear constellations can be obtained after applying the SSII cancellation technique (Fig. 7.14).

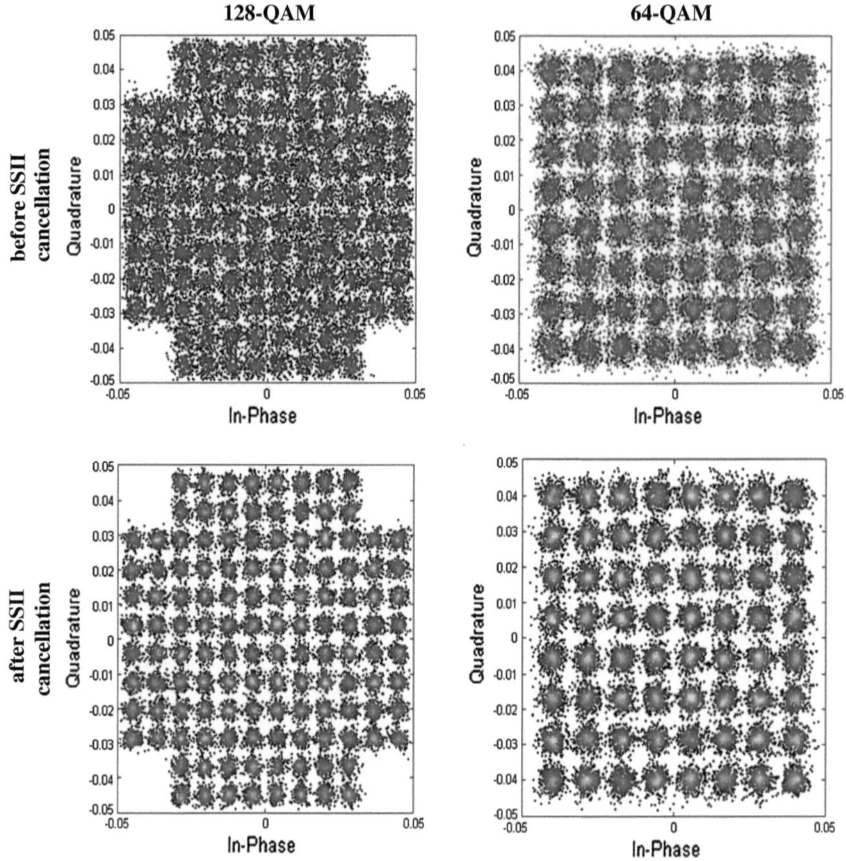

Fig. 7.15 The constellations of 64- and 128-QAM before and after SSII cancellation with 150-km SSMF transmission and the receiver power of −8 dBm

7.4 Conclusion

A novel nonlinear distortion compensation employing the SSII cancellation technique is proposed in an EAM-based OFDM-IMDD transmission system. Both EAM nonlinearity and dispersion-induced nonlinear distortion can be eliminated simultaneously for the first time, and accordingly, the OMI can be increased as optimizing the system. The experimental results show that the maximum SNR improvement of 7-dB and 6-dB can be accomplished in the highly-distorted cases at optical BtB and after 150-km uncompensated SSMF transmission, respectively. Moreover, for the 32-Gb/s OFDM signal, the proposed nonlinear compensation can reach the sensitivity improvement of 3.3 dB after 150-km uncompensated SSMF transmission.

References

1. T. Koonen, Fiber to the home/fiber to the premises: what, where, and when? Proc. IEEE **94**(5), 911–934 (2006)
2. J.I. Kani, F. Bourgart, A. Cui, A. Rafel, M. Campbell, R. Davey, S. Rodrigues, Next-generation PON-part I: technology roadmap and general requirements. IEEE Commun. Mag. **47**(11), 43–49 (2009)
3. F.J. Effenberger, J. Kani, Y. Maeda, Standardization trends and prospective views on the next generation of broadband optical access systems. IEEE J. Selected Areas Commun. **28**(6), 773–780 (2010)
4. R. Lin, Next generation PON in emerging networks, in *Paper OWH1, Optical Fiber Communication Conference (OFC)*, Mar 2008
5. P. Chanclou, F. Bourgart, B. Landousies, S. Gosselin, B. Charbonnier, N. Genay, A. Pizzinat, F. Saliou, B. Capelle, Q. Trung Le, F. Raharimanitra, A. Gharba, L. A. Neto, J. Guillory, Q. Deniel, B. L. Guyader, Technical options for NGPON2 beyond 10G PON, in *Paper We.9.C, European Conference and Exhibition on Optical Communication (ECOC)*, Sept 2011
6. F. Bourgart, NGPON2—where are the standards going, in *Paper NTu2F.3, National Fiber Optic Engineers Conference (NFOEC)*, Mar 2012
7. http://www.itu.int/dms_pub/itu-t/oth/06/13/T06130000200001PDFE.pdf
8. P.D. Townsend, G. Talli, C.W. Chow, E.M. MacHale, C. Antony, R. Davey, T. De Ridder, X.Z. Qiu, P. Ossieur, H.G. Krimmel, D.W. Smith, I. Lealman, A. Poustie, S. Randel, H. Rohde, Long reach passive optical networks, in *Invited paper ThW1, IEEE LEOS Annual Meeting*, Oct 2007
9. R.P. Davey, D.B. Grossman, M. Rasztovits-Wiech, D.B. Payne, D. Nesset, A.E. Kelly, A. Rafel, S. Appathurai, S.H. Yang, Long-reach passive optical networks. IEEE J. Lightwave Technol. **27**(3), 273–291 (2009)
10. K.Y. Cho, K. Tanaka, T. Sano, S.P. Jung, J.H. Chang, Y. Takushima, A. Agata, Y. Horiuchi, M. Suzuki, Y.C. Chung, Long-Reach Coherent WDM PON employing self-polarization-stablization technique. IEEE J. Lightwave Technol. **29**(4), 456–462 (2011)
11. D. Shea, J. Mitchell, A 10 Gb/s 1024-way-split 100-km long-reach optical-access network. IEEE J. Lightwave Technol. **25**(3), 658–693 (2007)
12. A. Lowery, L. Du, J. Armstrong, Performance of optical OFDM in ultralong-haul WDM lightwave systems. IEEE J. Lightwave Technol. **25**(1), 131–138 (2007)
13. I.B. Djordjevic, B. Vasic, Orthogonal frequency division multiplexing for high- speed optical transmission. Opt. Express **14**(9), 3767–3775 (2006)
14. J. Armstrong, OFDM for optical communications. IEEE J. Lightwave Technol. **27**(3), 189–204 (2009)
15. B.J.C. Schmidt, A.J. Lowery, J. Armstrong, Experimental demonstrations of electronic dispersion compensation for long haul transmission using direct-detection optical OFDM. IEEE J. Lightwave Technol. **26**(1), 196–203 (2008)
16. A.J. Lowery, J. Armstrong, Orthogonal-frequency-division multiplexing for dispersion compensation of long-haul optical systems. Opt. Express **14**(6), 2079–2084 (2006)
17. W. Shieh, X. Yi, Y. Ma, Y. Tang, Theoretical and experimental study on PMD-supported transmission using polarization diversity in coherent optical OFDM systems. Opt. Express **15**(16), 9936–9947 (2007)
18. N. Cvijetic, D. Qian, J. Hu, 100 Gb/s optical access based on optical orthogonal frequency-division multiplexing. IEEE Commun. Mag. **48**(7), 70–77 (2010)
19. Y.M. Lin, P.L. Tien, Next-Generation OFDMA-based passive optical network architecture supporting radio-over-fiber. IEEE J. Selected Areas Communication **28**(6), 791–799 (2010)
20. D. Qian, T.T.-O Kwok, N. Cvijetic, J. Hu, T. Wang, 41.25 Gb/s real-time OFDM receiver for variable-rate WDM-OFDMA-PON transmission, in *Paper PDPD9, Optical Fiber Communication Conference (OFC)*, Mar 2010

21. D. Qian, N. Cvijetic, J. Hu, T. Wang, 108 Gb/s OFDMA-PON with polarization multiplexing and direct detection. IEEE J. Lightwave Technol. **28**(4), 484–493 (2010)
22. L.A. Neto, A. Gharba, P. Chanclou, N. Genay, B. Charbonnier, M. Ouzzif, C. Aupetit-Berthelemot, J.L. Masson, High bit rate burst mode optical OFDM for next generation, in *Paper Tu. 3.B.5, European Conference and Exhibition on Optical Communication (ECOC)*, Sept 2010
23. D. Qian, N. Cvijetic, J. Hu, T. Wang, 108 Gb/s OFDMA-PON with polarization multiplexing and direct detection, in *Optical Fiber Communication Conference (OFC)*, Mar 2009
24. D.Z. Hsu, C.C. Wei, H.Y. Chen, J. Chen, M.C. Yuang, S.H. Lin, W.Y. Li, 21 Gb/s after 100 km OFDM long-reach PON transmission using a cost-effective electro-absorption modulator. Opt. Express **18**(26), 27758–27763 (2010)
25. D.Z. Hsu, C.C. Wei, H.Y. Chen, W.Y. Li, J. Chen, A cost-effective 33-Gbps intensity modulation direct detection multi-band OFDM LR-PON system employing a 10-GHz-based transceiver. Opt. Express **19**(18), 17546–17556 (2011)
26. D.Z. Hsu, C.C. Wei, H.Y. Chen, Y.C. Lu, J. Chen, A 40-Gbps OFDM LR-PON system over 100-km fiber employing an economical 10-GHz-based transceiver, in *Paper OW4B.2, Optical Fiber Communication Conference (OFC)*, Mar 2012
27. D.Z. Hsu, C.C. Wei, H.Y. Chen, Y.C. Lu, C.Y. Song, C.C. Yang, J. Chen, SSII cancellation in an EAM-based OFDM-IMDD transmission system employing a novel dynamic chirp model. Opt. Express **21**, 533–543 (2013)
28. H.Y. Chen, C.C. Wei, D.Z. Hsu, M.C. Yuang, J. Chen, Y.M. Lin, P.L. Tien, S.S.W. Lee, S.H. Lin, W.Y. Li, C.H. Hsu, J.L. Shih, A 40-Gbps OFDM PON system based on 10-GHz EAM and 10-GHz direct-detection PIN. IEEE Photon. Technol. Lett. **24**(1), 85–87 (2012)
29. A. Gharba, P. Chanclou, M. Ouzzif, J.L. Masson, L. A. Neto, R. Xia, N. Genay, B. Charbonnier, M. Hélard, E. Grard, V. Rodrigues, Optical transmission performance for DML considering laser chirp and fiber dispersion using AMOOFDM, in *International Congress on Ultra Modern Telecommunications and Control Systems and Workshops (ICUMT)*, pp. 1022–1026, Oct 2010
30. A. Gharba, M. Ouzzif, J. Le Masson, P. Chanclou, N. Genay, B. Charbonnier, E. Grard, V. Rodrigues, T.N. Duong, M. Hélard, Multi-band power allocation in AMOOFDM high data rate NG-PON downlink transmission direct modulation of linear laser, in *Paper JThA24, Optical Fiber Communication Conference (OFC)*, Mar 2010
31. L. Anet Neto, P. Chanclou, B. Charbonnier, A. Gharba, N. Genay, R. Xia, M. Ouzzif, C. Aupetit-Berthelemot, J. Le Masson, D. Erasme, E. Grard, V. Rodrigues, On the interest of chirped lasers for AMOOFDM transmissions in long distance PON networks, in *Paper OWK4, Optical Fiber Communication Conference (OFC)*, Mar 2011
32. C.C. Wei, Small-signal analysis of OOFDM signal transmission with directly modulated laser and direct detection. Opt. Lett. **36**(2), 151–153 (2011)
33. J. Wei, E. Hugues-Salas, R. Giddings, X. Jin, X Zheng, S Mansoor, J. Tang, Wavelength reused bidirectional transmission of adaptively modulated optical OFDM signals in WDM-PONs incorporating SOA and RSOA intensity modulators. Opt. Express **18**(10), 9791–9808 (2010)
34. K.Y. Cho, K. Tanaka, T. Sano, S.P. Jung, J.H. Chang, Y. Takushima, A. Agata, Y. Horiuchi, M. Suzuki, Y.C. Chung, Long-reach coherent coherent WDM PON employing self-polarization-stabilization technique. IEEE J. Lightwave Technol. **29**(4), 456–462 (2011) [WDM Coherent LR-PON]
35. G. Rizzelli, V. Ferrero, S. Straullu, S. Abrate, F. Forghieri, R. Gaudino, Optimization of uncooled RSOA parameters in WDM reflective PONs based on self-coherent or direct detection OLT receivers, in *Paper We.1.B.1, European Conference and Exhibition on Optical Communication (ECOC)*, Sept 2012
36. C.C. Wei, Analysis and iterative equalization of transient and adiabatic chirp effects in DML-based OFDM transmission systems. Opt. Express **20**(23), 25774–25789 (2012)

37. W. Yan, B. Liu, L. Li, Z. Tao, T. Takahara, J. C. Rasmussen, Nonlinear distortion and DSP-based compensation in metro and access networks using discrete multi-tone, in *Proceedings of ECOC, Mo.1.B.2* (2012)
38. T. Alves, J. Morgado, A. Cartaxo, Linearity improvement of directly modulated PONs by digital predistortion of coexisting OFDM-based signals, in *Proceedings of ANIC, AW4A.2* (2012)
39. Y. Bao, Z. Li, J. Li, X. Feng, B.-O. Guan, G. Li, Nonlinearity mitigation for high-speed optical OFDM transmitters using digital pre-distortion. Opt. Express **21**, 7354–7361 (2013)
40. D.Z. Hsu, C.C. Wei, H.Y. Chen, J. Chen, M.C. Yuang, S.H. Lin, W.Y. Li, 2.1-Tb/s•km OFDM long-reach PON transmission using a cost-effective electro-absorption modulator, in *Paper OMG2, Optical Fiber Communication Conference (OFC)*, Mar 2011

Part IV
Light Emitting Diodes

Chapter 8
Light Emitting Diodes

Chien-Chung Lin, Kuo-Ju Chen, Da-Wei Lin, Hau-Vie Han, Wei-Chih Lai, Jian-Jang Huang, Tien-Chang Lu, Shoou-Jinn Chang and Hao-Chung Kuo

Introduction: Light Emitting Diodes Light emitting diodes, or LEDs for short, invented in the mid-20th century, are one of the key technologies that are thriving in the past decade. With small volume and high wall-plug efficiency, LED is poised to take over the traditional light bulb invented by Thomas Edison in the next few decades. Besides indoor/outdoor lighting, the LED also plays a versatile role in traffic signal, display, automobile lights etc. The importance of LED technology is needless to say. In this chapter, the history of LED development and the essential features of the LED will be reviewed, and the critical results in efficiency improvement and light extraction can be revealed.

C.-C. Lin (✉)
Institute of Photonic System, National Chiao
Tung University, Tainan 711, Taiwan
e-mail: chienchunglin@faculty.nctu.edu.tw

K.-J. Chen · D.-W. Lin · H.-V. Han · T.-C. Lu · H.-C. Kuo
Department of Photonics and Institute of Electro-Optical Engineering,
National Chiao-Tung University, 1001 Ta-Hsueh Road, Hsinchu 300, Taiwan
e-mail: wind7535@yahoo.com.tw

D.-W Lin
e-mail: davidlin1006@hotmail.com

H.-V Han
e-mail: maurice5210@hotmail.com

T.-C Lu
e-mail: timtclu@mail.nctu.edu.tw

H.-C Kuo
e-mail: hckuo@faculty.nctu.edu.tw

W.-C. Lai
Department of Photonics, National Cheng Kung University,
No. 1, Daxue Road, East Dist., Tainan 70101, Taiwan
e-mail: weilai@mail.ncku.edu.tw

C.-C. Lee (ed.), *The Current Trends of Optics and Photonics*,
Topics in Applied Physics 129, DOI 10.1007/978-94-017-9392-6_8

8.1 LED History and Application

Even though the phenomenon of electroluminescence was discovered in the beginning of 20th century and revisited in semiconductors at both forward and reverse-biased point-contact diodes [1–4], it was not until the 1950–1960s, that the band theory of solids was applied and light emission from direct bandgap semiconductor was observed. In 1962, Nick Holonyak, Jr., at General Electric Company, developed the first practical visible-spectrum (red) LED [5, 6] using the concept of direct-bandgap alloys. Prof. Holonyak predicted from his previous data on GaAsP LED devices that a laser with such a material structure would operate in the visible red spectral region [7]. He and his colleagues demonstrated the first semiconductor injection laser operating in the visible in 1962 [8, 9]. Professor Holonyak is recognized as the "father of the light-emitting diode" due to his contribution to light-emitting diode and laser diode. In about 50 years, the development of direct-bandgap alloys in the III-V compound semiconductor system was the first step leading to the development of "an ultimate lamp."

Next, Allen in 1963 and Grimmeiss and Scholz in 1964 reported red LEDs with p-n junction fabricated by GaP [10]. Compared with GaAsP LED systems, GaP LED systems had higher efficiency; however, GaP's radiation efficiency was limited because of its indirect bandgap. In addition, the concept of alloy compound can be applied by doping N, Zn and Oiii, to the semiconductor materials. By changing the bandgap of the alloy materials, light with wavelength between red and green could be emitted. Later in 1972, a former graduate student of Prof. Holonyak, M. George Craford, invented the first yellow LED by N-doping to GaAsP/GaAs systems and improved the brightness of red and red-orange LEDs by a factor of ten [11, 12]. In 1976, T.P. Pearsall applied high-brightness, high-efficiency LEDs to optical fiber telecommunications using new semiconductor alloy materials (for example, $Ga_xIn_{1-x}As_yP_{1-y}$ by liquid-phase) with light emission in the optical fiber transmission wavelengths [13].

During 1980s, double heterostructure was applied to LED epitaxial structure. Due to the enhancement of carrier-confinement by double heterostructure, the rate of electron-hole recombination was improved. Thus, AlGaAs LED increased the efficiency of emission in the history of red-light LED. After

J.-J. Huang
Graduate Institute of Photonics and Optoelectronics, National Taiwan University,
1, Roosevelt Road, Taipei 106, Taiwan
e-mail: jjhuang@cc.ee.ntu.tw

S.-J. Chang
Institute of Microelectronics and Department of Electrical Engineering,
Advanced Optoelectronic Technology Center, Research Center for Energy Technology and Strategy, National Cheng Kung University, Tainan 70101, Taiwan
e-mail: changsj@mail.ncku.edu.tw

1985, Japanese researchers used AlGaInP system with double heterostructure of AlGaInP/GaInP as active layer as materials of visible-spectrum laser [14, 15]. By adjusting the proportion of the quaternary material AlGaInP, LED with 625 nm (red), 610 nm (orange), and 590 nm (yellow) spectral wavelength were successfully developed. In addition, compared with LED fabricated with GaAsP, AlGaInP LEDs had longer lifespan under higher temperature and humidity. In the 1990s, the efficiency was further improved by the development of processing technique of epitaxial growth and transfer transparent substrate technique [16, 17]. The sidewall of LED bare chips was tapered in order to improve light-extraction efficiency. At this stage, the development of red-light LED has become mature and stable.

Compared with the prosperity of the progress of red and green LEDs, the development and research of blue LEDs, which is the last color that comprises three primary colors, was rather slow. Utilizing GaN heteroepitaxial films grown on sapphire, Isamu Akasaki and Hiroshi Amano introduced stimulated emission of blue light in 1990 [18]. The process was known as "two-step MOCVD process", which included low-temperature AlN buffer layer and succeeding high-quality GaN epitaxial growth. For bright GaN p-n junction blue LEDs, these accomplishments are notable because of their pioneering works [19–21]. In 1994, Nichia Corporation demonstrated the first InGaN/GaN-based high-brightness blue LED [22–24]. The presence of blue LEDs and high-efficiency LEDs quickly led to the development of the first white LED, which employed a $Y_3Al_5O_{12}$:Ce, or "YAG", phosphor coating to mix down-converted yellow light with blue to produce white light [25].

Along with the development of LED technology, the efficiency and light output of LED rose exponentially, with a doubling occurring approximately every 36 months since the 1960s [26]. The advance is in parallel with development of other semiconductor technologies, optics and material science [26]. In particular, novel ideas such as semiconductor alloy, heterostructure, multiple quantum well, chip bonding, chip shaping, flip chip, surface texturing, etc. have been proposed and realized over the years to increase lumen output and efficiency. Without these technological innovations, it is impossible for us to reap the fruit of highly efficient light emitting devices today. In the following, we will review several key issues of a light emitting diode and many of the challenges we still need to face today.

8.2 Improvement of Droop and Internal Quantum Efficiency for GaN-Based LEDs by Epitaxial Technology

One of the most important characteristics of an LED is its quantum efficiency. Since its invention, most of the researches were actually put on the improvement of this factor. From epitaxial technology, from device design, the scientists will try everything they can to increase the efficiency to convert the electron-hole pairs into photons. For the GaN-based LEDs, this topic becomes even more important due to the infamous "efficiency droop" phenomenon, which inflicts this electron-to-photon conversion badly at high currents. In this section, in order to obtain an LED with high efficiency and low efficiency droop, several epitaxial technologies were proposed. For enhancing the readability, this chapter would be classified into two sections, including epitaxial template with nano-scale patterns and epitaxial structure design. In the first section, we will introduce SiO_2-based nanorod-array for patterned sapphire substrate (NAPSS), etched nano-pillars with embedded SiO_2 nano-masks, and a GaN epilayer with embedded air voids grown on Ar-implanted AlN/sapphire substrate. In the second section, we will review some epitaxial structure designs, including chirped multiquantum barrier and AlGaN staircase electron blocking layer. Our studies showed that these approaches can achieve high crystalline quality growth and effectively improve the internal quantum efficiency and efficiency droop behavior.

8.2.1 Introduction

To achieve the next generation applications in projectors, automobile headlights, and high-end general lightings, further improvements on the optical output power and the external quantum efficiency (EQE) of light-emitting diodes (LEDs) are required. Due to the difficulties and high cost to fabricate native GaN substrates, sapphire substrates still have been playing a major role for nitride-based LED production. The development of GaN-based LEDs has shown significant progress over the past decade, in particular for the growth of GaN on lattice-mismatched sapphire substrates using metal-organic chemical vapor deposition (MOCVD) [27, 28]. It has been shown that the epitaxial lateral overgrowth (ELO) using a micro-scaled SiN_x or SiO_x patterned mask on as-grown GaN seed crystal can effectively reduce the threading dislocation density (TDD) [29–31]. However, the two-step growth procedure and the need for a sufficient thickness for GaN coalescence are costly and time consuming. Meanwhile, high-quality GaN-based LEDs were demonstrated on a microscale patterned sapphire substrate (PSS) [32], where these patterns served as a template for ELO as well as the scattering centers for the guided light. Both the epitaxial crystal quality and the light extraction efficiency were improved by utilizing the PSS. On the other hand, with the advances in fabrication processes, nanoscale patterned substrates have become available and the improvements in LED performances are more significant than some of micron-scale techniques. Several studies indicated that the LEDs grown on the nanoscale PSS showed more enhancement in the EQE than those grown on the microscale PSS. However, the fabrication of nanoscale PSS generally required electron-beam lithography [33] or nanoimprinting techniques [34, 35], making it unfavorable for mass production.

In this chapter, we will review several nanotechnologies to fabricate a template for the nanoscale ELO (NELO) of GaN regrowth by MOCVD to produce high-efficiency GaN-based LEDs.

On the other hand, for lighting application, one often needs to inject a significantly larger current into these LEDs. These devices, however, inevitably suffer from a reduction in efficiency due to a high injection current administered [36]. In other words, the external quantum efficiency (EQE) of LED reaches its peak at a relatively low current density and monotonically decreases with a further increase in current. Although the origin of efficiency droop is still a topic of debate, it is generally recognized that the electron current leakage plays an important role in this regard [37]. In the following sections, we will further review some epitaxial structure designs, including chirped multiquantum barrier and AlGaN staircase electron blocking layer. These structures can improve the efficiency droop behavior of LED devices.

8.2.2 Nanoscale Epitaxial Patterned Template

8.2.2.1 Nanoscale Epitaxial Lateral Overgrowth of GaN-Based Light-Emitting Diodes on a SiO_2 Nanorod-Array Patterned Sapphire Template

One of the simplest methods to carry out nano-patterned substrate is to use the nano-scaled SiO_2 mask for lateral overgrowth. Previous researchers focused on micron-scaled masks [38, 39] have produced a robust technology that are widely applied in current LED industry. It is nature to shrink down the mask as the fabrication technique progresses. To form nano-patterns, self-assembled metal layer was considered for masking material [40].

GaN-based LEDs were grown on a 2 in. SiO_2 NAPSS using a low-pressure MOCVD system (Aixtron 2400 G). The preparation of the SiO_2 NAPSS template started with the deposition of a 200-nm-thick SiO_2 layer on a *c*-face (0001) sapphire substrate by plasma enhanced chemical vapor deposition, followed by the evaporation of a 10-nm-thick Ni layer, and the subsequent rapid thermal annealing with a flowing nitrogen gas at 850 °C for 1 min. The resulting self-assembled Ni clusters then served as the etch mask to form a SiO_2 nanorod array by etching the semiconductor layers in a reactive ion etch system for 3 min. Finally, the sample was dipped in a heated nitric acid solution (HNO_3) at 100 °C for 5 min to remove the residual Ni masks. As shown in Fig. 8.1a, the field-emission scanning electron micrograph (FESEM) indicated that the fabricated SiO_2 nanorods were approximately 100–150 nm in diameter with a density of 3 × 109 cm^{-2}. The spacing between nanorods was about 100–200 nm. Figure 8.1a also shows that the exposed sapphire surface was flat enough for epitaxy. As the deposition process began, localized and hexagonal island-like GaN nuclei were first formed from the sapphire surface to initiate GaN overgrowth, as shown in Fig. 8.1b. Figure 8.1c shows the cross-sectional

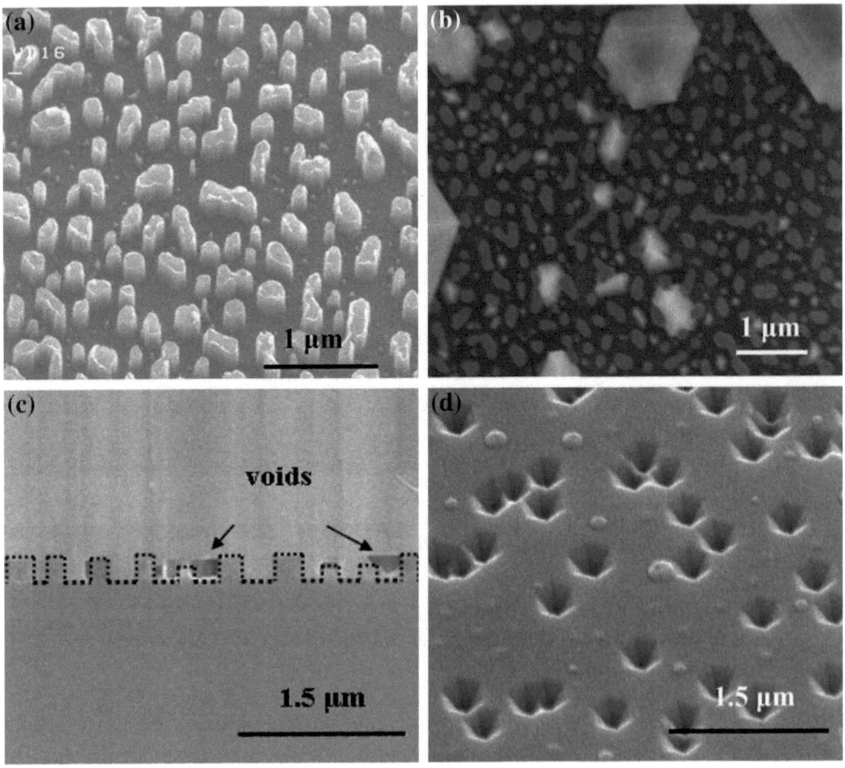

Fig. 8.1 FESEMs of **a** the fabricated SiO_2 nanorod array, **b** GaN nuclei on the SiO_2 NAPSS as growth seeds, **c** the GaN epilayer on a NAPSS in the cross-sectional view, and **d** the epitaxial pits on the *p*-GaN surface [42]

FESEM image of the GaN epi-layer, where voids with a size varying from 150 to 200 nm were observed between the highlighted SiO_2 nanorods. The existing of the voids between nanorods observed from the micrographs suggested that not all the exposed surface enjoyed the same growth rate. Hence, only the regions with higher growth rates, which might be originated from larger exposed surface, could play the role of a seedling layer, facilitating the lateral coalescence of GaN epitaxial growth. Lastly, the growth of a conventional LED structure, which consists of ten periods of InGaN/GaN multiple quantum wells and a 100-nm-thick *p*-GaN layer, was completed by MOCVD. The *p*-GaN layer of the NAPSS LED was grown at the relatively low temperature (800 °C), leading to the formation of hexagonal pits due to insufficient migration length of Ga atoms [41]. The FESEM image of the roughened *p*-GaN surface with randomly distributed pits is shown in Fig. 8.1d.

The TEM was employed to investigate the crystalline quality of GaN layers grown on a planar sapphire substrate and on a NAPSS. As shown in Fig. 8.2a, the

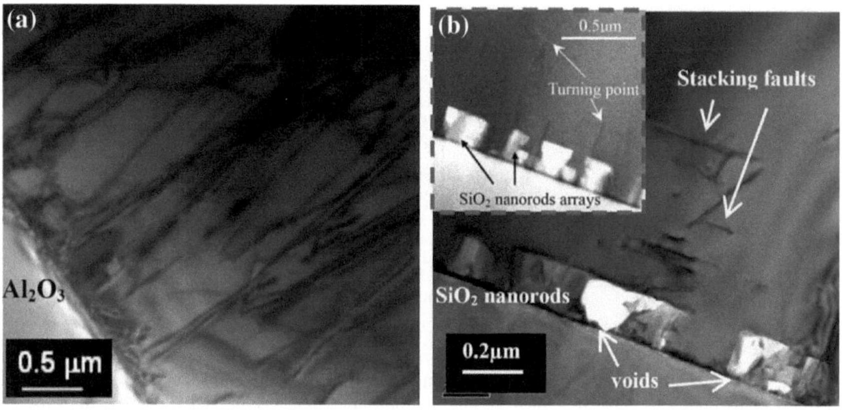

Fig. 8.2 The TEM images of the GaN/sapphire interface for the GaN epilayer grown on **a** a planar sapphire substrate and **b** on a NAPSS. The inset of (**b**) shows the dislocation bending phenomenon with visible turning points [42]

TDD of GaN on the planar sapphire substrate was greater than 10^{10} cm^{-2} due to both the large lattice mismatch (13 %) and the high thermal coefficient incompatibility (62 %) between sapphire and GaN. On the other hand, the crystalline quality of GaN epi-layer on a NAPSS was drastically improved from that grown on a planar sapphire substrate, as shown in Fig. 8.2b. We found that a number of stacking faults often occurred above the voids between SiO_2 nanorods, where visible threading dislocations (TDs) were rarely observed in the vicinities. It is believed that the presence of stacking faults could block the propagation of TDs [43]. Moreover, the TDs of the GaN layer on a NAPSS mainly originated from exposed sapphire surface, which could be bent due to the lateral growth of GaN. The inset of Fig. 8.2b shows the TEM image of the dislocation bending with visible turning points.

The completed epitaxial structure then underwent a standard four-mask LED fabrication process with a chip size of 350 × 350 μm^2 and packaged into TO-18 with epoxy resin on top. The schematic of a fabricated NAPSS LED is shown in the inset of Fig. 8.3a. The current-voltage (*I*-*V*) characteristics of the NAPSS LED and a conventional LED with the same chip size were measured at room temperature, as shown in Fig. 8.3a. The forward voltages at 20 mA were 3.27 V for the conventional LED and 3.31 V for the NAPSS LED. The nearly identical *I*-*V* curves indicate that the nanoscale roughness on the *p*-GaN surface had little impact on the *I*-*V* characteristics. Moreover, the NELO of GaN did not deteriorate in the electrical properties.

Figure 8.3b shows the measured light-output power versus the forward continuous dc current (*L*-*I*) for the NAPSS and conventional LEDs. At an injection current of 20 mA, the light-output power was approximately 22 and 14 mW for NAPSS- and conventional LEDs, respectively. The output power of the NAPSS LED was enhanced by a factor of 52 % compared to that of the conventional

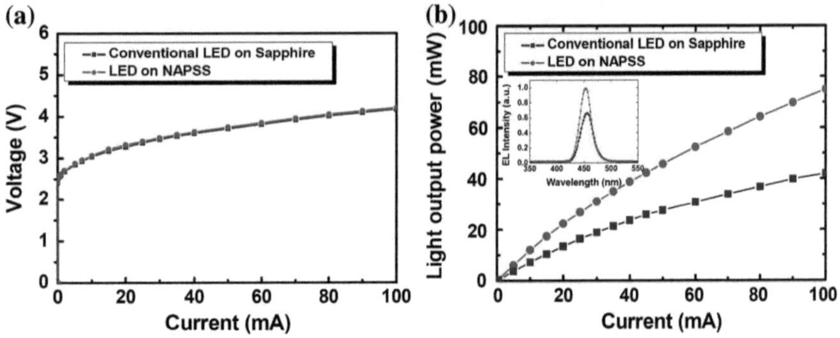

Fig. 8.3 Electrical and optical properties of a NAPSS and a conventional LED: **a** the current-voltage (*I-V*) curves, where the inset shows a schematic of a NAPSS LED, and **b** the current-output power (*L-I*) curves, where the inset shows the electroluminescence spectra for both devices at a driving current of 20 mA [42]

LED. The inset shows the normalized electroluminescence spectra for both devices at an injection current of 20 mA. A minor wavelength blue-shift of ~2 nm was observed for the NAPSS LED, attributed to the partial strain release by adopting the NELO scheme [44]. The EQE of the NAPSS LED was calculated to be ~40.2 %, which is an increase of 56 % when compared to that of the conventional LED (~25.7 %.) We believe that the 56 % enhancement in EQE originates from the improved internal quantum efficiency and the enhanced extraction efficiency. The SiO_2 NAPSS-assisted NELO method effectively suppresses the dislocation densities of GaN-based LEDs, which increases the internal quantum efficiency. Moreover, the embedded SiO_2 nanorods in the GaN epilayer contribute to the enhanced light extraction efficiency due to the scattering at the interfaces of different refractive indices.

8.2.2.2 Highly Efficient and Bright LEDs Overgrown on GaN Nanopillar Substrates

While nano-scale masks can improve the quality of LEDs, a more elaborated method based on this technology can be used, i.e., the GaN nano-pillar array substrate. The formation of nano-pillar is one more step than the previous two methods: dry etch to the saphhire surface. The benefit of this method is to maximize the dislocation bending because of stronger lateral growth, and a larger air-void array embedded in the substrate to scatter the light and increase the light-extraction efficiency.

The preparation procedures of the GaN epilayer with embedded micro voids and SiO_2 nanomask are discussed as follows. First, we deposited a 200-nm-thick SiO_2 layer on a 2-μm thick undoped GaN layer by plasma enhanced chemical vapor deposition (PECVD), followed by the evaporation of a 10-nm-thick Ni layer, and subsequent RTA process in the nitrogen gas ambient at 850 °C for 1 min. The resulting self-assembled Ni clusters then serve as etch masks to form SiO_2 nanorod arrays in an RIE system for 20 min to reach the sapphire surface. Finally, the sample was dipped in a heated nitric acid solution (HNO_3) at 100 °C for 5 min to remove the residual Ni. The 2-μm-tall GaN nanorods (NRs) with SiO_2 nanomask are produced as shown in Fig. 8.4a. It can be seen that GaN NRs with an average diameter of about 250–500 nm were formed. It was also found that GaN NR's density was around 3.3×10^8 cm^{-2} as shown in Fig. 8.4b. Next, we deposited a GaN-based LED structure on this GaN NRs template by a low pressure MOCVD (Veeco D75) system, denoted as NR-LEDs. At the same time, the identical GaN-based LED structure was also grown on sapphire without GaN NRs for comparison, denoted as conventional LEDs (i.e., C-LEDs). During the growth, trimethylgallium (TMGa), trimethylindium (TMIn) and ammonia (NH_3) were used as the sources for gallium, indium, and nitrogen, respectively. Silane (SiH_4) and biscyclopentadienyl magnesium (CP_2Mg) were used as the *n*-dopant and *p*-dopant sources, respectively. The epitaxial structure of the GaN-based LED overgrowth on GaN NRs template consists of 3 μm *n*-doped GaN (*n*-GaN), 10-pairs InGaN/GaN multi-quantum wells (MQWs), and 0.2 μm *p*-doped GaN (*p*-GaN) cap layer. Figure 8.4c shows the cross-sectional SEM

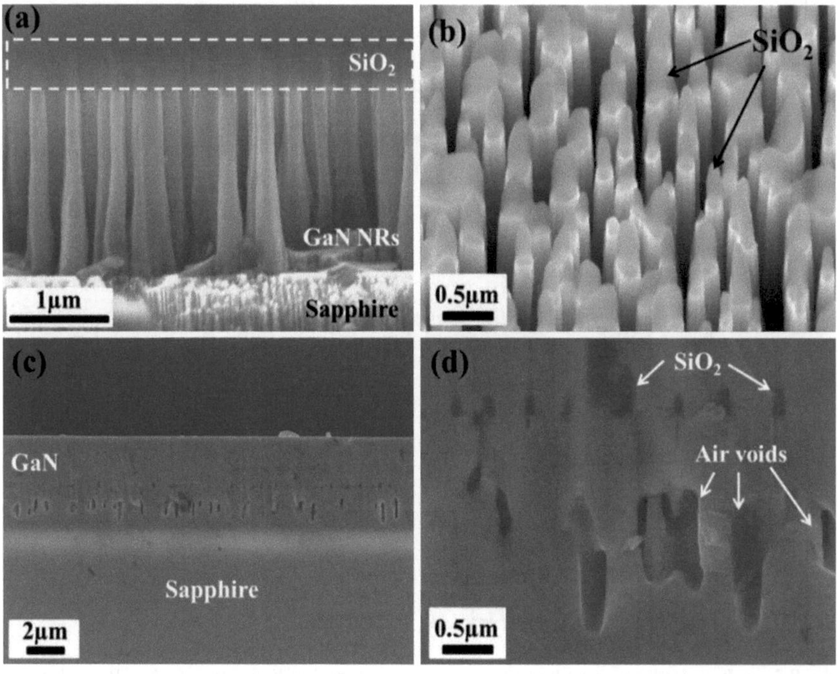

Fig. 8.4 **a** Cross-sectional; **b** tilted SEM image of GaN NRs template. **c** and **d** Cross-sectional SEM image of GaN epilayer [40]

image of GaN/InGaN epitaxial layers with voids and SiO_2 nanomasks after all the growth is completed. The estimated diameter of these voids ranges from 0.5 to 1 μm. These embedded voids and SiO_2 nanomasks (as shown in Fig. 8.4d) shall be able to increase the LEE due to extra light scattering [45].

Figure 8.5a–c show the mechanisms of the air gap formation by nanoscale epitaxial lateral overgrowth (NELOG) techniques on top of SiO_2 nanomask. First, GaN NRs with the SiO_2 nanomask were formed on sapphire substrate by top-down methods in Fig. 8.5a. As the GaN NRs grows upwards, there is also the lateral growth on the sidewall of individual NRs. Such lateral growth eventually narrows the gap between columns and forms holes with embedded air pockets as shown in Fig. 8.5b. Our pervious study showed this growth process adding extra m-plane $(10\bar{1}0)$ GaN on the side walls of the etched pillars and inclined r-plane facets $(1\bar{1}02)$ close to the top of nano-pillars [46]. Frajtag et al. [47] also reported that the semi-polar planes coalesce first due to their higher growth rates relative to the growth rates on the nonpolar side faces of nanowires. All of these growth mechanisms help the formation of voids in between nanorods. The final step consists of a planar epitaxial GaN overgrowth, and air-voids and SiO_2 nanomasks were encapsulated as show in Fig. 8.5c. To analyze the detailed epitaxial layer quality, we took the TEM pictures of

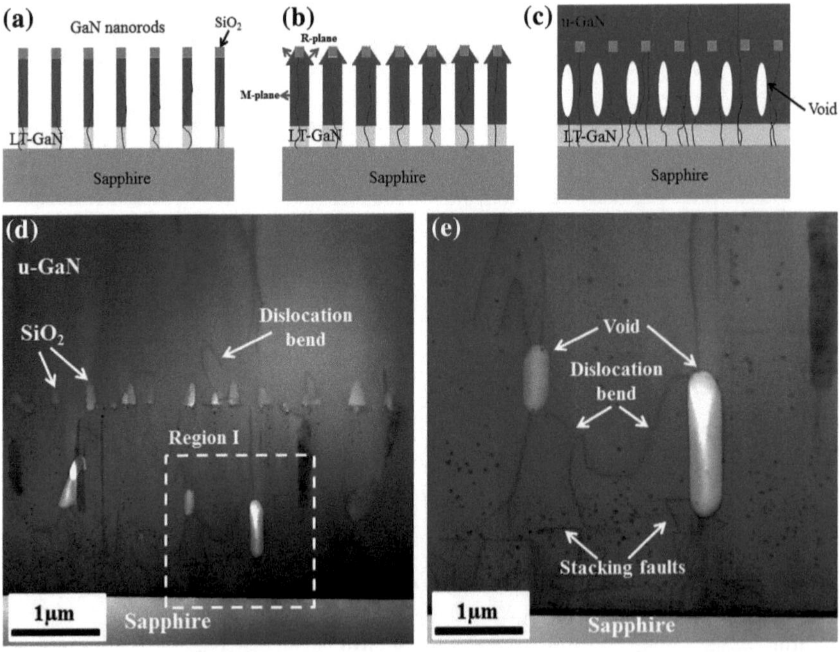

Fig. 8.5 **a**, **b** and **c** The procedure of the air-voids formation between a GaN NRs and u-GaN epitaxial layer. **d** TEM image of GaN epilayer overgrowth on GaN NRs, e HRTEM image of region I in (**d**). The diffraction condition is $g = 0002$ [40]

GaN epi-layers overgrowth on GaN NRs shown in Fig. 8.5d, and calculate the dislocation densities. Fewer TDDs are observed within the range in view. The dislocation density on the top of u-GaN is calculated to be around 5×10^7 cm^{-2}. Meanwhile, we found TDDs were bent near SiO_2 nanomasks. The behavior is similar to those occurred in the NELOG method on a SiO_2 nanorod-array patterned sapphire substrate [42]. The reduction of TDDs can be attributed to the misfit (mainly perpendicular to the *c*-axis) and dislocation bending occurred just above the voids, as shown in the inset of Fig. 8.5e.

Figure 8.6a displays the typical power-current-voltage (L-I-V) characteristics of NR-LEDs and C-LEDs. With an injection current of 20 mA, the forward voltage is 3.37 and 3.47 V, and the output power is 21.6 and 13.1 mW, for NR-LEDs and C-LEDs, respectively. The enhanced *L-I-V* characteristics can be attributed to the following factors. First, the TDD reduction of epitaxial layers leads to much fewer non-radiative recombination events and increases the photon generation efficiency. Second, more light can be extracted from the LED because of the light scattering effect from the embedded micro/nano-scale voids and SiO_2. In addition, at the reverse bias, the leakage current of the NR-LEDs is smaller than C-LEDs, as shown in Fig. 8.6b. Several types of dislocations can contribute to the reverse-bias

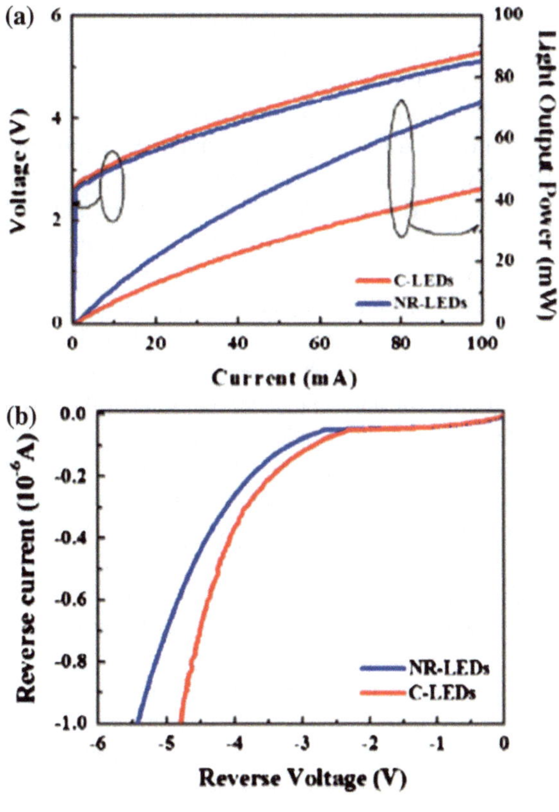

Fig. 8.6 **a** Forward I–V characteristics of all fabricated LEDs. **b** Reverse I–V characteristics of all fabricated LEDs [40]

leakage current, and one of the dominant types is the screw dislocation [48]. The reduction of screw dislocations can certainly help to reduce the reverse-bias current [48], and our measurement indicates a better crystal quality of NR-LEDs, which is confirmed by the TEM results.

8.2.2.3 Gallium Nitride-Based Light-Emitting Diodes with Embedded Air Voids Grown on Ar-implanted AlN/Sapphire Substrate

In this study, an approach aimed at achieving a light scattering layer beneath the active layer of GaN-based LEDs was demonstrated to improve the light-extraction efficiency (LEE). The AlN nucleation prepared by radio-frequency (RF) sputtering was initially deposited on sapphire substrate, and then a selective area Ar ion implantation was performed to periodically create damage regions on the AlN layer. The Ar-implanted AlN/sapphire (AIAS) substrates were served as templates. Next, the AIAS templates were loaded into MOVPE reactor to grow GaN-based LEDs without the request of a conventional low temperature nucleation layer [49–51]. Consequently, a series of air voids were formed around the sapphire/AlN/GaN interfaces to achieve a light scattering structure beneath the active layer for improving the LEE of LEDs. This technique features the fact that the surface morphology of AIAS was substantially flat in contrast to the surface of conventional PSS with concaves and convex. Detailed processing procedures and related results are discussed subsequently.

A 25 nm-thick ex situ sputtered AlN nucleation layer was deposited onto the sapphire substrates to serve as templates. During the RF sputtering process, AlN targets were used to deposit AlN thin film [50]. The typical growth rate and temperature of AlN thin films were 5 nm/min and 750 °C, respectively. Before the growth of LED epitaxial structures, Ar ion implantation, in which the dosage and energy were $1 \times 10^{16}/cm^2$ and 100 keV, respectively, was selectively performed on the AlN/sapphire templates. For the creation of periodically implanted regions, a 2 μm-thick photoresistor layer served as the mask layer. Circular openings on the PR mask layer with a diameter of 3 μm were defined using standard photolithography. In addition, a 90 nm-thick SiO_2 layer was then deposited in the opening, so that the implanted Ar ions would accumulate next to the surface of the AlN/sapphire templates. After the Ar-implantation process and removal of the mask layers, the diameter of the Ar-implanted regions and the spacing between the circular implanted regions were 3 μm. Figure 8.7a shows a schematic illustration in cross-section view of the AIAS substrate. Next, the AIAS substrates were loaded into the MOVPE reactor to grow LED epitaxial structures; these were labeled as LED-I. LED epitaxial structures grown on the AlN/sapphire templates, without the Ar ion implantation, were also prepared for comparison and labeled as LED-II. The epitaxial layers of the LEDs included a AlN nucleation ex situ layer prepared by RF sputtering, a 3 μm-thick unintentionally doped GaN (u-GaN) grown at 1,050 °C, a 1.7 μm-thick Si-doped n-GaN layer grown at 1,050 °C, a ten-pair $In_{0.3}Ga_{0.7}N$/GaN multi-quantum-well (MQW) structure

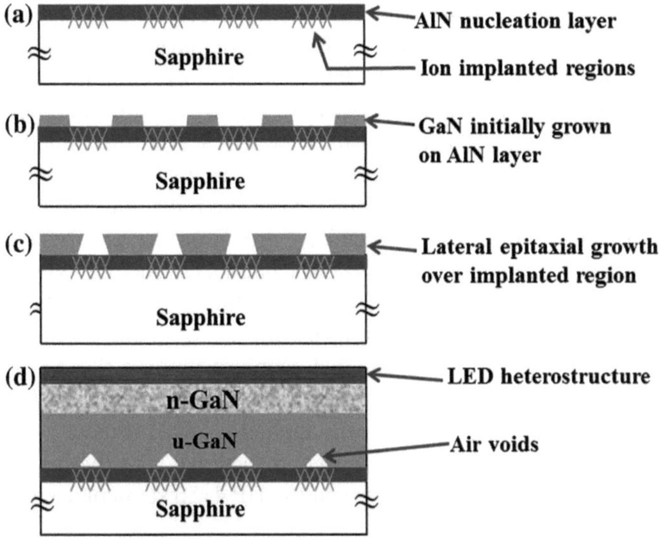

Fig. 8.7 Schematic illustration of **a** Ar-implanted AlN/sapphire template, **b** u-GaN layer grown on the AIAS, **c** lateral epitaxial growth occurs over the Ar-implanted regions on the AIAS, and **d** air voids embedded in the u-GaN layer [55]

grown at 750 °C, a 0.05 μm-thick Mg-doped p-$Al_{0.15}Ga_{0.85}N$ electron blocking layer, and a 0.2 μm-thick Mg-doped p-GaN top contact layer grown at 1,000 °C. The carrier concentrations of the n-GaN and p-GaN layers were around $8 \times 10^{18}/cm^3$ and $5 \times 10^{17}/cm^3$, respectively. Afterwards, a heavily Si-doped InGaN top layer was grown on the p-GaN contact layer [52]. During the growth of the InGaN/GaN LED epitaxial structure on the AIAS substrates, the u-GaN epitaxial layer was initially grown on the implantation-free regions, whereas the deposition of the u-GaN layer on the Ar-implanted regions did not occur because of the difference in lattice constants between the implanted and implantation-free regions. Consequently, selective growth occurred during the growth of the u-GaN epitaxial layer on the AIAS substrates to form a series of hexagonal concaves on the u-GaN layer, as shown in Fig. 8.7b, c. Next in sequential, u-GaN layer grew laterally over the implanted area to form air voids in the u-GaN layer under the MQW active layer, corresponding to Fig. 8.7d. Figure 8.8a shows the typical top-view micrograph of u-GaN layer grown on AIAS substrates obtained through scanning electron microscopy (SEM). Periodic holes on the SEM image indicate that selective growth took place during the growth of the u-GaN layer. This result could be attributed to the relatively lower growth rate of GaN at the Ar-implanted regions than at the implantation-free regions. The difference in the growth rates of the two aforementioned regions resulted from the difference in their lattice constants. In general, the crystal structure of semiconductor undergoing high-dose and/or high energy ion bombardment produces an amorphous layer [53]. In this study, the implanted Ar ions created a damage layer with a depth of approximately 50 nm from the sapphire surface estimated by simulation. According to a previous report, critical doses for the amorphization of sapphire or nitride semiconductors were shown to be around 10^{16}/cm^2, and the lattice constant of AIAS increased with the increase of implantation dose and/or energy [53, 54]. Figure 8.8a indicates concaves existing in the u-GaN layer grown on the AIAS. After forming the concaves in the u-GaN layer, the lateral coalesce eventually results in a continuous layer and leaves voids over the Arimplanted regions. Next, the remaining layers of a typical GaN-based LED structure were sequentially grown on the u-GaN layer with air voids embedded at the sapphire/AlN/u-GaN interfaces. Accordingly, GaN-based LEDs with embedded air voids beneath the InGaN/GaN MQW active layer could be formed using the AIAS substrates, as shown in Fig. 8.8b. On the other hand, GaN-based LEDs grown on conventional sapphire without the Ar-implantation process exhibited a flat and void-free sapphire/u-GaN interface (i.e., LED-II). For device process, all the LED wafers were applied to general mesa and electrodes formation procedures. Light output powers characteristics of the bare-chip LEDs were measured with an integral sphere and the HP 4156 semiconductor parameter analyzer, respectively.

Figure 8.9a shows the relative output power-current (L-I) characteristics of LED-I and LED-II. With a 20 mA current injection, the output power of LED-I has an enhancement in magnitude of 25 % compared with the LED-II. For the LED-I, embedded air voids resulted in a light scattering layer at the sapphire/AlN/u-GaN interfaces. The voids induced index steps at the sapphire/AlN/u-GaN interfaces that led to an increased probability of the photons escaping from the LED. In other words, light emitted from the active layer of LED-I experienced a significant reflection and redirection around the air voids, and thereby resulted in a shorter average path before the photons escaped into the free space, as shown in Fig. 8.9b [56]. Figure 8.9c shows the micrograph taken from the LED-I under an operation current of 10 mA. A periodic feature resulting from the light scattering at the embedded air voids could be observed on the near-field image, as indicated in the inset of Fig. 8.9c.

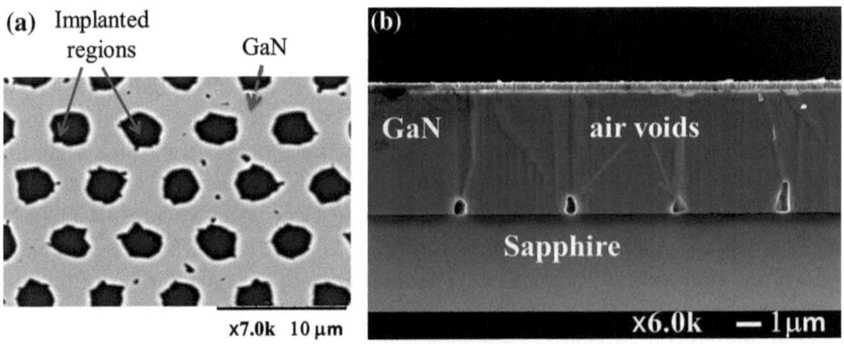

Fig. 8.8 SEM images (**a**) typical top-view micrograph from u-GaN epitaxial layer grown on an AIAS, (**b**) cross-section view obtained from a representative LED grown on an AIAS [55]

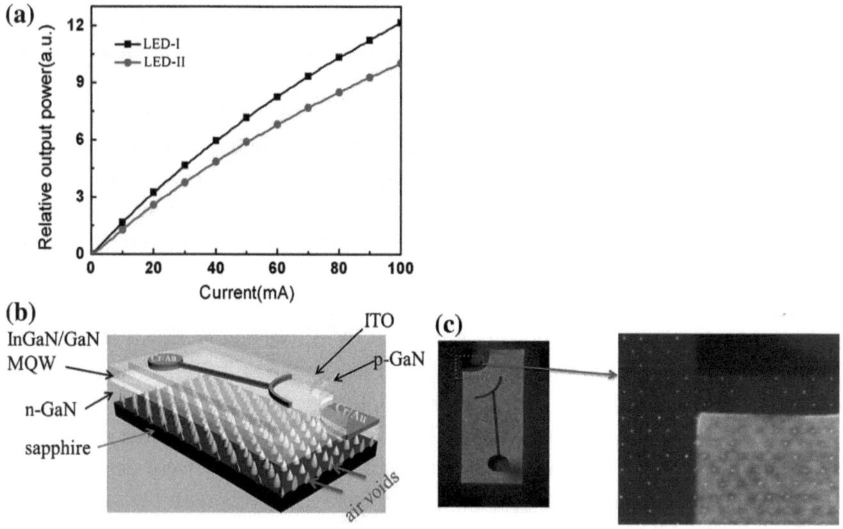

Fig. 8.9 **a** Relative output power of the LED-I and LED-II as a function of injection current, **b** illustration of photons paths in the LED-I, and **c** nearfield image from the LED-I driven under a current of 10 mA [55]

8.2.3 Low Efficiency Droop Epitaxial Structures

8.2.3.1 GaN-Based LEDs with a Chirped Multiquantum Barrier Structure

In general, the conventional method used to reduce electron current leakage is to insert a p-AlGaN electron blocking layer (EBL) between the multiquantum well (MQW) active region and the p-GaN cap layer [57]. With a large bandgap introduced, this EBL provides a potential barrier which can reduce the probability for the injected electrons to overflow from the MQW active region into the p-GaN cap layer. One possible method to further reduce electron current leakage is to use a multiquantum barrier (MQB) structure [58, 59]. MQB structure consists of alternating layers of narrow and wide bandgap semiconductor materials. Similar to crystal lattice, such an arrangement forms the allowed and forbidden carrier states over a very short range. By properly design the MQB structure, it is possible to achieve an effective barrier height which is much larger than classical barrier. Although MQB structure has already been used extensively in AlInGaP-based LEDs and laser diodes, only few reports on GaN-based LEDs with MQB structure can be found in the literature [59–61]. Furthermore, it has been shown that chirped MQB (CMQB) structure can provide an even larger electron blocking efficiency for AlInGaP-based LEDs, as compared to uniform MQB (UMQB) [62]. It should be noted that thicknesses of each narrow and wide bandgap semiconductor materials were kept the same for UMQB structure. In contrast, thicknesses of each narrow and wide bandgap semiconductors were different from each other for CMQB structure. A structure such as CMQB has never been applied to GaN-based LEDs. In this study, we report the fabrication of GaN-based blue LEDs with a CMQB structure. The electro-optical properties of the fabricated LEDs are also reported.

Samples used in this study were all grown on c-plane sapphire substrate by metalorganic chemical vapor deposition (MOCVD). Details of the growth can be found elsewhere [63, 64]. For comparison, samples with the three different EBLs, i.e., a conventional AlGaN, UMQB and CMQB were prepared. The structure of the conventional sample consists of a 50-nm-thick GaN nucleation layer grown at 550 °C, a 3-μm-thick n-type Si-doped GaN layer with electron concentration of 5×10^{18} cm^{-3} grown at 1,050 °C, a unintentionally doped MQW active region grown at 770 °C, a 45-nm-thick p-type Mg-doped $Al_{0.15}Ga_{0.85}N$ EBL grown at 1,050 °C and a 15-nm-thick p-type Mg-doped GaN cap layer with hole concentration of 1.2×10^{19} cm^{-3} grown at 1,050 °C and a Si-doped n+-InGaN/GaN short-period-superlattice (SPS) tunnel contact structure [65]. The electron and hole concentrations were determined by Hall measurement from the n-type and p-type epitaxial layers with the same growth conditions. The MQW active region consists of eight 2-nm-thick $In_{0.12}Ga_{0.88}N$ well layers separated by nine 15-nm-thick GaN barriers and the SPS structure consists of four pairs of 5-Å-thick $In_{0.23}Ga_{0.77}N$ layers and 5-Å-thick GaN layers.

The structures of UMQB LED and CMQB LED were identical to that of the conventional LED, except the UMQB and CMQB structures were used in place of AlGaN EBL for UMQB and CMQB LEDs, respectively. The UMQB structure consists of four 50-nm-thick GaN layers separated by five 5-nm-thick $Al_{0.15}Ga_{0.85}N$ layers. On the other hand, the CMQB structure consists of five $Al_{0.15}Ga_{0.85}N$ layers (7.5, 33.75, 60, 86.25, 112.5 Å, arranged in sequence from n-side to p-side) and four GaN layers (65.625, 46.875, 28.125, 9.375 Å, arranged in sequence from n-side to p-side) with different thicknesses. It should be noted that thicknesses of the five $Al_{0.15}Ga_{0.85}N$ layers were increased monotonically while thicknesses of the four GaN layers were decreased monotonically. It should also be noted that the total thicknesses of the UMQB and CMQB structures were each maintained at 45 nm, which is identical to that of $Al_{0.15}Ga_{0.85}N$ EBL used in conventional LED. Figure 8.10

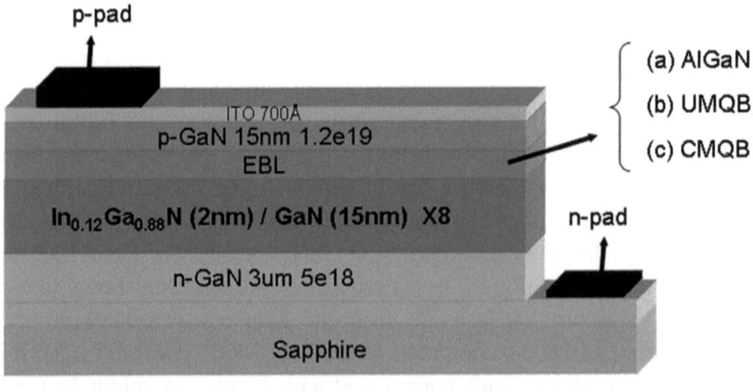

Fig. 8.10 The diagram of three different fabricated devices [66]

demonstrates the cross-sectional diagram of the samples used in this study. The as-grown samples were then annealed in the furnace at 750 °C in N2 ambient to activate Mg in the p-layers. Standard procedures were subsequently used to fabricate the 200 × 200 μm LED chips with indium-tin-oxide (ITO) p-contact and Ti–Al–Ti–Au n-contact. The current-voltage (I-V) characteristics of the fabricated devices were then measured at room temperature using an HP4156 semiconductor parameter analyzer. The chips were subsequently packaged as LED lamps. The intensity-current (L-I) characteristics of the packaged lamps were measured with the molded LEDs attached to an integrating sphere detector.

In order to minimize joule heating effects, electro-optical properties of the fabricated LEDs were all measured with pulse current injection. During these measurements, pulse current with 20 μs width and 1 % duty cycle was injected into these devices. Figure 8.11 shows the pulsed L-I characteristics measured from these three LEDs. From the L-I curves, it was found that output power observed from the CMQB LED was the largest, followed by the UMQB LED while the output power of the LED with a conventional AlGaN EBL was the smallest. With 20 mA injection current administered, it was found that the LED output powers were 27.5, 27.2 and 25.4 mW for CMQB LED, UMQB LED and LED with a conventional AlGaN EBL, respectively. When compared with the conventional LED with AlGaN EBL, an 8.3 and 7.1 % enhancement in output power from CMQB and UMQB LEDs were respectively achieved. The improvement was in fact obtained due to a larger effective barrier height provided by the MQB structure. It should be noted that output power measured from the CMQB LED was larger than that of the UMQB LED. It has been pointed out by Fujii et al. [67] that the reflection characteristics of MQB structure for electrons are analogous to the reflection characteristics of multilayer coating for light wave. Although UMQB could provide a high peak reflectivity, its bandwidth for electron wavelength or equivalently the electron energy is narrower. In other words, only electrons with energy either too high or too low could still overflow from the MQW active region into the p-cladding. In contrast, the thickness of the AlGaN and GaN layer

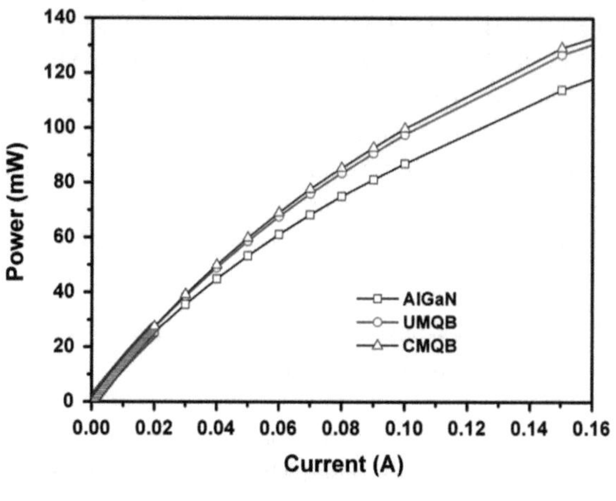

Fig. 8.11 Experimental L-I performance curves of the three different fabricated devices [66]

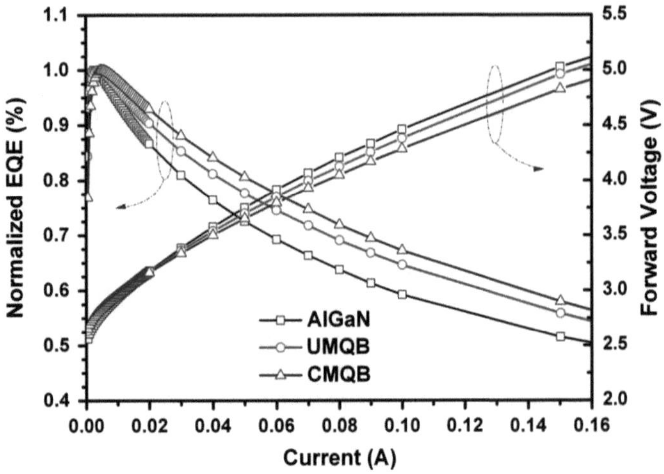

Fig. 8.12 Experimental EQE and forward voltage performance curves of the three different fabricated devices [66]

changes monotonically for the CMQB structure. Such a design could provide a larger bandwidth with a highly reflective energy band extended toward the both ends of the reflection spectrum [67]. Consequently, more electrons are expected to be reflected and a larger output power is obtained from the CMQB LED in return.

Figure 8.12 shows the normalized external quantum efficiency (EQE) as a function of injection current for these three LEDs. In this figure, the EQE was normalized to their respective maximum values. It can be seen clearly that we can indeed effectively mitigate the drooping effect by replacing the AlGaN EBL with UMQB structure or CMQB structure. As we increased the injection current to 150 mA, it was found that normalized EQE had reduced to 57.9, 55.8 and 51.5 % for CMQB LED, UMQB LED and LED with conventional AlGaN EBL, respectively. Figure 8.12 also shows I-V characteristics measured from these devices. With 20 mA current injection administered, it was found the forward voltages for CMQB LED, UMQB LED and LED with conventional AlGaN EBL were 3.163, 3.174 and 3.183 V, respectively. With MQB structure (i.e., either UMQB or CMQB), it is possible that some of the injected holes are confined and subsequently spread out in the in-plane direction before they enter the MQW active region. As a result, a better current spreading and thus a lower forward voltage can be achieved. These values also suggest that we can simultaneously improve the electrical property of the LEDs by using the CMQB structure.

8.2.3.2 InGaN-based Light Emitting Diodes with an AlGaN Staircase Electron Blocking Layer

In this study, we employed polarization engineering to prepare $Al_xGa_{1-x}N$ staircase EBLs in InGaN-based MQWs LEDs. Furthermore, we discuss herein the theoretical mechanisms in comparison with those of conventional $Al_xGa_{1-x}N$ EBLs (with a constant composition of Al), in terms of their external quantum efficiencies (EQEs) and calculated energy band diagrams.

All the InGaN-based blue LEDs were grown on c-plane (0001) sapphire substrates using an atmospheric pressure metal-organic chemical vapor deposition (AP-MOCVD) system (Taiyo Nippon Sanso: SR2000). A 25-nm-thick GaN nucleation layer, a 1-μm-thick undoped GaN buffer layer, and a 3-μm-thick Si-doped n-GaN layer, were sequentially deposited on the sapphire substrate. Next, the active region, consisting of five pairs of 3-nm-thick $In_{0.17}GaN$ wells and 12-nm-thick GaN barriers with emission peak around 450 nm, was deposited on the n-GaN layer. Three samples with different EBLs were prepared: an LED featuring a conventional p-$Al_{0.21}Ga_{0.79}N$ EBL having a thickness of 30 nm and two others with specifically designed $Al_xGa_{1-x}N$ staircase EBLs, in which the EBL was divided into three parts, with Al compositions (from low to high) of 0.07, 0.14, and 0.21 in one and (from high to low) of 0.21, 0.14, and 0.07 in the other. By adjusting the trimethylaluminum (TMAl) flow rate and growth time, the thickness of each staircase was controlled at 10 nm, meaning that the total thickness of each EBL was approximately 30 nm, similar to that of the conventional p-$Al_{0.21}Ga_{0.79}N$ EBL, and the p-type doping level for all EBLs are 1×10^{20} cm^{-3}. Finally, a 100-nm-thick p-GaN layer (doping level: 5×10^{19} cm^{-3}) was deposited on the EBLs in all samples. Schematic structures of the three LEDs are shown in Fig. 8.13. All the LED chips (dimensions: 1×1 mm^2) were fabricated using the standard process. All electrical characteristics were measured; however, using a semiconductor parameter analyzer (Agilent B1500A), with the output power measured using an integrating sphere (photodetector: Si for use as photoreceiver). To avoid the effect of heat on the EQE, a pulsed mode (width: 1 ms; duty cycle: 0.1 %) was employed to measure the output power. Moreover, SiLENSe software was employed to simulate the band structures of the LEDs.

Figure 8.14 displays the EQEs and forward voltages of the LEDs plotted with respect to the current density (up to 100 A cm^{-2}) for the conventional $Al_xGa_{1-x}N$

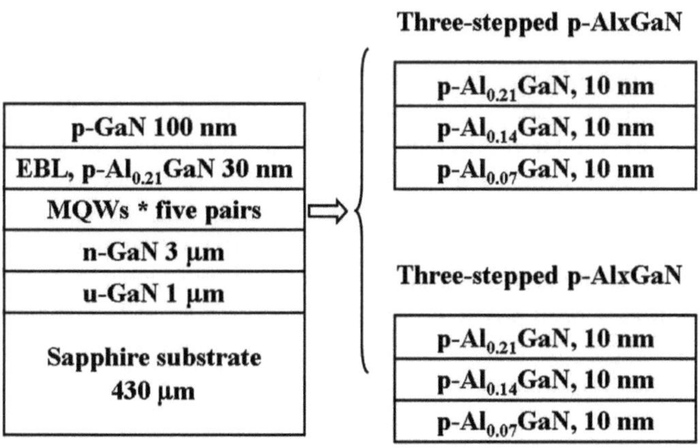

Fig. 8.13 Schematic representations of the reference structure and three-stepped p-AlxGaN EBLs [68]

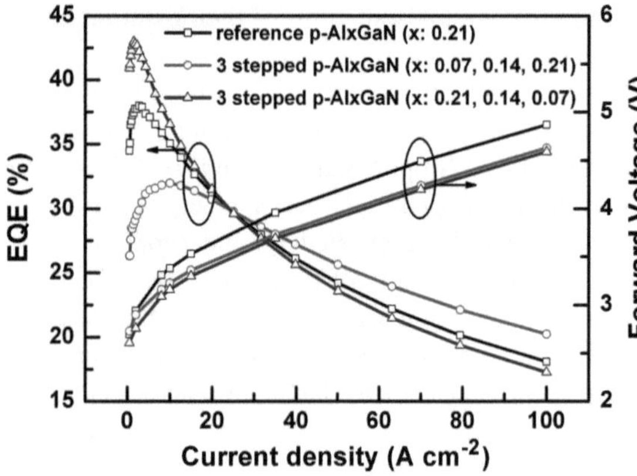

Fig. 8.14 EQEs and forward voltages, plotted with respect to current density, for the reference p-Al_xGaN EBL LED and the three-stepped p-Al_xGaN EBL LEDs [68]

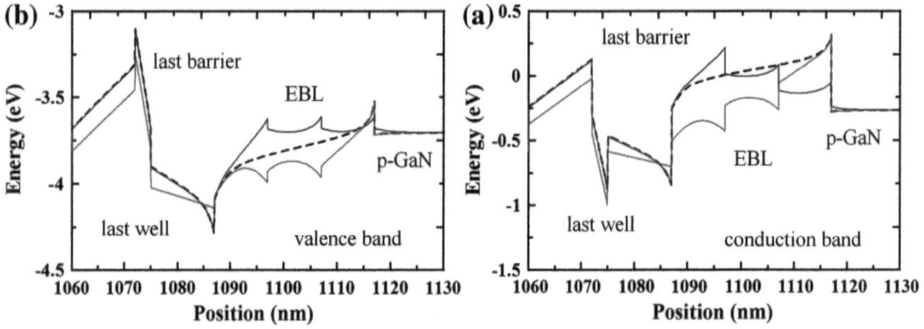

Fig. 8.15 Calculated **a** conduction and **b** valence band diagrams of the reference p-Al_xGaN EBL LED (*black dashed line*), the three-stepped p-Al_xGaN EBL LED (x: 0.07, 0.14, 0.21; *red line*), and the three-stepped p-Al_xGaN EBL LED (x: 0.21, 0.14, 0.07; *blue line*) at a current density of 2 A cm^{-2} [68]

EBL and the two $Al_xGa_{1-x}N$ staircase EBLs. The forward voltages for the two $Al_xGa_{1-x}N$ staircase-EBL LEDs were lower than those of the conventional $Al_xGa_{1-x}N$ EBL LED over the entire range of current densities, presumably because the lower average Al contents in the two staircase-EBL LEDs decreased the potential barriers at the MQWs-EBL interfaces. Furthermore, the LED incorporating the composition-stepped p-$Al_xGa_{1-x}N$ EBL (x: 0.21, 0.14, 0.07) exhibited the highest peak EQE, up to 43 % at a current density of 1.5 A cm^{-2}. In contrast, the LED featuring the composition-stepped p-$Al_xGa_{1-x}N$ EBL (x: 0.07, 0.14, 0.21) exhibited the lowest peak EQE, up to 32 %, at a current density of 10 A cm^{-2}, which the peak delay was significantly improved than those of the conventional $Al_xGa_{1-x}N$ EBL LED and composition-stepped p-$Al_xGa_{1-x}N$ EBL (x: 0.21, 0.14, 0.07).

To determine the mechanisms behind these phenomena, Fig. 8.15 presents the calculated band diagrams for these samples at a current density of 2 A cm^{-2}.

In contrast to the LED featuring the composition-stepped p-$Al_xGa_{1-x}N$ EBL (x: 0.07, 0.14, 0.21), the other two LEDs exhibited a sharp notch at the interface between the last barrier and the EBL, due to the stronger piezoelectric field of GaN relative to AlGaN. Meanwhile, the accumulated holes at the notch formed between EBL and p-GaN can easily tunnel into MQW region via EBL with the assistance of intermediate states in EBL, even though EBL appears to block holes in the valence band [69]. For the LED incorporating the composition-stepped p-$Al_xGa_{1-x}N$ EBL (x: 0.21, 0.14, 0.07), which had the sharpest V-shaped notch, thereby resulting in the higher EQEs at low current densities. The LED with the composition-stepped p-$Al_xGa_{1-x}N$ EBL (x: 0.07, 0.14, 0.21), however, exhibited lower EQEs at low current densities. Upon increasing the current density to 25 A cm^{-2} (see Fig. 8.14), the EQEs for the LED featuring the composition-stepped p-$Al_xGa_{1-x}N$ EBL (x: 0.07, 0.14, 0.21) began to surpass those of the other LEDs. Moreover, at a current density of 35 A cm^{-2}, we estimated the efficiency droops for the LEDs incorporating the composition-stepped p-$Al_xGa_{1-x}N$ EBL (x: 0.21, 0.14, 0.07), the reference LED (x = 0.21), and the LED featuring the composition-stepped p-$Al_xGa_{1-x}N$ EBL (x: 0.07, 0.14, 0.21) to be 17, 11, and 4 %, respectively, relative to their peak values.

Furthermore, Fig. 8.16 reveals interesting phenomena in the calculated band diagrams for the three LEDs at a current density of 100 A cm^{-2}. The band-bending of the conduction band (Fig. 8.16a) became more severe than that at a current density of 2 A cm^{-2} (Fig. 8.15a), presumably because of the larger electric field when the current density reached 100 A cm^{-2}. According to the calculated conduction band diagram for our particular EBL design, the LED incorporating the composition-stepped p-$Al_xGa_{1-x}N$ EBL (x: 0.07, 0.14, 0.21) exhibited the strongest electron confinement, followed by the reference LED (x = 0.21), with the LED featuring the composition-stepped p-$Al_xGa_{1-x}N$ EBL (x: 0.21, 0.14, 0.07) displaying the poorest electron confinement. These results can be explained by considering the degrees of electron leakage across the EBLs (simulation data not shown here). Moreover, Fig. 8.16b reveals severe downward band-bending at the interface

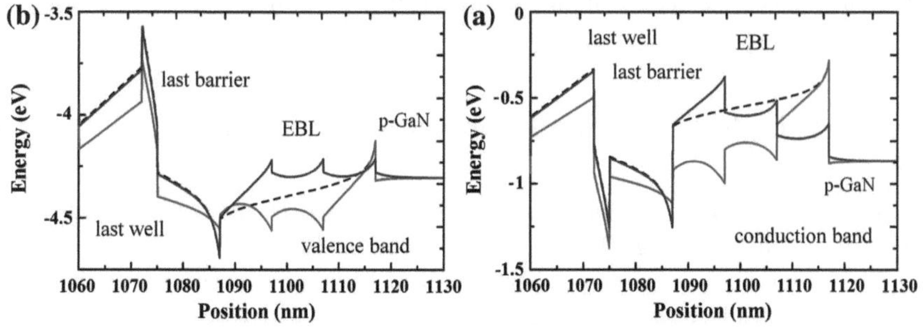

Fig. 8.16 Calculated **a** conduction and **b** valence band diagrams of the reference p-Al_xGaN EBL LED (*black dashed line*), the three-stepped p-Al_xGaN EBL LED (x: 0.07, 0.14, 0.21; *red line*), and the three-stepped p-Al_xGaN EBL LED (x: 0.21, 0.14, 0.07; *blue line*) at a current density of 100 A cm^{-2} [68]

between the last barrier and the EBL for both the reference LED (x = 0.21) and the LED incorporating the composition-stepped p-$Al_xGa_{1-x}N$ EBL (x: 0.21, 0.14, 0.07). In the LED featuring the composition-stepped p-$Al_xGa_{1-x}N$ EBL (x: 0.07, 0.14, 0.21), the last barrier against the first-step EBL was p-$Al_{0.07}Ga_{0.93}N$, the lower Al composition of which meant that less lattice mismatch existed between the last GaN barrier and the AlGaN EBL. Therefore, the piezoelectric field at the interface would decrease, thereby, mitigating the degree of band-bending. While at high current density, the tunneling process of holes can be negligible. As discussed above, the LED incorporating the composition-stepped p-$Al_xGa_{1-x}N$ EBL (x: 0.07, 0.14, 0.21) possessed a flat valence band and a much lower value of $\triangle E_V$ at the interface between the last barrier and the EBL; as a result, holes could pass more readily through the interface into the active region. Therefore, the recombination rate was enhanced at high current densities.

Theoretically, the amount of the piezoelectric polarization in the direction of the c-axis can be determined using the equation [70]:

$$P_{PE} = 2\frac{a - a_0}{a_0}(e_{31} - e_{33}\frac{C_{13}}{C_{33}}),$$

where a_0 is the equilibrium value of the lattice parameter, e_{31} and e_{33} are piezoelectric coefficients, and C_{13} and C_{33} are elastic constants. Because the value of $[e_{31} - e_{33}(C_{13}/C_{33})]$ was less than zero for AlGaN over the entire range of compositions, the piezoelectric polarization was negative for tensile strain (i.e., from GaN to AlGaN) and positive for compressive strain. In addition, for the Ga-face heterostructure, spontaneous polarization is negative, directing to the substrate as well [71]. Therefore, the total polarization (piezoelectric polarization plus spontaneous polarization) would induce the opposite electric field toward p-GaN and resist the transport of holes to the active layer. According to the formula above, the low lattice mismatch (i.e., from GaN to $Al_{0.07}Ga_{0.93}N$), relative to those of the reference LED (x = 0.21) and the LED incorporating the composition-stepped p-$Al_xGa_{1-x}N$ EBL (x: 0.21, 0.14, 0.07), would result in a lower degree of piezoelectric polarization between the last barrier and the EBL. Thus, the hole injection efficiency would increase, resulting in higher EQE performance.

8.3 LED Light Extraction Improvement

In recent years, LED dies have been extensively used in application about back light of monitor and lighting. In the application, higher brightness is always required. And therefore higher optical power of LED die is necessary. In the ideal LED die, each electron produces one photon in the active region and all photons emitted by the active region are also emitted into free space. However, in a real LED die, not all photons emitted from the active region are emitted into free space. Some photons may never leave the LED die to result in lower optical power. For example, in Fig. 8.17a photons generated in the active region tend

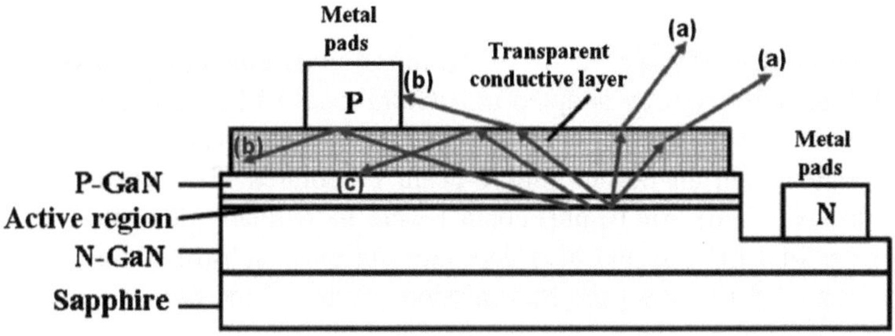

Fig. 8.17 Schematic drawings of **a** light partially absorbed, **b** light hindered, **c** light of total reflection in GaN-based LED die

to be partially absorbed by transparent conductive layer, which is on top side of epitaxial layer in GaN-based LED die [72]. And in GaN-based LED die, there are usually two top metal pads for wire bonding and top metal fingers for current spreading. However, in Fig. 8.17b such metal pads and fingers will hinder the extraction of light radiating from the active region. Furthermore the phenomenon of total internal reflection in the interface between LED die and air space in Fig. 8.17c, also referred to as the trapped light phenomenon, reduce the ability of the light to escape from the LED die [73].

The light extraction efficiency is defined as

$$\eta_{\text{extraction}} = \frac{\text{number of photons emitted into free space per second}}{\text{number of photons emitted from free active region per second}} \quad (8.1)$$

Thus higher optical power of LED die could be achieved by increasing light extraction efficiency in LED die.

8.3.1 Improvement in Transparent Conductive Layer with ITO Material and Patterned Structure

Once upon a time, most commercial nitride-based LED dies used semitransparent metal layer (i.e., Ni–Au) as transparent conductive layer. But the transmittance of the metal layer for LEDs is an important issue to consider since photons generated in the active region tend to be partially absorbed by this metal layer. It has been reported that the transmittance of conventional p-metal layer (i.e., Ni–Au) is only around 60–75 %. Hence, by increasing the transmittance of p-metal layer, it is expected that the LED output intensity should have been enhanced accordingly. Thus the semitransparent Ni–Au can be replaced by the transparent indium-tin-oxide (ITO) as the p-contact material. It is well known that ITO is a hard and chemically inert transparent material with high electrical conductivity and low optical absorption coefficient. Therefore, it is considered to be a good candidate for using as transparent contacts to many optoelectronic devices [74]. All of these advantages render ITO an attractive material for fabricating nitride-based LEDs. In fact, ITO has already been adopted in AlGaInP-based LEDs and AlInGaAs vertical-cavity surface-emitting lasers as transparent current spreading layers [75, 76].

For comparison, an 80-nm-thick e-beam evaporated ITO film and a thermal evaporated Ni(5 nm)–Au(10 nm) contact were individually deposited on the epitaxial layer of LED die and also deposited on glass substrates for transmission studies. Figure 8.18 shows the transmission spectra of the two different kinds of contact layers. In this figure, the transmittance of each film was normalized with respect to the transmittance of the glass substrate. It can be seen clearly that ITO was substantially more transparent than Ni–Au. At 465 nm, the transmittance of Ni–Au is only 63.7 % while the transmittance of ITO is able to reach 93.2 %; the results suggest that ITO indeed is an optical material suitable for use as upper contacts for nitride-based LEDs.

In the inset of Fig. 8.19, the electroluminescence (EL) spectra obtained for LEDs with ITO and Ni–Au individually on epitaxial layer, all measured at 20 mA,

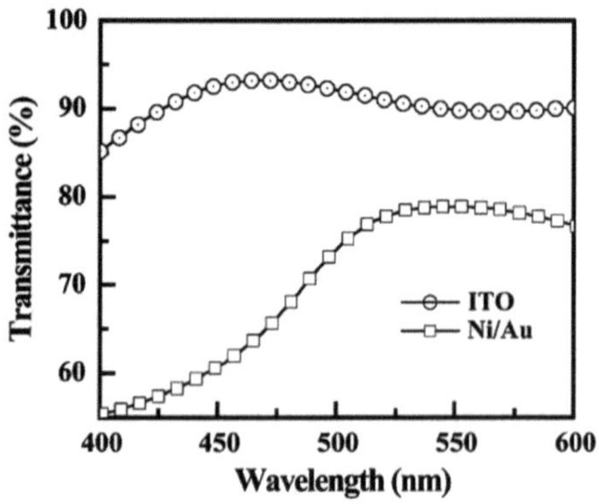

Fig. 8.18 Normalized optical transmittance as a function of wavelength for Ni(5 nm)–Au(10 nm) and ITO(80 nm) [72]

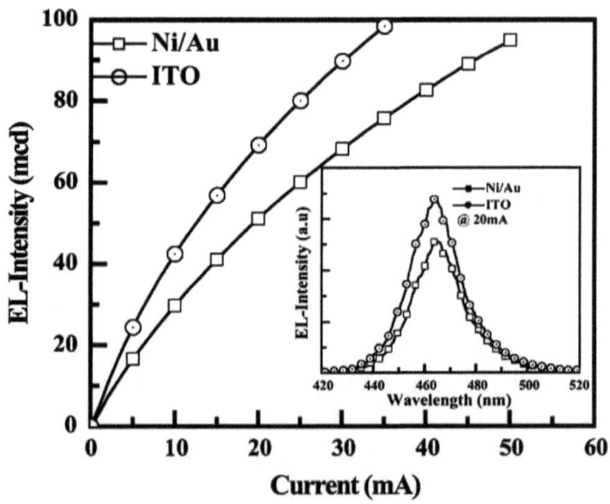

Fig. 8.19 L-I characteristics of LEDs with ITO and Ni–Au individually on epitaxial layer. The inset shows 20-mA EL spectra of LEDs with ITO and Ni–Au individually on epitaxial layer [72]

are depicted. It is clearly shown that even though the EL peak positions of these two LEDs were nearly identical, the EL intensity of the LED with ITO on epitaxial layer was about 30 % larger than that of the LED with Ni–Au on epitaxial layer. In Fig. 8.19, the intensity-current (L-I) characteristics of these two LEDs were compared. Again, that EL intensity of the LED with ITO on epitaxial layer was still larger than that of LED with Ni–Au on epitaxial layer. When a 20-mA current injection was administered, the EL intensity of LED with Ni–Au on epitaxial layer was measured to be 52 mcd. In contrast, the EL-intensity reaches 69 mcd for the LED with ITO on epitaxial layer. The output power of lamped LEDs with ITO p-contact was found to be 8.4 mW and at 20 mA, which is substantially larger than the other normal Ni–Au LEDs (i.e., 4.1 mW). A significant enhancement in EL intensity can be explained by the highly transparent nature of ITO, as compared to Ni–Au. As a result, more photons are emitted from the devices [72].

However, it has also been shown that good ohmic contact is difficult to achieve for ITO deposited on p-GaN [77–79]. In some study [80], the as-deposited 250 nm ITO films onto p-GaN in situ annealed in the RF sputtering chamber at 250 °C for 30 min in vacuum has improved electrical and optical property. Figure 8.20 shows optical transmittance as a function of wavelength of these contacts. It was also found that the in situ annealing could further increase the transmittance of the deposited ITO films. At 450 nm, the transmittance of in situ annealed ITO could reach 97.8 %. The high 97.8 % transmittance observed from the in situ annealed ITO films suggests ITO is indeed suitable optically to serve as the upper p-contact for nitride-based LEDs. The electrical properties of the ITO layers are also important. By using the Lehighton contactless measurement system, we found that the sheet resistance of as-deposited ITO films and in situ annealed ITO films was 25 Ω/□ and 15 Ω/□, respectively. Such a result suggests that post-deposition in situ annealing could not only improve the optical transmittance of ITO film but also could reduce its sheet resistance, which is important for the current spreading of LEDs.

Figure 8.21 shows the current-voltage (I-V) characteristic of the LEDs with different p-contacts. It can be seen that the LED forward voltage measured with a 20 mA current injection was 3.16, 5.74 and 4.28 V for the LEDs with Ni/Au,

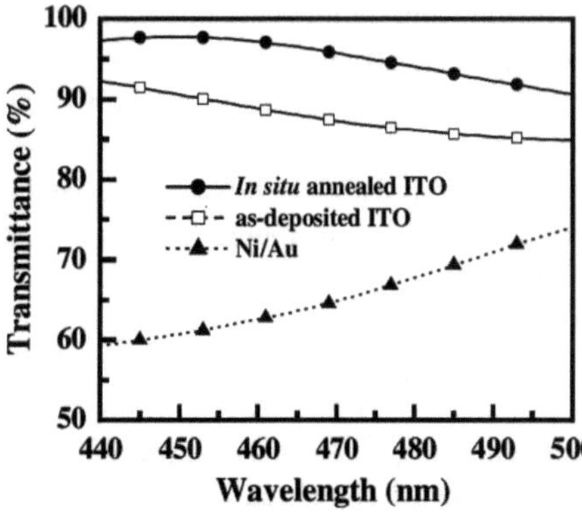

Fig. 8.20 Optical transmittance as a function of wavelength for Ni(5 nm)/Au(10 nm), as-deposited ITO(250 nm) and in situ annealed ITO(250 nm) films. In this figure, the transmittance of each film was corrected taking into account the absorption of the glass substrate [81]

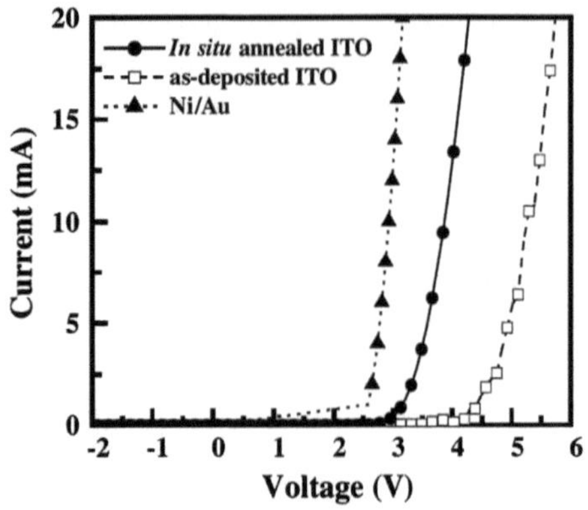

Fig. 8.21 I–V characteristics of LEDs with Ni/Au, as-deposited ITO and in situ annealed ITO p-contacts [81]

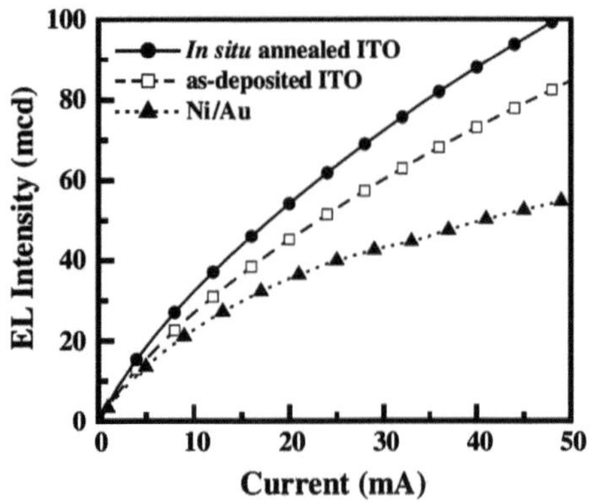

Fig. 8.22 L-I characteristics of LEDs with Ni/Au, as-deposited ITO and in situ annealed ITO p-contacts [81]

as-deposited ITO and in situ annealed ITO p-contact layer, respectively. It was found that we could effectively reduce the LED operation voltage by post deposition in situ annealing of ITO. It is believed such a decrease is due to the barrier height lowing of ITO on p-GaN after in situ annealing.

Figure 8.22 shows the EL intensity as a function of injection current of nitride-based LEDs with Ni/Au, as-deposited ITO and in situ annealed ITO p-contacts. With the same amount of injection current, it can be seen that the LED with in situ annealed ITO p-contact has the largest output EL intensity. With a 20 mA current injection, it was found that the EL intensity of LED with in situ annealed ITO p-contact was 55.4 mcd, which was much larger than the 35.4 mcd EL intensity observed from LED with Ni/Au p-contact. Such a significant increase in EL intensity can again be attributed to the more transparent nature of the in situ annealed ITO upper p-contact layer.

Besides, growing an n -InGaN-GaN short-period-superlattice (SPS) tunneling contact on top of the p-GaN cap layer could further reduce forward voltage with ITO contact [72]. It was found that we could achieve a nonohmic contact with a rectifying property by depositing ITO directly onto p-GaN. However, a rectifying contact can be converted into an ohmic contact with a specific contact resistance of 1.6 × $10^{-3}\Omega$ cm^2, if a Si-doped n -InGaN-GaN SPS structure is inserted between ITO and p-GaN layers. Even though the specific contact resistance of ITO on n -InGaN-GaN SPS structure is still higher than that of Ni–Au on n -SPS structure; nevertheless, this combination is still electrically plausible to be used as an upper contact to the nitride-based LEDs. Figure 8.23 depicts the I-V characteristic of the fabricated nitride-based LEDs. It was found that the operation voltage measured from the LED with ITO contact on p-GaN was much higher than those measured from the other three LEDs. It was also learned that inserting an n -SPS structure could significantly reduce the LED operation voltage, particularly for devices with ITO contact. The 20-mA forward voltage measured from LEDs with ITO on p-GaN, Ni–Au on p-GaN, ITO on n -SPS, and Ni–Au on n -SPS upper contacts was 6.01, 3.85, 3.24, and 3.07 V, respectively. When compared to the LED with Ni–Au on n -SPS upper contact, a slightly larger forward voltage at 20 mA was measured from the LED with ITO on n -SPS upper contact. However, a better electrical performance can still be obtained from LEDs with ITO on n -SPS upper contacts, in comparison with the conventional nitride-based LEDs with Ni–Au on p-GaN upper contacts.

According to Snell's law, the critical angles of total reflection at ITO/GaN interface and air/ITO interface are around 57° and 28°, respectively. Thus, a significant amount of photon will be reflected at the sample surface for conventional ITO LEDs, as shown in Fig. 8.24a. With the textured ITO surface, photons generated in the multi—quantum—well region should experience multiple scattering at the sample surface and could escape from the device easily [81], as shown in Fig. 8.24b. Thus, we can achieve larger output powers from the LEDs with imprinted ITO electrode.

Compared with conventional photolithography, imprint lithography is simple, low cost, and mass producible [82]. Chang et al. performed imprint lithography to pattern the ITO layer [83]. Detailed proceeding steps for imprint lithography are shown in Fig. 8.25. Subsequently, dilute hydrochloric acid (HCl) is used to

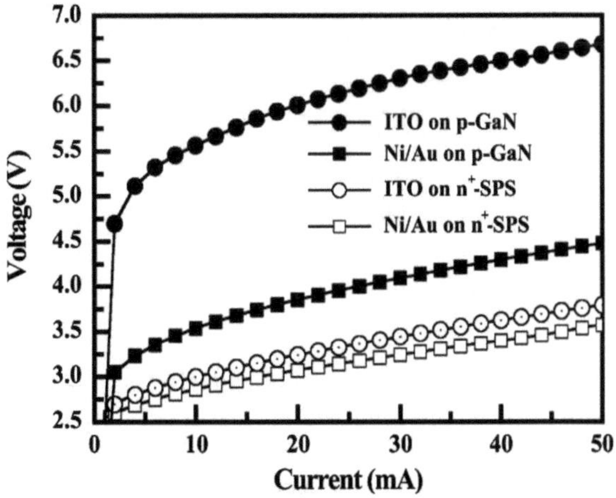

Fig. 8.23 I–V characteristic of the fabricated nitride-based LEDs with different upper contacts [72]

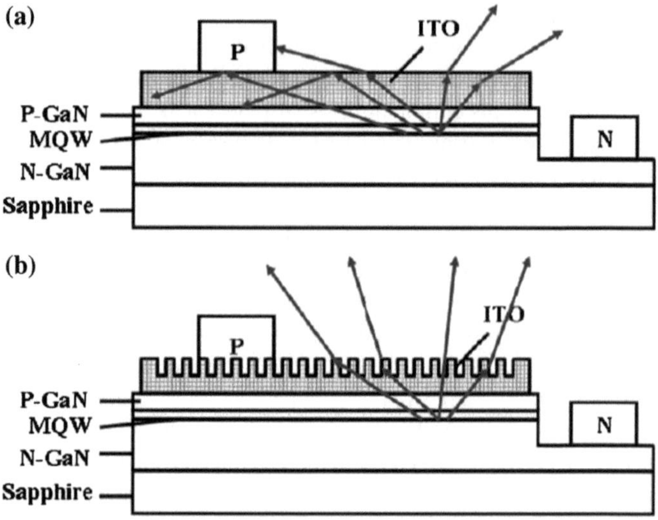

Fig. 8.24 Schematic drawings of **a** conventional ITO LED and **b** LED with imprinted ITO electrode [84]

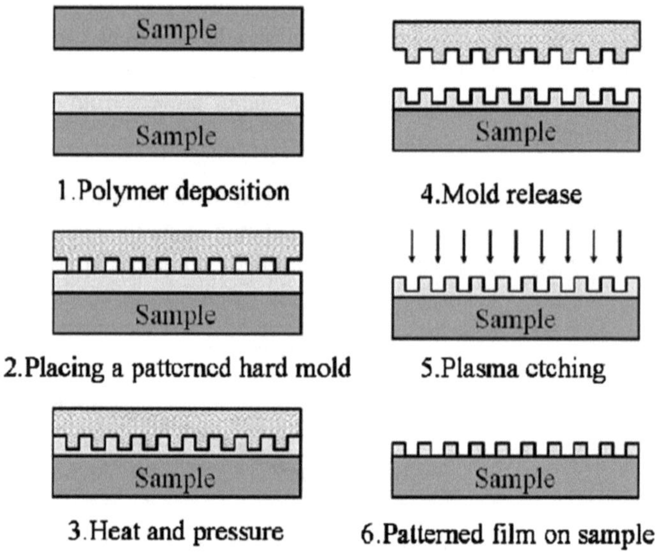

Fig. 8.25 Detailed proceeding steps for the imprint lithography [84]

partially etch the exposed ITO. Finally, acetone was used to remove the PMMA (Polymer) on the patterned ITO layer. Two circular patterns were used for imprint lithography. For the first pattern, the circular hole diameter and the spacing between holes of the first pattern were 1.75 and 1 μm, respectively. For the second

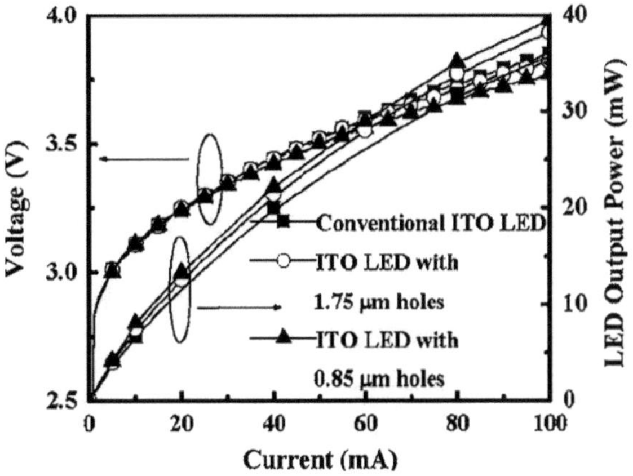

Fig. 8.26 Measured L-I-V characteristics of the fabricated LEDs [84]

pattern, the circular hole diameter and the spacing between holes were 0.85 and 0.5 μm, respectively.

Figure 8.26 shows measured intensity-current-voltage L-I-V. characteristics of the fabricated LEDs. Under 20 mA current injection, it was found that forward voltages were 3.24, 3.25, and 3.24 V for the conventional ITO LED, ITO LED patterned with 1.75 μm holes, and ITO LED patterned with 0.85 μm holes, respectively. The almost identical forward voltages can be attributed to the same epitaxial layers used in these devices. It also indicates that imprint lithography will not degrade the electrical properties of the LEDs. It was also found that output power increased with the injection current and no intensity saturation was observed up to 100 mA for all these three LEDs. Furthermore, it was found that we achieved the largest output power from the ITO LED patterned with 0.85 μm holes, followed by the ITO LED patterned with 1.75 μm holes while the output power of the conventional ITO LED was the smallest. With 20 mA injection current, it was found that the LED output powers were 11.7, 12.6, and 13.3 mW for the conventional ITO LED, ITO LED patterned with 1.75 μm holes, and ITO LED patterned with 0.85 μm holes, respectively. In other words, output powers could be enhanced 12 % by using the simple imprint lithography to pattern 0.85 μm holes on the ITO electrode.

8.3.2 Enhanced Light Extraction Efficiency Using Flip-Chip Structure

For top emitting III-N LED die with upper contacts, a significant amount of photons will be obscured by the bonding pads and metal fingers of the devices. One effective way to further increase LED output power is to use flip-chip technology [84–86]. As shown in Fig. 8.27, we should be able to achieve a larger

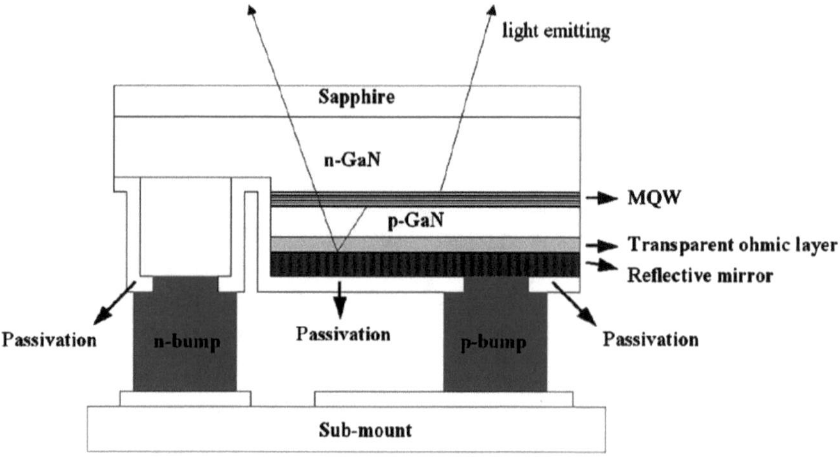

Fig. 8.27 Schematic diagrams of flip-chip LED with transparent ohmic contact and reflective mirror [88]

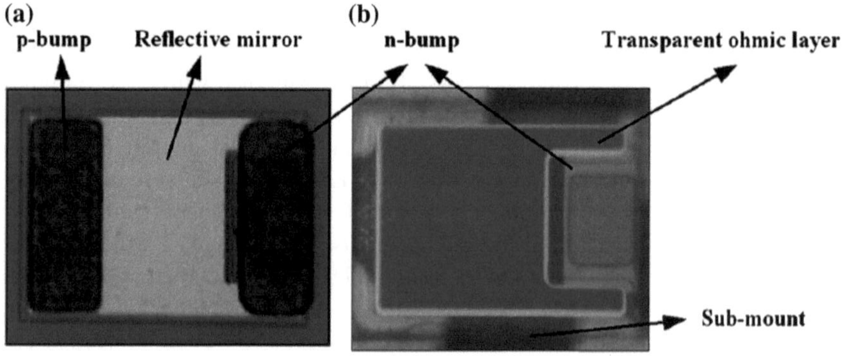

Fig. 8.28 Photographs of the chip **a** after electroplating and lift-off, and **b** after flip-chip [88]

output power with flip-chip technology since no bonding pads or wires exist on top of the devices so that photons could be emitted freely from LED die to air [87]. Figure 8.28a shows photograph of the entire chip after electroplating and lift-off. As shown in this figure, the two Sn/Au bumps were well defined after lift-off. Figure 8.28b shows photograph of the LED after flip-chip. And ITO/Ni (LEDI Flip-Chip), ITO (LEDII Flip-Chip), Ni (LEDIII Flip-Chip) were used as transparent ohmic layer and Ag layer was used as reflective mirror in nitride-based LED die. For comparison, conventional LED die with ITO (LEDIV Non-Flip-Chip) and Ni/Au (LEDV Non-Flip-Chip) upper contacts were also fabricated.

Figure 8.29 shows measured intensity-current-voltage L-I-V characteristics of the fabricated devices [88]. With 20 mA current injection, it was found that output powers were 21.0, 14.3, 16.0, 9.1, and 5.6 mW for LEDI, LEDII, LEDIII, LEDIV

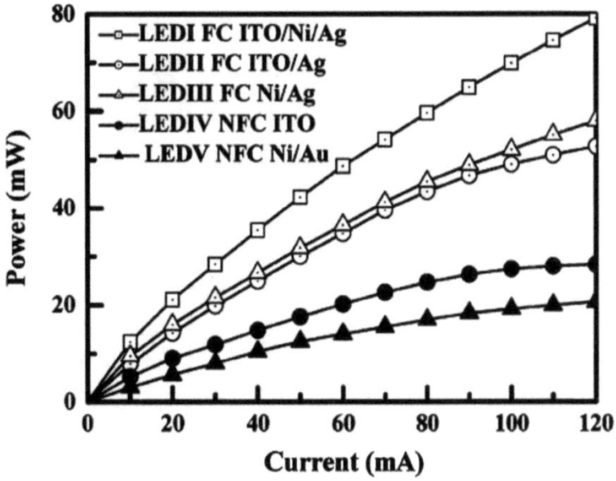

Fig. 8.29 Measured L-I characteristics of the fabricated devices [88]

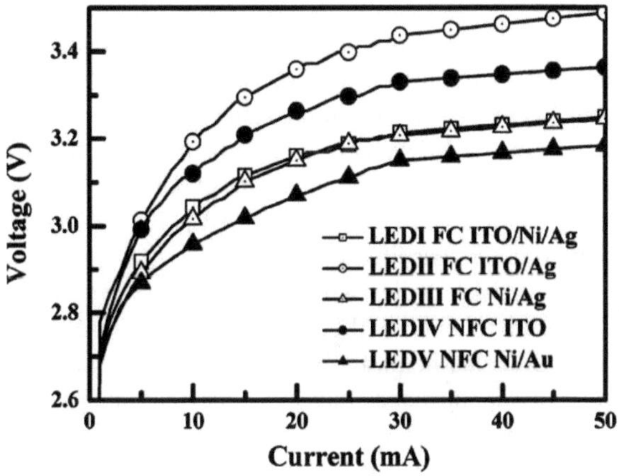

Fig. 8.30 Measured I-V characteristics of the fabricated LEDs [88]

and LEDV, respectively. In other words, FC LEDI with ITO/Ni/Ag is 3.75 times brighter than the conventional NFC LEDV with Ni/Au upper contact.

Figure 8.30 shows measured I-V characteristics of the fabricated LEDs. With 20-mA current injection, it was found that forward voltages were 3.16, 3.36, 3.15, 3.26, and 3.07 V for LEDI, LEDII, LEDIII, LEDIV and LEDV, respectively. It can be seen that the FC LED with ITO/Ni/Ag (i.e., LEDI) exhibited smaller operation voltage than NFC LED with ITO upper contact (i.e., LEDIV) and the voltage differences would be more obvious at higher curret injection. This is because FC LED with a large Ni-containing p-contact layer of lower sheet resistances, covering the entire p-type mesa can indeed provide more uniform current spreading and lower contact resistances.

Figure 8.31 shows life tests of relative luminous intensities measured from the fabricated LEDs, normalized to their respective initial readings. During life test, all five LEDs were driven by 30 mA current injection at room temperature. After

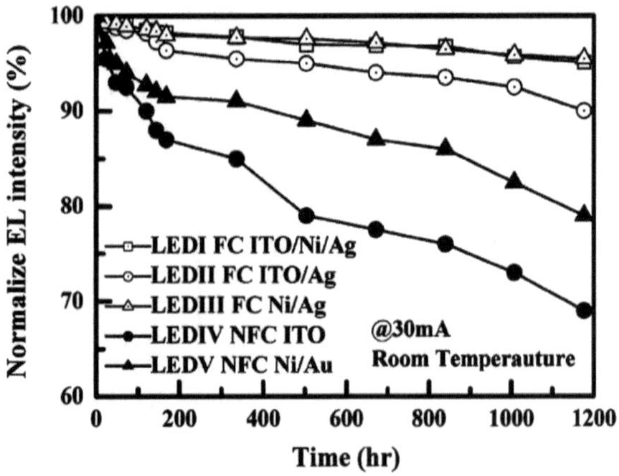

Fig. 8.31 Life tests of relative luminous intensities measured from the fabricated LEDs, normalized to their respective initial readings. During life test, all five LEDs were driven by 30-mA current injection at room temperature [88]

1,200 h, it was found that luminous intensities of LEDI, LEDII, LEDIII, LEDIV and LEDV decreased by 5.0, 10.0, 4.5, 29, and 20 %, respectively. It was found that FC LED dies were more reliable. It is known that the most of the heat is generated in the MQW active region. Since thermal path between the MQW active region and heat sink is much shorter for FC LED dies, most heat could flow away more easily, compared with NFC LED dies. With less thermal effect, FC LED dies achieved a longer lifetime.

For FC LED dies with flat surface, a significant portion of the light emitted from active region will be reflected at sapphire/air interface due to total reflection. This issue can be partially solved by either shaping or texturing the backside surface of sapphire substrate [89–92]. To shape or texture sapphire substrate, however, one needs to either use an inductively coupled plasma (ICP) etcher for dry etching or immerse sapphire into $H_2SO_4 + H_3PO_4$ solution at high temperatures for wet etching. It is also necessary to deposit a hard mask layer onto the sample and perform photolithography prior to etching. These processes are complex, costly, and time consuming. But we could use easier process of grinding to thin the sapphire substrate without polishing to rough the backside surface of sapphire substrate [93].

For the backside roughened FC LEDs (i.e., LED-I), sapphire substrate was thinned to around 90 μm by a SHUWA SGM-8000 grinding system. It should be noted that no polishing was performed after grinding. The conventional process of lapping and polishing were used to prepare FC power LEDs with flat backside surface (i.e., LED-II). For comparison, NFC power LEDs with Al backside reflector (i.e., LED-III) were also prepared. Figure 8.32a, b show schematic diagrams of LED-I and II, respectively.

Figure 8.33 shows measured output power as a function of injection current for these three LEDs. Under the same injection current, it was found that output powers observed from the two FC LEDs (i.e., LED-I and II) were larger than that observed from the NFC LED (LED-III), particular under high-current injections.

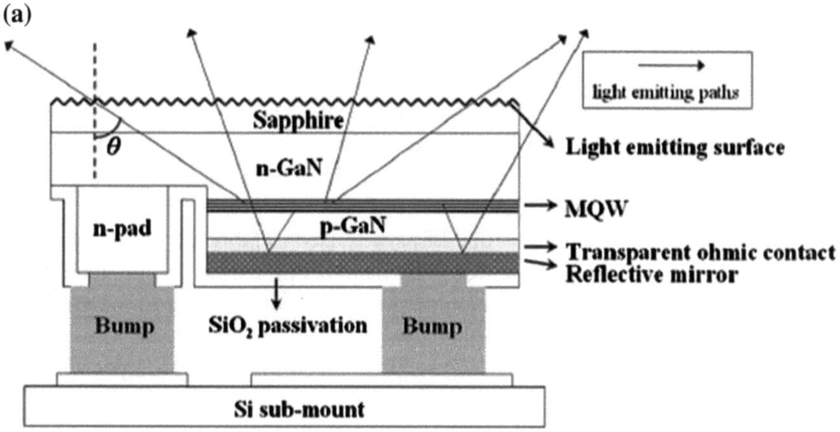

Fig. 8.32 Schematic diagrams of **a** LED-I with rough sapphire backside surface and **b** LED-II with conventional flat sapphire backside surface [94]

Fig. 8.33 Measured output power as a function of injection current for these three LEDs [94]

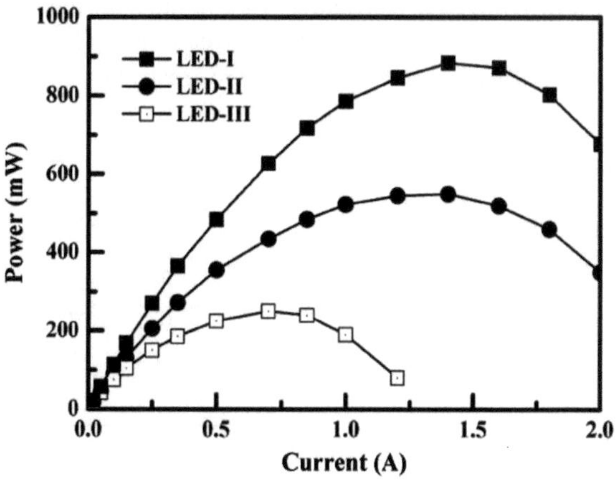

It should be noted that the maximum output intensity occurred at 1,400 mA for the two FC LEDs while occurred at 700 mA for the NFC LED. These results should be attributed to the much better thermal property of the FC LEDs. Under 350 mA current injection, it was found that output powers were 366.5, 271.8, and 185.1 mW for LED-I, II, and III, respectively. In other words, we can increase the output power of the power FC LEDs by about 35 % by roughening the backside surface of the sapphire substrate.

Further, technology of double-side PSSs was used to fabricate nitride-based FC LEDs to achieve larger output power [94]. Samples processed by this technology were all grown by metal-organic chemical vapor deposition. Prior to the growth of epitaxial layers, we first patterned the sapphire substrates by an inductively coupled plasma (ICP) etcher. During ICP etching, a 100-nm-thick Ni layer was used as the etching mask while Cl, BCl, and Ar were used as the etching gases. We designed our mask and controlled the etching parameters so as to achieve a periodic hole pattern with hole etching depth, hole diameter, and spacing of 1, 3, and 3 μm, respectively. We then removed the Ni mask to complete the preparation of PSS. After the growth of epitaxial layers and process, the epitaxial wafers were then lapped down to 150 m. We then used photolithography and ICP etching again to define textures on the backside surface of sapphire substrates. Using the same Ni mask and etching parameters, we patterned the sapphire backside surface with hole etching depth, hole diameter, and spacing of 1, 3, and 3 m, respectively. It should be noted that the hole patterns on the backside surface were not intentionally aligned with the PSS hole patterns. Schematic diagram of the fabricated FC LED with double-side PSS is shown in Fig. 8.34. For comparison, conventional FC LED without any patterning and FC LED prepared on PSS were also fabricated using exactly the same epitaxial layers.

Figure 8.35 shows room-temperature intensity-current-voltage (L-I-V) characteristics of the fabricated LEDs. Under the same current injection, it was also found that we achieved the largest output intensity from the FC LED with double-side PSS, followed by the FC LED on PSS. On the other hand, output power of the conventional FC LED was the smallest among the three devices. With 350-mA injection current, it was found that the LED output powers were 79.3, 98.1, and 121.5 mW for the conventional FC LED, FC LED prepared on PSS, and FC LED with double-side PSS, respectively. It should be noted that the peak

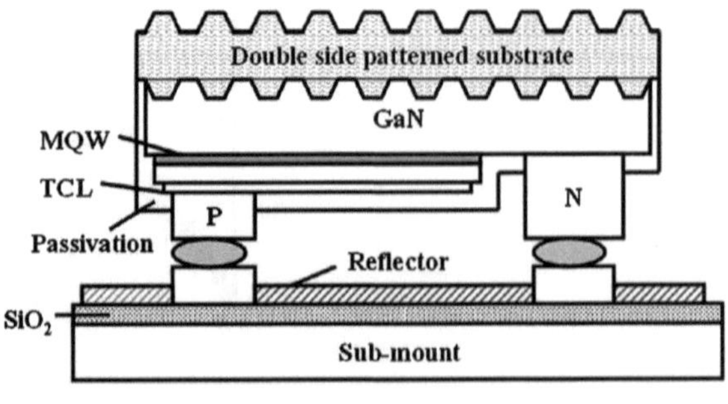

Fig. 8.34 Schematic diagram of the fabricated FC LED with double-side PSS [92]

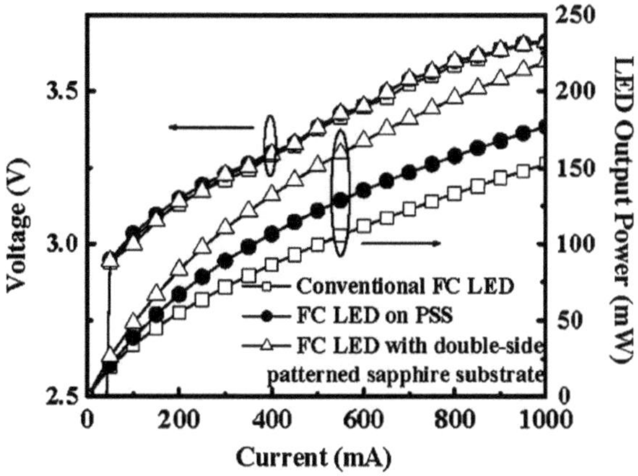

Fig. 8.35 Room-temperature L-I-V characteristics of the fabricated LEDs [92]

electroluminescence (EL) wavelength of the three LEDs all occurred at around 458 nm. Compared with conventional FC LED without any patterning, the 24 % EL intensity enhancement observed from the FC LED on PSS should be attributed to reduced dislocation density in the epitaxial layer and to enhanced light scattering at GaN-sapphire interface [95–98]. On the other hand, we can further improve the LED output intensity by texturing the sapphire bottom surface so that photons can experience multiple scattering and escape easily from the roughened sapphire surface, as shown in Fig. 8.35. It should be noted that the 121.5-mW output intensity observed from the FC LED with double-side PSSs is 53 % larger than that observed from the conventional FC LED. It should be noted that no optimization was performed in the current study. By optimizing the etching geometry (i.e., hole etching depth, hole diameter, and spacing) on both sides of the sapphire substrate, we should be able to further enhance the LED output intensity.

8.3.3 High-Efficiency LED Chip with High-Voltage Structure

There is conventionally one p-n junction in one LED die. This LED die has size of 500 μm × 1,000 μm and has 3 V at 120 mA for example, shown in Fig. 8.36. However longer path of light emitted frequently results in total internal reflection and photons absorbed by epitaxial layer, shown in Fig. 8.37. And relatively large chip size also results in the phenomenon of current crowding, especially in high power LED die with larger chip size.

Therefore a novel chip structure of High-Voltage type is studied, shown in Fig. 8.38a. High-Voltage LED die is separated into three p-n junctions for example and three p-n junctions are connected electrically by metal layer. This High-Voltage LED die has chip size of 500 μm × 1,000 μm and has 9 V at 40 mA. Figure 8.38b shows photons could be emitted into free space from the side wall of

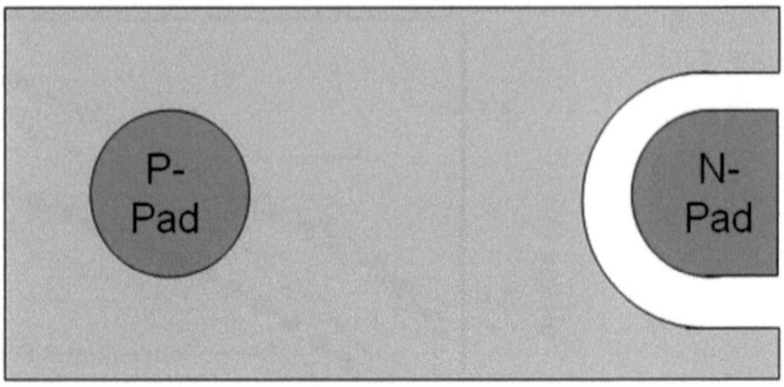

Fig. 8.36 Top-view diagram of convenional LED die

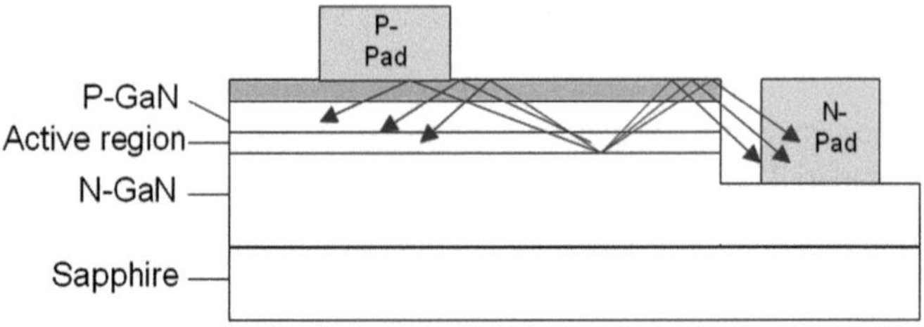

Fig. 8.37 Schematic drawings of conventional LED die

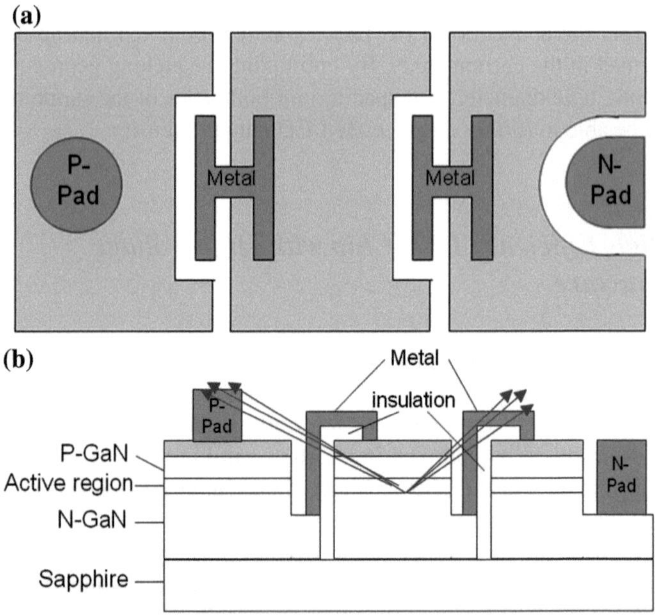

Fig. 8.38 **a** Top-view diagram of high-voltage LED die. **b** Schematic drawings of high-voltage LED die

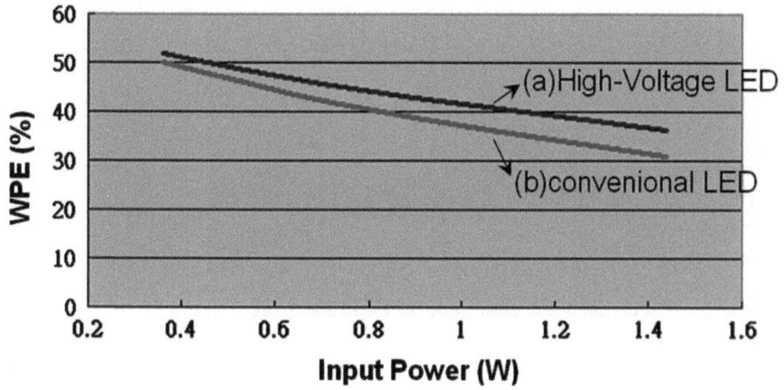

Fig. 8.39 WPE (%) of **a** high-voltage LED and **b** conventional LED at various input power

p-n junction due to shorter path of lighting. Thus the phenomenon of total internal reflection and photons absorbed by epitaxial layer will be avoided and light extraction efficiency will increases. Besides smaller size of every p-n junction will decrease deviation of current flowing path to let current spread more uniformly. It could lower operation voltage.

Figure 8.39 shows Wall Plug Efficiency (WPE) of High-Voltage LED die and conventional LED die at various input power. At 0.36 W of input power, High-Voltage LED and conventional LED have WPE of 52 and 50 %, individually and WPE of High-Voltage LED is higher 4 %, compared to conventional LED. We could also find conventional LED has droop effect more seriously at higher input power than High-Voltage LED. It is due to High-Voltage LED has lower operation voltage, higher light extraction efficiency, and more uniform current spreading.

In application, less IC drivers, converters, and transformers are needed to package with High-Voltage LED dies due to high-voltage property of LED die. On the other hand, because p-n junctions of High-Voltage LED are connected by deposited metal of semiconductor process, less processes of wire bonding are needed to package LED. Therefore High-Voltage LED chips could help the LED package house decrease cost of products.

8.4 LED Package for Better CCT, UV LED and QDs Application

In the modern LED system, one important but often overlooked characteristics is its package. The technologies applied for packages can affect the final cost of the module, and the end-user experiences. So it is important to develop an easy, cost-effective, and efficient packaging platform. As the LEDs are evolving into various applications, so are their package technologies. In a current LED system, as shown in Fig. 8.40, there are several parts that need to be addressed: (a) the lens for encapsulation and focusing (b) phosphor materials (c) LED chip (d) Submount

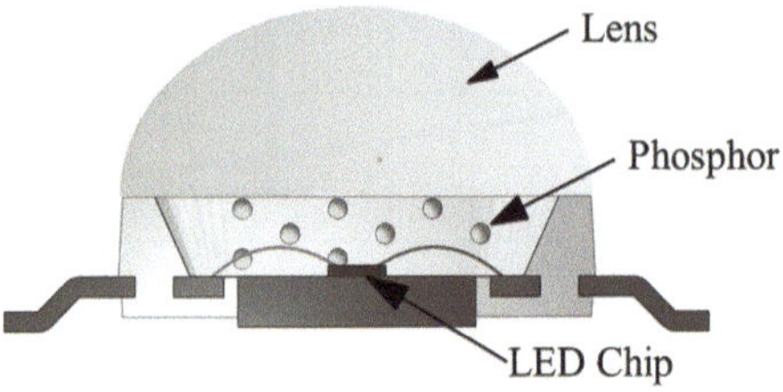

Fig. 8.40 The cross-sectional view of a generic LED package

and others. The LED chip issues are discussed in previous sections, and the remaining parts belong to the realm of packaging. In this section, several advanced concepts about LED packages will be reviewed, including the application of nanoparticles for light scattering, the elimination of non-uniform color temperature, and the introduction of highly efficient quantum dots as the packaging substances. Compared to traditional packaging concept, the progress of the nano-scale materials is really the true thrust of the next generation of packaging, and it marks another quantum jump of this field. In the following, several novel designs of the LED packaging technologies will be reviewed and discussed for their implication towards the actual module performances.

8.4.1 General Introduction of a LED Package

Recently, white light-emitting diodes (LEDs) have been regarded as the next-generation light sources due to the small size, environmental friendly process and high luminous efficiency, [99–101]. Generally, there are three methods to fabricate the white light LEDs including individual red, green and blue LEDs mixing, UV-LEDs exciting red, green and blue phosphors and blue-LEDs converting yellow phosphors, as shown in Fig. 8.41 [102]. Although the first two methods could fabricate the high CRI value white light, the high cost and low conversion efficiency become the serious disadvantages for high power white LEDs. Therefore, among different ways to produce the white light, combining the blue LED chip with the yellow phosphor is the most common method [103, 104].

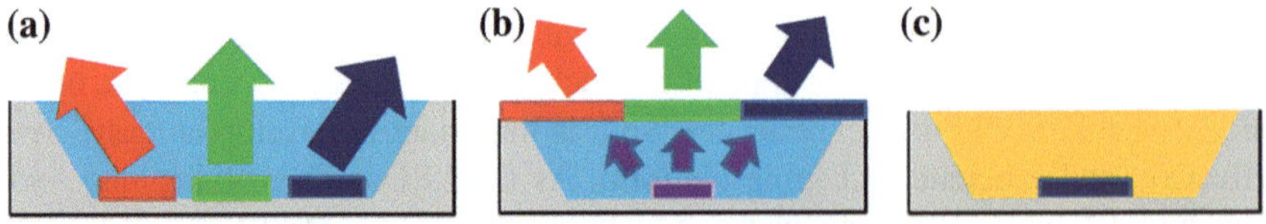

Fig. 8.41 Schematic diagrams of **a** individual *red*, *green* **b** UV-LEDs exciting *red*, *green* and **c** *blue*-LEDs converting *yellow* phosphors

The significant progress in phosphor-converted white LEDs had been strongly motivated by the advances in III-Nitride LEDs [105–113] serving as excitation sources. There are many new approaches of III-Nitride LEDs to reduce the charge separation issues in active region [105–107], efficiency-droop [108, 109], and the dislocation density in materials [109–113]. Furthermore, for the phosphors, there are also some novel materials developed to use in the solid-state lighting such as oxyfluoride [114], nitride [115], boride [116] and phosphide [117] hosts. In addition to the chip and material issues, the packaging engineering is becoming more and more important. From the package perspective, several issues can also directly affect the output of the white light LED. Among these issues, high luminous efficiency and uniformity of angular-dependent correlated color temperature (CCT) for white LEDs are two major challenges in order to meet the practical need in the solid state lighting [118].

Figure 8.42 shows the generic diagrams of dispense, conformal, remote and dispense with les phosphor LEDs. For dispense phosphor package, the phosphor was uniformly mixed with the silicone, and then filled in the lead-frame by dispensing technique. This approach is the most common method to use in the industry because of the easy fabrication, low cost and time. For remote phosphor packages, the phosphor layer is separated away from the chip, and this arrangement could effectively reduce the backscattering and exhibit higher conversion efficiency. To obtain high luminous efficiency, there are many examples demonstrated before such as the scattered photon extraction package and the ring-remote structure [119–121]. However, the disadvantages of remote phosphor structure are the concave surface and the non-uniform angular CCT. Therefore, the patterned structure of remote phosphor structure was proposed to improve the uniformity of CCT [122]. Conversely, conformal phosphor structure can get uniform color distribution and improve the distribution of angular CCT [123, 124]. However, according to the study, there are nearly 60 % re-emitted light from the phosphor layer

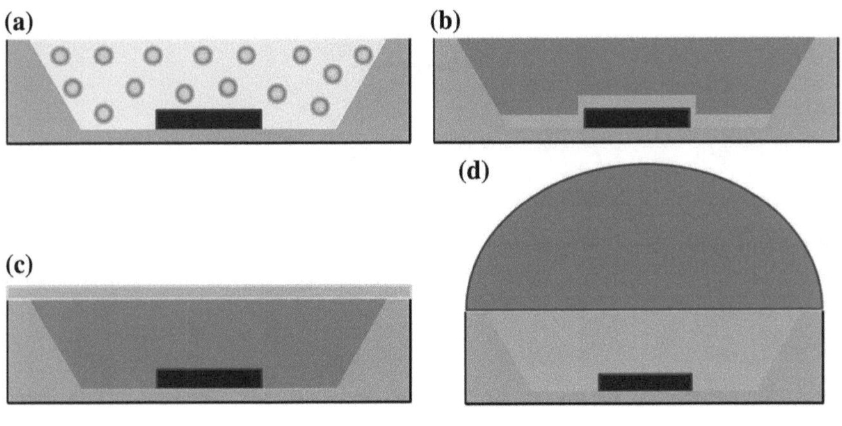

Fig. 8.42 Schematic diagrams of **a** dispense, **b** conformal and **c** remote **d** dispense with lens phosphor structure

are reflected backward, which influences the scattering and reflection characteristics of the phosphor particles [125]. Therefore, in the conformal phosphor structure, the lower light output power could be attributed to the large amount of light reflected back and forth and eventually lost inside the package. In addition, higher light extraction could be achieved by adding the design of a half-spherical lens on the LED package.

8.4.2 Improvement in Uniformity of Emission by ZrO_2 Nano-Particles for White LEDs

For the LED packages, the remote phosphor structure has higher luminous efficiency compared with the conventional dispensing phosphor structure due to the reduction of the backscattering light. However, there are still some disadvantages for the remote phosphor structure such as the concave encapsulant surface, which leads to non-uniform phosphor thickness during the fabrication. This phenomenon, called the yellow ring effect, could be attributed to the different optical paths of blue and yellow light, which leads to the inhomogeneous distribution of emission.

Here, the ZrO_2 nano-particles were used to improve the uniformity of correlated color temperature (CCT) distribution in remote phosphor WLEDs [126]. The superior scattering capability for the light can be provided by these ZrO_2 nano-particles. As a result, the intensity of blue light in large angle could be enhanced and thus the uniformity of CCT improves as well. In addition, because of the refractive index gradient between air and phosphor layers, the luminous flux also could be enhanced by ZrO_2 layer.

Figure 8.43 shows the process flow charts of the experiment. The following steps are the detail descriptions of the fabrication of the samples with remote

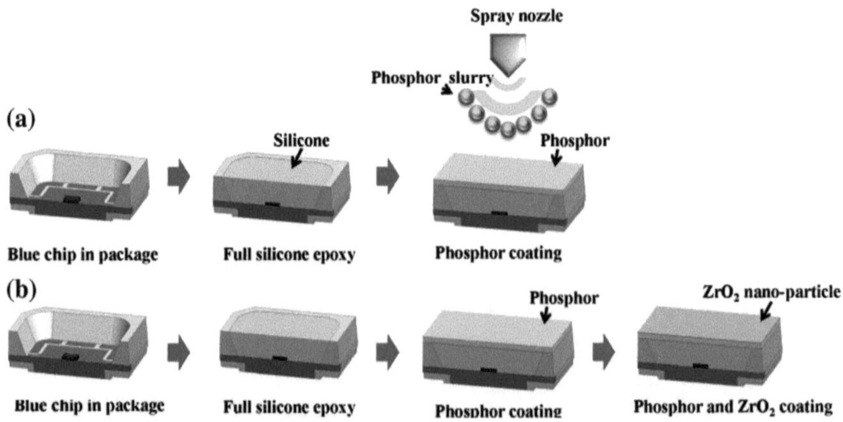

Fig. 8.43 Schematic diagram of process flow charts **a** without; **b** with ZrO_2 nano-particles remote phosphor structure [126]

phosphor structures: (1) the 24 mils square size blue LED chips with 450 nm peak emission wavelength are placed in the plastic lead-frame package. When the driving current is 120 mA, the radiant fluxes of bare blue LED chips were 95 mW. (2) The lead frame filled the transparent silicone by dispensing and baked at 150 °C for 1 h. (3) The pulse spray method was employed to spray phosphor slurry which is made of phosphor powders, silicone binder and alkyl-based solvent. The composition of the phosphor powders is $Y_3Al_5O_{12}$ (YAG) phosphors and the particle sizes are about 12 μm. Then, the conventional remote phosphor structures are formed when the phosphor slurry is sprayed onto the surface of transparent silicone, as shown in Fig. 8.43a. In Fig. 8.43b, the final step is fabricating the ZrO_2 nano-particles remote phosphor structure. The concentration of ZrO_2 nano-particles in phosphor slurry are 5 %. The ZrO_2 nano-particles are mixed with silicone binder and an alkyl-based solvent and sprayed onto the surface of the phosphor layer.

For comparing, when the driving current is 120 mA, the LED is picked with the same color temperature and color chromaticity coordinate. In the cross-sectional scanning electron microscopic (SEM) image, the ZrO_2 nano-particle dimension is around 300 nm, as shown in Fig. 8.44a. In Fig. 8.44b image, we also used the Energy Dispersive Spectrometer (EDS) to analyze the element of ZrO_2 nano-particles with silicone encapsulant. The existence of Zr and O elements in the silicone encapsulant can be confirmed by the EDS analysis.

In general, we can define the angular-dependent CCT uniformity by the difference between the max and min of CCT. In order to optimize the CCT deviation, different weights of ZrO_2 nano-particles with silicone layer on the remote phosphor structure are fabricated, as shown in Fig. 8.45a. When the weight of ZrO_2 nano-particles is 10 mg/cm^2, the CCT deviations is the lowest. Moreover, this result has 58 % improvement compared to the conventional remote phosphor structure. The far field images for conventional and ZrO_2 nano-particles remote phosphor structures are shown in the inset pictures of Fig. 8.45a. In Fig. 8.45b, the angular-dependent CCT of the conventional and 10 mg/cm^2 ZrO_2 nano-particles remote phosphor structures are measured. In the range of −70–70° of angular-dependent CCT results, the CCT deviation of ZrO_2, nano-particles and

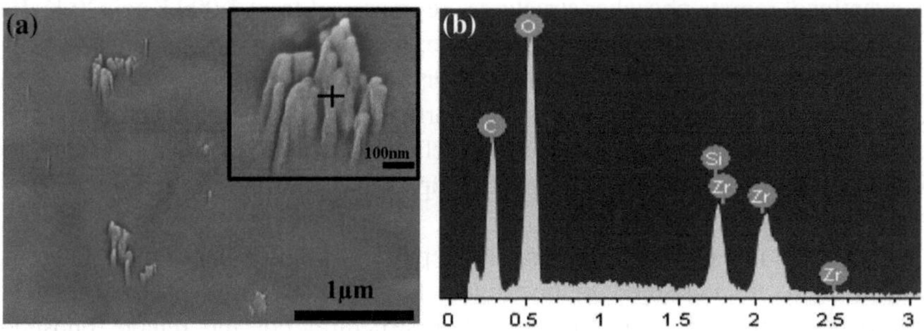

Fig. 8.44 **a** A cross sectional view of SEM image of the ZrO_2 nano-particles in silicone encapsulant. **b** The energy dispersive spectrometer (*EDS*) was taken by a JEOL JEM-2100F system [126]

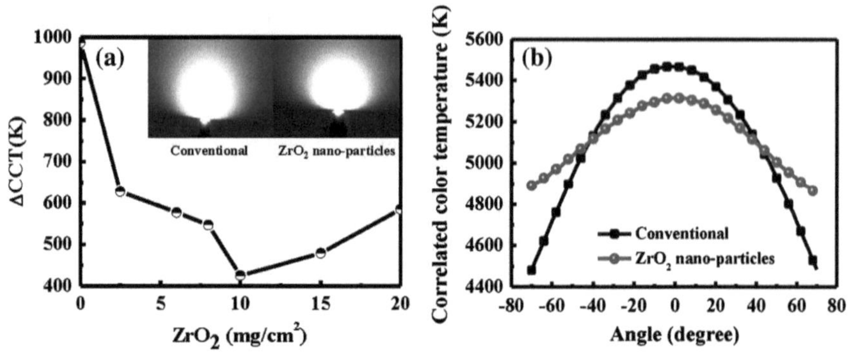

Fig. 8.45 **a** The CCT deviation of different weight of ZrO_2 layer in remote phosphor. **b** The angular-dependent correlated color temperature of conventional and ZrO_2 nano-particles remote phosphor [126]

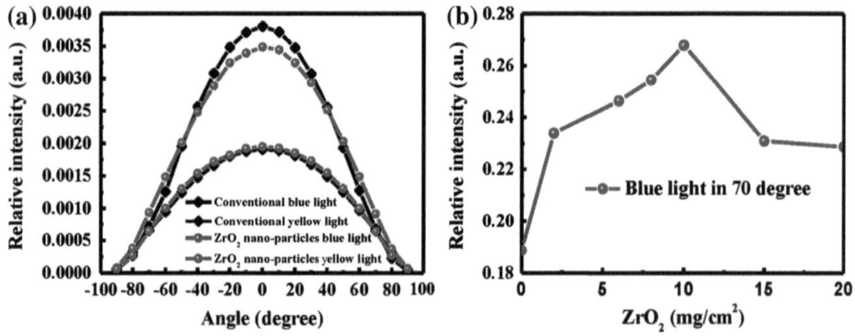

Fig. 8.46 **a** The intensity of *blue* and *yellow* light of conventional and ZrO_2 nano-particles remote phosphor structures, **b** the weight-dependent ZrO_2 nano-particles versus relative intensity of *blue light* in the 70° [126]

conventional remote phosphor structures are improved from 1,000 K to 420 K. For the conventional remote phosphor structure, the blue light at a large angle being trapped and reflected in the phosphor layer cause the inferior angular-dependent CCT deviation. Because the ZrO_2 nano-particles could provide an effective scattering capability to improve the ratio of yellow to blue lights in large angles, the sample with ZrO_2 nano-particles on the top of the phosphor layer has smoother angular-dependent CCT distribution.

As shown in Fig. 8.46a, the angular-dependent relative intensity of blue and yellow light are measured to investigate how the scattering of nano-particles affects the emission of remote phosphor structure. For the remote phosphor structure sample with ZrO_2 nano-particles, the blue light in normal direction is reduced and the divergent angle of blue light is larger than that of the conventional structure. Because the scattering effect by ZrO_2 nano-particles strongly influence the optical path of blue light, the CCT deviation is improved. However, the yellow-light distribution of the conventional and ZrO_2 nano-particles remote phosphor structures are almost the same. Moreover, the dependent-wavelength haze ratio is analyzed to understand this phenomenon. For the 10 mg/cm^2 sample, the haze is around 30 % for yellow light (around 600 nm), and 35 % for blue light (around 450 nm). The higher haze ratio indicates that the stronger scattering effect of photons. Thus, the yellow photons are scattered much less than the blue ones, which might be beneficial for color mixing and cause less variation on CCT for ZrO_2 samples. In the remote phosphor structure, the CCT deviation is mainly associated with the divergent angle of blue light. Moreover, we also measured the weight-dependent relative intensity of blue light with different weights of ZrO_2 nano-particles in the 70°, in Fig. 8.45b. The ZrO_2 nano-particles of 10 mg/cm^2 has the optimized condition because the highest intensity of blue light at large angle.

8.4.3 Enhanced Luminous Efficiency of WLEDs Using a Dual-Layer Structure of the Remote Phosphor Package

In the following experiments, in order to increase light output power, we use a dual-layer phosphor structure in a remote phosphor package, and the result is better than the conventional remote phosphor structure at the same CCT [127]. It can be shown both in experiment and simulation that a dual-layer phosphor structure yielded higher light transmission than a conventional phosphor structure. The TFCalc32 simulation tool is employed to demonstrate that the power intensity was enhanced between the silicone layer and the phosphor layers.

In order to create a uniform phosphor layer, we use the pulse spray coating method to fabricate a dual-layer structure in remote phosphor package structure. The phosphor powder was YAG:Ce^{3+} with the particle size of 13 μm. A peak emission wavelength at 450 nm InGaN-based blue LED was bonded on silver glue with gold wire in a lead-frame package. Figure 8.47 shows the schematic cross-sectional

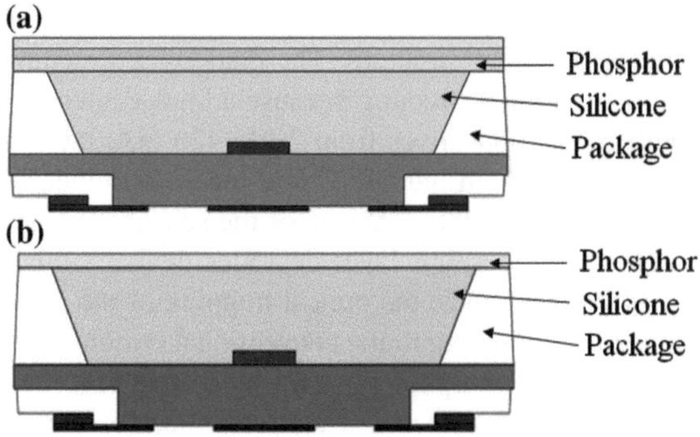

Fig. 8.47 Schematic cross-sectional view of a dual-layer and b conventional remote phosphor structures [127]

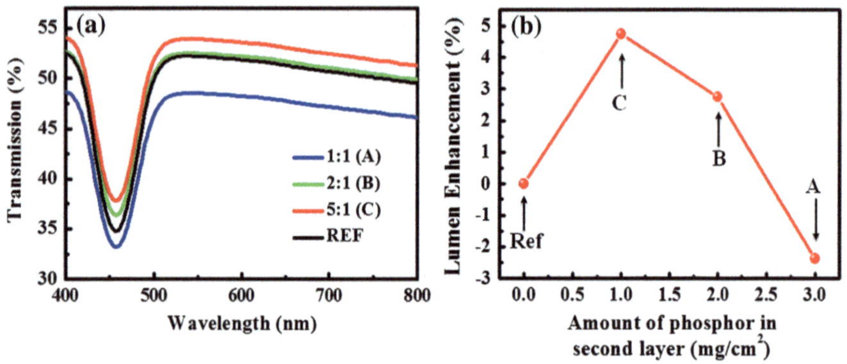

Fig. 8.48 **a** The lumen enhancement of the amount of phosphor in second layer. **b** The transmission of the different ratio with the wavelength [127]

view of a dual-layer remote phosphor structure. For comparing to different samples, the total density of the phosphor in dual-layer is set up as 6 mg/cm^2 for all the samples. Meanwhile, the spray coating step can control the density of phosphor at approximately 1.0 mg/cm^2 in each layers. The thickness of the top and bottom phosphor layers was adjusted to obtain the better luminous efficiency of the devices. Thus, the ratio of the sample A, B, and C represent the different density of phosphor layer in the dual-layer structure, which are 1:1, 2:1 and 5:1 for the top and bottom phosphor layers, respectively.

In Fig. 8.48a, it shows that the wavelength-dependent transmission of different ratios of the first and the second layers of the phosphor structure. In all of samples, sample C produced greater transmission than the other samples and we believe this is the optimized ratio. The high transmission means that the LED device can deliver more photons out of the package and increase the light output. However, the transmission curves have the intensity drop at 460 nm, this phenomenon can be attributed to the phosphor absorption. Figure 8.48b is a comparison of the lumen enhancement of the different samples. There is 5 % lumen enhancement of Sample C over the conventional remote phosphor structure. Increasing light extraction is critical to improving the luminous efficiency of LED structures.

Figure 8.49a shows the emission spectra of Sample C and the conventional remote phosphor structure. The dual-layer structure had higher light output than the conventional structure because a higher intensity in blue and yellow spectrum. With the current drives from 20 to 420 mA, the luminous flux and the luminous efficiency of both Sample C and the conventional remote phosphor structure are shown in Fig. 8.48b. Because of the self-absorption of phosphor, the optical trapping in the phosphor layer degrades device luminous efficiency. The dual-layer structure can reduce the optical trapping of the phosphor layer and increase luminous flux comparing to the conventional remote phosphor structure.

When the phosphor pumped by a blue ray, the direction of yellow rays will scatter to random ways. Therefore, most of the downward rays are lost in the

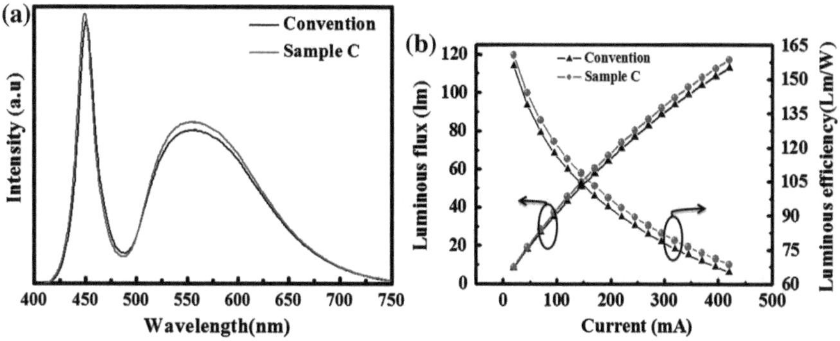

Fig. 8.49 **a** The emission spectra. **b** Luminous flux and luminous efficiency of the dual-layer and the conventional remote phosphor structures driven at currents from 20 to 420 mA [127]

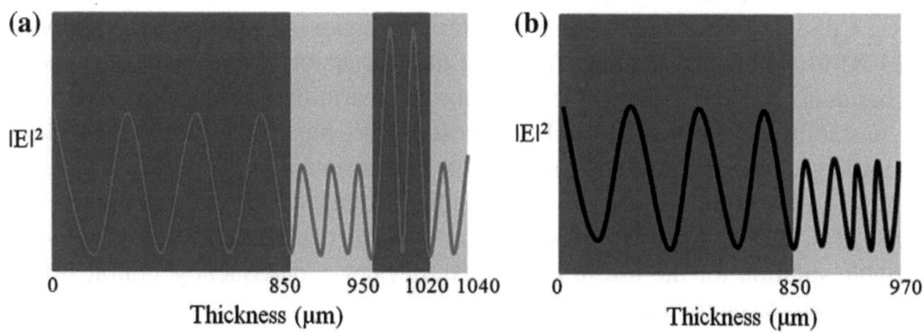

Fig. 8.50 Thickness-dependent $|E|^2$ of **a** the dual-layer and **b** the conventional LEDs by TFCalc32 simulation [127]

package and light output is reduced. Conversely, a thin silicone layer with a low refractive index is inserted into the phosphor layer in the dual-layer phosphor structure. The refractive indices of the phosphor and silicone encapsulants used in this study were 1.8 and 1.4, respectively. The total reflection angle needed to be increased despite of the small difference between the refractive indices of the phosphor and silicone. In order to increase the probability of phosphor excitation, increasing the reflection of blue ray is necessary. Therefore, we use the TFCalc32 to simulate the actual effect of blue photons coupling to the phosphor layer.

In the simulation model, the thickness of the first and second silicone layers were approximately 850 and 70 μm, respectively. On the other hand, the lengths of the first and second phosphor layers were approximately 100 and 20 μm. In the conventional phosphor structure, the silicone layer was approximately 850 μm, and the phosphor layer was roughly 120 μm. In Fig. 8.50, it shows that the electric field intensity for the different thicknesses of dual-layer and conventional phosphor structures. The electric field intensity in the second silicone layer was higher than in the conventional phosphor structure. Therefore, the advantage of the dual-layer structure is that the incident blue ray can be trapped in the lower refractive index medium, increasing the absorption of the phosphor layer and transferring more yellow rays than the conventional structure.

8.4.4 Resonant-Enhanced Full-Color Emission of Quantum-Dot-Based Display Technology Using a Pulsed Spray Method

Recently, colloidal CdSe/ZnS quantum dots (QDs) are considered as possible candidates in solid-state lighting [128, 129]. Figure 8.51 shows the emission and absorption spectrum of blue, green, yellow, orange, and red QDs. The high-photoluminescence efficiency, wide absorption spectrum and size-tunable band-gap of QDs have become very attractive features [130, 131]. With these available features, CdSe/ZnS QDs can be used as color conversion nanophosphors to improve the CRI for solid-state lighting. Although the conversion efficiency of phosphor is higher than QDs, the main problems for the rare-earth-based phosphors are chemical unstability, low efficiency and high cost when using in the white LEDs [132]. Accordingly, several studies focused on the enhancement of the quantum yield for QDs by multi-shell structure, thus suitable light source can be realized by quantum dots [133]. More importantly, the various diameters of CdSe/ZnS QDs offer the variable solution for the use in the display devices. Consequently, colloidal QDs have recently been used as.

In this study, the large-area and small footprint QDLEDs were fabricate by pulsed spray coating method. Furthermore, HfO_2/SiO_2 DBR with a stopband centered at 400 nm and a full width of approximately 60 nm was used to enhance the use of UV pumping photons, thereby increasing RGB emission intensity. Therefore, such RGB QDs with highly reflective DBR exhibits high color purity and provides an alternative method for display applications.

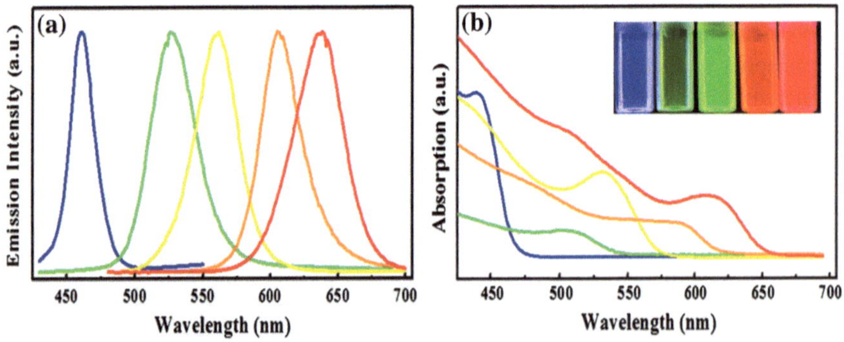

Fig. 8.51 The emission and absorption spectrum of *blue*, *green*, *yellow*, *orange*, and *red* QDs

The process flowcharts of the pixelated RGB QD arrays and non-patterned white light source on a 2-in substrate, as shown in Fig. 8.52. The concentrations of the RGB QD were approximately 1 mg/ml. The pixelated arrays are fabricated using the following steps: (1) an 11-pairs HfO_2/SiO_2 DBR was evaporated in an ion-assisted e-gun system on the top of the glass; (2) the mask was placed on the top of the substrate for alignment; and (3) the RGB QDs were sprayed onto the surface of the glass in the sequence of green, blue, and red. Similar steps can be followed for a 2-in. wafer white light source; however, no mask alignment is used.

The mask in our experiment was composed of an aluminum plate with an array of square (2 mm × 2 mm) holes. The interval of the each pixel was 5.22 mm. The alignment between the various colors can be achieved by visual inspection and XY translational stage adjustment. After PS QD deposition, AFM images were measured to inspect the surface, as shown in Fig. 8.53. The QDs aggregated and clustered after layer-by-layer deposition. The AFM measurement of the blue, green,

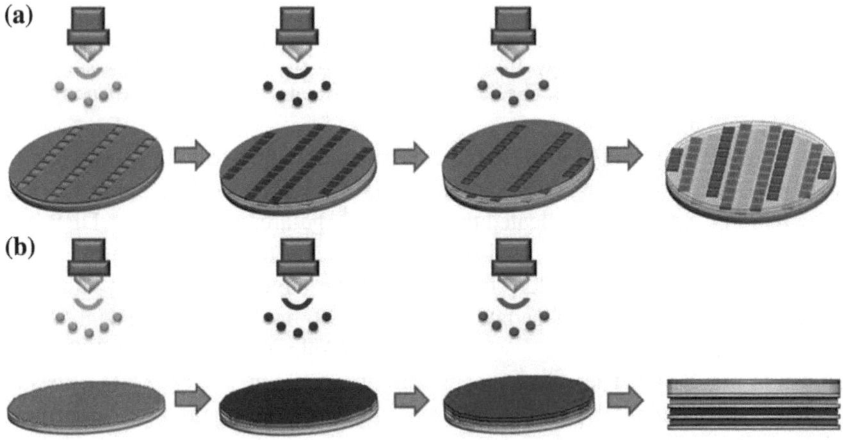

Fig. 8.52 **a** UV–Visible absorbance (*red*) and photoluminescence (*blue*) spectra of CdS QDs measure in toluene. The PLE spectrum was taken at the maximum of PL intensity (~470 nm). For the PL spectrum, the sample was excited by a light beam with 365 nm. The inset is the CdS quantum dot solution under UV excitation. **b** The measured reflectance spectra for QD-coated, No-QD coated, and AR-coated solar cells [134]

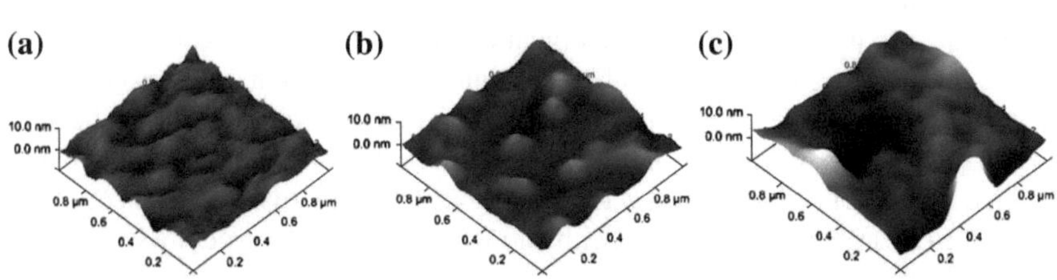

Fig. 8.53 AFM image of the**a** *blue*, **b** *green* and **c** *red* QD layer by pulsed spray coating method [134]

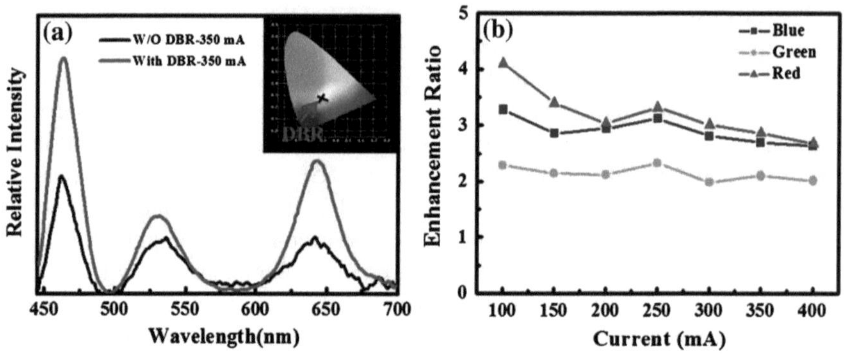

Fig. 8.54 **a** The relative intensity of the large-area samples with and without DBR operated under 350 mA. The inset shows the CIE color coordinates of the large-area samples with DBR. **b** Enhancement ratio of the intensity of large-area samples with and without DBR under various currents from 100 to 400 mA [134]

and red QD layer revealed a root-mean-square (RMS) roughness of 1.18, 0.816, and 0.788 nm, respectively. The RMS results indicate that uniform dispensing is possible with this technology.

The relative intensity of the large-area samples with DBR and the reference without DBR under a injection current of 350 mA for a UV pumping LED, as shown in Fig. 8.54a. It is indicated that that the large-area samples with DBR have higher intensity in red, blue, and green components than the reference without DBR. Moreover, the CIE color coordinates of the large-area samples with DBR, which provide a white light output, are (0.29, 0.29) and the enhancement of each color from the DBR sample over the non-DBR sample from 100 to 400 mA is shown in Fig. 8.54b. The enhancement ratio of red, green, and blue QD emissions is steady with the increase of the inject current. Moreover, the red emission shows higher enhancement than that of the other colors, which can be attributed to the proximity of red QD layers to the UV source. Conversely, the green QDs exhibited a weaker enhancement in general. This may have been caused by the substantial aggregation of green QDs when they were sprayed and dried on the substrate.

Figure 8.55 shows the image of the pixelated structure and 2-in. full-wafer sample under UV excitation. This result demonstrate the feasibility of this novel approach. If we focus on the control the volume of each color, it is possible to obtain any mixture of color on the chromaticity map via this QD technology.

To make this QD spray possible for lighting application, it is necessary to demonstrate the same capability in the regular LED package level, as shown in Fig. 8.56a. In this structure, QDLED is composed with the quantum dots/PDMS layer, a highly reflective DBR layer and a UV LED. Furthermore, The electroluminescence (EL) spectra of QDLED with and without the DBR structure is measured in Fig. 8.56b. It is obviously that three emission peaks occurred at 460 nm (blue band), 530 nm (green band), and 640 nm (red band), which were contributed by the RGB QDs. This indicated that EL with the DBR has stronger visible emission

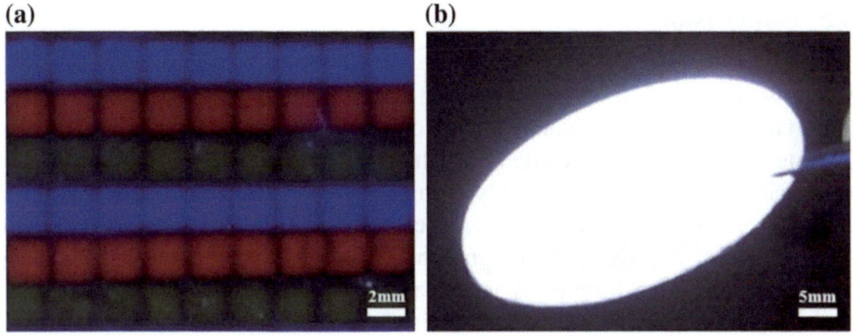

Fig. 8.55 Image of the **a** pixel pattern, **b** a 2-in. full wafer structure under UV excitation [134]

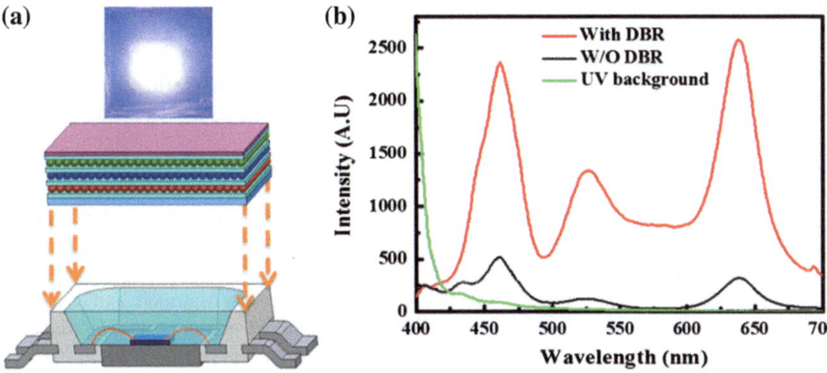

Fig. 8.56 Image of the **a** pixel pattern, **b** a 2-in. Full wafer structure under UV [134]

than that without the DBR structure. A good mixed white light can be generated via proper QD ratio, as can be seen in the inset of Fig. 8.56a.

8.4.5 Summary

In this section, we reviewed several advanced ideas for the LED packages. A successful package design should have a good lumen/watt efficiency and tolerable costs. The emission characteristics such as uniformity of CCT might be important for different applications. Some issues, such as the "yellow ring", can be solved by the nanoparticle doping and the extra scattering. Some of the novel materials, such as colloidal quantum dots, can be introduced to provide more variety of the delivered color. The target for LED package can be versatile and remain to be one of the key components of the overall system.

8.5 Conclusion

Since its first introduction half a century ago, light emitting diodes have been an important part of scientific community. Nowadays, the prevalence of the solid state lighting and the quest of eco-friendly technology have become crucial for our survival and the role of LED is quickly changing and not just a lab toy any more. As we could see from this chapter, the development of LED followed closely with the progress of material epitaxial growth, and the improvement of its efficiency has been experienced from every technical aspect. Accompany with the advanced packaging efforts, we believe this device is and will be important for our daily life.

References

1. V. Lossev, Behavior of contact detectors; the effect of temperature on the generating contact. Telegrafia I Telefonia Bez Provodov **18**, 45–62 (1923). (In Russian)
2. V. Lossev, Oscillating crystals. Wireless World Radio Rev. **271**, 93–96 (1924)
3. V. Lossev, Luminous carborundum detector and detection effect and oscillations with crystals. Phil. Mag. **5**, 1024–1044 (1928)
4. E.E. Loebner, Subhistories of the light emitting diode. IEEE Trans. Electron Dev. **23**, 675–699 (1976)
5. N. Holonyak, *Lemelson-MIT Prize Winner*. Lemenson-MIT Program, 13 Aug 2007 (2004)
6. N. Holonyak Jr, S.F. Bevacqua, Coherent (visable) light emission from Ga(As_{1-x} Px) junctions. Appl. Phys. Lett. **4**, 82–83 (1962)
7. N. Jr HolonyakJr, D.C. Jillson, S.F. Bevacqua, Halogen vapor transport and growth of epitaxial layers of intermetallic compounds and compound mixtures, in *Metallurgy of Semiconductor Materials*, vol. 15, ed. by J.B. Schroeder (Wiley, New York, 1961), pp. 49–59
8. N. Holonyak, Jr., Private Communication, Sept 2002
9. N. Holonyak Jr, S.F. Bevacqua, Coherent (visible) light emission from Ga(AsP) junctions. Appl. Phys. Lett. **1**, 82–83 (1962)
10. E. Fred Schubert, *Light-Emitting Diodes* (Cambridge University Press, Cambridge, 2003), pp. 8–9
11. T.S. Perry, M. George Craford [biography]. IEEE Spectr. **32**, 52–55 (1995)
12. B. Biography–Holonyak, D. Craford, Technology Administration. Retrieved 30 May 2007
13. T.P. Pearsall, B.I. Miller, R.J. Capik, K.J. Bachmann, Efficient, lattice-matched, double heterostructure LEDs at 1.1 mm from $Ga_xIn_{1-x}As_yP_{1-y}$ by liquid-phase epitaxy. Appl. Phys. Lett. **28**, 499 (1976)
14. M. Ikeda, K. Nakano, Y. Mori, K. Kaneko, N. Watanabe, Mocvd growth of algainp at atmospheric-pressure using triethylmetals and phosphine. J. Cryst. Growth **77**, 380–385 (1986)
15. M. Ikeda, Y. Mori, H. Sato, K. Kaneko, N. Watanabe, Room-temperature continuous-wave operation of an algainp double heterostructure laser grown by atmospheric-pressure metalorganic chemical vapor-deposition. Appl. Phys. Lett. **47**, 1027–1028 (1985)
16. K.H. Huang, J.G. Yu, C.P. Kuo, R.M. Fletcher, T.D. Osentowski, L.J. Stinson, M.G. Craford, A.S.H. Liao, Twofold efficiency improvement in high-performance algainp light-emitting-diodes in the 555–620 nm spectral region using a thick gap window layer. Appl. Phys. Lett. **61**, 1045–1047 (1992)
17. R.M. Fletcher, C.P. Kuo, T.D. Osentowski, K.H. Huang, M.G. Craford, V.M. Robbins, The growth and properties of high-performance algalnp emitters using a lattice mismatched gap window layer. J. Electron. Mater. **20**, 1125–1130 (1991)

18. H. Amano, T. Asahi, I. Akasaki, Stimulated emission near ultraviolet at room temperature from a GaN film grown on sapphire by MOVPE using an AlN buffer layer. Jpn. J. Appl. Phys. **29**, L205–L206 (1990)
19. H. Amano, N. Sawaki, I. Akasaki, Y. Toyoda, Metalorganic vapor-phase epitaxial-growth of a high-quality GaN film using an AlN buffer layer. Appl. Phys. Lett. **48**, 353–355 (1986)
20. I. Akasaki, H. Amano, K. Hiramatsu, N. Sawaki, High efficiency blue LED utilizing GaN film with AlN buffer layer grown by MOVPE, in *Gallium Arsenide and Related Compounds 1987*. Proceedings of 14th International Symposium, pp. 633–636 (1988)
21. H. Amano, M. Kito, K. Hiramatsu, I. Akasaki, P-type conduction in Mg-doped GaN treated with low-energy electron-beam irradiation (LEEBI). Jpn J. Appl. Phys. Part 2-Lett. **28**, L2112–L2114 (1989)
22. S. Nakamura, M. Senoh, T. Mukai, Highly p-typed Mg-doped GaN films grown with GaN buffer layers. Jpn J. Appl. Phys. Part 2-Lett. **30**, L1708–L1711 (1991)
23. S. Nakamura, T. Mukai, M. Senoh, High-power GaN p-n-junction blue-light-emitting diodes. Jpn J. Appl. Phys. Part 2-Lett. **30**, L1998–L2001 (1991)
24. S. Nakamura, T. Mukai, M. Senoh, N. Iwasa, Thermal annealing effects on p-type Mg-doped GaN films. Jpn J. Appl. Phys. **31**, L139–4215 (1992)
25. M.R. Krames, O.B. Shchekin, R. Mueller-Mach, G.O. Mueller, L. Zhou, G. Harbers, M.G. Craford, Status and future of high-power light-emitting diodes for solid-state lighting. J. Dis. Technol. **3**, 160–175 (2007)
26. R. V. Steele, The story of a new light source. Nat. Photonics **1**, 25-26 (2007)
27. Y. Narukawa, I. Niki, K. Izuno, M. Yamada, Y. Murazaki, T. Mukai, Phosphor-conversion white light emitting diode using InGaN near-ultraviolet chip. Japan. J. Appl. Phys. Part 2-Lett. **41**, L371–L373 (2002)
28. E.F. Schubert, *Light Emitting Diodes*, 1st edn. (Cambridge University Press, Cambridge, England, 2003)
29. A. Sakai, H. Sunakawa, A. Usui, Defect structure in selectively grown GaN films with low threading dislocation density. Appl. Phys. Lett. **71**, 2259–2261 (1997)
30. O.H. Nam, M.D. Bremser, T.S. Zheleva, R.F. Davis, Lateral epitaxy of low defect density GaN layers via organometallic vapor phase epitaxy. Appl. Phys. Lett. **71**, 2638–2640 (1997)
31. D.S. Wuu, W.K. Wang, K.S. Wen, S.C. Huang, S.H. Lin, S.Y. Huang, C.F. Lin, R.H. Horng, Defect reduction and efficiency improvement of near-ultraviolet emitters via laterally overgrown GaN on a GaN/patterned sapphire template. Appl. Phys. Lett. **89** (2006)
32. D.S. Wuu, W.K. Wang, K.S. Wen, S.C. Huang, S.H. Lin, R.H. Horng, Y.S. Yu, M.H. Pan, Fabrication of pyramidal patterned sapphire substrates for high-efficiency InGaN-based light emitting diodes. J. Electrochem. Soc. **153**, G765–G770 (2006)
33. A. Xing, M. Davanco, D.J. Blumenthal, E.L. Hu, Fabrication of InP-based two-dimensional photonic crystal membrane. J. Vac. Sci. Technol. B **22**, 70–73 (2004)
34. H.W. Huang, C.H. Lin, C.C. Yu, C.H. Chiu, C.F. Lai, H.C. Kuo, K.M. Leung, T.C. Lu, S.C. Wang, B.D. Lee, Enhanced light output from a nitride-based power chip of green light-emitting diodes with nano-rough surface using nanoimprint lithography. Nanotechnology **19**, 18530 (2008)
35. S. Li, A. Waag, GaN based nanorods for solid state lighting. J. Appl. Phys. **111**, 071101 (2012)
36. Y.L. Li, R. Huang, Y.H. Lai, Efficiency droop behaviors of InGaN/GaN multiple-quantum-well light-emitting diodes with varying quantum well thickness. Appl. Phys. Lett. **91**, 181113 (2007)
37. M.-H. Kim, M.F. Schubert, Q. Dai, J.K. Kim, E.F. Schubert, J. Piprek, Y. Park, Origin of efficiency droop in GaN-based light-emitting diodes. Appl. Phys. Lett. **91**, 183507 (2007)
38. T.S. Zheleva, O.H. Nam, W.M. Ashmawi, J.D. Griffin, R.F. Davis, Lateral epitaxy and dislocation density reduction in selectively grown GaN structures. J. Cryst. Growth **222**, 706–718 (2001)
39. A. Usui, H. Sunakawa, A. Sakai, A.A. Yamaguchi, Thick GaN epitaxial growth with low dislocation density by hydride vapor phase epitaxy. Japan. J. Appl. Phys. Part 2-Lett. **36**, L899–L902 (1997)

40. C.H. Chiu, M.H. Lo, T.C. Lu, P. Yu, H.W. Huang, H.C. Kuo, S.C. Wang, Nano-processing techniques applied in GaN-Based light-emitting devices with self-assembly Ni nano-masks. J. Lightwave Technol. **26**, 1445–1454 (2008)
41. Y.J. Lee, H.C. Kuo, T.C. Lu, S.C. Wang, High light-extraction GaN-based vertical LEDs with double diffuse surfaces. IEEE J. Quantum Elect. **42**, 1196–1201 (2006)
42. C.H. Chiu, H.H. Yen, C.L. Chao, Z.Y. Li, P. Yu, H.C. Kuo, T.C. Lu, S.C. Wang, K.M. Lau, S.J. Cheng, Nanoscale epitaxial lateral overgrowth of GaN-based light-emitting diodes on a SiO(2) nanorod-array patterned sapphire template. Appl. Phys. Lett. **93**, 081108 (2008)
43. Z.H. Feng, Y.D. Qi, Z.D. Lu, K.M. Lau, GaN-based blue light-emitting diodes grown and fabricated on patterned sapphire substrates by metalorganic vapor-phase epitaxy. J. Cryst. Growth **272**, 327–332 (2004)
44. K. Kusakabe, A. Kikuchi, K. Kishino, Characterization of overgrown GaN layers on nano-columns grown by RF-molecular beam epitaxy. Japan. J. Appl. Phys. Part 2-Lett. **40**, L192–L194 (2001)
45. E.H. Park, J. Jang, S. Gupta, I. Ferguson, C.H. Kim, S.K. Jeon, J.S. Park, Air-voids embedded high efficiency InGaN-light emitting diode. Appl. Phys. Lett. **93**, 191103 (2008)
46. M.H. Lo, Y.J. Cheng, H.C. Kuo, S.C. Wang, Enhanced electron-hole plasma stimulated emission in optically pumped gallium nitride nanopillars. Appl. Phys. Lett. **98**, 121101 (2011)
47. P. Frajtag, J.P. Samberg, N.A. El-Masry, N. Nepal, S.M. Bedair, Embedded voids formation by overgrowth on GaN nanowires for high-quality GaN films. J. Cryst. Growth **322**, 27–32 (2011)
48. C.H. Chiu, H.H. Yen, C.L. Chao, Z.Y. Li, P. Yu, H.C. Kuo, T.C. Lu, S.C. Wang, K.M. Lau, S.J. Cheng, Nanoscale epitaxial lateral overgrowth of GaN-based light-emitting diodes on a SiO(2) nanorod-array patterned sapphire template. Appl. Phys. Lett. **93**, 081108 (2008)
49. P. Puech, F. Demangeot, J. Frandon, C. Pinquier, M. Kuball, V. Domnich, Y. Gogotsi, GaN nanoindentation: a micro-Raman spectroscopy study of local strain fields. J. Appl. Phys. **96**, 2853–2856 (2004)
50. X.H. Wu, P. Fini, E.J. Tarsa, B. Heying, S. Keller, U.K. Mishra, S.P. DenBaars, J.S. Speck, Dislocation generation in GaN heteroepitaxy. J. Cryst. Growth **189**, 231–243 (1998)
51. C.H. Yen, W.C. Lai, Y.Y. Yang, C.K. Wang, T.K. Ko, S.J. Hon, S.J. Chang, GaN-based light-emitting diode with sputtered AlN nucleation layer. IEEE Photonics Tech. Lett. **24**, 294–296 (2012)
52. D. Hanser, E.A. Preble, T. Clites, T. Stephenson, R. Jacobs, T. Johnson, T. Paskova, K.R. Evans, CS Mantech Conference, Tampa, Florida, USA, pp. 18–21 (2009)
53. J.K. Sheu, J.M. Tsai, S.C. Shei, W.C. Lai, T.C. Wen, C.H. Kou, Y.K. Su, S.J. Chang, G.C. Chi, Low-operation voltage of InGaN/GaN light-emitting diodes with Si-doped In0.3Ga0.7N/GaN short-period superlattice tunneling contact layer. IEEE Electron Device Lett. **22**, 460–462 (2001)
54. C. Liu, B. Mensching, M. Zeitler, K. Volz, B. Rauschenbach, Ion implantation in GaN at liquid-nitrogen temperature: structural characteristics and amorphization. Phys. Rev. B **57**, 2530–2535 (1998)
55. J.K. Sheu, S.J. Tu, Y.H. Yeh, M.L. Lee, W.C. Lai, Gallium nitride-based light-emitting diodes with embedded air voids grown on Ar-implanted AlN/sapphire substrate. Appl. Phys. Lett. **101**, 151103 (2012)
56. J.K. Sheu, M.L. Lee, C.J. Tun, C.J. Kao, L.S. Yeh, S.J. Chang, G.C. Chi, Characterization of Si implants in p-type GaN. IEEE J. Sel. Top. Quantum Electron. **8**, 767–772 (2002)
57. S.-J. Tu, M.-L. Lee, Y.-H. Yeh, F.-W. Huang, P.-C. Chen, W.-C. Lai, C.-W. Chen, G.C. Chi, J.-K. Sheu, Improved output power of InGaN LEDs by lateral overgrowth on Si-implanted n-GaN surface to form air gaps. IEEE J. Quantum Electron. **48**, 1004–1009 (2012)
58. S. Grzanka, G. Franssen, G. Targowski, K. Krowicki, T. Suski, R. Czernecki, P. Perlin, M. Leszczynski, Role of the electron blocking layer in the low-temperature collapse of electroluminescence in nitride light-emitting diodes. Appl. Phys. Lett. **90**, 103507 (2007)

59. K. Iga, H. Uenohara, F. Koyama, Electron reflectance of multiquantum barrier (MQB). Electron. Lett. **22**, 1008–1010 (1986)
60. S.N. Lee, S.Y. Cho, H.Y. Ryu, J.K. Son, H.S. Paek, T. Sakong, T. Jang, K.K. Choi, K.H. Ha, M.H. Yang, O.H. Nam, Y. Park, E. Yoon, High-power GaN-based blue-violet laser diodes with AlGaN/GaN multiquantum barriers. Appl. Phys. Lett. **88**, 111101 (2006)
61. N. Tzer-En, W. Jen-Cheng, S. Hui-Tang, W. Ya-Fen, Effect of multiquantum barriers on performance of InGaN/GaN multiple-quantum-well light-emitting diodes. J. Appl. Phys. **102**, 033101–033107 (2007)
62. H. Hirayama, Y. Tsukada, T. Maeda, N. Kamata, Marked enhancement in the efficiency of deep-ultraviolet AlGaN light-emitting diodes by using a multiquantum-barrier electron blocking layer. Appl. Phys. Express **3**, 031002 (2010)
63. C.S. Chang, Y.K. Su, S.J. Chang, P.T. Chang, Y.R. Wu, K.H. Huang, T.P. Chen, High-brightness AlGaInP 573-nm light-emitting diode with a chirped multiquantum barrier. IEEE J. Quantum Electron. **34**, 77–83 (1998)
64. C.S. Chang, S.J. Chang, Y.K. Su, C.T. Lee, Y.C. Lin, W.C. Lai, S.C. Shei, J.C. Ke, H.M. Lo, Nitride-based LEDs with textured side walls. IEEE Photonics Tech. Lett. **16**, 750–752 (2004)
65. S.J. Chang, C.H. Chen, P.C. Chang, Y.K. Su, P.C. Chen, Y.D. Jhou, H. Hung, S.M. Wang, B.R. Huang, Nitride-based LEDs with p-InGaN capping layer. IEEE Trans. Electron Device **50**, 2567–2570 (2003)
66. Y.-Y. Lin, R.W. Chuang, S.-J. Chang, S. Li, Z.-Y. Jiao, T. Ko, S.J. Hon, C.H. Liu, GaN-based LEDs with a chirped multiquantum barrier structure. Photonics Tech. Lett., IEEE. **24**(18), 1600–1602 (Sep 15, 2012)
67. H. Fujii, K. Endo, H. Hotta, improvement of multiquantum-barrier effect by layer-thickness modulation. Appl. Phys. Lett. **64**, 3479–3481 (1994)
68. S.J. Chang, S.F. Yu, R.M. Lin, S. Li, T.H. Chiang, S.P. Chang, C.H. Chen, InGaN-based light-emitting diodes with an AlGaN staircase electron blocking layer. IEEE Photonics Tech. Lett. **24**, 1737–1740 (2012)
69. C.L. Reynolds Jr, A. Patel, Tunneling entity in different injection regimes of InGaN light emitting diodes. J. Appl. Phys. **103**, 086102 (2008)
70. O. Ambacher, J. Smart, J.R. Shealy, N.G. Weimann, K. Chu, M. Murphy, W.J. Schaff, L.F. Eastman, R. Dimitrov, L. Wittmer, M. Stutzmann, W. Rieger, J. Hilsenbeck, Two-dimensional electron gases induced by spontaneous and piezoelectric polarization charges in N- and Ga-face AlGaN/GaN heterostructures. J. Appl. Phys. **85**, 3222–3233 (1999)
71. F. Bernardini, V. Fiorentini, D. Vanderbilt, Spontaneous polarization and piezoelectric constants of III-V nitrides. Phys. Rev. B **56**, 10024–10027 (1997)
72. S.J. Chang, C.S. Chang, Y.K. Su, R.W. Chuang, W.C. Lai, C.H. Kuo, Y.P. Hsu, Y.C. Lin, S.C. Shei, H.M. Lo, J.C. Ke, J.K. Sheu, Nitride-based LEDs with an SPS tunneling contact layer and an ITO transparent contact. IEEE Photonics Tech. Lett. **16**, 1002–1004 (2004)
73. E. Fred Schubert, *Light-Emitting Diodes*, 2nd edn. (Cambridge University Press, Cambridge, 2006)
74. Y.K. Su, S.J. Chang, C.H. Chen, J.F. Chen, G.C. Chi, J.K. Sheu, W.C. Lai, J.M. Tsai, GaN metal-semiconductor-metal ultraviolet sensors with various contact electrodes. IEEE Sens. J. **2**, 366–371 (2002)
75. J.F. Lin, M.C. Wu, M.J. Jou, C.M. Chang, B.J. Lee, Y.T. Tsai, Highly reliable operation of indium tin oxide AlGaInP orange light-emitting-diodes. Electron. Lett. **30**, 1793–1794 (1994)
76. C.L. Chua, R.L. Thornton, D.W. Treat, V.K. Yang, C.C. Dunnrowicz, Indium tin oxide transparent electrodes for broad-area top-emitting vertical-cavity lasers fabricated using a single lithography step. IEEE Photonics Tech. Lett. **9**, 551–553 (1997)
77. C.L. Chua, R.L. Thornton, D.W. Treat, V.K. Yang, C.C. Dunnrowicz, Indium tin oxide transparent electrodes for broad-area top-emitting vertical-cavity lasers fabricated using a single lithography step. IEEE Photonics Tech. Lett. **9**, 551–553 (1997)

78. Y.C. Lin, S.J. Chang, Y.K. Su, T.Y. Tsai, C.S. Chang, S.C. Shei, S.J. Hsu, C.H. Liu, U.H. Liaw, S.C. Chen, B.R. Huang, Nitride-based light-emitting diodes with Ni/ITO p-type ohmic contacts. IEEE Photonics Tech. Lett. **14**, 1668–1670 (2002)
79. T. Margalith, O. Buchinsky, D.A. Cohen, A.C. Abare, M. Hansen, S.P. DenBaars, L.A. Coldren, Indium tin oxide contacts to gallium nitride optoelectronic devices. Appl. Phys. Lett. **74**, 3930–3932 (1999)
80. S.R. Jeon, Y.H. Song, H.J. Jang, G.M. Yang, S.W. Hwang, S.J. Son, Lateral current spreading in GaN-based light-emitting diodes utilizing tunnel contact junctions. Appl. Phys. Lett. **78**, 3265–3267 (2001)
81. C.S. Chang, S.J. Chang, Y.K. Su, Y.C. Lin, Y.P. Hsu, S.C. Shei, S.C. Chen, C.H. Liu, U.H. Liaw, InGaN/GaN light-emitting diodes with ITO p-contact layers prepared by RF sputtering. Semicond. Sci. Tech. **18**, L21–L23 (2003)
82. S.J. Chang, L.W. Wu, Y.K. Su, Y.P. Hsu, W.C. Lai, J.A. Tsai, J.K. Sheu, C.T. Lee, Nitride-based LEDs with 800 °C grown p-AlInGaN-GaN double-cap layers. IEEE Photonics Tech. Lett. **16**, 1447–1449 (2004)
83. T. Borzenko, M. Tormen, G. Schmidt, L.W. Molenkamp, H. Janssen, Polymer bonding process for nanolithography. Appl. Phys. Lett. **79**, 2246–2248 (2001)
84. S.J. Chang, C.F. Shen, W.S. Chen, C.T. Kuo, T.K. Ko, S.C. Shei, J.K. Sheu, Nitride-based light emitting diodes with indium tin oxide electrode patterned by imprint lithography. Appl. Phys. Lett. **91** (2007)
85. S.J. Chang, C.S. Chang, Y.K. Su, C.T. Lee, W.S. Chen, C.F. Shen, Y.P. Hsu, S.C. Shei, H.M. Lo, Nitride-based flip-chip ITO LEDs. IEEE Trans. Adv. Packag. **28**, 273–277 (2005)
86. J.R. Lee, S.I. Na, J.H. Jeong, S.N. Lee, J.S. Jang, S.H. Lee, J.J. Jung, J.O. Song, T.Y. Seong, S.J. Park, Low resistance and high reflectance Pt/Rh contacts to p-type GaN for GaN-based flip chip light-emitting diodes. J. Electrochem. Soc. **152**, G92–G94 (2005)
87. J. Song, D.S. Leem, J.S. Kwak, O.H. Nam, Y. Park, T.Y. Seong, Low resistance and reflective Mg-doped indium oxide-Ag ohmic contacts for flip-chip light-emitting diodes. IEEE Photonics Tech. Lett. **16**, 1450–1452 (2004)
88. S.J. Chang, W.S. Chen, Y.C. Lin, C.S. Chang, T.K. Ko, Y.P. Hsu, C.F. Shen, J.M. Tsai, S.C. Shei, Nitride-based flip-chip LEDs with transparent ohmic contacts and reflective mirrors. IEEE Trans. Adv. Packag. **29**, 403–408 (2006)
89. S.J. Chang, W.S. Chen, S.C. Shei, T.K. Ko, C.F. Shen, Y.P. Hsu, C.S. Chang, J.M. Tsai, W.C. Lai, A.J. Lin, Highly reliable high-brightness GaN-based flip chip LEDs. IEEE Trans. Adv. Packag. **30**, 752–757 (2007)
90. C.E. Lee, H.C. Kuo, Y.C. Lee, M.R. Tsai, T.C. Lu, S.C. Wang, C.T. Kuo, Luminance enhancement of flip-chip light-emitting diodes by geometric sapphire shaping structure. IEEE Photonics Tech. Lett. **20**, 184–186 (2008)
91. D.-S. Han, J.-Y. Kim, S.-I. Na, S.-H. Kim, K.-D. Lee, B. Kim, S.-J. Park, Improvement of light extraction efficiency of flip-chip light-emitting diode by texturing the bottom side-surface of sapphire substrate. IEEE Photonics Tech. Lett. **18**, 1406–1408 (2006)
92. C.F. Shen, S.J. Chang, W.S. Chen, T.K. Ko, C.T. Kuo, S.C. Shei, Nitride-based high-power flip-chip LED with double-side patterned sapphire substrate. IEEE Photonics Tech. Lett. **19**, 780–782 (2007)
93. S.-H. Huang, R.-H. Horng, K.-S. Wen, Y.-F. Lin, K.-W. Yen, D.-S. Wuu, Improved light extraction of nitride-based flip-chip light-emitting diodes via sapphire shaping and texturing. IEEE Photonics Tech. Lett. **18**, 2623–2625 (2006)
94. S.-J. Chang, W.S. Chen, S.C. Shei, C.T. Kuo, T.K. Ko, C.F. Shen, J.M. Tsai, W.-C. Lai, J.-K. Sheu, A.J. Lin, High-brightness InGaN–GaN power flip-chip LEDs. IEEE J. Lightwave Technol. **27**, 1985–1989 (2009)
95. K. Tadatomo, H. Okagawa, Y. Ohuchi, T. Tsunekawa, T. Jyouichi, Y. Imada, M. Kato, H. Kudo, T. Taguchi, High output power InGaN ultraviolet light plus emitting diodes fabricated on patterned substrates using metalorganic vapor phase epitaxy. Physica Status Solidi a-Appl. Res. **188**, 121–125 (2001)

96. W.K. Wang, D.S. Wuu, S.H. Lin, P. Han, R.H. Horng, T.C. Hsu, D.T.C. Huo, M.J. Jou, Y.H. Yu, A.K. Lin, Efficiency improvement of near-ultraviolet InGaN LEDs using patterned sapphire substrates. IEEE J. Quantum Electron. **41**, 1403–1409 (2005)
97. Y.J. Lee, J.M. Hwang, T.C. Hsu, M.H. Hsieh, M.J. Jou, B.J. Lee, T.C. Lu, H.C. Kuo, S.C. Wang, Enhancing the output power of GaN-based LEDs grown on wet-etched patterned sapphire substrates. IEEE Photonics Tech. Lett. **18**, 1152–1154 (2006)
98. S.J. Chang, Y.C. Lin, Y.K. Su, C.S. Chang, T.C. Wen, S.C. Shei, J.C. Ke, C.W. Kuo, S.C. Chen, C.H. Liu, Nitride-based LEDs fabricated on patterned sapphire substrates. Solid-State Electron. **47**, 1539–1542 (2003)
99. S. Pimputkar, J.S. Speck, S.P. DenBaars, S. Nakamura, Prospects for LED lighting. Nat. Photonics **3**, 179–181 (2009)
100. E.F. Schubert, J.K. Kim, Solid-state light sources getting smart. Science **308**, 1274–1278 (2005)
101. K.J. Chen, H.C. Chen, M.H. Shih, C.H. Wang, M.Y. Kuo, Y.C. Yang, C.C. Lin, H.C. Kuo, The influence of the thermal effect on CdSe/ZnS quantum dots in light-emitting diodes. IEEE J. Lightwave Technol. **30**, 2256–2261 (2012)
102. H.C. Chen, K.J. Chen, C.C. Lin, C.H. Wang, C.C. Yeh, H.H. Tsai, M.H. Shih, H.C. Kuo, Improvement of lumen efficiency in white light-emitting diodes with air-gap embedded package. Microelectron. Reliab. **52**, 933–936 (2012)
103. H.T. Huang, Y.P. Huang, C.C. Tsai, Planar lighting system using array of blue LEDs to excite yellow remote phosphor film. J. Disp. Technol. **7**, 44–51 (2011)
104. H.C. Chen, K.J. Chen, C.H. Wang, C.C. Lin, C.C. Yeh, H.H. Tsai, M.H. Shih, H.C. Kuo, T.C. Lu, A novel randomly textured phosphor structure for highly efficient white light-emitting diodes. Nanoscale Res. Lett. **7**, 1–5 (2012)
105. R.M. Farrell, E.C. Young, F. Wu, S.P. DenBaars, J.S. Speck, Materials and growth issues for high-performance nonpolar and semipolar light-emitting devices. Semicond. Sci. Tech. **27** (2012)
106. H. Zhao, G. Liu, J. Zhang, J.D. Poplawsky, V. Dierolf, N. Tansu, Approaches for high internal quantum efficiency green InGaN light-emitting diodes with large overlap quantum wells. Opt. Express **19**, A991–A1007 (2011)
107. J. Zhang, N. Tansu, Improvement in spontaneous emission rates for InGaN quantum wells on ternary InGaN substrate for light-emitting diodes. J. Appl. Phys. **110**, 113110 (2011)
108. H. Zhao, G. Liu, R.A. Arif, N. Tansu, Current injection efficiency induced efficiency-droop in InGaN quantum well light-emitting diodes. Solid-State Electron. **54**, 1119–1124 (2010)
109. C.H. Wang, S.P. Chang, P.H. Ku, J.C. Li, Y.P. Lan, C.C. Lin, H.C. Yang, H.C. Kuo, T.C. Lu, S.C. Wang, C.Y. Chang, Hole transport improvement in InGaN/GaN light-emitting diodes by graded-composition multiple quantum barriers. Appl. Phys. Lett. **99**, 171106 (2011)
110. Y.-K. Ee, J.M. Biser, W. Cao, H.M. Chan, R.P. Vinci, N. Tansu, Metalorganic vapor phase epitaxy of III-nitride light-emitting diodes on nanopatterned AGOG sapphire substrate by abbreviated growth mode. IEEE J. Sel. Top. Quantum Electron. **15**, 1066–1072 (2009)
111. Y.J. Lee, C.H. Chiu, C.C. Ke, P.C. Lin, T.C. Lu, H.C. Kuo, S.C. Wang, Study of the excitation power dependent internal quantum efficiency in InGaN/GaN LEDs grown on patterned sapphire substrate. IEEE J. Sel. Top. Quantum Electron. **15**, 1137–1143 (2009)
112. Y.-K. Ee, X.-H. Li, J. Biser, W. Cao, H.M. Chan, R.P. Vinci, N. Tansu, Abbreviated MOVPE nucleation of III-nitride light-emitting diodes on nano-patterned sapphire. J. Cryst. Growth **312**, 1311–1315 (2010)
113. Y. Li, S. You, M. Zhu, L. Zhao, W. Hou, T. Detchprohm, Y. Taniguchi, N. Tamura, S. Tanaka, C. Wetzel, Defect-reduced green GaInN/GaN light-emitting diode on nanopatterned sapphire. Appl. Phys. Lett. **98**, 151102 (2011)
114. W.B. Im, N. George, J. Kurzman, S. Brinkley, A. Mikhailovsky, J. Hu, B.F. Chmelka, S.P. DenBaars, R. Seshadri, Efficient and color-tunable oxyfluoride solid solution phosphors for solid-state white lighting. Adv. Mater. **23**, 2300 (2011)

115. S.E. Brinkley, N. Pfaff, K.A. Denault, Z. Zhang, H.T. Hintzen, R. Seshadri, S. Nakamura, S.P. DenBaars, Robust thermal performance of Sr2Si5N8:Eu2+: an efficient red emitting phosphor for light emitting diode based white lighting. Appl. Phys. Lett. **99**, 241106 (2011)
116. Y. Zhang, L. Wu, M. Ji, B. Wang, Y. Kong, J. Xu, Structure and photoluminescence properties of KSr4(BO3)(3):Eu3+ red-emitting phosphor. Opt. Mater. Express **2**, 92–102 (2012)
117. H. Li, H.K. Yang, B.K. Moon, B.C. Choi, J.H. Jeong, K. Jang, H.S. Lee, S.S. Yi, Tunable photoluminescence properties of Eu(II)- and Sm(III)-coactivated Ca9Y(PO4)(7) and energy transfer between Eu(II) and Sm(III). Opt. Mater. Express **2**, 443–451 (2012)
118. Y. Shuai, N.T. Tran, F.G. Shi, Nonmonotonic phosphor size dependence of luminous efficacy for typical white LED emitters. IEEE Photonics Tech. Lett. **23**, 552–554 (2011)
119. J.K. Kim, H. Luo, E.F. Schubert, J.H. Cho, C.S. Sone, Y.J. Park, Strongly enhanced phosphor efficiency in GaInN white light-emitting diodes using remote phosphor configuration and diffuse reflector cup. J. Appl. Phys. Part 2-Lett. Express Lett. **44**, L649–L651 (2005)
120. M.T. Lin, S.P. Ying, M.Y. Lin, K.Y. Tai, S.C. Tai, C.H. Liu, J.C. Chen, C.C. Sun, Ring Remote phosphor structure for phosphor-converted white LEDs. IEEE Photonics Tech. Lett. **22**, 574–576 (2010)
121. H.T. Huang, C.C. Tsai, Y.P. Huang, A direct-view backlight with UV excited trichromatic phosphor conversion film. J. Dis. Technol. **6**, 128–134 (2010)
122. H.C. Kuo, C.W. Hung, H.C. Chen, K.J. Chen, C.H. Wang, C.W. Sher, C.C. Yeh, C.C. Lin, C.H. Chen, Y.J. Cheng, Patterned structure of remote phosphor for phosphor-converted white LEDs. Opt. Express **19**, A930–A936 (2011)
123. M.R. Krames, O.B. Shchekin, R. Mueller-Mach, G.O. Mueller, L. Zhou, G. Harbers, M.G. Craford, Status and future of high-power light-emitting diodes for solid-state lighting. J. Dis. Technol. **3**, 160–175 (2007)
124. C. Sommer, J.R. Krenn, P. Hartmann, P. Pachler, M. Schweighart, S. Tasch, F.P. Wenzl, The Effect of the phosphor particle sizes on the angular homogeneity of phosphor-converted high-power white LED light sources. IEEE J. Sel. Top. Quantum Electron. **15**, 1181–1188 (2009)
125. N. Narendran, Y. Gu, J.P. Freyssinier-Nova, Y. Zhu, Extracting phosphor-scattered photons to improve white LED efficiency. Phys. Status Solidi **202**, R60–R62 (2005)
126. H.-C. Chen, K.-J. Chen, C.-C. Lin, C.-H. Wang, H.-V. Han, H.-H. Tsai, H.-T. Kuo, S.-H. Chien, M.-H. Shih, H.-C. Kuo, Improvement in uniformity of emission by ZrO_2 nano-particles for white LEDs. Nanotechnology **23** (2012)
127. K.-J. Chen, H.-C. Chen, M.-H. Shih, C.-H. Wang, H.-H. Tsai, S.-H. Chien, C.C. Lin, H.-C. Kuo, Enhanced luminous efficiency of WLEDs using a dual-layer structure of the remote phosphor package. IEEE J. Lightwave Technol. **31**, 1941–1945 (2013)
128. P. Reiss, J. Bleuse, A. Pron, Highly luminescent CdSe/ZnSe core/shell nanocrystals of low size dispersion. Nano Lett. **2**, 781–784 (2002)
129. I.L. Medintz, H.T. Uyeda, E.R. Goldman, H. Mattoussi, Quantum dot bioconjugates for imaging, labelling and sensing. Nat. Mater. **4**, 435–446 (2005)
130. J. Lee, V.C. Sundar, J.R. Heine, M.G. Bawendi, K.F. Jensen, Full color emission from II-VI semiconductor quantum dot-polymer composites. Adv. Mater. **12**, 1102 (2000)
131. R. Xie, D. Battaglia, X. Peng, Colloidal InP nanocrystals as efficient emitters covering blue to near-infrared. J. Am. Chem. Soc. 129, 15432– (2007)
132. X. Wang, W. Li, K. Sun, Stable efficient CdSe/CdS/ZnS core/multi-shell nanophosphors fabricated through a phosphine-free route for white light-emitting-diodes with high color rendering properties. J. Mater. Chem. **21**, 8558–8565 (2011)
133. E. Jang, S. Jun, H. Jang, J. Llim, B. Kim, Y. Kim, White-light-emitting diodes with quantum dot color converters for display backlights. Adv. Mater. **22**, 3076–3080 (2010)
134. K.-J. Chen, H.-C. Chen, K.-A. Tsai, C.-C. Lin, H.-H. Tsai, S.-H. Chien, B.-S. Cheng, Y.-J. Hsu, M.-H. Shih, C.-H. Tsai, H.-H. Shih, H.-C. Kuo, Resonant-enhanced full-color emission of quantum-dot-based display technology using a pulsed spray method. Adv. Funct. Mater. **22**, 5138–5143 (2012)

Part V
Solar Cells

Chapter 9
Solar Cells

Ching-Fuh Lin

Symbols

kW	10^3 Watt
TW	10^{12} Watt
kWh	Kilowatt hour
B	Boron
ν	Frequency of light
$h\nu$	Photon energy
E	Energy
eV	Electron volt
nm	Nanometer
I	Electric current
I_0	Dark current
I_{pc}	Photo current
exp	Exponential
k	Boltzmann constant
T	Temperature
V_{oc}	Open-circuit voltage
ln	Nature log
I_{sc}	Short-circuit current
R	Resistor
I-V	Current-voltage
V	Voltage
P_m	Maximum power

C.-F. Lin (✉)
Innovative Photonics Advanced Research Center, Graduate Institute of Photonics and Optoelectronics, Department of Electrical Engineering, National Taiwan University, Taiwan, No. 1, Sec. 4, Roosevelt Rd, Taipei 10617, Taiwan
e-mail: lincf@ntu.edu.tw

C.-C. Lee (ed.), *The Current Trends of Optics and Photonics*,
Topics in Applied Physics 129, DOI 10.1007/978-94-017-9392-6_9

I_m	Maximum current
V_m	Maximum voltage
FF	Fill Factor
η	Efficiency
P_{in}	Power input
Ec	Conduction band
S	Ground state
S^*	Excited state
h	Planck constant
μm	Micrometer
I^-	Iodide ion
I_3^-	Triiodide ion
e^-	Electron
Pb	Lead

9.1 Introduction

Since 1765, the world had been reshaped with the steam engine invented by James Watt. This is generally known as the first industrial revolution. After then, human labors and animal power were replaced with machines and fossil energy became the major resources to power modern industry. Energy consumption from fossil fuels that include coal, oil, and natural gases grows almost 40 times from 1900 to 2000. No reverse trend is foreseen in the near future.

The use of fossil energy for more than one hundred years had created three crises. The first is certainly the possible depletion of fossil energy. The second is the global warming due to the emission of greenhouse gases. The emission of carbon dioxide reached the historical record of 30,600 million tons in 2010. In comparison, it was less than 15,000 million tons in 1970. Even the worse, after 2004, non-OECD countries generated more carbon dioxide than OECD countries. Such fact certainly worsens the emission of carbon dioxide. Many experts attribute the global worming to the emission of greenhouse gases like carbon dioxide. A recent report from NASA, USA, states that the global temperature has an obvious increase since 1980. As compared to the average temperature between 1951 and 1980, the global temperature has increased 0.5–0.6 °C. The north hemisphere has even more increase. Many glaciers disappeared. The worst part is that, on Greenland, the ice melts to dilute and lower the salinity of the Arctic sea and then significantly influences the conveyed ocean currents.

The global warming had further led to the third crisis of dramatic weather change. Many places have record rainfalls or extreme wind speed in recent years. For example, the Mississippi River in the United States had huge floods in April and May 2011 because two major storms created record levels of rainfall. Such flood spread over several states, including Missouri, Illinois, Kentucky, Tennessee, Arkansas, Mississippi, and Louisiana. In the coming year, October, 2012, hurricane Sandy also severely damaged east coast of the United States. Its wind diameter is nearly 2,000 km, covering a broad range of area, from eastern part of the US and Canada. It affected twenty-four states, including south state of Florida and the north state of Maine. Although it is controversial to directly attribute the dramatic weather change to global warming, such phenomena cannot be experimented in the laboratory. The coincidence between the temperature rise and the detrimental weathers in recent years give us a strong warning. It is time for us to reduce global warming. Therefore, using other energy resources that do not emit greenhouse gases to replace fossil energy is an important issue.

Solar, hydropower, ocean wave, tide, biomass, wind, and geothermal energies are all included in renewable energies. Hydropower has a long history of being used by human beings and still generates about 15 % of global electricity. Unfortunately, hydropower relies deeply on large rivers with steady water current. Ocean waves and tides need to take into account the corrosion of salty water and are still not mature enough for practical implementation. Biomass needs huge lands to grow plants and could possibly cause food crisis due to its land competition with agricultural crops. Geothermal also relies on particular areas that are closely linked to earthquake zones. Thus, among all renewable energies, wind and solar are probably the two mostly developed technologies that have the potential to generate significant electricity worldwide. The global wind power has the potential of approximately 1.3×10^{16} kW. This amount is about 850 times of global power demand nowadays. Some estimate that about 72–170 TW is extractable in a cost-competitive manner. This amount is still much larger than human needs. However, the wind power is usually not stable and dramatically influenced by geographical conditions.

In comparison with other renewable energies, solar energy has more advantages. First, it is almost everywhere. For the most populated areas, from equator to 40° of latitude, sunshine is abundant. Second, solar power is 174,000 TW. The human demand of power is about 16 TW, so solar energy is about 11,500 times of human needs. With the solar intensity of 1 kW/cm^2 and 15 % of efficiency, to generate 16 TW of power for global consumption, we need an area of about 1×10^5 km^2, which is only 0.07 % of total land area on earth surface. For a regular house with 100 m^2 roof, even with only 10 % efficiency, using solar panels on the roof will give 10 kW of power. If the house is in the region that has 3.5 h of equivalent sunshine, it will have 35 kWh of electricity each day. Then it will generate energy of over 1,000 kWh per month. Such energy can sufficiently provide the need for a usual household.

Because solar energy has the above potential, this chapter will focus on the topic of solar cells. We will start with the operation principles. Then several types of solar cells will be discussed in depth.

9.2 Principles of Solar Cells

9.2.1 Semiconductor p-n Junction Solar Cells

The first solar cell had been invented for more than 70 years. In 1940, R. Ohl discovered that the silicon p-n junction was light sensitive [1]. He patented it in 1941. Then D. Chapin, C. S. Fuller and G. Pearson reported 4.5 and 6 % from a silicon solar cell [2] in 1954. This solar cell was also based on p-n junction. The first energy crisis in 1973 had sparked the fast development of solar cells, so less than 15 years later the first Si solar cell beyond 20 % was reported in 1985. The Si-based solar cells are still the main stream of commercial use. Other semiconductor types of solar cells are also rapidly developed. They all have similar operation principles that are based on p-n junction. Therefore, the operation principle of semiconductor p-n junction solar cells will be first discussed here.

Before discussing p-n junction, we shall review p-type and n-type semiconductors briefly and take Si as an example of semiconductors to understand their characteristics. A silicon semiconductor that is not doped with any impurity is called intrinsic. When an intrinsic Si semiconductor is doped with impurity atoms that have one more electron than Si in the outer electronic orbit, e.g. P, then there will be more electrons than holes. This extra electron can move freely in the semiconductor. Because electron is negatively charged, such doped semiconductor is called n-type semiconductor. However, the n-type semiconductor is not negatively charged. The atom with one more electron than Si becomes positively charged when this extra electron leaves it. The extra electrons and the positively charged atoms are equal in numbers, so the overall charge of the semiconductor is neutral.

Similarly, when an intrinsic Si semiconductor is doped with impurity atoms that have one electron less than Si in the outer electronic orbit, e.g. B (Boron), there will be less electrons than holes. A hole represents that the orbit is lack of an electron. Thus a hole is positively charged and such semiconductor is thus called p-type. The electron in the neighboring locations can move to this orbit that is lack of an electron. This is equivalent to hole moving to another location. Therefore a p-type semiconductor has more holes than electrons. Also, the p-type semiconductor is not positively charged. Because the atom with one electron less than Si becomes negatively charged when another electron comes to fill up this orbit. The numbers of the holes and the negatively charged atoms are equal, so the overall charge of the semiconductor is neutral.

A p-n junction is formed when a p-type semiconductor and an n-type semiconductor are brought together to have a close contact. Figure 9.1 shows a schematic of the p-n junction. Near the junction, the holes in the p-type semiconductor diffuse to the n-type semiconductor because the hole concentration there is higher than the n-type semiconductor. As holes move to the n-type semiconductor, the left-behind atoms become negatively charged ions. Similarly, electrons in the n-type semiconductor diffuse to the p-type semiconductor, so the atoms there become positively charged. Although the positively charged ions and the

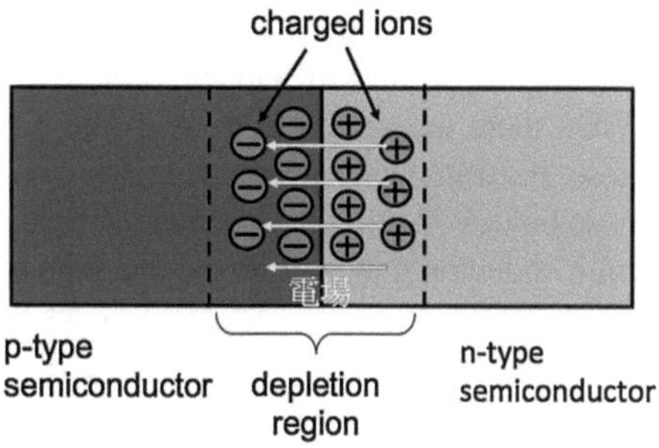

Fig. 9.1 A schematic of the semiconductor p-n junction

negatively charged ions are close near the p-n junction, they cannot move because those ions are located at the lattice sites in the semiconductor crystal. However, the positively charged ions and the negatively charged ions will create a huge electric field, called built-in field. The region that is occupied with the charged ions is called depletion because there are very few electrons and holes in this region. Thus the built-in field is in the depletion region. The direction of the field is indicated by the arrow signs in Fig. 9.1.

This built-in electric field in the depletion region is very important for the operation of semiconductor p-n junction solar cells. First, let's assume that an extra electron-hole pair is generated in the depletion region no matter how it is generated. Because this field is very strong, the electron and the hole will be pulled away and transport toward opposite directions. Electron will be forced to move to the n-type semiconductor and the hole is pulled toward the p-type semiconductor.

Next, let's discuss how the extra electron-hole pair is generated. One of the possibilities is that the p-n junction is illuminated by light. In this case, light is usually viewed as a collection of many photons. The energy of a photon is proportional to the light frequency. If the frequency of light is ν, then the corresponding photon energy is equal to $h\nu$, i.e., $E = h\nu$. If the energy of photon is larger than the bandgap of semiconductors, it can be absorbed and cause the electron in the valence band to transit to the conduction band. That is, an electron in the conduction band and a hole in the valence band are generated. From this point of view, the bandgap of semiconductor is an important factor that influences the absorption of light in the semiconductors. For Si, its bandgap is 1.12 eV, so light with energy $E = h\nu > 1.12$ eV will be absorbed. In other words, if the wavelength of light is less than 1,100 nm, it can be absorbed by Si.

The photo-generated electron-hole pair can form an exciton, which is a particle similar to a hydrogen atom with the negatively charged electron orbiting around the positively charged hole. The exciton has an energy less than an electron-hole pair that is not bonded. This energy is called binding energy of exciton. In semiconductors, the exciton binding energy is small. For example, in Si, it is only 15 meV. This energy is much smaller than thermal energy, so most of the excitons in Si will dissociate and become freely moving electrons and holes. Then those electrons and holes will move according to what is described previously.

On the other hand, if extra electron-hole pairs are generated outside of the depletion region, what will happen? Let us take n-type semiconductor as an example. When there are more electrons in the n-type semiconductors at a certain location, those electrons will diffuse. Similarly, when there are more holes than background, those holes will diffuse. Nonetheless, because holes are minority carriers in the n-type semiconductor, they will recombine with electrons sooner or later, depending the diffusion length or carrier lifetime of holes therein. If those holes are able to diffuse to the depletion region, they will not be recombined. Instead, the built-in electric field will pulled them toward the p-type semiconductor. If the distance between the depletion region and the location that extra holes are generated is less than the diffusion length, the extra holes will very likely move to the depletion region before being recombined with electrons. As a result, the carrier lifetime or diffusion of minority carriers is a very important parameter for semiconductors. Similarly, as extra electron-hole pairs are generated in the p-type semiconductors, extra electrons will be recombined if they cannot diffuse to the depletion due to recombination with holes. On the contrary, if the location that generates extra electrons is within the diffusion length from the depletion region, the electrons are most likely able to move to the depletion region and then pulled by the electric field toward the n-type semiconductor. From another point of view, if the minority carriers, i.e., electrons in the p-type semiconductor or holes in the n-type semiconductor, are generated at the locations that are far from the depletion region, more than the diffusion distance, they will mostly recombine with majority carriers and disappear, so cannot move to the opposite side of the p-n junction.

As mentioned previously, if extra electrons or holes are generated in the depletion region or at a location not far from the depletion region, they will possibly move to the opposite side of the p-n junction and then accumulate there. Now if the p-n junction is connected through an external circuit, as shown in Fig. 9.2, the accumulated electrons in the n-type semiconductor can flow through the circuit to the p-type semiconductors and then recombine with holes there. The accumulated holes in the p-type semiconductor recombine with electrons and disappear. This process gives us the current in the external circuit, which can be utilized for practical applications. This current is generated from the light illumination and so called photo current, usually indicated as Ipc.

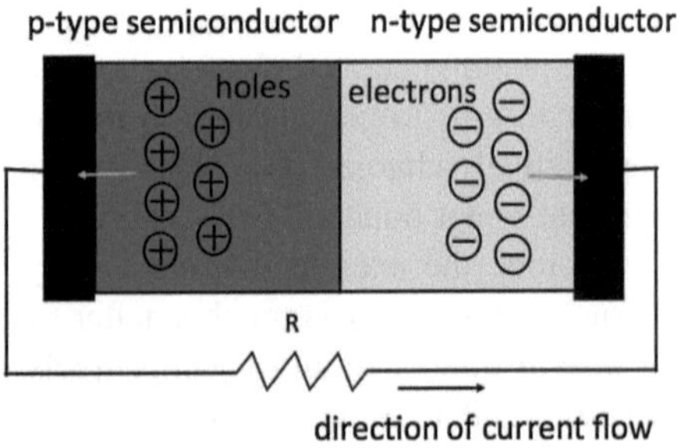

Fig. 9.2 A p-n junction is connected through an external circuit

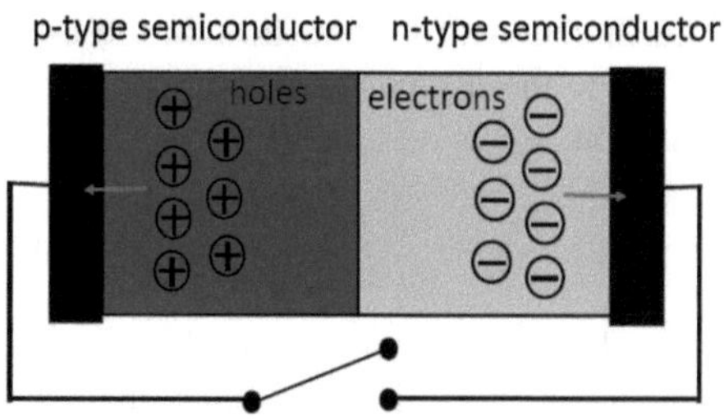

Fig. 9.3 A p-n junction under open circuit

However, if the p-n junction is not connected externally, it becomes open circuit, as shown in Fig. 9.3. Then there will be no current flowing through the external circuit. As a result, electrons and holes that transport to the n-type and p-type semiconductors, respectively, will accumulate there. Because electrons and holes are charged particles, they will give rise to another field and a measurable voltage from the external circuit. This field can counter-balance the built-in field in the depletion region. Thus, when the forces from these two fields are equal, electrons and holes will not move further to cross the p-n junction. The voltage between the p-type semiconductor and the n-type semiconductor is given implicitly in the following formula.

$$I = I_0[\exp(\frac{eV}{kT}) - 1] - I_{pc} = 0 \tag{9.1}$$

where I_0 is the dark current.

In (9.1), the current consists of two parts. One is the photo current. The other is the current due to the voltage between the p-type semiconductor and the n-type semiconductor. As explained previously, when light is illuminated on the semiconductor, electrons will move toward the n-type semiconductor and vice versa for holes. Hence the photo current flows from the n-type semiconductor to the p-type semiconductor. However, for the current due to the voltage between the two types of semiconductors, because positively charged holes accumulate in the p-type semiconductor, the voltage is higher at this side than the other side of n-type semiconductor. The induced current has the direction from the p-type semiconductor to the n-type semiconductor. Therefore, the two currents have the opposite directions and the total is then given by (9.1). When the external circuit is open, because there is no current passing through the external, the total current is equal to zero. This leads to the open-circuit voltage given as follows.

$$V_{oc} = \frac{kT}{e} \ln\left(\frac{I_{pc}}{I_0} + 1\right) \tag{9.2}$$

On the other hand, if the resistor in Fig. 9.2 has zero resistance, it becomes short-circuit. Then the voltage across the p-type semiconductor and the n-type semiconductor is zero. The current is represented as Isc. According to (9.1), this current is given by

$$I = I_{sc} = I_{pc} \tag{9.3}$$

The direction of current flow in the external circuit is from the p-type semiconductor to the n-type semiconductor. Substituting (9.3) to (9.2) gives the following equation.

$$V_{OC} = \frac{kT}{e} \ln\left(\frac{I_{SC}}{I_0} + 1\right) \tag{9.4}$$

Thus the short-circuit current, Isc, and the open-circuit voltage, Voc, are not independent. In typical applications, the solar cell is connected to some appliance in the external circuit. This appliance can be represented by a resistor, R, as shown in Fig. (9.2). Then the relation between the voltage and the current is given by

$$I = I_0[\exp(\frac{eV}{kT}) - 1] - I_{sc} \tag{9.5}$$

In (9.5), the direction of current is defined as from the p-type semiconductor to the n-type semiconductor inside the p-n junction, so its direction is opposite to the arrow sign shown in Fig. 9.2. The current-voltage (I-V) curve is schematically shown in Fig. 9.4. This curve has an interception with the horizontal axis at Voc and an interception with the vertical axis at Isc. The actual operation point depends on the resistor in the external circuit. The relation of voltage and the current of the resistor is given by

$$\mathrm{V = IR} \tag{9.6}$$

Here the direction of current I is the same as the arrow sign shown in Fig. 9.2. To make the two equations (9.5) and (9.6) have the same direction of current flow, (9.6) should be replaced with (9.7).

$$\mathrm{V = -IR} \tag{9.7}$$

The operation point is the interception of the two curves from (9.5) and (9.7), as shown in Fig. 9.5. This point will certainly vary with the resistance. At one extreme, if the resistance is zero, the curve of V = IR is the vertical axis. Then the interception is at I = Isc and V = 0. Because V = 0, there will be no output power. For another extreme, if the resistance is infinite, then I = 0, so the curve of V = IR is the horizontal axis. The interception becomes I = 0 and V = Voc. There will be no output power, either, because there is no output current.

The above two cases are extreme situations. In practical conditions, the curve of V = −IR, which is actually a straight line, locates between the vertical axis and

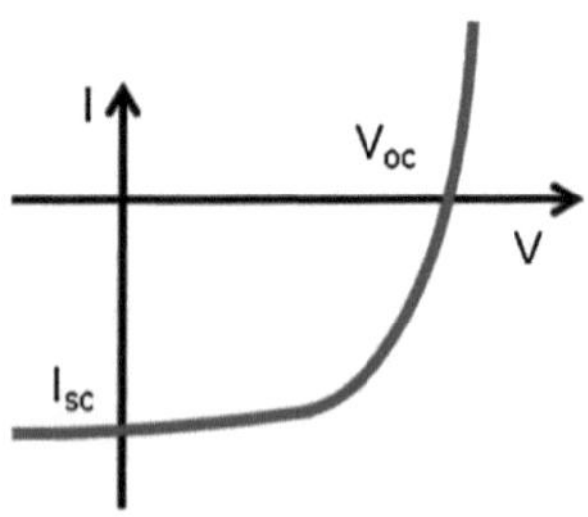

Fig. 9.4 Current-voltage curve of a solar cell under light illumination

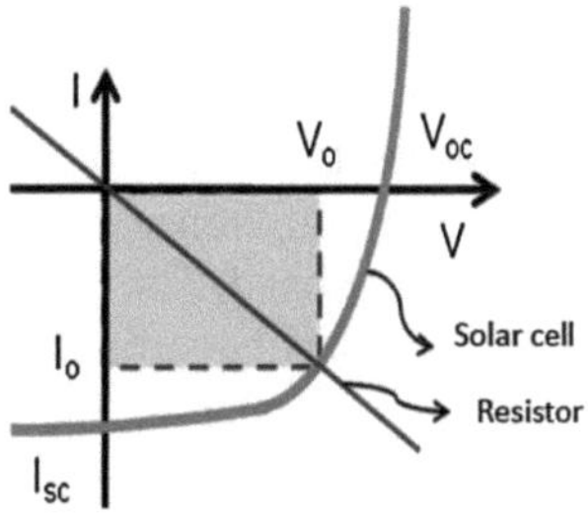

Fig. 9.5 Current-voltage curve of a solar cell intercepts with the curve of an external resistor

the horizontal axis. As schematically shown in Fig. 9.5, the output voltage is Vo and the output current is Io. The output power is given by $P_o = I_oV_o$. As we vary the resistance R, the output power will vary and has a maximum, denoted as P_m. Its corresponding voltage and current are V_m and I_m, respectively.

$$P_m = I_m \times V_m \tag{9.8}$$

The maximum power is less than the product of Isc and Voc. The ratio of Pm to the product of Isc and Voc is named as Fill Factor (FF).

$$FF = \frac{P_m}{I_{sc}V_{oc}} = \frac{I_mV_m}{I_{sc}V_{oc}} \tag{9.9}$$

Therefore, the output power Pm can be written as Pm = Isc × Voc × FF.

The efficiency of a solar cell, η, is then given by

$$\eta = \frac{P_m}{P_{in}} = \frac{I_mV_m}{P_{in}} \tag{9.10}$$

where P_{in} is the incident power of sunshine. Because solar intensity varies with the latitude and weather, it is not a constant value. To make the evaluation of solar cells trustful, a standard solar intensity is defined as 100 mW/cm^2 or 1 kW/m^2. Hence, the evaluation of solar cells is usually taken from a unit area.

9.2.2 Organic Solar Cells

Organic solar cells had been developed before 1983 [3]. The first noticeable device of organic solar cells was the two-layer structure with 1 % efficiency discovered by Tang [4] and then 5 % device invented in 2000 [5]. By convention, organic solar cells are made of organic materials that are conductive. Organic materials differ from usual inorganic semiconductors in several aspects. First, conductive organic materials have much smaller mobility than inorganic semiconductors. Second, the binding energy of excitons in organic materials is much larger than that in inorganic semiconductors. Third, the lifetime and the diffusion length of excitons in organic materials are much shorter than the lifetime and the diffusion length of carriers in inorganic semiconductors. Such differences make the operation principle of organic solar cells different from semiconductor p-n junction solar cells. However, there is one similarity between these two types of materials. They have a bandgap. For inorganic semiconductors, the bandgap is the energy difference between the edge of the conduction band and the edge of the valence band. For organic materials, the bandgap is between the lowest unoccupied molecular orbital (LUMO) and the highest occupied molecular orbital (HOMO). Similar to the edge of the conduction band, LUMO is the lowest energy level of a group of closely spaced energy levels. Also, HOMO is the highest energy level of a group of closely spaced energy levels. Between LUMO and HOMO, there are no energy levels.

There are also two types of organic materials, named donor and acceptor. Donor is the kind of molecules that are easier to give away electrons than accepting electrons when they are close to another type of molecules. On the contrary, acceptor is the kind of molecules that easily receive electrons. Whether a molecule is easy to give away or to receive electrons is determined by the electronegativity. The larger the electronegativity of a molecule is, the more it is likely to attract electrons. Thus a donor has a smaller electronegativity than an acceptor.

Because the binding energy of excitons in organic materials is large, it is unlikely that electrons and holes are unbounded. Hence there are not many freely moving electrons and holes in organic materials. Therefore, when the donor and the acceptor are in touch with one another, electrons and holes do not move from one side to anther like in semiconductors. Therefore, there is no depletion region. As a result, there is no built-in field to pull electrons from the donor to the acceptor and to pull holes from the acceptor to the donor. Then how excitons dissociate and release electrons and holes? Also, how electrons and holes are separated and move toward the opposite directions to form current flow in the external circuit?

First, the organic solar cells have two electrodes of anode and cathode with different work functions. Usually anode has a higher work function than the cathode. In equilibrium, the work function will be forced to be at the same energy, so causing the energy levels to tilt, as shown in Fig. 9.6a. The tilted energy levels make electrons tend to move toward the cathode and the holes toward anode. The tiled energy field also indicates that there is an electric field between the anode and the cathode. However, this field is not as strong as the built-in field in the depletion region of the p-n junction in semiconductors. Next, when a photon incident on the donor, it will be absorbed and cause the electron below the HOMO level transit to the energy level above the LUMO level, as shown in Fig. 9.6b, leaving the empty energy level near the HOMO level, which is equivalent to a hole. Then in a very short period of time, the electron and hole will form an exciton. The electron in the exciton corresponds to an energy level lower than the LUMO level, as shown in Fig. 9.6c. An exciton has neutral charge, so it is actually not influenced by the tilted energy level. The excitons may move toward all directions. Here we focus on those that move toward the interface between the donor and the acceptor, as shown in Fig. 9.6d. If the energy level of the electron in the exciton is larger than the LUMO level of the acceptor, this electron will transit to the LUMO level of acceptor, as shown in Fig. 9.6e. At this point, the electron and the hole will separate. That is

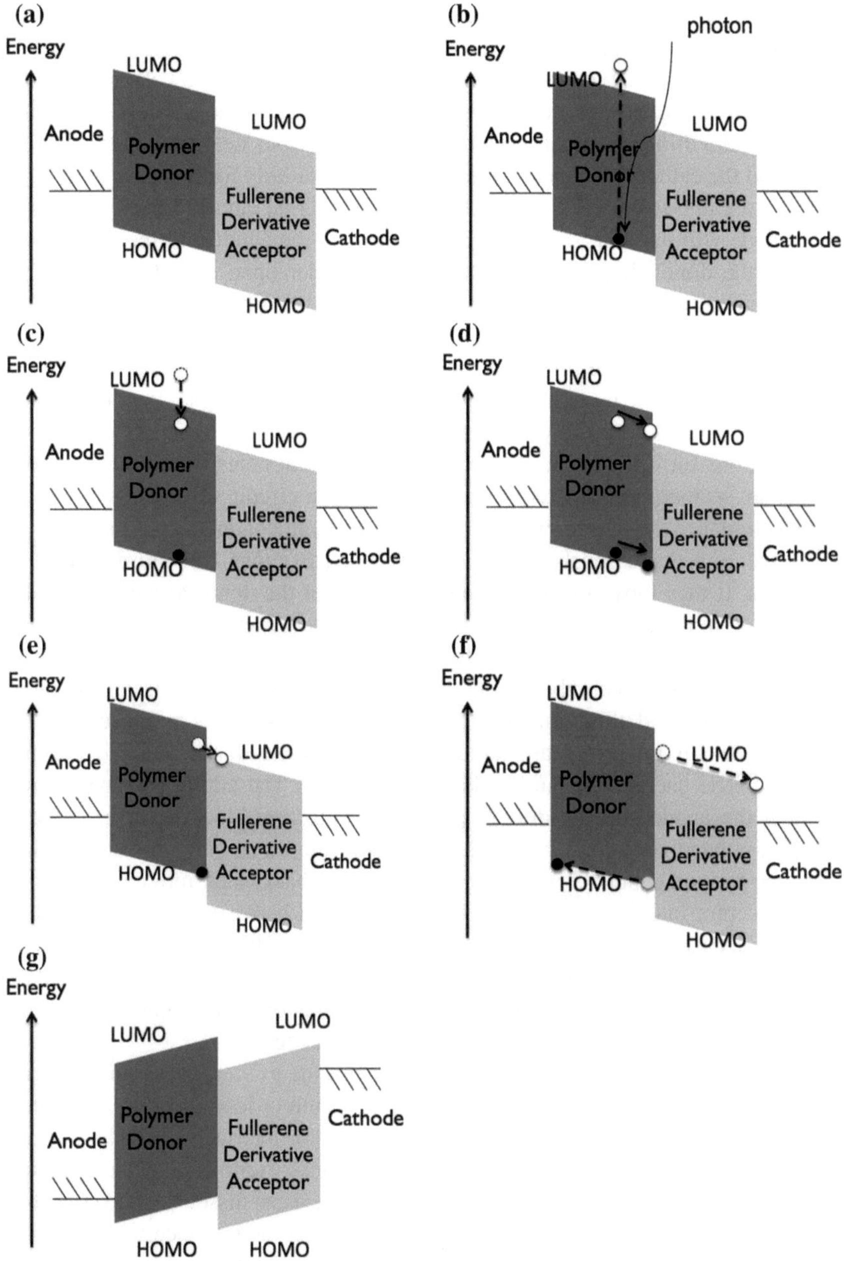

Fig. 9.6 **a** Schematic energy levels of an organic solar cell; **b** incidence of a photon causing the electron under the HOMO level to transit to the high energy level; **c** the electron at the high energy level relax to another energy level lower than the LUMO level of donor-; **d** the electron and the hole forms an exciton and move toward the interface of the donor and the acceptor; **e** at the interface, the electron transit to the LUMO level of the acceptor; **f** the separated electron and hole transport toward the opposite directions; **g**) under open circuit, electrons and holes accumulate in the acceptor and the donor-, respectively, causing the energy levels to reverse the tilting direction

the exciton dissociate at the interface of the donor and the acceptor. As long as the electron and the hole are separated, they will move to the opposite electrodes, as shown in Fig. 9.6f, due to two reasons. One is the tilted energy level or the electric field that pulls the electron to the cathode and the hole to the anode. The other is caused by the diffusion because there are more electrons near the interface of the donor and the cathode than near the electrode and similarly for holes.

If the organic solar cell is connected with an external circuit, the electrons near the cathode will pass through this external route to the anode and recombine with holes in the donor. If the organic solar cell is under open circuit, electrons and accumulate in the acceptor and holes will accumulate in the donor. Because of the negative charge of electrons and the positive charge of holes, they will build an electric field that has the direction opposite to the one set up by the cathode and anode. In the extreme case, it will be even larger than the original one, making the energy levels tilted oppositely, as shown in Fig. 9.6g. Such tilted energy levels will stop electrons further move to the acceptor and prevent holes from moving toward the anode, so there will be zero current at the open-circuit voltage. Its I-V curve will be similar to the one shown in Fig. 9.4 for the semiconductor p-n junction.

As mentioned previously, the photo-generated excitons can move toward all directions. If they move toward the anode instead of the donor-acceptor interface, the electron in the exciton will relax to the lower energy levels in anode, too, leading the dissociation of exciton, while holes will also move to anode. Then electrons and holes will recombine quickly in the anode, so there will be no current generated. To avoid this situation, a sandwiched structure is proposed. In this structure, two more layers are added to sandwich the donor and the acceptor materials. The energy diagram is shown in Fig. 9.7. The additional layer placed between the anode and the donor is called the electron-blocking layer. It has a conduction-band edge higher than the LUMO of donor, so electrons are blocked. Also, the exciton cannot dissociate and is bounced back, then moving toward the donor-acceptor interface. Similarly, the layer inserted between the cathode and the acceptor has a valance-band edge much lower than the HOMO of the acceptor, so holes are blocked. It is hence called hole-blocking layer. This layer is particularly important if the acceptor also absorbs light and exciton can be generated in this layer. The process is similar to the one described for the donor.

The above description is still too ideal. Because the exciton of organic materials has very short diffusion length, typically around 10 nm or less, most of the excitons will disappear before they move to the donor-acceptor interface if the thickness of the donor is larger than 20 nm. As a result, only those excitons generated approximately within 10 nm from the donor-acceptor interface can dissociate and contribute to the photo current. Those that are generated far from the interface will disappear. If the layer of donor is thick and light is incident from the donor side, most of light will be absorbed by the donor material near the anode. Although they will generate excitons, but very few are able to move to the interface without recombination. To increase the ratio of excitons able to reach the donor-acceptor interface, the layer of the donor has to be thin. However, a thin donor- layer cannot absorb much light, so not much light is converted to electricity. This becomes a contradictory condition.

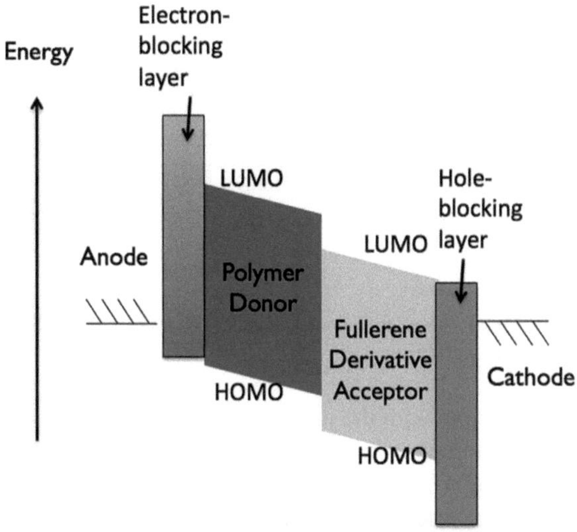

Fig. 9.7 Band diagram with two more layers added to sandwich the donor- and the acceptor materials

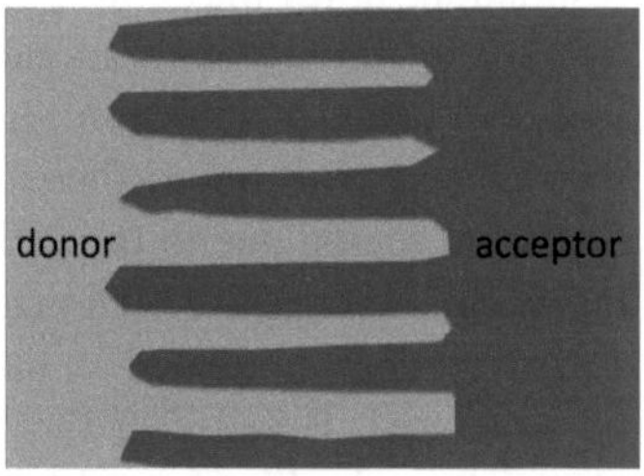

Fig. 9.8 Inter-penetrated nano-network of the donor- and the acceptor

To overcome this deadlock, a bulk heterojucntion structure is proposed. A schematic of the structure is shown in Fig. 9.8. In this structure, the donor and the acceptor inter-penetrate into each other with many fingers. Ideally, each finger has the cross section of less than 20 nm, so the photo-generated excitons can meet the donor–acceptor interface quickly to provide fast dissociation of excitons. In the meantime, the effective thickness of the donor- and the acceptor is increased, so the light absorption is also increased. The structure in Fig. 9.8 is usually called inter-penetrated nano-network because those penetrated fingers are in the nanometer scale.

9.2.3 Hybrid Heterojunction Solar Cells

Hybrid heterojunction solar cells include several types. Here we focus on the type that consists of an inorganic semiconductor and an organic material. The heterojunction formed by an inorganic semiconductor and an organic material is different from the semiconductor p-n junction. Figure 9.9a schematically shows the relative energy levels of the two materials before they are in contact. Here we use

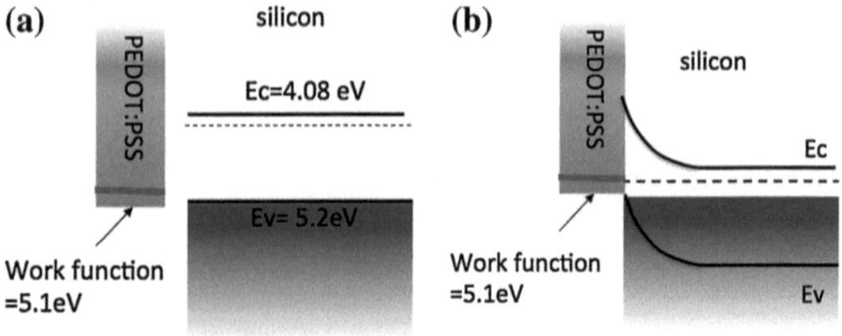

Fig. 9.9 **a** The relative energy levels of Si and PEDOT:PSS; **b** The band diagram when the system is in equilibrium

the example with the junction formed from the n-type silicon and poly(3,4- ethylenedioxythiophene):poly(styrenesulfonate) (PEDOT:PSS). When the two materials are brought to contact one another, they will interact and then achieve equilibrium. Because the work function of PEDOT:PSS is very close to the valence-band edge of Si, the excitons in PEDOT: PSS can possibly dissociate and the corresponding holes may transport to Si. As the holes move to n-type Si, they will quickly recombine with an electron. As a result, the bands of Si near the interface will bend, as shown in Fig. 9.9b. Because the electrons near the interface are recombined with holes, there will be much less electrons. There will be also a depletion region due to the lack of electrons. There will also be a built-in field in the depletion region, which mainly locates in the silicon side.

As light is illuminated on the Si, it will be absorbed and electron-hole pairs will be generated. Because of the band bending, electrons will be forced toward the electrode at the Si side and holes will move to PEDOT:PSS and then to the electrode connected with PEDOT:PSS. If the solar cell is connected with an external circuit, electrons will pass through it, then enter the PEDOT:PSS, and recombine with holes there. If the solar cell is under open circuit, electrons will accumulate in Si, while holes will accumulate in PEDOT:PSS. The accumulate electrons and holes at the opposite sides will create another field to balance the original built-in field, leading to the flat band and so a measurable open-circuit voltage.

9.2.4 Dye-Sensitized Solar Cells

The conventional device structure of a dye-sensitized solar cell (DSSC) consists of the mesoporous TiO film, the dye adsorbed on the TiO_2 film, the iodine based redox electrolyte ($I^-I_3^-$ electrolyte), and the counter electrode, as shown in Fig. 9.10.

The mesoporous TiO film is usually coated on a fluorine-doped SnO transparent electrode (FTO) as the photoelectrode. The process starts from coating TiO_2 paste. The paste is then sintered at 450–500 °C. It produces a TiO_2 film with thickness of ~10 μm. Then the dye is adsorbed on the surface of the porous structure of the TiO_2 film. Because this mesoporous film is composed of TiO_2 nanoparticles

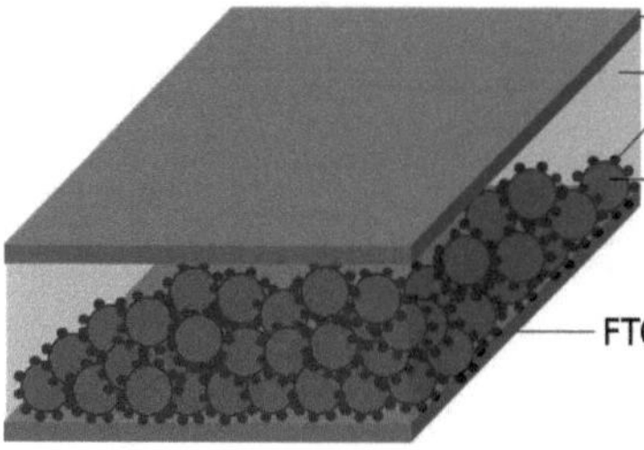

Fig. 9.10 The conventional device structure of DSSC

(~10–30 nm), it has huge surface area and the amount of dye adsorbed is very large. The amount of dye adsorbed significantly influences the light harvesting efficiency. Practically, in order to scatter the incident light effectively to increase the light absorption, this TiO_2 film normally contains large TiO_2 particles (~250–300 nm). Theoretically, the larger thickness of TiO_2 film is expected to adsorb more dyes. However, a thick film may increase the charge recombination between the injected electrons and the dye due to the numerous grain boundaries between TiO_2 nanoparticles and the low electron diffusion coefficient in the TiO_2 film [6]. Thus the TiO_2 film should be controlled to near the optimized thickness [7].

For the adsorption of dye, the dye is attached to the TiO_2 surface through the carboxyl groups (COO^- and COOH moieties). It can be verified by Raman spectroscopy [8]. Under the illumination of light, in the anode, the excited electron is injected from the dye to the TiO_2 photoelectrode. For the typical Ru complex dye, the metal-to-ligand charge-transfer (MLCT) transition is observed [9]. Furthermore, in the cathode, the transition of electrons between the dye and the counter electrode is conducted by the electrolyte. Typical electrolyte used in the DSSC contains I^-/I_3^- redox ions. Pt-coated TCO substrate is usually used as the counter electrode.

The operation principle of the DSSC and the energy band diagram are illustrated in Fig. 9.11.

Under the illumination, the dye is excited from the ground state (S) to the excited state (S^*).

$$S + h\nu \rightarrow S^*$$

The excited electrons are injected into the conduction band of the TiO_2 photoelectrode, resulting in the oxidation of the dye. Injected electrons in the conduction band of TiO_2 are transported between TiO_2 nanoparticles with diffusion toward FTO.

$$S^* \rightarrow S^+ + e^-(TiO_2)$$

The oxidized dye (S^+) accepts electrons from I^- ion mediator (redox electrolyte), regenerating the ground state (S). I^- is oxidized to the oxidized state, I_3^-.

$$2S^+ + 3I^- \rightarrow 2S + I_3^-$$

The oxidized redox mediator, I_3^-, diffuses toward the counter electrode and is reduced to I^- ions.

$$I_3^- + 2e^- \rightarrow 3I^-$$

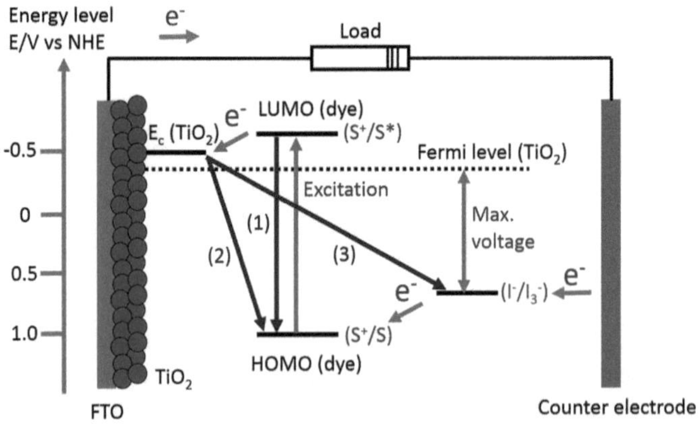

Fig. 9.11 Energy-band diagram and operation principle of a typical DSSC. The electron transfer processes are illustrated with blue arrows. Unfavorable back-reaction processes are illustrated with red arrows

Thus, the electric power is generated. The obtained photocurrent is determined by the energy gap of the dye. The open-circuit voltage produced is determined by the energy gap between the Fermi level of TiO_2 and the potential of electrolyte. In theory, the smaller LUMO-HOMO energy gap of dye is, the larger photocurrent will be achieved. However, the LUMO and the HOMO energy levels of the dye need to be appropriately designed with the consideration of the energy levels of TiO_2 and electrolyte. For the efficient charge injection, it requires a driving force of about 100–200 meV [10]. In order to inject electrons effectively, the LUMO level of dye must be sufficiently negative in comparison with the conduction band of TiO_2. Also, in order to accept electrons effectively, the HOMO level of dye must be sufficiently positive in comparison with the potential of electrolyte.

Unfavorable back-reaction processes in a typical DSSC include (1) the relaxation of the excited state of dye, (2) the charge recombination between the injected electrons and the dye cations, and (3) the charge recombination between the injected electrons and the electrolyte [11], as shown with red arrows in Fig. 9.11. Electron injection from the dye into the TiO_2 photoelectrode typically happens in the femtosecond to picosecond time scale [12]. The process is much faster than the charge recombination process, which typically happens in the micro to millisecond time scale. In addition, the back reaction of a conduction band electron to the oxidized dye is much slower than the reduction of the oxidized dye by the electrolyte. Thus this cell structure gives the electrons sufficient time to be collected by the contacts.

9.2.5 Tandem Solar Cells

The above solar cells all use single junction. The single-junction solar cells all have their output voltage limited by the bandgap of materials because of

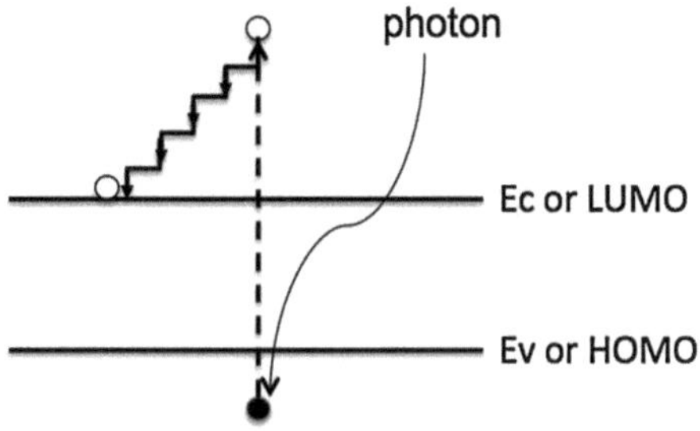

Fig. 9.12 Illustration of electron-phonon scattering process that causes electron to lose energy and relax to the band edge or the LUMO level

significant thermal loss. As illustrated in Fig. 9.12, as the photon energy is much larger than the bandgap, the electron will transit to an energy level much higher than the conduction-band edge of the semiconductors or the LUMO level of organic material. This high-energy electron will then experience very fast electron-phonon scattering. The energy of electron will be transferred to phonons until its energy reaches the band edge or it becomes bounded in the exciton. The holes will experience similar process, too. Such process is very fast, only a few picoseconds. Therefore, before electrons and holes transport to the external circuit, they have already relaxed to the band edges or form excitons. As a result, lots of energy will be lost. For example, if the photon energy is 2.24 eV, being absorbed by Si, the thermal loss will be 1.12 eV because Si has a bandgap of 1.12 eV. The thermal loss will be as large as 50 %.

To reduce the thermal loss, one can use large bandgap materials. However, those materials can then only absorb light with energy larger than the bandgap. Sunlight covers a wide range of optical spectrum. As shown in Fig. 9.13 [13], the solar spectrum covers from 300 nm up to 4000 nm. If the entire solar spectrum is converted to current, there will be a large current density of around 62 mA/cm^2. Nonetheless, large bandgap materials can absorb only a small portion of light. As a result, there is a conflicting reality. Large bandgap will lead to small photo current, while small bandgap will lead to small output voltage due to the thermal loss. Consequently, the output voltage and the output current are contradictory. A tradeoff has to be made. The evaluated maximum device efficiency is about 31 % from a single junction solar cell.

To surpass this limitation due to the tradeoff, multiple junctions are proposed. Each junction is only responsible for absorbing a short range of solar spectrum. To absorb a broad spectrum of the sunlight, each junction is certainly made of one type of semiconductor that has different bandgap from other semiconductors used for other junctions. The layout is schematic shown in Fig. 9.14. Solar cell I is made of the material that has the largest bandgap. Solar cell II is made of the material that has the second largest bandgap. Others follow the same order.

Each junction of solar cell in the multi-junction structure is similar to the single junction described previously. When they are made of tandem cells, the neighboring cells have their p-side and n-side connected together. As a result, the holes from the p-side will recombine with the electrons in the n-side of another solar

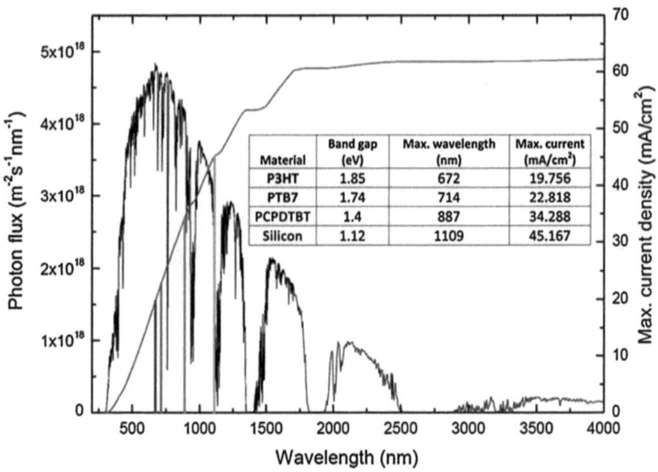

Fig. 9.13 solar spectrum [13]

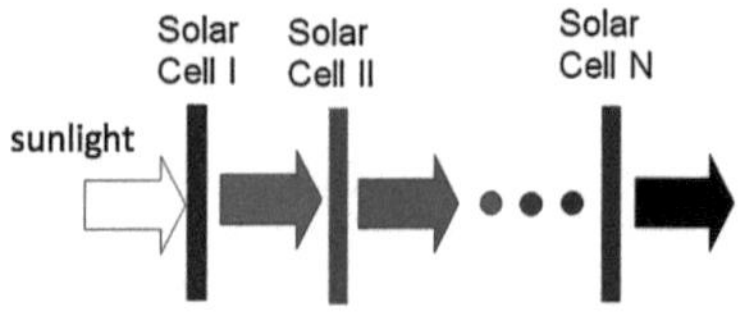

Fig. 9.14 Layout of multi-junction solar cells

cell. It has to be one to one correspondence. That is, each hole will recombine with one electron. If there are more holes or electrons, they will transport to the neighboring cell and eliminate the holes or electrons generated there, leading to the reduction of the photo current. This will result in the waste of photo-generated carriers in certain cells and so the power conversion efficiency. To maximize the power conversion efficiency, each cell should have the same current. Therefore, the tandem cells need very delicate design. Each cell should absorb the same photon number of sunlight and generate the same amount of electron-hole pair.

9.3 Progress of Various Types of Solar Cells

In this section, we will introduce various types of solar cells and their recent progress.

9.3.1 Crystalline Si Solar Cells

Crystalline Si solar cells are the most conventional type and are the main stream of solar-cell industry. The fabrication starts from the formation of raw Si materials, which are then further refined to a high purity. The high-quality Si are made as ingots, which are sliced to Si wafers using diamond saws. Afterwards, standard semiconductor processing steps are applied to form p-n junction, metal contacts, and then cells. The p-n junction will have a built-in electrical field, so as electrons and holes are generated, they are pulled away and move to opposite electrodes, following the principles described previously. Currently the best efficiency of single-crystalline Si solar cells and the multiple-crystalline solar cells is 25 and 20.4 %, respectively [14]. The overall cost of using solar cell to harvest solar energy and to convert to electricity is still larger than using fossil energies, so quite many efforts are still required to make solar energy using crystalline Si solar cells become a significant portion of future energy supplies.

9.3.2 Amorphous Si Solar Cells

Amorphous Si (a-Si) solar cells are usually with the p-i-n structure. The p-layer and the n-layer provide the function of creating a built-in electrical field. The i-layer is sandwiched between the p-layer and the n-layer. It is responsible for light absorption. The thickness of this layer is around 0.2–0.5 um. Because the a-Si has a larger bandgap than crystalline, it absorbs photons with energy larger than 1.5–1.8 eV, depending on the fine structure of a-Si. This energy is larger than 1.1 eV, bandgap of Si, so the absorption spectrum of a-Si is narrower than crystalline Si, but it has a much larger absorption coefficient, 10^4–10^5 cm^{-1}. Increasing the thickness of the i-layer can enhance the light absorption, but will also increase the probability of electron-hole recombination and hence decrease the photo-current and the open-circuit voltage.

The p-, n-, and i-layers are usually deposited using plasma-enhanced chemical vapor deposition (PECVD). They can be placed on stainless-steel foils, glass substrates, or plastics. If the flexible substrates like stainless-steel foils or plastics are used, the fabrication can be done with roll-to-roll production, which will have great advantages in mass-production, shipping, and implementation on the house roofs and racks in the fields. The a-Si solar cells still use transparent conducting oxide for light transmission and current conduction. Also they highly rely on huge vacuum processing. Thus the overall cost is only slightly lower than crystalline solar cells. In addition, the best efficiency of a-Si solar cells is only slightly larger than 10 % [14], making it less preferable than crystalline Si solar cell for practical implementation.

9.3.3 III-V Semiconductor Solar Cells

The III-V semiconductor solar cells are made of compound semiconductors that consist of group III and group V chemical elements in the periodic table. Typical

III-V compound semiconductors used for solar cells are based on GaAs, InP, and GaN. Because they are also semiconductors, the operation principle is similar to crystalline Si solar cells, as described previously. A great advantage of III-V compound semiconductors is that the bandgap can be engineered using different ratio of III-V chemical elements. For example, AlGaAs has the bangap between AlAs and GaAs, depending on the chemical ratio between Al and Ga. Also, most of the III-V compound semiconductors are direct-bandgap materials, so their light absorption coefficient is much larger than that of Si. It can be as large as 10^5 cm^{-1}, so 95 % of light can be absorbed with only 1 um of thickness. The disadvantage of the III-V semiconductor solar cells is that those semiconductors are usually grown using metalorganic vapour phase epitaxy (MOCVD) or molecular beam epitaxy (MBE), which is very expensive. Therefore, the cost of III-V semiconductor materials is very high. To reduce the material cost, III-V semiconductor solar cells are usually used in concentrated solar system, which a lens of concentrator is applied to collect a large area of sunlight to be focused on the chip of the III-V semiconductor solar cells.

On the other hand, the epitaxial growth and the possibility of bandgap engineering make the III-V semiconductor solar cells most likely formed in tandem structures, so extremely high efficiency is possible. So far, the highest efficiency is achieved from the III-V tandem solar cells in a concentrated solar system [14].

9.3.4 CIGS Thin-Film Solar Cells

CIGS thin-film solar cells are made of compound semiconductors that consist of four different chemical elements: copper, indium, gallium and selenium. The n-type or p-type of CIGS compound is controlled from the ratio of Cu_2Se and In_2Se_3. It is also a direct-bandgap semiconductor, so has a large absorption coefficient, ~10^5 cm^{-1} and about 1-um of thin film could absorb about 95 % of sun light. CIGS can be deposited on flexible substrates, so roll-to-roll production is possible, making it very attractive for relatively easy mass-production, shipping, and implementation on the house roofs and racks in the fields. In addition, a very good efficiency of 19.6 % [14] has been demonstrated, so it attracts significant attention recently. However, it also faces some challenges. It uses toxic chemical of Cd and Se. In addition, the storage of In on earth is small, limiting the possible expansion of CIGS solar cells to a very large scale. The processing steps and controls to obtain good CIGS crystal for high efficiency solar cells are complicated and involve expensive vacuum apparatuses.

9.3.5 CdTe Thin-Film Solar Cells

CdTe thin-film solar cells are formed from p-type CdTe compound semiconductor and the n-type CdS compound semiconductor. The p-type CdTe is about

2 um, while the n-type CdS is only 0.1 um. Therefore, light is mainly absorbed by the CdTe layer, which has absorption coefficient slightly less than 10^5 cm^{-1}. Its bandgap is about 1.45 eV, so it is able to absorb light from UV to about 855 nm. Because either the p-type CdTe layer or the n-type CdS layer consists of only two chemical elements, the material processing is simpler than CIGS. Its best efficiency is also 19.6 % [14]. Now some portion of commercial solar panels is made of CdTe thin-film solar cells. However, there are still some concerns. First, it uses Cd, which is toxic, so recycling Cd material is important to avoid environmental contamination. Second, the storage of Te element on earth is small, so imposing its limitation on possible expansion to a very large scale.

9.3.6 Dye-Sensitized Solar Cells

In dye-sensitized solar cells (DSSCs), dye molecules are used for light absorption. However, dye is not a conductive material, so the layer of dye has to be very thin, so photo-generated electrons and holes can be transferred to their neighboring conducting layers. Because the layer of dye is very thin, it absorbs very small amount of light. To increase the light absorption, porous structure of electron-conducting materials, e.g., TiO_2, is used and dye molecules are adsorbed on its surface. As the overall thickness of the porous electron-conducting material is thick enough, significant absorption of light can be achieved. Fortunately, quite mature technologies had already developed to fabricate the porous electron-conducting material (TiO_2) and to adsorb dye molecules. Also, to achieve good contact of dye with the hole transportation material, liquid electrolyte containing I^- and I_3^- ions is used. Such DSSCs exhibit efficiency of 11.9 % [15]. The fabrication steps are simple. Roll-to-roll production is also possible. Thus it has been deemed as a very promising technique. However, the use of liquid electrolyte makes it inconvenient to use. As a result, solid-state electrolytes become an attractive direction of improvement. Solid-state DSSCs now could have the best efficiency of 10.2 %. As long as the efficiency is further increased from the solid-state DSSCs, it should be pretty promising for practical applications. Nonetheless, its efficiency does not have significant progress over these years.

Some research groups replace the dye molecules with the perovskite materials in the similar structure. Because the perovskite material has better conductivity than dye molecules, no porous structure is required. Hence, its fabrication is even simpler than conventional DSSCs. Perovskite solar cells can be fabricated using solution process or evaporation. The solution-processed ones can have the efficiency of 11.4 % and 19.3 % [16, 17], while the evaporated ones have the efficiency of 15.4 % [18]. The only concern is the use of heavy metal Pb in perovskite, giving some risk of environmental contamination and increased cost of recycling.

9.3.7 Organic Solar Cells

Organic solar cells have the advantages of low cost, light weight, simple process, and flexibility. Compared to semiconductor types, organic solar cells can be fabricated with solution processes such as spin-coating, spray, screen-printing, dip-coating, blade coating, and so on. They can certainly be fabricated using roll-to-roll production. However, due to the short diffusion length of excitons in organic polymers, some nano-morphology of inter-link configuration is necessary. It is not easy to control. Currently the best efficiency of organic solar cells is around 10 % [14], but is limited to small areas. Large-area organic solar cells do not have high efficiency although they have great potential of low lost. Also, the easy interaction of organic material with air and moisture makes organic solar cells short lifetime, so encapsulation will be necessary.

References

1. M. Riordan, L. Hoddeson, The origins of the pn junction. IEEE Spectr. **34**, 46–51 (1997)
2. D. Chapin, C. Fuller, G. Pearson, A new silicon p-n junction photocell for converting solar radiation into electrical power. J. Appl. Phys. **25**, 676–677 (1954)
3. G.A. Chamberlain, Organic solar cells: a review. Solar Cells **8**, 47–83 (1983)
4. C.W. Tang, Two-layer organic photovoltaic cell. Appl. Phys. Lett. **48**, 183–185 (1986)
5. J.H. Schön, Ch. Kloc, B. Batlogg, Efficient photovoltaic energy conversion in pentacene-based heterojunctions. Appl. Phys. Lett. **77**, 2473–2475 (2000)
6. F. Gao, Y. Wang, D. Shi, J. Zhang, M. Wang, X. Jing, R. Humphry-Baker, P. Wang, S.M. Zakeeruddin, M. Grätzel, Enhance the optical absorptivity of nanocrystalline TiO_2 film with high molar extinction coefficient ruthenium sensitizers for high performance dye-sensitized solar cells. J. Am. Chem. Soc. **130**, 10720–10728 (2008)
7. S. Ito, S.M. Zakeeruddin, R. Humphry-Baker, P. Liska, R. Charvet, P. Comte, M.K. Nazeeruddin, P. Péchy, M. Takata, H. Miura, S. Uchida, M. Grätzel, High-efficiency organic-dye-sensitized solar cells controlled by nanocrystalline-TiO_2 electrode thickness. Adv. Mater. **18**, 1202–1205 (2006)
8. C. Pérez León, L. Kador, B. Peng, M. Thelakkat, Characterization of the adsorption of Ru-bpy dyes on mesoporous TiO_2 films with UV-Vis, Raman, and FTIR spectroscopies. J. Phys. Chem. B **110**, 8723–8730 (2006)
9. J.F. Endicott, H.B. Schlegel, M.J. Uddin, D.S. Seniveratne, MLCT excited states and charge delocalization in some ruthenium-ammine-polypyridyl complexes. Coord. Chem. Rev. **229**, 95–106 (2002)
10. http://www.diss.fu-berlin.de/diss/servlets/MCRFileNodeServlet/FUDISS_derivate'_000000002568/02_2.pdf?hosts=
11. M.V. Martínez-Díaz, M. Ince, T. Torres, Phthalocyanines: colorful macroheterocyclic sensitizers for dye-sensitized solar cells. Monatshefte für Chemie-Chemical Monthly **142**, 699–707 (2011)
12. D. Kuciauskas, J.E. Monat, R. Villahermosa, H.B. Gray, N.S. Lewis, J.K. McCusker, Transient absorption spectroscopy of ruthenium and osmium polypyridyl complexes adsorbed onto nanocrystalline TiO_2 photoelectrodes. J. Phys. Chem. B **106**, 9347–9358 (2002)
13. M. Wright, A. Uddin, Organic-inorganic hybrid solar cells: a comparative review. Solar Energy Mater. Solar Cells **107**, 87–111 (2012)

14. M.A. Green, K. Emery, Y. Hishikawa, W. Warta, E.D. Dunlop, Solar cell efficiency tables (version 42). Progress in Photovoltaics **21**, 827–837 (2013)
15. I. Chung, B. Lee, J. He, R.P.H. Chang, Mercouri G. Kanatzidis, All-solid-state dye-sensitized solar cells with high efficiency. Nature **485**, 486–490 (2012)
16. G.E. Eperon, V.M. Burlakov, P. Docampo, A. Goriely, H.J. Snaith, Morphological control for high performance, solution-processed planar heterojunction perovskite solar cells. Adv. Funct. Mater. **24**, 151–157 (2014)
17. H. Zhou, Q. Chen, G. Li, S. Luo, T.-B. Song, H.-S. Duan, Z. Hong, J. You, Y. Liu, Y. Yang, Interface engineering of highly efficient perovskite solar cells. Science **345**, 542–546 (2014)
18. M. Liu, M.B. Johnston, H.J. Snaith, Efficient planar heterojunction perovskite solar cells by vapour deposition. Nature **501**, 395–398 (2013)

Part VI
Liquid Crystal Technology

Symbols

A	Absorbance
Δn	Birefringence
d	Cell gap
T_c	Clearing Temperature
DR	Dichroic Ratio
$\Delta\varepsilon$	Dielectric Anisotropy
$\check{n}$	Director Axis
n_{eff}	Effective Index
n_e	Extraordinary Refractive Index
$\vec{q}$	Grating vector
n	Index of Refraction
S	Order Parameter
n_o	Ordinary Refractive Index
Γ, δ	Phase Retardation
V_{th}	Threshold Voltage
ϕ	Twisted Angle
$\vec{k}$	Wave vector
λ	Wavelength

Chapter 10
Physics of Liquid Crystals

Chi-Yen Huang

10.1 Brief History

Liquid crystals (LCs) are matter in a state that has properties between those of conventional liquid and those of solid crystals. The discovery of LCs can be track back to the end of the 19th century [1, 2]. In 1888, an Austrian botanical physiologist, Friedrich Reinitzer examined the physic-chemical properties of cholesterol derivatives. He found that cholesteryl benzoate has two melting points. At 145.5 °C it melts into an intermediate cloudy liquid, and at 178.5 °C the cloudy liquid becomes clear. With the help of Otto Lehmann and von Zepharovich, they identified that the intermediate "fluid" was crystalline. After his accidental discovery, Reinitzer did not pursue studying LCs further. The research was continued by Lehmann. Eventually he realized that the cloudy liquid was a new state of matter and coined the name "liquid crystal", illustrating that it was something between a liquid and a solid, sharing important properties of both. However, LCs were not popular among scientists and the material remained a pure scientific curiosity for about 80 years.

LCs became a hot research topic for the flat panel since 1962 at RCA Laboratories. In 1970, the twisted nematic field effect in LCs was filed for patent by Hoffmann-LaRoche in Switzerland, with Wolfgang Helfrich and Martin Schadt listed as inventors. James Fergason, while working with Sardari Arora and Alfred Saupe at liquid crystal institute, Kent State University, filed an identical patent in the United States in 1971. Then the company of Fergason ILIXCO (now LXD Incorporated) produced the first LCDs based on the TN-effect. The invention of TN LCDs then opened a new era for flat panel displays.

C.-Y. Huang (✉)
Department of Physics and Graduate Institute of Photonics, National Changhua University of Education, Changhua, Taiwan
e-mail: chiyen@cc.ncue.edu.tw

C.-C. Lee (ed.), *The Current Trends of Optics and Photonics*,
Topics in Applied Physics 129, DOI 10.1007/978-94-017-9392-6_10

10.2 LC Phases

The LC phases (called mesophases) occurs between the crystalline (solid) and isotropic (liquid) state [3]. The constituents are elongated rod-like (calamitic) or disk-like (discotic) organic molecules, as shown in Fig. 10.1. The size of the molecules is typically a few nanometers (nm). The ratio between the length and the diameter of the rod-like molecules, or the ratio between the diameter and the thickness of the disk-like molecules is about 5 or larger. In mesophases, the molecules of the organic compound have not only the orientational order like crystal solids but also fluidity like isotropic liquids.

According to the formation methods, LCs can be categorized as lyotropic and thermotropic types. Thermotropic phases are occurred in a certain temperature range. If the temperature is raised too high, thermal motion destroys the cooperative ordering of the LC phase, pushing the material into a conventional isotropic liquid phase. If the temperature is too low, most LC materials will become a conventional crystal. Many thermotropic LCs exhibit various phases as temperature is changed. For instance, a particular rod-like LC molecule (called mesogen) may exhibit various smectic following nematic (and finally isotropic) phases as temperature is increased, as shown in Fig. 10.2. Most thermotropic LCs are composed of rod-like molecules and applied extensively on displays and electro-optical devices. Figure 10.2b, c illustrate the smectic phases with one dimensional translation order and orientation order. Figure 10.2d shows the nematic phase in which only a long range orientational order of the molecular axes exists. The cholesteric phase is also a nematic phase except that it is composed of chiral molecules. Therefore, the structures acquire a spontaneous twist about a helical axis normal to the director, as shown in Fig. 10.3. Notably, in each plane normal to the helical axis, the LC is in a nematic phase. A smectic LC possesses the higher degree of order, and thus is close in structure to the solid crystals. However, it is the nematic and cholesteric phases that have the greatest number of electro-optical applications.

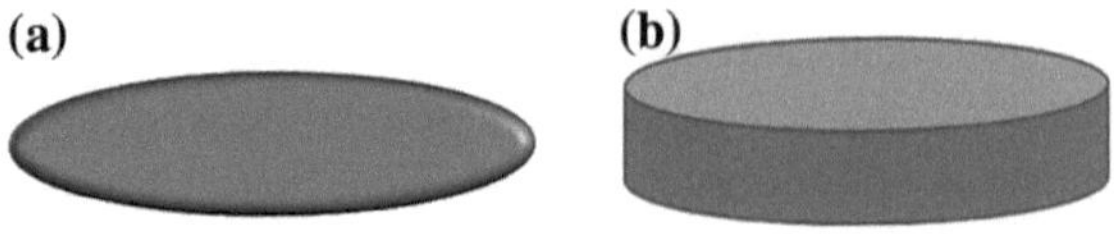

Fig. 10.1 **a** Rod-like molecule, and **b** disk-like molecule

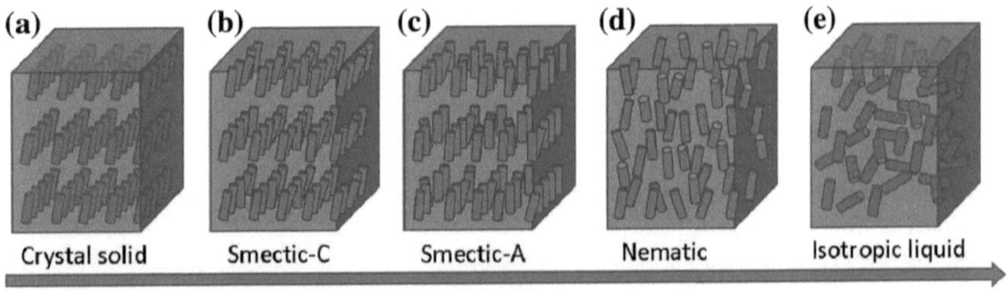

Fig. 10.2 Schematic representations of phases of rode-like LC molecules

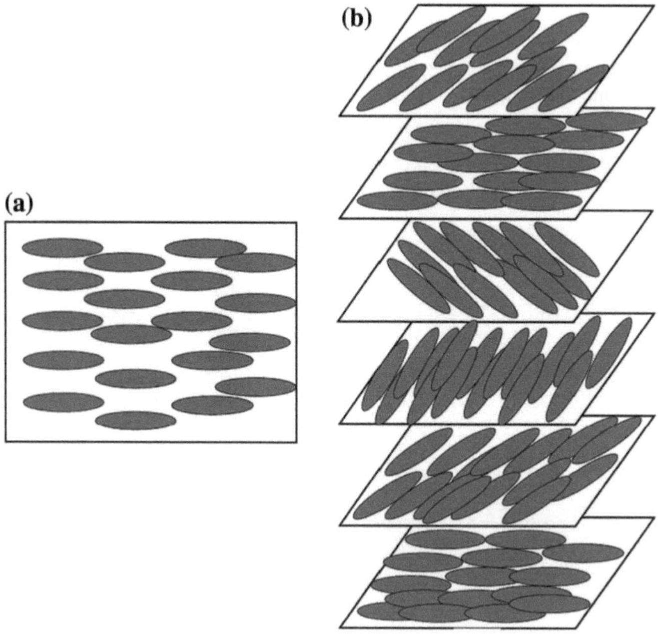

Fig. 10.3 **a** Nematics, **b** cholesterics

A lyotropic LC consists of two or more components that exhibit liquid-crystalline properties in certain concentration ranges. A compound that has two immiscible hydrophilic and hydrophobic parts within the same molecule is called an amphiphilic molecule. Many amphiphilic molecules show lyotropic liquid-crystalline phase sequences depending on the volume balances between the hydrophilic part and hydrophobic part. Soap is an everyday example of a lyotropic LC.

10.3 Nematic LC

10.3.1 Order Parameter

Figure 10.4 depicts the collective arrangement of the rod-like mesogenic molecules in the nematic phase. They have long-range orientational order but lack of positional order. On the average, they have the tendency to parallel to a certain direction called director, $\hat{n}$. In the figure, θ is the angle between the director $\hat{n}$ and the long axis $\hat{L}$ of the single molecule.

The order of the collective LC molecules can be described by a quantity, called order parameter S, which is defined as

$$S = \frac{1}{2}\left\langle 3\cos^2\theta - 1\right\rangle, \tag{10.1}$$

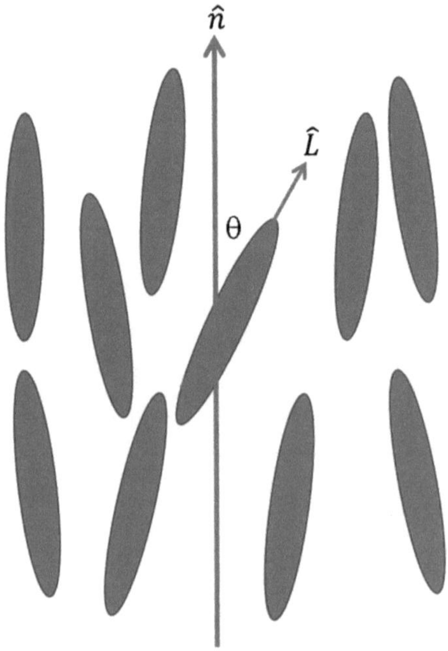

Fig. 10.4 Schematic representation of the collective arrangement of rod-like mesogenic molecules

where bracket “< >” denotes the average of the entire molecules. The order parameter represents the directional order of a certain kind of material macroscopically. In theory, the order parameter S of the material is in between 0 and 1, because the order parameter of the isotropic liquid (out of order) is 0, but that of the crystalline solid is 1 (in perfect order). The order parameter of the nematic LCs is approximately between 0.4 and 0.6, and decreases with increasing temperature. The following equation describes the order parameter, S, with ambient temperature T,

$$S = \left(1 - \frac{T}{T_C}\right)^{\beta}, \tag{10.2}$$

where T_C is the clearing temperature and β is the material constant. The order is associated with a variety of physical properties, including refraction index, dielectric, magnetic, elastic and viscosity constants of the LCs.

10.3.2 Birefringence

LC is an optical anisotropic substance; its index of refraction is dependent on the polarization and propagation direction of the incident light [4]. In optics, nematic LC is uniaxial crystal with optical axis parallel with the director of the

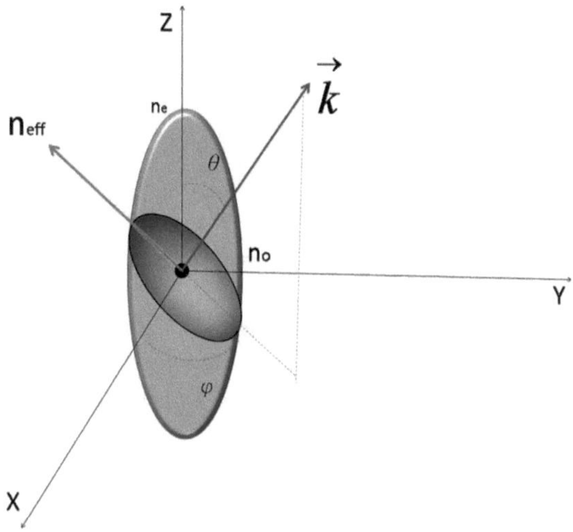

Fig. 10.5 Schematic representation of the index ellipsoid for effective refraction index calculation

LC molecules. As shown in Fig. 10.5, when the incident light propagates in the $\vec{k}$ direction and have angle of θ on the y-z plane, the refraction index of the light in the y-z plane become the effective index of refraction, $n_{eff}(\theta)$. The formula for $n_{eff}(\theta)$ can be written as,

$$\frac{\cos^2\theta}{n_o^2} + \frac{\sin^2\theta}{n_e^2} = \frac{1}{n_{eff}^2(\theta)}, \tag{10.3}$$

The birefringence can then written as,

$$\Delta n(\theta) = n_{eff}(\theta) - n_o, \tag{10.4}$$

where n_o is the ordinary index of refraction of the LC molecules.

10.3.3 Dielectric Anisotropy

The dielectric anisotropy means the difference of the dielectric constants along the long axis and the short axis of the LC director. When the applied electric field is parallel (perpendicular) to the director of the LC molecules, the measured dielectric constant is $\varepsilon_{//}(\varepsilon_{\perp})$. The dielectric anisotropy is defined as $\Delta\varepsilon = \varepsilon_{//} - \varepsilon_{\perp}$. The main cause of the dielectric anisotropy is that the dielectric displacement and the induced polarization of the substance are not parallel with the direction of the applied electric field.

When an external electric field E is applied to a nematic cell, the electric displacement $\vec{D}$ can be written as

$$\begin{aligned}\vec{D} &= \varepsilon_{//}\vec{E}_{//} + \varepsilon_{\perp}\vec{E}_{\perp} \\ &= \varepsilon_{\perp}\vec{E} + (\varepsilon_{//} - \varepsilon_{\perp})(\hat{n}\cdot\vec{E})\hat{n}, \\ &= \varepsilon_{\perp}\vec{E} + \Delta\varepsilon(\hat{n}\cdot\vec{E})\hat{n}\end{aligned} \tag{10.5}$$

where $\vec{E} = \vec{E}_{//} + \vec{E}_{\perp}$, in which $\vec{E}_{//}$ is the electric field parallel with, and $\vec{E}_{\perp}$ is the electric field perpendicular to the long axis of the LC director. The electric contribution to the free energy density f_{elec} of the nematics can be written as

$$f_{elec} = -\frac{1}{2}\varepsilon_o\Delta\varepsilon(\hat{n}\cdot\vec{E})^2, \tag{10.6}$$

10.3.4 Elastic Continuum Theory of Nematic LCs

Similar to liquids and solids, LCs possess the curvature elasticity. The elastic constants of LCs govern the restoring torque that arises when the LC system is deformed from its equilibrium state. In elastic continuum theory, the basic deformations of the LC molecules can be divided as splay, twist and bend, as shown in Fig. 10.6. The elastic free energy density of the nematic molecules is [5],

$$f_e = \frac{1}{2}\left[k_{11}\left(\nabla\cdot\vec{n}\right)^2 + k_{22}\left(\vec{n}\cdot\nabla\times\vec{n}\right)^2 + k_{33}\left(\vec{n}\times\nabla\times\vec{n}\right)^2\right], \tag{10.7}$$

where k_{11}, k_{22}, and k_{33} are the splay, twist and bend elastic constants of nematic LCs, respectively. LCs can be deformed easily by external fields, such as electric field, magnetic field or pressure. For most LC materials, the elastic constants are in the range of 3–25 p N (10^{-12} N).

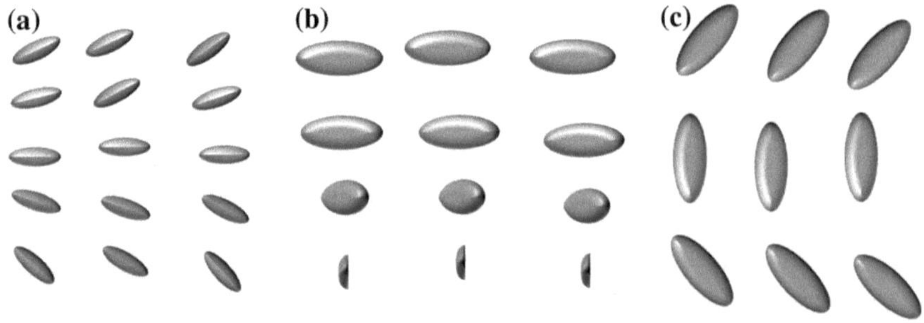

Fig. 10.6 Schematic drawing of **a** splay, **b** twist, and **c** bend deformations of nematics

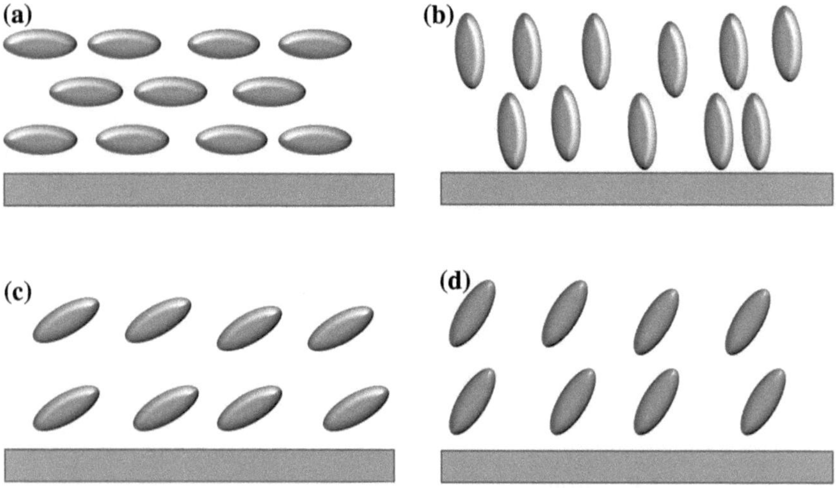

Fig. 10.7 Fundamental molecular alignments of nematics on substrate surface. **a** Parallel, **b** perpendicular, **c** small tile from parallel, **d** small tilt from perpendicular

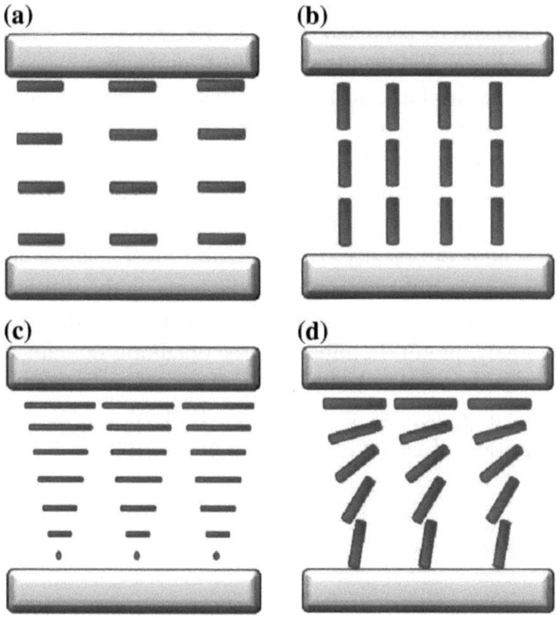

Fig. 10.8 Fundamental molecular alignments of nematic cells. **a** Homogeneous, **b** homeotropic, **c** twisted, and **d** hybrid alignments

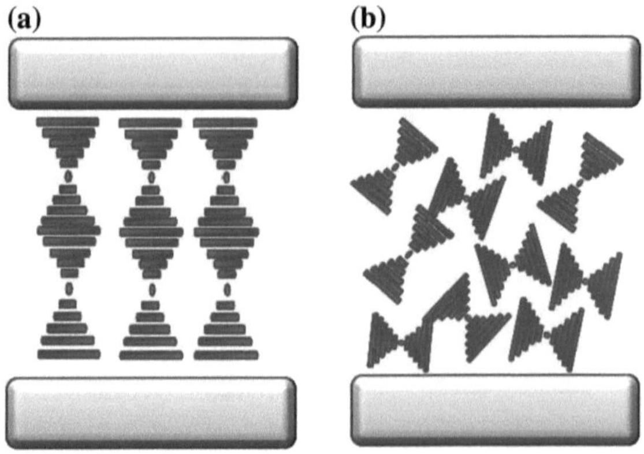

Fig. 10.9 Fundamental molecular alignments of cholesteric cells. **a** Planar state, and **b** focal conic state

10.4 Surface Alignment

The orientation of LCs plays an essential role in electro-optical application of the LC devices, in which uniform and well pre-defined orientation of the LC molecules is required. Without special surface treatment, the LCs will generally have many domains, each with different orientation direction. Surface treatment of the employed substrate has been one of the convenient methods to control the alignment of the LCs. The factors that affect the alignment of LCs including dipolar interactions, chemical bonding, van der Waals interactions, steric factors, surface topologies and the elasticity of LC molecules.

The two basic surface alignments of the LCs on the substrate surface are parallel and homeotropic alignments [6], as shown in Fig. 10.7a, b. Rubbing the substrate surface with linen cloth and lens tissue has been a simple way to promote parallel alignment. The etched glass surface causes the LCs to have the homeotropic alignment. Some crystalline cleavages (Al_2O_3, $LiNbO_3$) also orient nematics homeotropically. However, the most popular technique for the homeotropic alignment is the coating of surfactants on the substrate surface. In practice, a small tilt from the parallel and perpendicular conditions, as, shown in Fig. 10.7c, d, namely, pretilt, is important for obtaining domain free orientation under external fields. In nematic LC cells, there are several commonly used combinations of surface alignment as shown in Fig. 10.8. As for the cholesteric LC cell, the two fundamental alignments are shown in Fig. 10.9.

References

1. Liquid Crystal. http://en.wikipedia.org/wiki/Liquid_crystal
2. History and Properties of Liquid Crystals. http://www.nobelprize.org/educational/physics/liquid_crystals/history/index.html
3. D.K. Yang, S.T. Wu, *Fundamentals of Liquid Crystal Devices* (Wiley, Chichester, 2006)

4. E. Hecht, *Optics* (Addison-Wesley, Reading, 2002)
5. P.G. Gennes, J. Prost, *The Physics of Liquid Crystals* (Clarendon Press, Oxford University Press, Oxford, 1993)
6. B. Bahadur, *Liquid Crystals: Applications and Uses* (World Scientific, Singapore, 1991)

Chapter 11
Photo-Alignment Technology

Tsung-Hsien Lin

LCDs are used in many everyday devices, including mobile phones, notebook computers and wide screen televisions. A typical LCD is composed of two glass substrates between which is a gap that is filled with LC. The arrangement of the LC is controlled by the boundary condition of an alignment film. A long-range orientational interaction causes the preferable alignment of a LC by the surface boundary to extend into the LC bulk. The alignment layer is typically rubbed with a special cloth to induce a unidirectional alignment of the LC. In the first devices based on twisted LCs, these LCs were aligned only by rubbing transparent electrodes. The rubbing technique became much more reliable when the glass substrate was covered with a rubbed polymer layer. The rubbing of polymers is a very simple technique and it is now extensively utilized in both small scientific laboratories and large LCD factories.

Despite the extensive use of rubbing in LCD manufacturing, it has some serious shortcomings that arise from the contact involved in rubbing. These critically affect the production of the most recent generation of LCDs and miniature LC photonic devices. First, rotational rubbing for the polymer surface causes the accumulation of static charges and the formation of fine dust particles. The accumulation of static electricity may destroy the transistors or diodes in the LCD. The second shortcoming is the effect of decreasing in manufacturing yield. Rubbing for large and high-resolution active matrix LCDs (AMLCDs) involve additional difficulties that are associated with the precise control of the rubbing characteristics. Other problems are the difficulty of producing local structures with the required LC director alignment in the selected areas of the device. Finally, this method cannot be used to align LCs in closed volumes.

For all of the above reasons, new alignment solutions and approaches are sought. There are some alternative methods of rubbing include laser writing, micro-rubbing, ion beam irradiation and photoalignment. Of these methods, the

T.-H. Lin (✉)
Department of Photonics, National Sun Yat-sen University, Kaohsiung, Taiwan
e-mail: jameslin@faculty.nsysu.edu.tw

C.-C. Lee (ed.), *The Current Trends of Optics and Photonics*,
Topics in Applied Physics 129, DOI 10.1007/978-94-017-9392-6_11

photoalignment method has the greatest potential for industry. Photoalignment has been studied and developed for a long time. To perform photoalignment, a substrate must be coated with a layer of photosensitive material whose molecules can orientate in an ordered manner under irradiation by light. Anisotropic interfacial interaction between the alignment film and the LC enable the alignment of the LCs to be controlled by the action of light on the aligning layer.

The photoalignment technique has not only display applications but also potential advanced applications in telecommunications and organic electronics, in which rubbing is very problematic. Accordingly, the synthesis of photoaligning materials and the development of different photoalignment approaches have become some of the most important topics in applied LCs science. Over the last 20 years, a variety of photoalignment techniques have been developed and improvements have been made. The required materials are now readily commercially available.

Truly revolutionary events concerning the industrial use of photoalignment have recently occurred. In 2009, Sharp Corporation of Japan announced the development of a new LCD, the Ultraviolet-induced multi-domain Vertical Alignment (UV^2A) display. One year later, this technology was used at Sharp's largest factory anywhere in the world, their Sakai Plant, to produce tenth-generation LCD panels. The company is now implementing a complete changeover from Advanced Super View (ASV) mode panels in its televisions. Some other applications of photoalignment are also close to practical realization, and even current applications of photoalignment provide new incentives for research in this field. This chapter introduces several main photoalignment mechanisms. These are (i) photo-isomerization in azo-compound-containing polymer/dye films [1–5]; (ii) bulk-mediated photo-isomerization and adsorption effects [6–9]; (iii) photo-crosslinking in cinnamic side-chain polymers [10–14], and (iv) photo-degradation in polyimide films [15–17].

11.1 Photo-Isomerization in Azo-Compound-Containing Polymer/Dye Films

Of the four aforementioned photoaligning mechanisms, azobenzene-based methods support the tuning of alignment direction owing to a reversible *trans*-*cis* transformation of azobenzene. Photoalignment using photo-responsive layers is more thermally stable than bulk-mediated photoalignment. Surface treatment of the substrate is the key to aligning mesogenic molecules. The surface (layer), which comprises photochromic molecules, and modifies the molecular orientation of the LCs, is generally called the "command surface" [18]. Ichimura et al. were the first to propose the idea of using photochromic monolayers to regulate the LC arrangement [5]. Under irradiation by ultraviolet (UV) light (365 nm), the azo-molecules changed from the *trans*- to *cis*- isomers, causing planar alignment. They changed back to the *trans*-conformation, and a vertical alignment, under visible light (440 nm). Both *trans*- and *cis*- states of the photochemical material were thermally stable. Generally, *trans*-azobenzene is dichroic, absorbing light whose polarization is parallel to the long axes of the dye molecules. When the azo-dye is optically excited by polarized light, the increase in internal energy forces the molecules to reorient in cycles of *trans-cis-trans* isomerization to release that energy. Rotational motions consequently occur till the absorption oscillator of the azobenzene is orthogonal to the incident polarization. Concurrently, vibrational motions may also be established in the process of the release of internal energy but these are not thought to influence the photoalignment. Herein, most of the internal energy is transferred by isomerization and the thermal reactions are negligible. Finally, almost all of the azo-molecules rotate until the absorption axes are perpendicular to the direction of polarization of the pump light, yielding a uniform arrangement. This claim can be confirmed by measuring the absorption spectra of a dye layer before and after exposure to polarized UV as the probing light. Owing to the geometry of the *trans*-isomer, the interfacial mesogenic molecules line up with the azo-compound, and the elastic continuum will simultaneously lead to a planar-aligned LC. Surprisingly, the resultant alignment seems to be stable and not to undergo any thermal disturbance. According to Chigrinov et al, who studied the birefringence relaxation of a photoaligned LC film after the pumping light was switched off, the order parameter firstly dropped slightly and thereafter remained constant [19]. The direction of alignment can be altered by exposing the LC again to UV.

Owing to such features of photoalignment as the absence of ion and impurity problems, rewritability, usability on flexible and curved surfaces, ease of patterning, the spatially distributed optical axis orientation and control of the pretilt angle and anchoring energy, it has been extensively applied to fabricate LC-based optical devices that cannot be formed by conventional rubbing [20]. Here we introduce some applications of surface-mediated photoalignment. An optical axis grating is a polarization grating whose optical axis has an orientation that varies periodically in the direction parallel to the substrate [21]. It can be regarded as a periodic arrangement of multiple waveplates. LC polarization gratings are able to be simply fabricated by two-beam holographic photoalignment. In general, one beam interferes with the other, whose polarization is orthogonal to that of the first, on the aligning film, creating a spatially distributed vector field. The orientation of these molecular axes on the surface is able to determine the arrangement of the LC in the bulk. A mesogen-based polarization grating has been demonstrated to be a promising diffractive optical element, which typically exhibits insensitivity to wavelength and high diffraction efficiency (>97 %) [22, 23]. Such polarization gratings are used to develop polarimeters [24] and imaging systems [23, 25]. Combining two optical axis gratings can produce various interesting diffraction outcomes [23]. The Q-plate is another type of waveplate whose optical axes are distributed as a function of the topological charge "q" [26]. This device is usually used for shaping wavefronts by spin-to-orbital angular momentum conversion. A mesogen-based q-plate can be realized by photoalignment. In the fabrication setup, a linear polarizer, a "V" mask and a plano-convex cylindrical lens are positioned between the pumping light source and the photoalignment film [27]. The *q* plate can be formed with any desired semi-integer *q* value by control of the rates of rotation of the polarizer and the aligning film in the photoaligning process. In practice, an electric field is applied to adjust the effective indices to satisfy the phase retardation requirements. Therefore, not only is the manufacturing procedure simplified, but also the conversion efficiency can be increased to 99 % [27]. A micropolarizer array is now an important component of an imaging system, because polarization analysis provides valuable information about the (physical, chemical, geometric and other) characteristics of objects and the removal of unpolarized light eliminates the ambient scattering noise [28, 29]. A mesogen-based micropolarizer normally consists of a linear polarizer and a patterned, micron-sized waveplate array that is composed of twisted and untwisted nematic pixels and/or electrically controlled birefringence pixels [27, 29, 30]. The micron-sized pattern of LC alignment is difficult to achieve by rubbing but easy to achieve by photoaligning. Surface-mediated photoalignment is also effective for the fabrication of flexible and rewritable displays [31], gradient-index lenses [32], and other devices [33].

To enhance the thermal stability of the alignment film and its durability against exposure to strong light after the aligning process, Takada et al. synthesized a new azobenzene material SDA-1 [33, 34], which is typically co-doped with a small amount of thermo-polymerization initiator V-65 into an organic solvent. The photoaligning and subsequent thermal polymerization processes yield a thermally stable azo-polymer at ambient temperatures of up to 250 °C. SDA-1 can also intact at a UV intensity of 175 MJ/m^2. Notably, the absorption band shifts to the UV range, following polymerization, enhancing its stability under exposure to visible light. SDA-2 was designed by replacing the COOH substituent with CF3 to prevent the damp-induced hydrolysis and/or any effect of the polarity of water on the rotational reorientation of LC [33, 35].

11.2 Bulk-Mediated Photo-Isomerization and Adsorption Effects

Voloschenko et al. [36] were the first to identify the adsorption effect in homogeneously aligned azo-dye doped LC films. They found that the absorption band of methyl red (MR) resulted in the photoalignment of LC on an isotropic surface of a polyninylcinnamate (PVCN) film, as depicted in Fig. 11.1. The authors posited that phototransformation of the MR molecules close to the surface significantly changed their adsorption capacity. Komitov et al. suggested that *trans*-isomers near the substrates absorbed light and transformed to the *cis*- form,

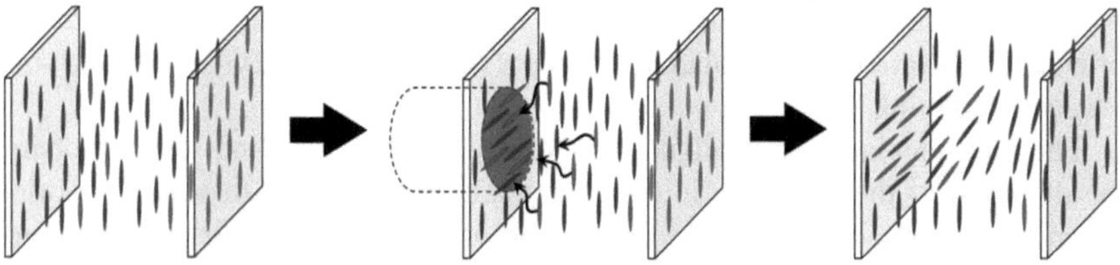

Fig. 11.1 Process of bulk-mediated photoalignment: when the sample is pumped by linearly polarized light, azobenzene molecules adsorb on the illuminated substrate and reorient parallel/perpendicular to the polarization leading to the alignment of the LC

which adsorbed at the surface [37, 38]. Since the MR molecules are anisotropic and dichroic and are in an anisotropic matrix, light-induced adsorption should be anisotropic and sensitive to the polarization of the exciting beam. Magyar et al. observed the planar alignment of nematic LC 5CB on a fused quartz surface upon illumination by polarized UV light [39]. The axis of easy orientation was perpendicular to the polarization of the incident light. The authors suggested that the observed phenomenon was caused by the light-induced rearrangement of the LC layer that was adsorbed on the quartz surface. The authors also suggested that the absorption dichroism of LC molecules caused the anisotropic desorption of these molecules from the surface, making the adsorbed layer anisotropic with the long molecular axis perpendicular to the direction of polarization. Ouskova et al. [40, 41] showed that photoalignment of LCs in an MR-doped LC system depends strongly on the intensity of the pumping beam. They attributed the photoalignment effect to two mechanisms—adsorption and desorption, which dominate in strong- and weak-intensity regimes, respectively. These mechanisms tend to reorient LC molecules perpendicular to the direction of polarization of the pump beam. The observation of the LC alignment perpendicular to the polarization in the cells with pure LC, supposedly owing to the light-induced desorption of LC molecules, suggested the possibility of an analogous effect in a dye-doped LC. Subsequent studies demonstrated that in LC that is doped with MR, light-induced desorption from the spontaneously adsorbed MR layer results in the alignment of the LC perpendicular to the polarization upon exposure to light of weak intensity.

Lee et al. [42] suggested that the surface morphology of the adsorbed substrate depends considerably on not only the intensity but also the duration of the pumping. In the weak-intensity regime, the homogeneous and fine layer of adsorbed dyes and the layer with microgrooves dominate in the early and late stages, respectively, tending to cause the LCs to reorient perpendicular and parallel to the direction of polarization of the pump beam. The inhomogeneous ribbon-like and rough adsorbed layer dominates in the strong intensity regime, perhaps severely disturbing the orientation of the LCs. Dark adsorption does not occur on substrates that are not coated with PVCN. Additionally, the authors found that long exposure produces ripple structures with periods of the order of 100 nm and microgrooves oriented in the direction of polarization in polyvinyl-alcohol (PVA) films and some vertically aligning layers.

The bulk-mediated photoalignment technique can be used not only to control the direction of the easy axis in the azimuthal plane but also to adjust the orientation of the axis in the zenithal plane. The pretilt angle is achieved by controlling the parameters of the ripple structure in the cell with initially vertically aligned LCs [43]. The combination of the fixed alignment force that is exerted by the homeotropically aligning layer and the variable one that is induced by light-induced ripple structures supports a method for generating various LC pretilt angles from 24° to 63.5°. Additional control of the pretilt angle by field-assisted photoalignment in a vertically aligned LC that is doped with a dichroic azo dye has also been proposed [44]. Sharp switching from homeotropic alignment to planar alignment is possible in LC cells that are doped with MR and oligomeric nanoparticles [45]. Light-induced adsorption of MR dopant changes the homeotropic alignment that is induced by these particles to planar.

The basic mechanism of the alignment effect is the light-induced reorientation of azobenzene chromophores, which is promoted by *trans-cis* isomerization cycles. When light irradiates dye-doped LC films, azo dye molecules undergo *trans-cis* isomerization, reorientation, diffusion, and adsorption toward the irradiated surface. Based on the model of Fedorenko et al. [46], the evolution of light-induced anchoring involves competition between light-induced desorption and adsorption of the dye molecules on the aligning surface, and the total light irradiation dose determines the final anchoring. Firstly, this evolution causes the drift of the initial easy axis away from the polarization owing to light-induced desorption of dye molecules from the surface at low exposures. Then, the light-induced adsorption begins to dominate and longer exposures result in the drift of the easy axis toward the polarization of the incident light.

The morphology of the adsorbed MR layers PVA films observably changed upon long exposures [47]. SEM and AFM were used to identify smooth layers of adsorbed molecules on the surfaces following exposures that caused the reorientation of the directors perpendicular to the direction of polarization and layers with well-oriented ripples upon exposures that caused alignment toward the direction of polarization [42, 48]. The adsorbed layer consists of MR molecules with their long axes perpendicular to the ripples. The ripples are parallel to the polarization and contribute to the final alignment. The ripples are associated with the interference between the exciting beam and the light that is scattered from the surface. The ripple structures were not observed on the PVCN surfaces. The doses to which the cells with PVCN surfaces were exposed were supposedly too low to form significant microgrooves. A well-established layer of spontaneously adsorbed MR molecules is also likely to be necessary for the formation of the ripple structure; however, the MR molecules don't adsorb on PVCN surface spontaneously. It is suggested that excited MR molecules close to the surface undergo *trans-cis*

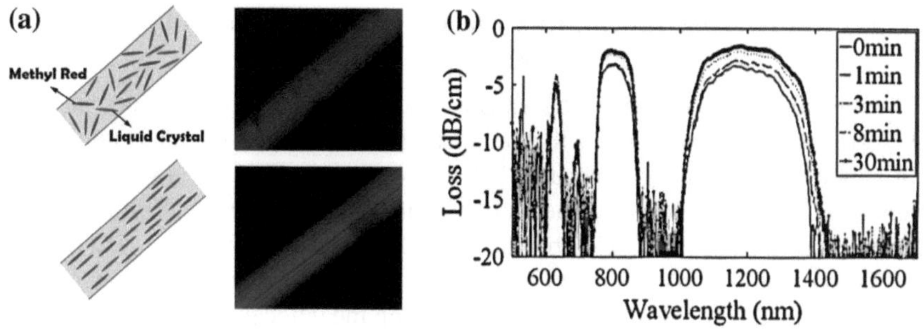

Fig. 11.2 **a** Schematic depictions and microscopic photographs of a methyl red-doped photonic liquid crystal fiber before and after photoalignment. **b** Transmission spectra of a photonic liquid crystal fiber photoaligned at pump intensity of 80, 160 and 320 mW/cm^2

isomerization, reorientation, diffusion and finally adsorption onto the substrate with their long axes perpendicular to the directions of both the polarization and the propagation of the impinging light.

The proposed photoalignment scheme is highly effective in fabricating LCDs and photonic devices. Lin et al. developed optically rewritable LCDs [49, 50] in which an image is written, erased, and rewritten by bulk-mediated photoalignment. Jau et al. demonstrated a photo-rewritable transmissive flexible-LCD that was based on the aligning effect of the photo-induced adsorption of azo dye on a flexible indium zinc oxide/polycarbonate (IZO/PC) substrate. A polarizer-free, electrically switchable and optically rewritable display that is based on dye-doped polymer-dispersed LCs has been developed [51]. Owing to bulk-mediated alignment, the adsorbed dye in the illuminated region aligns the LC homeotropically and thereby generates a transparent pattern on the scattering film without any polarizer. The adsorbed dyes can be erased and readsorbed by thermal and optical treatments, respectively. The combination of bulk-mediated hybrid photoalignment and homeotropic alignment in a single pixel supports the fabrication of LCDs with a controllable viewing-angle, [52] on which simultaneously displays two images that can be seen from different viewing angles. Other applications in photonic devices, such as optical gratings [53–55] and Fresnel lenses have also been investigated [56–60]. Photoalignment is a unique tool for achieving LC alignment in mechanically inaccessible volumes. As displayed in Fig. 11.2, it has been successfully applied to align LCs in photonic crystal fibers so as to reduce scattering loss and improve their optical performance [61, 62]. LC configuration can be controlled by varying the photoirradiation. A line-shaped and linearly polarized beam is applied onto a rotated homogeneous dye-doped LC cell to fabricate axially symmetric photoalignment devices for helical twisting power measurement and modulations of polarizations and beam profiles [63–70].

11.3 Photo-Crosslinking in Cinnamic Side-Chain Polymers

LC molecules can be aligned by causing them interact with the side-chains of photo-crosslinkable polyimides that are coated on the surface of a substrate. These polyimides can be small molecules [71–73], polymers [10, 74, 75], or polymer blends [76]. Linear photo-polymerization results in the preferred depletion of the cinnamic side-chain molecules by the [2 + 2] cycloaddition that is induced by irradiation with linearly polarized UV light. In the cycloaddition reaction, cyclobutane is formed from the double bonds following the delocalization of π electrons. The long axis of cyclobutane molecules preferably aligns perpendicular to the direction of the polarized UV light ($\vec{P}$). The van der Waals' (or dispersion) interaction causes the LC molecules to align perpendicular to $\vec{P}$.

The [2 + 2] cycloaddition is a general reaction that is undergone by cinnamate [10, 74, 75], coumarin [12, 77–79], chalconyl [11], tetrahydrophthalimide [13] and maleimide [13, 14] photosensitive polyimides. The [2 + 2] cycloaddition of cinnamate and chalcone species, is accompanied by *trans-cis* isomerization [80, 81]. However, materials with a rigid molecular structure, such as coumarin, tetrahydrophthalimide and maleimide, cannot undergo *trans-cis* isomerization [77]. One can fabricate materials containing several photosensitive fragments of different nature. The materials combining the fragments that undergo *trans-cis* isomerization and polymerization [82, 83], *trans-cis* isomerization and [2 + 2] cycloaddition [84, 85], and [2 + 2] cycloaddition and polymerization [72, 86] have been demonstrated. Although mechanisms of photoalignment are not well established for these materials, the superior thermal stability, extraordinary optical and dielectrical properties make them very promising for some applications.

The photoaligning molecular structure is designed to enable the direction of the easy axis on its irradiated surface to be easily predicted. Reznikov et al. [87] found that the direction of the easy axis in the azimuthal plane depends upon the position of the cinnamate fragments in the side polymer chains. Owing to the [2 + 2] cycloaddition of the cinnamate moieties of the polymer, the direction of alignment is the direction of the axis of geometric anisotropy of the crosslinked fragments. Obi et al. [88] and Kawatsuki et al. [89] identified other cinnamate polymers that exhibited "parallel" LC alignment. These polymers, as well as some of cinnamate monomers [72], gave both "parallel" and "perpendicular" alignment, depending on the degree of exposure [77].

The structure of photosensitive molecules can be manipulated to adjust the direction of the easy axis in the *zenithal* plane. The pretilt angle can be controlled by attaching hydrophobic groups such as fluorine, alkyl, alkoxy, fluoroalkyl, fluoroalkoxy and others to photosensitive molecules. These groups can be attached as individual side chains [90, 91], as terminal groups of photosensitive side chains [72, 92] or fragments of main polymer chains. They can also be blended with the photoaligning materials [76, 93, 94]. The method by which photosensitive fragments are connected to the polymer chain is important. For example, the position of azobenzene fragments in the main chain can promote the photosensitivity and thermal stability of alignment. The attachment of the fragments to a side chain increases the reaction rate and the photoinduced ordering. It also enables LC pretilt angles to be controlled over a wide range [95]. According to Akiyama et al.'s study [96], the attachment of azobenzene in the *ortho* position provides alignment with much greater photo- and thermal stability of the than that in the *para* position. Therefore, the molecular structure of the photoaligning materials determines their photoaligning characteristics.

Owing to their irreversible photochemistry and highly constrained molecular motions, photocrosslinking materials may combine excellent photosensitivity with high resistance of the LC alignment to heat and light. Actually, the photoaligning materials that are available commercially are such photocrosslinking materials.

11.4 Photo-Degradation in Polyimide Films

Photo-degradation includes photodissociation, which is the breakup of molecules into smaller pieces by the absorption of light. It also includes the irreversible shape alteration of molecules. Photo-degradable organic films comprise materials (commonly polymers) that are susceptible to irreversible photo-destruction, such as by chain scission, photooxidation or other processes. The targets of these reactions are normally aromatic fragments of main or side polymer chains. Apart from PIs [17, 93, 97, 98], polysilanes [99, 100], polystyrene, polyesters and some other aromatic polymers [101] undergo photoalignment upon photodestruction. Photo-destroyable PIs usually have a sufficiently high anchoring energy of 10^{-5} J/m^2 [102, 103]. Characteristics of these materials include pronounced image sticking and display flicker.

Photosensitive fragments of organic photoaligning materials have aromatic groups, which frequently π conjugate with other parts of the fragments, determining their absorption spectra. Hence, the rate of photodegradation of photoaligning materials is rather sensitive to the structure of the aromatic fragments. The manner in which photosensitive fragments are connected to the polymer chain is important. For example, the position of azobenzene fragments in the main chain can enhance the photo- and thermal stability of LC alignment. The attachment of such fragments to a side chain increases the reaction rate and the photo-induced ordering. This position also increases the range of controllable LC pretilt angles [95]. The molecular mass of the photoaligning polymer also importantly affects photoalignment properties. Increasing the molecular weight often improves the thermal stability of photoalignment by reducing the mobility of the polymer chains. Molecular weight may also affect the rate of photoreactions and photo-ordering [104].

Since PI materials have high thermal stability and are generally used in the LCD industry, their modification for photoalignment applications is highly desirable. The first efforts so to do revealed that polarized light at $\lambda = 257$ nm can induce LC alignment perpendicular to the direction of the incident polarized UV light [17]. Accordingly, the direction of the maximum density of unbroken PI chains, which is perpendicular to the direction of the polarization of the light, is the direction of LC alignment. The photoaligning direction can be changed by varying the direction of polarization of the UV light [97]. Therefore, before UV exposure, PI chains in the film are randomly oriented. PI chains parallel to the direction of polarization of the UV light to which they are exposed are selectively decomposed, and the corresponding photoproducts become randomly relocated in the PI film. The residual PI chains that are perpendicular to the polarization of the UV remain unaffected, causing the anisotropic van der Waals' forces to align the LC molecules along the optical axis of the film [105].

Numerous applications depend on large pretilt angles. For example, larger pretilt angles cause the bend mode to be more stable at a zero bias voltage [106]. Large angles are also required to establish the so-called stressed splay-twist (SST) mode, which may be useful for rapid LC switching and field-sequential-color LCD [107]. Although pretilt angles of close to 90° for in the vertical alignment nematic mode is possible, angles in the range of 10°–80° are difficult to achieve.

The mechanism of the generation of such a large pretilt angle is particularly interesting. The photoalignment mechanism involves preferential bond breaking. The deep-UV light breaks bonds in some preferred direction, inducing alignment at the surface. However, the control of the pretilt angle depends on an additional mechanism of competition between the horizontal and vertical PIs. The pretilt angle of the LC molecules depends on the deep-UV dosage in the vertical PI layer owing to and is governed by controlled photo-degradation [108]. For a pure vertical PI that is not exposed to DUV, an angle of 90° is necessarily obtained, but exposure to deep-UV reduces the anchoring strength by degrading the material. As the deep-UV radiation destroys the vertical alignment, the pretilt angle is reduced. In fact, control of the dosage of light enables any pretilt angle between 0° between 90° to be achieved [94]. This method of generating large pretilt angles under deep-UV radiation is quite useful and universal. However, since deep-UV must be used, the light source is more difficult to control. Furthermore, the exploitation of photodegradation is not desirable as it may cause the alignment surface to become contaminated.

References

1. S.R. Nersisyan, N.V. Tabiryan, D.M. Steeves et al., Study of azo dye surface command photoalignment material for photonics applications. Appl. Optics. **49**, 1720–1727 (2010)
2. K. Ichimura, H. Akiyama, N. Ishizuki et al., Command surfaces, 6 Azimuthal orientation of liquid crystals photo-controlled by an azobenzene pendent polymer. Die Makromolekulare Chemie, Rapid Communic. **14**, 813–817 (1993)
3. H. Akiyama, T. Kawara, H. Takada et al., Synthesis and properties of azo dye aligning layers for liquid crystal cells. Liq. Cryst. **29**, 1321–1327 (2002)
4. S. Lucht, D. Neher, T. Miteva et al., Photoaddressable polymers for liquid crystal alignment. Liq. Cryst. **30**, 337–344 (2003)

5. K. Ichimura, Y. Suzuki, T. Seki et al., Reversible change in alignment mode of nematic liquid crystals regulated photochemically by command surfaces modified with an azobenzene monolayer. Langmuir **4**, 1214–1216 (1988)
6. T.H. Lin, H.C. Jau, S.Y. Hung et al., Photoaddressable bistable reflective liquid crystal display. Appl. Phys. Lett. **89**, 021116–021113 (2006)
7. B. Umanskii, N. Novoseletskii, S. Torgova et al., Azodyes as Photo-Alignment Materials for Liquid Crystals. Mol. Cryst. Liq. Crys. **412**, 313–319 (2004)
8. J.L. West, L. Su, Y. Reznikov, Photo-alignment using adsorbed dichroic molecules. Mol. Cryst. Liq. Crys. A **364**, 199–210 (2001)
9. V. Kozenkov, S. Iudin, E. Katyshev et al., Photoinduced optical anisotropy in multilayer Langmuir films. Pisma Zhurnal Tekhnischeskoi Fiziki **12**, 1267–1272 (1986)
10. M. Schadt, K. Schmitt, V. Kozinkov et al., Surface-induced parallel alignment of liquid-crystals by linearly polymerized photopolymers. Jpn. J. Appl. Phys. **1**(31), 2155–2164 (1992)
11. Y. Makita, T. Natsui, Kimura Si et al., Photo alignment materials with high sensitivity to near UV light. J. Photopolym. Sci. Technol. **11**, 187–192 (1998)
12. M. Schadt, H. Seiberle, A. Schuster, Optical patterning of multidomain liquid-crystal displays with wide viewing angles. Nature **381**, 212–215 (1996)
13. LO. Vretik, VG. Syromyatnikov, VV. Zagniy et al, Problem of photoalignment in the LCD's development: Synthetic routes in its solving. Mol Cryst Liq Crys 486:57/(1099)–1065/(1107) (2008)
14. H. Murai, T. Nakata, T. Goto, Liquid crystal photo-alignment layers made from aromatic bismaleimides. Liq. Cryst. **29**, 669–673 (2002)
15. J. Chen, D.L. Johnson, P.J. Bos et al., Model of liquid crystal alignment by exposure to linearly polarized ultraviolet light. Phys. Rev. E **54**, 1599–1603 (1996)
16. J.H. Kim, S. Kumar, S.D. Lee, Alignment of liquid crystals on polyimide films exposed to ultraviolet light. Phys. Rev. E **57**, 5644–5650 (1998)
17. M. Hasegawa, Y. Taira, Nematic homogeneous photoalignment by polyimide exposure to linearly polarized UV. J. Photopolym. Sci. Technol. **8**, 241–248 (1995)
18. K. Ichimura, Photoalignment of liquid-crystal systems. Chem. Rev. **100**, 1847–1874 (2000)
19. V. Chigrinov, S. Pikin, A. Verevochnikov et al., Diffusion model of photoaligning in azo-dye layers. Phys. Rev. E **69**, 061713 (2004)
20. V. Chigrinov, H.S. Kwok, I. Nishiyama et al., 63.1: invited paper: liquid crystal devices based on photo-alignment and photo-patterning materials. SID Int. Symp. Dig. Tec. **42**, 939–941 (2011)
21. B.R. Kimball, N.V. Tabiryan, D.M. Steeves et al., Optical axis gratings in liquid crystals and their use for polarization insensitive optical switching. J Nonlinear Opt. Phys. **18**, 1–47 (2009)
22. C. Oh, M.J. Escuti, Achromatic diffraction from polarization gratings with high efficiency. Opt. Lett. **33**, 2287–2289 (2008)
23. S.R. Nersisyan, N.V. Tabiryan, L. Hoke et al., Polarization insensitive imaging through polarization gratings. Opt. Express **17**, 1817–1830 (2009)
24. M.W. Kudenov, M.J. Escuti, E.L. Dereniak et al., White-light channeled imaging polarimeter using broadband polarization gratings. Appl. Optics. **50**, 2283–2293 (2011)
25. H. Sarkissian, B. Park, N. Tabirian et al., Periodically aligned liquid crystal: potential application for projection displays. Mol. Cryst. Liq. Crys. **451**, 1–19 (2006)
26. L. Marrucci, C. Manzo, D. Paparo, Optical spin-to-orbital angular momentum conversion in inhomogeneous anisotropic media. Phys. Rev. Lett. **96**, 163905 (2006)
27. S. Slussarenko, A. Murauski, T. Du et al., Tunable liquid crystal q-plates with arbitrary topological charge. Opt. Express **19**, 4085–4090 (2011)
28. C.K. Harnett, H.G. Craighead, Liquid-crystal micropolarizer array for polarization-difference imaging. Appl. Optics. **41**, 1291–1296 (2002)

29. X. Zhao, A. Bermak, F. Boussaid et al., High-resolution photoaligned liquid-crystal micropolarizer array for polarization imaging in visible spectrum. Opt. Lett. **34**, 3619–3621 (2009)
30. SR. Nersisyan, NV. Tabiryan Polarization imaging components based on patterned photoalignment. Mol Cryst Liq Crys 489:156/(482)–168/(494) (2008)
31. V. Chigrinov, A. Murauski, Q. Yu et al., 15.1: optically rewritable liquidcrystal technology: a new green epaper approach. SID Int. Symp. Dig. Tec. **41**, 195–198 (2010)
32. M.C. Tseng, F. Fan, C.Y. Lee et al., Tunable lens by spatially varying liquid crystal pretilt angles. J. Appl. Phys. **109**, 083109–083105 (2011)
33. V. Chigrinov, H.S. Kwok, H. Takada et al., Photo-aligning by azo-dyes: physics and applications. Liq. Cryst. Today **14**, 1–15 (2005)
34. H. Takada, H. Akiyama, H. Takatsu et al., P-104: aligning layers using azo dye derivatives for liquid crystal devices. SID Int. Symp. Dig. Tec. **34**, 620–623 (2003)
35. A.D. Kiselev, V. Chigrinov, D.D. Huang, Photoinduced ordering and anchoring properties of azo-dye films. Phys. Rev. E **72**, 061703 (2005)
36. D. Voloshchenko, A. Khyzhnyak, Y. Reznikov et al., Control of an easy-axis on nematic-polymer interface by light action to nematic bulk. Jpn. J. Appl. Phys. **1**(34), 566–571 (1995)
37. L. Komitov, C. Ruslim, Y. Matsuzawa et al., Photoinduced anchoring transitions in a nematic doped with azo dyes. Liq. Cryst. **27**, 1011–1016 (2000)
38. L. Komitov, K. Ichimura, A. Strigazzi, Light-induced anchoring transition in a 4,4'-disubstituted azobenzene nematic liquid crystal. Liq. Cryst. **27**, 51–55 (2000)
39. G. Magyar, J. West, Y. Reznikov et al., Light induced LC alignment on the isotropic non-photosensitive surface. Mol. Cryst. Liq. Crys. A **329**, 71–79 (1999)
40. E. Ouskova, D. Fedorenko, Y. Reznikov et al., Hidden photoalignment of liquid crystals in the isotropic phase. Phys. Rev. E **63**(2 Pt 1), 021701 (2001)
41. E. Ouskova, Y. Reznikov, S.V. Shiyanovskii et al., Photo-orientation of liquid crystals due to light-induced desorption and adsorption of dye molecules on an aligning surface. Phys. Rev. E **64**, 051709 (2001)
42. C.R. Lee, T.L. Fu, K.T. Cheng et al., Surface-assisted photoalignment in dye-doped liquid-crystal films. Phys. Rev. E **69**, 031704 (2004)
43. Andy Y.-G. Fuh, C.K. Liu, K.T. Cheng et al., Variable liquid crystal pretilt angles generated by photoalignment in homeotropically aligned azo dye-doped liquid crystals. Appl. Phys. Lett. **95**, 161104–161103 (2009)
44. K.Y. Yang, W. Lee, Voltage-assisted photoaligning effect of an azo dye doped in a liquid crystal with negative dielectric anisotropy. Opt. Express **18**, 19914–19919 (2010)
45. W.Z. Chen, Y.T. Tsai, T.H. Lin, Photoalignment effect in a liquid-crystal film doped with nanoparticles and azo-dye. Appl. Phys. Lett. **94**, 201114–201113 (2009)
46. D. Fedorenko, E. Ouskova, V. Reshetnyak et al., Evolution of light-induced anchoring in dye-doped nematics: Experiment and model. Phys. Rev. E **73**, 031701 (2006)
47. Andy Y.-G. Fuh, C.C. Liao, K.C. Hsu et al., Laser-induced reorientation effect and ripple structure in dye-doped liquid-crystal films. Opt. Lett. **28**, 1179–1181 (2003)
48. C.R. Lee, T.S. Mo, K.T. Cheng et al., Electrically switchable and thermally erasable biphotonic holographic gratings in dye-doped liquid crystal films. Appl. Phys. Lett. **83**, 4285–4287 (2003)
49. T.H. Lin, H.C. Jau, S.Y. Hung et al., Photoaddressable bistable reflective liquid crystal display. Appl. Phys. Lett. **89**, 021116 (2006)
50. C.T. Wang, H.C. Jau, T.H. Lin, Optically controllable bistable reflective liquid crystal display. Opt. Lett. **37**, 2370–2372 (2012)
51. Andy Y.-G. Fuh, C.-C. Chen, C.K. Liu et al., Polarizer-free, electrically switchable andoptically rewritable displays based on dye-dopedpolymer-dispersed liquid crystals. Opt. Express **17**, 7088–7094 (2009)

52. Andy Y.-G. Fuh, J.C. Chen, S.Y. Huang et al., Binary liquid crystal alignments based on photoalignment in azo dye-doped liquid crystals and their application. Appl. Phys. Lett. **96**, 051103–051103 (2010)
53. S.Y. Huang, S.T. Wu, Andy Y.-G. Fuh, Optically switchable twist nematic grating based on a dye-doped liquid crystal film. Appl. Phys. Lett. **88**, 041104–041103 (2006)
54. W.Y. Wu, T.-S. Mo, Andy Y.-G. Fuh, Polarization characteristics of diffracted beams from twisted nematic gratings fabricated by the photoalignment effect in dye-doped liquid-crystal films. J. Opt. Soc. Am. B **23**, 1737–1742 (2006)
55. W.Y. Wu, Andy Y.-G. Fuh, Rewritable liquid crystal gratings fabricated using photoalignment effect in dye-doped poly(vinyl alcohol) film. Jpn. J. Appl. Phys. Part **1**(46), 6761–6766 (2007)
56. K.T. Cheng, C.K. Liu, C.L. Ting et al., Electrically switchable and optically rewritable reflective Fresnel zone plate in dye-doped cholesteric liquid crystals. Opt. Expr **15**, 14078–14085 (2007)
57. L.C. Lin, H.C. Jau, T.H. Lin et al., Highly efficient and polarization-independent Fresnel lens based on dye-doped liquid crystal. Opt. Express **15**, 2900–2906 (2007)
58. Andy Y.-G. Fuh, J.C. Chen, K.T. Cheng et al., Polarization-independent and electrically tunable liquid-crystal Fresnel lenses based on photoalignment in dye-doped liquid crystals. J. Soc. Inf. Display **18**, 572–576 (2010)
59. Andy Y.-G. Fuh, J.C. Chen, K.T. Cheng et al., Polarization-independent and electrically tunable liquid-crystal Fresnel lenses based on photoalignment in dye-doped liquid crystals. J. Soc. Inf. Disp. **18**, 572–576 (2010)
60. Y.H. Huang, S.W. Ko, S.C. Chu et al., High-efficiency Fresnel lens fabricated by axially symmetric photoalignment method. Appl. Optics. **51**, 7739–7744 (2012)
61. C.H. Chen, C.H. Lee, T.H. Lin, Loss-reduced photonic liquid-crystal fiber by using photoalignment method. Appl. Opt. **49**, 4846–4850 (2010)
62. C.H. Lee, C.H. Chen, C.L. Kao et al., Photo and electrical tunable effects in photonic liquid crystal fiber. Opt. Express **18**, 2814–2821 (2010)
63. S.W. Ko, Y.Y. Tzeng, C.L. Ting et al., Axially symmetric liquid crystal devices based on double-side photo-alignment. Opt. Express **16**, 19643–19648 (2008)
64. Y.Y. Tzeng, S.W. Ke, C.L. Ting et al., Axially symmetric polarization converters based on photo-aligned liquid crystal films. Opt. Express **16**, 3768–3775 (2008)
65. S.W. Ko, S.H. Huang, Andy Y.-G. Fuh et al., Measurement of helical twisting power based on axially symmetrical photo-aligned dye-doped liquid crystal film. Opt. Express **17**, 15926–15931 (2009)
66. S.W. Ko, C.L. Ting, Andy Y.-G. Fuh et al., Polarization converters based on axially symmetric twisted nematic liquid crystal. Opt. Express **18**, 3601–3607 (2010)
67. Andy Y.-G. Fuh, S.W. Ko, S.H. Huang et al., Polarization-independent liquid crystal lens based on axially symmetric photoalignment. Opt. Express **19**, 2294–2300 (2011)
68. Y.H. Huang, S.W. Ko, S.C. Chu et al., High-efficiency Fresnel lens fabricated by axially symmetric photoalignment method. Appl. Opt. **51**, 7739–7744 (2012)
69. S.W. Ko, T.H. Lin, Y.H. Huang et al., Electrical control of shape of laser beam using axially symmetric liquid crystal cells. Appl. Opt. **51**, 1540–1545 (2012)
70. Y.H. Huang, S.W. Ko, M.S. Li et al., Modulation of shape and polarization of beam using a liquid crystal q-plate that is fabricated via photo-alignment. Opt. Express **21**, 10954–10961 (2013)
71. D. Shenoy, K. Grueneberg, J. Naciri et al., Photo-dimerized monolayer (PDML) versus rubbed polyimide (RPI): A comparison of electro-optic properties. Jpn. J. Appl. Phys. **2**(37), L1326–L1329 (1998)
72. O. Yaroshchuk, L.G. Cada, M. Sonpatki et al., Liquid-crystal photoalignment using low-molecular-weight photo-cross-linkable composites. Appl. Phys. Lett. **79**, 30–32 (2001)
73. C.H. Zhang, Z.H. Yang, M.X. Ding, Liquid crystal alignment generated by linearly polarized UV light on photoactive low molecular mass compounds. Liq. Cryst. **30**, 65–69 (2003)

74. A.G. Dyadyusha, T.Y. Marusiĭ, Y.A. Reznikov et al., Orientational effect due to a change in the anisotropy of the interaction between a liquid crystal and a bounding surface. JETP Lett. **56**, 17–21 (1992)
75. A. Dyadyusha, V. Kozinkov, T. Marusii et al., Light-induced planar alignment of nematic liquid crystal on anisotropic surface without microgrooves. Ukr. Fiz. Zh. **36**, 1059–1061 (1991)
76. S.J. Sung, K.Y. Cho, J.K. Park, Photo-induced liquid crystal alignment of poly(vinyl cinnamate) and fluorinated polyimide blends. Mat. Sci. Eng. C-bio. S **24**, 181–184 (2004)
77. P.O. Jackson, M. O'Neill, W.L. Duffy et al., An investigation of the role of cross-linking and photodegradation of side-chain coumarin polymers in the photoalignment of liquid crystals. Chem. Mater. **13**, 694–703 (2001)
78. M. Schadt, H. Seiberle, Optical patterning of multidomain LCDs. SID Int. Symp. Dig. Tec. **5**, 367–370 (1997)
79. M. Obi, S. Morino, K. Ichimura, The reversion of photoalignment direction of a liquid crystal induced by a polymethacrylate with coumarin side chains. Macromol. Rapid Comm. **19**, 643–646 (1998)
80. K. Ichimura, Y. Akita, H. Akiyama et al., Photoreactivity of polymers with regioisomeric cinnamate side chains and their ability to regulate liquid crystal alignment. Macromolecules **30**, 903–911 (1997)
81. B. Sapich, J. Stumpe, I. Gerus et al., Photoinduced anisotropy and LC photoalignment properties of polyvinylcinnamate films. Mol. Cryst. Liq. Crys. **352**, 443–452 (2000)
82. V. Chigrinov, A. Muravski, H.S. Kwok et al., Anchoring properties of photoaligned azo-dye materials. Phys. Rev. E **68**, 061702 (2003)
83. V. Kyrychenko, L. Vretik, O. Yaroshchuk et al Highly stable photoalignment of liquid crystals by the layers of photocrosslinkable azopolymers. In: *Abstracts of 23rd International Liquid Crystal Conference*, Krakow, Poland, 11–16 July 2010
84. J.Y. Kim, T.H. Kim, T. Kimura et al., Surface relief grating and liquid crystal alignment on azobenzene functionalized polymers. Opt. Mater. **21**, 627–631 (2003)
85. H. Kang, D. Kang, J.C. Lee, Liquid crystal alignment property of polystyrene derivatives containing dual photoreactive side groups. Polymer **50**, 2104–2112 (2009)
86. D.H. Kim, S.K. Park, S.B. Kwon et al., P-173: newly developed cellulose-based photopolymer with high anchoring energy and low-image-sticking. SID Int. Symp. Dig. Tec. **37**, 867–870 (2006)
87. Y. Reznikov, O. Yaroshchuk, I. Gerus et al., Relationship between molecular structure of photosensitive fragments and aligning properties of polysiloxanes. Mol. Cryst. Liq. Crys. C **9**, 333–342 (1998)
88. M. Obi, S. Morino, K. Ichimura, Reversion of photoalignment direction of liquid crystals induced by cinnamate polymer films. Jpn. J. Appl. Phys. **2**(38), L145–L147 (1999)
89. N. Kawatsuki, H. Takatsuka, T. Yamamoto et al., Photoregulated liquid crystal alignment on photoreactive side-chain liquid-crystalline polymer. Jpn. J. Appl. Phys. **1**(36), 6464–6469 (1997)
90. L. Vretik, L. Paskal, V. Syromyatnikov et al New photoalignment materials in LCD's development: Liquid crystal pretilt angle variations by using fluoroalkylmethacrylates. Mol Cryst Liq Crys 468:173/(525)-179/(531) (2007)
91. A. Dyadyusha, T. Marusii, Y. Reznikov et al., Oblique photoalignment of a nematic liquid crystal by modified polyvinylcynnamates. Proc. SPIE **273**, 151–154 (1995)
92. A. Dyadyusha, A. Khizhnyak, T. Marusii et al., An oblique orientation of nematic liquid crystals on a photosensitive aligning polymer. Mol. Cryst. Liq. Crys. A **263**, 399–413 (1995)
93. B. Park, K.J. Han, Y. Jung et al., Director tilting of liquid crystals on photoisomerizable polyimide alignment layers doped with homeotropic surfactant. J. Appl. Phys. **86**, 1854–1859 (1999)
94. J.Y.L. Ho, V.G. Chigrinov, H.S. Kwok, Variable liquid crystal pretilt angles generated by photoalignment of a mixed polyimide alignment layer. Appl. Phys. Lett. **90**, 243506 (2007)

95. B. Park, Y. Jung, H.H. Choi et al., Thermal and optical stabilities of photoisomerizable polyimide layers for nematic liquid crystal alignments. Jpn. J. Appl. Phys. **1**(37), 5663–5668 (1998)
96. H. Akiyama, K. Kudo, K. Ichimura, Command surfaces. 10. Novel Polymethacrylates with laterally attached Azobenzene groups displaying photoinduced optical anisotropy. Macromol. Rapid Comm. **16**, 35–41 (1995)
97. J.L. West, X. Wang Y. Li et al Polarized UV-exposed polyimide films for liquid-crystal alignment. SID Int. Symp. Dig. Tec .26:703–705 (1995)
98. J.H. Kim, B.R. Acharya, S. Kumar et al., A method for liquid crystal alignment using in situ ultraviolet exposure during imidization of polyimide. Appl. Phys. Lett. **73**, 3372–3374 (1998)
99. O. Yaroshchuk, A. Kadashchuk, Liquid crystal photoalignment properties of polymethylphenylsilane. Appl. Surf. Sci. **158**, 357–361 (2000)
100. A. Dyadyusha, S. Nespurek, Y. Reznikov et al., Liquid crystal alignment on photoinduced species in poly[methyl(phenyl)silylene] film. Mol. Cryst. Liq. Crys. **359**, 387–397 (2001)
101. M. Hasegawa, Key molecular structure determination of photoalignment materials from the effects of linearly polarized deep UV light on several polymers. Jpn. J. Appl. Phys. **1**(39), 1272–1277 (2000)
102. M. Nishikawa, T. Kosa, J.L. West, Effect of chemical structures of polyimides on unidirectional liquid crystal alignment produced by a polarized ultraviolet-light exposure. Jpn. J. Appl. Phys. **2**(38), L334–L337 (1999)
103. C.J. Newsome, M. O'Neill, Spectroscopic analysis of photoinduced degradation of polyimide for liquid crystal alignment: linear and nonlinear effects. J. Appl. Phys. **88**, 7328–7333 (2000)
104. N. Kawatsuki, T. Shiraku, E. Uchida, Influence of molecular weight on photoinduced reorientation of 4-methoxyazobenzene-containing polymethacrylate films using 633 nm He-Ne laser. Mol. Cryst. Liq. Crys. **441**, 163–171 (2005)
105. M. Nishikawa, B. Taheri, J.L. West, Mechanism of unidirectional liquid-crystal alignment on polyimides with linearly polarized ultraviolet light exposure. Appl. Phys. Lett. **72**, 2403–2405 (1998)
106. F.S.Y. Yeung, H.S. Kwok, Fast-response no-bias-bend liquid crystal displays using nanostructured surfaces. Appl. Phys. Lett. **88**, 063505 (2006)
107. Y.W. Li, L. Tan, F.S.Y. Yeung et al., 12.4: distinguished student paper: passive-matrix-driven field-sequential-color LCD. SID Int. Symp. Dig. Tec. **38**, 154–157 (2007)
108. F.S. Yeung, J.Y. Ho, Y.W. Li et al., Variable liquid crystal pretilt angles by nanostructured surfaces. Appl. Phys. Lett. **88**, 051910 (2006)

Chapter 12
Liquid Crystal Display—Present Status and Emerging Technology

Ko-Ting Cheng

Nowadays, in your daily lives, the applications of LCs, such as LCD TVs, computer/laptop monitors, mobile phones, navigators, digital cameras, and so on, are widely used worldwide. Even, wherever you are whatever you do, you can take your electronics out of your pocket to check your e-mails, listen to music, watch movies, find your way, and so many others. LCDs provide high performance in regard to the weight, volume, contrast ratio, viewing angle, low power consumption, etc. That's why the high-quality applications of LCDs are indispensable and paid much attention by scientists and electronics industries in the recent decade. Undoubtedly, LCDs have acted as a leader in the field of flat panel displays, in not only small-sized displays, but also large-sized ones. Refer to cathode ray tube (CRT) monitors; LCDs have almost completely replaced CRT monitors in the display market because of their bulky and high power consumption, despite their good response time, viewing angle, image quality. However, in addition to the present status of LCD applications, some emerging technologies, such as wide viewing angle displays, fast response displays, blue phase LCDs, flexible displays, 3D displays, high resolution displays, in-cell touch panel, electronic papers, tablets, etc., have also attracted much attention. In this chapter, the introductions of the present status of LCD industry, as well as the emerging technology for improving the performances of LCDs will be given.

12.1 Liquid Crystal Display Modes

Regarding the commercial products of LCDs [1–3], their operating principles for modulating light intensity and/or light polarization include polarization rotation, phase retardation, scattering, absorbance, etc. LCs possess, as described in

K.-T. Cheng (✉)
Department of Optics and Photonics, National Central University, Zhongli, Taiwan
e-mail: chengkt@dop.ncu.edu.tw

C.-C. Lee (ed.), *The Current Trends of Optics and Photonics*,
Topics in Applied Physics 129, DOI 10.1007/978-94-017-9392-6_12

Chap. 10, several anisotropic properties, such as refractive index (birefringence, Δn), dielectricity (dielectric anisotropy, $\Delta\varepsilon$), elastic constants, susceptibility, and others. Briefly, without applying any external field, LCs are aligned along the direction corresponding to the alignment layers. Because of the properties of dielectric anisotropy and flow of LCs, LCs can be oriented by applying an external field, resulting in light modulation based on phase retardation, polarization rotation, light reflection and transmission, etc. Additionally, if the applied external field is removed, LCs can recover to its initial orientation state due to the elasticity and viscosity. Notably, LCs in LCDs do not emit light, they only control whether light gets through LCs or not, so that back-light/front-light units are needed for LCDs according to transmissive or reflective modes. In terms of the operating mechanisms, the commercially adopted LCDs are divided into the following modes [1–3], including twisted nematic (TN), super twisted nematic (STN), mixed-mode twisted nematic (MTN), vertical alignment (VA), optically compensated bend (OCB, π-cell), in-plane-switching (IPS), ferroelectric LC (FLC), guest-host (G-H), polymer dispersed LC (PDLC), cholesteric reflection, LC on silicon (LCoS), reflective, transflective modes, and so on. Typically, for a LCD, each pixel is divided into three (red, green and blue) or four (red, green, blue and yellow/white) sub-pixels with the color based on additive color principle. For another kind of LCDs, called field sequential color (FSC), or color sequential displays, all of the pixels display the image for only one color of RGB at the one third frame time, indicating that its mechanism is based on separating color temporally not spatially. In this section, we will focus on several modes those attach importance to LCD industry recently.

(a) polarization rotation mode

The physical mechanism for LCDs using polarization rotation mode is based on waveguiding effect [4, 5], showing that the polarization direction of the incident light beam follows the twist of LC director in a LC cell. Hence, the polarizations of incident and output light are two points to be discussed. Briefly, if the polarization direction of the incident light is selected to be parallel to the LC director near the incident layer, the polarization state will be rotated to the direction of the LC director close to the output layer. However, strictly speaking, the polarization rotation effect is valid with the limitation, called Mauguin condition,

$$\phi \ll \frac{2\pi d \Delta n}{\lambda}, \tag{12.1}$$

where ϕ, d, Δn and λ are, respectively, the twisted angle, thickness of LC cell, birefringence, and wavelength of light [4]. In LC industry, the representative is twisted nematic (TN) LCD, extensive commercially used in the recent decade. It was first proposed by Schadt and Helfrich in 1971 [5]. For example, in a typical 90° TN cell, the polarization direction of the incident linearly polarized light is rotated by 90°. In other word, the polarization direction of the output light remains linearly polarized if the condition of Mauguin limitation is satisfied. Otherwise, the polarization of the output beam will be changed to elliptic polarization, indicating that the transmission will be reduced under cross- or parallel- polarizers. For a special case of 90° TN cell ($\phi = 90°$), the Mauguin condition can be reduced to

$$\frac{\lambda}{4} \ll d\Delta n, \tag{12.2}$$

Here, we do not talk about its details of electromagnetic wave, but operating principle of 90° TN-LC cell will be shown. First of all, a LC cell having a pair of orthogonal rubbing directions, resulting in a total twisted angle of 90°, is considered. In brief, the polarization of linearly polarized light can be rotated by 90°, indicating that the polarization will follow the twist of the local director. It should be noted that in order to ensure the unidirectional rotation of the LCs, a little fraction of chiral dopant will be doped into the TN-LC cells. A typical 90° TN-LC cell can be divided into two kinds of operating mode; they are normally white (NW) and normally black (NB) modes. The word "normally" means that the state without applying any voltage. Hence, the transmissions of NW and NB modes at voltage-off state are, respectively, bright and dark. According to the layout, the only difference between NW and NB modes is the transmission axes of polarizer and analyzer. As shown in Fig. 12.1a, b (Fig. 12.1c, d), it is the NW (NB) mode that the transmission axes of polarizer and analyzer are set perpendicular (parallel) to each other. Considering the NW mode TN-LC cell, it is bright state if no voltage is applied (Fig. 12.1a). However, when a voltage is applied onto the TN-LC cell, the LCs ($\Delta\varepsilon > 0$) will be aligned along the direction of electric field if the voltage is high enough. At this moment, the TN-LC cell changes to homeotropic LC cell so that zero transmission can be obtained under cross-polarizers (Fig. 12.1b).

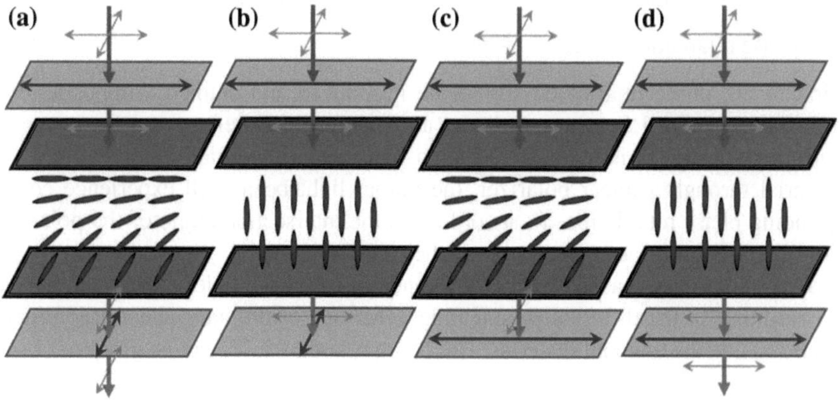

Fig. 12.1 Schematic diagrams of twisted nematics, **a** *bright* (field-off) and **b** *dark* states (field-on) of normally *white* TN LCDs; **c** *dark* (field-off) and **d** *bright* (field-on) states of normally *black* TN LCDs

Moreover, if the amplitude of the applied voltage is between zero and that to align LCs homeotropically, gray scales can be obtained. Here, we give the formula describing the transmittance of a NW 90° TN-LC cell.

$$T = \cos^2 X + \left(\frac{\Gamma}{2X}\cos 2\beta\right)^2 \sin^2 X, \tag{12.3}$$

where β is the angle between the polarization axis and the front LC director, ϕ the twisted angle, $X = \sqrt{\phi^2 + (\Gamma/2)^2}, \Gamma = 2\pi d\Delta n/\lambda$, d, Δn and λ are, respectively, cell gap, birefringence, and wavelength. For a special case, cosX = ±1, i.e. X = mπ, m is an integer; T is independent of β and the Gooch-Tarry condition can be obtained [4].

$$\frac{d\Delta n}{\lambda} = \sqrt{\left(m^2 - \frac{1}{4}\right)}, \tag{12.4}$$

For m = 1, it's the Gooch-Tarry first minimum. Notably, 90° TN-LC cell has been widely used for many kinds of displays, such as transmissive, projective and reflective LCDs, using two polarizers. However, the narrow viewing angle and gray scale inversion, caused by optical anisotropy of LCs, off-axis light leakage from cross-polarizers, and others, are the major drawback of TN-LC cells even the performances are acceptable for notebook computer applications. Therefore, compensated films are employed to enhance their performance.

The mixed twisted nematic (MTN) mode is developed based on TN mode. The main difference between TN mode and MTN mode LCDs are the setting of polarizers and the total twisted angles. For MTN mode LCD, the angle of intersection of polarizer axis and rubbing is set to be non-zero angle, and the total twisted angle of 90° is unnecessary. The mixed mode means the combination of polarization rotation and phase retardation. Usually, it should be noted that MTN mode can be adopted to fabricate a reflective LCD, and will be introduced in next section [1–4].

(b) phase retardation mode

First of all, phase retardation should be introduced and defined. Supposing one uniaxial crystal or LC, whose ordinary and extraordinary refractive indexes are n_o and n_e, respectively, is prepared. When a light beam is normally incident into the material through a linear polarizer, the output light beam will experience corresponding phase retardation, Γ, according to the intersection angle of the polarization direction and the optical axis of the uniaxial material. The formation of the phase retardation is caused by the difference of refractive indexes between n_o and n_e, resulting in the different propagating velocities of ordinary and extraordinary beams inside the uniaxial material. The phase retardation, Γ, can be given as [1–4]

$$\Gamma = \frac{2\pi d(n_e - n_o)}{\lambda} = \frac{2\pi d\Delta n}{\lambda}, \tag{12.5}$$

where d is LC cell gap, Δn is birefringence and λ is the wavelength of incident light. For two special cases, polarization direction of the incident light parallel or perpendicular to optical axis of the uniaxial material, the phase retardation equals zero. Otherwise, the effective phase retardation causes the transformation of polarization of the output beam. For example, just like half-waveplates, if the effective phase retardation is $\pi/2$, the incident linearly polarized light will be transferred to elliptically polarized light (be rotated by 2θ, θ: intersection angle). Accordingly, if a uniaxial LC layer is sandwiched between two cross polarizers, the transmission can be electrically tuned because of the variation of Δn by applying voltage, or the so-called electrically controlled birefringence (ECB) effect. Many types of LCDs, including vertical alignment, homogeneous alignment, hybrid alignment, in-plane-switching (IPS) modes, and others, are operated using this ECB effect.

In principle, the relationship between transmittance and phase retardation under crossed ($T_\perp$) and parallel ($T_{//}$) polarizers can be given as,

$$T_\perp = \sin^2 2\theta \ \sin^2\left(\frac{\Gamma}{2}\right), \quad T_{//} = 1 - \sin^2 2\theta \ \sin^2\left(\frac{\Gamma}{2}\right), \tag{12.6}$$

where θ is the angle between the polarizer direction and the LC director.

(c) scattering mode

In scattering mode for modulating light intensity, LCs and polymers are two candidates for use in such kind of LC devices, including displays, light shutters, diffusers, etc. Initially, they will be homogeneously mixed with each other, and then, will be phase separated. Two famous catalogs—polymer-dispersed LCs (PDLCs) and polymer-stabilized LCs (PSLCs) are developed according to their morphologies.

Regarding PDLCs, the micron- or sub-micron-size LC droplets, isolated from one another, will be formed and dispersed in polymer matrix. The incident light will be scattered because of the mismatch of refractive indexes of LCs and polymers at zero applied voltage. Notably, the concentration of polymer and LCs are comparable with each other. In PSLCs, divided into scattering and non-scattering devices, the concentration of LCs is much higher than that of the polymer. Here, the scattering devices, where exist multi-domain LC structures, are discussed only. The size of LC domain is comparable to the scattered light. Notably, the size of LC domains in both of PDLCs and PSLCs can be controlled during phase separation processes. Here, a brief introduction of the operation mechanism for scattering mode LC devices will be shown following. As depicted in Fig. 12.2a, the material of polymer matrix is optically isotropic with its refractive index, n_p, and the LC droplets are randomly oriented in polymer matrix in the absence of an applied voltage. Regarding the boundary of two materials with discontinued refractive indexes, the refractive indexes mismatch between the LC droplets and polymer matrix leads to strong scattering in the voltage-off state (scattering state). On the other hand, the director of LCs with $\Delta\varepsilon > 0$ are aligned along the applied electric filed in the voltage-on state (transparent). A very important key is that the ordinary refractive index (n_o) of the used LCs should be as close as n_p for the application of light shutters. Accordingly, the unidirectional director provides almost insignificant differences ($n_o = n_p$) in the refractive indexes of neighboring polymers such that the PDLC appears perfectly transparent, as shown in Fig. 12.2b. However, the polymer matrix walls produce a strong surface

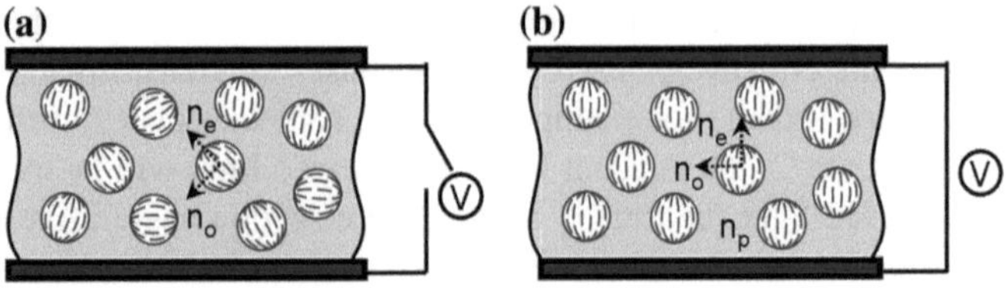

Fig. 12.2 Schematic diagrams of polymer-dispersed LCs; **a** scattering (field-off) state and **b** transparent (field-on) state

anchoring effect that increases driving voltages, indicating that the operated voltage of a traditional PDLC scattering mode light shutter is high [6, 7]. Accordingly, many scientists have recently paid much attention to such approaches as surface rubbing effect [8], dichoric dyes [9], and others, to reduce the operated voltage of PDLCs. Next, a newly proposed approach, particular thermally-induced phase separation (TIPS), is introduced [6]. Fuh et al. reported particular TIPS method that involves a combination of dissolution process of LCs and poly(N-vinylcarbazole) (PVK), and TIPS. Briefly, a nematic LC cell fabricated by two substrates coated with uniform PVK films is heated and then cooled, generating the rough PVK layers onto the surfaces of the substrates. The LC cell having rough PVK layers produces micron-sized, multiple domains of disordered LCs that can scatter incident light. The fabricated LC light shutter possessed the advantages of low driving voltage, fast response in the order of milliseconds, independent of polarization, high contrast ratio, and being polarizer free.

Moreover, scattering mode light shutters can also be used to fabricate many LC devices, such as holography [10], smart windows [11], etc.

(d) absorbance mode

So far, we have introduced several physical mechanisms, such as polarization rotation, phase retardation and scattering, for modulating the transmissive light through a LC cell. In the following section, we'd like to discuss another mode for modulating light intensity, absorbance mode. The famous LCD type using absorbance mode is guest-host (G-H) display that does not need any polarizer. To introduce G-H LCD, dichroic dyes should be paid attention to firstly. The LC and dyes play the roles of host and guest, respectively, because of the properties of electrical switching of LCs and anisotropic absorbance of dichroic dyes. Usually, the LC host is highly transparent in visible wavelength, and the selected dyes should strongly absorb one polarization direction of incident light, and unabsorb the other one. Importantly, the rod-like dichroic dyes are aligned along the long axis of LCs. Thus, the mechanism of G-H LCDs is electrically controllable light absorption. About the dichroic dyes, dichroic ratio, DR, the most important parameter in determining the performance of a G-H LCD, is defined as

$$DR \equiv \frac{A_{\parallel}}{A_{\perp}} = \frac{1+2S}{1-S}, \tag{12.7}$$

where $A_{//}$ and $A_{\perp}$ are the absorbances when the polarization direction of the incident light is, respectively, parallel and perpendicular to the optical axis of the dyes. S is the order parameter [12]. According to the absorbance performance, the contrast ratio, defined as the transmittance ratio at voltage on and off states, of G-H LCDs increases with increasing DR. Additionally, the contrast ratio is also dependent on the cell thickness and concentration of dyes, but the prices are the increases of operating voltage, response time, etc. Notably, powder dyes exhibit extremely low solubility into LCs. Several LC dyes are developed to enhance the concentration of dyes [13, 14].

Next, we'd like to introduce the most commonly used G-H LCDs—White-Taylor cell, which does not need any polarizer [15]. The materials used in the cell are nematic LC ($\Delta\varepsilon > 0$), about 2 wt% absorbing dyes and about 1 wt% chiral dopant. Due to the doping chiral dopant, the dye-doped chiral nematic LCs present twisted structures, in which the twisted aligned dyes can absorb incident unpolarized light at voltage-off state (dark state, Fig. 12.3a). Moreover, at voltage-on state (bright state, Fig. 12.3b), LCs and dyes are aligned along the direction of electric field, resulting in the low absorbance. Since no polarizer is required, the brightness is excellent, which is very suitable for fabricating reflective mode LCDs. Additionally, other kinds of G-H LCDs, which need polarizers or quarter waveplates, using dye-doped nematic LCs have also been developed [16]. Also, Toshiba group demonstrated a prototype of reflective color LCDs using 3-layer guest-host mode [17].

(e) cholesteric reflection mode

In recent decade, cholesteric LCs (CLCs), also known as chiral nematic LCs, have been studied by many scientists [18–23]. CLCs usually consist of nematics and a chiral dopant, as well as have two stable textures, planar (P) and focal conic (FC) textures. The P textures (Fig. 12.4a) present reflective state due to the Bragg reflection. Briefly, circularly polarized light with the same handedness as that of the P textures of CLCs is selectively reflected. The central reflection wavelength (λ_c) of the selective reflection band by P textures can be described by p and $\bar{n}$, and can be expressed as $\lambda_c = \bar{n}\, p\cos\theta$ ($\theta = 0$ for normally incident light, $\bar{n} = (n_{\|} + n_{\perp})/2$). Pitch length ($p$) is determined by the concentration (c) of the chiral dopant and the relative helical twisting power (HTP) according to their relationship $HTP = 1/pc$. Positive (negative) values of HTP

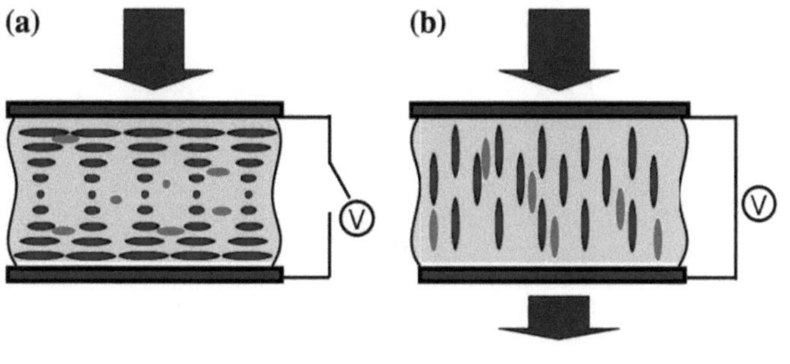

Fig. 12.3 White-Taylor guest-host cell; **a** *dark* (field-off) state and **b** *bright* (field-on) state

represent right- (left-) handed helix. Moreover, the FC textures (Fig. 12.4b) show strong light scattering because of the randomly arranged helical axes. Such various potential bistable (reflective planar and scattering focal conic states) LCD technologies have been developed because they present low power consumption [22, 23].

Moreover, P textures can be switched to FC ones by applying a voltage with relatively low amplitude. In this state, the Bragg reflection can be eliminated mostly, and the incoming light is scattered. The FC textures are transformed into homeotropic texture, which is a transparent state, if a relative high voltage is applied. When the applied high voltage is switched off rapidly, CLCs in homeotropic textures can be relaxed back into P textures. Notably, the homeotropic textures can also be switched back to FC textures by reducing the high applied voltage to a low bias voltage, or removing the high voltage gradually. Figure 12.5 presents the possible texture transitions of CLCs. Importantly, both of P and FC textures are stable states, and can be used to achieve bistability with applying any external voltage to maintain them. Accordingly, cholesteric reflection modes LCDs are widely used because of their low power consumption, bistability, readable under sun, good viewing angle, etc. Also, polymer-stabilized cholesteric textures, full-color reflective LCDs, and others, are developed to improvement their performance.

(f) LC on silicon (LCOS) mode

LC on silicon mode, which is a reflective LCD, has been widely adopted for developing projection systems and virtual display (micro-display). Such a kind of reflective LCDs is fabricated by only one glass substrate (front substrate) and a silicon wafer (back substrate), containing the corresponding circuits for addressing each pixel. Briefly, the pixel electrodes of LCOS displays are coated with a flat and highly reflective aluminum mirror and an alignment layer, as shown in Fig. 12.6. LCOS can be adopted to achieve high-resolution LCDs because of the

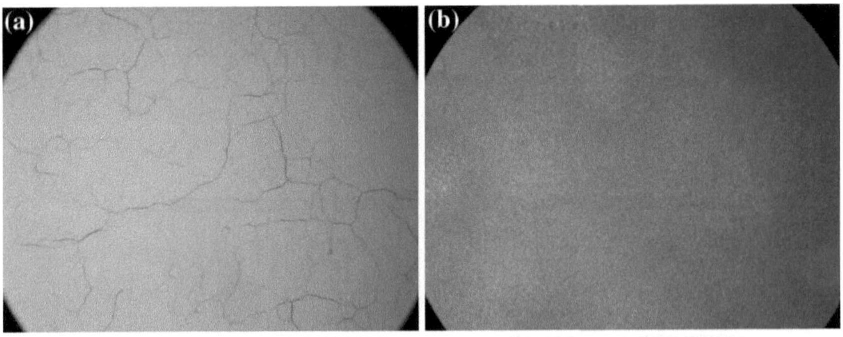

Fig. 12.4 Observations of cholesterics under cross-polarized optical microscope; **a** planar and **b** focal conic textures

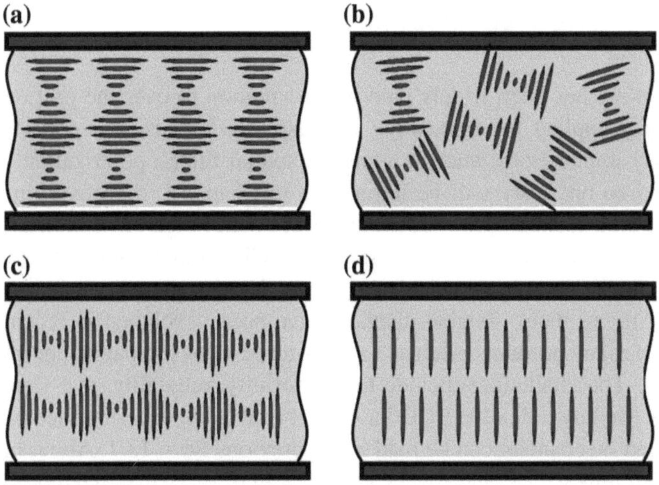

Fig. 12.5 Schematic diagrams of cholesterics; **a** planar texture, **b** focal conic texture, **c** fingerprint texture and **d** homeotropic textures

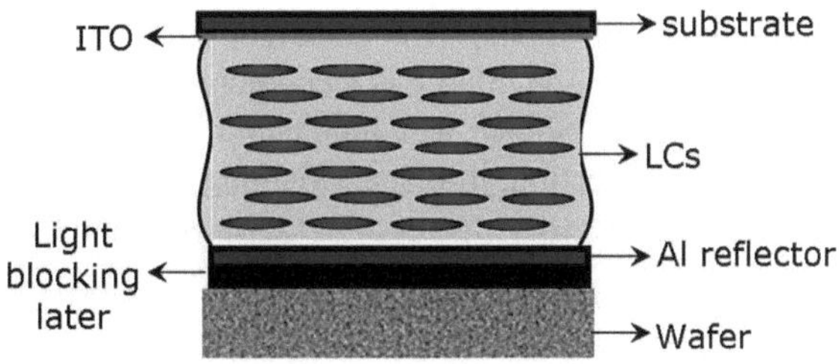

Fig. 12.6 Basic configurations of LC on silicon cell

existing silicon technology. Moreover, considering the color display using LCOS technology, three methods, including color pixellization, multiple LCOS panels, and field sequential color, can be used to achieve color. However, because of a small size of LCOS display, the second and third approaches are commonly used in projectors and virtual displays, respectively. It should be noted that the mechanism of field sequential color (FSC), or the so-called color sequential display, is based on separating color temporally rather than spatially (sub-pixels), so that the switching time of LC cell is extremely important.

However, LCOS mode still has some drawbacks, resulting from the temperature. The stability of surface alignment layer will be reduced due to the high temperature inside the LCOS system so that the LC director alignment will be disturbed. Accordingly, usually, the used materials to form the alignment layers are inorganic materials, such as SiOx. Additionally, color breakup is another issue should be solved in color sequential display.

12.2 Emerging Technology

LC technology has been widely paid much attention to over the past decade. LCs are not only applied for fabricating flat displays, but also for developing many non-display applications, such as gratings, spatial filters, polarization convertors, lenses, and so on. They will be introduced in Chap. 14. Moreover, in LC fields, some emerging technologies, such as flexible LCDs, high resolution LCDs, green technologies, reflective LCDs, transflective LCDs, three-dimension (3D) LCDs, fast response LCDs, touch panels, and others, have attracted considerable interest recently. Among them, flexible displays have been developed in recent years for applying in many products, such as smart cards, e-books, e-papers, mobile phones, and others. The challenges include the uniformity, reliability, etc. Commercially, customers can buy a flexible LCD based on bistable cholesteric reflective displays. The detailed mechanism can be read in the previous Sect. 12.1. About high resolution, the famous 4K2K and 8K LCDs (ultra high definition, UHD) are believed to dominate the mainstream market. To achieve such high resolution, the technical processes, including high aperture ratio TFT pixel design, high accurate process equipments, optimization of the processes, and so on, are the main keys. Moreover, green technology in LC fields can be extended to multi-stability, low power consumption, simple fabrication processes, and so on. The techniques of polymer stabilizations, LED backlight units, frontier LC materials are developed to achieve the issue. 3D LCD technologies will be described in detail in Chap. 13. Touch panel, combined with high definition flat display, is another challenge in LC fields. In this section, in-plane-switching (IPS)/fringing field switching (FFS), blue phase, and reflective/transflective LCDs will be introduced.

(a) in-plane-switching (IPS)/fringing field switching (FFS) LCD

Conventionally, as shown above, the direction of the applied voltage to drive LCs is perpendicular to the substrates (parallel to the normal direction of the substrates). With the applied voltage, the electrically controllable birefringence results in proper phase retardation, and can be adopted to control gray scales with polarizers. However, the phase retardations obtained from right and left sides are different due to the directional asymmetry of tilted LC. Such a shortage causes narrow viewing angle and asymmetric transmittance, especially along the vertical directions. To overcome the problems, it is in-plane-switching (IPS) technique that was proposed in 1970s by Soref et al. [24, 25]. IPS mode is, in brief, a driving scheme by a fringing electric field, which keeps the directors of LCs lying parallel to the surface of substrates during the treatment of electrical driving. Restated, the directors of LCs are rotated in the plane parallel to the direction of the applied transverse electric field, so that the viewing angle and asymmetric transmittance can be improved. In IPS mode, to provide a transverse electric field, the periodically interdigital electrodes are fabricated onto one of the substrates to generate the fringing field in the transverse plane. It should be noted that the distance (l) between the electrodes should be larger than the cell gap (d) ($l > d$) and the width of the electrode. Refer to the previous Sect. 12.1(b), the phase retardation mode can be achieved in both of homogeneous (planar) and homeotropic (vertical) alignments using longitudinal fields. In the IPS mode, both of them can also be switched by transverse fields [26]. Especially, in homeotropic alignment, LCs having either positive or negative dielectric anisotropy can be adopted. The famous electronics, including iphone, ipad, LG smart phone, etc., are employed IPS mode as their displays.

The first design of the IPS cell, using positive dielectric anisotropy LCs in homogeneous alignment cell, was proposed by Hitachi [27, 28]. Such a LC cell, as shown in Fig. 12.7, shows normally black mode under cross-polarizers, indicating that in the field-off state, the LCs are aligned by the homogeneous alignment layer in parallel to the direction of the polarizer, so that no phase retardation can be experienced. Finally, the incident light is thus blocked by the analyzer. Moreover, in the field-on state, the LCs are electrically rotated so that the phase retardation modulates the polarization of the incident light. Hence, the transmittance can be electrically switched. Additionally, IPS modes using positive dielectric anisotropy LCs in homeotropic alignment layers were also developed by Hyundai [29] and Samsung [30]. To overcome the dark lines, existing between the electrodes, Hyundai and Samsung groups proposed different approaches to achieve four-domain configurations. The viewing angle can be improved further.

Furthermore, to improve the performance of LCDs, such as aperture ratio, viewing angle, small-size high-resolution, large-size ultra-hugh-resolution, etc., Hydis group developed fringing field switching (FFS), which also utilizes the transverse field to switch LC directors [31]. The configuration of the FFS mode is very similar to that of the IPS mode. The main difference between these two modes is the electrode gap. In the FFS mode, the electrode gap (l) is smaller than the cell gap (d) ($l < d$), so that the fringing field can reorient the LCs just above the electrodes. Hence, the high transmittance can be achieved because of the absence of the dead zones. Also, LCs with both positive and negative dielectric anisotropy

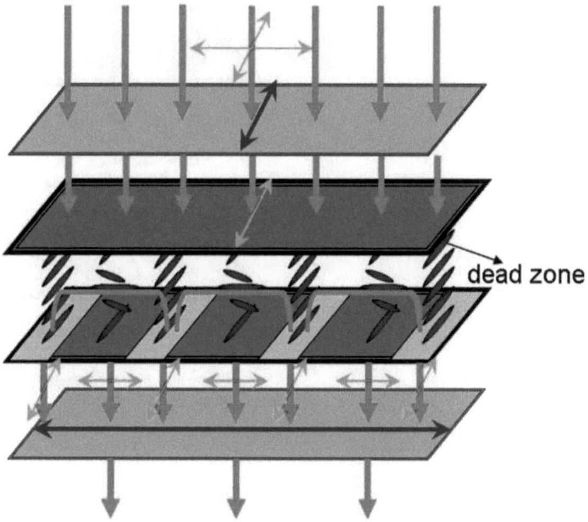

Fig. 12.7 Basic configuration of the IPS mode LCD in voltage-on state with a homogeneous alignment LCs ($\Delta\varepsilon > 0$) parallel to the stripes of electrodes

can be used in the FFS mode. Figure 12.8 shows the basic configuration of the FFS mode LCD with a homogeneous alignment parallel to the stripes of electrodes. The used LCs are with positive dielectric anisotropy. Clearly, the fringing field distributes over the electrodes and their gaps, resulting in high transmittance. The LCDs, produced by the FFS mode, are called Super-LCDs. Recently, the Hydis group also developed Advanced FFS (AFFS) technique, providing superior performance, low power consumption, wide viewing angle, high transmittance, fast response, and color gamut with high luminosity, as well as the improvement of outdoor readability [32]. Additionally, high aperture FFS has also been developed for small size LC panels [33]. AUO also proposed several techniques, including VA-IPS with positive dielectrically anisotropic LCs, VA-FFS using inverse electric fields, etc., to improve the performances of LCDs.

In summary, IPS and FFS modes provide some advantages, such as wide viewing angle, high contrast, and others, resulting from the principle of in-plane reorientation of LCs. The incoming light from any directions experiences the short axis of LCs so that the viewing angle can be widened and symmetric. Additionally, IPS and FFS modes LCDs are hard screen mode LCD, suitable for the applications of touch panel displays. However, the yields and the cost are the main keys for IPS and FFS modes to compete with the general LCD panels with wide viewing angle.

(b) blue phase LCD

BP-LCs, including PBI (body-centered cubic symmetry), BPII (simple cubic symmetry) and BPIII (amorphous structures), are LC phases that appear in a narrow

Fig. 12.8 Basic configuration of the FFS mode LCD in voltage-on state with a homogeneous alignment parallel to the stripes of electrodes

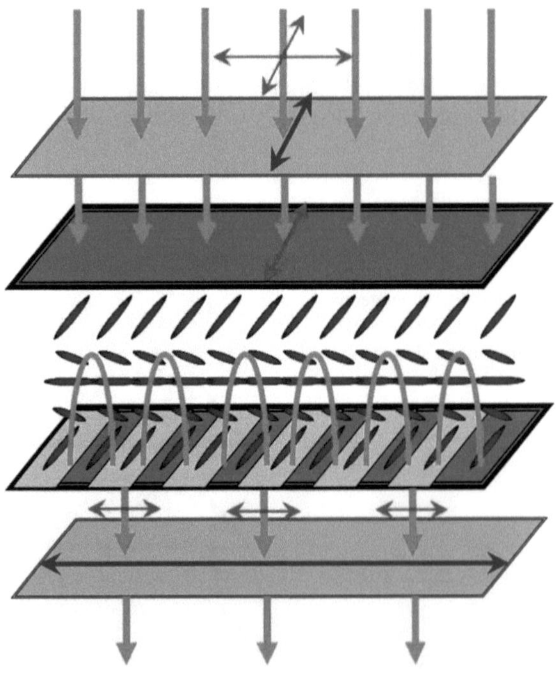

temperature range, usually less than a few K, between chiral nematic and isotropic phases. Unlike the cholesteric LCs with single twisted structures, BP-LCs are assembled by double twisted structures. Such two twisted structures are crossed with each other and generate double twisted cylinders (DTCs), as shown in Fig. 12.9. BP-LCs exhibit selective Bragg reflections due to their three-dimensional cubic structures with lattice periods in the order of 100 nm. Moreover, observing BP-LCs under polarized optical microscope, BPI and BPII present colorful platelet textures, while BPIII fog-like textures. [34]

For application in LCDs, because BP-LCs present extremely fast response (μs), wide viewing angle, alignment layer free process, cell gap insensitivity, etc., they are much benefit for the next generation of display and electro-optical devices if the temperature range can be broadened. Briefly, the isotropic sphere will be elongated or flattened to be ellipsoid, depending on whether the LCs are positive or negative dielectric anisotropic materials, when electric field is applied. In 2008, Samsung announced its prototype TFT-LCD panel with 240 Hz image frame rate based on blue phase LCs (BP-LCs) at 2008 SID (Society for Information Display) [35]. Moreover, not only the direct-view, but also the projection displays based on Kerr effect using BP-LCs high have potential for LCDs. Regarding projection displays, field-sequential, or the so-called color-sequential, projection LCDs with several advantages of high optical efficiency, low cost, and others, are newly and famous displays in the recent decades. However, color breakup, resulting from

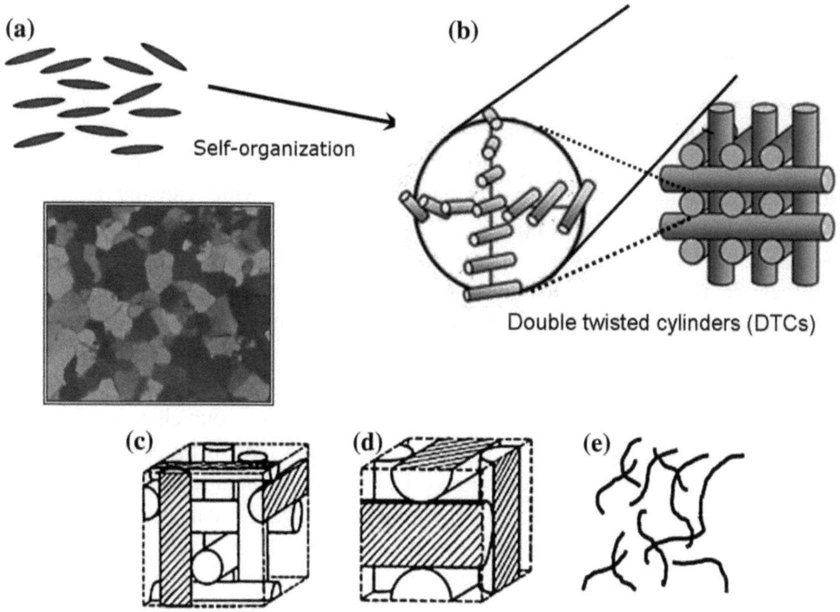

Fig. 12.9 Blue phase LCs, **a** observation under cross-polarized optical microscopy; **b** double twisted cylinders (DTCs); **c** PBI (body-centered cubic symmetry); **d** BPII (simple cubic symmetry) and **e** BPIII (amorphous structures)

slow response time, should be paid much attention. Hence, BP-LCs are the promising candidates for such display applications due to their fast response [36–41].

Regarding to the temperature range, recently, it is demonstrated that the polymer can be used to stabilize BP-LCs to extend the temperature range more than 60 K, including room temperature [42]. Also, other approaches were developed to broaden the temperature range of BP-LCs [43]. Accordingly, the stabilized BP-LCs at room temperature possess the electro-optical switching with extremely fast response (μs).

Under the application of electric field, the optically isotropic LCs can be changed to anisotropic LCs with the optical anisotropy parallel to the applied field. It indicates that the described IPS or FFS modes are adopted to produce birefringence for BP-LCDs. Notably, the optically anisotropy follows the Kerr Effect [37]. In order to increase the phase retardation, the birefringence can be increased with the increase of the applied field until the saturated birefringence appears. However, the extremely high operating voltage of BP-LCDs based on IPS mode is not suitable for practical application. To reduce such high operating voltage, several modified structures of IPS electrodes, such as protrusion electrodes [44], wall-shaped electrodes [45], double-penetrating fringe fields [38], and corrugated electrodes [36], are developed. Also, Kim et al. and Wu et al. also proposed vertical field switching (VFS), whose electric field is in longitudinal direction and uniform, in BP-LCs [40, 46, 47]. However, although BP-LCs were discovered a 100 years ago and were also improved technically, they were almost failed to be really applied for fabricating practical devices due to the described instinct shortages.

(c) transflective mode LCD

Nowadays, transmissive LCDs have been widely developed in most of the display electronics. As described above [1–4], the mechanisms for operating principle of light modulations include polarization rotation, phase retardation, scattering, absorbance, and so on. The performances, including wide viewing angle, fast response, low power consumption, high contrast ratio, etc., are also improved. However, an inevitable issue for transmissive LCDs is the continued "on" backlight source as long as the display is turned on. Restated, the relatively high power consumption cannot be reduced according to this point. Of course, light emitting diodes are great for reducing the power consumption of backlight unit. However, it is worth to be noticed. Additionally, if you bring your electronics out of doors under sunlight, the contrast of the displayed image onto your electronics can be reduced by the sunlight. So, it is reflective LCDs that have been developed [1]. It is clear to understand that the reflective LCDs do not include built-in backlights so that the undesired ambient light for transmissive LCDs can be used to be the "front light" to reflective LCDs for displaying information. Accordingly, comparing reflective LCDs with transmissive ones, not only the outdoor readability can be improved, but also the power consumption can be reduced. But, how about bring your electronics with reflective LCDs into a dark room? Reflective LCDs do not work there. Combing transmissive and reflective LCDs can keep their advantages in one LCD, which is named transflective LCDs.

As described above, transflective LCDs, whose light sources are the ambient light or backlight units, are developed to be used under both bright and dark ambient, and can simultaneously or independently display images based on the transmissive and reflective modes. Hence, the transflective LCDs can be applied onto the electronics, which can be used indoors and outdoors, such as mobile phones, watches, digital cameras, navigators, and so on. Briefly, transflective LCDs can be divided into several categories according to the operating mechanisms [48]; they are absorption, scattering, reflection and phase retardation modes. Among them, the phase retardation mode, requiring two polarizers usually, has higher potential for commercial applications than the others, due to their higher contrast ratio, lower driving voltage, and so forth. Here, we will introduce two main types of transflective LCDs based on phase retardation mode, one is dual-cell-gap transflective LCDs, and the other one is single-cell-gap transflective LCDs. The key point is the identical phase retardation from the transmissive region (T-region) and reflective region (R-region) after the light leave the LCs regions when any amplitude of the electric field is applied.

(1) dual-cell-gap transflective LCDs

Figure 12.10 shows the schematic configuration of a simple dual-cell-gap transflective LCD with the homogeneous aligned LCs in both the T- and R-regions. The cell gaps of these two region, d_T and d_R are different, and the relationship between d_T and d_R is $d_T = 2d_R$. Notably, the LC alignments in T- and R-regions are the same so that the complicated alignment process is unnecessary. Regarding the waveplates in the configuration, a half- and a quarter-wave plate are setup on both sides of T- and R-regions. It should be noted that the combination of these tow waveplates with a linear polarizer equals a broadband circular polarizer [49], and can be employed to achieve a normally white mode transflective LCD. Restated, good dark state can be obtained, and is not too sensitive to the variation of cell gap in T- and R-regions. Although the mechanism for such a dual-cell-gap transflective

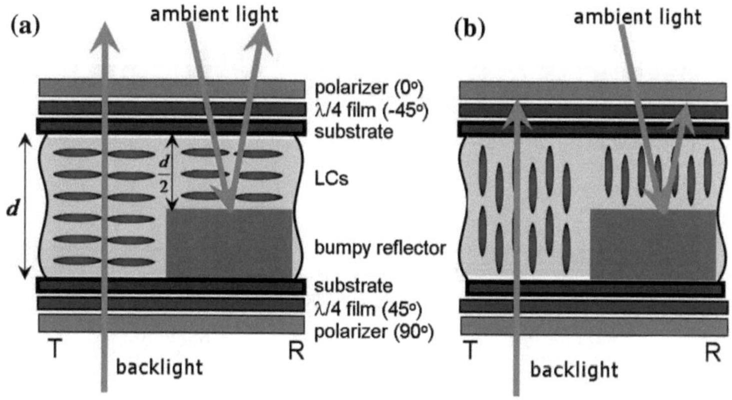

Fig. 12.10 Schematic configuration of a simple dual-cell-gap transflective LCD with the homogeneous aligned LCs in both the T- and R-regions. **a** *bright* and **b** *dark* states

LCD is very simple, several problems, including the different response time in T- and R-regions, narrow viewing angle, complicated fabrication process, etc., should be considered [50]. Fortunately, so far, the manufacturing processes are completely compatible with the existing technique, so the dual-cell-gap transflective LCDs are the first candidate for the commercial transflective LCD products.

Furthermore, many scientists worldwide have also proposed several kinds of dual-cell-gap transflective LCDs, such as dual-cell-gap transflective vertical alignment LCD [51], dual-cell-gap transflective hybrid alignment LCD [52], dual-cell-gap transflective FFS LCD [53, 54], and so on. The detailed operating principles of these kinds of transflective LCDs follow the electrically controllable birefringence LCDs, as describe above.

(2) single-cell-gap transflective LCDs

It is clear to understand that the difference between dual-cell-gap and single-cell-gap transflective LCDs. An identical cell gap in T- and R-regions of single-cell-gap transflective LCDs is made. Ideally, similar to the dual-cell-gap transflective LCDs, the electro-optical properties, including the dynamic responses, viewing angles, gray-scale controls, and others, in T- and R-regions, are identical to each other. The backlight travels the LC film once in the T-regions, while the ambient light twice in the R-regions. General speaking, single-cell-gap transflective LCDs are more favorable because they are easier to fabricate, but the matching phase retardation between T- and R-regions is difficult to achieve. In recent decade, many single-cell-gap transflective LCDs has been developed, for example, transflective TN, transflective STN, transflective MTN, transflective HA/hybrid, transflective BP, transflective IPS LCDs, and so on.

For instance, Fig. 12.11 depicts that the dual LC alignments, comprising the homogeneous (HA) and hybrid (HB) alignments, sandwiched between two crossed

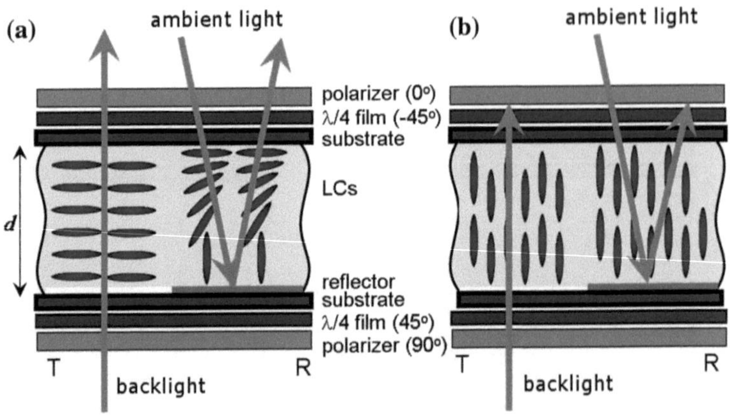

Fig. 12.11 Schematic configuration of a single-cell-gap transflective LCD with the dual LC alignment of homogeneous and hybrid alignments both the T- and R-regions, respectively. **a** *bright* and **b** *dark* states

polarizers. Such a binary configuration can be used to examine the electro-optical characteristics of the transmissive and reflective pixels in one kind of single-cell-gap transflective LCDs. [55]. Theoretically, the optical path of the reflective pixel (hybrid alignment) is twice as long as that of the transmissive pixel (homogeneous alignment) by a reflector, and the phase retardation for the bright state in the T-region should be set as $\lambda/2$ ($=\pi$), while that in the R region $\lambda/4$ ($=\pi/2$). Therefore, the quarter-wave plate is set between the substrate and polarizer to optimize the bright and dark states. Moreover, the angles between the transmission axes of the polarizers and the rubbing direction (along the y-axis) were $+45°$ and $-45°$. The same phase retardation ($2\pi d\Delta n/\lambda$) in the T- and R- (with a reflector) regions can be essential in achieving high optical performance of the transflective LCDs. Hence, it schematically depicts the designed single-cell-gap transflective LCD. In this design, according to the electrically controlled birefringence (ECB) mode LCD, the transmittance observed under the cross-polarizers can be given by $\sin^2 2\beta \sin^2(\delta/2)$, where β and δ denote the angle between the polarization direction of the incident light and the rubbing direction, and the phase retardation $[2\pi d(n\text{-}n)/\lambda]$, respectively [2–4]. In this case, the transmittance observed under the cross-polarizers equals $\sin^2(\delta/2)$ when β is set to 45°. Therefore, the maximum (minimum) transmittance implies that the phase retardation equals $(2n + 1)\pi$ $[(n\pi)]$, where n represents an integer. Additionally, the phase retardation in T- and R-regions decreases when the applied voltage increases, ultimately reaches nearly zero. The ratio of phase retardation in the R-region to that in T-region at an applied voltage should be optimized to be around 0.5. The bright, dark, and grayscale states that originated from the phase retardation between 0 and π, therefore, can be obtained by applying an AC voltage in the LCDs.

In summary, as discussed herein, the single-cell-gap transflective LCDs based on dual LC configuration require either dual-rubbing process or complicated electrode designs, so such kind of transflective LCDs have not been commercialized yet.

References

1. S.T. Wu, D.K. Yang, *Reflective Liquid Crystal Displays* (John Wiley, Chichester, 2001)
2. E. Lueder, *Liquid Crystal Displays* (John Wiley, Chichester, 2001)
3. D.K. Yang, S.T. Wu, *Fundamentals of Liquid Crystal Devices* (John Wiley, Chichester, 2006)
4. P. Yeh, C. Gu, *Optics of Liquid Crystal Displays* (John Wiley, Chichester, 1999)
5. M. Schadt, W. Helfrich, Voltage-dependent optical activity of a twisted nematic liquid crystal. Appl. Phys. Lett. **18**, 127–128 (1971)
6. Y.D. Chen, Andy Y.-G. Fuh, K.T. Cheng, Particular thermally induced phase separation of liquid crystal and poly(N-vinyl carbazole) films and its application. Opt. Express **20**, 16777–16784 (2012)
7. Andy Y.-G. Fuh, C.C. Chen, C.K. Liu et al., Polarizer-free, electrically switchable and optically rewritable displays based on dye-doped polymer-dispersed liquid crystals. Opt. Express **17**, 7088–7094 (2009)
8. G.Z. Liu, D.L. Xia, W.J. Yang et al., The surface rubbing effect on morphologies of LC droplets and electro-optic properties of flexible PDLC films. Sci. China Ser. Biol. Chem. **52**, 2329–2335 (2009)

9. K.J. Yang, D.Y. Yoon, Electro-optical characteristics of dye-doped polymer dispersed liquid crystals. J. Ind. Eng. Chem. **17**, 543–548 (2011)
10. R.L. Sutherland, V.P. Tondiglia, V.L. Natarajan, Electrically switchable volume gratings in polymer-dispersed liquid crystals. Appl. Phys. Lett. **64**, 1074 (1994)
11. http://www.polytronix.com/index.html
12. A.C. Lowe, R.J. Cox, Order Parameter and the performance of nematic guest-host displays. Mol. Cryst. Liq. Cryst. **66**, 309–318 (1981)
13. S.T. Wu, J.D. Margerum, M.S. Ho et al., Liquid crystal dyes with high solubility and large dielectric anisotropy. Appl. Phys. Lett. **64**, 2191 (1994)
14. N. Tabiryan, U. Hrozhyk, S. Serak, Nonlinear refraction in photoinduced isotropic state of liquid crystalline azobenzenes. Phys. Rev. Lett. **93**, 113901 (2004)
15. D.L. White, G.N. Taylor, New absorptive mode reflective liquid crystal display device. J. Appl. Phys. **45**, 4718 (1974)
16. T. Uchida, H. Seki, C. Shishido et al., Bright dichroic guest–host LCDs without a polarizer. Proc. Soc. Inf. Disp. **22**, 41–46 (1981)
17. K. Taira, H. Iwanaga, A. Hotta et al, Optical and color design of the reflective three-layer guest-host color LCD. in SID Tech Digest 1996 international workshop on active-matrix liquid-crystal displays in conjunction with IDW'96, pp. 333–336 (1996)
18. D.K. Yang, L.C. Chien, J.W. Doane, Cholesteric liquid crystal/polymer dispersion for haze-free light shutters. Appl. Phys. Lett. **60**, 3102–3104 (1992)
19. M. Xu, D.K. Yang, Dual frequency cholesteric light shutters. Appl. Phys. Lett. **70**, 720–722 (1997)
20. K.T. Cheng, C.K. Liu, C.L. Ting et al., Electrically switchable and optically rewritable reflective Fresnel zone plate in dye-doped cholesteric liquid crystals. Opt. Express **15**, 14078–14085 (2007)
21. D.K. Yang, J.L. West, L.C. Chien et al., Control of reflectivity and bistability in displays using cholesteric liquid crystals. J. Appl. Phys. **76**, 1331–1333 (1994)
22. Andy Y.-G. Fuh, Z.H. Wu, K.T. Cheng et al., Direct optical switching of bistable cholesteric textures in chiral azobenzene-doped liquid crystals. Opt. Express **21**, 21840–21846 (2013)
23. Y. Wang, Q. Li, Light-driven chiral molecular switches or motors in liquid crystals. Adv. Mater. **24**, 1926–1945 (2012)
24. R.A. Soref, Transverse field effect in nematic liquid crystals. Appl. Phys. Lett. **22**, 165 (1973)
25. R.A. Soref, Field effect in nematic liquid crystals obtained with interdigital electrodes. J. Appl. Phys. Lett. **45**, 5466 (1974)
26. M. Ohe, M. Yoneya, K. Kondo, Switching of negative and positive dielectro-anisotropic liquid crystals by in-plane electric fields. J. Appl. Phys. Lett. **82**, 528 (1997)
27. M. Ohe, M. Ohta, S. Arantani et al., Principle and characteristics of lelctro-optical behavior with in-plane switching mode. Asia Disp. **95**, 577 (1995)
28. M. Ohta, M. Ohe, K. Kondo, Development of super-TFT-LCDs with in-plane switching mode. Asia Disp. **95**, 707 (1995)
29. S.H. Lee, H.Y. Kim, I.C. Park et al., Rubbing-free, vertically aligned nematic liquid crystal display controlled by in-plane field. Appl. Phys. Lett. **71**, 2851 (1997)
30. K.H. Kim, S.B. Park, J.U. Shim et al., New LCD modes for wide-viewing-angle applications. SID Tech. Digest. **29**, 1085 (1998)
31. S.H. Lee, S.L. Lee, H.Y. Kim, Electro-optic characteristics and switching principle of a nematic liquid crystal cell controlled by fringe-field switching. Appl. Phys. Lett. **73**, 2881 (1998)
32. K.H. Lee, H.Y. Kim, K.H. Park et al., A novel outdoor readability of portable TFT-LCD with AFFS technology. SID Symposium Digest. Tech. Pap. **37**, 1079–1082 (2006)
33. S. Choi, S. Sun, K. Kim et al., Novel gray-toneless technology for mask reduction in high aperture rate FFS mode. SID Symposium Digest. Tech. Pap. **44**, 1126–1129 (2013)
34. S. Meiboom, J.P. Sethna, P.W. Anderson et al., Theory of the blue phase of cholesteric liquid crystals. Phys. Rev. Lett. **46**, 1216–1219 (1981)

35. http://www.samsung.com/us/aboutsamsung/news/newsIrRead.do?news_ctgry=irnewsrelease&news_seq=8351
36. M. Jiao, Y. Li, S.T. Wu, Low voltage and high transmittance blue-phase liquid crystal displays with corrugated electrodes. Appl. Phys. Lett. **96**, 011102 (2010)
37. J. Yan, H.C. Cheng, S. Gauza et al., Extended Kerr effect in polymer-stabilized blue-phase liquid crystals. Appl. Phys. Lett. **96**, 071105 (2010)
38. L. Rao, H.C. Cheng, S.T. Wu, Low voltage blue-phase LCDs with double-penetrating fringe fields. J. Disp. Technol. **6**, 287–289 (2010)
39. J. Yan, S.T. Wu, Polymer-stabilized blue phase liquid crystals: a tutorial. Opt. Mater. Express **1**, 1527–1535 (2011)
40. H.C. Cheng, J. Yan, T. Ishinabe et al., Wide-view vertical field switching blue-phase LCD. J. Disp. Technol. **8**, 627–633 (2012)
41. H.C. Cheng, J. Yan, T. Ishinabe et al., Blue-phase liquid crystal displays with vertical field switching. J. Disp. Technol. **8**, 98–103 (2012)
42. H. Kikuchi, M. Yokota, Y. Hisakado et al., Polymer-stabilized liquid crystal blue phases. Nat. Mater. **1**, 64–68 (2002)
43. F. Castles, F.V. Day, S.M. Morris et al., Blue-phase templated fabrication of three dimensional nanostructures for photonic applications. Nat. Mater. **11**, 599–603 (2012)
44. L. Rao, Z. Ge, S.T. Wu et al., Low voltage blue-phase liquid crystal displays. Appl. Phys. Lett. **95**, 231101 (2009)
45. M. Kim, M.S. Kim, B.G. Kang et al., Wall-shaped electrodes for reducing the operation voltage of polymer-stabilized blue phase liquid crystal displays. J. Phys. D Appl. Phys. **42**, 235502 (2009)
46. Y.H. Kim, S.T. Hur, K.W. Park et al, A vertical-field-driven polymer-stabilized blue phase liquid crystal displays. in SID Symposium Digest **42**, pp. 298–301 (2011)
47. H.C. Cheng, J. Yan, T. Ishinabe et al., Vertical field switching for blue-phase liquid crystal devices. Appl. Phys. Lett. **98**, 261102 (2011)
48. X. Zhu, Z. Ge, T.X. Wu et al., Transflective liquid crystal displays. J. Disp. Technol. **1**, 15–29 (2005)
49. S. Pancharatnam, Achromatic combinations of birefringent plates: part I. An achromatic circular polarizer. Proc. Indian Acad. Sci. **41**, 130–136 (1995)
50. M. Shibazaki, Y. Ukawa, S. Takahashi et al., Transflective LCD with low driving voltage and wide viewing angle. SID Dig. Tech. Pap. **34**(1), 90–93 (2003)
51. H.D. Liu, S.C. Lin, A novel design wide view angle partially reflective super multi-domain homeotropically aligned LCD. in SID Digest of Technical Papers, pp. 558–561 (2002)
52. C.L. Yang, Electro-optics of a transflective liquid crystal display with hybrid-aligned liquid crystal texture. Jpn. J. Appl. Phys. **43**, 4273–4275 (2004)
53. T.B. Jung, J.C. Kim, S.H. Lee, Wide-viewing-angle transflective display associated with a fringe-field driven homogeneously aligned nematic liquid crystal display. Jpn. J. Appl. Phys. **42**, L464–L467 (2003)
54. T.B. Jung, J.H. Song, D.S. Seo et al., Viewing angle characteristics of transflective display in a homogeneously aligned liquid crystal cell driven by fringe-field. Jpn. J. Appl. Phys. **43**, L1211–L1213 (2004)
55. Andy Y.-G. Fuh, C.Y. Huang, C.K. Liu et al., Dual liquid crystal alignment configuration based on nanoparticle-doped polymer films. Opt. Express **19**, 18525–18531 (2011)

Chapter 13
Liquid Crystal 3D Displays

Yi-Pai Huang

3D display was expected to be the dominant next-generation display technology for producing more natural images. For autostereoscopic 3D technologies, observers could perceive 3D images with the naked eyes, yet still has the degradation of image quality for a given autostereoscopic 3D display. In this chapter, various LC lens (LC-lens) technologies for improving the issues are illustrated, followed by the future developments of LC-lens on 3D applications.

13.1 Introduction of 3D Display

Three-dimensional display technologies can be categorized into stereoscopic type and auto-stereoscopic type, as listed in Fig. 13.1. The stereoscopic 3D displays require the viewers to wear glasses or special devices such as active shutter glasses type, passive polarizer type, or head-mounted display (HMD), to perceive 3D images [1–8]. Thus, viewers perceive correct left and right images for the corresponding eyes through the special devices. Moreover, the stereo technique only provides the same two-view stereo image that there are no perspective images for arbitrary positions. Hence, the auto-stereoscopic (glasses-free) 3D displays are definitely the future trend.

The glasses-free 3D displays include holographic type [9–12], volumetric type [13, 14], multi-plane type [16–18], integral imaging and multiplexed-2D type [19–27]. The holographic displays use the concept of interference to reconstruct the 3D scene through reconstruction of wave intensity and phase. However, this method now is too complicated to reconstruct good dynamic and full color 3D scene due to the special light source, complicated image content, and high

Y.-P. Huang (✉)
Department of Photonics, National Chiao Tung University, Hsinchu, Taiwan
e-mail: boundshuang@mail.nctu.edu.tw

C.-C. Lee (ed.), *The Current Trends of Optics and Photonics*,
Topics in Applied Physics 129, DOI 10.1007/978-94-017-9392-6_13

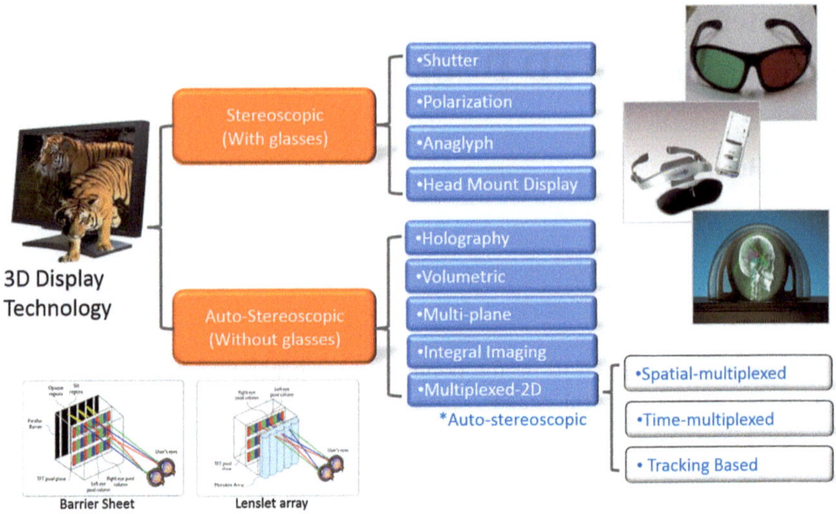

Fig. 13.1 Classification of 3D technologies

quality display requirement. The volumetric displays and multi-plane displays are good approaches while they require huge space. Therefore, recently, researchers are more attracted on the compact multiplexed-2D displays for 3D applications because it can be achieved with flat panel displays.

The spatial-multiplexed 3D displays generally includes two components, optical device and display panel. The optical device, such as the lenslet array or barrier sheet, is used to project different images to the corresponding directions. And the display panel is used to display the images. Two common optical devices, for example, used for these 3D displays are fixed barrier [21, 28] and lenticular lens sheet [25–27, 29–31]. This method provides different information to the corresponding left and right eyes simultaneously. Hence, human brain fuses this different information and then 3D sensation can be obtained. Besides, the two-view or multi-view 3D displays can be achieved by generating two or more view information to be the image content. However, current spatial-multiplexed method will much decreases the resolution of 3D images, especially can't be used for reading the text contents. Therefore, liquid crystal lens (LC-lens) was proposed for 2D/3D switching function. Except 2D/3D switching, the 3D rotation and 3D scanning function of LC-lens were also been reported to extend the application of 3D displays.

13.2 LC Lens for 3D Display

In 1979, the LC material was first employed for lens applications [32]. The LC is injected into the space between the plano and concave glass substrates, or plano and convex glass substrates, respectively. The transparent electrodes are coated on the glasses, as shown in Fig. 13.2. Applying the driving voltages on the electrodes, the LCs is re-aligned according to the potential distribution within the cells. Thus, the optical powers of the concave and convex lenses can be tuned without physically change the lens shape. The polarized light is also needed in order to meet effective refractive index changing of the LC lens (i.e., the un-polarized light cannot meet the index changing that the incident light path cannot be changed after passing through the LC cell).

13.2.1 2D/3D Switching Function of LC-Lens

For 2D/3D switching LC-lens, three approaches, polarization active micro-lens, Active LC lenticular lens, Electric-field-driven LC(ELC) lenticular lens, had been reported and described in detail in this session.

13.2.1.1 Polarization Active Micro-Lens

The polarization activated micro-lens was proposed by G.J. Woodgate, and J. Harroldas in 2003 [33, 34], which the basic structure is shown in Fig. 13.3. The polarization of incident light is not changed after encountering the polarization switching cell (TN cell) when applying driving voltage on the TN cell. Thus, the 2D mode can be obtained when the polarization of polarized light do not be changed. On the other hand, the polarization of polarized light is changed that the light is focused after passing through micro-lens layer without applying driving voltage on the TN cell. Hence, the 3D mode is achieved.

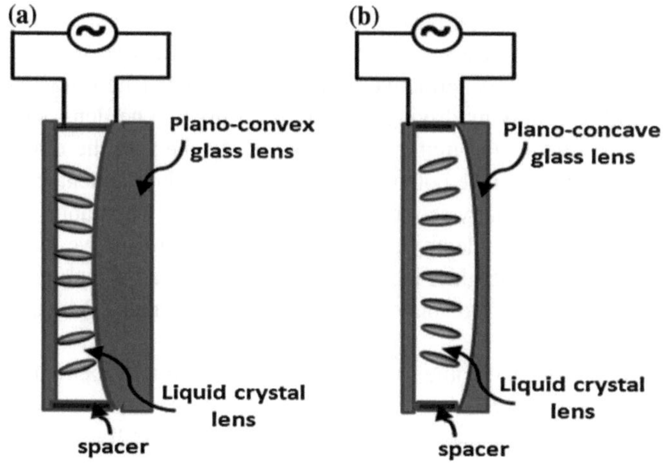

Fig. 13.2 The first proposed LC lenses in 1979. **a** Plano-convex cell, and **b** plano-concave cell

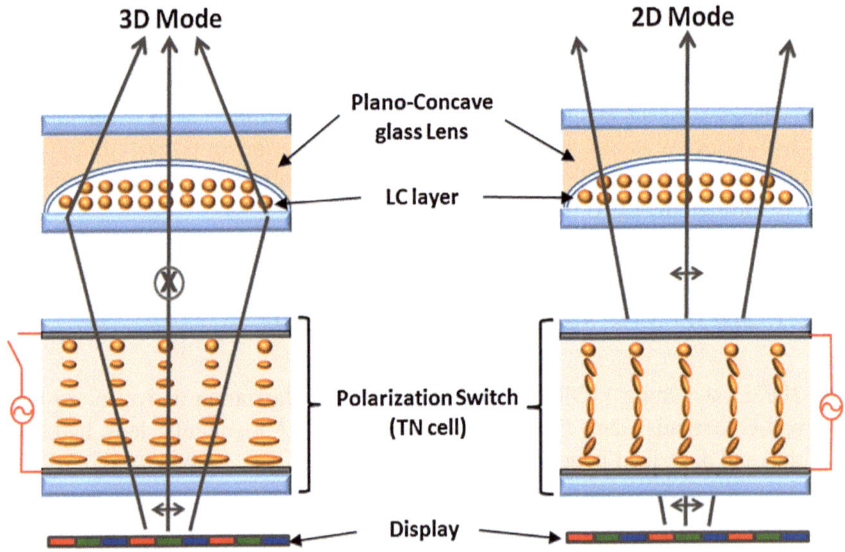

Fig. 13.3 On and off state of polarization active micro-lens

13.2.1.2 Active LC Lenticular Lens

Just like the structure of inhomogeneous LC-lens, S.T. de Zwart et al., developed an active LC lenticular lens for 2D/3D switchable display [35, 36]. As shown in Fig. 13.4, the LC molecules (positive dielectric anisotropy)are aligned perpendicular to the plane. The refractive index of the concave glass lens, n_g, is chosen equal to ordinary refractive index, n_o. The incident polarized light from the underlying LCD is with polarization perpendicular to the plane (i.e., extra-ordinary light). Hence, the polarized light encounter a difference of refractive index distribution from n_e (the extra-ordinary refractive index of LCs) to n_o (the refractive index of glass lens) without applying driving voltage. It results in a net lens-action to form a 3D mode, i.e., the spatial-multiplexed type 3D display; on the other hand, the LCs reorient according to the potential distribution thus the incident extra-ordinary light encounters the ordinary refractive index of LCs, n_o, when applying the driving voltage on the planar electrodes. Since the ordinary refractive index matched the refractive index of the glass lens, the LC lens acts as a normal glass plate and the light is not refracted. A display which can display normal 2D images could be achieved.

Nevertheless, there are some drawbacks of the above two methods, active LC lenticular lens and polarization active micro-lens, such as mismatching of LC alignment at the interface and the tough fabricating process for the concave lens.

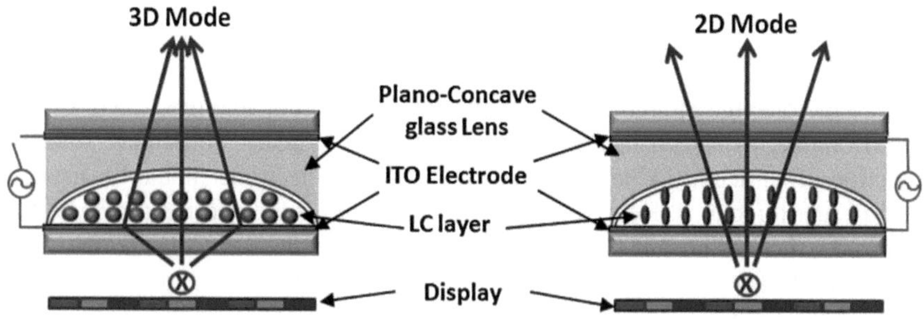

Fig. 13.4 Active LC lenticular lens

Hence, the simplest LC lens, homogeneous LC layer and homogeneous glass structure, for 3D displays was invented to overcome these issues.

13.2.1.3 Electric-Field-Driven LC (ELC) Lenticular Lens

For the Electric-field-driven LC (ELC) lens [37–40], the electrodes are designed to be the symmetric structure with respect to the centerline, and the different driving voltages are connected to the electrodes which are located at different layers, as shown in Fig. 13.5. The electrode layers are fabricated on the side of bottom substrate, and an insulator layer (SiO_2 layer) is sandwiched between these electrode layers. Applying specific different voltages on the different electrode layers, the phase retardation profile, which approach to the parabolic curve, of the LC layer can be obtained. The ELC lens also can be utilized for switching between 2D and 3D mode by changing the driving voltage on and off. However, the assembly and response time are two issues of this kind of LC lens.

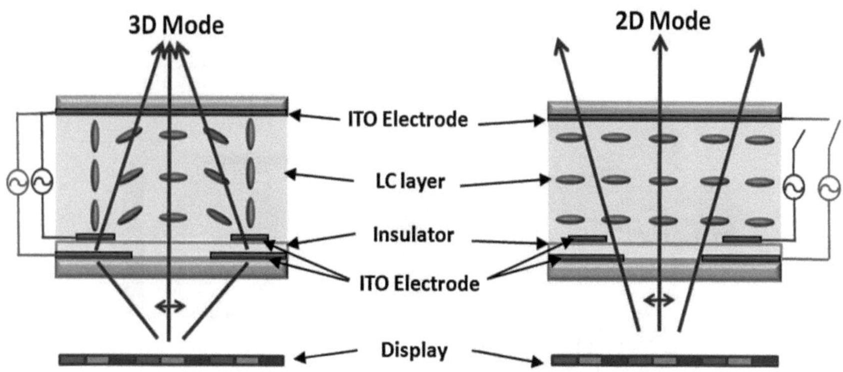

Fig. 13.5 Principle of electric-field-driven LC lens [20–24]

13.2.2 Rotation Function for Mobile Application

Due to the raising requirement of portable device, the rotatable function of display on 3D mode, as shown in Fig. 13.6, also become popular. To achieve rotatable function of lenticular LC lens, the two sets of stripe electrode should be arranged in different directions. However, when we apply the voltage to the top or bottom electrode, there is no full flat electrode on the other side. According to the structures and electrode field of LC cell, the LC cell won't be generated as a lenticular lens. To form the shape of lenticular lens, the variation of electrode field should be change only in one direction. Therefore the function of high resistance layer can be used to overcome this issue.

According to the switchable function of electrode, Chang et al. [41] proposed rotatable lenticular LC lens which is coated high resistance on both inner electrodes. In order to form lenticular lens in two directions, the top electrodes and the bottom electrodes should almost perpendicular to each other. To achieve switchable function in two directions, top and bottom electrodes are coated with high-resistance material, as shown in Fig. 13.7. When all bottom electrodes are connected to Vcom and top electrodes are driven by interlaced voltage, VD and Vcom, the LC lenticular lens will arrange in horizontal direction. Due to the same driving voltage, the bottom high resistance layer becomes a full electrode with ground voltage. On the other hand, distribution of voltage on the top side becomes a gradient and continuous distribution, because of the function of high resistance layer. Alternatively, in vertical direction, the bottom electrodes are driven by VD and Vcom individually, and whole electrodes on the top are operated by Vcom.

Except crossed electrodes, M. Kashiwagi and S. Uehara optimized the LC alignment direction (LCAD) and Electrode array direction (EAD) to realize the rotatable 2D/3D display with wider vertical and horizontal viewing angles in both landscape

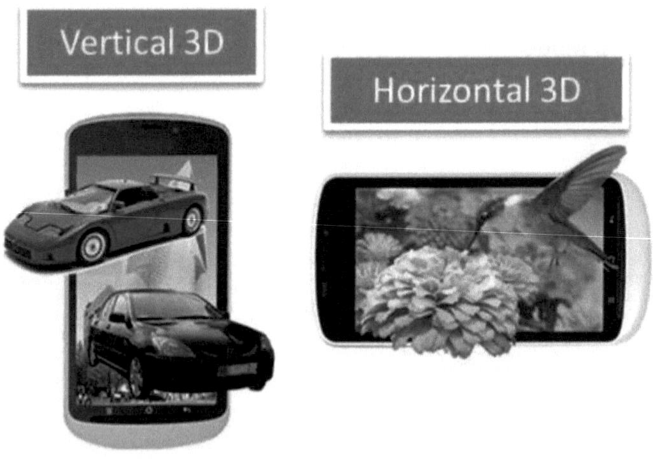

Fig. 13.6 The schematic plot of rotatable 3D display for mobile application

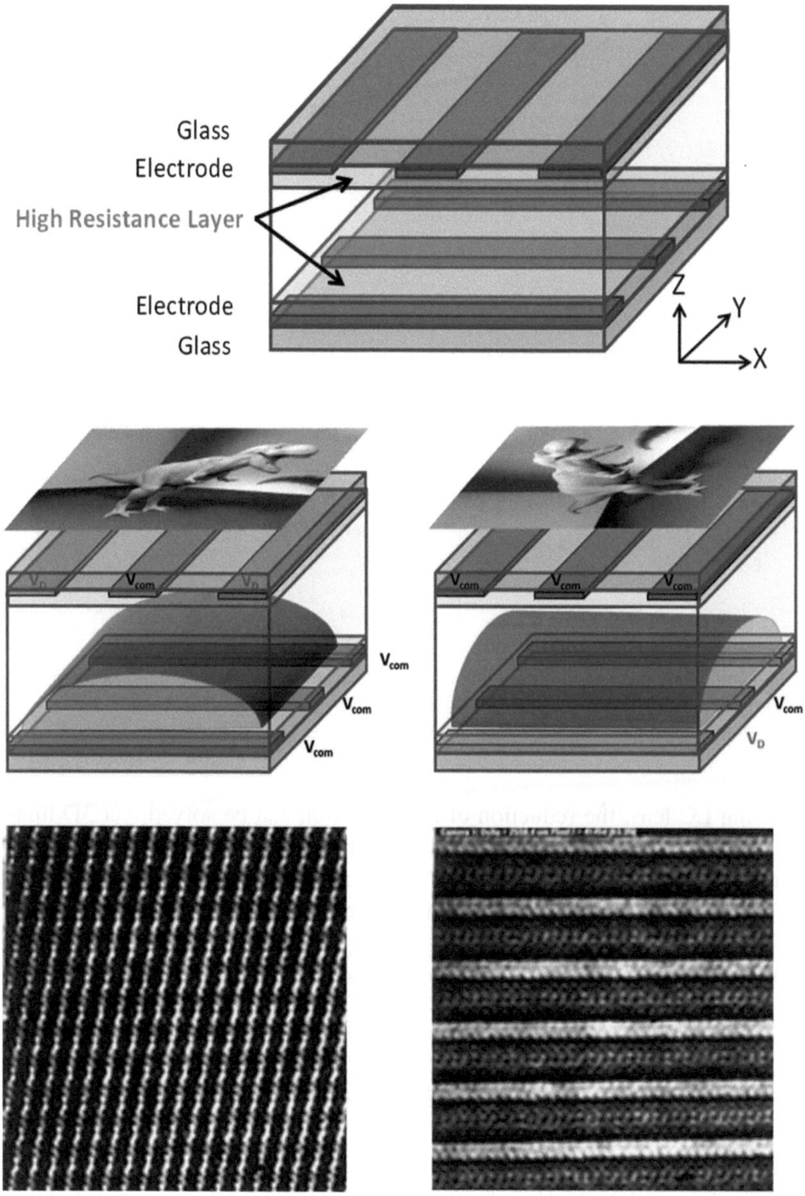

Fig. 13.7 The schematic plot of rotatable LC lens and the focused pattern on vertical and horizontal direction

and portrait orientations [42]. The conditions of two electrode directions and LC alignment direction are both less than 50° of θ_R, the results show that 3D viewing azimuth angle is over 30° in both landscape and portrait orientations (Fig. 13.8).

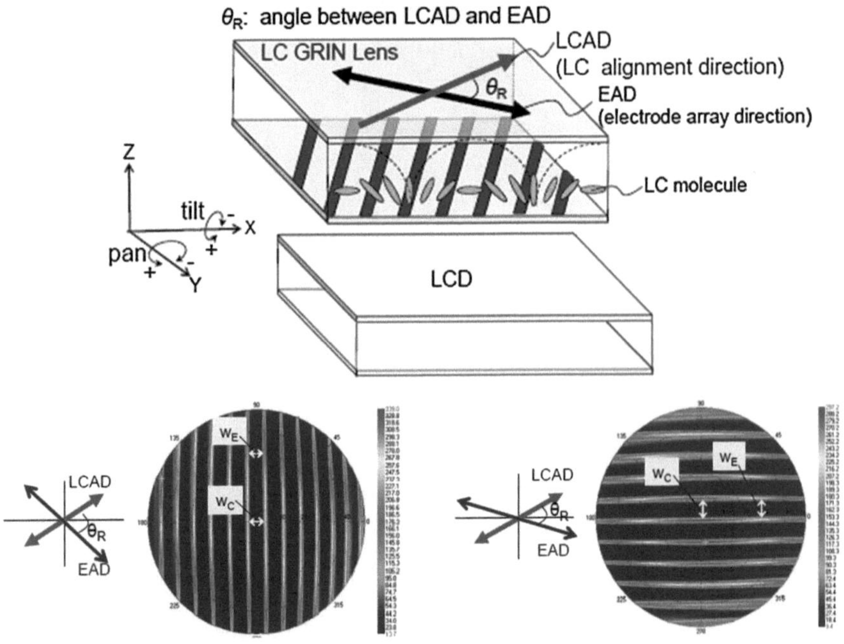

Fig. 13.8 Optical model and luminance measurement results of non-orthogonal electrodes array system proposed by Uehara et al. [42]

By using LC lens, the reduction of 2D resolution can be solved, yet 3D image isn't. For having more freedom of views for multi-users, the view-points have to be increased. The pixel resolution (spatial domain) for 3D image has to be sacrificed for yielding more views to widen the viewing angle (Spatial-multiplexed). Consequently, in order to have a full resolution 3D display, "temporal domain" or "time-multiplexed" technologies have to be considered.

13.2.3 3D Scanning LC-Lens for Full Resolution 3D Image

Resolution decreasing is always a major issue of auto-stereo 3D display. To improving the resolution. Huang, et al., proposed the active scanning Multi-electrode Driven LC lens (MeD-LC lens) [43, 44]. The MeD-LC lens can not only formed ideal parabolic lens curvature for 2D/3D switch with low crosstalk, but also can scan the electrode voltage to move the lens for multi-view points.

At the voltage on state, local electric fields are formed by the gradient distribution of electric field. Hence, the lenticular MeD-LC lens array can be obtained by making repeated patterns as well. In addition, by applying the operating voltage sequentially on each electrode, the LC in MeD-LC lens reorients and builds

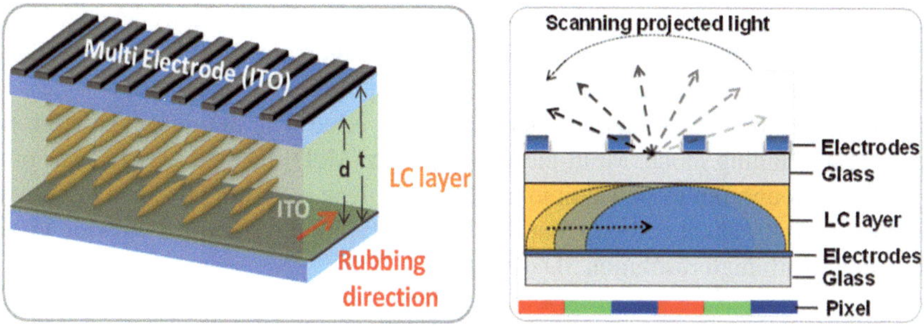

Fig. 13.9 Multi-electrode structure and scanning of LC-lens

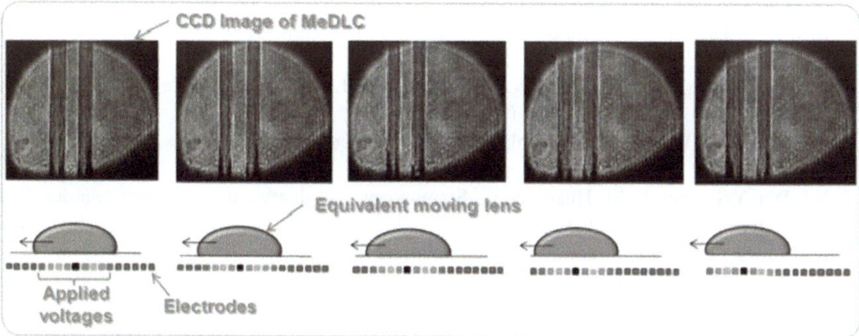

Fig. 13.10 Experiment results of scanning effect

in a lens-like shape, thus the lens could move (shift) sequentially on the horizontal direction to project the images to different viewing angle, as t he schematic shown in Fig. 13.9.

Due to the multi-electrodes are periodic, the scanning function could be operated by changing the voltage sequentially. The captured image with scanning MeD-LC lens is demonstrated in Fig. 13.10. Currently, however, the response time of MeD-LC lens is still not adequate for super time-multiplexed 3D display.

13.3 Conclusions and Future Developments

For various types of the 2D/3D switching displays, the liquid-crystal (LC) lens was one of the most favorable designs for such display applications. The LC lens would focus and just pass the incident light with and without supplying the operating voltage.

Three types of LC lens, polarization active micro-lens, Active LC lenticular lens, and Electric-field-driven LC (ELC) lenticular lens, had been reported and successfully demonstrated. To further using crossed electrode coated by high resistance material or controlling the LC alignment direction (LCAD) and Electrode array direction (EAD), 3D rotation display can be obtained.

However, the resolution decreasing is still an issue with 3D images. Therefore, temporal scanning mode, Multi-electrode Driven (MeD) Fresnel LC-lens was proposed. Although the response time of above LC-lens were not adequate in current stage due to the thick cell-gap, but illustrated a future development on wide viewing angle with high resolution 3D display.

References

1. N. Holliman, 3D Display System (2002)
2. L. Hill, A. Jacobs, 3-D liquid crystal displays and their applications. Proc. IEEE **94**, 575–590 (2006)
3. A.K. Srivastava, J.L. deB, D.L. Tocnaye, L. Dupont, Liquid crystal active glasses for 3D cinema. J. Disp. Technol. **6**, 522–530 (2010)
4. S. Pastoor, M. Wopking, 3-D displays: A review of current technologies. Displays **17**, 100–110 (1997)
5. Y.J. Wu, Y.S. Jeng, W.M. Huang et al., Stereoscopic 3D display using pattern retarder. Proc. SID Symp. Dig. **39**, 260–263 (2008)
6. S.M. Jung, I.J. Chung et al., A novel polarizer glasses-type 3D displays with an active retarder. Proc. SID Symp. Dig. **40**, 348–351 (2009)
7. M.G. Tomilin, Head-mounted displays. J. Opt. Technol. **66**, 528–533 (1999)
8. K. Keller, A. State, H. Fuchs, Head mounted displays for medical use. J. Disp. Technol. **4**, 468–472 (2008)
9. B. Javidi, E. Tajahuerce, Three dimensional object recognition using digital holography. Opt. Lett. **25**, 610–612 (2000)
10. J.Y. Son, B. Javidi, S. Yano, K.H. Choi, Recent developments in 3D image technologies. J. Disp. Technol. **6**, 394–403 (2010)
11. Y. Takaki, N. Okada, Reduction of image blurring of horizontally scanning holographic display. Opt. Express **18**, 11327–11334 (2010)
12. Y. Takaki, M. Yokouchi, N. Okada, Improvement of grayscale representation of the horizontally scanning holographic display. Opt. Express **18**, 24926–24936 (2010)
13. P. Soltan, J.A. Trias, M. McDonald et al., Laser based 3-D volumetric display system. Nav. Eng. J. **107**, 233–243 (1995)
14. H.H. Refai, Static volumetric three-dimensional display. J. Disp. Technol. **5**, 391–397 (2009)
15. M. Gately, L. Sawalha et al., A three-dimensional swept volume display based on LED arrays. J. Disp. Technol. **7**, 503–514 (2011)
16. S. Suyama, S. Ohtsuka, H. Takada, K. Uehira, S. Sakai, Apparent 3-D image perceived from luminance-modulated two 2-D images displayed at different depths. Vis. Res. **44**, 785–793 (2004)
17. H. Kuribayashi, M. Date, S. Suyama, H. Tatada, A method for reproducing apparent continuous depth in a stereoscopic display. J. Soc. Info. Disp. **14**, 493–498 (2006)
18. C.Y. Hsu, Y.P. Huang, Y.C. Chang, Wide viewing angle 3D depth-fused display (DFD) system. Proc. SID Symp. Dig. **39**, 1655–1658 (2008)
19. C.V. Berkel, J.A. Clarke, Characterization and optimization of 3D-LCD module design. Proc. SPIE **3012**, 179–187 (1997)
20. S.J. Young, B. Javidi, Three-dimensional image methods based on multi-view image. J. Disp. Technol. **1**, 125–140 (2005)

21. C.H. Chen, Y.P. Huang, H.P.D. Shieh et al., Liquid crystal panel for high efficiency barrier type autostereoscopic three-dimensional displays. Appl. Opt. **48**, 3446–3454 (2009)
22. W. Maphepo, Y.P. Huang, H.P.D. Shieh, Enhancing the brightness of parallax barrier based 3D Flat panel mobile displays without compromising power consumption. J. Disp. Technol. **6**, 60–62 (2010)
23. M. Tsuboi, S. Kimura, Y. Takaki, T. Horikoshi, Design conditions for attractive reality in mobile-type 3-D display. J. Soc. Info. Disp. **18**, 698–703 (2010)
24. Y. Takaki, Multi-view 3-D display employing a flat-panel display with slanted pixel arrangement. J. Soc. Info. Disp. **18**, 476–482 (2010)
25. Y. Takaki, K. Tanaka, J. Nakamura, Super multi-view display with a lower resolution flat-panel display. Opt. Express **19**, 4129–4139 (2011)
26. N.S. Holliman, N.A. Dodgson, G.E. Favalora, L. Pockett, Three-dimensional displays: a review and applications analysis. IEEE Tran. Broadcast. **57**, 362–371 (2011)
27. J. Nakamura, T. Takahashi, C.W. Chen, Y.P. Huang, Y. Takaki, Analysis of longitudinal viewing freedom of reduced-view super multi-view display and increased longitudinal viewing freedom using eye-tracking technique. J. Soc. Info. Disp. **20**, 228–234 (2012)
28. G. Hamagishi, Analysis and improvement of viewing conditions for two-view and multi-view display. Proc. SID Symp. Dig. **40**, 340–343 (2009)
29. H.J. Im, S.M. Jung, H.N. Shin et al., Mobile 3-D display based on a LTPS 2.4-in. VGA LCD panel attached with lenticular lens sheets. Proc SID Symp. Dig. **39**, 256–259 (2008)
30. B. Javidi, F. Okano, *Three-Dimensional Television, Video, and Display Technologies* (Springer, New York, 2002)
31. J.Y. Son, B. Javidi, Three-dimensional image methods based on multi-view images. J. Disp. Technol. **1**, 125–140 (2005)
32. S. Sato, Liquid-crystal lens–cells with variable focal length. Jpn. J. Appl. Phys. **18**, 1679–1684 (1979)
33. G.J. Woodgate, A new architecture for high resolution autostereoscopic 2D/3D displays using free-standing liquid crystal microlenses. Proc. SID Symp. Dig. **36**, 378–381 (2005)
34. G.J. Woodgate, High efficiency reconfigurable 2D/3D autostereoscopic display. Proc. SID Symp. Dig. **34**, 394–397 (2003)
35. O.H. Willemsen, S.T. De Zwart, 2D/3D switchable displays. J. Soc. Inf. Disp. **14**(8), 715–722 (2006)
36. C.M. Krijn, S.T. de Zwart, 2D/3D displays based on switchable lenticulars. J. Soc. Inf. Disp. **16**(8), 847–855 (2008)
37. Y. Li, S.T. Wu, Polarization independent adaptive microlens with a blue-phase liquid crystal. Opt. Express **19**, 8045–8050 (2011)
38. H.W. Ren, S.T. Wu, Getting in tune: electronically controlled liquid crystal yields tunable-focal-length lenses. SPIE's optics express magazine (2004), pp. 24–27
39. A.F. Naumov et al., Liquid-crystal adaptive lenses with modal control. Opt. Lett. **23**(13), 992–994 (1998)
40. M. Ye, S. Sato, Low-voltage-driving liquid crystal lens. Jpn. J. Appl. Phys. **49**, 100204-1–100204-3 (2010)
41. Y.C. Chang, T.H. Jen, Y.P. Huang et al., High-resistance liquid-crystal lens array for rotatable 2D/3D autostereoscopic display. Opt. Express **22**, 2714–2724 (2014)
42. S. Uehara, M. Baba et al., LC GRIN lens mode with wide viewing angle for rotatable 2D/3D tablet. Proc. SID Symp. Dig. **42**, 154–157 (2013)
43. Y.P. Huang, L.Y. Liao, C.W. Chen, 2D/3D switchable autostereoscopic display with multi-electrically driven liquid crystal (MeD-LC) lenses. J. Soc. Inf. Disp. **18**(9), 642–646 (2010)
44. Y.P. Huang, C.W. Chen, Superzone fresnel liquid crystal lens for temporal scanning auto-stereoscopic display. J. Disp. Technol. **8**(11), 650–655 (2012)

Chapter 14
Liquid Crystals for Non-display Applications

Andy Ying-Guey Fuh

As discussed in the previous Sections, LCs are highly birefringent. Also, their birefringence can be electrically controllable, the so-called Electrically Controllable Birefringence (ECB) effect [1]. The voltage required to change the birefringence is significantly low. Moreover, LCs is highly transparent in the visible and near-infrared wavelength regions [2]. All these are the main features that LCDs (LCDs) are based on. No doubt, these features also make LCs be applied for many non-display photonic devices, for example, LC gratings [3–28], spatial filters [29–33], polarization converters [34–54], lenses [55–73], q-plate for shaping laser beams [47–51, 74–93], and lasers [94–100]. The advantages of these LC devices are electrically switchable due to the ECB effect of LCs. In this Section, some of these non-display devices are presented. Moreover, devices fabricated by doping azo dyes in LC will be introduced.

14.1 Liquid Crystal Gratings

Among the LC devices, a diffraction grating is the simplest one to fabricate. Optical diffraction gratings play a central role in current photonic technologies such as in multiplexing and processing optical signals (optical interconnection, beam steering, and sensor array interrogation). Various types and structures of LC gratings can be made with the uses of pure nematic LCs (NLCs), NLC doped with azo dye [3–9], doped with nano-particles, mainly the carbon 60, carbon

A.Y.-G. Fuh (✉)
Department of Physics, National Cheng Kung University, Tainan, Taiwan
e-mail: andyfuh@mail.ncku.edu.tw

C.-C. Lee (ed.), *The Current Trends of Optics and Photonics*,
Topics in Applied Physics 129, DOI 10.1007/978-94-017-9392-6_14

nano-tube [3, 10, 11], cholesteric LCs [12–18], and polymer-LC composites [19, 20], etc. The doping of azo dye, carbon 60 or carbon nano-tube induces the grating effect through the photo-refractive effect in the sample. A persisting grating can be formed with these doping materials adsorbed periodically onto the substrate inner surfaces, which gives rise to the formation of binary LC alignment configurations. The PDLC gratings are most fabricated using the holographic method (discuss below). The polymerization preferentially initiates in the high-intensity regions. The consumption of monomers in these regions thus lowers their chemical potential. This gives rises to the diffusion of monomers from the low-intensity regions towards the high-intensity ones. On the other hand, as the LCs are not consumed, their chemical potential increases in the high-intensity regions over the low-intensity ones due to the consumption of monomers. Hence, there is a diffusion of LC molecules from the high-intensity regions towards the low-intensity ones to equalize the chemical potential in the writing area. The detailed dynamics of such a PDLC-grating formation can be found in the [21]. Apparently, PDLC films can be used to record the interference patterns of the light beams [22], and holograms [23].

Basically, there are two main approaches to fabricate LC gratings. One is to form alternate binary domains using, for example, the patterned electrodes, or patterned alignments [24–26]. The other is the holographic method [6]. Two coherent laser beams are applied to overlap on the sample at an intersecting angle. Among these two methods, the latter one has been more commonly used due to its flexibility to adjust the grating spacing. Thus, LC gratings fabricated by the holographic method are presented here.

Referred to Fig. 14.1, two coherent plane waves $\vec{E}_a$ and $\vec{E}_b$, are incident on the recording material in x-z plane with an intersection angle of α.

$$\vec{E}_j(\mathbf{r},\mathbf{t}) = \mathrm{Aj}\,\exp(\mathrm{i}\,\vec{k}_j\,\mathrm{r} - \mathrm{i}\omega\mathrm{t}); \quad \mathrm{j} = \mathrm{a\,or\,b} \tag{14.1}$$

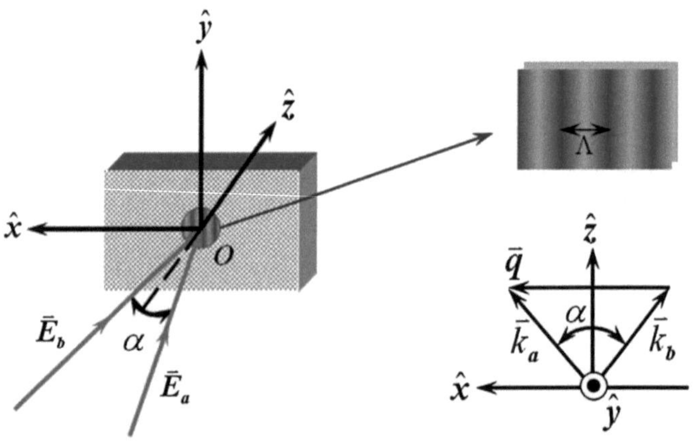

Fig. 14.1 Schematic drawing of interference produced by two coherent plane waves

Their wave vectors, amplitudes are $\vec{k}_a$ and $\vec{k}_b$, $\vec{A}_a$ and $\vec{A}_b$, respectively. The grating vector of the grating pattern produced by the interference is given by

$$\vec{q} = \pm\left(\vec{k}_a - \vec{k}_b\right) = \pm 2k_x\hat{x}. \tag{14.2}$$

With reference to Fig. 14.1, q satisfies $\left|\vec{q}\right| = q = 2\left|\vec{k}_a\right| \sin\left(\frac{\alpha}{2}\right) = \frac{4\pi \sin(\frac{\alpha}{2})}{\lambda}$, and the spatial period of the two interfering beams, called the grating spacing, is

$$\Lambda = \frac{2\pi}{q} = \frac{\lambda}{2\sin\left(\frac{\alpha}{2}\right)}, \tag{14.2}$$

where λ and α are the wavelength of the writing beam and the intersection angle, respectively.

Generally, when two coherent waves with arbitrary polarizations interfere, the amplitude of the intensity modulation and the state of polarization of the total field varies spatially in both magnitude and direction, as determined by the phase variations between these two waves. The final form of the interference region has various contributions from the intensity and/or polarization distribution, which are determined by the polarization states of these two interfering waves. Therefore, the gratings that exhibit an intensity (polarization) distribution are called intensity (polarization) gratings. In most experiments, the angle of interference, the so-called intersection angle, α, is very small. This condition causes the intensity modulation ΔI to be zero when $\vec{A}_a \perp \vec{A}_b$, such as for $(\hat{A}_a, \hat{A}_b)$ of (s, p), (+45°, −45°) and (RCP, LCP) polarizations. RCP and LCP are right- and left-circularly polarized light, respectively. In the three cases of interest herein, the interfering patterns generate pure polarization gratings [9, 27, 28]. For (s, s), (p, p), (+45°, +45°), (−45°, −45°), (RCP, RCP) and (LCP, LCP), the interfering patterns produce pure intensity gratings. The total interference field pattern, as shown in Table 14.1, is determined for one cycle for a small angle of interference, α.

14.2 Spatial Filters

A Spatial filter is commonly used in the Fourier optical signal process. Controlling the spatial distribution of optical information is very important. Such control has been practically applied in the fields of photographic image enhancement, pattern recognition and image-inversion, using a highly nonlinear dye-doped NLC film as a phase-modulation element. Another form of selective spatial frequency reconstruction was demonstrated using functionalized mesogenic composites with holographic capability. Both the linear and nonlinear spatial filters can be fabricated using LC cells.

Kato et al. presents a nonlinear spatial filtering using an azo dye-doped liquid-crystal cell [29]. Photoisomerization of azo-dye molecules leads to local alignment changes in the LC molecules because of interaction among the molecules. Consequently the polarization of transmitted light is modulated, depending on

Table 14.1 The total field vector distribution for different polarization combination of the interfering waves

$\delta \equiv \frac{2\pi x}{\Lambda}$		Total field interference pattern in the x-y plane					Comments
Polarization of interfering wave $\left(\vec{E}_a, \vec{E}_b\right)$	δ	0	π/2	π	3π/2	2π	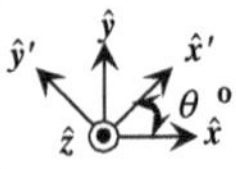
	x	0	Λ/4	Λ/2	3Λ/4	Λ	
(s, s)		↕	↕	·	↕	↕	Intensity grating (arb. α)
(p, p) in x-y plane		↔	↔	·	↔	↔	Intensity grating (arb. α)
(p, p) in x-z plane		↔	↺	↕	↻	↔	Polarization grating (large α)
(s, p)		⤢	↺	⤡	↻	⤢	Polarization grating (arb. α)
(+45°, +45°)		⤢	⤢	·	⤢	⤢	Intensity grating (small α)
(+45°, −45°)		↕	↺	↔	↻	↕	Polarization grating (small α)
(RCP, RCP)		↺	↺	·	↺	↺	Intensity grating (small α)
(RCP, LCP)		↕	⤢	↔	⤡	↕	Polarization grating (small α)

The spatial variations of intensity or polarization in x-direction dependent on different polarization combinations of the interfering waves can cause so-called intensity or polarization gratings

the induced light intensity. By selection of suitable polarization of the transmitted light self-aligned filtering is realized. Fuh et al. reported the fabrication of a few linear spatial filters. An electrically switchable spatial filter using a polymer-dispersed LC (PDLC) cell [30] is demonstrated based on the fact that the size of the LC droplet formed in a PDLC film is inversely proportional to the intensity of curing. Controlling the driving voltage on the PDLC sample can filter particular spatial frequencies in the Fourier optical signal process. A polarization controllable spatial filter based on azo-dye-doped LC film [31] is realized. The fabrication of such a polarization controllable spatial filter relies on the fact that the various intensities of the diffracted orders are responsible for various changes of the polarization state induced by the photo-aligned DDLC film. Particular spatial orders in the Fourier optical signal process can be filtered using an analyzer placed behind the sample by controlling the polarization state of the diffracted orders. Later, a transflective spatial filter based on azo-dye-doped cholesteric LC (DDCLC) films is presented [32]. The mechanism is due to the fact that the various intensities of the diffracted orders are responsible for various degree of transparency induced by the photoisomerized DDCLC film. High- and low-pass images in the Fourier optical signal process can be observed simultaneously through the reflected and transmitted signals, respectively. Another novel electro- and photo-controllable LC spatial filter is fabricated in a LC film with a photoconductive layer [33]. Such a spatial filter is formed because of the controllability of the photoelectro-induced screen effect of the space charge in the LC cell. An applied *DC* voltage or incident pumped intensity can be controlled to enable different spatial distributions of the diffraction pattern of the target object to be selected for filtering by the LC cell, such that various reconstructed images can be obtained.

14.3 Polarization Converters

LCs are commonly employed as polarization-converting materials because its director, which is an unit vector along the common axis of LC molecular orientation, can be modulated by an external field. Numerous methods have been realized. The first approach utilizes an LC spatial light modulator to cause axial polarization [34]. The second method is to control LC directors by concentrically circular rubbing [35]. Recently, Wu et al. presented a polarization converter using a sheared polymer network LC and LC gel [36, 37]. Some common concerns about the aforementioned approaches are their complex fabrication processes, and flexibility.

Tzeng et al. [38] reported the fabrication of LC polarization converters, which can be used to produce axially symmetrical polarization beams using the photo-alignment method. Photo-alignment techniques are discussed in Chap. 11. As seen, one of the photo-alignment methods is based on the azo compounds. Suitable LC alignments can be obtained via two approaches; one is to firstly spin-coat a layer of azo material on the substrates and followed by the light exposure, the other is the doping of azo material in LCs and then the cell is photo-excited by a suitable laser beam [38, 39]. Figure 14.2a shows the obtained homogeneous-radial sample, which can convert a linearly polarized beam into a radially polarizaed one. Figure 14.2b shows the image of the homogeneous-radial LC film under a crossed-polarized optical microscope (POM). The result is reasonable, and expected with the radially polarized emerging light. In the POM setup, the polarizer axis is parallel to the front LC director. Hence, the region from which the emerging beam is polarized perpendicular (parallel) to the analyzer axis is dark (bright), and a brightness gradient exists between these two directions. If the incident linearly polarized light with its polarization is set perpendicular to the front LC director (Fig. 14.2c), then the linearly polarized beam is converted to an azimuthally polarized beam. The image of the homogeneous-radial LC film under a crossed POM becomes that in Fig. 14.2d, as expected.

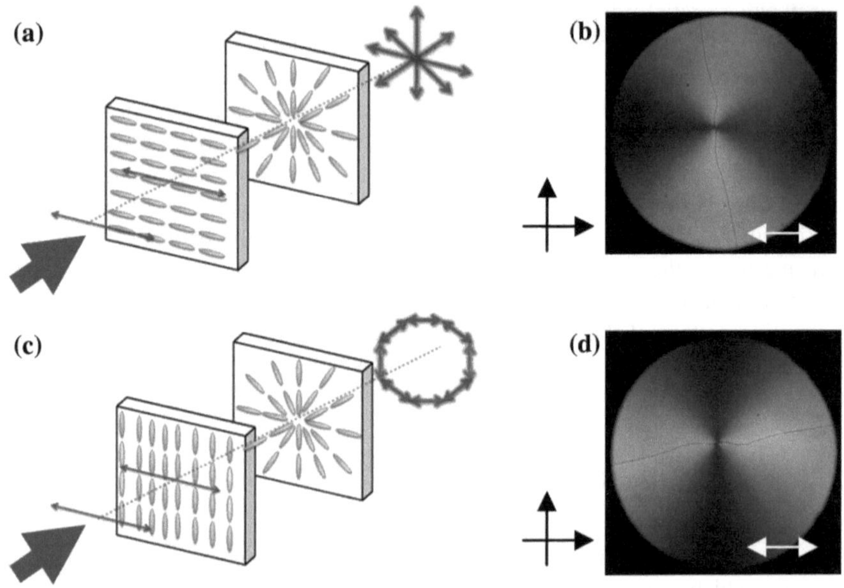

Fig. 14.2 Generations of **a** radially and **c** azimuthally polarized light by a homogeneous-radial LC film; **b**, **d** show images of **a** and **c**, respectively, under the crossed polarized optical microscope. *R* rubbing direction. *P* polarizer, *A* analyzer

Axially symmetric polarized beams have attracted much interest in recent years. They can be adopted in diffractive optics and imaging systems [34]. An important property of these beams is their unique cylindrical symmetry polarization, which enables the amplification of axially polarized beams to high power and good beam quality [40]. Radially and azimuthally polarized beams can be generated inside a laser cavity [41–43]. The adopted approaches, however, are based on complex resonator configurations, or require special fabricating techniques. Alternatively, radially or azimuthally polarized beams can be obtained directly from a Gaussian beam, outside the laser cavity, with specially designed mode converters. Such a conversion can be performed using various interferometric arrangements by interferometric superposition of linearly polarized beams [44, 45] or the use of a spatially varying retarder [46].

Ko et al. further reported the fabrication of the azimuthal-azimuthal (Fig. 14.3a) and radial-radial (Fig. 14.3b) samples using the so-called double-side photo-alignment method [39]. To fabricate a hybrid azimuthal-radial LC cell, two substrates were disassembled from both azimuthal and radial samples were combined to fabricate a new azimuthal-radial cell, as presented in Fig. 14.3c. Figures 14.3d, e show images of the axially symmetric azimuthal and radial LC samples, respectively, under crossed POM.

The hybrid azimuthal-radial cell can be used as a polarization converter. Figure 14.4a presents the polarization converter effects based on azimuthal-radial

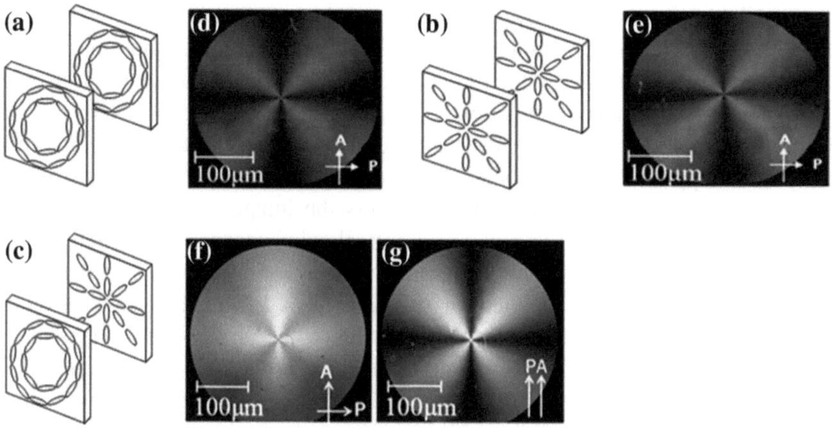

Fig. 14.3 Axially symmetric **a** azimuthal, **b** radial and **c** twisted nematic LC structures. Images of axially symmetric **d** azimuthal, **e** radial and **f** and **g** twisted nematic LC devices under a polarized optical microscope. *P* polarizer, *A* analyzer

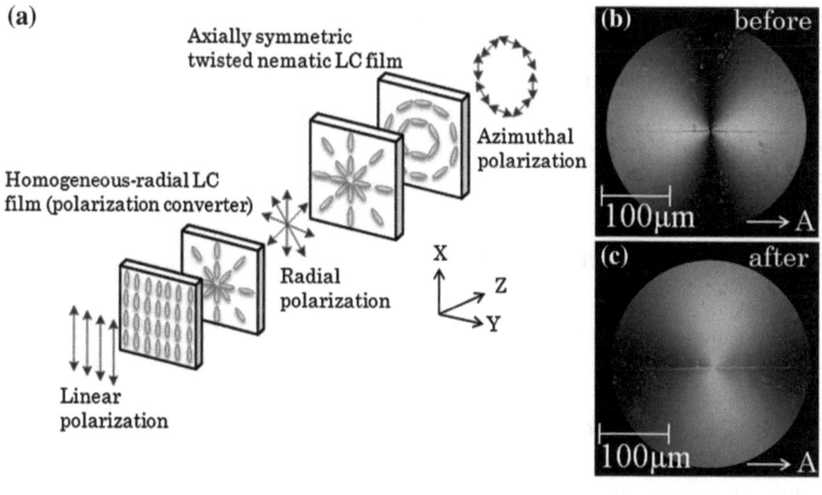

Fig. 14.4 **a** Setup for converting polarization of linearly polarized beam using axially symmetric LC devices; **b** radially polarized light and **c** azimuthally polarized light analyzed using an analyzer (y-axis) under POM

LC cell in an axially symmetric optical system. Initially, a homogeneous-radial LC film (polarization converter) [38] was adopted to transform linearly polarized light into radially polarized light. The radial polarization was analyzed using an analyzer with transmission along the y-axis; Fig. 14.4b presents the corresponding image. The region in which the polarization is perpendicular to the analyzer is in the dark state. When an azimuthal-radial LC cell was placed behind the homogeneous-radial LC film, it converted radially polarized light to azimuthally polarized light.

Figure 14.4c presents the analyzed (using a y-axis analyzer) image. This image is the opposite of the image of radial polarization (Fig. 14.4b) in transmission.

The radial-radial (Fig. 14.3b) LC cell can be utilized to modulate laser-beam shape. Laser-beam shape modulation is a rapidly developing field, which is driven by both technological improvements and the ever-increasing demand for its use in optical trapping (OT) and other applications [47–51]. Figure 14.5a presents the experimental setup. Figures 14.5b presents the image of an axially symmetric LC sample under a cross polarizing optical microscope. A Gaussian He–Ne laser beam passes through a polarizer and is incident onto the center of the radial-radial device, to which is applied an AC voltage (1 kHz). The diameter of the laser beam is ~1.5 mm. The upper figure in Fig. 14.5c presents the shapes of the beams that emerge from the sample at voltages of 0 V (donut-shaped) and 17 V (Gaussian), while the bottom figure gives their cross sectional irradiance distributions. Notably, the conversion efficiency from the incident linearly polarized Gaussian beam to donut-shaped beam was measured to be ~86.9 %. The donut shape is caused by the phase retardation that is provided by the axially symmetric LC cell [52–54], when no external voltage is applied. The transmitted beam undergoes destructive interference at its center when it passes through the axially symmetrically distributed LC molecules. When the applied voltage is sufficiently (~17 V), the LC molecules are re-oriented almost perpendicular to the substrates. Therefore, the transmitted laser retains its original Gaussian shape. Figure 14.6 presents the images of the beam shape which is formed by a radial-radial cell with varying applied voltage.

14.4 Liquid Crystal Lenses

As mentioned above, LCs possess a high birefringence, and a low control voltage. Thus, the use of LCs for adaptive systems has significantly increased in the last decade. Much research on the use of LCs is focused on the switchable lenses. Basically, the LC lenses are mainly based two designs; one is to establish a LC distribution that leads to a LC gradient index (GRIN) lens; the other one is through the use of a LC Fresnel plate. The former usually uses the patterned electrodes, and thus focal length and/or intensity are electrically controllable. Such a LC lens is first proposed by Sato [55] with the focal length switched between 150 and 230 mm. Several electrode designs have been presented, for examples, the stacked planar electrode geometry [56] or forming a certain electric field [57–60] by spherical electrode, the high resistance electrode [61], the hexagonal-hole-patterned electrode [62], double LC layers [63], four sub-electrodes [64] and hole and ring electrodes [65]. Different LC materials such as ferroelectric material [66], blue-phase LCs [67] were also proposed for improving the electro-optics properties. Various LC Fresnel lenses have been developed, the polymer-dispersed LC (PDLC) [68], polymer-stabilized LC [69], polarization-insensitive LC (PILC) [70] and dye-doped LC (DDLC) [71].

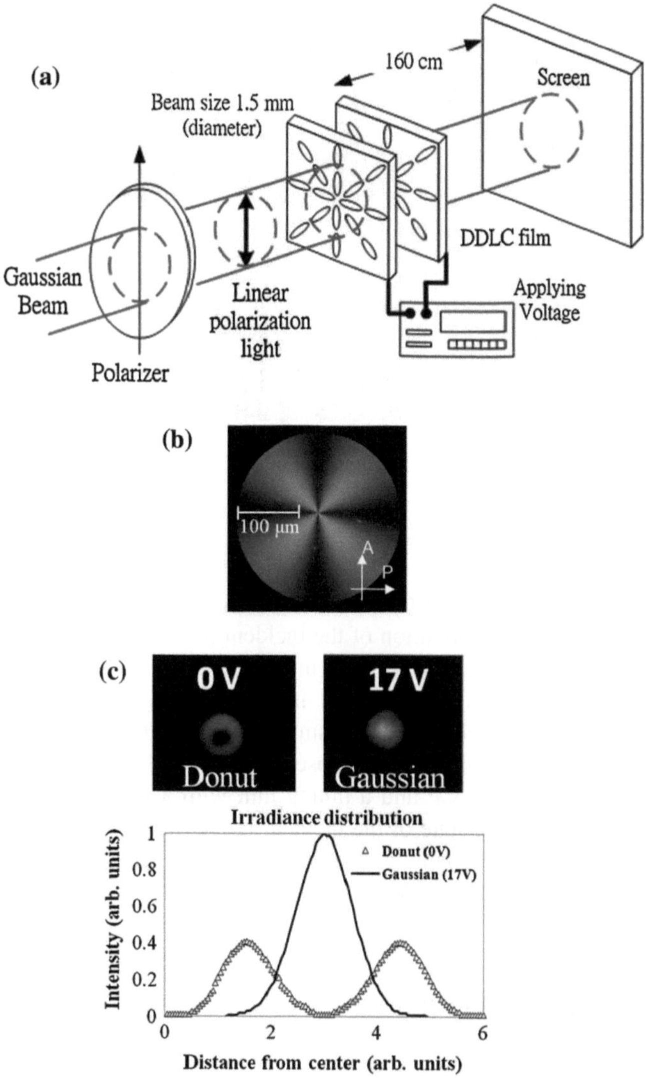

Fig. 14.5 **a** Experimental setup, **b** the images of a radial LC device under a cross-polarizer optical microscope, **c** the beam shapes (*upper*) and the irradiance distributions (*bottom*) of the output donut-shaped beam (0 V) and the Gaussian beam (17 V). *P* polarizer, *A* analyzer

Notably, the focal effects of all above-mentioned LC lenses are electrically controllable. Huang et al. demonstrates the feasibility of a microlens array (MLA) with a focal intensity that can be optically tuned by controlling the polarization of incident light [72]. The proposed MLA has a focusing unit based on birefringent liquid crystalline polymer (LCP), and a tuning unit with a photo-alignment layer for controlling the polarization state of incident laser light. The optically

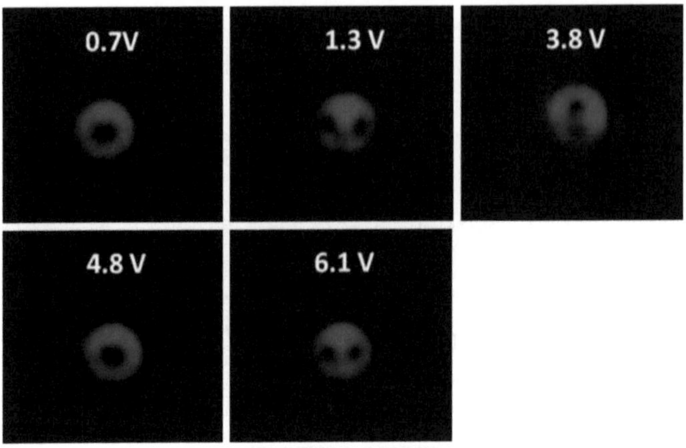

Fig. 14.6 Various beam shapes obtained using an ASDDLC device under various applied voltages

variable refractive indices of LCP allow a positive or negative MLA to be realized by controlling the polarization of the incident light. Further, the same authors demonstrate a liquid crystalline polymer microlens array (LCP MLA) with an all-optically tunable and multi-stable focal intensity through photochemical phase transition [73]. The operational mechanism of the optical tuning is associated with the photo-isomerization effect. The proposed LCP MLA device has a focusing unit based on a birefringence LCP and a tuning unit with a light responsive material to control the polarization state of the incident probe beam. The optically variable refractive indices of LCP enable a positive or negative MLA that can control the polarization of incident light to be realized.

14.5 Liquid Crystal Q-Plates

Gaussian beams (in transverse electromagnetic wave (TEM_{00}) mode) are the most common used laser beam-profiles. Their particular distribution is preserved as the beams propagate, and such beams can be focused into a diffraction-limited spot. In recent decades, the manipulation of the shapes of laser beams has rapidly developed because of technological improvements in beam-shaping devices and the ever-increasing demand for their applications [47–51]. The profile intensity of a laser-beam can be shaped by modulating the amplitude [74, 75] or the phase [76, 77] of the beam. Although amplitude modulation is effective, it usually wastes much of the energy of the laser beam. When properly performed, phase modulation can almost losslessly reproduce the intensity of a beam. Hence, phase modulation methods are more efficient in practice. A novel device, called a q-plate (QP),

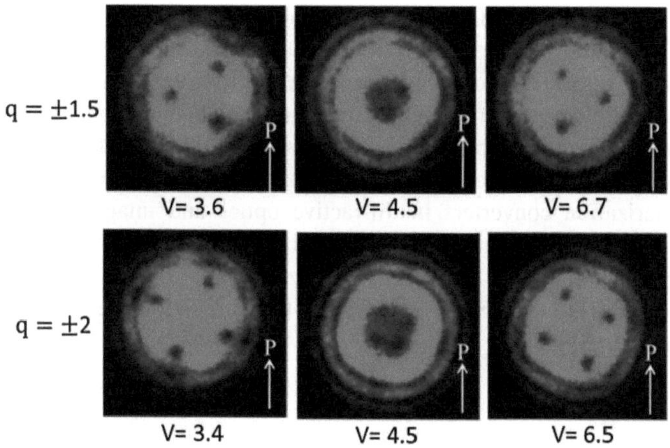

Fig. 14.7 Various beam shapes were obtained by modulation by LC QP with q values of ±1.5 and ±2 when suitable voltages were applied

which is based on the phase modulation of LC, has been developed [78]. It can be patterned with a specific transverse topology and contains a well-defined integer or semi-integer topological charge at its center [79–81]. This device has been applied to convert a Gaussian into a beam with orbital angular momentum [79, 82] and it has been used in various areas in the field of optics [83–85]. Vector vortex beams that are produced by the modulating circularly polarized Gaussian beams are commonly coincident with corresponding singular points in the optical phase. Vortex beams are strongly correlated to singular optics with the optical phase singular point and can be used in optical tweezers [86, 87], imaging [88, 89], atomic trapping [90], and quantum information [91].

Huang et al., recently, present LC QPs with positive or negative integer and semi-integer q values, using the axially symmetric photo-alignment method [92]. The shift in the phase-retardation that is caused by fabricated LC QP devices when a voltage is investigated and the fabricated QPs are shown to be useful for the electrical tuning the beam shapes. Moreover, the polarization distribution of a linearly Gaussian beam after it has been modulated by an LC QP is analyzed. Numerical simulation was performed to confirm the LC alignment structures on the substrates and for comparison of the results with observations made under a crossed-POM. The results of the simulation are consistent with the experimental results. The LC QPs were utilized to modulate a linearly polarized Gaussian laser beam and the distributions of linear polarization in three modes upon the application of a suitable voltage were studied. The LC QP device can be utilized as a beam shape modulator and a spatial polarization converter in diffractive optics and imaging systems.

Figure 14.7 shows the images of the beams that were modulated by LC QPs with q values of ±1.5 and ±2 under suitable applied voltages. LC QPs with the same absolute q values identically modulate a linearly polarized TEM_{00} Gaussian beam, as revealed in the POM images. As the applied voltage is varied, each LC QP has three modes, which are a donut-shaped mode and two beam structures with 2q dark spots. The dark spots are resulting from the phase singularities that caused by far-field beam interference, and can be calculated using the Fraunhofer diffraction [93] for the propagation of a linearly polarized Gaussian beam through the center of LC QP. These devices can be used as beam-shaped modulators and spatial polarization converters in diffractive optics and imaging systems. Also, the proposed beam-shape manipulation can be used in an advanced optical tweezers system. The intensity profile and polarization distribution were simulated by MATLAB and 1D-DIMOS. The results of the simulation were consistent with experimental findings.

References

1. D.K. Yang, S.T. Wu, *Fundamentals of Liquid Crystal Devices* (Wiley, Chichester, 2006)
2. I.C. Khoo, *Liquid Crystals Physical Properties and Nonlinear Optical Phenomena*. (Wiley, New York, 1995)
3. I.C. Khoo, Holographic grating formation in dye- and fullerene C60-doped namatic liquid-crystal film. Opt. Lett. **20**, 2137–2139 (1995)
4. L. Frey, M. Kaczmarek, J.M. Jonathan et al., Analysis of gratings induced in azo-dye doped liquid crystals. Opt. Mat. **18**, 91–94 (2001)
5. Andy Y.-G. Fuh, C.C. Liao, K.C. Hsu et al., Dynamic studies of holographic gratings in dye-doped liquid crystal films. Opt. Lett. **26**, 1767–1769 (2001)
6. Andy Y.-G. Fuh, C.C. Liao, K.C. Hsu et al., Laser-induced reorientation effect and ripple structure in dye-doped liquid crystal films. Opt. Lett. **28**, 1179–1181 (2003)
7. S.Y. Huang, S.T. Wu, Andy Y.-G. Fuh, Optically switchable twist nematic grating based on a dye-doped liquid crystal film. Appl. Phys. Lett. **88**, 041104 (2006)
8. S.T. Wu, Y.S. Chen, J.H. Guo et al., Fabrication of twist nematic gratings using polarization hologram based on azo dye-doped liquid crystals. Jpn. J. Appl. Phys. **45**, 9146–9151 (2006)
9. W.Y. Wu, M.S. Li, H.C. Lin et al., Two-dimensional holographic polarization grating formed on azo-dye-doped PVA films. J. Appl. Phys. **103**, 083119–083124 (2008)
10. W. Lee, C.S. Chiu, Observation of self-diffraction by gratings in nematic liquid crystals doped with carbon nanotubes. Opt. Lett. **26**, 521–523 (2001)
11. P.M. Ajayan, Carbon nanotubes, in *Nanostructured Materials and Nanotechnology*, ed. by H.S. Nalwa (Academic Press, San Diego, 2002), pp. 329–360
12. D. Subacius, P.H. Bos, O.D. Lavrentovich, Switchable diffractive cholesteric gratings. Appl. Phys. Lett. **71**, 1350 (1997)
13. S.N. Lee, L.C. Chien, S. Sprunt, Polymer-stabilized diffraction gratings from cholesteric liquid crystals. SID Symp. Tech. Digest. **29**, 834–837 (1998)
14. S.W. Kang, L.C. Chien, Field-induced and polymer-stabilized two-dimensional cholesteric liquid crystal gratings. Appl. Phys. Lett. **90**, 221110 (2007)
15. Andy Y.-G. Fuh, C.H. Lin, M.F. Hsieh et al., Cholesteric gratings doped with a dichroic dye. Jpn. J. Appl. Phys. **40**, 1334–1338 (2001)
16. Andy Y.-G. Fuh, C.H. Lin, C.Y. Huang, Dynamic pattern formation and beam-steering characteristics of cholesteric gratings. Jpn. J. Appl. Phys. **41**, 211–218 (2002)
17. H.C. Jau, T.H. Lin, R.X. Fung et al., Optically-tunable beam steering grating based on azobenzene doped cholesteric liquid crystal. Opt. Express **18**, 17498–17503 (2010)
18. H.C. Jau, T.H. Lin, Y.U. Chen et al., Direction switching and beam steering of cholesteric liquid crystal gratings. Appl. Phys. Lett. **100**, 131909 (2012)

19. R.L. Sutherland, V.P. Tondiglia, L.V. Natarajan et al., Electrically switchable volume grating in PDLCs. Appl. Phys. Lett. **64**, 1074–1076 (1994)
20. Andy Y.-G. Fuh, M.S. Tsai, L.J. Huang et al., Optically switchable grating based on polymer-dispersed liquid crystal films doped with a guest-host dye. Appl. Phys. Lett. **74**, 2572–2574 (1999)
21. Andy Y.-G. Fuh, T.C. Ko, M.S. Tsai et al., Dynamical studies of grating formed in polymer-dispersed liquid crystal films. J. Appl. Phys. **83**, 679–683 (1998)
22. Andy Y.-G. Fuh, C.Y. Huang, M.S. Tsai et al., Polymer-dispersed liquid crystal films: Interference pattern recording and its application. Jpn. J. Appl. Phys. **35**, 630–633 (1996)
23. Andy Y.-G. Fuh, M.S. Tsai, C.Y. Huang et al., Polymer-dispersed liquid crystal films for storing optical holographic images. Opt. Quant. Electron. **28**, 1535–1541 (1996)
24. C.J. Yu, D.W. Kim, J. Kim et al., Polarization-invariant grating based on a photoaligned liquid crystal in an oppositely twisted binary configuration. Opt. Lett. **30**, 1995–1997 (2005)
25. Andy Y.-G. Fuh, J.C. Chen, S.Y. Huang et al., Binary liquid crystal alignments based on photoalignment in azo dye-doped liquid crystals and their application. Appl. Phys. Lett. **96**, 051103 (2010)
26. W. Hu, A. Srivastava, F. Xu et al., Liquid crystal gratings based on alternate TN and PA photoalignment. Opt. Express **20**, 5384–5391 (2012)
27. G.P. Crawford, J.N. Eakin, M.D. Radcliffe et al., Liquid-crystal diffraction gratings using polarization holography alignment techniques. J. Appl. Phys. **98**, 123102 (2005)
28. C. Oh, M. Eskuti, Achromatic diffraction from polarization grating with high efficiency. Opt. Lett. **33**, 2287–2289 (2008)
29. J. Kato, I. Yamaguchi, H. Tanaka, Nonlinear spatial filtering with a dye-doped liquid-crystal cell. Opt. Lett. **21**, 767–769 (1996)
30. Andy Y.-G. Fuh, T.H. Lin, Electrically switchable spatial filter based on polymer-dispersed liquid crystal film. J. Appl. Phys. **96**, 5402–5404 (2004)
31. T.H. Lin, Andy Y.-G. Fuh, Polarization controllable spatial filter based on azo-dye-doped liquid crystal film. Opt. Lett. **30**, 1390–1392 (2005)
32. T.H. Lin, Andy Y.-G. Fuh, Transflective spatial filter based on dye-doped cholesteric liquid crystal film. Appl. Phys. Lett. **87**, 011106 (2005)
33. S.Y. Huang, T.C. Wung, Andy Y.-G. Fuh et al., Electro- and photo-controllable spatial filter based on a liquid crystal film with a photoconductive layer. Appl. Phys. B **97**, 749–752 (2009)
34. J.A. Davis, D.E. McNamara, D.M. Cottrell et al., Two-dimensional polarization encoding with a phase-only liquid-crystal spatial light modulator. Appl. Opt. 39, 1549 (2000)
35. R. Yamaguchi, T. Nose, S. Sato, Liquid crystal polarizers with axially symmetrical properties. Jpn. J. Appl. Phys. 28, 1730 (1989)
36. Y.H. Wu, Y.H. Lin, H. Ren et al., Axially-symmetric sheared polymer network liquid crystals. Opt. Express **13**, 4638 (2005)
37. H. Ren, Y.H. Lin, S.T. Wu, Linear to axial or radial polarization conversion using a liquid crystal gel. Appl. Phys. Lett. **89**, 051114 (2006)
38. Y.Y. Tzeng, S.W. Ke, C.L. Ting et al., Axially symmetric polarization converters based on photo-aligned liquid crystal films. Opt. Express **16**, 3768–3775 (2008)
39. S.W. Ko, Y.Y. Tzeng, C.L. Ting et al., Axially symmetric liquid crystal devices based on double-side photo-alignment. Opt. Express **16**, 19643–19648 (2008)
40. I. Moshe, S. Jackel, A. Meir, Production of radially or azimuthally polarized beams in solid-state lasers and the elimination of thermally induced birefringence effects. Opt. Lett. **28**, 807 (2003)
41. R. Oron, S. Blit, N. Davidson et al., The formation of laser beams with pure azimuthal or radial polarization. Appl. Phys. Lett. **77**, 3322 (2000)
42. T. Moser, M.A. Ahmed, F. Pigeon et al., Generation of radially polarized beams in Nd:YAG lasers with polarization selective mirrors. Laser Phys. Lett. **1**, 234 (2004)
43. Y. Kozawa, S. Sato, Generation of a radially polarized laser beam by use of a conical Brewster prism. Opt. Lett. **30**, 3063 (2005)
44. S.C. Tidwell, G.H. Kim, W.D. Kimura, Efficient radially polarized laser beam generation with a double interferometer. Appl. Opt. **32**, 5222 (1993)

45. N. Passily, R.S. Denis, K. Ait-Ameur et al., Simple interferometric technique for generation of a radially polarized light beam. J. Opt. Soc. Am. A **22**, 984 (2005)
46. G. Machavariani, Y. Lumer, I. Moshe et al., Efficient extracavity generation of radially and azimuthally polarized beams. Opt. Lett. **32**, 1468 (2007)
47. C. Bustamante, Z. Bryant, B.S. Smith, Ten years of tension: single- molecule DNA mechanics. Nature **421**, 423–427 (2003)
48. S.M. Block, D.F. Blair, H.C. Berg, Compliance of bacterial flagella measured with optical tweezers. Nature **338**, 514–518 (1989)
49. C. Selhuber-Unkel, I. Zins, O. Schubert et al., Quantitative optical trapping of single gold nanorods. Nano Lett. **8**, 2998–3003 (2008)
50. R. Agarwal, K. Ladavac, Y. Roichman et al., Manipulation and assembly of nanowires with holographic optical traps. Opt. Express **13**, 8906–8912 (2005)
51. F. Hajizadeh, N.S. Reihani, Optimized optical trapping of gold nanoparticles. Opt. Exp. **18**, 551–559 (2010)
52. A. Jesacher, S. Fürhapter, S. Bernet et al., Diffractive optical tweezers in the Fresnel regime. Opt. Express **12**, 2243–2250 (2004)
53. S.C. Chu, K. Otsuka, Doughnut-like beam generation of Laguerre-Gaussian mode with extremely high mode purity. Opt. Commun. **281**, 1647–1653 (2008)
54. V.V. Kotlyar, A.A. Kovalev, R.V. Skidanov et al., Simple optical vortices formed by a spiral phase plate. J. Opt. Technol. **74**, 49–58 (2007)
55. S. Sato, Liquid-crystal lens-cells with variable focal length. Jpn. J. Appl. Phys. **18**, 1678–1684 (1979)
56. M. Ye, B. Wang, S. Sato, Liquid crystal lens with focal length variable from negative to positive values. IEEE Photon Tech. Lett. **18**, 78–81 (2006)
57. H. Ren, Y.H. Fan, S. Gauza et al., Tunable-focus flat liquid crystal spherical lens. Appl. Phys. Lett. **84**, 4789–4791 (2004)
58. T. Nose, S. Sato, A liquid crystal microlens obtained with a nonuniform electric field. Liq. Cryst. **5**, 1425–1433 (1989)
59. B. Wang, M. Ye, S. Sato, Experimental and numerical studies on liquid crystal lens with spherical electrode. Mol. Cryst. Liq. Cryst. **433**, 217–227 (2005)
60. B. Wang, M. Ye, M. Honma et al., Liquid crystal lens with spherical electrode. Jpn. J. Appl. Phys. **41**, L1232–L1233 (2002)
61. P.J.W. Hands, A.K. Kirby, G.D. Love, Adaptive modally addressed liquid crystal lenses. in *Proceedings of SPIE 5518: Liq Cryst VIII*, pp. 136–143 (2004)
62. M. Ye, S. Hayasaka, S. Sato, Liquid Crystal Lens Array with Hexagonal-Hole-Patterned Electrodes. Jpn. J. Appl. Phys. **43**, 6108–6111 (2004)
63. B. Wang, M. Ye, S. Sato, Liquid crystal lens with stacked structure of liquid-crystal layers. Opt. Commun. **250**, 266–273 (2005)
64. M. Ye, B. Wang, S. Sato, Liquid crystal lens with focus movable in focal plane. Opt. Commun. **259**, 710–722 (2006)
65. C.W. Chiu, Y.C. Lin, C.P. Chao Paul et al., Achieving high focusing power for a large aperture liquid crystal lens with novel hole-and ring electrodes. Opt. Express **16**, 19277–19285 (2008)
66. N.A. Clark, M.A. Handschy, S.T. Lagerwall, Ferroelectric liquid crystal electro-optic using the surface stabilized structure. Mol. Cryst. Liq. Cryst. **94**, 213–234 (1983)
67. Y.H. Lin, H.S. Chen, H.C. Lin et al., Polarizer-free and fast response microlens arrays using polymer-stabilized blue phase liquid crystals. Appl. Phys. Lett. **96**, 113505 (2010)
68. H. Ren, Y.H. Fan, S.T. Wu, Tunable Fresnel lens using nanoscale polymer-stabilized liquid crystals. Appl. Phys. Lett. **83**, 1515–1517 (2003)
69. Y.H. Fan, H. Ren, S.T. Wu, Switchable Fresnel lens using polymer-stabilized liquid crystals. Opt. Express **11**, 3080–3086 (2003)
70. D.W. Kim, C.J. Yu, H.R. Kim et al., Polarization-insensitive liquid crystal Fresnel lens of dynamic focusing in an orthogonal binary configuration. Appl. Phys. Lett. **88**, 203505–203507 (2006)

71. T.H. Lin, Y. Huang, Andy Y.-G. Fuh et al., Polarization controllable Fresnel lens using dye-doped liquid crystals. Opt. Express **14**, 2359–2364 (2006)
72. S.Y. Huang, T.C. Tung, C.L. Ting et al., Polarization-dependent optical tuning of focal intensity of liquid crystal polymer microlens array. Appl. Phys. B **104**, 93–97 (2011)
73. S.Y. Huang, T.C. Tung, H.C. Jau et al., All-optical controlling of the focal intensity of a liquid crystal polymer microlens array. Appl. Opt. **50**, 5883–5888 (2011)
74. P. Srinivasan, M.K. Poutous, Z.A. Roth et al., Spatial and spectral beam shaping with space-variant guided mode resonance filters. Opt. Express **17**, 20365–20375 (2009)
75. Z. Tian, M. Nix, S.H. Yam, Laser beam shaping using a single-mode fiber abrupt taper. Opt. Lett. **34**, 229–231 (2009)
76. W.W. Simmons, G.W. Leppelmeier, B.C. Johnson, Optical beam shaping devices using polarization effects. Appl. Opt. **13**, 1629–1632 (1974)
77. S.W. Ko, T.H. Lin, Y.H. Huang et al., Electrical control of shape of laser beam using axially symmetric liquid crystal cells. Appl. Opt. **51**, 1540–1545 (2012)
78. S. Slussarenko, A. Murauski, T. Du et al., Tunable liquid crystal q-plates with arbitrary topological charge. Opt. Express **19**, 4085–4090 (2011)
79. L. Marrucci, C. Manzo, D. Paparo, Optical spin-to-orbital angular momentum conversion in inhomogeneous anisotropic media. Phys. Rev. Lett. **96**, 163905 (2006)
80. L. Marrucci, E. Karimi, S. Slussarenko et al., Spin-to-orbital conversion of the angular momentum of light and its classical and quantum applications. J. Opt. **13**, 064001 (2011)
81. E. Nagali, F. Sciarrino, F.D. Martini et al., Polarization control of single photon quantum orbital angular momentum states. Opt. Express **17**, 18745–18759 (2009)
82. L. Marrucci, C. Manzo, D. Paparo, Pancharatnam-Berry phase optical elements for wave front shaping in the visible domain: switchable helical mode generation. Appl. Phys. Lett. **88**, 221102 (2006)
83. E. Karimi, S. Slussarenko, B. Piccirillo, Polarization-controlled evolution of light transverse modes and associated pancharatnam geometric phase in orbital angular momentum. Phys. Rev. A **81**, 053813 (2010)
84. E. Nagali, L. Sansoni, F. Sciarrino et al., Optimal quantum cloning of orbital angular momentum photon qubits through hong-ou-mandel coalescence. Nat. Photon **3**, 720–723 (2009)
85. S. Slussarenko, B. Piccirillo, V. Chigrinov et al., Liquid crystal spatial-mode converters for the orbital angular momentum of light. J. Opt. **15**, 025406 (2013)
86. S.H. Tao, X.C. Yuan, J. Lin et al., Residue orbital angular momentum in interferenced double vortex beams with unequal topological charges. Opt. Express **14**, 535–541 (2006)
87. C.H.J. Schmitz, K. Uhrig, J.P. Spatz et al., Tuning the orbital angular momentum in optical vortex beams. Opt. Express **14**, 6604–6612 (2006)
88. F. Tamburini, G. Anzolin, G. Umbriaco et al., Overcoming the Rayleigh Criterion Limit with Optical Vortices. Phys. Rev. Lett. **97**, 163903 (2006)
89. C. Maurer, A. Jesacher, S. Furhapter et al., Tailoring of arbitrary optical vector beams. New J. Phys. **9**, 78 (2007)
90. S. Franke-Arnold, J. Leach, M.J. Padgett et al., Optical ferris wheel for ultracold atoms. Opt. Express **15**, 8619–8625 (2007)
91. G. Molina-Terriza, J.P. Torres, L. Torner, Twisted photons. Nat. Phys. **3**, 305–310 (2007)
92. Y.H. Huang, S.W. Ko, M.S. Li et al., Modulation of shape and polarization of beam using a liquid crystal q-plate that is fabricated via photo-alignment. Opt. Express **21**, 10954–10961 (2013)
93. J.W. Goodman, *Introduction to Fourier Optics*. (Roberts & Company, Atlanta, 2004), (Chapter 4)
94. I.P. Il'chishin, E.A. Tikhonov, V.G. Tishchenko et al., Generation of a tunable radiation by impurity cholesteric liquid crystals. JETP Lett. **32**, 24–27 (1980)
95. V.I. Kopp, Z.Q. Zhang, A.Z. Genack, Lasing in chiral photonic structures. Prog. Quantum Electron. **27**, 369–416 (2003)
96. J. Schmidtke, W. Stille, H. Finkelmann et al., Laser emission in a dye doped cholesteric polymer network. Adv. Mater. **14**, 746–749 (2002)

97. S. Furumi, S. Yokoyama, A. Otomo et al., Phototunable photonic bandgap in a chiral liquid crystal laser device. Appl. Phys. Lett. **84**, 2491–2493 (2004)
98. S.M. Morris, A.D. Ford, M.N. Pivnenko et al., Enhanced emission from liquid-crystal lasers. J. Appl. Phys. **97**, 023103 (2005)
99. T.H. Lin, Y.J. Chen, C.H. Wu et al., Cholesteric liquid crystal laser with wide tuning capability. Appl. Phys. Lett. **86**, 161120 (2005)
100. C.R. Lee, S.H. Lin, H.C. Yeh et al., Color cone lasing emission in a dye-doped cholesteric liquid crystal with a single pitch. Opt. Express **17**, 12910–12921 (2009)

Chapter 15
Liquid Crystals for Bio-medical Applications

Yi-Hsin Lin

Thermotropic nematic LCs can modulate light in phases, amplitudes, and polarizations [1]. Many photonic applications based LCs have developed, such as displays, electro-optical switches, lenses, optical vortex generators, variable optical attenuations, solar cells and so on. Thermotropic nematic LCs have great potential in bio-medical applications. In this session, we mainly introduce two the bio-medical applications based on thermotropic nematic LCs: biosensors and ophthalmic lenses. In biosensors, the key is interfacial interactions between biosamples and LC molecules. We will introduce a LC and polymer complex system, the LC and polymer composite film (LCPCF), whose surface free energy is electrically switchable. LCPCF can help to test the sperm quality and high-density-lipoprotein (HDL). In ophthalmic applications, we will introduce the challenges and requirements for design of ophthalmic lenses. A polarizer-free LC lens with large aperture size is also introduced.

15.1 Biosensors Using Nematic Liquid Crystals

In general, interfacial interactions between LCs and bio-samples (or sensing targets) are the key to design a biosensor using nematic LCs [2–7]. Such an interfacial interaction often causes the re-orientations of LC molecules at the interface. As a result, optical properties of LC change and the change of the orientation of LC can be read out by observing under a polarizing microscopy. LC can optically amplify the biological binding events and help us to study the biology. The main approach is to immobilize thermotropic nematic LC onto a solid surface as a sensing interface for bio-samples. 5CB is a commonly used a LC material for biosensing. In the literatures, 5CB has been proved for sensing many bio-samples,

Y.-H. Lin (✉)
Department of Photonics, National Chiao Tung University, Hsinchu, Taiwan
e-mail: yilin@mail.nctu.edu.tw

C.-C. Lee (ed.), *The Current Trends of Optics and Photonics*,
Topics in Applied Physics 129, DOI 10.1007/978-94-017-9392-6_15

such as biotin/antibiotin-IgG, phosphopeptide, DNA, proteins, phospholiqid, bacteria and viruses, human embryonic stem cells, and protein-peptide binding events [6, 8–15]. The thermotropic nematic LC materials are soft condensed materials which have to be sustained in reservoirs and it takes times to sense the biosamples (30 min ~ few hours). In order to sustain thermotropic nematic LC materials, LC and polymer complex systems are great candidates to turn the soft LC materials into film-like structures which are easy to apply voltage to further stimulate the LC orientations for biosensing. Recently, we develop a LC and polymer composite film (LCPCF) whose wettability (or surface free energy) is electrically tunable due to the orientation of LC molecules on LCPCF [16–21]. LCPCF is a film fabricated through photoinduced phase separation process. The ingredients are thermotropic nematic LC E7 (Merck), a liquid crystalline monomer (4-(3-Acryloyloxypropyloxy)-benzoic acid 2-methyl-1, 4-phenylene ester) and a photoinitiator. LCPCF is a platform for realizing biosensing devices on a basis of droplet manipulation induced by interfacial interaction between anisotropic LC molecules and bio-materials.

The physical mechanism of electrically tunable wettability of LCPCF is based on the electrically tunable surface free energy of LCPCF. Figure 15.1 illustrates a typical LCPCF device. Generally speaking, a surface free energy of a smooth solid surface can be determined by placing a fluidic droplet on the solid surface and the surface free energy is described according to the Young's contact angle, a result of a balance of three phases: vapor, liquid and solid. The surface free energy is re-modified by the roughness according to the Wenzel's magnification of wettability. When the solid surface consists of several materials or so-called chemical heterogeneity, the surface free energy can be further re-modified according to the Cassie's linear composition of interfacial energies of materials. The Young's

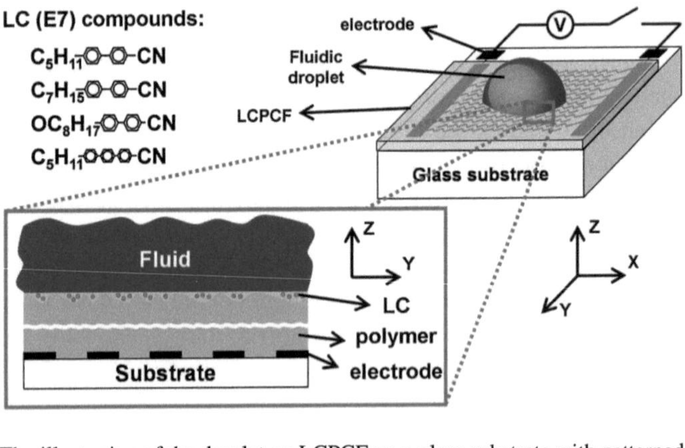

Fig. 15.1 The illustration of the droplet on LCPCF on a glass substrate with patterned electrodes. The magnification of the interface is exaggeratedly depicted in the inlet. The four chemical structures of LC (E7) compounds of LCPCF are listed as well. The monomer is a liquid crystalline monomer (4-(3-Acryloyloxypropyloxy)-benzoic acid 2- methyl-1, 4-phenylene ester) [21]

balance of three phases, Wenzel's magnification of wettability and Cassie's chemical heterogeneity are the factors that can help us to describe the wetting properties or surface free energy of a surface. However, one more factor, molecular orientations, can also be considered when we design an actively switchable surface, LCPCF for example. The surface free energy of the LCPCF in the air ($\gamma_{\text{LCPCF-air}}$) with an unit of J/m^2 can be expresses in 15.1 according to the Chibowski's film pressure model [21]:

$$\gamma_{LCPCF-air}(\phi(V)) = \frac{\gamma_{L-air} \times (1 + \cos\theta_a(V))^2}{2 + \cos\theta_r(V) + \cos\theta_a(V)}, \tag{15.1}$$

where $\gamma_{\text{L-air}}$ is the surface free energy of the testing fluid (L) in the air, θ_a is advancing angle of a fluidic drop on the LCPCF, θ_r is a receding angle of a fluidic drop, V is an applied voltage, and ϕ is the average tilt angle of LC molecules with respect to x-axis in Fig. 15.1. The advancing angle (θ_a) and the receding angle (θ_r) are the angles at which the contact line changes when the volume of the fluidic drop increases and decreases, respectively. In experiments, the surface free energy of the LCPCF ($\gamma_{\text{LCPCF-air}}$) increases from 36×10^{-3} to 51×10^{-3} J/m^2 with the applied pulsed voltage [21]. According to Cassie's model, the surface free energy of LCPCF in the air is a linear composition of the surface free energies of polymer ($\gamma_{\text{p-air}}$) and the surface free energy of LC ($\gamma_{\text{LC-air}}$). The surface free energy of LCPCF can also be expressed as:

$$\gamma_{LCPCF-air}(\phi(V)) = R_w \times [f_{LC} \times \gamma_{LC-air}(V) + f_p \times \gamma_{p-air}], \tag{15.2}$$

where R_w is the roughness factor defined as the ratio of the actual surface area to the geometric surface area, f_{LC} is the fraction of LC and f_p is the fraction of polymer. The LC materials (E7) of LCPCF consisting of 4 compounds can simply be divided by three parts: the alkyl/alkoxy chain, the cyano (CN) group, and the phenyl/terphenyl part. The surface free energy of the LCs anchored among the polymer grains on LCPCF can be expressed as: $\gamma_{LC-air}(\phi(V)) = \gamma_{CN} \times \sin^2\phi(V) + \gamma_{ph-ter} \times \cos^2\phi(V)$, where $\gamma_{\text{ph-ter}}$, and $\gamma_{\text{alky-alko}}$ are the surface free energy of phenyl/terphenyl part, alkyl chain, and cyano group, respectively. $\gamma_{\text{ph-ter}}$ is ~29×10^{-3} J/m^2, $\gamma_{\text{alky-alko}}$ is ~29×10^{-3} J/m^2, and γ_{CN} is ~154×10^{-3} J/m^2 [21]. The averaged LC molecules tilt up from 0 to 32° with the applied pulsed voltage under calculations from experiments.

The detail physical mechanism near the surface of LCPCF is also depicted in Fig. 15.2a, b. At voltage-off state, the strong anchoring force which is provided by polymer grains results in the LC molecules aligned along x-direction. The surface free energy of LCPCF results from the interaction between phenyl/terphenyl part of LC (E7) and the testing water. When the applied pulsed voltage overcomes the anchoring force of the polymer grains and elastic properties of LC materials, the LC molecules change the orientations gradually. The CN-group tilts up with the applied pulsed voltage because of the fringing electric field, and the strong interaction between polar CN-group and polar fluid

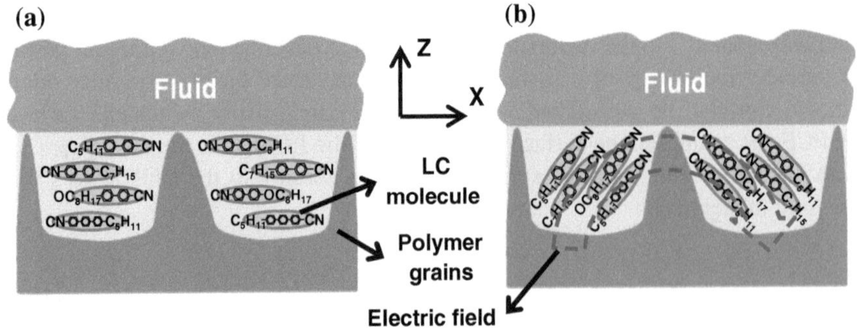

Fig. 15.2 The physical mechanism of electrically tunable free energy of LCPCF. **a** At voltage-off state, the LC molecules are anchored among the polymer grains. **b** At voltage-on state, the LC molecules are reoriented by the electric field and the cyano groups of LC materials tilt toward the interface between the polar fluid and LC materials [21]

(de-ionized water). Figures 15.3a–d show scanning electron microscope images of the surface of LCPCF at different magnification. The surface show elongated aggregation of polymer grains and void holes for LC molecules. The size of the

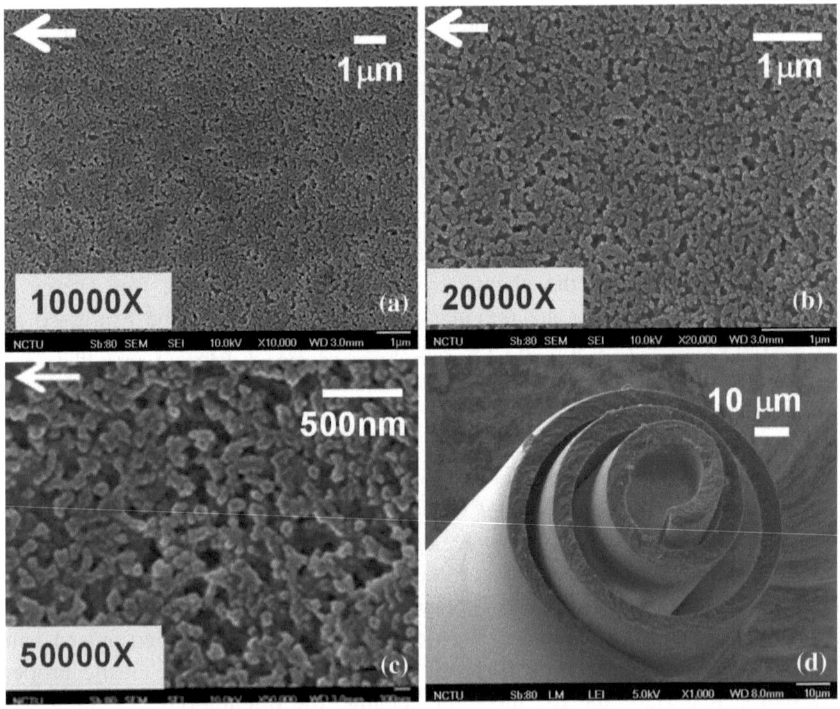

Fig. 15.3 **a**, **b**, **c** SEM images of the surface of LCPCF at different magnification. **d** SEM image of the *side view* of rolled LCPCF. The *arrows* indicate the direction of LC anchored the polymer grains

LC domains is around 100–200 nm. The root-mean-square (RMS) roughness of the LCPCF is ~30 nm. The domain size and roughness can be adjusted by phase separation process. Figure 15.3d shows SEM image of the side-view of a rolled LCPCF. LCPCF is a thin film and the thickness is around 4–30 μm. The wetting properties of LCPCF indicate the potentials of alignment layers to LC molecules for photonic applications. However, LCPCF is thicker, the applied voltage is higher.

To manipulate a polar droplet on LCPCF, the two regions of the interdigitated chevron electrodes are patterned identically, as depicted in Fig. 15.4a. At V = 0, the contact angles on both sides of the droplet are identical. When we apply a voltage in the left interdigitated region, the left region of LCPCF is more hydrophilic because of the tilts of LC molecules anchored to the polymer grains, so the droplet experiences an imbalanced Young's force [22] and the droplet is forced to move toward the left, as depicted in Fig. 15.4b. Figure 15.5a shows the dynamics of water droplet when we applied ~250 V_{rms} square pulses (f = 1 kHz) to the left electrodes in Fig. 15.4a for a time duration of 600 ms. Figure 15.5b shows the water contact angle on the left and right of the droplet as a function of time. The contact angles on both sides of the droplet are the same at V = 0. When we turned on the pulsed voltage in the left region in Fig. 15.5a at 0.6 s, this region became more hydrophilic due to the reorientation of the LC directors. The contact angle on the left (~56.77°) became smaller than that on the right (~79.54°). The droplet then experienced an imbalanced net force (~98 μN) causing the droplet to move toward left. The net force (F) can be calculated according to: $F = \pi \times r_o \times \gamma_{LV}(\cos(\theta_{right}) - \cos(\theta_{left}))$ where γ_{LV} represents the surface free energy of the liquid–vapor interface (~72.5×10^{-3} J/m^2), r_o is the radius of the drop (~1.18 mm for a 3 μL water drop), and θ_{right} and θ_{left} stand for the local contact angles on the right and on the left of the droplet. Between 0.6 and 0.65 s, the contact angles on the left and on the right of the water droplet oscillate to overcome the resistive force of the surface resulting from the hysteresis of LCPCF, as shown in Fig. 15.5c. The hysteresis of LCPCF,

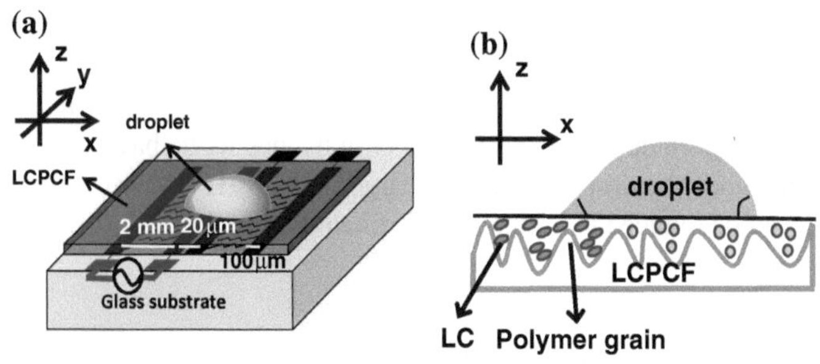

Fig. 15.4 **a** The droplet manipulation on LCPCF, and **b** the magnification of the surface of LCPCF near two adjacent electrode regions. At voltage-off, *rod-like* LC molecules anchored among the polymer grains are aligned along y-direction. At voltage-on states, LC molecules tilt up by the electric fields. The tilted LC molecules result in more hydrophilic surface properties of LCPCF [20]

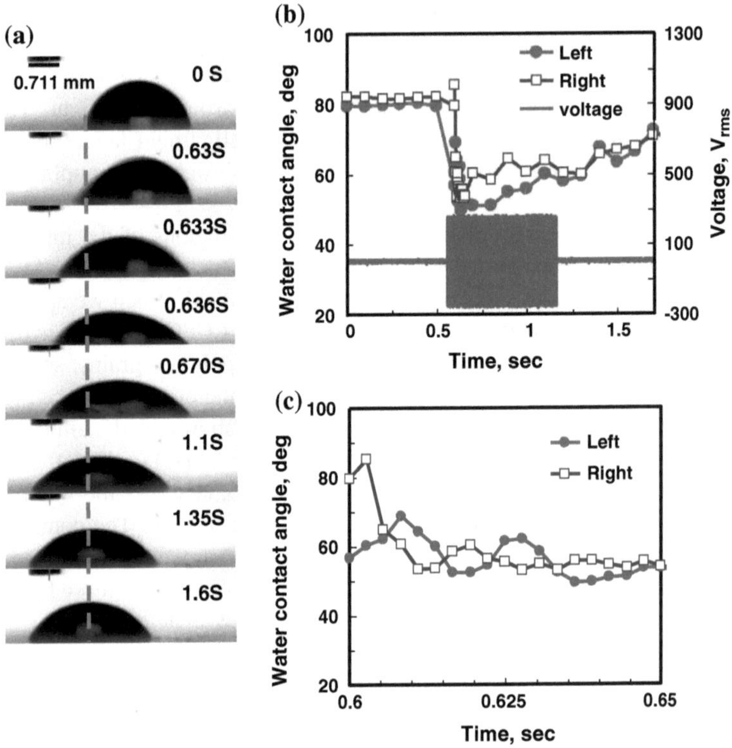

Fig. 15.5 **a** The photos of de-ionized water droplet manipulation at different times. **b** The water contact angle on the *left* and *right* of the droplet as a function of time under a *squared* pulsed voltage (250 V_{rms}) with 600 ms time duration. **c** The water contact angle as a function of time in (**b**) between 0.6 and 0.65 s. Droplet volume was ~3 μl

the difference between advancing (~64°) and receding angle (~52°), is ~12°. After 0.65 s, the water droplet slides toward the smaller contact angle on the left. At 1.2 s, the voltage is turned off and the droplet contracts, with high contact angles on both sides. The reason why we used periodic electric fields is mainly to overcome the LCPCF's hysteresis and keep the droplet moving forward. The net force is around 50–150 μN for droplet volume 1–9 μL. Gravity begins to have an effect on the water droplet when the droplet volume is larger than 6 μL. The high driving voltage (250 V_{rms}) can be reduced by reducing the thickness of LCPCF.

15.2 Sperm Testing Devices Using Droplet Manipulation

The great motivation in developing sperm testing devices is that I suffered from a difficulty for having a baby and the sperm testing device is an intermediate amazement during the process I was figuring out solutions. The quality of human spermatozoa, such as concentration, motility, and morphology, is one of the factors for infertility which is a global human issue. Figure 15.6 shows the microscopic image of human spermatozoa (or sperm). Most of men are not comfortable to do the sperm test at a hospital. Up to now, no handy testing devices on the market test sperm quality. Manipulating orientations of LC molecules can help to test sperm quality. When a water droplet is placed on LCPCF, the water droplet can be manipulated utilizing the surface wettability gradient induced by an inhomogeneous reorientation of LC molecules among polymer grains. The LCPCF can sense the semen through the motion of a semen droplet which in turn depends on the information of spermatozoa in the semen. We first demonstrated a sperm testing device by either a back-and-forth stretch or a collapse of a semen droplet on LCPCF [20].

The sperm sensing mechanism of LCPCF is depicted in Fig. 15.7. Based on the experimental results, two semen drop motions: the back-and-forth stretch (Figs. 15.7a–c) and the collapse (Figs. 15.7d–g) of semen drops, are related to the sperm quality. The semen drop simply consists of infertile sperms (gray), fertile sperms (black), and seminal plasma, a fluidic medium for sperms. In the beginning, the LC molecules anchored among the polymer grains are aligned along y-direction and all the sperms uniformly disperse in the semen drop (Fig. 15.7a). When we apply the electric field at the left region of LCPCF, the LC molecules tilt up. The cyano terminal group of LC molecules is more hydrophilic; as a result, the left side of the LCPCF is more hydrophilic as well. The semen drop experiences a net Young's force resulting from the change of the wettability. A fluidic flow (or the flow of seminal plasma) inside the semen drop is then induced. (The white arrow in Fig. 15.7b) All the sperms are then flushed by such a fluidic flow. However, the fertile sperms swim upstream against the fluidic flow because of the nature of the fertile sperms [23]. (Figure 15.7b) When we turn off the voltage, the LC molecules reorientate back along y-direction owing to the elastically essential properties of LC and the anchoring force of polymer grains. The wettability of LCPCF goes back. The fertile and infertile sperms then dispersed inside the semen drop (Fig. 15.7c). Therefore, we observe the back-and-forth stretch of a semen drop when the voltage is on and off periodically. The high concentration, high motility, better morphology, the high percentage of sperms swimming in a linear trajectory represent that more sperms can swim upstream against the

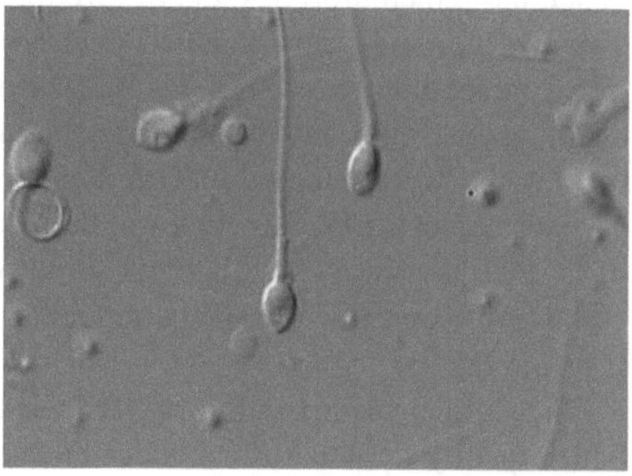

Fig. 15.6 Microscopic image of human spermatozoa (sperms)

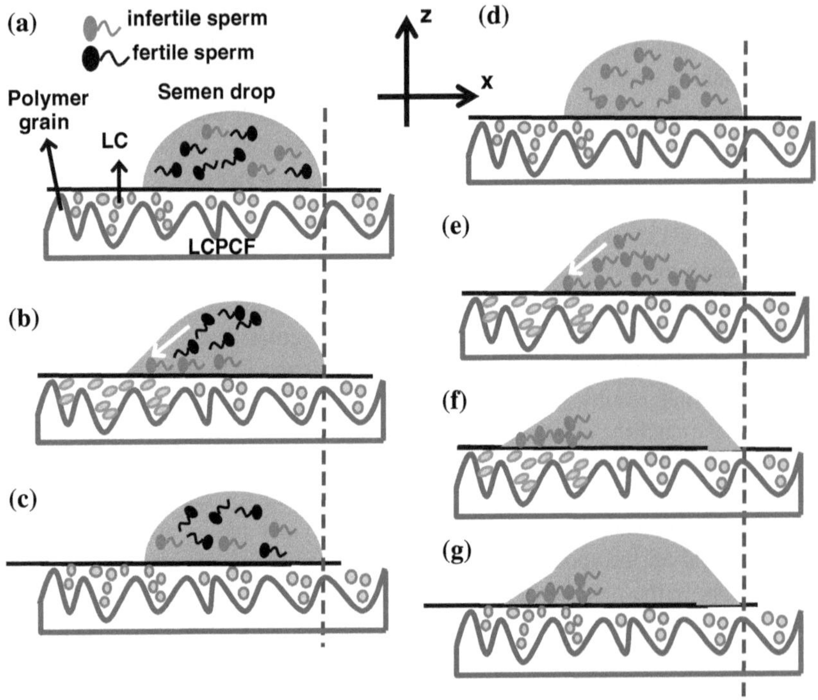

Fig. 15.7 **a**, **b**, **c** The schematic mechanism of stretches of a semen droplet. **a** At V = 0, LC molecules are aligned along y-direction. The fertile (*black*) and infertile (*gray*) sperms swim randomly. **b** When the voltage is applied to the *left* of LCPCF, the LC molecules tilt up and the *left* LCPCF is more hydrophilic. The induced fluidic flow (*white arrow*) washes away all the sperms meanwhile the fertile sperms swim upstream against the flow. **c** When the voltage turns off, the wettability of LCPCF is same as (**a**) and the semen drop stretches back. **d** At V = 0, the infertile sperms disperse in the semen drop. **e** When the voltage is applied to the *left* region of LCPCF, the induced fluidic flow (*white arrow*) washes away all the sperms. **f** and **g** The infertile sperms are deposited on the *left region* [20]

flow current of seminal plasma induced by the wettability gradient of LCPCF. The mechanism of the collapse of the semen drop is also illustrated in Figs. 15.7d–g. When the semen drop has lots of infertile sperms, the infertile sperms are washed away by the fluidic flow induced by the wettability gradient due to the orientation of LC molecules. (Figure 15.7e) Because the infertile sperms have weak ability to swim against the flow, the infertile sperms are attracted and trapped by the field of LCs and weak fringing electric fields on the surface of LCPCF. (Figure 15.7f) The surface of LCPCF is then re-modified. As a result, the switchable properties of LCPCF are invalid by the deposition of the sperms on LCPCF. Therefore, the collapse of semen drop and the aggregation near the voltage region are observed (Fig. 15.7g).

The droplet manipulation on LCPCF can realize a sperm testing device. The better quality spermatozoa results in back-and-forth stretches of a semen drop on LCPCF; otherwise, the semen drop collapses. From the experimental results, stretching of a semen droplet indicates the high concentration (>100 million/mL), good morphologies (>15 %), good motility (>50 %) of spermatozoa, and more than 30 % of spermatozoa swimming jointly forward. The stretching distance is linearly proportional to the motility of sperm. In fact, human fertilization involves not only sperm quality, but also sperm-oocyte processes. However, good sperm quality is still a significant factor for fertility potential. The experimental results indicate the great capabilities of LCPCF to be used as a sperm quality tester. Potential applications for this device include sperm testers and microfluidic devices for Assisted Reproductive Technology.

15.3 Sensing for High Density Lipoprotein in Human Serum

High-density lipoprotein (HDL) is one of the major carriers of cholesterol in the human blood for carrying cholesterol back from tissues/organs to the live. In the clinics, the high-density lipoprotein cholesterol (HDL-C), the cholesterol easters inside HDL, are usually measured to indicate the risk of diseases, such as cardiovascular diseases, breast and lung cancers, non-Hodgkin's lymphoma, and reducing the risk of Alzheimer's disease and dementia [24]. In fact, HDL concentration is proportional to a HDL-C concentration. Actually, HDL has high surface polarity compared to other particle existed in human serum because of special apolipoproteins [25]. LCPCF is suitable to sense polar fluids. By manipulating the orientations of LC molecules, the surface of LCPCF can be switched from less polar surface to polar surface when the biphenyl/terphenyl parts of LC molecules facing to LC/fluids interface flip to cyano (CN) group of LC molecules. The sensing mechanism of HDL is illustrated in Fig. 15.8 [26]. The LC molecules of LCPCF consist of three part: cyano(CN)-group (terminal group), alkyl-group (terminal group), and biphenyl/terphenyl part (core part). When a HDL drop is placed on LCPCF, the interface between HDL and LCPCF is illustrated in Figs. 15.8a, b. At a null voltage (i.e.V = 0), LC molecules are aligned along x-direction. The biphenyl/terphenyl of LC molecules contacts with the HDL/LC interface and LC molecules have a relative weak interaction with HDL compared to the case of the titling CN-group (i.e. the right figure in Fig. 15.8a) When the applied voltage V exceeds threshold voltage (V_{th}), the CN-group (or one of the terminal groups) of LC molecules tilts up toward the HDL because of the polar-polar interaction between HDL and LC molecules. In addition, the polarity of HDL concentration increases as the HDL concentration increases [25]. As a result, the biosensor is designed based on HDL droplet manipulation on LCPCF, as illustrated in Fig. 15.8b. A HDL drop is placed on LCPCF and the glass substrate is patterned by two regions of ITO electrodes. At V = 0, HDL drop is placed on the right region of electrodes. In the right figure in Fig. 15.8b, the HDL drop collapses to the left as we apply pulsed voltages on the left region of electrodes because the left region of LCPCF is more

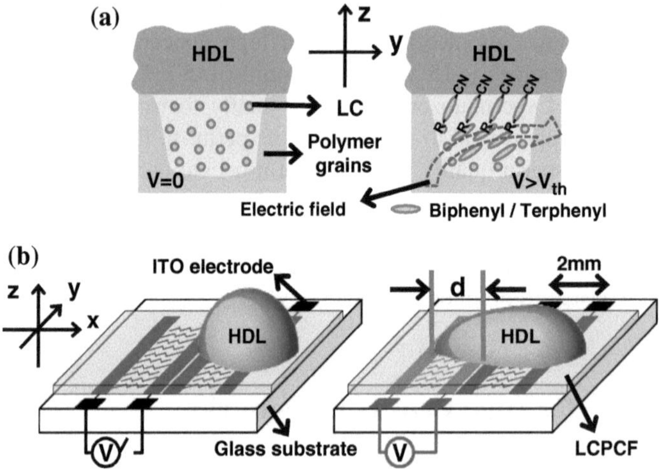

Fig. 15.8 **a** *Left* LC molecules anchored between polymer grains are along x direction at V = 0. *Right* LC molecules change the tilt angle due to anchoring force from polymer grains, the external electric field and interaction between HDL and LC when $V \gg V_{th}$. CN and R stand for the cyano group and the alkyl/alkoxy chain, respectively. **b** *Left* The HDL drop is placed on LCPCF without voltage. *Right* The voltage is applied at the *left region* of the electrodes. The HDL drop moves toward *left* and the collapse distance is d. The collapse distance depends on the concentration of HDL [26]

hydrophilic to HDL resulting from the tilting CN-groups of LC molecules in the left region of LCPCF. As a result, the HDL drop experiences a net Young's force to move the HDL drop. In addition, the viscosity of HDL drop is high and then the HDL drop collapses instead of doing the translation motion. When HDL concentration is higher, the polar-polar interaction at HDL/LC interface is larger and then the region applied electric fields on LCPCF is more hydrophilic. Thus, the collapse distance d is larger as HDL concentration is higher. By droplet manipulation on LCPCF, a biosensor for HDL can be realized.

In order to prove the LCPCF indeed can sense HDL of human serum, we collected the blood samples directly from human beings and removed the blood cells to obtain human serum. Human serums consist of water, proteins and lipoproteins. The lipoproteins also consists of many particles, such as chylomicron (CM), very low density lipoprotein (VLDL), low density lipoprotein (LDL), and high density lipoprotein (HDL). The parameters of the samples are pre-measured and we selected the samples of human serum with different HDL-C concentrations (i.e. the cholesterol easters inside HDL) ranging from 18 to 92 mg/dL, but other parameters are controlled in the relatively small variation range of individual parameter, such as such as LDL-C ranging from 103 to 153 mg/dL, triglyceride (TG) ranging from 102 to 165 mg/dL, total protein (i.e. Albumin + Globulin) ranging from 6.1 to 7.6 mg/dL, albumin ranging from 3.7 to 4.9 mg/dL, and A/G (i.e. the ratio of albumin to globin) ranging from 1 to 1.9. LCPCF is actually used

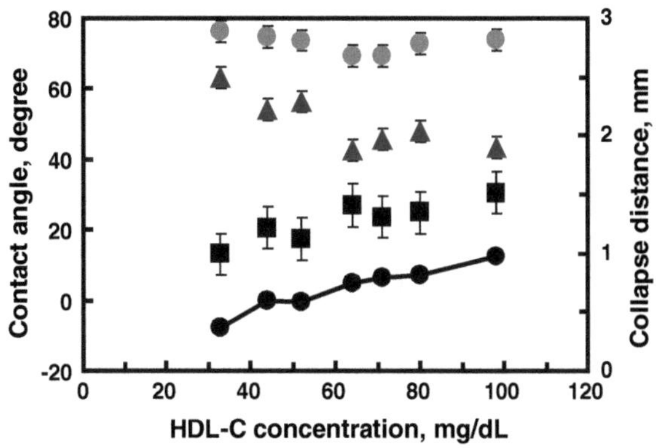

Fig. 15.9 The contact angle as a function of the HDL-C concentration in human serum at V = 0 V_{rms} (*grey dots*), at V = 200 V_{rms} (*triangles*). *Black squares* stand for the contact angle difference between 0 and 200 V_{rms}. The *solid dotted line* represents the collapse distance at different HDL-C concentrations in human serum. The concentration of HDL-C is proportional to the concentration of HDL in human serum

for sensing HDL in human serum, not HDL-C. Since we could not measure the HDL concentration directly, we used an indirect parameter, HDL-C concentration which can be measured in a standard process in clinics, to show the sensing effect of LCPCF for sensing HDL concentration. Then we performed the experiments of contact angles at V = 0 and V = 200 V_{rms}, and collapse distance at different HDL-C concentrations, as shown in Fig. 15.9. The contact angles at V = 0 and V = 200 V_{rms} decrease with HDL-C concentrations. The contact angle difference between 0 and 200 V_{rms} increases with HDL-C concentrations. The collapse distance increases with HDL-C concentration. This means LCPCF can sense HDL which embraces HDL-C. The polar HDL affects the orientations of LC near the interface. Actually, the collapse distance can also be affected by other particles of liproproteins, such as LDL, VLDL, and CM. We can also design sensing device for LDL, VLDL and CM as well as long as exhibit the polarity.

15.4 Ophthalmic Lenses Using Nematic Liquid Crystals

Human eyes are delicate imaging systems. The incident light passes through a cornea, a crystalline lens and retina of a human eye to form an image. The performances of eyes degrade due to refractive anomalies and then result in myopia, hypermetropia, and presbyopia. Myopia and hypermetropia mainly result from anomaly of the length of eyes or anomaly of focusing power of cornea and the crystalline lens. However, presbyopia mainly originates from the age-related degradation of the crystalline lens and then affects the eye's ability to focus, so-called amplitude of accommodation [27]. Declination of the amplitude of accommodation means that the difference between the farthest vision and the nearest vision is decreased or become shorter. Such an amplitude of accommodation decreases linearly with the age and then turns out static (around 0–2 D diopter, or m^{-1}) after age of 50 [28, 29]. The crystalline lenses of eyes are actually tunable-focusing lenses whose lens powers or focal lengths change with the curvature of the lens surfaces [30, 31]. By adopting an extra-artificially tunable focusing lens using LCs or a LC lens with a tunable accommodation, the degradation of the crystalline lens of eye can be compensated. In this way, the visional malfunction resulting from an aging crystalline lens can be corrected. Even though the natural accommodations of eyes disappear, elderly can still have their accommodations in vision. In addition, people who suffer from myopia and presbyopia need two pairs of eyeglasses. The solutions on the market are bifocal eyeglasses, two focuses in two areas of an ophthalmic lens, and progressive eyeglasses, multiple focuses in an ophthalmic lens. However, it is not easy for people to get used to bifocal or progressive eyeglasses. The tunable LC lenses possessing both adjustable positive and negative lens powers are suitable for the tunable eyeglasses of myopia-presbyopia.

The illustration of the image system of a human eye with an ophthalmic LC lens can be simplified as depicted in Fig. 15.10. The system consists of a LC lens (ophthalmic lens), a cornea, a crystalline lens and a retina as an image sensor. The LC lens and the crystalline lens of the eye are tunable-focusing lenses. On a basis of the image formations, the distance between the solid lens and object (d_o) can be expressed as [32]:

$$d_o = \frac{1}{P_{LC}(V) + S(P_{cryst})}, \tag{15.3}$$

where P_{LC} is the voltage(V)-dependent lens power of the LC lens, P_{cryst} is the lens power of the crystalline lens in the eye, and $S(P_{cryst})$ is the effective lens power of eyes which satisfies (15.4).

$$S(P_{cryst}) = \frac{-1}{d_g - \frac{1}{S'(P_{cryst})}}, \tag{15.4}$$

where d_g is the distance between the LC lens and cornea, and $S'(P_{cryst})$ satisfies (15.5).

$$S'(P_{cryst}) = P_c - \frac{1}{\frac{d_1}{n} - \frac{1}{P_{cryst} - \frac{n}{d_2}}}, \tag{15.5}$$

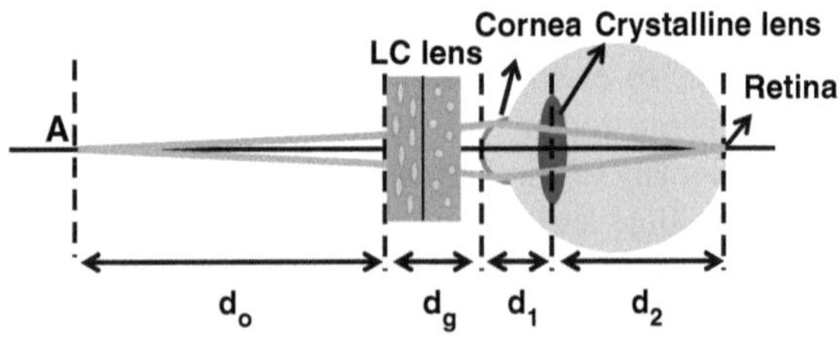

Fig. 15.10 Illustration of LC lens in a human eye image system [32]

where d_1 is the distance between the cornea and the lens in the eye, d_2 is the distance between the retina and the lens in the eye. n is the refractive index of the eye ball (~1.333 in average), and P_c is the lens power (or the inverse of the focal length) of the cornea (~42.735 m^{-1}) [31]. In (15.3), the objective distance (d_o) can be adjusted to see nearer or further by adjusting the lens power of crystalline lens and the lens power of LC lens. LC lens power could be positive and negative depending on the orientation of LC directors. Therefore, the LC lens could be an ophthalmic lens for myopia-presbyopia. The lens power of the LC lens ($P_{LC}(V)$) can be expressed as [33]:

$$P_{LC}(V) = \frac{2 \cdot \delta n(V) \cdot d_{LC}}{r^2}. \tag{15.6}$$

where d_{LC} is the thickness of the LC layer, r is the radius of the aperture of the LC lens, and $\delta n(V)$ is the difference of the refractive indices between the rim and the middle of the aperture. By design proper electrodes to generate an inhomogeneous electric field, the proper orientational distribution of LC directors results in lens-like phase profile or spatial optical phase difference of a lens. P_{LC} can also be positive or negative. The concept of wavefront conversion of LC lens is illustrated in Fig. 15.11. From optical theory, the idea is simple. However, to realize LC lenses as an ophthalmic lenses is not easy. The main challenges of the ophthalmic lenses using LC are: polarizer-free, large aperture (>20 mm), large tunable range of the lens power (i.e. an inverse of focal length), image quality, scattering-free, and aspherical phase profile.

A LC lens usually requires a polarizer which decreases the light efficiency at least of 50 %. To remove the requirement of a polarizer, a polarization independent LC phase modulation is required. Many polarization independent LC phase modulations are demonstrated and proposed [35–43]. Table 15.1 lists the performances of polarization-independent phase modulations in the literatures, such as phase shift, driving voltage and response time. Among the proposed polarization independent LC phase modulations, the double-layered type LC phase modulation with large phase shift is more suitable for designing ophthalmic lenses. The optical mechanism of the polarization independency of the double-layered type LC phase modulation is that two eigen-polarizations of incident light experience the same phase shift contributed by two orthogonal orientations of LC layers. In the double-layered type LC phase modulation, the phase shift is proportional to the

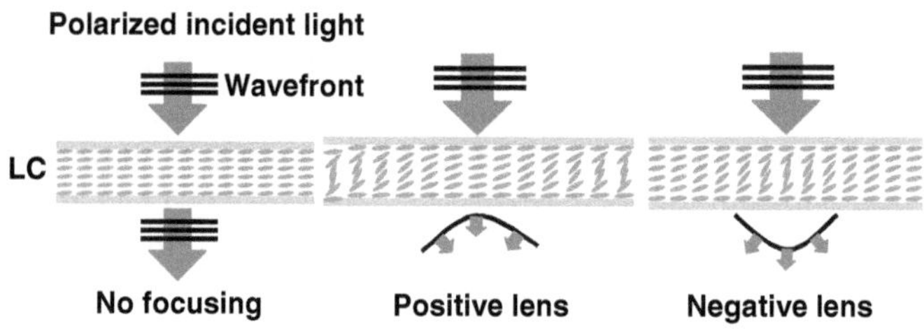

Fig. 15.11 Concept of wavefront conversion of LC lens [34]

Table 15.1 The list of performance of polarizer-free LC phase modulations [36–43]

Mode	Phase shift ($\Delta\varphi$), radians (π)	Driving voltage (V_{rms})	Response time	References
Double-layered LC	8.1	40	$\tau_{total} = 300$ ms	Ren and Wu [35]
Double-layered LC gel	1.1	180	$\tau_{rise} = 0.2$ ms $\tau_{decay} = 0.5$ ms	Lin et al. [37]
Homeotropic LC gel	0.08	180	$\tau_{rise} = 590$ μs $\tau_{decay} = 150$ μs	Ren et al. [38]
PDLC	0.09	60	$\tau_{rise} = 0.8$ ms $\tau_{decay} = 1.9$ ms	Ren et al. [39]
PSCT	0.025	160	$\tau_{rise} = 75$ μs $\tau_{decay} = 793$ μs	Ren et al. [40]
SP-PDLC	0.1	40	$\tau_{total} = 3.8$ ms	Lin et al. [41]
T-PNLC	0.28	30	$\tau_{total} = 1.6$ ms	Lin and Tsou [42]
PSBP-LC	1 π	150	$\tau_{total} = 3.0$ ms	Lin et al. [36]

birefringence of LC and the thickness of the LC layer. The double-layered type LC phase modulation shows lensing effect after we apply an inhomogeneous electric field to generate a parabolic phase profile for a good image quality.

For the large aperture size (>10 mm), the LC lenses on a basis of diffractive Fresnel lenses have been demonstrated in ophthalmic applications of presbyopia [44]. Such LC lenses require complex Fresnel electrodes and they have only two steps switches, on (focus) and off (no focus). As a result, delicate Fresnel patterns of such lenses need to be customized individually for different people. To further enlarge the aperture size (>20 mm), the limitation is fabrication process. The most important is that the image quality of any types of Fresnel lens can not be good enough for vision care. Without Fresnel lenses, it is still difficult to realize ophthalmic lenses using LC materials.

In order to maintain the uniform thickness of the LC layer for the consideration of uniform response time, the inhomogeneous electric fields must be applied to the LC layer. When the aperture size of the LC lens is large than 10 mm, the inhomogeneous electric fields of the hole-patterned electrode are difficult to apply to the LC layer for generating the parabolic phase profile (or spatial phase difference). In addition, when the tunable range of the lens power of the LC lenses is up to 10 D for the ophthalmic applications, the LC layer has to be thick (>1 mm). In this way, and the scattering is unavoidable and the response time is slow. By using LCPCF, we can separate the LC thickness into several thin layers as a multilayer structure. In this way, the aperture size can be enlarged to >10 mm and the phase increases with the number of layers. Figure 15.12 shows the proposed double-layered structure of polarizer-free LC lens [33, 45]. LCPCF is the uniaxial polymeric layer in Fig. 15.12. Without applied voltages, the lens power of the LC lens is zero. When the applied voltage $V_2 > V_1$, the LC lens is a negative lens because the LC directors around the rim of the aperture are parallel to the glass substrates and the LC directors inside the aperture are more perpendicular to the

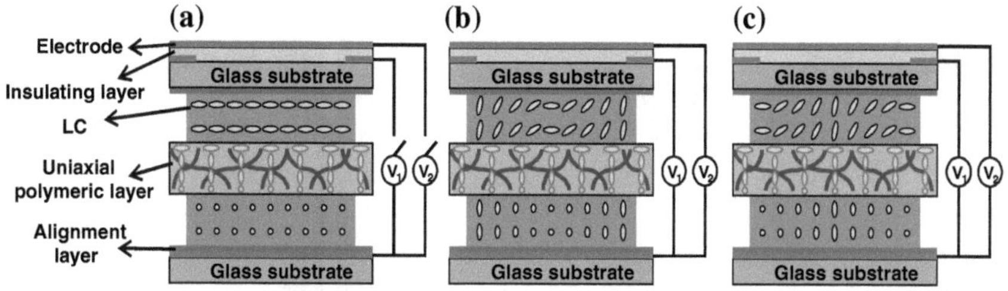

Fig. 15.12 The structure of the polarization independent LC lens. **a** When $V_1 = V_2 = 0$, the LC lens has no lensing effect. **b** When $V_1 > V_2$, the LC lens is a positive lens. **c** When $V_1 < V_2$, the LC lens is a negative lens [33]

glass substrates. As a result, an incident plane wave is converted to a diverging parabolic wave by the LC lens. Similarly, the LC lens is a positive lens when the applied voltage $V_2 < V_1$. An incident plane wave is then converted to a converging parabolic wave by the positive LC lens. By changing the magnitude of the voltages, the focal length or lens power of the LC lens is electrically switchable. The functions of LCPCF (the uniaxial polymeric layer) are cell separators and alignment layer. By proper design, LCPCF is optically isotropic and does not contribute an extra focusing property to the LC lens.

Figure 15.13a shows the image performance of the double-layered structure of polarizer-free LC lens with aperture size of 6 mm for the object charts placing at 50, 100, and 360 cm away from the LC lens. The thickness of two LC layers (Merck, MLC-2070) is 50 μm. The average lens power of the LC lens switches from −1.17 D (Diopter or m^{-1}) to +1.99 D, which means the total tunable lens power is 3.16 D for the positive lens. Without the LC lens power (i.e. $P_{LC} = 0$ D), the image system is set to mimic a myopic eye and the farthest point of vision is at 100 cm. When $P_{LC} = -1$ D, the farthest point the eye can see is extended at 360 cm. When $P_{LC} = +2$ D, the nearest point the eye can see can be closer to 50 cm. In fact, we are also developing the double layered LC lens with different aperture sizes of 6, 10 and 20 mm, as show in Fig. 15.13b. The tunable ranges of the lens powers of the LC lenses are: −1 to +2 D for the aperture of 6 mm, −2.5 to +3 D for the aperture of 10 mm, and −0.4 to +0.8 D for the aperture of 20 mm. The tunable range of the lens power of the LC lens can be improved by adding LC layers, increasing the thickness of the LC layer, and adopting LC materials with high birefringence. We are keeping working on polarizer-free LC lens with large aperture size and large tunable range of the lens power for ophthalmic lens applications.

Besides ophthalmic lens, the LC lenses have many applications, such as the imaging systems of portable devices, pico-projection systems, holographic projection systems, solar cells, ophthalmic lenses for myopia/presbyopia and endoscopic systems [33, 46–51]. In the imaging system, LC lenses can help to realize auto-focusing system and optical zoom system for the portable devices, such as cell phones and cameras [46]. As to the pico-projection systems, attached a LC lens to a pico-projection system can help to electrically adjust the focusing properties of the projected image without mechanically adjusting position of a projection lens

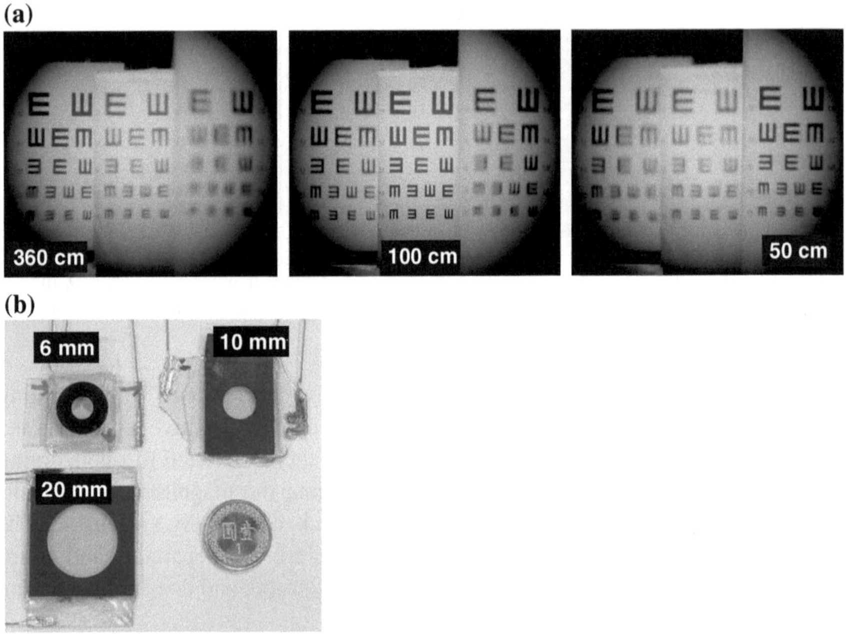

Fig. 15.13 The image performance of the double-layered LC lens. **a** The clear object is at 360 cm when $P_{LC} = -1$ D, at 100 cm when $P_{LC} = 0$ D, and at 50 cm when $P_{LC} = 2$ D. The aperture size is 6 mm. **b** The photo of the developed polarizer-free LC lenses with different aperture size [45]

[47]. For the holographic projection system, a LC lens can help to correct the mismatch of chromatic image size which is important for the full-color holographic projection system [49]. As to the endoscopic system, a LC lens can be adopted to electrically enlarge the depth-of-field of the endoscopic system [50]. LC lenses can also be used as a concentrator and a sun tracker in a concentrating photovoltaic (CPV) system to stabilize the output power of CPV system [48, 51].

An active eyewear device is a future trend. Vision care has great market in the world. LC is a great material to modulate light with mature development of technology. Up to now, LC lenses for ophthalmic application still have lots to improve for practical application. We can foresee that active eyewears based on LC ophthalmic lenses can be a platform for integrated multi-functions, such as function of displayed images, image captures, light shutters, attenuations, CPV system for charging, and optical zoom for superman's vision and so on.

References

1. D.K. Yang, S.T. Wu, *Fundamentals of Liquid Crystal Devices* (John Wiley, Chichester, 2006)
2. G.T. Stewart, Liquid crystals in biology I. Historical, biological and medical aspects. Liq. Cryst. **30**(5), 541–557 (2004)
3. G.T. Stewart, Liquid crystals in biology II. Origins and processes of life. Liq. Cryst. **31**(4), 443–471 (2004)

4. S.J. Woltman, G.P. Crawford, G.D. Jay, *Liquid Crystals: Frontiers in Biomedical Applications* (World Scientific, Hackensack, 2007)
5. S.J. Woltman, G.D. Jay, G.P. Crawford, Liquid-crystal materials find a new order in biomedical applications. Nat. Mater. **6**, 929–938 (2007)
6. A. Hussain, A.S. Pina, A.C.A. Roque, Bio-recognition and detection using liquid crystals. Biosens. Bioelectron. **25**, 1–8 (2009)
7. J.P.F. Lagerwall, G. Scalia, A new era for liquid crystal research: applications of liquid crystals in soft matter nano-, bio- and microtechnology. Curr. Appl. Phys. **12**, 1387–1412 (2012)
8. J.M. Brake, M.K. Daschner, Y.Y. Luk et al., Biomolecular interactions a phospholpid-decorated surfaces of liquid crystals. Science **302**, 2094–2097 (2003)
9. Y.Y. Luk, M.L. Tingey, K.A. Kickson et al., Imaging the binding ability of proteins immobilized on surfaces with different orientations by using liquid crystals. J. Am. Chem. Soc. **126**, 9024–9032 (2004)
10. M.L. Tingey, S. Wilyana, E.J. Snodgrass et al., Imaging of affinity microcontact printed proteins by using liquid crystals. Langmuir **20**, 6818–6826 (2004)
11. B.H. Clare, N.L. Abbott, Orientations of nematic liquid crystals on surfaces presenting controlled densities of peptides: amplification of protein-peptide binding events. Langmuir **21**, 6451–6461 (2005)
12. N.A. Lockwood, J.C. Mohr, L. Ji et al., Thermotropic liquid crystals as substrates for imaging the reorganization of matrigel by human embryonic stem cells. Adv. Funct. Mater. **16**, 618–624 (2006)
13. A.D. Price, D.K. Schwartz, DNA hybridization-induced reorientation of liquid crystal anchoring at the nematic liquid crystal/Aqueous interface. J. Am. Chem. Soc. **130**, 8188–8194 (2008)
14. S. Sivakumar, K.L. Wark, J.K. Gupta et al., Liquid crystal emulsions as the basis of biological sensors for the optical detection of bacteria and viruses. Adv. Funct. Mater. **19**, 2260–2265 (2009)
15. C.H. Chen, K.L. Yang, Detection and quantification of DNA adsorbed on solid surfaces by using liquid crystals. Langmuir **26**(3), 1427–1430 (2010)
16. Y.H. Lin, H. Ren, Y.H. Wu et al., Electrically tunable wettability of liquid crystal/polymer composite films. Opt. Express **16**(22), 17591–17598 (2008)
17. Y.P. Chiu, C.Y. Shen, Y.H. Lin, Characteristics of electrically switchable wettability surfaces of liquid crystal and polymer composite films. Jpn. J. Appl. Phys. **49**, 071604 (2010)
18. Y.P. Chiu, C.Y. Shen, W.C. Wang, Electrically surface-driven switchable wettability of liquid crystal/polymercomposite film. Appl. Phys. Lett. **96**, 131902 (2010)
19. Y.H. Lin, J.K. Li, T.Y. Chu et al., A bistable polarizer-free electro-optical switch using a droplet manipulation on a liquid crystal and polymer composite film. Opt. Express **18**(8), 10104–10111 (2010)
20. Y.H. Lin, T.Y. Chu, W.L. Chu et al., A sperm testing device on a liquid crystal and polymer composite film. J. Nanomedic. Nanotechnol. **S9**, 001 (2011)
21. Y.H. Lin, T.Y. Chu, Y.S. Tsou et al., An electrically switchable surface free energy on a liquid crystal and polymer composite film. Appl. Phys. Lett. **101**, 233502 (2012)
22. P.G. De Gennes, F.B. Wyart, D. Quere, *Capillarity and Wetting Phenomena Drops, Bubbles, Pearls, Waves* (Springer, Berlin, 2004)
23. D. Seo, Y. Agca, Z. Feng, J. Critser, Development of sorting, aligning, and orienting motile sperm using microfluidic device operated by hydrostatic pressure. Microfluid. Nanofluid. **3**, 561–570 (2007)
24. D. McGrowder, C. Riley, E.Y.S.A. Morrison et al., The role of high-density lipoproteins in reducing the risk of vascular diseases, neurogenerative disorders, and cancer. Cholesterol **2011**, 496925 (2011)
25. I.D. Cruzadoa, S. Songa, S.F. Crouseb et al., Characteriazation and quantification of the apoproteins of high-density lipoprotein by capillary electrophoresis. Anal. Biochem. **243**, 100–109 (1996)
26. Y.H. Lin, K.H. Chang, W.L. Chu, A biosensor of high-density lipoprotein of human serum on a liquid crystal and polymer composite film. Proc. SPIE **8828**, 88280I (2013)

27. L. Wemer, F. Trindade, F. Pereira et al., Physiology of Accommodation and presbyopia. Arq. Bras. Oftalmol. **63**, 503–509 (2000)
28. E. Hermans, M. Dubbelman, R.V.D. Heijde RVD et al., The shape of the human lens nucleus with accommodation. J. Vis. **7**(10), 16.1–16.10 (2007)
29. G.Y. Yoon, D.R. Williams, Visual performance after correcting the monochromatic and chromatic aberrations of the eye. J. Opt. Soc. Am. **19**, 266–275 (2002)
30. M. Jalie, *Opthalmic lenses and dispensing*, 3rd edn. (Elsevier Health Sciences, London, 2008)
31. D.A. Atchison, G. Smith, in *OPTICS of the HUMAN EYES*. (Elsevier Science Limited, Amsterdam, 2002)
32. Y.H. Lin, H.S. Chen, Electrically tunable-focusing and polarizer-free liquid crystal lenses for ophthalmic applications. Opt. Express **21**(8), 9428–9436 (2013)
33. H.C. Lin, M.S. Chen, Y.H. Lin, A review of electrically tunable focusing liquid crystal lenses. Trans. Electr. Electron. Mater. **12**, 234–240 (2011)
34. M. Ye, S. Sato, Optical properties of liquid crystal lens of any size. Jpn. J. Appl. Phys. **41**, L571–L573 (2002)
35. H. Ren, S.T. Wu, *Introduction to Adaptive Lenses* (Wiley, Hoboken, 2012)
36. Y.H. Lin, H. Ren, Y.H. Wu et al., Polarization-independent liquid crystal phase modulator using a thin polymer-separated double-layered structure. Opt. Express **13**(22), 8746–8752 (2005)
37. Y.H. Lin, H.S. Chen, H.C. Lin et al., Polarizer-free and fast response microlens arrays using polymer-stabilized blue phase liquid crystals. Appl. Phys. Lett. **96**(11), 113505 (2010)
38. H. Ren, Y.H. Lin, S.T. Wu, Polarization independent and fast-response phase modulators using double-layered liquid crystal gels. Appl. Phys. Lett. **88**(6), 061123 (2006)
39. H. Ren, Y.H. Lin, C.H. Wen, S.T. Wu, Polarization-independent phase modulation of a homeotropic liquid crystal gel. Appl. Phys. Lett. **87**(19), 191106 (2005)
40. H. Ren, Y.H. Lin, Y.H. Fan et al., Polarization-independent phase modulation using a polymer-dispersed liquid crystal. Appl. Phys. Lett. **86**(14), 141110 (2005)
41. Y.H. Lin, H. Ren, Y.H. Fan et al., Polarization-independent and fast-response phase modulation using a normal-mode polymer-stabilized cholesteric texture. J. Appl. Phys. **98**, 043112 (2005)
42. Y.H. Lin, Y.S. Tsou, A polarization independent liquid crystal phase modulation adopting surface pinning effect of polymer dispersed liquid crystals. J. Appl. Phys. **110**, 114516 (2011)
43. Y.H. Lin, M.S. Chen, W.C. Lin et al., A polarization-independent liquid crystal phase modulation using polymer-network liquid crystals in a 90° twisted cell. J. Appl. Phys. **112**, 024505 (2012)
44. G.Q. Li, D.L. Mathine, P. Valley et al., Switchable electro-optic diffractive lens with high efficiency for ophthalmic applications. Proc. Natl. Acad. Sci. U.S.A. **103**(16), 6100–6104 (2006)
45. H.S. Chen, M.S. Chen, Y.H. Lin, in Electrically tunable ophthalmic lenses for myopia and presbyopia using liquid crystals. In: *Optics of Liquid Crystals*. (Honolulu, Hawaii, 29 Sept–4 Oct 2013
46. H.C. Lin, Y.H. Lin, A fast response and large electrically tunable-focusing imaging system based on switching of two modes of a liquid crystal lens. Appl. Phys. Lett. **97**, 063505 (2010)
47. Y.H. Lin, M.S. Chen, A pico projection system with electrically tunable optical zoom ratio adopting two liquid crystal lenses. J Disp. Technol. **8**, 401–404 (2012)
48. Y.S. Tsou, Y.H. Lin, A.C. Wei, Concentrating photovoltaic system using a liquid crystal lens. IEEE Photon. Technol. Lett. **24**(24), 2239–2242 (2012)
49. N. Collings, Y.H. Lin, H.C. Lin et al., Tunable liquid crystal lens for a holographic projection system. Proc. SPIE **8828**, 88281B (2013)
50. Y.H. Lin, H.S. Chen, An endoscopic system adopting a liquid crystal lens with an electrically tunable depth-of-field. Opt. Express **21**(15), 18079–18088 (2013)
51. Y.S. Tsou, K.H. Chang, Y.H. Lin, A droplet manipulation on a liquid crystal and polymer composite film as a concentrator and a sun tracker for a concentrating photovoltaic system. J. Appl. Phys. **113**, 244504 (2013)

Part VII
Advanced Trends of Nanophotonics

Wei Ting Chen and Din Ping Tsai

Introduction

Nanophotonics is the study of the behavior of light-matter interaction at the nanometer scale. By adding the dimensions of optical devices and components to sub-wavelength scale, nanophotonics provides new opportunities for fundamental science and practical applications. One of the goals of nanophotonics development is to manipulate light at the nanoscale, which may not be limited by the chemical composition of natural materials and the diffraction limit of electromagnetic wave. Nanophotonics has several advantages with such diffraction-unlimited properties for functional applications: (i) nanoscale footprints-smaller components and devices; (ii) photon-electron process in nanoscale—faster processing speed, and (iii) nanoscale confinement of optical radiation and electromagnetic fields—enhancing the light-matter interactions and dramatically reducing the optical energy consumption. The characterization of drastic optical localization within such components strongly enhances the typically weak interaction between light and matter, which increases the energy efficiency to obtain desired effects and phenomena. This chapter covers two major parts of the latest trends of nanophotonics, plasmonics, and metamaterials. Several cutting-edge approaches harvested from the extraordinary properties of nanophotonics, which are conducted to advanced trends relate to: Micro/nano-lasers with the smallest plasmonic nanolaser, theoretical models of the micro/nano-cavity, and semiconductor micro-lasers with tuning functions on a flexible substrate (Chap. 16, 17 and 19), nanostructures light-emitting diode (LED) with better light extraction and reduced piezoelectric field induced by strain (Chap. 18, 24), one-dimensional photonic crystal nanowires with small footprints and ultrahigh Q-factors (Chap. 20), nano-structured waveguides with slow light effect (Chap. 21), negative refraction index generated by glancing angle deposition (Chap. 22), improving the light harvest of solar cell with anti-reflective nanostructures (Chap. 23), high-sensitive plasmonic biosensors with Fano-like resonance (Chap. 25) are addressed.

Plasmonics, the coherent electrons oscillation of noble materials driven by photons, has abilities to confine electromagnetic field to be smaller than the wavelength of incident electromagnetic wave. There are two branches of plasmonics: "surface plasmon polaritions" (SPP) and "localized surface plasmons" (LSP). Surface plasmon polaritions are supported by metallic thin film, and can be usually excited by either grating coupling or total internal reflection to provide additional wave vector to match the phase difference between free space propagating wave and surface plasmon polaritions. On the contrary, the localized surface plasmons of a given metallic nanoparticles can be directly excited by free space propagating wave, and its resonance frequency can be tailored by its geometrical dimensions and the refraction index of environment as well as polarization state of incident electromagnetic wave. At both surface plasmon and localized surface plasmon resonance, the electromagnetic field is strongly enhanced in the immediate vicinity, and therefore the light-matter interaction is enhanced. As a result, plasmonics show a wide range of potential applications—nanocavity, high-sensitive bio-sensing, LED, nano-laser, etc.

Artificial materials with sub-wavelength structure, which are so-called metamaterials, have attracted a lot of attention. The optical properties of metamaterials are determined by their artificial structures rather than their material composition. The Greek word "meta" is translated as "beyond," which means that the central concept is to construct new materials with those optical properties that are not found or hardly observed in nature. For instance, the negative refraction is the typical example of metamaterials that reword the formula of Snell's law. Moreover, the function of metamaterials as artificial atoms or molecules provides an entirely new route to further enhance the capability to design and create novel material properties. While through the near-field interaction between metamaterials, they can offer and generate more fascinating physical and optical properties unavailable in nature or chemically synthesized materials. Metamaterials therefore enable us to tailor the propagation of electromagnetic wave, even more transformation optics and optical cloaking. Till date, the research agenda on metamaterials is shifting from fundamental researches to functionalities for practical applications, such as tunable, switchable, and biosensing devices. It will also pave a new way to integrate photonics with electronics in not only telecommunication systems but also opto-electron circuit devices and applications.

Finally, I hope this chapter proves to be a useful guideline for both current and future researchers, and inspires people toward cutting-edge breakthroughs in the field of nanophotonics. Any comment and suggestion related to this chapter is highly appreciated.

Chapter 16
Diffraction-Unlimited Plasmonic Nanolaser

Yu-Jung Lu, Jisun Kim, Hung-Ying Chen, Chihhui Wu, Nima Dabidian, Charlotte E. Sanders, Chun-Yuan Wang, Ming-Yen Lu, Bo-Hong Li, Xianggang Qiu, Wen-Hao Chang, Lih-Juann Chen, Gennady Shvets, Chih-Kang Shih and Shangjr Gwo

Symbols

λ Wavelength
n Effective index

16.1 Main Text

In 2003, it was proposed that utilizing plasmonic cavities can break this limit—a new class of lasers based on surface plasmon amplification by stimulated emission of radiation (spaser) has been proposed by Bergman and Stockman [2]. In such cavities, surface plasmon polaritons (SPPs) excited in noble metal structures

Y.-J. Lu · H.-Y. Chen · C.-Y. Wang · S. Gwo (✉)
Department of Physics, National Tsing-Hua University, Hsinchu 30013, Taiwan
e-mail: gwo@phys.nthu.edu.tw

J. Kim · C. Wu · N. Dabidian · C.E. Sanders · G. Shvets · C.-K. Shih
Department of Physics, The University of Texas at Austin, Austin, TX 78712, USA

C.-K Shih
e-mail: shih@physics.utexas.edu

M.-Y. Lu · L.-J. Chen
Department of Materials Science and Engineering, National Tsing-Hua University, Hsinchu 30013, Taiwan

B.-H. Li · X. Qiu
Beijing National Laboratory for Condensed Matter Physics and Institute of Physics, Chinese Academy of Sciences, Beijing 100190, China

W.-H. Chang
Department of Electrophysics, National Chiao-Tung University, Hsinchu 30010, Taiwan

C.-C. Lee (ed.), *The Current Trends of Optics and Photonics*,
Topics in Applied Physics 129, DOI 10.1007/978-94-017-9392-6_16

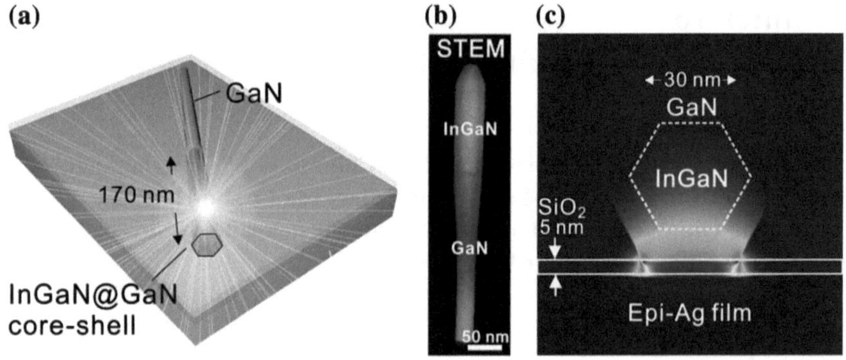

Fig. 16.1 **a** Schematic of metal-oxide-semiconductor (MOS) structure plasmonic laser, contain a single InGaN@GaN core-shell nanorod on a SiO_2-covered epitaxial Ag film (28 nm). **b** Scanning transmission electron microscopy (STEM) image of a single-crystalline InGaN@GaN core-shell nanorod. **c** The calculated energy-density distribution, the sandwich structure is utilized to confine SPPs within the low-*k* dielectric nanogap [1]

adjacent to gain media shrink the optical mode volume and provide the necessary feedback mechanism for a spaser. However, it remains to be seen whether one indeed can overcome the high losses in three-dimensional (3D)-confined, deep-subwavelength plasmonic cavities with the currently available noble metals and semiconductor gain materials (Fig. 16.1).

In this work, an important breakthrough is reported in the growth of atomically smooth, epitaxial Ag films on Si substrates, which can be used as the platform material to fabricate low-loss plasmonic cavities due to their superior material properties. In particular, by using single, shape-controlled InGaN/GaN core–shell nanorods via plasma-assisted molecular beam epitaxy (PAMBE) growth as the gain media with a large gain coefficient in the green spectral region, we were able to realize diffraction-unlimited nanolasers that can be operated under continuous-wave (CW) conditions with an ultralow threshold power above liquid nitrogen temperature. Nearly 100 % polarized lasing mode was experimentally observed. According to the numerical simulations, a large proportion of energy emissions are transformed into in-plane directional and coherent surface plasmon in the near filed (a spaser) which is in agreement with the experimental results of measured temporal coherence signature. These nanolasers, with record-small cavity and mode volumes are orders of magnitude smaller than the 3D diffraction limit and the feature sizes are comparable with that of the state-of-the-art metal-oxide-semiconductor (MOS) transistors in nanoelectronics (Fig. 16.2).

Size mismatches between electronics and photonics have been a huge barrier to realize on-chip optical communications and computing systems. The recent surge of research interest in nanoplasmonics has been largely due to its capability to break the diffraction limit. However, the lossy nature of conventional plasmonic

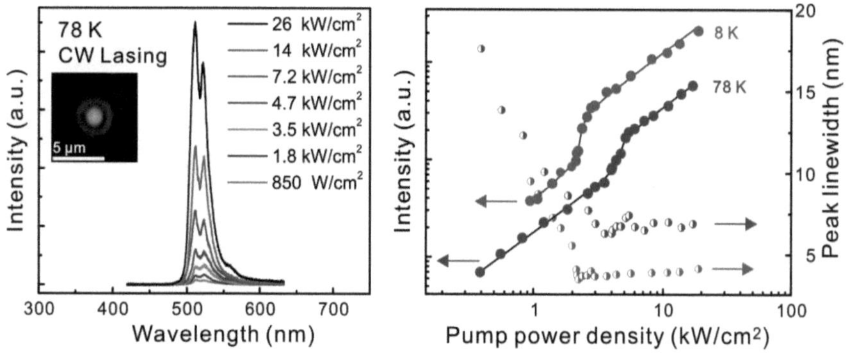

Fig. 16.2 CW lasing spectra from a single nanorod at 78 K under varying optical power densities excited by a CW semiconductor diode laser at 405 nm. It also revealed the temperature dependent lasing signatures, including the concurrent lasing thresholds of output intensity kink and spectra linewidth narrowing

materials requires a remedy to overcome this difficulty. Most plasmonic devices are based on granular polycrystalline Ag films where surface roughness and grain boundaries lead to scattering losses of surface plasmons [3]. This means that atomically smooth plasmonic structures with single-crystalline structures are highly desirable building blocks for low-loss plasmonic applications. On the other hand, the growth of high-quality nitride semiconductor nanorods by using PAMBE [4, 5] offers several advantages, including broadband-tunable light emissions in the full visible region and amphoteric doping [5] (*n*- and *p*-type GaN) for electrically injected devices.

In summary, the present all-epitaxial approach opens a scalable platform for low-loss, active nanoplasmonics. In particular, the nitride semiconductor-based plasmonic nanolasers are expected to be compatible with the existing multi-functional on-chip devices (e.g., integrating these nanolasers with state-of-the-art CMOS electronics) or future integrated plasmonic circuits for advanced applications.

References

1. Y.-J. Lu et al., Plasmonic nanolaser using epitaxially grown silver film. Science **337**, 450 (2012)
2. D.J. Bergman, M.I. Stockman, Surface plasmon amplification by stimulated emission of radiation: quantum generation of coherent surface plasmons in nanosystems. Phys. Rev. Lett. **90**, 027402 (2003)
3. R.F. Oulton et al., Plasmon lasers at deep subwavelength scale. Nature **461**, 629 (2009)
4. C.-Y. Wu et al., Plasmonic green nanolaser based on a metal–oxide–semiconductor structure. Nano Lett. **11**, 4256 (2011)
5. Y.-J. Lu, H.-W. Lin, H.-Y. Chen, Y.-C. Yang, S. Gwo, Single InGaN nanodisk light emitting diodes as full-color subwavelength light sources. Appl. Phys. Lett. **98**, 233101 (2011)

Chapter 17
Modeling of Micro and Nanolaser Cavities

Shu-Wei Chang

Symbols

Ω	Computation domain
Ω_a	Active region
$\mathbf{r}$	Position vector
$\mathbf{E}(\mathbf{r})$	Electric field
$\mathbf{J}_s(\mathbf{r})$	Current-density source
ω	Oscillation frequency of the field and source
c	Speed of light in vacuum
ε_0	Vacuum permittivity
μ_0	Vacuum permeability
$\overline{\overline{\varepsilon_r}}(\mathbf{r}, \omega)$	Relative permittivity tensor
n	Label of basis modes
$\mathbf{f}_n(\mathbf{r}, \omega)$	Modal profile of basis mode n
$\mathbf{j}_{s,n}(\mathbf{r}, \omega)$	Current-density source of basis mode n
$\Delta\varepsilon_{r,n}(\omega)$	Permittivity variation related to basis mode n
$U(\mathbf{r})$	Indicator function of the active region
ω_n	Natural oscillation frequency of basis mode n
V_a	Volume of the active region Ω_a
$P(\omega)$	White-source power spectrum
FWHM	Full width at half maximum
$\Delta\omega_n$	Full-width-at-half-maximum linewidth of basis mode n

S.-W. Chang (✉)
Research Center for Applied Sciences, Academia Sinica, 128,
Section 2, Academia Rd., Nankang, Taipei 11529, Taiwan
e-mail: swchang@sinica.edu.tw

S.-W. Chang
Department of Photonics, National Chiao-Tung University, 1001,
University Rd., Hsinchu 30010, Taiwan

C.-C. Lee (ed.), *The Current Trends of Optics and Photonics*,
Topics in Applied Physics 129, DOI 10.1007/978-94-017-9392-6_17

Q_n	Quality factor of basis mode n
$g_{\mathrm{th},n}$	Threshold gain of basis mode n
G_n	Multiplication rate of basis mode n
$\tau_{\mathrm{p},n}$	Photo lifetime of basis mode n
$\Gamma_{\mathrm{E},n}$	Energy confinement factor of mode n
$v_{\mathrm{g,a}}(\omega_n)$	Group velocity in the active material
$n_{\mathrm{g,a}}(\omega_n)$	Group index of the active material
$n_{\mathrm{a}}(\omega)$	Refractive index of the active material
$\varepsilon_{\mathrm{g,a}}(\omega)$	Group permittivity of the active material
$\Gamma_{\mathrm{E},n}^{(\mathrm{old})}(\Omega)$	Conventional energy confinement factor
$\overline{\overline{\varepsilon}}_{\mathrm{g}}(\mathbf{r},\omega)$	Group permittivity tensor
$V_{\mathrm{eff},n}$	Effective mode volume in the rate equation
$V_{\mathrm{QM},n}$	Effective mode volume for the Purcell effect and cavity quantum electrodynamics
WGM	Whispering gallery mode
TE_r	Radially transverse-electric
R	Radius of the dielectric sphere
L	Angular momentum quantum number of whispering gallery modes
M	Magnetic quantum number of whispering gallery modes
θ	Polar angle
ϕ	Azimuthal angle
$Y_{LM}(\theta,\phi)$	Spherical harmonic
$f_L(r)$	Radial modal profile
k_0	Propagation constant in vacuum
k_{a}	Propagation constant in the active sphere
$h_L^{(1)}(k_0 r)$	Spherical Hankel function of the first kind
$j_L(k_{\mathrm{a}} r)$	Spherical Bessel function of the first kind
$\Delta\varepsilon_{\mathrm{r},L}(\omega)$	Permittivity variation of the whispering gallery mode
$\hbar\omega_L$	Resonance photon energy of the whispering gallery mode
Q_L	Quality factor of the whispering gallery mode
$Q_{\mathrm{rad},L}$	Quality factor of the whispering gallery mode due to radiation loss
$Q_{\mathrm{mat},L}$	Quality factor of the whispering gallery mode due to material loss
$g_{\mathrm{th},L}$	Threshold gain of the whispering gallery mode
$\Gamma_{\mathrm{E},L}$	Energy confinement factor of the whispering gallery mode
$V_{\mathrm{eff},L}$	Effective modal volume of the whispering gallery mode
$V_{\mathrm{QM},L}$	Quantum mechanical effective modal volume of the whispering gallery mode
R_{b}	Integration radius of the spherical computation domain

17.1 Introduction

The laser cavity, in addition to the gain medium, plays an important role in active photonic devices. While the intrinsic photon emissions and multiplications originate from gain materials, the spectral (resonance) properties, modal characteristics, and couplings (communication) to other optical components, which can be systematically refined or engineered for specific purposes, are determined by laser cavities. Proper cavity designs usually boost up the power efficiency of lasers and bring about various functionalities aiming at different goals. Such requirements/concerns are commonly encountered in practical applications of quantum electronics and active optoelectronic/photonic devices. In views of the small laser devices nowadays, these issues will become even more critical as laser cavities shrink down to the micro and nanoscale [1–13] since device characteristics become quite sensitive to the size and geometry variations of cavities.

To understand a laser and its basic features before the real device fabrication, some calculations are often helpful in the evaluations and optimizations of its performance under the existing fabrication constraints and system specifications. A typical computation tool adopted in investigations of laser cavities is the finite-difference-time-domain (FDTD) method [14–17]. In this scheme, the electric and magnetic fields are solved by directly propagating them through the time-dependent Maxwell's equations. The advantages of this approach are the easiness of numerical implementations, small memory usages, and simultaneous searches of spectral resonances (cavity modes). Its disadvantages, however, include the stability issue related to sizes of time steps [15], source-dependent outcomes, long computation times for modes with high quality (Q) factors, being less capable of dispersions beyond Debye or Lorentz-Drude types [18–22], and so on. In addition, for some device parameters such as confinement factors, threshold gain, and modal volumes, which are indicative to the performance of ultrasmall active photonic devices, the FDTD method cannot provide them directly. Usually, some post processings or numerical estimations using the evolved fields have to be performed in order to obtain these parameters [23, 24], which turn out to be unreliable in quite a few cases of ultrasmall lasers.

In this contributed chapter, we will introduce an alternative computation scheme aimed at laser cavities [25]. The approach originates from viewpoints of gain and lasing of cavity modes in Maxwell's equations, namely, active and self-induced oscillations rather than passive resonances. The formulation is based on a generalized eigenvalue (GE) problem for lasing mode [25–30], in which the eigenvalue is the amount of permittivity variation. Contrary to time evolutions based on the FDTD method, the memory usage in this type of GE problems is the main concern. However, this approach does have its own advantages. Since the GE problem is carried out in the frequency domain, in principle, any frequency dispersions that obey the causality but are not necessarily limited to the Debye or Lorentz-Drude types can be incorporated in the formalism. In addition, except for the intrinsic degeneracy of various cavity modes, the GE problem does not mix different categories of modes together, and hence the source-dependent phenomena in the FDTD method is prevented. This approach has been applied and compared to real experimental data [31, 32] with decent agreements. In the following, we shall also show that merely from the information of the eigenvalue, a few important parameters of laser cavities can be directly evaluated rather than ambiguously estimated [33].

17.2 Formulation Based on Generalized Eigenvalue Problem

In Fig. 17.1, we show the computation domain of a generic laser cavity and denote it as Ω. The cavity contains a so-called active region Ω_a, in which the photon amplification due to the populated-inverted gain medium takes place. Since the coherent radiation of the laser can be effectively thought of as being generated by a finitely-sized source, the fields far away from the cavity have to be outgoing waves. Therefore, in practical computations, perfectly-matched layers (PMLs) are usually utilized to avoid artificial reflections from the truncated computation domain.

We now consider the frequency-domain wave equation of the electric field $\mathbf{E}(\mathbf{r})$ in the presence of a source $\mathbf{J}_s(\mathbf{r})$ which oscillates at a frequency ω and only exists in the active region Ω_a:

$$\nabla \times \nabla \times \mathbf{E}(\mathbf{r}) - \left(\frac{\omega}{c}\right)^2 \overline{\overline{\varepsilon}}_r(\mathbf{r}, \omega)\mathbf{E}(\mathbf{r}) = i\omega\mu_0 \mathbf{J}_s(\mathbf{r}), \tag{17.1a}$$

$$\mathbf{J}_s(\mathbf{r}) = \mathbf{0} \quad \forall\, r \notin \Omega_a. \tag{17.1b}$$

where c is the speed of light in vacuum; μ_0 is the vacuum permeability; and $\overline{\overline{\varepsilon}}_r(\mathbf{r}, \omega)$ is the relative permittivity tensor of the cold cavity, that is, no gain is present in Ω_a yet. To expand the source $\mathbf{J}_s(\mathbf{r})$ and field $\mathbf{E}(\mathbf{r})$ at least partially, we need some set of basis functions. For this purpose, we choose a simple linear relation for the mode profile $\mathbf{f}_n(\mathbf{r}, \omega)$ and associated current source $\mathbf{j}_{s,n}(\mathbf{r}, \omega)$ of a basis labeled by n:

$$\nabla \times \nabla \times \mathbf{f}_n(\mathbf{r}, \omega) - \left(\frac{\omega}{c}\right)^2 \overline{\overline{\varepsilon}}_r(\mathbf{r}, \omega)\mathbf{f}_n(\mathbf{r}, \omega) = i\omega\mu_0 \mathbf{j}_{s,n}(\mathbf{r}, \omega), \tag{17.2a}$$

$$\mathbf{j}_{s,n}(\mathbf{r}, \omega) = -i\omega\varepsilon_0 \Delta\varepsilon_{r,n}(\omega) U(\mathbf{r})\mathbf{f}_n(\mathbf{r}, \omega), \tag{17.2b}$$

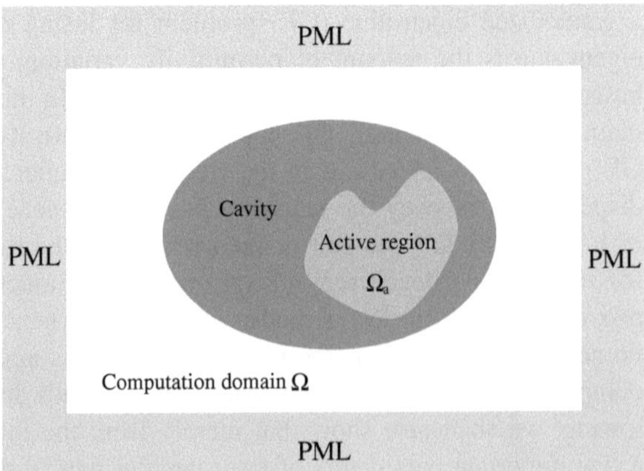

Fig. 17.1 The computation domain Ω of a generic laser cavity. The cavity contains an active region Ω_a. PMLs are inserted on the rims of Ω to avoid field reflections

where ε_0 is the vacuum permittivity; $\Delta\varepsilon_{r,n}(\omega)$ is a proportional constant at this stage; and $U(\mathbf{r})$ is an indicator function ensuring that the source is only present in Ω_a:

$$U(\mathbf{r}) = \begin{cases} 1, & \mathbf{r} \in \Omega_a, \\ 0, & \text{otherwise.} \end{cases} \tag{17.3}$$

After substituting (17.2b) into (17.2a), we obtain the following GE problem in the differential form:

$$\nabla \times \nabla \times \mathbf{f}_n(\mathbf{r},\omega) - \left(\frac{\omega}{c}\right)^2 \overline{\overline{\varepsilon}}_r(\mathbf{r},\omega)\mathbf{f}_n(\mathbf{r},\omega) = \left(\frac{\omega}{c}\right)^2 \Delta\varepsilon_{r,n}(\omega)U(\mathbf{r})\mathbf{f}_n(\mathbf{r},\omega), \tag{17.4}$$

where the parameter $\Delta\varepsilon_{r,n}(\omega)$ now plays the role of eigenvalues. The physical meaning of $\Delta\varepsilon_{r,n}(\omega)$ can be understood by moving the right-hand side (RHS) of (17.4) to its left-hand side (LHS). In this way, the parameter $\Delta\varepsilon_{r,n}(\omega)$ is interpreted as the amount of permittivity variation in Ω_a that is required for the self oscillation (lasing) of basis mode n at a given frequency ω. The real part $\mathrm{Re}[\Delta\varepsilon_{r,n}(\omega)]$ shifts the oscillation frequency of mode n to ω while the imaginary part $\mathrm{Im}[\Delta\varepsilon_{r,n}(\omega)]$, which is a *negative* number in passive cavities, characterizes the corresponding threshold gain at this frequency. We note that the frequency ω is not the natural resonance frequency ω_n of mode n. However, through the dependence of eigenvalue $\Delta\varepsilon_{r,n}(\omega)$ on ω, we can determine this frequency.

In the following sections, we will describe how to obtain useful information of mode n from the frequency-dependent parameter $\Delta\varepsilon_{r,n}(\omega)$.

17.3 White-Source Spectrum and Spectral Property

In addition to the interpretation of the required permittivity variation in the active region for self oscillations, the parameter $\Delta\varepsilon_{r,n}(\omega)$ is also connected to the source $\mathbf{j}_{s,n}(\mathbf{r}, \omega)$ through (17.2b). Alternatively, the mode $\mathbf{f}_n(\mathbf{r}, \omega)$ may be regarded as the response field to the driving source $\mathbf{j}_{s,n}(\mathbf{r}, \omega)$ at a frequency ω.

Suppose that a specific source $\mathbf{J}_s(\mathbf{r})$ has a spectrum proportional to $\mathbf{j}_{s,n}(\mathbf{r}, \omega)$ at all the positions in Ω_a, namely,

$$\mathbf{J}_s(\mathbf{r}) = a(\omega)\mathbf{j}_{s,n}(\mathbf{r},\omega), \tag{17.5}$$

where $a(\omega)$ is the strength at ω. Let us impose the *white-source* condition to $\mathbf{J}_s(\mathbf{r})$ by requiring that the integral of the square magnitude $|\mathbf{J}_s(\mathbf{r})|^2$ in Ω_a is frequency- independent:

$$\int_{\Omega_a} d\mathbf{r} |\mathbf{J}_s(\mathbf{r})|^2 = \vartheta^2 V_a, \tag{17.6}$$

where ϑ is an average strength; and V_a is the volume of the active region. We then substitute (17.2b) into (17.6) and obtain the following constraint

$$|a(\omega)|^2 = \frac{\vartheta^2 V_a}{\varepsilon_0^2 \omega^2 \left|\Delta\varepsilon_{r,n}(\omega)\right|^2} \left[\int_{\Omega_a} d\mathbf{r} |\mathbf{f}_n(\mathbf{r}, \omega)|^2 \right]^{-1}. \quad (17.7)$$

Using this relation and (17.2b) again, the power spectrum $P(\omega)$ generated by the white source $\mathbf{J}_s(\mathbf{r})$ has a simple form as follows:

$$P(\omega) = -\frac{1}{2} \int_{\Omega_a} d\mathbf{r} \mathrm{Re}\left[\mathbf{E}(\mathbf{r}) \cdot \mathbf{J}_s^*(\mathbf{r})\right] = -\frac{|a(\omega)|^2}{2} \int_{\Omega_a} d\mathbf{r} \mathrm{Re}\left[\mathbf{f}_n(\mathbf{r}, \omega) \cdot \mathbf{j}_{s,n}^*(\mathbf{r}, \omega)\right]$$
$$= \frac{\vartheta^2 V_a}{2\varepsilon_0} \mathrm{Im}\left[\frac{1}{\omega \Delta\varepsilon_{r,n}(\omega)}\right]. \quad (17.8)$$

Note that the white-source condition in (17.6) is meant to provide an equal footing on the driving source at each frequency ω. In this way, if the power spectrum $P(\omega)$ is frequency-dependent, this dependence should be the intrinsic nature of the mode rather than from the magnitude of the source.

Equation (17.8) indicates that the spectrum $P(\omega)$ may be significant if the denominator $\omega\Delta\varepsilon_{r,n}(\omega)$ inside the imaginary part nearly vanishes. Motivated by this fact, we define the natural resonance frequency ω_n of mode n as the one minimizing the magnitude of this denominator, namely,

$$\left|\omega_n \Delta\varepsilon_{r,n}(\omega_n)\right| = \min\left[\left|\omega \Delta\varepsilon_{r,n}(\omega)\right|\right]. \quad (17.9)$$

We then expand the white-source power spectrum $P(\omega)$ around the resonance frequency ω_n. Note that the perturbation of $\omega\Delta\varepsilon_{r,n}(\omega)$ around $\omega = \omega_n$ has to lead or lag $\omega_n\Delta\varepsilon_{r,n}(\omega_n)$ by $\pi/2$ in the phase angle (or simply proportional to $-i$) so that $|\omega_n\Delta\varepsilon_{r,n}(\omega_n)|$ is an extreme value [25]:

$$\omega \Delta\varepsilon_{r,n}(\omega) \approx \omega_n \Delta\varepsilon_{r,n}(\omega_n) \left[1 - i\frac{(\omega - \omega_n)}{(\Delta\omega_n/2)}\right], \quad (17.10a)$$

$$\Delta\omega_n \equiv -2i\left[\omega_n \Delta\varepsilon_{r,n}(\omega_n)\right] \left\{ \left. \frac{\partial[\omega \Delta\varepsilon_{r,n}(\omega)]}{\partial\omega} \right|_{\omega=\omega_n} \right\}^{-1} \in IR^+ \quad (17.10b)$$

where $\Delta\omega_n$ is a *real* parameter related to the frequency derivative of $\Delta\varepsilon_{r,n}(\omega)$ at ω_n. If we apply this expansion to (17.8), the spectrum $P(\omega)$ around ω_n, aside of some proportional constant, is approximately

$$P(\omega) \sim \frac{\Delta\omega_n/2}{(\omega - \omega_n)^2 + (\Delta\omega_n/2)^2} - \frac{\mathrm{Re}[\Delta\varepsilon_{r,n}(\omega_n)]}{\mathrm{Im}[\Delta\varepsilon_{r,n}(\omega_n)]} \frac{(\omega - \omega_n)}{(\omega - \omega_n)^2 + (\Delta\omega_n/2)^2}. \quad (17.11)$$

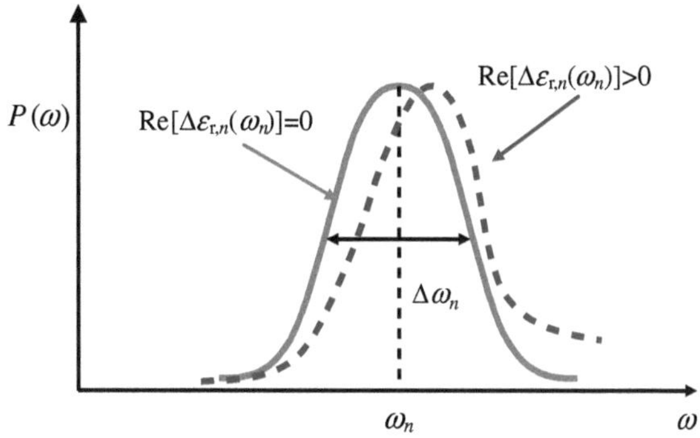

Fig. 17.2 The lineshape of the white-source power spectrum $P(\omega)$ as $\text{Re}[\Delta\varepsilon_{r,n}(\omega_n)] = 0$ (*red solid*, Lorentzian) or $\text{Re}[\Delta\varepsilon_{r,n}(\omega_n)] > 0$ (*blue dashed*, Fano lineshape tilted toward the high-energy side)

For modes with sufficiently high Q factors, we usually have $|\text{Re}[\Delta\varepsilon_{r,n}(\omega_n)]| \ll |\text{Im}[\Delta\varepsilon_{r,n}(\omega_n)]|$, namely, these cavity modes are nearly *resistive*, or equivalently, slightly capacitive/inductive. In that case, the white-source power spectrum $P(\omega)$ is dominated by the symmetric Lorentzian [the first term at RHS of (17.11)], as shown in Fig. 17.2, and the parameter $\Delta\omega_n$ has an interpretation of the full-width-at-half- maximum (FWHM) linewidth. From this linewidth, we then obtain the Q factor of mode n as follows:

$$Q_n = \frac{\omega_n}{\Delta\omega_n} = \frac{i}{2\Delta\varepsilon_{r,n}(\omega_n)} \left.\frac{\partial[\omega\Delta\varepsilon_{r,n}(\omega)]}{\partial\omega}\right|_{\omega=\omega_n}. \qquad (17.12)$$

Equations (17.10b) and (17.12) reveal that once the frequency behavior of $\Delta\varepsilon_{r,n}(\omega)$ around ω_n is known, the FWHM linewidth and Q factor can be obtained through the frequency derivative of this parameter at resonance.

For lossy cavity modes or those which exhibit additional coupling to some continuum modes, the magnitude of the ratio $\text{Re}[\Delta\varepsilon_{r,n}(\omega_n)]/\text{Im}[\Delta\varepsilon_{r,n}(\omega_n)]$ may be comparable to unity. Under such circumstances, the spectrum $P(\omega)$ become the skew-symmetric Fano lineshape, as indicated in Fig. 17.2. The peak of the lineshape may tilt toward the blue (red) side of ω_n if $\text{Re}[\Delta\varepsilon_{r,n}(\omega_n)]$ is positive (negative). Although this asymmetric lineshape is derived under the white-noise condition, it can be still present on the emission spectrum once a physically meaningful model of the source $\mathbf{J}_s(\mathbf{r})$ is utilized in calculations of the emission spectrum.

17.4 Threshold Gain, Confinement Factor, and Modal Volume

We can utilize the permittivity variation $\Delta\varepsilon_{r,n}(\omega_n)$ of mode n to estimate the corresponding threshold gain $g_{th,n}$. For simplicity, let us assume that the active region is locally homogeneous and isotropic with a relative permittivity $\varepsilon_{r,a}(\omega)$ ($\text{Re}[\varepsilon_{r,a}(\omega)] > 0$ for $\omega \sim \omega_n$). The threshold gain is proportional to the difference between the imaginary parts of the complex propagation constants at ω_n before and after the gain insertion:

$$g_{\mathrm{th},n} = -2\left(\frac{\omega_n}{c}\right)\mathrm{Im}\left[\sqrt{\varepsilon_{\mathrm{r,a}}(\omega_n) + \Delta\varepsilon_{\mathrm{r},n}(\omega_n)} - \sqrt{\varepsilon_{\mathrm{r,a}}(\omega_n)}\right]. \tag{17.13}$$

For typical cavity modes, the real part $\mathrm{Re}[\varepsilon_{\mathrm{r,a}}(\omega_n)]$ of the permittivity in Ω_a is the dominant one among various permittivity terms in (17.13). In this way, we can use the binomial expansion for the two square roots in (17.13) and derive an approximate expression for the threshold gain $g_{\mathrm{th},n}$ as

$$g_{\mathrm{th},n} \approx -\left(\frac{\omega_n}{c}\right)\frac{\mathrm{Im}[\Delta\varepsilon_{\mathrm{r},n}(\omega_n)]}{\sqrt{\mathrm{Re}[\varepsilon_{\mathrm{r,a}}(\omega_n)]}}. \tag{17.14}$$

Equations (17.13) and (17.14) indicate that once the permittivity variation $\Delta\varepsilon_{\mathrm{r},n}(\omega_n)$ at resonance is known, the threshold gain $g_{\mathrm{th},n}$ of mode n is immediately accessible, in contrast to the indirect approach based on FDTD calculations.

The expressions of the Q factor Q_n (17.12) and threshold gain $g_{\mathrm{th},n}$ [(17.13) and (17.14)] enable us to calculate the confinement factor directly using the gain-loss balance of the photon density in the laser rate equation. Without the consideration of spontaneous emissions, the steady-state multiplication rate G_n of the photons in mode n, which is the inverse of the photon lifetime $\tau_{\mathrm{p},n}$, has to balance the loss rate:

$$\frac{1}{\tau_{\mathrm{p},n}} = \frac{1}{\Delta\omega_n} = \frac{\omega_n}{Q_n} = G_n, \tag{17.15a}$$

$$G_n = \Gamma_{\mathrm{E},n} v_{\mathrm{g,a}}(\omega_n) g_{\mathrm{th},n}, \tag{17.15b}$$

where $\Gamma_{\mathrm{E},n}$ is the energy confinement factor of mode n; and $v_{\mathrm{g,a}}(\omega_n)$ is the group velocity of the active material at resonance. The group velocity $v_{\mathrm{g,a}}(\omega_n) = c/n_{\mathrm{g,a}}(\omega_n)$ is inversely proportional to the group index $n_{\mathrm{g,a}}(\omega_n)$, which can be calculated through the frequency derivative of the refractive index $n_\mathrm{a}(\omega) \approx \mathrm{Re}[\varepsilon_{\mathrm{r,a}}(\omega)]^{1/2}$ in the active region and is approximated as [34]

$$n_{\mathrm{g,a}}(\omega_n) = \left.\frac{\partial[\omega n_\mathrm{a}(\omega)]}{\partial\omega}\right|_{\omega=\omega_n} \approx \frac{\varepsilon_{\mathrm{g,a}}(\omega_n) + \mathrm{Re}[\varepsilon_{\mathrm{r,a}}(\omega_n)]}{2n_\mathrm{a}(\omega_n)}, \tag{17.16}$$

where $\varepsilon_{\mathrm{g,a}}(\omega) = \partial\,\mathrm{Re}[\omega\varepsilon_{\mathrm{r,a}}(\omega)]/\partial\omega$ is called the group permittivity in the active region. Equating (17.15a) to (17.15b), we derive an expression of the inverse of the confinement factor of mode n as follows:

$$\frac{1}{\Gamma_{\mathrm{E},n}} = \frac{2Q_n\mathrm{Im}[\Delta\varepsilon_{\mathrm{r},n}(\omega_n)]}{\varepsilon_{\mathrm{g,a}}(\omega_n) + \mathrm{Re}[\varepsilon_{\mathrm{r,a}}(\omega_n)]}. \tag{17.17}$$

The conventional expression of the inverse of the energy confinement factor is often written as the ratio between two field integrals:

$$\frac{1}{\Gamma_{\mathrm{E},n}^{(\mathrm{old})}(\Omega)} = \frac{\int_{\Omega} d\mathbf{r}\mathbf{f}_n^*(\mathbf{r},\omega_n)\cdot\frac{\varepsilon_0}{4}\{\overline{\overline{\varepsilon}}_{\mathrm{g}}(\mathbf{r},\omega_n)+\mathrm{Re}[\overline{\overline{\varepsilon}}_{\mathrm{r}}(\mathbf{r},\omega_n)]\}\mathbf{f}_n(\mathbf{r},\omega_n)}{\int_{\Omega_\mathrm{a}} d\mathbf{r}\mathbf{f}_n^*(\mathbf{r},\omega_n)\cdot\frac{\varepsilon_0}{4}\{\varepsilon_{\mathrm{g,a}}(\omega_n)+\mathrm{Re}[\varepsilon_{\mathrm{r,a}}(\mathbf{r},\omega_n)]\}\mathbf{f}_n(\mathbf{r},\omega_n)}, \quad (17.18)$$

where the superscript "old" means that the expression is due to the conventional definition; and $\overline{\overline{\varepsilon}}_{\mathrm{g}}(\mathbf{r},\omega) = \partial\mathrm{Re}[\overline{\overline{\varepsilon}}_{\mathrm{r}}(\mathbf{r},\omega)]/\partial\omega$ is called the group permittivity tensor. The factor $\Gamma_{\mathrm{E},n}^{(\mathrm{old})}(\Omega)$ can be intuitively thought of as the ratio between the electromagnetic energies of mode n in the active region Ω_a and whole computation domain Ω. For the two integrands in the numerator and denominator of (17.18), the parts related to $\overline{\overline{\varepsilon}}_{\mathrm{g}}(\mathbf{r},\omega_n)$ and $\varepsilon_{\mathrm{g,a}}(\omega_n)$ originate from the electric energies corresponding to the (dispersive) media in Ω and Ω_a, respectively. On the other hand, the two counterparts related to $\overline{\overline{\varepsilon}}_{\mathrm{r}}(\mathbf{r},\omega_n)$ and $\varepsilon_{\mathrm{r,a}}(\omega_n)$, after some approximations, reflect the magnetic energies. Ideally, the energy confinement factor is a physical parameter and should be fixed once the active region and cavity structure have been known. However, the definition of $\Gamma_{\mathrm{E},n}^{(\mathrm{old})}(\Omega)$ in (17.18), in fact, depends on the computation domain Ω. The reason is that far-field magnitudes of any lasing modes in the free space behave asymptotically as

$$\lim_{r\to\infty}|\mathbf{f}_n(\mathbf{r},\omega)| \sim \frac{1}{r}, \quad (17.19)$$

where $r = |\mathbf{r}|$ is the distance between a far observation point and the laser cavity roughly located at the origin. With this characteristic, if the computation domain Ω is a ball with a volume V, the numerator in (17.18) goes as $V^{1/3}$ as the ball becomes sufficiently large. Hence, the conventional expression $\Gamma_{\mathrm{E},n}^{(\mathrm{old})}(\Omega)$ is not invariant to Ω, which is nonphysical. The correct energy confinement factor is $\Gamma_{\mathrm{E},n}$ in (17.17), and it is manifestly invariant to any computation domains. In fact, one can show that $\Gamma_{\mathrm{E},n}$ is closely related to the conventional one $\Gamma_{\mathrm{E},n}^{(\mathrm{old})}(\Omega)$ as follows [33]:

$$\frac{1}{\Gamma_{\mathrm{E},n}} = \frac{1}{\Gamma_{\mathrm{E},n}^{(\mathrm{old})}(\Omega)} + [\text{counter terms}], \quad (17.20)$$

where "counter terms" are composed of some volume and surface integrals of $\mathbf{f}_n(\mathbf{r}, \omega)$ in Ω and on its surface, respectively. These counter terms do depend on the computation domain Ω, but the two dependencies of $\Gamma_{\mathrm{E},n}^{(\mathrm{old})}(\Omega)$ and counter terms on Ω cancel each other, and the net effect is the energy confinement factor $\Gamma_{\mathrm{E},n}$ that is invariant to Ω.

With the correct energy confinement factor $\Gamma_{\mathrm{E},n}$, we can then write down well-defined expressions for modal volumes of mode n. There are two modal volumes commonly encountered in active nanophotonics. One is the modal volume $V_{\mathrm{eff},n}$ used in the rate equation for the photon density of the lasing mode, which can be connected to $\Gamma_{\mathrm{E},n}$ as

$$V_{\mathrm{eff},n} = \frac{V_{\mathrm{a}}}{\Gamma_{\mathrm{E},n}} = \frac{V_{\mathrm{a}}}{\Gamma_{\mathrm{E},n}^{(\mathrm{old})}(\Omega)} + [\text{counter terms}], \tag{17.21}$$

where V_{a} is the volume of the active region; the counter terms here are different from those in (17.20) but still maintain the invariance of $V_{\mathrm{eff},n}$ to Ω; and the remaining term other than counter terms in the last equality is the conventional but ill-defined expression of $V_{\mathrm{eff},n}$. The other one, $V_{\mathrm{QM},n}$, is related to the quantum nature of photons associated with a particular cavity mode ("QM" refers to quantum-mechanical) and is commonly utilized in the Purcell effect and cavity quantum electrodynamics [35]. Its expression is

$$\begin{aligned} V_{\mathrm{QM},n} &= \frac{\frac{1}{\Gamma_{\mathrm{E},n}}\int_{\Omega_{\mathrm{a}}} d\mathbf{r}\mathbf{f}_n^*(\mathbf{r},\omega_n)\cdot\frac{\varepsilon_0}{4}\{\varepsilon_{\mathrm{g,a}}(\omega_n)+\mathrm{Re}[\varepsilon_{\mathrm{r,a}}(\mathbf{r},\omega_n)]\}\mathbf{f}_n(\mathbf{r},\omega_n)}{\mathbf{f}_n^*(\mathbf{r}_{\mathrm{p}},\omega_n)\cdot\frac{\varepsilon_0}{4}\{\overline{\overline{\varepsilon}}_{\mathrm{g}}(\mathbf{r}_{\mathrm{p}},\omega_n)+\mathrm{Re}[\overline{\overline{\varepsilon}}_{\mathrm{r}}(\mathbf{r}_{\mathrm{p}},\omega_n)]\}\mathbf{f}_n(\mathbf{r}_{\mathrm{p}},\omega_n)} \\ &= \frac{\int_{\Omega} d\mathbf{r}\mathbf{f}_n^*(\mathbf{r},\omega_n)\cdot\frac{\varepsilon_0}{4}\{\overline{\overline{\varepsilon}}_{\mathrm{g}}(\mathbf{r},\omega_n)+\mathrm{Re}[\overline{\overline{\varepsilon}}_{\mathrm{r}}(\mathbf{r},\omega_n)]\}\mathbf{f}_n(\mathbf{r},\omega_n)}{\mathbf{f}_n^*(\mathbf{r}_{\mathrm{p}},\omega_n)\cdot\frac{\varepsilon_0}{4}\{\overline{\overline{\varepsilon}}_{\mathrm{g}}(\mathbf{r}_{\mathrm{p}},\omega_n)+\mathrm{Re}[\overline{\overline{\varepsilon}}_{\mathrm{r}}(\mathbf{r}_{\mathrm{p}},\omega_n)]\}\mathbf{f}_n(\mathbf{r}_{\mathrm{p}},\omega_n)} \\ &\quad + [\text{counter terms}], \end{aligned} \tag{17.22}$$

where $\mathbf{r}_{\mathrm{p}}$ is the location at which the field strength is maximal; and the rest is the conventional expression of $V_{\mathrm{QM},n}$. Again, the corresponding counter terms also make $V_{\mathrm{QM},n}$ invariant to Ω. We note that the modal volume $V_{\mathrm{eff},n}$, which is inversely proportional to $\Gamma_{\mathrm{E},n}$, usually reflects how well a lasing mode overlaps with the gain medium. On the other hand, the modal volume $V_{\mathrm{QM},n}$ is more indicative of physical sizes of modes.

17.5 Example: Whispering Gallery Modes

We take whispering gallery modes (WGMs) in a dielectric sphere as a demonstration of the presented approach in previous sections. For simplicity, only the fundamental radially transverse-electric (TE_r) modes are considered. These modes do not have the radial electric field and can be regarded as the standing waves confined near a certain equatorial surface of the dielectric sphere, as shown in the inset of Fig. 17.3a. They are characterized by the angular momentum mode number L, twice of which is just the number of wave packets in their standing wave patterns. In this example, we set the radius R of the dielectric sphere to 2.5 μm. In addition to the cavity, the sphere also plays the role of the active region Ω_{a}. Its relative permittivity $\varepsilon_{\mathrm{r,a}}$ is assumed to be nondispersive but complex with a value $(1.5 + 0.0005i)^2$. In this way, the dielectric sphere is intrinsically absorptive. Its surrounding is air, of which the relative permittivity is unity.

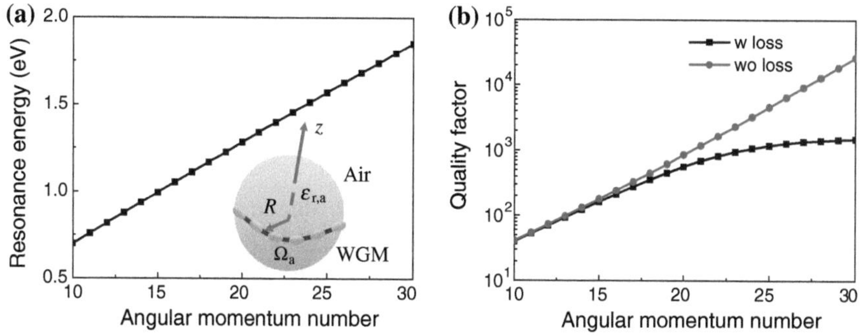

Fig. 17.3 **a** The resonance energies $\hbar\omega_L$ of the fundamental TE$_r$ *WGMs* as a function of L. The *inset* shows the standing-wave behavior of these modes near a certain equator of the sphere. **b** The Q factors Q_L of the same WGMs versus L (*black squares*). The counterparts in the lossless sphere (*red circles*) are also shown for comparisons

The mode profiles of these TE$_r$ modes depend on the polar angle θ and azimuthal angle ϕ through the spherical harmonic $Y_{LM}(\theta, \phi)|_{M=L}$ and its partial derivatives. After the separation of the θ and ϕ dependencies, these modes are characterized by some functions $f_L(r)$ which only spatially depend on the radial coordinate r. These functions $f_L(r)$ can be written in terms of the spherical Hankel function of the first kind $h_L^{(1)}(k_0 r)$ and spherical Bessel function of the first kind $j_L(k_a r)$ outside and inside the sphere, respectively, where $k_0 = \omega/c$ is the propagation constant in the air, and $k_a = k_0\sqrt{\varepsilon_{r,a} + \Delta\varepsilon_{r,L}(\omega)}$ is the propagation constant in the sphere (active region) with the permittivity variation $\Delta\varepsilon_{r,L}(\omega)$ taken into account. After demanding the continuity of the electric and magnetic fields transverse to the radial direction ($\hat{r}$), we can write down the transcendental equation for $\Delta\varepsilon_{r,L}(\omega)$ as

$$\sqrt{\varepsilon_{r,a} + \Delta\varepsilon_{r,L}(\omega)}\left\{\frac{1}{xj_L(x)}\frac{d[xj_L(x)]}{dx}\right\}\Bigg|_{x=k_aR} = \left\{\frac{1}{xh_L^{(1)}(x)}\frac{d[xh_L^{(1)}(x)]}{dx}\right\}\Bigg|_{x=k_0R}. \tag{17.23}$$

The permittivity variation $\Delta\varepsilon_{r,L}(\omega)$ in (17.23) is then solved self-consistently via the transcendental equation.

Figure 17.3a shows the resonance photon energies $\hbar\omega_L$ of various fundamental TE$_r$ modes obtained through the criterion in (17.9). The spectral spacings between neighboring resonance photon energies labeled by different L are nearly identical, which reflects the standing-wave characteristic of these WGMs. In Fig. 17.3b, we show the Q factors Q_L of these WGMs as a function of L. The counterparts without any absorption loss ($\varepsilon_{r,a} = 1.5^2$) are shown for comparisons. The Q factors in both cases increase with L because the radiation loss of these WGMs is suppressed as L becomes larger. However, for modes in the lossy sphere, while the corresponding Q factors nearly coincide with those in absence of the absorption at the low-L side, they gradually saturate at the large-L side, in contrast to the exponentially-increasing counterparts of the lossless sphere. The clamping of Q_L at the large-L side for the lossy sphere is a result of the material absorption. From a hand-waving argument on the constituents of Q factors:

$$\frac{1}{Q_L} = \frac{1}{Q_{\mathrm{rad},L}} + \frac{1}{Q_{\mathrm{mat},L}}, \tag{17.24}$$

where $Q_{\mathrm{rad},L}$ and $Q_{\mathrm{mat},L}$ are the components of Q_L due to the radiation and material losses, we can see that the exponentially-increasing $Q_{\mathrm{rad},L}$ with L would ultimately have no influence on Q_L, and the relatively L-independent component $Q_{\mathrm{mat},L}$ should take over. This explains the gradual saturation of Q_L as L turns larger. We note that the resonance energies $\hbar\omega_L$ and the radiation Q factors $Q_{\mathrm{rad},L}$ of these modes calculated through the presented approach are nearly identical to their counterparts obtained through the asymptotic expansion of Bessel functions [36, 37].

In Fig. 17.4, we show the threshold gain $g_{\mathrm{th},L}$ of these fundamental TE$_r$ WGMs as a function of L. The counterparts in the lossless sphere are also shown for comparisons. In both cases, the high threshold gains at the low-L side are mainly caused by the radiation loss, and therefore they are nearly identical. In accordance to the trends of Q factors, the threshold gains of the lossy sphere stop decreasing at the high-L side due to the material loss. On the other hand, the counterparts of the lossless sphere decrease exponentially with L, reflecting that the radiation loss is the only loss mechanism under such circumstances.

From the Q factors Q_L and threshold gains $g_{\mathrm{th},L}$ of the WGMs, we can calculate their energy confinement factors $\Gamma_{\mathrm{E},L}$ and modal volumes $V_{\mathrm{eff},L}$ and $V_{\mathrm{QM},L}$ through (17.18), (17.21), and (17.22), respectively. The results are shown in Fig. 17.5a, b, where the two modal volumes are presented in the unit of the volume of the active region $V_{\mathrm{a}} = 4\pi R^3/3$. For most of the modes, the energy confinement factor $\Gamma_{\mathrm{E},L}$ is an increasing function of L, indicating that the WGM with the larger L overlaps better with the active region. We note that the energy confinement factor $\Gamma_{\mathrm{E},L}$ calculated from the presented formulation is always smaller than unity, which is physical because some fractions of the WGMs are always distributed

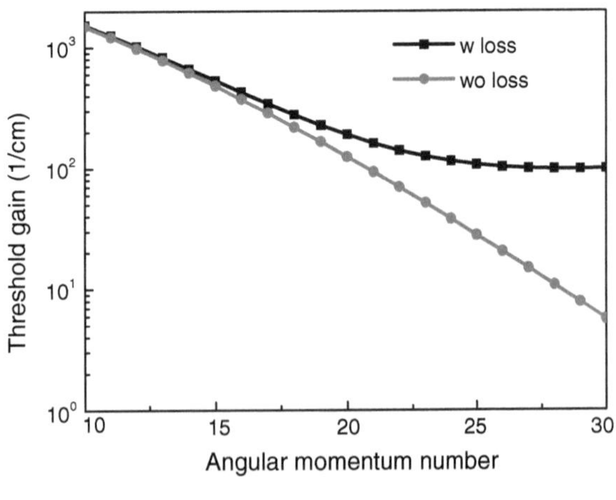

Fig. 17.4 The threshold gains $g_{\mathrm{th},L}$ of the fundamental TE$_r$ WGMs in the lossy sphere versus L (*black squares*). The counterparts in the lossless sphere (*red circles*) are shown for comparisons

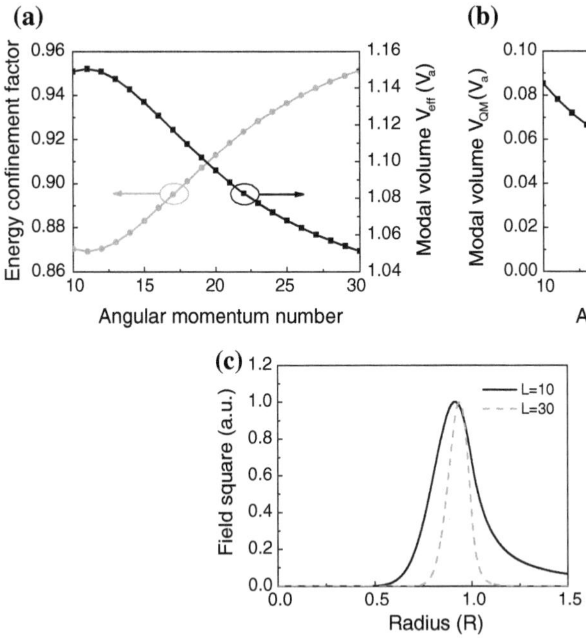

Fig. 17.5 **a** The energy confinement factors $\Gamma_{E,L}$ (*red circles*) and modal volumes $V_{eff,L}$ (*black squares*) of the fundamental TE_r WGMs versus L. **b** The modal volume $V_{QM,L}$ of the same modes as a function of L. Note that the modal volumes are presented in the unit of V_a. **c** The square magnitudes $|f_L(r)|^2$ for two fundamental TE_r WGMs at $L = 10$ (*black solid*) and $L = 30$ (*red dashed*)

outside the active region, and hence the modes cannot overlap perfectly with the gain medium. On the other hand, since the whole dielectric sphere is the active region, the confinement factor $\Gamma_{E,L}$ is quite high and can be close to unity at the high-L side. Corresponding to $\Gamma_{E,L} < 1$, the modal volume $V_{eff,L}$ is always larger than V_a but decreases with L in most of the cases. As mentioned above, the modal volume $V_{eff,L}$ is inversely proportional to $\Gamma_{E,L}$ and mainly indicates how well a particular WGM overlap with the active region. It does not really reflect the size of the mode. The other modal volume $V_{QM,L}$, in fact, does a better job than $V_{eff,L}$ does in this sense. As shown in Fig. 17.5b, in addition to the decreasing behavior with L, the modal volumes $V_{QM,L}$ are only a few percents of V_a. These small fractions actually make sense since the mode profiles of these WGMs are well confined near the rim of the sphere at a sufficiently large L. These field profiles do not occupy the whole sphere. Such a tight confinement becomes more prominent as L increases and can be observed by comparing the square magnitudes $|f_L(r)|^2$ of the two modes at $L = 10$ and $L = 30$ in Fig. 17.5c. The mode at $L = 30$ is much less extended into the sphere and free space than its counterpart at $L = 10$ is.

As the last illustration, let us demonstrate the uncertainty hidden in the conventional but ill-defined expressions of modal volumes. Let us set the computation

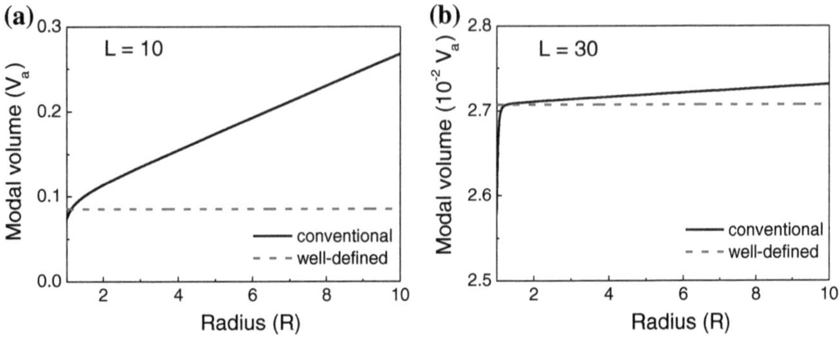

Fig. 17.6 **a** The conventional but ill-defined modal volume corresponding to $V_{QM,L}$ at $L = 10$ as a function of the integration radius R_b (shown in unit of R), and **b** the counterpart at $L = 30$. Although the two modal volumes increase with R_b at different rates, they ultimately diverge to the infinity

domain Ω as a ball which has a radius $R_b > R = 2.5\ \mu m$ and is concentric to the dielectric sphere. In Fig. 17.6a, b, we show the outcomes as calculated from the conventional definition corresponding to $V_{QM,L}$ [see the expression other than the counter terms in (17.22)] as a function of the integration boundary R_b for the WGMs at $L = 10$ and $L = 30$, respectively. For the low-Q mode ($L = 10$) shown in Fig. 17.6a, since its leaky field is quite significant outside the sphere, the corresponding conventional expression intercepts with the correct value $V_{QM,L} = 10$ at $R_b \gtrsim R$ but soon deviates from it significantly. The conventional expression would ultimately approach the nonphysical infinity as R_b increases, and the divergence is more prominent for modes with the lower Q factor. On the other hand, for the counterpart of the high-Q mode ($L = 30$) in Fig. 17.6b, the corresponding divergence seems to be much milder as R_b increases. This phenomenon indicates that for high-Q modes, the conventional expression of the modal volume might provide a reasonable estimations if the computation domain Ω is decently but not excessively big. However, one has to keep in mind that the divergence of the conventional expressions of modal volumes as the size of the computation domain increases is destined to occur sooner or later, no matter the target mode has a low or high Q factor.

17.6 Conclusion

In summary, we have presented an alternative approach to model the lasing modes in micro and nanolaser cavities from viewpoints of active photonic devices. This approach is based on a frequency-domain GE problem which incorporates sources in the form of permittivity variations and can take any material frequency dispersions constrained by the causality into account. In addition, with this formulation, many important parameters specific to the active photonic devices can be directly evaluated. No post processing via field integrations is required in this case. The parameters obtained in this way do not bear any uncertainties and ambiguities as compared with those brought by the indirect estimations through field integrations. The presented method here can be helpful for the performance evaluation of ultra small lasers, in which the variations due to the geometry and material become critical.

References

1. M.T. Hill, Y.S. Oei, B. Smalbrugge, Y. Zhu, T. de Vries, P.J. van Veldhoven, F.W.M. van Otten, T.J. Eijkemans, J.P. Turkiewicz, H. de Waardt, E.J. Geluk, S.H. Kwon, Y.H. Lee, R. Notzel, M.K. Smit, Lasing in metallic-coated nanocavities. Nat. Photonics **1**, 589–594 (2007)
2. M.T. Hill, M. Marell, E.S.P. Leong, B. Smalbrugge, Y. Zhu, M. Sun, P.J. van Veldhoven, E.J. Geluk, F. Karouta, Y.S. Oei, R. Notzel, C.Z. Ning, M.K. Smit, Lasing in metal-insulator-metal sub-wavelength plasmonic waveguides. Opt. Express **17**, 11107–11112 (2009)
3. R.F. Oulton, V.J. Sorger, T. Zentgraf, R.M. Ma, C. Gladden, L. Dai, G. Bartal, X. Zhang, Plasmon lasers at deep subwavelength scale. Nature **461**, 629–632 (2009)
4. M.P. Nezhad, A. Simic, O. Bondarenko, B. Slutsky, A. Mizrahi, L. Feng, V. Lomakin, Y. Fainman, Room-temperature subwavelength metallo-dielectric lasers. Nat. Photonics **4**, 395–399 (2010)
5. K. Yu, A. Lakhani, M.C. Wu, Subwavelength metal-optic semiconductor nanopatch lasers. Opt. Express **18**, 8790–8799 (2010)
6. S. Kita, S. Hachuda, K. Nozaki, T. Baba, Nanoslot laser. Appl. Phys. Lett. **97**, 161108 (2010)
7. C.Y. Lu, S.W. Chang, S.L. Chuang, T.D. Germann, D. Bimberg, Metal-cavity surface-emitting microlaser at room temperature. Appl. Phys. Lett. **96**, 251101 (2010)
8. M. Nomura, Y. Ota, N. Kumagai, S. Iwamoto, Y. Arakawa, Zero-cell photonic crystal nanocavity laser with quantum dot gain. Appl. Phy. Lett. **97**, 191108 (2010)
9. B. Ellis, M.A. Mayer, G. Shambat, T. Sarmiento, J. Harris, E.E. Haller, J. Vuckovic, Ultralow-threshold electrically pumped quantum-dot photonic-crystal nanocavity laser. Nat. Photonics **5**, 297–300 (2011)
10. M.W. Kim, P.C. Ku, Semiconductor nanoring lasers. Appl. Phys. Lett. **98**, 201105 (2011)
11. C.Y. Lu, S.L. Chuang, A. Mutig, D. Bimberg, Metal-cavity surface-emitting microlaser with hybrid metal-DBR reflectors. Opt. Lett. **36**, 2447–2449 (2011)
12. Y.J. Lu, J. Kim, H.Y. Chen, C. Wu, N. Dabidian, C.E. Sanders, C.Y. Wang, M.Y. Lu, B.H. Li, X. Qiu, W.H. Chang, L.J. Chen, G. Shvets, C.K. Shih, S. Gwo, Plasmonic nanolaser using epitaxially grown silver film. Science **337**, 450–453 (2012)
13. J.Y. Suh, C.H. Kim, W. Zhou, M.D. Huntington, D.T. Co, M.R. Wasielewski, T.W. Odom, Plasmonic bowtie nanolaser arrays. Nano Lett. **12**, 5769–5774 (2012)
14. K. Yee, Numerical solution of initial boundary value problems involving maxwell's equations in isotropic media. IEEE Trans. Antennas Propag. **14**, 302–307 (1966)
15. A. Taflove, M. Brodwin, Numerical solution of steady-state electromagnetic scattering problems using the time-dependent Maxwell's equations. IEEE Trans. Microw. Theory Tech. **23**, 623–630 (1975)
16. A. Taflove, Application of the finite-difference time-domain method to sinusoidal steady-state electromagnetic-penetration problems. IEEE Trans. Electromagn. Compat **EMC-22**, 191–202 (1980)
17. A. Taflove, S.C. Hagness, *Computational Electrodynamics: The Finite-Difference Time-Domain Method*, 3rd edn. (Artech House, London, 2005)
18. R. Luebbers, F. Hunsberger, FDTD for Nth-order dispersive media. IEEE Trans. Antennas Propag. **40**, 1297–1301 (1992)

19. D. Kelley, R. Luebbers, Piecewise linear recursive convolution for dispersive media using FDTD. IEEE Trans. Antennas Propag. **44**, 792–797 (1996)
20. T. Kashiwa, I. Fukai, A treatment by the FD-TD method of the dispersive characteristics associated with electronic polarization. Microwave Opt. Technol. Lett. **3**, 203–205 (1990)
21. R.M. Joseph, S.C. Hagness, A. Taflove, Direct time integration of Maxwell's equations in linear dispersive media with absorption for scattering and propagation of femtosecond electromagnetic pulses. Opt. Lett. **16**, 1412–1414 (1991)
22. M. Okoniewski, M. Mrozowski, M. Stuchly, Simple treatment of multi-term dispersion in FDTD. Microwave Guided Wave Lett. IEEE **7**, 121–123 (1997)
23. C. Manolatou, F. Rana, Subwavelength nanopatch cavities for semiconductor plasmon lasers. IEEE J. Quantum Electron. **44**, 435–447 (2008)
24. A. Mock, First principles derivation of microcavity semiconductor laser threshold condition and its application to FDTD active cavity modeling. J. Opt. Soc. Am. B **27**, 2262–2272 (2010)
25. S.W. Chang, Full frequency-domain approach to reciprocal microlasers and nanolasers—perspective from Lorentz reciprocity. Opt. Express **19**, 21116–21134 (2011)
26. T. Søndergaard, B. Tromborg, General theory for spontaneous emission in active dielectric microstructures: example of a fiber amplifier. Phys. Rev. A **64**, 033812 (2001)
27. A.G. Vlasov, O.P. Skliarov, An electromagnetic boundary value problem for a radiating dielectric cylinder with reflectors at both ends. Radio. Eng. Electron. Phys. **22**, 17–23 (1977)
28. E. Smotrova, A. Nosich, Mathematical study of the two-dimensional lasing problem for the whispering-gallery modes in a circular dielectric microcavity. Opt. Quant. Electron. **36**, 213–221 (2004)
29. E. Smotrova, A. Nosich, T. Benson, P. Sewell, Cold-cavity thresholds of microdisks with uniform and nonuniform gain: quasi-3-D modeling with accurate 2-D analysis. IEEE J. Sel. Top. Quantum Electron. **11**, 1135–1142 (2005)
30. A.I. Nosich, E.I. Smotrova, S.V. Boriskina, T.M. Benson, P. Sewell, Trends in microdisk laser research and linear optical modelling. Opt. Quant. Electron. **39**, 1253–1272 (2007)
31. Y.G. Wang, S.W. Chang, C.C. Chen, C.H. Chiu, M.Y. Kuo, M.H. Shih, H.C. Kuo, Room temperature lasing with high group index in metal-coated GaN nanoring. Appl. Phys. Lett. **99**, 251111 (2011)
32. R.S. Moirangthem, P.J. Cheng, P.C.H. Chien, B.T.H. Ngo, S.W. Chang, C.H. Tien, Y.C. Chang, Optical cavity modes of a single crystalline zinc oxide microsphere. Opt. Express **21**, 3010–3020 (2013)
33. S.W. Chang, Confinement factors and modal volumes of micro- and nanocavities invariant to integration regions. IEEE J. Sel. Top. Quantum Electron. **18**, 1771–1780 (2012)
34. S.W. Chang, S.L. Chuang, Fundamental formulation for plasmonic nanolasers. IEEE J. Quantum Electron. **45**, 1014–1023 (2009)
35. J.M. Gerard, B. Gayral, Strong Purcell effect for InAs quantum boxes in three-dimensional solid-state microcavities. J. Lightwave Technol. **17**, 2089–2095 (1999)
36. S. Schiller, Asymptotic expansion of morphological resonance frequencies in Mie scattering. Appl. Opt. **32**, 2181–2185 (1993)
37. L.A. Weinstein, *Open Resonators and Open Waveguides* (Golem Press, Boulder, 1969)

Chapter 18
Nano Structure Light Emitting Devices

Yuh-Jen Cheng

Symbols

τ_r Radiative life time
τ_{nr} Non-radiative life time
τ_{PL} Photoluminescent life time
η Internal quantum efficiency (IQE)

Nano structure light emitting devices have gained great research interests in recent years. Conventional light emitting devices are often fabricated on 2D planar substrates. The device structure as well as device design follow the planar geometry. Recently, there are new trends in using 3D structure for light emitting device design. It potentially provides a new design dimension to improve device performance and may reveal new device physics. In recent developments, III-nitride based light emitting diode (LED) has gain significant progress in power efficiency and wavelength coverage. Its' high efficiency and long life time make it the choice of future lighting device. The large bandgap coverage from UV to near IR of the III-nitride semiconductor compound can potentially open up many useful new device applications. However, there are still many challenges waiting to be resolved. The current device structure is still based on a 2D planar design. Using nano structure can potentially offer opportunities to further increase its efficiency and also resolve some of the problems faced in 2D planar structure. There are many different possible nanostructure design depending on the material and method used to fabricate them. This section is devoted to report the fabrication and properties of III-nitride based crystalline nanopyramid LEDs.

Y.-J. Cheng (✉)
Research Center for Applied Sciences, Academia Sinica, 128 Sec. 2, Academia Rd, Nankang, Taipei 115, Taiwan
e-mail: yjcheng@sinica.edu.tw

C.-C. Lee (ed.), *The Current Trends of Optics and Photonics*,
Topics in Applied Physics 129, DOI 10.1007/978-94-017-9392-6_18

The band gap of InGaN semiconductor can be potentially varied from UV (3.4 eV) to near infrared (0.7 eV) [1]. In principle, such devices can cover the entire visible spectrum, thereby promising multiple color lighting applications without the use of down-converting phosphors which have significant Stokes shift energy loss. Conventional InGaN/GaN multiple quantum wells (MQWs) are often grown on a polar c-plane substrate. The typical emission wavelength is in the blue region, where the efficiency is optimal. To have emission wavelength in the green to red region, it is required to increase In incorporation in InGaN/GaN MQWs. However, the efficiency drops rapidly as In concentration increases [2, 3]. One of the primary causes is the large strain induced piezoelectric field due to the lattice mismatch among the sapphire template, GaN, and InGaN. The piezoelectric field can significantly reduce the spatial electron-hole wave function overlap and leads to low internal quantum efficiency (IQE) [4, 5]. One way to reduce this problem is to grow MQWs on other crystal planes with lower polarization field. Experimental demonstrations have shown improved emission efficiency for MQWs grown on semipolar and nonpolar substrates [6–8]. These substrates are, however, not readily available.

The semipolar and nonpolar crystal surfaces can be found in naturally formed crystalline structure using 3-dimensional epitaxial growth. Nano size hexagonal pyramids can be grown by selective area growth on the opening holes of a SiO_2 masked c-plane GaN substrate, which is more readily available [9]. The pyramid facets are typically {10-11} or {11-22} semipolar planes. In addition to the low polarization field of these crystal planes, the small footprint of 3D nano structure on a substrate can also provide better strain relaxation, resulting in lower defect density and the further reduction of the piezoelectric field. The small footprint can significantly relieve the lattice mismatch problem in planar hetero-epitaxy. There have been strong interests in using these semipolar pyramid facets for LED applications, in particular high In content LEDs [10, 11]. Photoluminescent (PL) studies of these nano scale facets have demonstrated significant reduction in the polarization field and better IQE [12, 13]. The reduced polarization field leads to better electron-hole wave function overlap and faster radiative recombination rate. The IQE enhancement is however not in full proportion to the enhancement of radiative recombination rate. The reason lies in the fact that the non-radiative recombination rate of semipolar InGaN QWs is also shorten, which has not been well investigated. Reports for electrically driven high In content nanopyramid LEDs are also very limited [14].

Here, we report the fabrication and properties of electrically driven nanopyramid LEDs. High In content MQWs for green, olivine, and amber color emission were fabricated. A thin p-GaN layer conformal to the nanopyramid geometry was grown on the nanopyramid facets. The time-resolved photoluminescence (TRPL) measurement shows fast sub-ns radiative recombination lifetime. The peak emission wavelengths of these samples monotonically decrease with increasing temperature, indicating less localized potential [15–18]. This in turn cause faster non-radiative lifetime, partly compromising the faster radiative lifetime. The internal quantum efficiencies (IQEs) of the green, olivine, and amber color LED were

30, 25, and 21 %, respectively, which are hard to achieve for c-plane MQWs in the similar wavelength range. The efficiency decreases with increasing wavelength is attributed to the increase of the residual piezoelectric field in MQWs as In content increases.

The fabrication steps are shown in Fig. 18.1a–d. Nanopillars were first fabricated by patterned top-down etching of an n-doped GaN substrate (Fig. 18.1a). The nanopillar side walls were coated with spin-on-glass (Fig. 18.1b), followed by a MOCVD regrowth to grow nanopyramids and InGaN/GaN MQWs on the tops of nanopillars (Fig. 18.1c). The spin-on-glass coating was to prevent GaN from growing on the side walls during the regrowth. A thin layer of p-GaN conformal to the pyramid facets was subsequently grown to form arrays of nanopyramid LEDs. A standard LED fabrication process was used to fabricate LED chips, where indium tin oxide (ITO) was used as a transparent conducting layer (Fig. 18.1d). Three LEDs with different In concentrations were fabricated. The peak emission wavelengths were 500, 550, and 600 nm, respectively. Their visual colors appear to be green, olivine, and amber color. For the convenience of reference, each is referred as G-, O-, and A-LED. The high resolution tunneling electron microscope (HRTEM) image taken by a JEOL JEM-2,100 electron microscope at 200 kV is shown in Fig. 18.1e, showing the MQWs grown on the inclined pyramid facets. The measured QW and barrier widths are 2 and 8 nm, respectively. The widths are uniform due to the crystalline facet growth. The darker color at the bottom pyramid coalescent boundaries implies possible higher In content and/or coalescent defects. There are also some coalescent defects at the apex region.

The photoluminescent properties of the nanopyramids were investigated by spectrally resolved cathodoluminescent (CL) microscopy, as shown in Fig. 18.2a–f. The images taken from the G-LED sample are shown

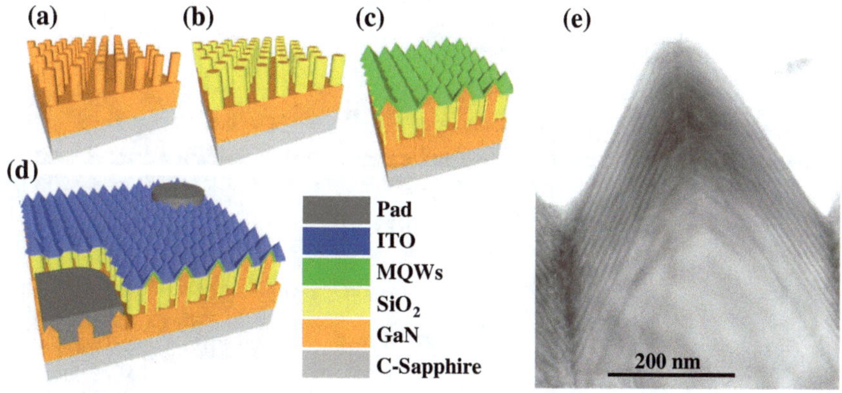

Fig. 18.1 The fabrication steps of a nanopyramid LED. **a** Nanopillars by patterned *top-down* etch. **b** Nanopillar sidewall coated with oxide. **c** Nanopyramids grown on the *top* of nanopillars, followed by MQW growth. **d** *Top* surface indium tin oxide (ITO) deposition and metal pad fabrication. **e** HRTEM cross-section image taken from a nanopillar sample *right* after MQW growth

here to illustrate the general characteristics. Figure 18.2a is the plane view scanning electron microscope (SEM) image, showing the nanopyramid structure. Figure 18.2b–f are the spectrally resolved CL images taken from 460 to 560 nm at every 20 nm step. The bright pattern indicates the emission location of the selected wavelength on the pyramid facets. The much lower CL intensity in Fig. 18.2f is attributed to defects in MQWs at the apex region. The bright triangle contour becomes smaller as the wavelength is scanned from 480 to 560 nm, as shown in Fig. 18.2b–f, indicating that it moves up the pyramids as the wavelength increases. The bright triangle pattern basically follows the pyramid height contour, as can be seen by comparing it with the pyramid SEM image Fig. 18.2a. This indicates that the MQW emission redshifts as location moves from the bottom to apex region on the pyramid facets. It is attributed to the increase of In incorporation as the region moves up the facet. The variation of In concentration in MQWs was confirmed by an energy dispersive spectrometer (EDS) Oxford X-Max. The same specimen used for the previous cross section HRTEM analysis was used for the EDS analysis. The e-beam was focused on the InGaN QW and scanned along the QW. The In content increases from 15 to 30 % as the QW is scanned from bottom to apex region. There is no noticeable QW thickness variation in the HRTEM image. We therefore attribute this redshift mainly to the In incorporation changes.

The internal quantum efficiencies (IQE) of these samples were investigated by the temperature dependent PL measurement. Figure 18.3a shows the integrated PL intensity over the emission spectrum versus temperature from 20 to 300 K. The IQE is assumed to be close to unity at low temperature because

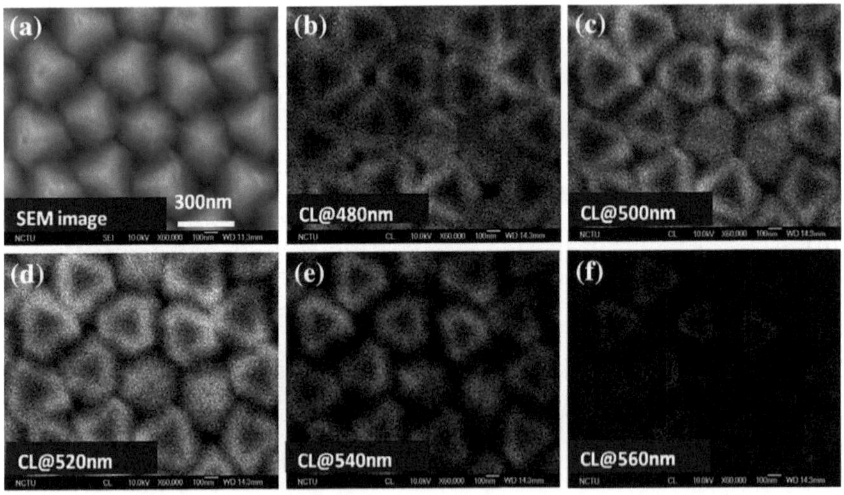

Fig. 18.2 **a** SEM *top view*. **b–f** Spectrally resolved CL images at various wavelength. The bright pattern follows the pyramid height contour, indicating a redshift of MQW emission as the location moves from the *bottom* to *top* of nanopyramids

non-radiative recombination is mostly suppressed at low temperature [19]. The IQE was obtained by normalizing the integrated PL intensity to its low temperature value. The measured room temperature IQEs of G-, O-, and A-LED are 30, 25, and 21 %, respectively. These values are hard to achieve for c-plane MQWs, e.g. 12 % was reported for c-plane MQWs at 570 nm [20]. The decrease in efficiency as the sample changes from G- to A-LED is because the increase of In content induces larger internal polarization field which results in larger electron-hole wave function separation. We have also measured the PL lifetimes by TRPL, as shown in Fig. 18.3b. The PL lifetime decreases with increasing temperature because the non-radiative process is thermally activated. The PL emission peak energy versus temperature is shown in Fig. 18.3c. The peak energy monotonically decreases with increasing temperature. It is different from the often observed red-blue-red shift (S-curve) in the c-plane MQWs, which is attributed to the localized states due to the fluctuations of In distribution in InGaN MQWs [16, 21]. The absence of S-curve implies a significant reduction of localized potentials for MQWs grown on the nanopyramid facets. This could be due to the better strain relaxation provided by the nanostructure and/or the growth property of the pyramid semipolar facets [10, 11].

The PL lifetime τ_{PL} is related to the radiative and nonradiative lifetime, τ_r and τ_{nr}, by $1/\tau_{PL} = 1/\tau_r + 1/\tau_{nr}$. The IQE η is defined as $\eta = \tau_{PL}/\tau_r$. From the measured τ_{PL} and η, we can obtain the temperature dependent τ_r and τ_{nr}, as shown in Fig. 18.4a–c. The τ_r remains constant at low temperature below ~50 K and increases about linearly at higher temperature. This relates to the well known behavior that the radiative lifetime is nearly constant for a localized exciton and increases linearly with temperature for a 2D confined exciton [22]. The measured result indicates that, at low temperature, the excitons are trapped in localized potentials due to inhomogeneous In composition in MQWs. As temperature increases, excitons are excited out of the traps and become free excitons in the 2D MQWs, resulting in approximately linearly increasing radiative lifetime with temperature. The rather low temperature transition implies small localized potentials. The measured room temperature (τ_r, τ_{nr}) for G-, O-, and A-LED are (0.92, 0.40),

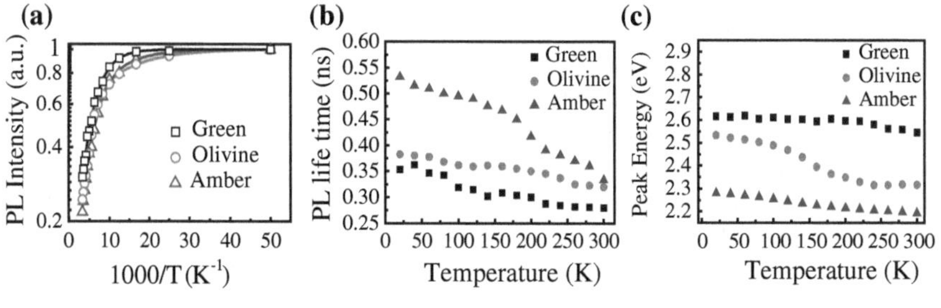

Fig. 18.3 **a** Integrated PL intensity versus temperature. The PL intensity is normalized to its value at 20 K. **b** Time resolved PL decay time constant versus temperature. **c** Emission peak energy versus temperature

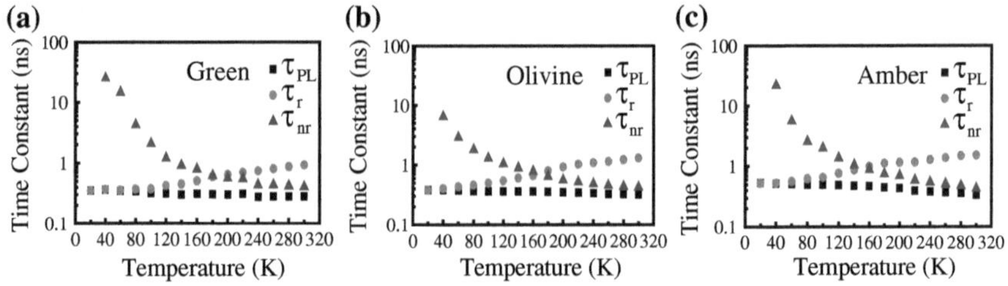

Fig. 18.4 The radiative and nonradiative lifetime versus temperature derived from measured PL lifetime and IQE for *green* (**a**), *olivine* (**b**), and *amber* (**c**) nanopyramid LED, respectively

(1.3, 0.42), and (1.5, 0.43) ns, respectively. These τ_r values are about two orders of magnitude smaller than the typical hundreds of ns value reported for c-plane MQWs [20, 25]. The faster radiative recombination is attributed to the lower polarization field in the nanopyramid facets. It is worth to note that the IQE is not determined by τ_r alone. τ_{nr} also plays an equally important role. The measured τ_{nr} is also significantly shortened as compared with the reported hundred ns lifetime [20]. This is attributed to the above observed less localized potential in semipolar MQWs. The localized potential has been considered as one of the important factors in preventing the capture of excitons by the threading dislocation defects [23, 24]. The suppression of localized potentials will result in the increase of the probability of the capture of excitons by the defects and therefore shorten τ_{nr}. As a result, the improvement of IQE by the large enhancement in τ_r is partly compromised by the faster τ_{nr}.

The three substrates were fabricated into 300 μm × 300 μm LED chips using standard LED fabrication steps. Figure 18.5a–c shows the optical microscope images of these LEDs under electrical injection along with the measured light-current-voltage (L–I–V) curves. The dark pats are the p- and n-metal contacts. The I–V curves show typical diode turn on behavior around 3 volts. The turn-on threshold becomes less distinct as the sample changes from G- to A-LED. Above turn-on, the voltage increases significantly up to 6 volts at 200 mA. This is higher than the normal c-plane LED driving voltage, indicating a higher serial resistance. It is attributed to the imperfection in the current spreading ITO layer on the corrugated nanopyramid surface. The L–I curves shows a slow turn on of light. It becomes worse in particular for A-LED, where the light does not turn on until the current reaches 40 mA. It implies a considerable current leakage. The leakage paths are likely at the pyramid coalescent boundaries and the apex region, where defects are high. The EL spectra versus injection current are shown in Fig. 18.5d–f. The peak emission wavelength blue shifts as the current increases. It is likely related the gradual turn on of the gradient In content MQWs from top to bottom. The blue shifts for G-, O-, and A-LED are about 13, 25, and 50 nm, respectively.

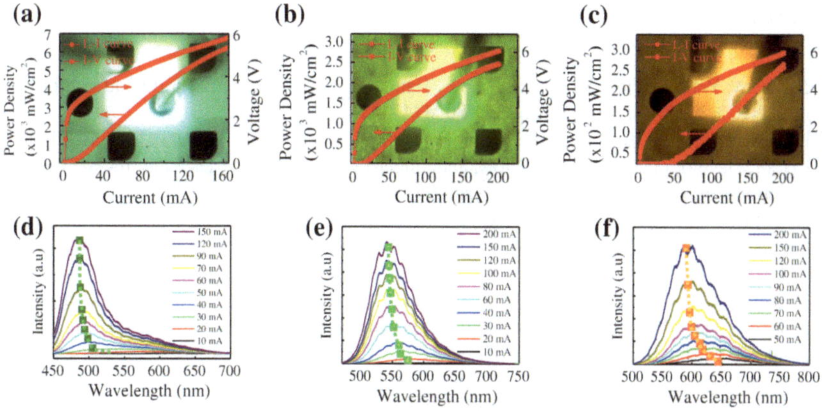

Fig. 18.5 **a–b** Optical microscope images of the G-, O-, and A-LED under electrical injection, showing *green*, *olivine*, and *amber* color. **d–f** EL spectrum versus injection current, showing the *blue* shift of peak emission as current increases due to the screening of polarization field

The increasing blue shift is due to the increasing polarization field in MQWs as In concentration increases. These blue shift values are, nevertheless, still relatively small compared with that in a c-plane MQW [10, 25]. The peak emission wavelengths at 100 mA are 500, 550, and 600 nm, respectively. Their emission colors appear to be green, olivine, and amber, respectively. The slightly redshifted color appearance with respect to their peak emission wavelength is due to the long wavelength tail in the emission spectrum. The full-width-at-half-maximum bandwidths are 60, 80, and 110 nm. The broad spectrum is due to the gradient In distribution in the MQWs. This could be a useful feature in phosphor free white light LED applications if the bandwidth can be further extended.

In summary, electrically driven green, olivine, and amber color nanopyramid LEDs were demonstrated. MQWs were grown on the nanopyramid semipolar facets. Temperature dependent PL measurement shows high IQEs despite the high In content in InGaN/GaN MQWs. The measured fast radiative lifetime indicates improved radiative recombination efficiency due to the low polarization field in semipolar MQWs. Low localized potential was also observed. It leads to reduced non-radiative lifetime, which partly compromises the enhancement of IQE by the faster radiative lifetime. The electrical injection demonstrates the promising potential of using nanopyramid design for high In content LED applications.

Acknowledgments The authors thank the support of the National Science Council of Taiwan contract NSC 100-2112-M-001-021-MY3. The authors also thank Dr. T. C. Hsu and M. H. Shieh of Epistar Corporation, Dr. Chung-Hsiang Lin of LuxtalTek Corporation, and the Industrial Technology Research Institute (ITRI) Taiwan for their technical support.

References

1. J. Wu, W. Walukiewicz, K. Yu, J. Ager, E. Haller, H. Lu, W. Schaff, Small band gap bowing in In1-xGaxN alloys. Appl. Phys. Lett. **80**, 4741–4743 (2002)
2. P.T. Barletta, E.A. Berkman, B.F. Moody, N.A. El-Masry, A.M. Emara, M.J. Reed, S.M. Bedair, Development of green, yellow, and amber light emitting diodes using InGaN multiple quantum well structures. Appl. Phys. Lett. **90**, 151109 (2007)
3. D. Fuhrmann, C. Netzel, U. Rossow, A. Hangleiter, G. Ade, P. Hinze, Optimization scheme for the quantum efficiency of GaInN-based green-light-emitting diodes. Appl. Phys. Lett. **88**, 071105 (2006)
4. Y.-L. Lai, C.-P. Liu, Y.-H. Lin, R.-M. Lin, D.-Y. Lyu, Z.-X. Peng, T.-Y. Lin, Effects of the material polarity on the green emission properties of InGaN/GaN multiple quantum wells. Appl. Phys. Lett. **89**, 151906 (2006)
5. F. Bernardini, V. Fiorentini, D. Vanderbilt, Spontaneous polarization and piezoelectric constants of III-V nitrides. Phys. Rev. B **56**, R10024–R10027 (1997)
6. N. Akopian, G. Bahir, D. Gershoni, M.D. Craven, J.S. Speck, S.P. DenBaars, Optical evidence for lack of polarization in (11-20) oriented GaN/(AlGa)N quantum structures. Appl. Phys. Lett. **86**, 202104 (2005)
7. M. Ueda, K. Kojima, M. Funato, Y. Kawakami, Y. Narukawa, T. Mukai, Epitaxial growth and optical properties of semipolar (11-22) GaN and InGaN/GaN quantum wells on GaN bulk substrates. Appl. Phys. Lett. **89**, 211907 (2006)
8. R. Sharma, P.M. Pattison, H. Masui, R.M. Farrell, T.J. Baker, B.A. Haskell, F. Wu, S.P. DenBaars, J.S. Speck, S. Nakamura, Demonstration of a semipolar (10-1-3) InGaN/GaN green light emitting diode. Appl. Phys. Lett. **87**, 231110 (2005)
9. K. Hiramatsu, Epitaxial lateral overgrowth techniques used in group III nitride epitaxy. J. Phys. Condens. Matter **13**, 6961–6992 (2001)
10. C. Liu, A. Satka, L.K. Jagadamma, P.R. Edwards, D. Allsopp, R.W. Martin, P. Shields, J. Kovac, F. Uherek, W. Wang, Light emission from InGaN quantum wells grown on the facets of closely spaced GaN nano-pyramids formed by nano-imprinting. Appl. Phys. Express **2**, 121002 (2009)
11. I.H. Wildeson, R. Colby, D.A. Ewoldt, Z. Liang, D.N. Zakharov, N.J. Zaluzec, R.E. García, E.A. Stach, T.D. Sands, III-nitride nanopyramid light emitting diodes grown by organometallic vapor phase epitaxy. J. Appl. Phys. **108**, 044303 (2010)
12. T. Kim, J. Kim, M.-S. Yang, S. Lee, Y. Park, U.-I. Chung, Y. Cho, Highly efficient yellow photoluminescence from 11–22 InGaN multiquantum-well grown on nanoscale pyramid structure. Appl. Phys. Lett. **97**, 241111 (2010)
13. H. Yu, L.K. Lee, T. Jung, P.C. Ku, Photoluminescence study of semipolar 10–11 InGaN/GaN multiple quantum wells grown by selective area epitaxy. Appl. Phys. Lett. **90**, 141906 (2007)
14. Y.-H. Ko, J.-H. Kim, L.-H. Jin, S.-M. Ko, B.-J. Kwon, J. Kim, T. Kim, Y.-H. Cho, Electrically driven quantum dot/wire/well hybrid light-emitting diodes. Adv. Mater. **23**, 5364–5369 (2011)
15. Y. Kawakami, K. Omae, A. Kaneta, K. Okamoto, Y. Narukawa, T. Mukai, S. Fujita, In inhomogeneity and emission characteristics of InGaN. J. Phys. Condens. Matter **13**, 6993–7011 (2001)
16. Y.-H. Cho, G.H. Gainer, A.J. Fischer, J.J. Song, S. Keller, U.K. Mishra, S.P. DenBaars, S-shaped temperature-dependent emission shift and carrier dynamics in InGaN/GaN multiple quantum wells. Appl. Phys. Lett. **73**, 1370–1372 (1998)
17. H.P.D. Schenk, M. Leroux, P. de Mierry, Luminescence and absorption in InGaN epitaxial layers and the van Roosbroeck-Shockley relation. J. Appl. Phys. **88**, 1525–1534 (2000)
18. A. Bell, S. Srinivasan, C. Plumlee, H. Omiya, F.A. Ponce, J.·Christen, S. Tanaka, A. Fujioka, Y. Nakagawa, Exciton freeze-out and thermally activated relaxation at local potential fluctuations in thick AlxGa1–xN layers. J. Appl. Phys. **95**, 4670–4674 (2004)

19. Y. Narukawa, S. Saijou, Y. Kawakami, S. Fujita, T. Mukai, S. Nakamura, Radiative and nonradiative recombination processes in ultraviolet light-emitting diode composed of an In0.02Ga0.98 N active layer. Appl. Phys. Lett. **74**, 558–560 (1999)
20. T. Li, A.M. Fischer, Q.Y. Wei, F.A. Ponce, T. Detchprohm, C. Wetzel, Carrier localization and nonradiative recombination in yellow emitting InGaN quantum wells. Appl. Phys. Lett. **96**, 031906 (2010)
21. P.G. Eliseev, P. Perlin, J. Lee, M. Osinski, Blue temperature-induced shift and band-tail emission in InGaN-based light sources. Appl. Phys. Lett. **71**, 569–571 (1997)
22. Y. Narukawa, Y. Kawakami, S. Fujita, S. Nakamura, Dimensionality of excitons in laser-diode structures composed of InxGa1-xN multiple quantum wells. Phys. Rev. B **59**, 10283–10288 (1998)
23. N.K. van der Laak, R.A. Oliver, M.J. Kappers, C.J. Humphreys, Role of gross well-width fluctuations in bright, green-emitting single InGaN/GaN quantum well structures. Appl. Phys. Lett. **90**, 121911 (2007)
24. A. Hangleiter, F. Hitzel, C. Netzel, D. Fuhrmann, U. Rossow, G. Ade, P. Hinze, Suppression of nonradiative recombination by v-shaped pits in GaInN/GaN quantum wells produces a large increase in the light emission efficiency. Phys. Rev. Lett. **95**, 127402 (2005)
25. S.-P. Chang, Y.-C. Chen, J.-K. Huang, Y.-J. Cheng, J.-R. Chang, K.-P. Sou, Y.-T. Kang, H.-C. Yang, T.-C. Hsu, H.-C. Kuo, C.-Y. Chang, Electrically driven nanopyramid green light emitting diode. Appl. Phys. Lett. **100**, 061106 (2012)

Chapter 19
Flexible Micro/Nano-lasers and Compact Optical Curvature Sensors

Min-Hsiung Shih and Kung-Shu Hsu

Symbols

λ_L	lasing wavelength
R	bending radius
n	refractive index
H_z	magnetic field in z component
Q	quality factor
m	azimuthal number
1/R	bending curvature
$\Delta\lambda$	wavelength shift
ΔR	bending radius variation
$\Delta\lambda/\Delta R$	sensitivity
M	1st Brillouin zone symmetry point in hexagonal lattice
K	1st Brillouin zone symmetry point in hexagonal lattice
Γ	1st Brillouin zone symmetry point in hexagonal lattice

19.1 Introduction

In recent decade, the semiconductor lasers have been widely study for light source and photonic integrated circuit. Focused to small size semiconductor laser, the microdisk lasers and photonic crystal lasers have been followed with interest. Semiconductor microdisk cavities have attracted a lot of attention for applications in photonic integrated circuits due to their promising and versatile optical

M.-H. Shih (✉)
Research Center for Applied Sciences, Academia Sinica, Taipei, Taiwan
e-mail: mhshih@gate.sinica.edu.tw

K.-S. Hsu
Department of Photonics, National Chiao Tung University (NCTU), Hsinchu, Taiwan

C.-C. Lee (ed.), *The Current Trends of Optics and Photonics*,
Topics in Applied Physics 129, DOI 10.1007/978-94-017-9392-6_19

functions, for instance, lasers [1], modulators [2] and sensors [3]. The microdisk cavity is a good choice for study of the photonic device due to high quality factor and small size. On the other way, the two-dimensional (2-D) photonic crystal (PhC) lasers become a promising technology as novel light sources in dense chip-scale optical systems due to high quality factor (Q) and small mode volume. A number of photonic crystal lasers have been reported in a suspended membrane [4] or on the dielectric substrates such as Si/SiO_2 [5], III-V compound semiconductor [6] and sapphire [7]. Such lasers exhibit excellent optical properties including small mode volume, high quality factors and ultra-low thresholds. To increase the functionality of the photonic crystal lasers, the tunable optical properties are necessary. A possible approach to achieve the wavelength-tunable lasers is to vary the environmental refractive index of the device [8]. Conversely, the index variation can be highly detected by photonic crystal lasers via measuring the lasing wavelength shift with high spectral resolution and excellent sensitivity. It has been studied extensively for applications in bio-chemical detections [9], gas sensing [10]. However, there are only few reports for wavelength-tunable lasers by deforming the lithographic laser geometry [11]. In particular, most of the photonic crystal lasers are formed in a suspended membrane [12] or fabricated on a hard substrate with low refractive index [7]. All of those reports, the device are fabricated on a hard substrate. With these configurations, it is hard to modify the laser geometry to make the optical properties tunable. Therefore, polymer or organic photonic devices are commonly used because of their flexible applications and low cost. Recent studies have reported on III-V materials-based light emitting diodes [13, 14] and compact lasers [15–17] on flexible and stretchable polymer/organic substrates. Because the optical properties of compact lasers are sensitive to its geometry, this organic or polymer-semiconductor hybrid platform allows fine-tuning of optical properties for small devices.

In this section, we have introduced different laser cavities on a flexible polydimethylsiloxane (PDMS) substrate. The tuning ability in lasing wavelength, threshold and lasing wavelength were investigated by bending the compact cavities to varied curvatures. With a flexible platform, those microdisk laser and photonic crystal laser can not only function as a micrometer size tunable light source, but also be applied for the laser/sensor array on the non-flat surfaces. And the optical curvature sensor had been studied widely with the long period fiber grating system [18]. This optical fiber technology can be applied in the curvature monitoring for larger structures such as bridges and buildings. However, these fiber gratings are not suitable for curvature sensing in either chip-scale integrated circuits or two-dimensional in-plane detection due to its structure size and special geometry.

19.2 Flexible Microdisk Cavity Laser

First, a compact optical curvature sensor with the InGaAsP microdisk laser on a polydimethylsiloxane (PDMS) substrate has been presented. Figure 19.1a shows the illustration of an InGaAsP microdisk cavity on a PDMS substrate. The

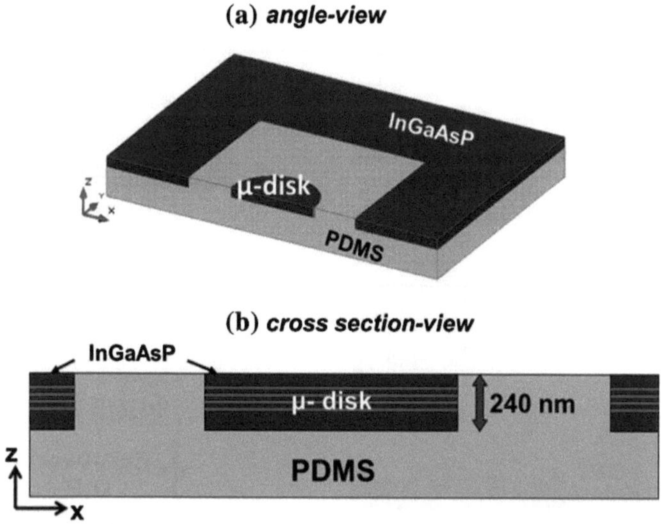

Fig. 19.1 The illustrations of an InGaAsP microdisk cavity on a PDMS substrate from **a** angle view and **b** cross section view

InGaAsP microdisk is embedded inside a flexible PDMS layer which is benefit to the bending ability of the cavity. Figure 19.1b shows the cross section-view of the InGaAsP microdisk cavity.

The low index (n = 1.41) of the PDMS also improve the vertical confinement of the optical mode in the disk, compare to microdisks on Si or GaAs substrates. With a flexible platform, this novel laser can function not only as a light source for the photonic integrated circuits on the non-flat surface, but also as a compact sensing device for the curvature of the bent substrate. The microdisk cavities were implemented in a 240 nm thick InGaAsP layer on the InP substrate. The InGaAsP layer consists of four 10 nm thick strained InGaAsP quantum wells (QWs) which is designed for the lasers operated near 1,550 nm wavelength. A silicon nitride (SiNx) layer and a polymethylmethacrylate (PMMA) layer are deposited subsequently for the dry etching processes and electron beam lithography. The microdisk patterns were defined by electron beam lithography followed by two dry etching steps with CHF_3/O_2 mixture and $CH_4/Cl_2/H_2$ mixture gas in the inductive couple plasma (ICP) system. The microdisk structures then flipped and mounted to a 250 μm thick substrate. The InP substrate was removed by HCl solution. The size of the fabricated microdisk array is from 0.5 to 10 μm in diameter for studying the compact lasers. Figure 19.2a shows the flow chat of the fabrication process and 19.2b shows the photograph of the InGaAsP microdisk cavity array on a PDMS substrate.

Figure 19.3a shows the SEM images of a microdisk array with varied diameters on the PDMS substrate. Figure 19.3b is a close-up view of a microdisk cavity with a diameter of 4.75 μm. Before characterizing the curvature sensing ability of the flexible lasers, the lasing action of the microdisk cavities has to be verified. The cavities were optically pumped at room temperature by an 850 nm wavelength

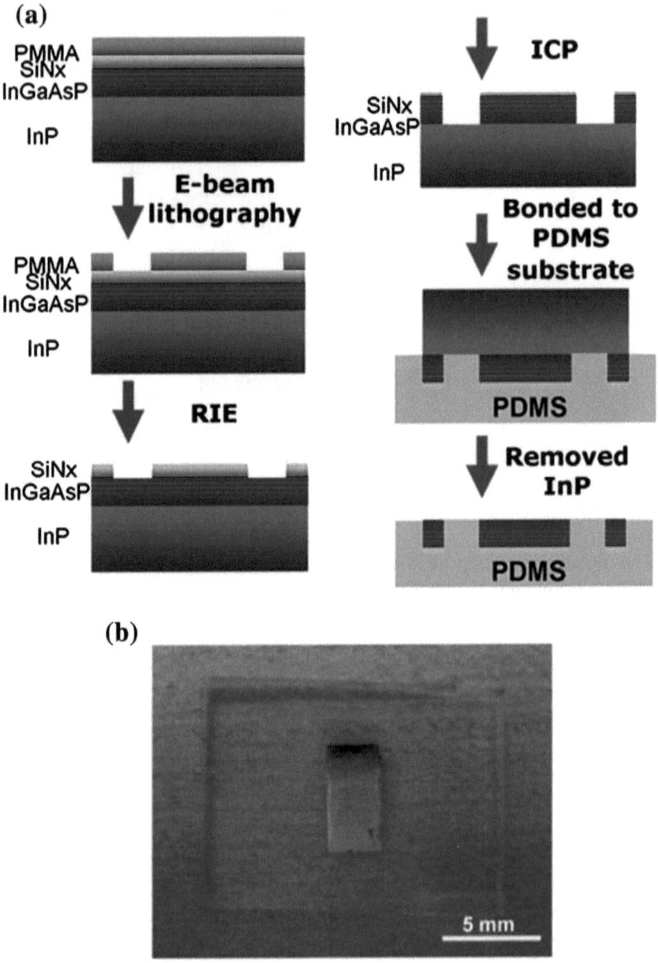

Fig. 19.2 **a** The flow chats of the fabrication process. **b** Photograph of a InGaAsP microdisk laser array on a PDMS substrate after fabrication procedure

diode laser at normal incidence with a 1.5 % duty cycle and a 30 ns pulse width. The pump beam was focused on the devices by a 100x objective lens. The pumping spot size is approximately 2 μm in diameter. The output power of the device was then collected by a multi-mode fiber connected to an optical spectrum analyzer (OSA) for spectrum characterization.

During experiment, the lasing action of the flexible microdisk cavities was achieved with the low threshold power. We also measured the photoluminescence (PL) spectra from non-patterned region of the InGaAsP QWs membrane before and after bonding it on the PDMS substrate. The PL spectrum shows no much difference before and after bonding process. This result implies the stress between InGaAsP layer and the PDMS substrate can be ignored for this device. Figure 19.4a

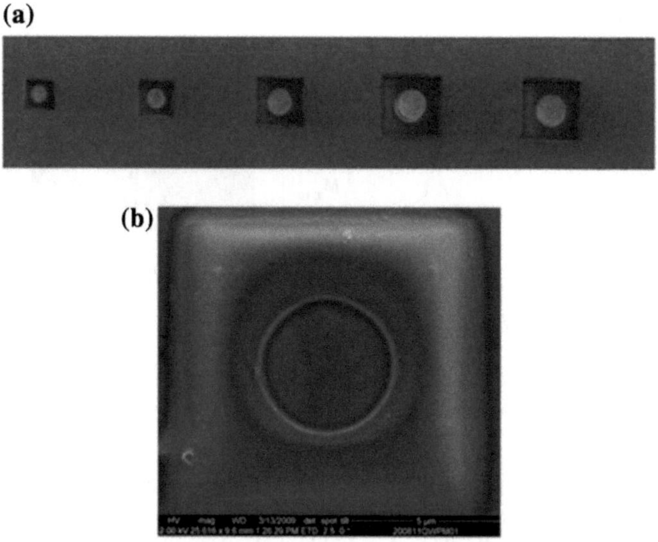

Fig. 19.3 **a** The SEM image of the different size microdisk cavity. **b** The close-up view of a microdisk cavity

Fig. 19.4 **a** A lasing spectrum from a microdisk laser with 4.75 μm diameter. **b** The light-in-light-out (L-L) curve of this laser

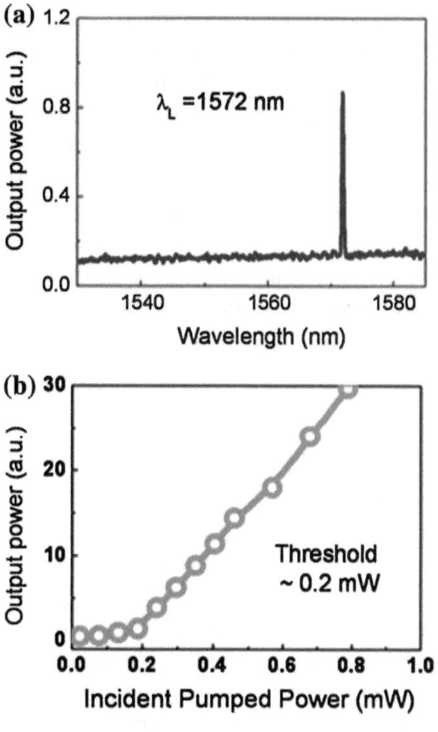

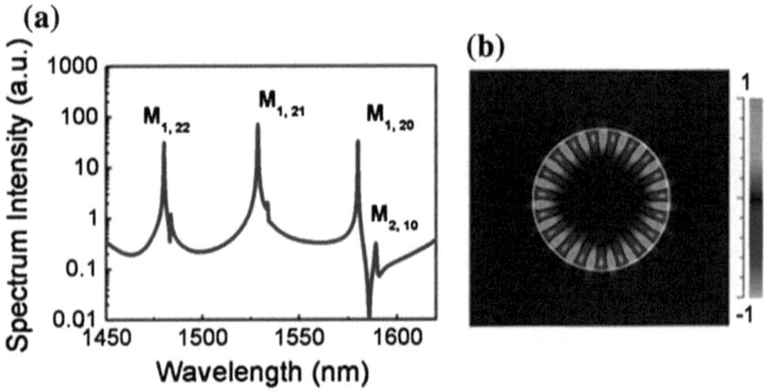

Fig. 19.5 **a** The simulated spectrum for the microdisk cavity on the PDMS substrate. **b** The calculated Hz components of the M1,20 mode of a 4.75 μm microdisk cavity

shows a lasing spectrum from a microdisk laser with 4.75 μm diameter. The lasing wavelength is around 1,572 nm. The light-in-light-out (L-L) curve of this laser is shown in the Fig. 19.4b. The threshold of the laser happened at 0.18 mW of incident pumped power. This corresponds to an effective threshold power is only 9.8 μW after estimating the material absorption, surface reflectivity of the cavity structure. This low threshold power of a single laser will minimize thermal effects of the devices, and benefit the further integration of the laser array.

In order to understand the cavity mode of the flexible microdisk cavity on the PDMS substrate, three-dimensional (3-D) finite-difference time-domain (FDTD) method was used to perform the simulation. Figure 19.5a shows the simulated spectrum for a 4.75 μm InGaAsP microdisk cavity on the PDMS substrate. There are three first-order whispering-gallery modes (WGMs) within the QWs gain region. Their azimuthal numbers (*m*) of WGMs are 22, 21, and 20 which are label with $M_{1,22}$, $M_{1,21}$, and $M_{1,20}$. Due to the index contrast of InGaAsP and PDMS, the higher order WGMs have lower response and lower quality factor (Q) values. Therefore the lasing action of higher order WGMs is not expected. By comparing the measured and simulated spectra, the lasing mode from the measured microdisk was verified to be the first-order mode $M_{1,20}$. The calculated H_z components of the $M_{1,20}$ mode of a 4.75 μm microdisk cavity is shown in Fig. 19.5b.

19.3 Optical Curvature Sensor with the Microdisk Laser

After the demonstration of the InGaAsP microdisk laser on a flexible polydimethylsiloxane (PDMS) substrate, a compact optical curvaturesensor with the flexible microdisk laser was presented in the section. The dimension of the sensor is less than 10 microns. This small size makes it suitable for monitoring the local curvature within few μm regions and for performing the curvature detection or mapping in two-dimensional planes with the microdisk sensor array. The 3-D FDTD was used to perform the simulation for a bent microdisk cavity.

To characterized curvature sensing ability of the flexible microdisk laser, the cavity was bended along the diameter of the disk on a bent metal surface. Figure 19.6 illustrates the microdisk cavity with the curvature 1/R after bending. We expected the lasing characteristics of the microdisk can be manipulated by varying the bending radius R, which could be applied as sensing parameters of a curvature sensor. The microdisk was mounted on the metal plates with different bending radius. These curved metal surfaces were formed by bending the metal plates in the slots with varied bending radii with a homemade curvature component. Figure 19.7a shows the real devices with the bending. In the experiment, the devices were placed on a bent metal surface at the bending stage, and the bending radius R was controlled by the micrometer. The stage is shown in the Fig. 19.7b. We also verify the surface curvature of the device with optical microscope and SEM images after the device is mounted on the bent metal surface. In the experiment, the variation of output power, lasing threshold and lasing wavelength were all observed by bending the microdisk cavity and fixing the pumped conditions. However the lasing wavelength is a better sensing parameter by considering convenient operation of the compact sensors. Therefore we demonstrated the optical curvature sensor based on the lasing wavelength of the flexible microdisk laser. Figure 19.8 shows the lasing spectra of a 4.75 μm microdisk laser at the varied bending radius (R = infinity, 45, 15 and 12.5 mm) under the same pumping

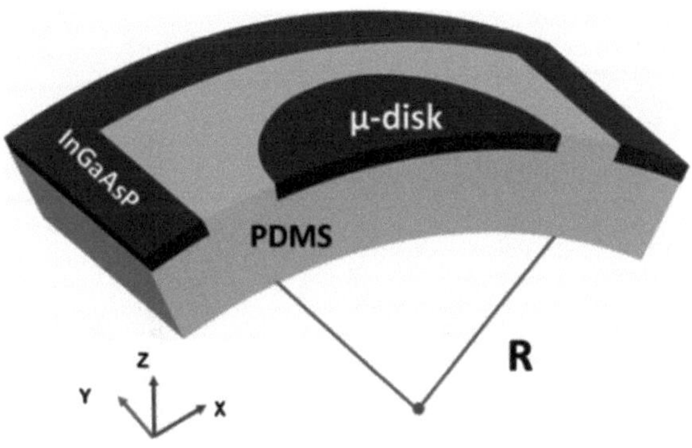

Fig. 19.6 The illustration of a microdisk cavity with a bending radius *R* during the experiment

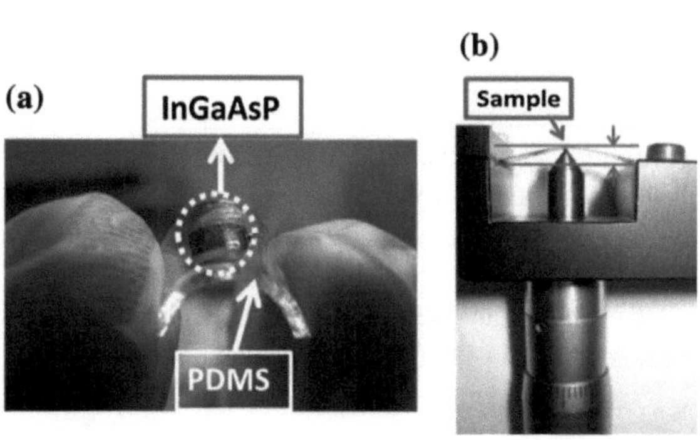

Fig. 19.7 **a** Bent photonic device. **b** Curved sample on a bent mental surface at bending stage

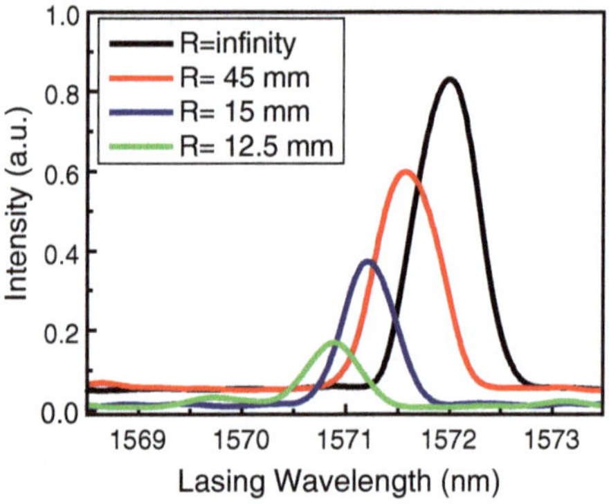

Fig. 19.8 The lasing spectra of a 4.75 μm microdisk laser at the varied bending radius (R = infinity, 45, 15 and 12.5 mm) under the same pumping conditions and fixed 2.0 mW incident pumped power

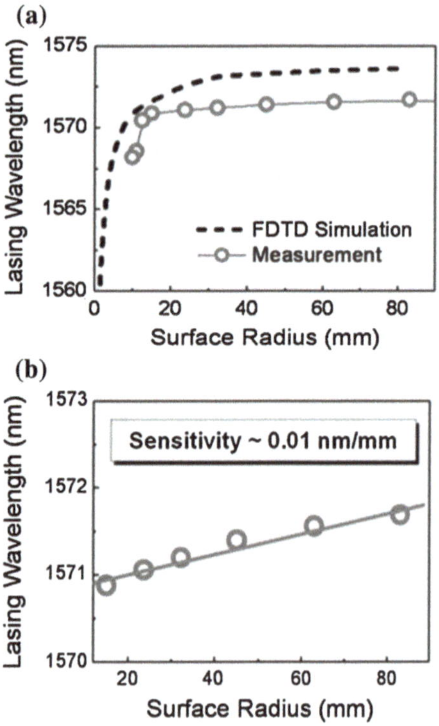

Fig. 19.9 **a** The comparison of the 3D-FDTD simulation and measurement at different curvature. **b** The curvature dependence of lasing wavelength in this linearly region

conditions and a fixed 2.0 mW incident pumped power. According to the data, the lasing wavelength reduces and output power decreases as the bending curvature increases. We have to note that the microdisk achieved lasing at all these bending radii.

To understand the relation between the lasing wavelength and bending radius, the 3-D FDTD method was also used to perform the simulation for the wavelength of the bent microdisk at varied bending curvatures. Figure 19.8 shows the comparison of the 3-D FDTD simulation and measurement at different radii. The black curve is the calculated lasing wavelength of the bended micro-disk cavity on a PDMS substrate at varied radii. The wavelength of the operated mode is blue-shifted linearly as the bending radius decreases from flat to 20 mm, and the wavelength decrease dramatically once the bending radius is smaller than 20 mm. The measured lasing wavelength is also plotted in Fig. 19.9a with the red open-circles. The experiment results agree with the FDTD simulation very well. The difference between the simulation and measurement in wavelength is approximately 2 nm, which is less than 0.2 %. This difference is believed to be caused by the small difference in refractive index used in the simulation and experiment. The lasing wavelength is varied linearly in a wide bending region between flat to 20 mm radius which is suitable for the curvature sensing. Figure 19.9b shows the curvature dependence of lasing wavelength in this linearly region. The sensitivity of the curvature sensor ($^{\Delta\lambda}/_{\Delta R}$) is verified to be approximately 0.01 nm/mm. Although it is not higher than other larger size curvature sensors [19], this value is good enough for the micrometer-size sensor to detect the local curvature variation within the few μm region. The micrometer size, flexible platform also stands on a vantage point for the high-density integration of the sensor arrays in a single chip.

19.4 Tunable Photonic Crystal Laser with on the PDMS Flexible Substrate

In this section, the lasing wavelength fine-tuning by manipulating the curvature of flexible PhC lasers have been demonstrated, as shown in Fig. 19.10a. Band-edge lasers exhibit single mode emission characteristics over large areas and high output powers [20–23].

PhC patterns were fabricated in a 240 nm thick InGaAsP layer on the InP substrate. The InGaAsP layer consisted of four strained InGaAsP QWs with an emission peak at 1,550 nm. A 240 nm thick silicon-nitride (SiNx) layer and a 300 nm thick polymethylmethacrylate (PMMA) resist were deposited on the epitaxial wafer for dry etching and electron-beam lithography. Triangular-lattice air holes were defined on PMMA using electron-beam lithography. After the RIE and ICP dry etching processes, the patterns were transferred to the SiNx layer. The QW layer was then bonded to a 500-μm-thick PDMS substrate. The structure was formed by removing the InP substrate with a solution of HCl. Figure 19.10b shows scanning electron microscope (SEM) images of fabricated PhCs. The area of each photonic crystal cavity is about 15 μm × 15 μm.

The devices were optically pumped at room temperature. The structure achieved lasing at a low threshold power. Figure 19.11a shows the lasing spectrum from the PhC band-edge laser with a lattice constant of 395 nm. The lasing

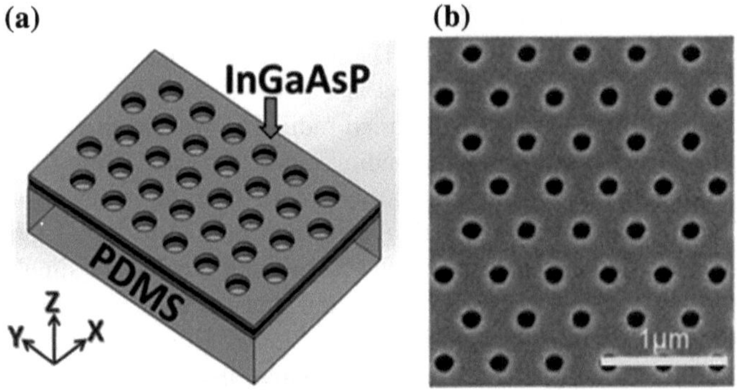

Fig. 19.10 **a** The illustration of a triangular-lattice photonic crystal band edge laser on a PDMS substrate. **b** The SEM image of the fabricated structure

wavelength is approximately 1,581 nm. Figure 19.11b shows the light-in light-out (L-L) curve for the laser with a threshold power of 160 μW, which is estimated by the thickness and absorption of the InGaAsP QWs. The linewidth (blue dots) narrowing effect dependence of pumping power was also plotted in Fig. 19.11b and the estimated quality factor (Q) at transparence in experiment is about 3,000. To confirm the operating lasing modes of the band-edge laser, structures with different lattice constants were optically-pumped and their lasing wavelengths were recorded. Figure 19.11c shows the lasing wavelengths from the PhC lattices with different lattice constants. Two groups of these data were identified as operation lasing modes with normalized frequencies of 0.249 and 0.264. These lasing wavelengths can cover most of the InGaAsP QWs gain region (the gray area). To understand the lasing modes, the corresponding band structure for the TE-like modes was calculated using the three-dimensional (3-D) plane-wave expansion (PWE) method. Figure 19.11d shows the band structure. Band-edge lasing modes are likely to occur near highly symmetrical band structure points. The flat dispersion curve near the band-edge indicates a low group velocity of light and strong localization. Compared to the simulation and measurements, the 0.249 normalized frequency lasing mode corresponds to the first M (M_1) band-edge point and the 0.264 mode corresponds to the first K (K_1) band-edge point. Other band-edge modes were not observed because they fell outside the gain spectrum.

The PhC devices were then bent along the Γ-M direction of the lattices by our homemade bending stage as shown in Fig. 19.7b. Figure 19.12a shows the illustration of bent PhC lattices with bending radius R along Γ-M direction and Fig. 19.12b shows the illustration of Γ-M and Γ-K direction of the flat PhC lattices. The maximal bending curvature reached an approximate bending radius of 20 mm. This is sufficiently large for most applications. At the same pumping conditions and pumping position, the fabricated structure achieved lasing at various bending curvatures.

During the bending, PhC lattices were extended along the Γ-K direction, which is shown in Fig. 19.13a. In real, the PhC lattices were extended to 0.15 % when

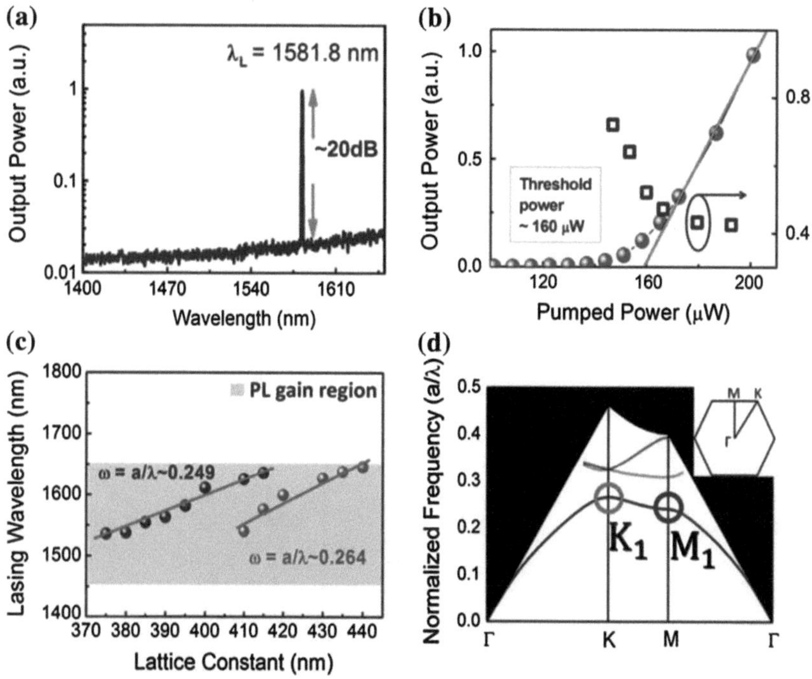

Fig. 19.11 **a** Lasing spectrum of the triangular-lattice PhC band-edge laser on a PDMS substrate. **b** The laser L-L curve. **c** The lasing wavelength versus the lattice constant of the band-edge laser and the InGaAsP QW PL gain region. The normalized frequencies of the lasing modes are approximately 0.249 and 0.264. **d** Band structure of the triangular-lattice PhC with a 0.25 r/a ratio

Fig. 19.12 **a** Illustration of the bent PhC laser. **b** The Illustration of the flst PhC laser

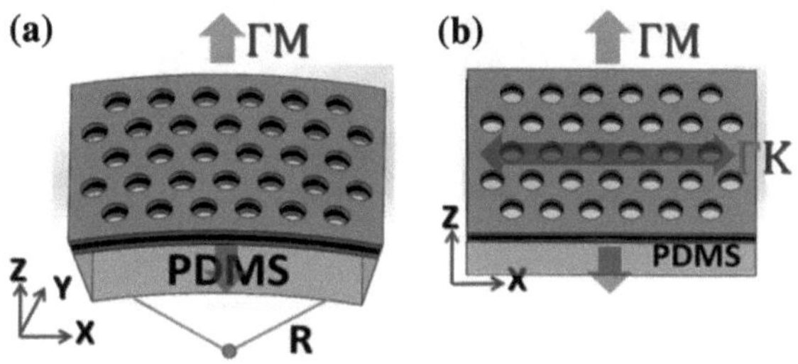

the bending radius reached 20 mm and the lattice extension percentage is linearly as the bending curvature from flat to 0.04 mm^{-1} which is shown in Fig. 19.13b. The extended percentage value was obtained through the real cavity with extended lattices and simulation (3-D PWE method). The lasing wavelength bending property was measured to ensure that mechanical wavelength tuning of PhC band-edge lasers on flexible substrates is repeatable and reliable. Figure 19.13c, d show lasing wavelengths of the K_1 and M_1 modes as the bending radius decreased from the

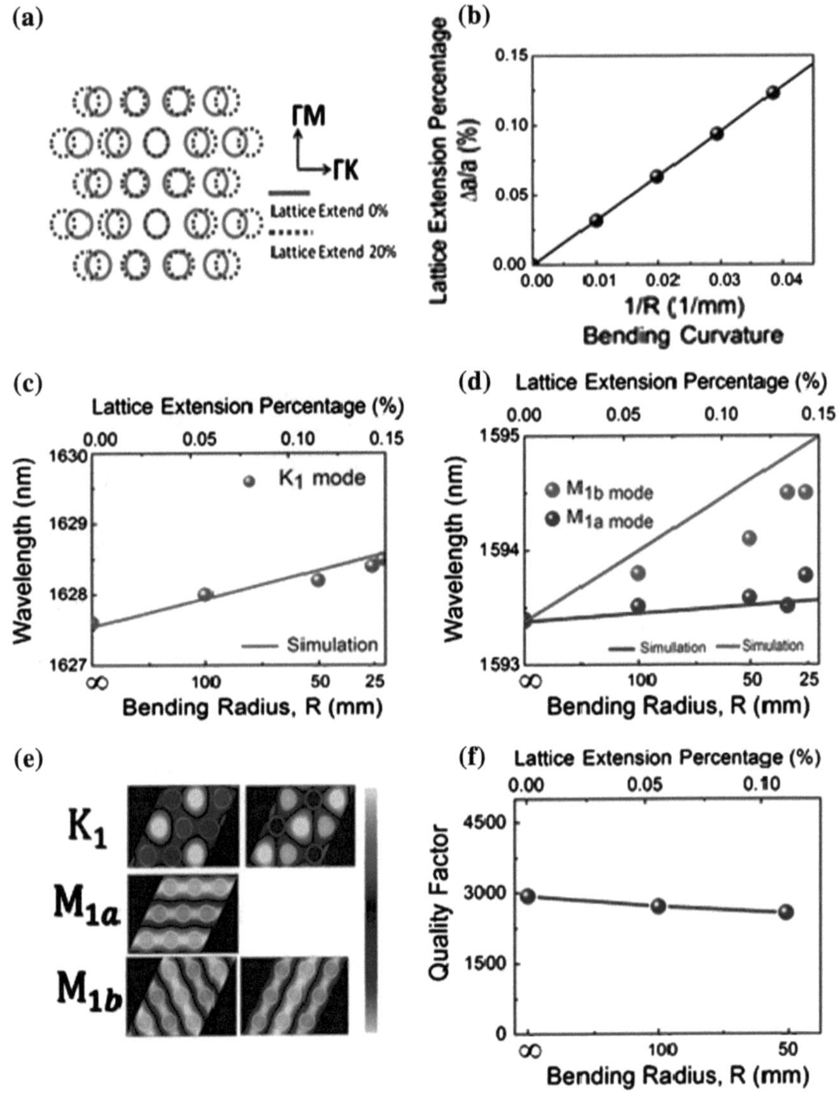

Fig. 19.13 **a** Diagram of PhC lattice distortion. **b** Photonic crystal lattices extended percentage varied with bending curvature **c** and **d** Comparison between measured lasing frequency (*dots*) and lattice distortion frequency (*lines*). **e** Calculated Hz mode profiles of the K1, M1a, and M1b band-edge modes. **f** The quality factor of the laser with different bending curvature

flat (R = ∞). The solid lines in the Fig. 19.13c, d are the calculated wavelengths of the lasing modes from the 3-D PWE simulation. The lasing wavelength was red-shift as the bending radius decreased. The K_1 and M_1 band-edge modes had different red-shift responses to the bending radius change. In Fig. 19.13d, the M_{1a}

and M_{1b} modes split from the M_1 mode when the laser was bent. On the other hand, the quality factor (Q) of laser in bending experiment is decreased around 15 % when the laser was bent to 50 mm which was shown in Fig. 19.13e. And the threshold pumped power raised about 20 % when the laser was bent.

The solid lines in Fig. 19.13c, d show the simulated wavelength of the band-edge modes by using the 3-D PWE method. The resonant frequencies of the K_1 and M_1 modes decreased with two distinct peak shifts and the M_1 mode split into two modes: M_{1a} and M_{1b}. Figure 19.13e shows the simulated H_z field mode profile corresponding to modes K_1, M_{1a}, and M_{1b}. The lasing oscillation of the M_1 band-edge mode exists in three M directions of the triangular lattices [22]. When the lattice was extended in the Γ-K direction, M_1 band-edge mode degeneracy was broken and split into two modes. The lasing oscillation of the M_{1a} band-edge mode occurs along the bending direction and the M_{1b} mode is composed of lasing oscillations in two other Γ-M directions. By contrast, the K_1 band-edge mode did not split significantly when the lattice is extended in the Γ-K direction. This is because the K_1 band-edge mode consists of three nonparallel wave vectors which form a closed loop [22], so the K_1 mode has less impact due to the lattice deformation in the Γ-K direction. A good agreement is observed between the simulation and measurement. The small difference between the measurement and simulation results is attributed to an inaccuracy in the PDMS index variation [24]. The wavelength tunability by bending of the K_1, M_{1a} and M_{1b} modes are 16.5, 9.6 and 28.9 nm/mm^{-1}, respectively. Since the propagation directions of the M_{1b} mode are more parallel to the Γ-K direction, and it has the stronger dependence to the lattice extension. Therefore the M_{1b} mode has a higher tunability, as expected. Figure 19.13f shows the experimental quality factor (Q) with different bending radius and the Q value was decreased around 10 % from flat to 50 mm in bending radius. The results indicate that the flexible PhC laser can serve as a wavelength-tunable light source by mechanically deforming the laser structure. It could also be used as a novel, highly sensitive, ultra-compact optical sensor for local geometry deformation.

In conclusion, we had discussed the flexible InGaAsP micro-/nano-lasers, and a chip-scale optical curvature sensor. Compact microdisk and photonic crystal structures were fabricated on a flexible polydimethylsiloxane (PDMS) substrate. The lasing of the flexible compact cavities was achieved with a low threshold power around 1,550 nm wavelength. The lasing wavelength fine tuning of the flexible microdisk and photonic crystal lasers were also demonstrated by bending the micro-cavities from flat to 10 mm bending radius. The lasing wavelength shift is attributed to cavity distortion. A good agreement between experiment and modeling was also obtained. We also demonstrated a compact optical curvature sensor with the flexible microdisk laser. The curvature dependence of lasing wavelength was characterized by bending the cavity at different bending radii. The measurements showed that the lasing wavelength decreases monotonously with an increasing bending curvature. The curvature sensitivity, which is lasing wavelength to the bending radius ($\Delta\lambda/\Delta R$) is approximately 0.01 nm/mm from the experiment.

References

1. S.L. McCall, A.F.J. Levi, R.E. Slusher, S.J. Pearton, R.A. Logan, Whispering-gallery mode microdisk lasers. Appl. Phys. Lett. **60**, 289–291 (1992)
2. K. Djordjev, S.J. Choi, S.J. Choi, P.D. Dapkus, Active semiconductor microdisk devices. J. Lightwave Technol. **20**, 105–113 (2002)
3. S.Y. Cho, N.M. Jokerst, A polymer microdisk photonic sensor integrated onto silicon. IEEE Photon. Technol. Lett. **18**, 2096–2098 (2006)
4. O. Painter, R.K. Lee, A. Scherer, A. Yariv, J.D. O'Brien, P.D. Dapkus, I. Kim, Two-dimensionaphotonic band-gap defect mode laser. Science **284**, 1819 (1999)
5. C. Monat, C. Seassal, X. Letartre, P. Regreny, P.R. Romeo, P. Viktorovitch, M.L.V. d'Yerville, D. Cassagne, J.P. Albert, E. Jalaguier, S. Pocas, B. Aspar, Modal analysis and engineering on InP-based two-dimensional photonic-crystal microlasers on a Si wafer. IEEE J. Quantum Electron. 39, 419 (2003)
6. T.B. Wang, W.C. Hsu, I.L. Chen, T.D. Lee, K.H. Su, H.P.D. Yang, C.H. Chiouc, Single-mode InGaAs photonic crystal vertical-cavity surface-emitting lasers emitting at 1170 nm. J. Electrochem. Soc. **154**, H351 (2007)
7. M.H. Shih, W. Kuang, T. Yang, M. Bagheri, Z.J. Wei, S.J. Choi, L. Lu, J.D. O'Brien, P.D. Dapkus, Experimental characterization of the optical loss of sapphire-bonded photonic crystal laser cavities. IEEE Photon. Technol. Lett. **18**, 535 (2006)
8. M. Lončar, A. Scherer, Y.M. Qiu, Photonic crystal laser sources for chemical detection. Appl. Phys. Lett. **82**(26), 4648–4650 (2003)
9. M.R. Lee, P.M. Fauchet, Two-dimensional silicon photonic crystal based biosensing platform for protein detection. Opt. Expr. **15**, 4530 (2007)
10. T. Sünner, T. Stichel, S.-H. Kwon, T.W. Schlereth, S. Höfling, M. Kamp, A. Forchel, Photonic crystal cavity based gas sensor. Appl. Phys. Lett. **96**, 261112 (2008)
11. S. Riechel, C. Kallinger, U. Lemmer, J. Feldmann, A. Gombert, V. Wittwer, U. Scherf, A nearly diffraction limited surface emitting conjugated polymer laser utilizing a two-dimensional photonic band structure. Appl. Phys. Lett. **77**, 2310 (2000)
12. P.T. Lee, J.R. Cao, S.J. Choi, Z.J. Wei, J.D. O'Brien, P.D. Dapkus, Room-temperature operation of VCSEL-pumped photonic crystal lasers. IEEE Photon. Technol. Lett. **14**, 435 (2002)
13. S.I. Park, Y. Xiong, R.H. Kim, P. Elvikis, M. Meitl, D.H. Kim, W. Jian, J. Yoon, C.J. Yu, Z. Liu, Y. Huang, K.C. Hwang, P. Ferreira, X. Li, K. Choquette, J.A. Rogers, Printed assemblies of inorganic light-emitting diodes for deformable and semitransparent displays. Science **325**, 977 (2009)
14. R.H. Kim, D.H. Kim, J. Xiao, B.H. Kim, S.I. Park, B. Panilaitis, R. Ghaffari, L. Yao, M. Li, Z. Liu, V. Malyarchuk, D.G. Kim, A.P. Le, R.G. Nuzzo, D.L. Kaplan, F.G. Omenetto, Y. Haung, Z. Kang, J.A. Rogers, Waterproof AlInGaP optoelectronics on stretchable substrates with applications in biomedicine and robotics. Nat. Mater. **9**, 929 (2010)
15. M.H. Shih, K.S. Hsu, Y.C. Wang, Y.C. Yang, S.K. Tsai, Y.C. Liu, Z.C. Chang, M.C. Wu, Flexible compact microdisk lasers on a polydimethylsiloxane (PDMS) substrate. Opt. Expr. **17**, 991 (2009)
16. M.H. Shih, K.S. Hsu, Wan Kuang, Y.C. Yang, Y.C. Wang, S.K. Tsai, Y.C. Liu, Z.C. Chang, M.C. Wu, Compact optical curvature sensor with a flexible microdisk laser on a polymer substrate. Opt. Lett. **34**, 2733 (2009)
17. S. Riechel, C. Kallinger, U. Lemmer, J. Feldmann, A. Gombert, V. Wittwer, U. Scherf, A nearly diffraction limited surface emitting conjugated polymer laser utilizing a two-dimensional photonic band structure. Appl. Phys. Lett. **77**, 2310 (2000)
18. S.W. James, R.P. Tatam, Optical fibre long-period grating sensors: characteristics and application. Meas. Sci. Technol. **14**, R49–R61 (2003)
19. Y. Liu, L. Zhang, J.A.R. Williams, I. Bennion, Optical bend sensor based on measurement of resonance mode splitting of long-period fiber grating. IEEE Photon. Technol. Lett. **12**, 531–533 (2000)

20. M. Imada, S. Noda, A. Chutinan, T. Tokuda, M. Murata, G. Sasaki, Coherent two-dimensional lasing action in surface-emitting laser with triangular-lattice photonic crystal structure. Appl. Phys. Lett. **75**, 316–318 (1999)
21. S. Noda, M. Yokoyama, M. Imada, A. Chutianan, M. Mochizuki, Polarization mode control of two dimensional photonic crystal laser by unit cell structure design. Science **293**, 1123–1125 (2001)
22. M. Notomi, H. Suzuki, T. Tamamura, Directional lasing oscillation of two-dimensional organic photonic crystal lasers at several photonic band gaps. Appl. Phys. Lett. **78**, 1325 (2001)
23. D. Ohnishi, T. Okano, M. Imada, S. Noda, Room temperature continuous wave operation of a surface emitting two-dimensional photonic crystal diode laser. Opt. Expr. **12**, 1562–1568 (2004)
24. K. Hoshino, I. Shimoyama, An elastic thin-film micolens array with a pneumatic actuator, *Proceedings of IEEE Conference on Micro Electro Mechanical Systems* (Institute of Electrical and Electronics Engineers, Interlaken, Switzerland, 2001), pp. 321–324

Chapter 20
Driving Lightwave in Nanopatterned Nanowire

Po-Tsung Lee and Tsan-Wen Lu

Symbols

Γ, γ_d	Confinement factor in dielectric
ε_{Si}	Dielectric constant of silicon
ε_{SiO2}	Dielectric constant of silicon-dioxide
λ	Wavelength
σ_ε	Emission cross section of the Er
a, a_n	Lattice constant
k_x	Wavevector number in x direction
n	Number of photonic crystal periods
n_{NB}	Refractive index of nanobeam
n_r	Refractive index of nanoring
r, r'	Air hole radius
t_{NB}	Thickness of nanobeam
t_{NFB}	Thickness of nano-fishbone
t_r	Thickness of nanoring
t_{SiO2}	Thickness of SiO_2 slot
w	Width of nanobeam and nano-fishbone
w_r	Width of nanoring,
E	Amplitude of electric field
E_x	Amplitude of electric field in x component
E_y	Amplitude of electric field in y component
E_z	Amplitude of electric field in z component

P.-T. Lee (✉) · T.-W. Lu
Department of Photonics, National Chiao Tung University,
1001 Ta-Hsueh Road, Hsinchu 30010, Taiwan
e-mail: potsung@mail.nctu.edu.tw

T.-W. Lu
e-mail: tsanwenlu@gmail.com

C.-C. Lee (ed.), *The Current Trends of Optics and Photonics*,
Topics in Applied Physics 129, DOI 10.1007/978-94-017-9392-6_20

E_{zSi} Amplitude of electric field in silicon in z component
E_{zSiO2} Amplitude of electric field in silicon-dioxide in z component
N_{Er} Concentration of the Er
Q Quality factor
Q_{lasing} Quality factor for lasing
Q_{N1} Quality factor of N_1 nanocavity
Q_{N2} Quality factor of N_2 nanocavity
R Bending radius
V_{eff} Effective mode volume

20.1 Background: Two-Dimensional Photonic Crystals on Slabs

Since the first show up [1, 2], photonic crystals (PhCs) have caused a revolution in manipulating lightwave in wavelength scale. In the past two decades, two-dimensional (2D) PhCs on thin slabs are the most widely-used structure owing to its feasibility in fabrication process. Engineering the photonic bands of 2D PhCs results in various novel effects of lightwave, for example, photonic band gap (PBG), slow light [3], negative refraction [4], and so on. Among these effects, PBG effect provides possibilities to further design PhC defects with different freedoms, for example, waveguides and nanocavities, which have long been the key components for constructing versatile devices in photonic integrated circuits (PICs). PhC waveguides can guide lightwave with very low loss, bend the light propagation in tight angles, produce slow light propagation [5], and even be used for optical manipulation [6]. And PhC nanocavities can confine the photon flows with very low loss (that is, ultrahigh quality factors (Qs)) in wavelength-comparable regions (that is, ultrasmall mode volumes (V_{eff})), which had obtained tremendous successes in nanolasers [7, 8], optical sensors [9, 10], and other functional passive components [11, 12]. Comparing with traditional micro-optical devices [13] based on total internal reflection (TIR), using 2D PhCs for realizing these of devices can provide new scenario in manipulating photons in wavelength scale.

Although 2D PhC devices have long been regarded as the superior candidate to control the photons on-demand in planar PICs [14], using 2D PhCs to realize nanocavities and waveguides has two drawbacks. First, for nanocavities, a huge amount of lattice periods is essential for maintaining their excellent properties and leads to very large device footprint even larger than the micro-optic resonators [13] (micro-disks, -toroids, -rings, and so on) based on TIR. Second, in addition to very large device footprint, using 2D PhCs to construct waveguides also increases the complexity in fabrication and restricts their integration abilities with other components because of PhC lattice geometry.

20.2 Minimizing Device Footprint: One-Dimensional Photonic Crystals on Nanowires

A feasible approach to overcome above drawbacks is to manage light by PhCs arranged in only one dimension, that is, using one-dimensional (1D) PhCs. This structure can be a traditional ridge waveguide with periodic air-holes (PhCs) at the center for engineering the propagating mode of the ridge waveguide [15], as shown in Fig. 20.1a, which is also known as 1D PhC nanowire or nanobeam (NB). With parameters defined in Fig. 20.1a, including NB width w = lattice constant a, NB thickness t_{NB} = a, air-hole radius r over a (r/a) ratio = 0.34, and refractive index (n_{NB}) of 3.4 (for InGaAsP), theoretic transverse-electric(TE)-like band diagram of 1D PhC NB via plane wave expansion (PWE) method is shown in Fig. 20.1b. The 1st and 2nd bands from low to high frequencies (curves

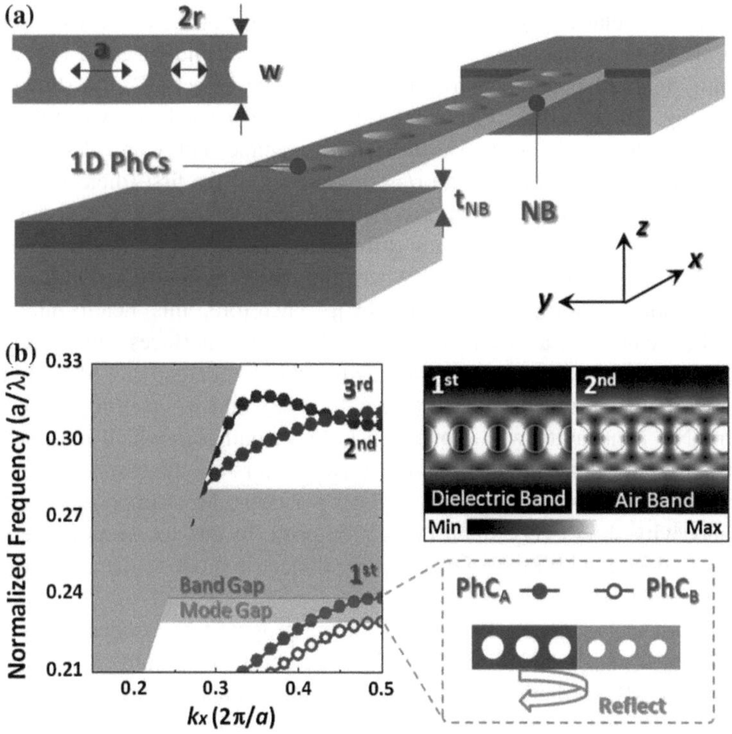

Fig. 20.1 **a** Scheme of a suspended NB with 1D PhCs and **b** its TE-like band diagram. Theoretic $|E|^2$ field distributions of the dielectric and air bands are shown as the inset of **b**. Illustration of forming mode-gap for the dielectric band by hetero-interface of PhC_A/PhC_B on NB is also shown in **b**

with solid circles) are the dielectric and air bands, whose electric fields at band edges ($k_x = \pi/a$) concentrate in dielectric and air regions respectively, shown as the insets of Fig. 20.1b. According to their field distributions, the dielectric band can provide good gain enhancement for an active light emitters and the air band is suitable for optical sensors. Based on this kind of 1D PhC NB, in the following sub-sections, we will show various nanocavity designs on different type NBs for different polarizations, which show small device footprints, highly flexible integration abilities, and various unique features.

20.3 Photonic Crystal Nanobeam: Efficient Nanolasers

To form a nanocavity and confine the propagation modes in 1D PhC NB, one can utilize the PhCs as the mirrors via PBG effect, as illustrated in Fig. 20.2a, which was a very common design in early stages. However, this kind of cavity design usually shows relatively low Q owing to the significant radiation loss caused by the abrupt PBG confinement [16]. The other approach illustrated in Fig. 20.2a is designing double hetero-PhCs to produce mode gap effect [17]. As shown in the inset of Fig. 20.1b, a PhC hetero-interface can be formed via two different 1D PhCs with different r/a on the same NB, for example, PhC_A and PhC_B with r/a_A and r/a_B, where r/a_A is larger than r/a_B. Frequency of the first (dielectric) band in PhC_B (red curve with hollow circles) is lower than that in PhC_A (red curve with solid circles) on NB. The frequency difference represented by the shadow region in Fig. 20.1b forms the mode gap, where the dielectric mode in PhC_A with frequency inside cannot propagate in PhC_B. Therefore, this hetero-interface of PhC_A/PhC_B can serve as a mirror and double hetero-interfaces can form a nanocavity. In addition to r/a ratio, similar cavities with mode gap can be designed via tuning waveguide width [18], effective index [19], and so on. Via proper design, 1D PhC NB nanocavity will show high Q and small V_{eff}, which mean low loss and strong light-matter interaction. These two factors are the essentials for highly efficient nanolasers with thresholdless lasing actions. Most importantly, 1D PhC NB nanocavity shows very small device footprint. In this sub-section, we show a nanocavity design with mode gap effect for dielectric mode to demonstrate highly efficient nanolasers.

As shown in Fig. 20.2b, PhC mirror of the proposed N_1 nanocavity is formed by n air holes with gradually-increased lattice constant on both sides [20]. From the center of NB, the lattice constant a_n is linearly increased by a 5 nm increment, where a_1 is 360 nm and r is fixed at 126 nm. In fabrication, tuning lattice constant to form the hetero-PhCs shows higher controllability and precision than changing air hole size. This kind of multi-step double-hetero PhCs will result in a gently varied frequency cave with mode gap effects, as shown in Fig. 20.2c. With n = 14, theoretic Q and V_{eff} of fundamental (0th-order) dielectric mode in N_1 nanocavity are 10^6 and $0.37(\lambda/n_{NB})^3$. Furthermore, in Fig. 20.2c, different high order (1st- and 2nd-order) modes also exist in this N_1 nanocavity, but with lower Qs and larger V_{eff}.

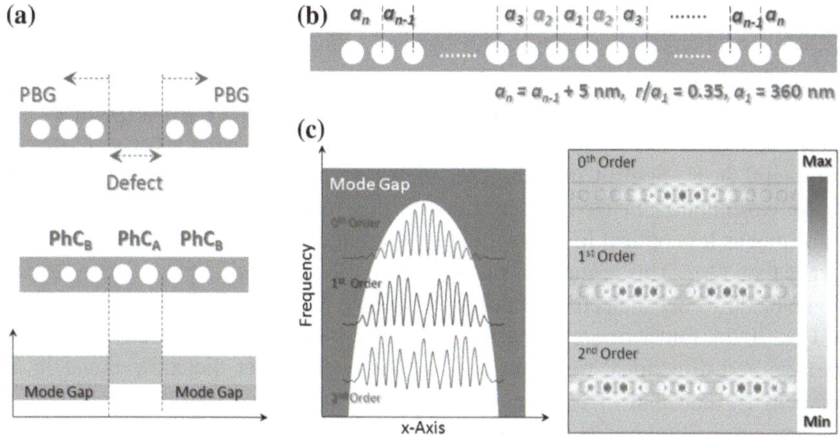

Fig. 20.2 **a** Using PBG (*top*) and mode gap (*bottom*) effects to form nanocavities for confining the propagation mode in 1D PhC NB. **b** Design of 1D PhC NB N_1 nanocavity with mode gap effect via lattice gradually-shifted PhC mirrors. **c** Different order dielectric modes confined in N_1 nanocavity with mode gap

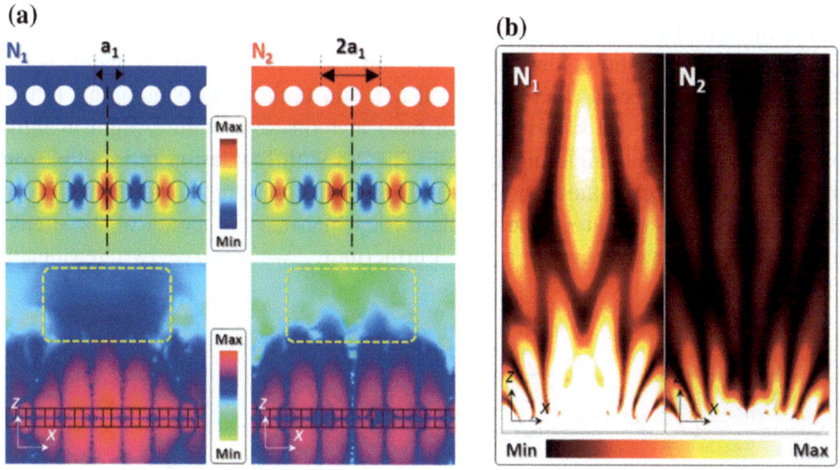

Fig. 20.3 **a** Theoretic profiles in E_y field in xy-plane and time-averaged power flow distributions in xz-plane of the 0th-order dielectric modes in N_1 and N_2 nanocavities. **b** Theoretic radiation patterns of the dielectric modes in N_1 and N_2 nanocavities in xz-plane

One significant benefit of PhC nanocavity is that the mode properties can be significantly altered via slight change in lattice parameters. To show this feature, we design a nanocavity with larger cavity size of $2a_1$ than that of N_1 nanocavity, named N_2 nanocavity, as shown in Fig. 20.3a. Unlike the dielectric vein locates at the center of N_1 nanocavity, the center of N_2 nanocavity is an air hole. In Fig. 20.3a, mode profiles in E_y field (dominant field) of the 0th-order dielectric modes confined in N_1 and N_2 nanocavities show even and odd symmetries respectively owing to their different cavity geometries, which lead to very different dielectric mode properties. For example, Q of the dielectric modes in N_1 nanocavity (Q_{N1}) is always smaller than that in N_2 nanocavity (Q_{N2}) under different n. With n = 14, Q_{N2} is 3×10^6. As shown in Fig. 20.3a), because the odd symmetry in E_y field of the 0th-order dielectric mode in N_2 nanocavity leads to mode cancellation in far-field, less power flow distributes above and below the slab (means less optical loss) than that in N_1 nanocavity, which makes Q_{N2} always larger than Q_{N1}. This argument is confirmed by the theoretic time-averaged power flows of 0th-order dielectric modes in N_1 and N_2 nanocavities in xz-plane, shown in Fig. 20.3a. Furthermore, the difference in field symmetry also leads to stronger vertical and directional radiation into the air of 0th-order dielectric mode in N_1 nanocavity than that in N_2 nanocavity. This is confirmed by theoretic radiation patterns in $|E^2|$ fields (20 μm above the slabs) of the 0th-order dielectric modes in these two nanocavities shown in Fig. 20.3b. Therefore, above results show the modal properties can be greatly changed via slightly tuning different lattice parameters, for example, the nanocavity symmetries shown here.

To demonstrate and verify above features, the designed N_1 and N_2 nanocavities are realized on the epitaxial structure consisted of compressively strained InGaAsP multi-quantum-wells (MQWs) on InP substrate via a series of electron-beam lithography with proximity correction, dry-, and wet-etching processes [21]. Figure 20.4a shows the tilted- and top-views of N_1 and N_2 nanocavities under scanning electron microscope (SEM). Via optical pulse pumping (15 ns pulse duration and 100 kHz repetition rate) at room temperature, light-in light-out (L-L) curves and single-mode lasing spectra with low lasing thresholds from the 0th-order dielectric modes in N_1 and N_2 nanocavities are shown in Fig. 20.4b. Both lasing emissions are linearly polarized, which are the nature of the dielectric modes dominated by E_y fields. Furthermore, we observe two reasonable

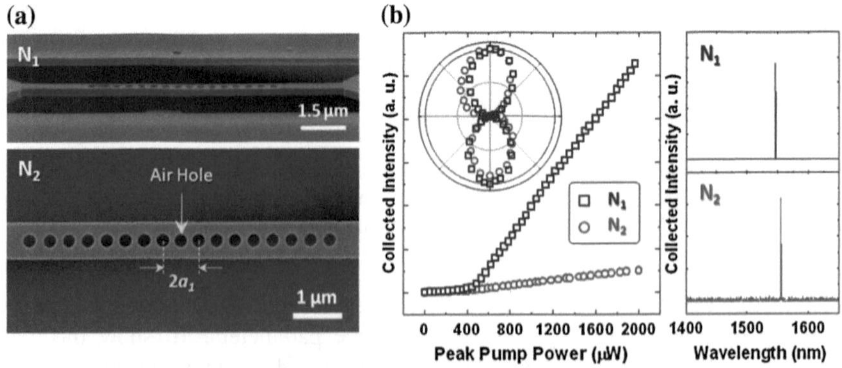

Fig. 20.4 **a** Tilted- and top-view SEM pictures of 1D PhC NB N_1 and N_2 nanocavities. **b** L-L curves, measured linear polarizations, and single-mode lasing spectra of the 0th-order dielectric modes in N_1 and N_2 nanocavities with the same lattice parameters

phenomenons from the L-L curves of N_1 and N_2 nanocavities with the same lattice parameters. First, the threshold of N_2 nanocavity is smaller than that of N_1 nanocavity, which is always observed in all samples. This is because Q_{N2} is always higher than Q_{N1} under the same lattice parameters, as we predicted. Second, the slope efficiencies of 0th-order dielectric modes in N_1 nanocavities are always larger than those in N_2 nanocavities, that is, the dielectric mode in N_1 nanocavity shows stronger vertically directional emission than that in N_2 nanocavity, which agrees with the prediction in Fig. 20.3b. These results show that the modal symmetries can greatly affect the lasing properties, including lasing thresholds and emission directions/intensities.

In addition, 1D PhC NB nanocavity is also a good candidate for the robust and efficient nanolasers on a flexible platform [22], owing to its simple architecture and very small cavity size. To verify this application, we first manufacture a 1D PhC single-defect nanocavity on a suspended InGaAsP NB [21] shown in the inset of Fig. 20.5, which is consisted of a central missing air hole, the tapered and outer PhC mirrors. The tapered PhC mirrors are designed for efficiently reducing the scattering loss, which have fixed r/a ratio and linearly varied lattice constant from cavity to the outer PhC mirror. Then the NB nanocavity is directly transferred onto a flexible polypropylene substrate via acrylic-based optical glue (refractive index ~ 1.48) stamping process [23] illustrated in Fig. 20.5.

As shown in Fig. 20.6a, the flexible-NB nanocavity is fixed on mounts with different bending radii R and pumped by strictly-fixed power of 1 mW. Lasing wavelengths and linewidths of the flexible-NB nanocavity show only 0.15 nm and 0.1 nm variations under different R from ∞ (without bending) to 2.5 mm. In addition, from the lasing spectra of flexible-NB nanocavity under $R = \infty$ and 2.5 mm shown in Fig. 20.6b, the laser emission intensity shows less than 4.8 % variation. These results guarantee the robust lasing properties of the flexible-NB

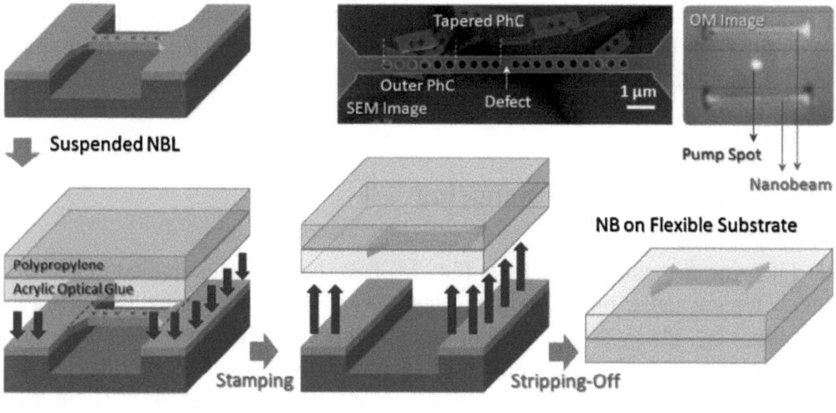

Fig. 20.5 Illustration of transferring the NB nanocavity onto a flexible substrate. The inset images are the SEM (*left*) and optical microscopic (OM) (*right*) pictures of NB nanocavity before and after transferring onto the flexible substrate

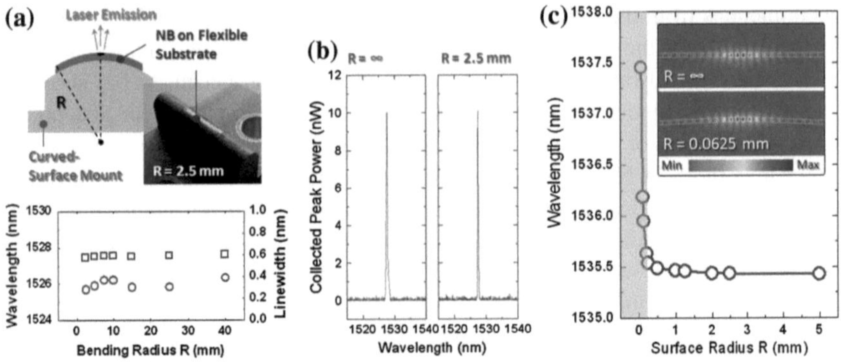

Fig. 20.6 **a** (*Top*) Scheme and picture of the flexible-NB nanocavity on the home-made mount surface with bending radius R. (*Bottom*) Lasing wavelengths and linewidths of the flexible-NB nanocavity under different R from ∞ to 2.5 mm. **b** Lasing spectra of flexible-NB nanocavity when R = ∞ and 2.5 mm. **c** Theoretic wavelength variation of the flexible-NB nanocavity under different R from 5.0 to 0.0625 mm via 3D FEM. The insets show the theoretic mode profiles in electric-field of the flexible-NB nanocavity when R = ∞ and 0.0625 mm

nanocavity in large experimental bending radius range, while there is only very slight deformation on the nanocavity because of its relatively small cavity size. This argument is verified by theoretic wavelengths of the 0th-order dielectric mode under different R from 5 to 0.0625 mm via three-dimensional (3D) finite-element method (FEM) shown in Fig. 20.6c. The lasing wavelength is almost invariant when R > 1.0 mm, which confirms the experimental results. Once R becomes smaller than 1.0 mm, the wavelength shows a significant red-shift, which is caused by the considerable nanocavity elongation owing to the NB bending. Theoretic mode profiles in electric-field of the 0th-order dielectric mode when R = ∞ and 0.0625 mm are shown as the insets of Fig. 20.6c. This kind of flexible-NB nanocavity with low threshold, small device footprint, and robust lasing properties under different bending radii will be very suitable for serving as an efficient nanolaser in future flexible PICs with ultrahigh component density.

20.4 Photonic Crystal Nanoring

2D PhC ring resonators (PhCRRs) [24, 25] have been widely applied in add-drop filters, logical units, optical buffers, and sensors for constructing versatile PICs via proper bus PhC waveguides. Comparing with traditional microring resonators, light can operate at slow light region [3] or recycle with low loss via PBG effect in 2D PhCRRs with compact ring sizes, which are also beneficial for efficient lasers in PICs. However, in addition to the drawback of large device footprint caused by 2D PhCs, using 2D PhC waveguides are the essential to efficiently lead the light into or out from the ring, while the PhC lattice geometry limits the waveguide

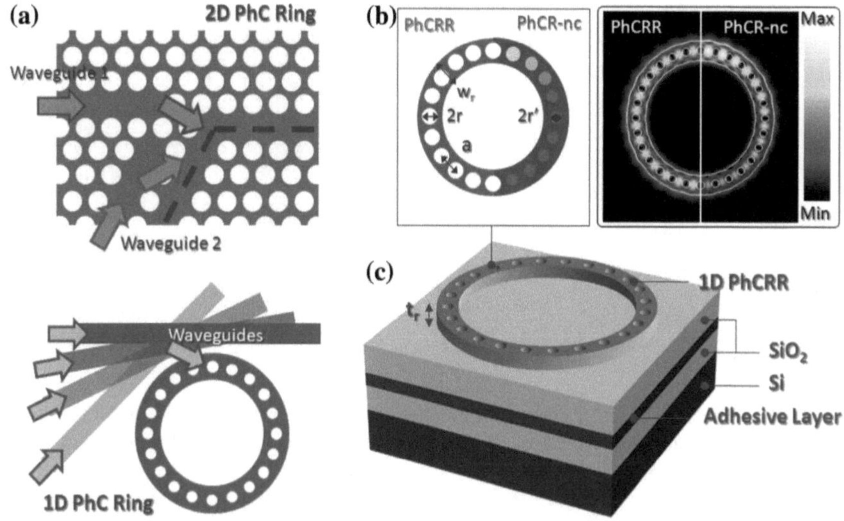

Fig. 20.7 **a** Differences in arrangements of coupling waveguides between (*top*) 2D and (*bottom*) 1D PhCRRs. **b** Parameters of 1D PhCRR and PhCR-nc and their theoretic dielectric mode profiles in electric field via 3D FEM. **c** Scheme of 1D PhCRR on SiO_2 with adhesive layer

arrangement and restricts the integration ability, as shown in Fig. 20.7a. To overcome these drawbacks, one can encircle 1D PhC NB to form a 1D PhCRR [26], which has small device footprint and high integration flexibility in PICs via ridge waveguide coupling without PhC lattice geometry restriction, that is, the ridge waveguides can be arranged in arbitrary directions, as illustrated in Fig. 20.7a.

Design of 1D PhCRR and its parameter definitions are shown in Fig. 20.7b along with theoretic dielectric mode profile in electric field, which is similar to that at the dielectric band edge in 1D PhC NB shown in Fig. 20.1b. This dielectric mode in 1D PhCRR also shows good field overlapping with the dielectric region (that is, high confinement factor γ_d, defined as the ratio of mode energy in dielectric region) and slow light effect for enhanced light-matter interactions in serving as an efficient laser. However, this 1D PhCRR cannot be realized on the widely used suspended slab structure owing to the lack of mechanical support. Therefore, as shown in Fig. 20.7c, an underlying SiO_2 substrate with low refractive index is used to provide mechanically stable supporting without inducing significant optical losses. Via this structure, 1D PhCRR can be applied in realizing InGaAsP/SiO_2 hybrid lasers. With a = 400 nm, refractive index of ring $n_r = 3.4$, thickness $t_r = 190$ nm, and total period number n of 28, theoretically, Q and γ_d will increase when r/a ratio decreases or ring width w_r increases, as shown in Fig. 20.8a. With r/a ratio = 0.27 and $w_r = 1.3a$, Q (~5,800) and γ_d (~0.460) of the dielectric mode, two important factors for lasing in a nanoresonator, are sufficient for lasing.

To realize this structure, first, 800 nm SiO_2 layers are deposited on InGaAsP MQWs and silicon (Si) wafers by plasma enhanced chemical vapor

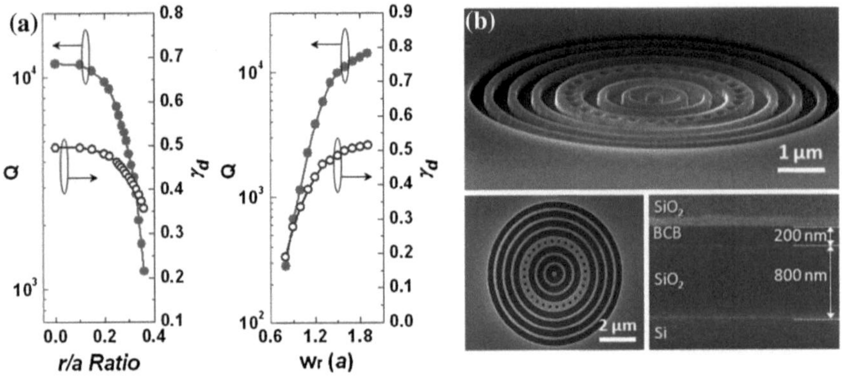

Fig. 20.8 **a** Theoretic Q and γ_d factors of the dielectric modes in 1D PhCRRs on SiO_2 as functions of (*left*) r/a ratio and (*right*) w_r. **b** Tilted-, top-views, and BCB bonding interface cross-sectional SEM pictures of 1D PhCRR on SiO_2

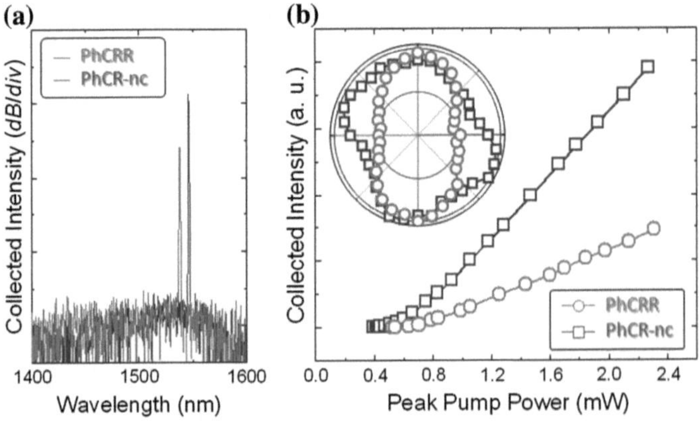

Fig. 20.9 **a** Single-mode lasing spectra, **b** L-L curves, and polarizations of the dielectric modes lasing in 1D PhCRR and PhCR-nc

deposition at 80 °C. These two wafers with SiO_2 are spin-coated diluted DVS-bis-benzocyclobutene (BCB) and then jointed under uniform pressure of 130 kPa at 250 °C for 2 h. Subsequently, the substrate and capping layer of MQWs are removed via diluted HCl wet etching at room temperature to leave MQWs on SiO_2. The MQWs are followed by electron-beam lithography and dry-etching to manufacture PhC nanostructures [21]. Tilted-, top-views, and BCB bonding interface cross-sectional SEM pictures of 1D PhCRR on SiO_2 are shown in Fig. 20.8b.

Under optical pulse pumping, single-mode lasing spectrum from the dielectric mode is obtained and shown in Fig. 20.9a), which has low lasing threshold of 0.57 mW (effective threshold ~ 36 μW) estimated from the L-L curve shown in Fig. 20.9b. Dissimilar to highly linear-polarized laser emissions of the dielectric modes in NB nanocavities shown in Fig. 20.4b, the emission from PhCRR exhibits a low polarized degree (PD) of 2.3 owing to the rotational symmetry of the ring, shown as the inset of Fig. 20.9b. Ideally, PD should be one because laser emission from the dielectric mode has identical E_x and E_y fields. Experimentally, PD > 1 is attributed to the fabrication imperfections of the characterized device.

Comparing with traditional microring resonators, an important feature of PhCRR is that the mode properties can be altered via purposely-tuned PhCs. To show this capability, the air-hole radii of 1D PhCRR is linearly shrunk from the top (r) to the side region (r′) under fixed a, as shown in Fig. 20.7b, which forms a PhC ring nanocavity (PhCR-nc) with mode gap illustrated earlier. In Fig. 20.7b, with Δr/a (r/a – r′/a) of 0.02 and the same parameters as for PhCRR, the dielectric mode in 1D PhCR-nc shows 1.4 times field enhancement in the cavity, which leads to a reduced V_{eff} of $1.7(\lambda/n_r)^3$ (for PhCRR, V_{eff} is $2.9(\lambda/n_r)^3$) without significant reductions in Q (~5,700) and γ_d (~0.456). In Figs. 20.9a and b, single-mode lasing with lower effective threshold of 29 μW, larger slope efficiency, and stronger emission than those in PhCRR are observed. With almost the same Q and γ_d of the dielectric modes in PhCRR and PhCR-nc, the reduced threshold is attributed to smaller V_{eff} in PhCR-nc than that in PhCRR because small V_{eff} leads to strong coupling between spontaneous emissions and lasing mode [27]. In addition, inducing a nanocavity breaks the rotational symmetry of the ring and changesfar-field profile of the dielectric mode, which leads to stronger vertical emissions and larger slope efficiency in PhCR-nc than those in PhCRR. These results mean that we can easily obtain two different nanolasers with planar or vertical emissions via controlling the symmetry of the PhC ring.

Moreover, SiO_2 layer below PhC ring has good thermal conductivity and plays the role of efficient heat sinker, where most heat flux is through, as shown by theoretic total heat flux as a function of time in Fig. 20.10a. Therefore, the nanolasers based on PhCRR and PhCR-nc on SiO_2 ought to have long-term thermal stabilities. In experiments, the devices are excited continuously by fixed pump power of 1.5 mW. After 6 h, both of them show almost invariant lasing linewidths and slight blue-shifts of 0.2 nm in lasing wavelength owing to the surface oxidation of InGaAsP, as shown in Fig. 20.10b. These results also confirm the thermal stability of the BCB adhesive layer.

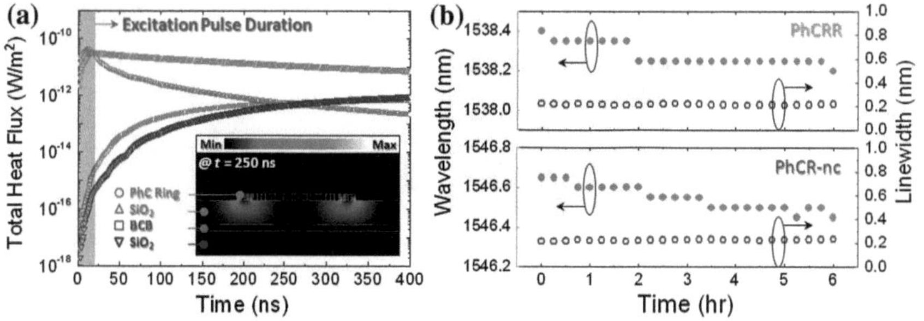

Fig. 20.10 **a** Theoretic total heat flux in PhC ring, *top* SiO_2, BCB, and *bottom* SiO_2 as a function of time. **b** Lasing wavelengths and linewidths of 1D PhCRR and PhCR-nc as a function of time

20.5 Photonic Crystal Nano-Fishbone: Ultralow Loss TM-Polarized Mode

So far, most studied modes in PhC nanocavities are TE-polarized, which had shown their capabilities of serving as various functional active and passive components [7–12, 20, 21, 23–26]. Although transverse-magnetic (TM)-polarized PhC nanocavities draw less attentions on the roadmap, they are still very important in various applications, especially for quantum cascade lasers (QCLs) [28], silicon lasers with horizontal gain slots [29], plasmonic nanophotonics [30], and so on. However, in 2D PhC system, TM-polarized PhC nanocavities are usually designed via dielectric rods for large TM-polarized PBG. In addition to large device footprint, there are two serious drawbacks. First, the structural discontinuity of dielectric rods makes current injection difficult when serving as an active light source. Second, the claddings and substrates with refractive indices larger than air are the essentials for supporting the discontinuous rods, which limit Qs of the nanocavities.

Actually, via properly choosing lattice parameters, TE-polarized 1D PhC NB shown in Sect. 20.3 can exhibit large TM-polarized PBG at the same time [31]. This continuous structure can simultaneously overcome the disadvantages of large device footprint, difficulty of current injection, and low Q in 2D TM-polarized PhC nanocavity based on dielectric rods. To achieve ultrahigh Q, we applied a nanocavity design based on Bloch mode index matching [19] on silicon (Si) NB, which has 10-period outer PhC mirrors and 12-period gradually-varied PhC mirrors with linearly varied lattice constant a_n (n = 1–13) with 5 nm increment under the fixed air-hole radius (r_n) over a_n (r_n/a_n) ratio of 0.32, as illustrated in Fig. 20.11a. With NB width w of 340 nm, theoretic Qs of the TM_{00} mode in nanocavities with different NB thickness t_{NB} are shown in Fig. 20.11b, which shows

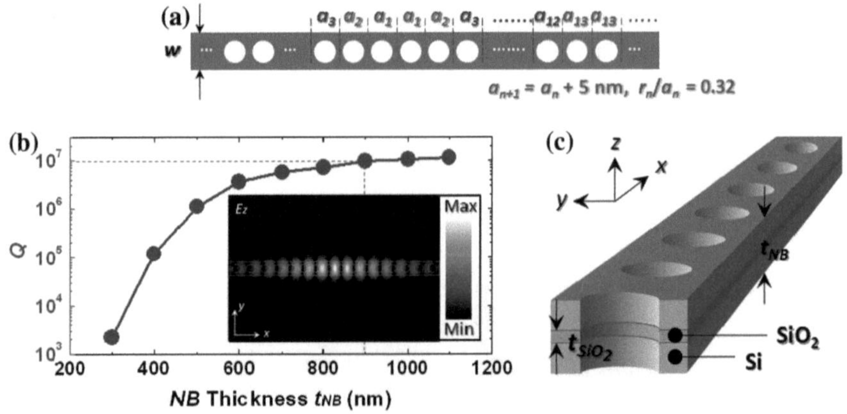

Fig. 20.11 **a** Design of 1D PhC NB nanocavity based on Bloch mode index matching and **b** Theoretic Qs of TM_{00} modes in 1D PhC NB nanocavities with different t_{NB} from 300 to 1,100 nm. TM_{00} mode profile for E_z-field in xy-plane confined in the nanocavity with t_{NB} = 900 nm is shown as the inset. **c** Scheme of 1D PhC NB nanocavity with a horizontal SiO_2 slot

a saturated value higher than 10^7 when $t_{NB} > 900$ nm. The simulated TM_{00} mode profile for E_z-field in xy-plane is shown as the inset in Fig. 20.11b.

Although this design shows an ultrahigh Q TM_{00} mode never seen before in PhC nanocavities, the mode volume is still quite large ($V_{eff} \sim 1.2(\lambda/n_{Si})^3$) owing to the gradually varied PhC mirrors. To reduce V_{eff} without degrading Q, we further design a horizontal SiO_2 slot with thickness t_{SiO2} and refractive index n_{SiO2} of 1.44 through the entire Si NB for providing TM_{00} mode an index discontinuity in z-direction, as shown in Fig. 20.11c. Theoretic TM_{00} mode profiles for E_z-field in xy-, yz-, and xz-planes when $t_{SiO2} = 20$ nm are shown in Fig. 20.12a, where the E_z field is strongly enhanced inside the SiO_2 slot. This is because the electric flux density has to satisfy Maxwell equation. The E_z field of TM_{00} mode will obey the relationship of $\varepsilon_{Si}E_{zSi} = \varepsilon_{SiO2}E_{zSiO2}$ [32], where ε_{Si}, ε_{SiO2}, E_{zSi}, and E_{zSiO2} represent the dielectric constants and E_z fields in Si and SiO_2. Thus, the E_z field in the SiO_2 slot shows $\varepsilon_{Si}/\varepsilon_{SiO2}$ times enhancement compared to that in Si, which leads to a greatly reduced V_{eff} of TM_{00} mode. With fixed t_{NB} of 900 nm, theoretic Q and V_{eff} of TM_{00} modes in 1D PhC NB nanocavities with SiO_2 slots under different t_{SiO2} are shown in Fig. 20.12b. Q increases to a optimized value of 1.5×10^7 when $t_{SiO2} = 20$ nm, which is even higher than that of nanocavity without the SiO_2 slot because the presence of low index SiO_2 slot can further fine tune Bloch mode index matching. In addition, the V_{eff} monotonically decreases with decreasing t_{SiO2}. When $t_{SiO2} = 10$ nm, V_{eff} and Q/V_{eff} are $0.176(\lambda/2n_{SiO2})^3$ and $5.9 \times 10^7(\lambda/2n_{SiO2})^{-3}$ respectively. This ultrahigh Q/V_{eff} cannot be achieved by most optical nanocavities, which is beneficial for realizing nanolasers and useful to the researches of quantum-electro dynamics and applications requiring strong light-matter interactions.

This horizontally-slotted PhC nanocavity shows several unique features and advantages. First, the horizontal slot formed on the entire NB efficiently reduces the V_{eff} and optimizes the Q simultaneously. Second, the most important advantage, the horizontal slot can be easily made via sputtering or chemical vapor deposition with nanometer scale precisions without needing the optimized electron beam lithography and dry etching processes used in PhC vertically-slotted nanocavities [33]. Third, the optical gain medium, for example, rare-earth doped SiO_2, can be placed in the

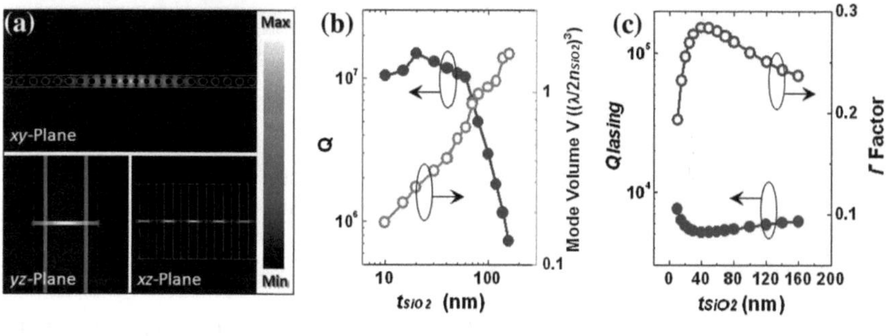

Fig. 20.12 **a** Theoretic TM_{00} mode profiles for E_z-field in xy-, yz-, and xz-planes when $t_{SiO2} = 20$ nm. **b** Theoretic Q, V_{eff}, **c** Q_{lasing}, and Γ factor of TM_{00} modes in 1D PhC NB nanocavities with SiO_2 slots, where $t_{SiO2} = 10$–160 nm

horizontal slot by co-sputtering or ion-implantation process during or after the slot formation to extend the radiative photon life time [34], which cannot be achieved in most reported PhC vertically-slotted nanocavities. Therefore, with the enhanced mode properties and easiness to insert the optical gain medium, we evaluate this horizontally-slotted nanocavity for realizing Si-based nanolasers via required minimum Q for lasing (Q_{lasing}), gain condition of the erbium (Er) dopant, and optical loss of the nanocavity, which are illustrated by the following equation [35]:

$$Q_{la\sin g} > \frac{2\pi}{\lambda} n_{slot} \frac{1}{\Gamma \sigma_{\varepsilon} N_{Er}} \tag{20.1}$$

where the terms σ_ε and N_{Er} represent the emission cross section and concentration of the Er, which are set as 4×10^{-21} cm^2 and 1×10^{22} ions/cm^3. The overlap Γ factor with the Er-doped SiO_2 slot is defined as the ratio of mode energy inside the slot. Theoretic Q_{lasing} and Γ factor under different t_{SiO2} are shown in Fig. 20.12c, which strongly correlates with each other. For $t_{SiO2} = 10$ nm, Γ factor is 0.193 and Q_{lasing} reaches the highest value of 7.6×10^3. Comparing with the results in Fig. 20.12b, Qs of nanocavities with present t_{SiO2} range are more than two to three orders higher than Q_{lasing}, which show they are sufficient for lasing.

Although we can get ultrahigh Q and Q/V_{eff} from the TM_{00} mode in above 1D PhC NB nanocavity with horizontal SiO_2 slot, these marvelous values only appear for sufficiently thick NB and nanocavity with sufficient large PhC periods. Such device size may be too massive for some applications, and also increases the complexity in manufacturing. Therefore, to further reduce the device volume, we show a different design named PhC nano-fishbone (NFB) nanocavity [36]. Unlike 1D PhC NB with periodic air-holes located at the center, 1D PhC NFB is consisted of periodic half air-holes on both sides of the ridge waveguide, as shown in Fig. 20.13a. For 1D PhC Si NFB with

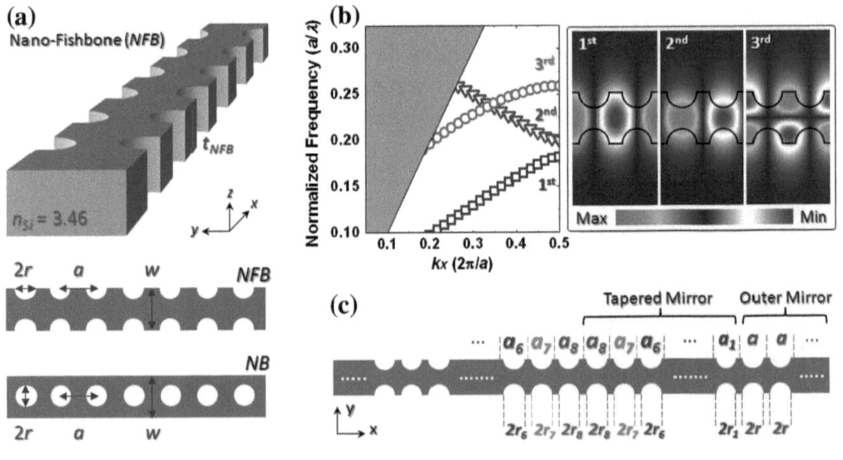

Fig. 20.13 **a** Schemes and lattice parameters of 1D PhC NFB and NB. **b** The first three TM-like bands in 1D PhC NFB and their theoretic mode profiles in E_z field at $k_x = 0.5(2\pi/a)$ via PWE method. **c** Design of 1D PhC NFB nanocavity with the tapered and outer PhC mirrors based on Bloch mode index matching

$w = t_{NFB} = a$ and $r/a = 0.34$, theoretic TM-like band diagram with E_z fields of the first three bands at the band edge ($k_x = 0.5(2\pi/a)$) are shown in Fig. 20.13b. Instead of the fundamental TM_{00} modes (1st and 2nd bands) with very small PBG region and weak slow light effect near the band edge, we focus on the odd-like TM_{10} mode (3rd band) with field concentrating in the regions between the half air-holes. To well confine the TM_{10} mode at the band edge, we propose a NFB nanocavity design with the outer and tapered PhC mirrors on both sides as illustrated in Fig. 20.13c, which is based on the same principle of Bloch mode index matching for designing NB nanocavity. However, the outer and tapered mirrors are formed by only 5-period PhCs with fixed a and 8-period gradually-varied PhCs with lattice constant $a_n = (1\text{–}0.02n)$ a ($n = 1\text{–}8$) and fixed r_n/a_n ratio of 0.34, which has less PhC periods than the NB nanocavity in Fig. 20.11a. With $t_{NFB} = 0.90a$ and $w = a$, theoretic TM_{10} mode profile in $|E_z|$ field in xy-plane in 1D PhC NFB nanocavity is shown in Fig. 20.14a, which has ultrahigh Q of 1.8×10^7 and small V_{eff} of $1.14(\lambda/n_{Si})^3$. In addition, as shown in Fig. 20.14b, with fixed $t_{NFB} = 0.90a$, Q is always larger than 10^6 in the studied ranges of $w = 0.85a\text{–}1.03a$ and $r/a = 0.29\text{–}0.36$, which shows very large tolerances of 1D PhC NFB nanocavity in design and fabrication.

For comparison, TM_{00} mode in NB nanocavity with the same parameters defined in Fig. 20.13c is shown in Fig. 20.14a, which has much lower Q of 900 and larger V_{eff} of $2.8(\lambda/n_{Si})^3$ than those of TM_{10} mode in NFB nanocavity. The differences in Q and V_{eff} between TM_{00} and TM_{10} mode can be understood by their cross-sectional $|E_z|$ field distributions along x-axis shown in Fig. 20.14a. TM_{10} mode in NFB nanocavity shows a Gaussian distribution with narrow linewidth and strong field enhancement, which are responsible for its small V_{eff} and high Q. However, for TM_{00} mode in NB nanocavity, thirteen PhC periods on each side of the nanocavity are insufficient to form a complete Gaussian distribution, thus leads

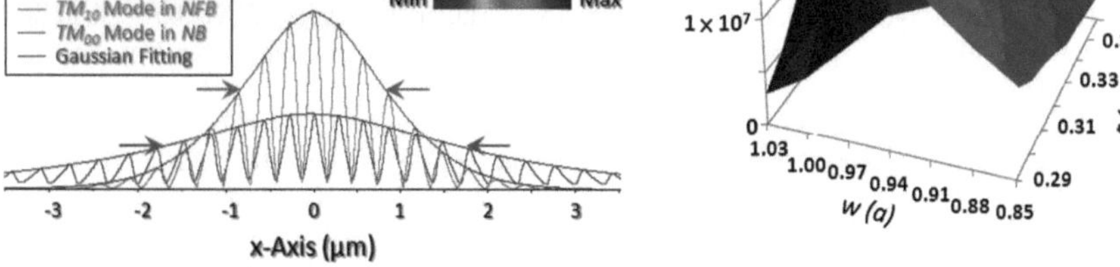

Fig. 20.14 **a** (*Top*) Theoretic mode profiles in $|E_z|$ field in xy-plane and (*Bottom*) cross-sectional $|E_z|$ field distributions along x-axis with Gaussian fittings of TM_{10} and TM_{00} modes confined in 1D PhC NFB and NB nanocavities with the same parameters. **b** Theoretic Q mapping of the TM_{10} mode in 1D PhC NFB nanocavity as functions of w and r/a

to considerable optical losses into the light cone and results in a relatively low Q and large V_{eff} when the device size is too compact.

Furthermore, in addition to the advantage in device size, another important feature of NFB nanocavity is the reduced total etched surface area. Minimizing etched surface area can reduce unnecessary carrier losses from non-radiative surface recombination in active optoelectronic devices. For comparison, under the same thickness of 0.90a and width of a, NFB nanocavities show 30 to 33 % reduction in total etched surface area than NB nanocavities in the r/a range of 0.29–0.36.

20.6 Summary

In summary, we have introduced a variety of ultrahigh Q nanocavities for TE and TM polarizations based on different types 1D PhC NBs. These devices not only have small device footprints, but also show good integration abilities without lattice geometry restriction in design. Most importantly, via slightly changing PhCs on purpose, we can greatly alter the mode properties. Comparing with the massive 2D PhC devices, this kind of 1D PhC NB devices can provide new scenarios as efficient active light sources or other functional passive devices in ultra-condensed planar PICs.

References

1. E. Yablonovitch, Inhibited spontaneous emission in solid-state physics and electronics. Phys. Rev. Lett. **58**, 2059–2062 (1987)
2. S. John, Strong localization of photons in certain disordered dielectric superlattices. Phys. Rev. Lett. **58**, 2486–2489 (1987)
3. T.F. Krauss, Slow light in photonic crystal waveguides. J. Phys. D Appl. Phys. **40**, 2666–2670 (2007)
4. E. Cubukcu, K. Aydin, E. Ozbay, S. Foteinopoulou, C.M. Soukoulis, Electromagnetic waves: Negative refraction by photonic crystals. Nature **423**, 604–605 (2003)
5. Y.A. Vlasov, M. O'Boyle, H.F. Hamann, S.J. McNab, Active control of slow light on a chip with photonic crystal waveguides. Nature **438**, 65–66 (2005)
6. P.T. Lin, P.T. Lee, All-optical controllable trapping and transport of subwavelength particles on a tapered photonic crystal waveguide. Opt. Lett. **36**, 424–426 (2011)
7. B. Ellis, M. Mayer, G. Shambat, T. Sarmiento, E. Haller, J.S. Harris, J. Vuckovic, Ultralow-threshold electrically pumped quantum-dot photonic-crystal nanocavity laser. Nat. Photonics **5**, 297–300 (2011)
8. T.W. Lu, P.T. Lin, K.U. Sio, P.T. Lee, Optical sensing of square lattice photonic crystal point-shifted nanocavity for protein adsorption detection. Appl. Phys. Lett. **96**, 213702 (2010)
9. S. Kita, S. Otsuka, S. Hachuda, T. Endo, Y. Imai, Y. Nishijima, H. Misawa, T. Baba, Super-sensitivity in label-free protein sensing using nanoslot nanolaser. Opt. Express **19**, 17683–17690 (2011)
10. T.W. Lu, P.T. Lee, Ultra-high sensitivity optical stress sensor based on double-layered photonic crystal microcavity. Opt. Express **17**, 1518–1526 (2009)
11. M. Bartha, O. Benson, Manipulation of dielectric particles using photonic crystal cavities. Appl. Phys. Lett. **89**, 253114 (2006)

12. K. Nozaki, A. Shinya, S. Matsuo, Y. Suzaki, T. Segawa, T. Sato, Y. Kawaguchi, R. Takahashi, M. Notomi, Ultralow-power all-optical RAM based on nanocavities. Nat. Photonics **6**, 248–252 (2012)
13. K.J. Vahala, Optical microcavities. Nature **424**, 839–846 (2003)
14. T. Baba, Photonic crystals remember the light. Nat. Photonics **1**, 11–12 (2007)
15. J.S. Foresi, P.R. Villeneuve, J. Ferrera, E.R. Thoen, G. Steinmeyer, S. Fan, J.D. Joannopoulos, L.C. Kimerling, H.I. Smith, E.P. Ippen, Photonic-bandgap microcavities in optical waveguide. Nature **390**, 143–145 (1997)
16. Y. Akahane, T. Asano, B.S. Song, S. Noda, High-Q photonic nanocavity in a two-dimensional photonic crystal. Nature **425**, 944–947 (2003)
17. B.S. Song, S. Noda, T. Asano, Y. Akahane, Ultra-high-Q photonic double-heterostructure nanocavity. Nat. Mater. **4**, 207–210 (2005)
18. M. Notomi, E. Kuramochi, T. Tanabe, Large-scale arrays of ultrahigh-Q coupled nanocavities. Nat. Photonics **2**, 741–747 (2008)
19. P.B. Deotare, M.W. McCutcheon, I.W. Frank, M. Khan, M. Lončar, High quality factor photonic crystal nanobeam cavities. Appl. Phys. Lett. **94**, 121106 (2009)
20. P.T. Lee, T.W. Lu, L.H. Chiu, Dielectric-band photonic crystal nanobeam lasers. J. Lightwave Technol. **31**, 36–42 (2013)
21. T.W. Lu, L.H. Chiu, P.T. Lin, P.T. Lee, One-dimensional photonic crystal nanobeam lasers on a flexible substrate. Appl. Phys. Lett. **99**, 071101 (2011)
22. J. Hu, L. Li, H. Lin, P. Zhang, W. Zhou, Z. Ma, Flexible integrated photonics: where materials, mechanics and optics meet. Opt. Mater. Express **3**, 1313–1331 (2013)
23. C. Jansen, S. Wietzke, V. Astley, D.M. Mittleman, M. Koch, Mechanically flexible polymeric compound one-dimensional photonic crystals for terahertz frequencies. Appl. Phys. Lett. **96**, 111108 (2010)
24. Z. Qiang, W. Zhou, R.A. Soref, Optical add-drop filters based on photonic crystal ring resonators. Opt. Express **15**, 1823–1831 (2007)
25. W.Y. Chiu, T.W. Huang, Y.H. Wu, Y.J. Chan, C.H. Hou, H.T. Chien, C.C. Chen, A photonic crystal ring resonator formed by SOI nano-rods. Opt. Express **15**, 15500–15506 (2007)
26. T.W. Lu, W.C. Tsai, C.Y. Wu, P.T. Lee, Laser emissions from one-dimensional photonic crystal rings on silicon-dioxide. Appl. Phys. Lett. **102**, 051103 (2013)
27. K. Nozaki, S. Kita, T. Baba, Room temperature continuous wave operation and controlled spontaneous emission in ultrasmall photonic crystal nanolaser. Opt. Express **15**, 7506–7514 (2007)
28. M. Bahriz, V. Moreau, R. Colombelli, O. Crisafulli, O. Painter, Design of mid-IR and THz quantum cascade laser cavities with complete TM photonic bandgap. Opt. Express **15**, 5948–5965 (2007)
29. T.W. Lu, P.T. Lin, P.T. Lee, Photonic crystal horizontally slotted nanobeam cavity for silicon-based nanolasers. Opt. Lett. **37**, 569–571 (2012)
30. A.M. Lakhani, M. Kim, E.K. Lau, M.C. Wu, Plasmonic crystal defect nanolaser. Opt. Express **19**, 18237–18245 (2011)
31. Y. Zhang, M.W. McCutcheon, I.B. Burgess, M. Lončar, Ultra-high-Q TE/TM dual-polarized photonic crystal nanocavities. Opt. Lett. **34**, 2694–2696 (2009)
32. V.R. Almeida, Q. Xu, C.A. Barrios, M. Lipson, Guiding and confining light in void nanostructure. Opt. Lett. **29**, 1209–1211 (2004)
33. S. Kita, K. Nozaki, S. Hachuda, H. Watanabe, Y. Saito, S. Otsuka, T. Nakada, Y. Arita, T. Baba, Photonic crystal point-shift nanolaser with and without nanoslots: design, fabrication, lasing and sensing characteristics. IEEE J. Sel. Top. Quantum Electron. **17**, 1632–1647 (2011)
34. C. Creatore, L.C. Andreani, M. Miritello, R. Lo, Savio, and F. Priolo, Modification of erbium radiative lifetime in planar silicon slot waveguides. Appl. Phys. Lett. **94**, 103112 (2009)
35. T.J. Kippenberg, J. Kalkman, A. Polman, K.J. Vahala, Demonstration of an erbium-doped microdisk laser on a silicon chip. Phys. Rev. A **74**, 051802 (2006)
36. T.W. Lu, P.T. Lee, Photonic crystal nano-fishbone nanocavity. Opt. Lett. **38**, 3129–3132 (2013)

Chapter 21
Slow Light in Nano-structured Waveguides

Chii-Chang Chen

Symbols

n_g Group index
c Velocity of the light in vacuum
k Wavenumber
ω Angular frequency of the light
v_g Group velocity
Δf Frequency bandwidth

An optical pulse can be decomposed by several different sinusoidal waves with different frequencies. The velocity of the sinusoidal waves is the phase velocity. The velocity of the optical pulse is so-called the group velocity. In vacuum, the velocity of the light with the different frequencies is identical to be 3×10^8 m/s. In non-vacuum environments, the phase velocity of the light could be different. This phenomenon is called the dispersion. The dispersion can be presented by different ways such as refractive-index versus frequency curve or frequency versus wavenumber curve, etc. The group index is expressed by $n_g = c dk/d\omega$ where c, k and ω are the velocity of the light in vacuum, the wavenumber and the angular frequency of the light, respectively. The group index can be simply obtained by the slope in the refractive-index versus frequency curve. The group velocity which is defined to be $v_g = d\omega/dk$ can also be obtained.

The devices to reduce or to control the propagation velocity of optical pulses can be used in the optical communication systems. The technique is called dispersion engineering. The devices are often called optical delay lines. The delay of optical pulses can be achieved by moving some mechanical mirrors to enlarge the propagation length of the optical pulse in free space or by tuning the refractive

C.-C. Chen (✉)
Department of Optics and Photonics, National Central University,
300 Chung-Da Rd, Chung-Li 32001, Taiwan
e-mail: trich@dop.ncu.edu.tw

C.-C. Lee (ed.), *The Current Trends of Optics and Photonics*,
Topics in Applied Physics 129, DOI 10.1007/978-94-017-9392-6_21

index of the materials where the pulses propagate through. Electromagnetically induced transparency is a method to delay the optical pulses using the transition between the atomic systems in dilute gasses [1]. The group velocity has been reduced to the speed of bike around 17 m/s [1]. However, the size of the experimental setup is hard to be integrated to the practical use. An alternative method to provide the large delay time for optical pulses is to use the materials or structures with a large dispersion property. The phenomenon in which the optical pulses can be significantly delayed is named slow light.

To provide the large delay time and to miniaturize the devices, it is found that the nano-structures such as the photonic crystals can produce the large dispersion. Figure 21.1a, b show the nano-structure and the schematic dispersion relation of the waveguide mode in the photonic band structure, respectively. The nano-structure consists of a photonic crystal waveguide formed by a line defect of missing air holes. In Fig. 21.1b, the slope of the waveguide mode multiplied by a factor of 2π is the group velocity ($d\omega/d\mathrm{k}$). A more flattened curve means a lower group velocity within the corresponding frequency (vertical axis). This phenomenon can be often observed in the waveguide mode of the photonic band structure. Since the size of the devices is in micrometers, the devices might be integrated into the circuit of integrated optics or the devices of optics communications. Therefore, the nano-structure waveguides to obtain the slow light have been intensively investigated [2–5]. To measure the group index, the Fabry-Pérot resonance or Mach-Zehnder interference can be used in frequency domain [2]. In time domain, the measurement of the delayed phase or time of the transmitted light can also be implemented [3]. In [2] the slow-light photonic crystal waveguide is fabricated in silicon membrane. The group index has been experimentally obtained to be 300 [2].

In Fig. 21.1b, we can observe that the bandwidth with low group velocity is narrow. However, a pulse signal in time domain corresponds to a signal consisting of wide bandwidth light source in frequency domain. Therefore, the uniformity of the group velocity within the bandwidth of the light source which is expressed by group velocity dispersion (GVD) $d^2\mathrm{k}/d^2\omega$ is important. The group velocity is uniform as the value of GVD is close to zero. With wide-bandwidth uniform group velocity, the pulse will keep the form in time domain. In opposite, in the case with large GVD, the pulse in time domain will be spread during the propagation in the slow light material or structure. In this case, the dispersion compensation structure should be applied to solve the problem of pulse deformation. Figure 21.2 illustrates schematically the problem of large GVD.

Therefore, the capacity to slow down an optical pulse is decided by the delay time which is controlled by the group index and also by the frequency bandwidth in which the optical pulse can be delayed with the similar group index (low GVD). The product $\mathrm{n_g}\Delta\mathrm{f/f}$ where $\Delta\mathrm{f}$ is the frequency bandwidth becomes an important factor. In [3], as the group index is 233, the product $\mathrm{n_g}\Delta\mathrm{f/f}$ is 0.257. The result in [4] is also in the same order of magnitude. It seems to have a bottleneck to improve the performance.

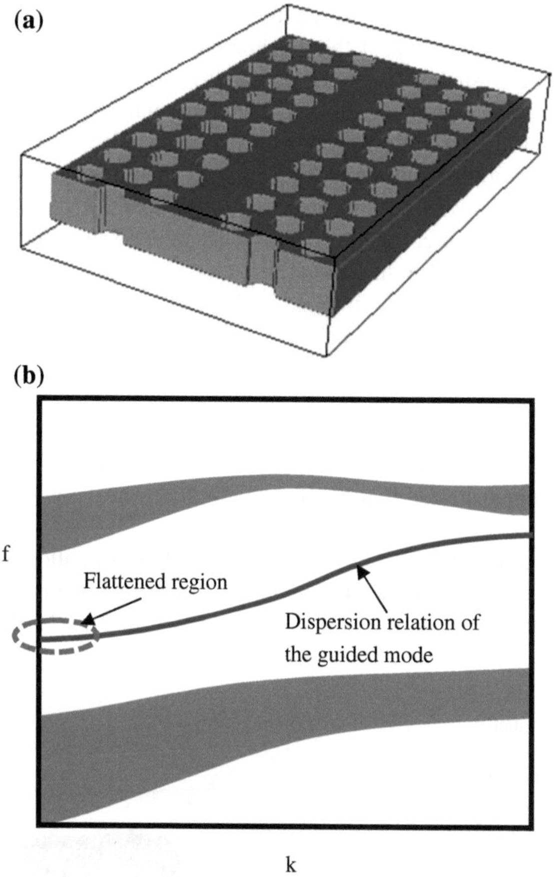

Fig. 21.1 **a** nano-structure waveguide in photonic crystals, **b** schematic dispersion relation of the waveguide mode

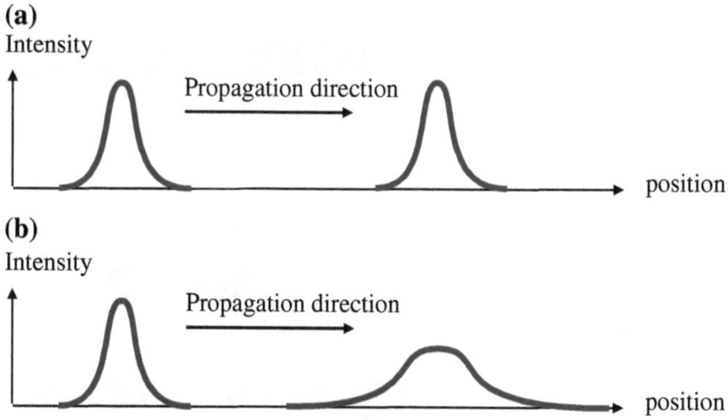

Fig. 21.2 Schematic drawing of the deformation of a pulse after the propagation in **a** zero dispersion materials or structures **b** large GVD materials or structures

To enhance the light-matter interaction, the slow or even stop light might be the solutions. The application of this phenomenon is to ameliorate the efficiency of the active devices such as lasers or photodiodes [5–7]. However, for the applications of the optical buffers or delayed lines, the enhanced light-matter interaction could induce the strong absorption of the light pulse or wavelength change due to high-order optical non-linearities. One of the solutions is to use the hollow waveguides. The idea has been proposed in [8] in which the radiation pressure was studied in an omnidirectional-reflector waveguide. The semiconductor hollow waveguides formed by omnidirectional reflectors (SHOW-ODR) have been realized by two methods as shown in Fig. 21.3 [9, 10]. The omnidirectional reflector is formed by periodically deposited Si and SiO_2 thin films. The first method uses the wafer bonding technique. The propagation loss is around 1 dB/cm for both the TE and TM modes which is close to the specification of commercial use. However, the end polishing is required [9]. The second method uses the photoresist as the sacrificing layer [10]. The end polishing is waived. The experimental result has also demonstrated that the 90° bending loss of the waveguide can be as low as 1 dB.

An alternative structure of the hollow waveguide is so-called doughnut waveguides [11]. The structure is formed by tori as illustrated in Fig. 21.4a. The light is confined in the center of the structures as shown in Fig. 21.4b. The usage of

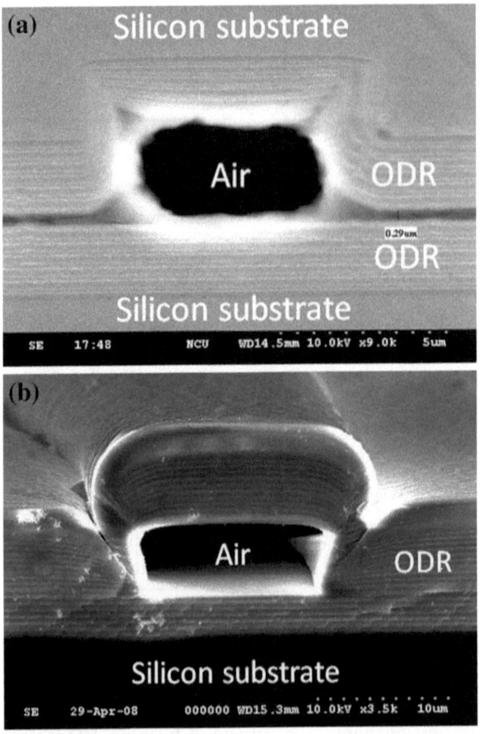

Fig. 21.3 Semiconductor hollow waveguide formed by omnidirectional reflectors **a** by wafer bonding [9], **b** by using sacrificing layer [10]

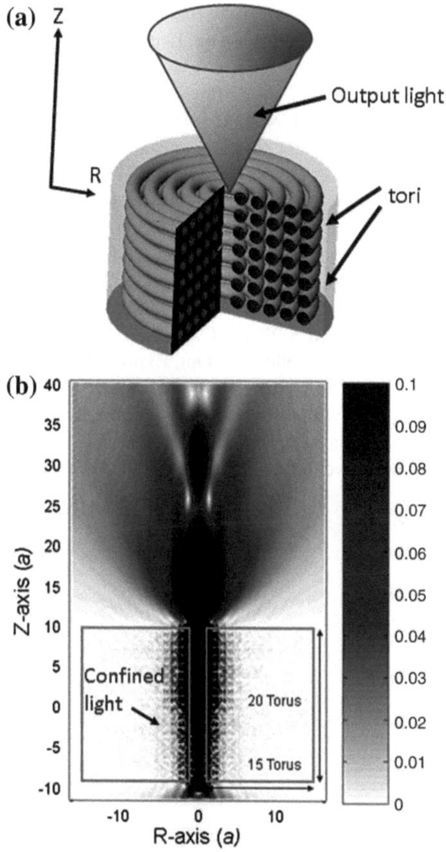

Fig. 21.4 **a** Schematic structure of doughnut waveguide. **b** The optical field distribution confined in the doughnut waveguide [9]

the dielectric materials of the doughnut waveguide might be less than the SHOW-ODR leading to lessen the problem of material absorption and non-linearities.

In summary, the device with slow light effect has been demonstrated recently. The device size is in the scale of micrometer. The feature can facilitate the integration of the devices into optical communication system. The enhancement of the product $n_g\Delta f/f$ of the device is required. To reduce the problem of the absorption and the non-linearities problem, the hollow waveguides may be one of the solutions.

References

1. L. Vestergaard Hau, S.E. Harris, Z. Dutton, C.H. Behroozi, Light speed reduction to 17 meters per second in an ultracold atomic gas. Nature **397**, 594–598 (1999)
2. Y.A. Vlasov, M.O'Boyle, H.F. Hamann1, S.J. McNab, Active control of slow light on a chip with photonic crystal waveguides. Nature **438**, 65–69 (2005)

3. S. Kubo, D. Mori, T. Baba, Low-group-velocity and low-dispersion slow light in photonic crystal waveguides. Opt. Lett. **32**, 2981–2983 (2007)
4. H. Wu, D.S. Citrin, L.Y. Jiang, X.Y. Li, Polarization-independent slow light in annular photonic crystals. Appl. Phys. Lett. **102**, 141112 (2013)
5. J.K. Yang, H. Noh, M.J. Rooks, G.S. Solomon, F. Vollmer, H. Cao, Lasing in localized modes of a slow light photonic crystal waveguide. Appl. Phys. Lett. **98**, 241107 (2011)
6. R. Hayakawa, N. Ishikura, H.C. Nguyen, T. Baba, Two-photon-absorption photodiodes in Si photonic-crystal slow-light waveguide. Appl. Phys. Lett. **102**, 031114 (2013)
7. W.C. Lai, S. Chakravarty, Y. Zou, Y. Guo, R.T. Chen, Slow light enhanced sensitivity of resonance modes in photonic crystal biosensors. Appl. Phys. Lett. **102**, 041111 (2013)
8. M.L. Povinelli, M. Ibanescu, S.G. Johnson, J.D. Joannopoulos, Slow-light enhancement of radiation pressure in an omnidirectional-reflector waveguide. Appl. Phys. Lett. **85**, 9 (2004)
9. S.S. Lo, M.S. Wang, C.C. Chen, Semiconductor hollow optical waveguides formed by omnidirectional reflectors. Opt. Express **12**, 6589–6593 (2004)
10. H.K. Chiu, F.L. Hsiao, C.H. Chan, C.C. Chen, Compact and low-loss bent hollow waveguides with distributed Bragg reflector. Opt. Express **16**, 15069–15073 (2008)
11. Y.L. Tsai, K.C. Hsu, C.C. Chen, C.Y. Chen, J.Y. Chang, Optical confinement using a doughnut waveguides. J. Phys. D Appl. Phys. **43**, 245103 (2010)

Chapter 22
Obliquely Deposited Negative Index Film

Yi-Jun Jen

Symbols

θ_v	The angle of incoming vapor flux
ε	Equivalent permittivity
μ	Equivalent permeability
GLAD	Glancing angle deposition
LPM	Longitudinal plasmon mode
NRA	Nanorod array
SWF	Sandwiched film
TPM	Transverse plasmon mode

22.1 Introduction

Metamaterials with subwavelength structures have been demonstrated to have novel optical properties, including negative refractive indices. Technological development has led to the reduction of the scale of metamaterials, which now exhibit a negative refractive index from microwave to visible wavelengths. Since a typical metamateiral is a kind of metal-dielectric composite, its optical performance is always affected by the absorption problem, preventing such applications as perfect lens and cloaking [1]. However, between the discovery of metamaterials and the development of metadevices, the mass-produce of metamaterials is an important issue. Apart from lithography, glancing angle deposition has been demonstrated to be a simple method of coating a nano-structured thin film with a negative refractive index at visible wavelengths. Basically, the glancing angle

Y.-J. Jen (✉)
Department of Electro-Optical Engineering, National Taipei University of Technology, Taipei 106, Taiwan
e-mail: jyjun@ntut.edu.tw

C.-C. Lee (ed.), *The Current Trends of Optics and Photonics*,
Topics in Applied Physics 129, DOI 10.1007/978-94-017-9392-6_22

deposition (GLAD) is a technique that involves tilting the substrate such that the incoming vapor flux makes an angle θ_v with respect to the substrate normal, causing the shadowing effect, which ensures that incident atoms join the nuclei that have already formed on the surface [2]. The shadowing effect reveals that a larger deposition angle results in the formation of a film with a larger porosity. Therefore, the fraction of metal in the film can be controlled by varying the deposition angle. In 2007, GLAD was used applied to control the effective refractive index of a dielectric thin film [3]. The tunable refractive indices of a five-layered structure realize broadband and wide angle antireflection. The birefringence of the dielectric thin film was also exploited to form a multilayered structure that performs achromatic phase retardation.

22.2 A Nanostructured Thin Film with Negative Index

Metal nanorod arrays (NRAs) can also be prepared using GLAD. Owing to the property of metal, the fabrication of a metal nanorod array requires a very large deposition angle between the surface normal and the direction of the vapor flux. Slanted silver nanorods were grown at around 70° to the surface normal when the deposited vapor flux was incident at a deposition angle of 85°. The non-unity permeability must be considered in determining the equivalent electromagnetic parameters, because when the fraction of metal exceeds about 30 %, the traditional effective medium approximation fails to apply in the prediction of the refractive index of the film. When the equivalent permittivity $\varepsilon = \varepsilon' + i\varepsilon''$ and permeability $\mu = \mu' + i\mu''$ satisfy the inequality $\varepsilon'\mu'' + \varepsilon''\mu' < 0$, the real part of the refractive index becomes negative.

A slanted silver NRA exhibits negative refractive index at visible wavelengths. When it is illuminated by p-polarized light (with the direction of the oscillating electric field along the rods) and s-polarized light (the direction of the oscillating electric field is perpendicular to the rods), the longitudinal plasmon mode (LPM) and the transverse plasmon mode (TPM) are excited, respectively. In the LPM, the film exhibits high absorption over a broad range of wavelengths. The TPM typically applies at wavelengths that are shorter than the wavelengths to which the LPM applies. The slanted Ag NRA exhibits both an anisotropic refraction index and extinction coefficient. The huge phase retardation, measured normally from the Ag NRA, suggests that one of the plenary refractive index is negative real while the other is positive. The strong positive to negative birefringence is demonstrated by measuring the polarization-dependent refractive index and impedance. A walk-off interferometer is used to measure the transmission and reflection coefficients, which are used to calculate the equivalent permittivity and permeability. The associated optical parameters including refractive index and impedance can then be derived from the permittivity and permeability using the equations. A slanted Ag NRA with average diameter of 108 nm average separation of 191 nm, length of 336 nm and tilt angle of 76° with respect to surface normal exhibits negative real refractive index when the NRA is normally illuminated with polarization along the rods [4]. The negative real part is measured to be around −0.6 in the green light and −0.9 for the red light.

22.3 Interference Observation from a Low Loss Film with Negative Index

In recent work, a sandwiched Al-SiO_2-Al upright nanorod array was fabricated by rotating a substrate rapidly during deposition. Such a sandwiched NRA exhibits a polarization-independent negative refractive index and equivalent permeability over visible wavelengths. The loss of the sandwiched film (SWF) is low enough to allow light to propagate through, and to be reflected back from the film producing the interference effect. Besides exhibiting negative real refractive index, the extinction coefficient is reduced to be around 0.1. The figure of merit defined as the ratio of real part to imaginary part of refractive index becomes 50 at red light. Notably, this film has low reflectance over the visible regime. The antireflection that is caused by destructive interference from the film observed and demonstrated herein. The negative phase change that is caused by the backward propagation of a wave in a medium with a negative refractive index contributes to the destructive interference between the first two orders of reflected waves. Figure 22.1 presents the first two reflected waves with a wavelength of 676 nm traced through a sandwiched Al-SiO_2-Al film with average diameter of 161 nm, average separation of 323 nm, average Al thickness of 195 nm and average SiO_2 thickness of 50 nm.

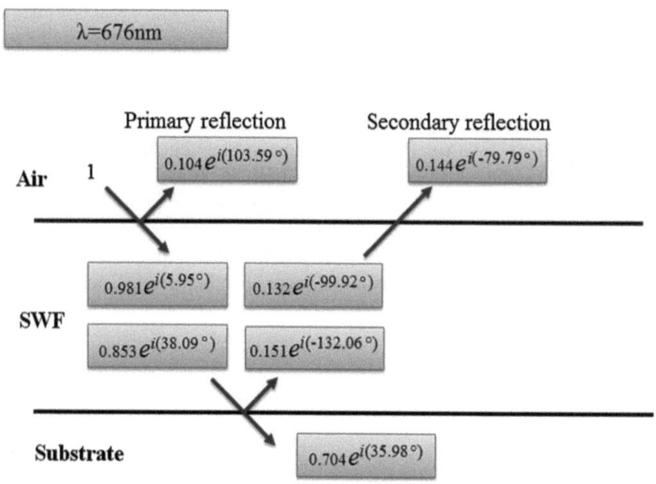

Fig. 22.1 Wave tracing for the SWF normally illuminated by light with an electric field amplitude of unity at wavelength of 676 nm

The bianisotropic parameters including equivalent permittivity, equivalent permeability and bianisotropic parameter are measured using walk-off interferometer [5]. The negative phase change that is associated with the backward wave propagation is approximately 180°, causing the first two reflected waves to cancel each other out in complete antireflection.

22.4 Summary

Newly designed optical coatings are expected to comprise multilayered structures that comprise one or more films with negative indices. More compact structures will can replace traditional multilayered structures, providing the same optical responses such as antireflection, high reflection and absorption [6]. Thin films with negative refractive indices represent a new alternative for use in optical coating design. Additionally, the mutually independent impedance and refractive index also give designers of optical coatings one more parameter to control. Tailoring a nanostructured thin film with a specific impedance and refractive index is an important issue, and depends on accurate control of the near-field electric and magnetic field distributions within the nanostructure in the future.

References

1. C.M. Soukoulis, M. Wegener, Past achievements and future challenges in the development of three-dimensional photonic metamaterials. Nat. Photonics **5**, 523–530 (2011)
2. I.J. Hodgkinson, Q.H. Wu, *Birefringent Thin Films and Polarizing Elements* (World Scientific, Singapore, 1997)
3. J.-Q. Xi, M.F. Schubert, J.K. Kim, E.F. Schubert, M. Chen, S.-Y. Lin, W. Liu, J.A. Smart, Optical thin-film materials with low refractive index for broadband elimination of Fresnel reflection. Nat. Photonics **1**, 176–179 (2007)
4. Y.-J. Jen, C.-H. Chen, C.-W. Yu, Deposited metamaterial thin film with negative refractive index and permeability in the visible regime. Opt. Lett. **36**, 1014–1016 (2011)
5. C.É. Kriegler, M.S. Rill, S. Linden, M. Wegener, Bianisotropic photonic metamaterials. IEEE J. Sel. Top. Quantum Electron. **16**, 367–375 (2010)
6. K. Aydin, V.E. Ferry, R.M. Briggs, H.A. Atwater, Broadband polarization-independent resonant light absorption using ultrathin plasmonic super absorbers. Nat. Commun. **2**, 517 (2011)

Chapter 23
Antireflective Nanostructures for Solar Cells

Hsin-Ping Wang, Jr-Hau He and Hung-Chih Chang

Symbols

n_{eff}	Effective refractive index
n_0	Refractive indices of air
n_s	Refractive indices of substrate
n_i	Refractive indices of intermediate films
R	Reflectance
ARC	Antireflection coating
AOI	Angle of incident
EMT	Effective medium theory

The AR coatings (ARCs) are utilized to suppress undesired Fresnel reflection between different media, directly enhancing amount of photons to enter the devices. The development of ARCs is a long process with multiple stages.

23.1 Thin Film AR

The concept of ARCs can be proposed by Lord Rayleigh in the 19th century when he found that the glass with tarnished surface can increase the transmittance of incident light [2]. Scientists later realized the AR effects can be achieved

H.-P. Wang · J.-H. He (✉) · H.-C. Chang
Department of Electrical Engineering, Institute of Photonics and Optoelectronics,
National Taiwan University, No. 1, Sec. 4, Roosevelt Road, Taipei 10617, Taiwan
e-mail: jhhe@cc.ee.ntu.edu.tw

H.-P Wang
e-mail: chicken19860813@hotmail.com

H.-C Chang
e-mail: b94505047@ntu.edu.tw

C.-C. Lee (ed.), *The Current Trends of Optics and Photonics*,
Topics in Applied Physics 129, DOI 10.1007/978-94-017-9392-6_23

by the idea of light destructive interference, and the first ARC was demonstrated by Fraunhofer [3] who used the etched surface to suppress reflection. After a long period of development, now the most widely used ARCs in industry are the quarter-wavelength-thick dielectric layers, due to their simplicity and high fabrication rate.

23.1.1 Single Layer ARC

Inthe design of quarter-wavelength ARCs, the single antireflective layer is assumed to be sandwiched by two semi-infinite media, which are usually the air and the substrate. For this simple layer structure, the refractive index of the intermediate film should be equal to $\sqrt{n_0 n_s}$ in order to attain the minimized reflectance, as dictated by the following equation: [4–6]

$$R = \left(\frac{n_0 n_s - n_i^2}{n_0 n_s + n_i^2} \right)^2 \tag{23.1}$$

where R is reflectance, n_0, n_s, and n_i are the refractive indices of air, substrate, and intermediate films, respectively. To fulfill the requirement of Eq. (23.1), the material selected for quarter-wavelength ARCs are usually the dielectrics with proper n_i, such as ZnO, ITO, TiO_2, SiO_2, Si_3N_4, etc. For example, on Si-based solar cells, where the substrate is of $n_s \sim 4.2$ at the wavelength of 528 nm, the ARC made of SiN_x with $n_i \sim 2.05$ gives near zero reflectance at 528 nm. However, quarter-wavelength ARCs only render the minimized reflectance at a narrow wavelength range, as shown in Fig. 23.1a [6, 7]. Therefore, this type of ARC may not be suitable for broadband AR applications.

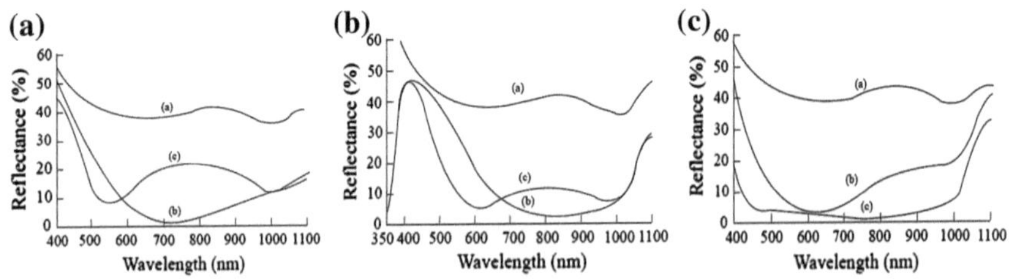

Fig. 23.1 **a** Reflection of (*a*) uncoated silicon (*b*) coated SiO single quarter wavelength optical thickness (*c*) coated MgF_2 + SiO double layer with equal quarter wavelength thickness. **b** Reflection of (*a*) uncoated silicon (*b*) coated CeO_2 single quarter wavelength optical thickness (*c*) coated MgF_2 + CeO_2 double layer with equal quarter wavelength thickness. **c** Reflection of (*a*) uncoated silicon (*b*) coated ZnS single quarter wavelength optical thickness (*c*) coated MgF_2 + ZnS double layer with equal quarter wavelength thickness. Reprinted with permission from [7]

23.1.2 Multi-layer ARC

To solve the disadvantages in single ARCs, double or multi-layer ARCs are fabricated to suppress reflection over a wide range of wavelengths. The broadband AR abilities of a multi-layer structure are attained via the destructive interferences among the waves reflected from layer interfaces. For double ARCs, the two dielectric layers are usually of identical thicknesses and have the refractive indices that follow the rule: $n_0 n_s = n_1 n_2$, where n_1 and n_2 are the indices of the two layers. Figure 23.1a–c show the effectiveness of three different types of double ARC, which are made of MgF_2 ($n = 1.38$), SiO ($n = 1.85$), CeO_2 ($n = 2.2$), and ZnS ($n = 2.3$) [7]. Compared with the bare surface and the single-layer ARCs, all of the double ARCs exhibit two valleys in reflectance. The positions of two reflectance minima can be controlled through suitable selection of layer thicknesses and refractive indices. Although multi-layer ARCs can increase the range of working wavelength, it should be noted that multi-layer ARCs only function properly at nearly normal incident angles. The reflections generally exhibit gradual increase for the angles away from normal incidence. To eliminate the reflectance in wide ranges of wavelengths and incident angles, many groups have developed the ARCs with subwavelength nanostructures. For instance, Xi et al. utilized the oblique-angle deposition technique to create a graded refractive index by controlling the porosity of the deposited TiO_2 and SiO_2 films, as shown in Fig. 23.2a. The oblique-angle nanostructures exhibit nearly zero reflectance not only in a very wide range of wavelengths, but also in a wide range of incident angles, which can be seen in Fig. 23.2b, c [8].

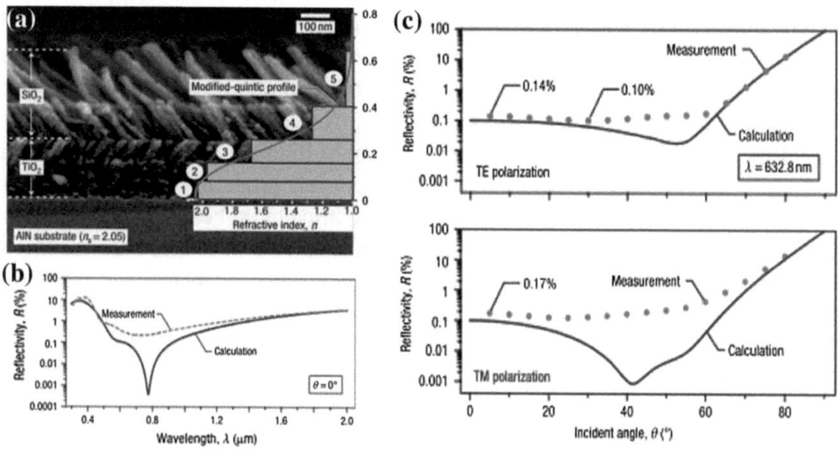

Fig. 23.2 **a** ARC with Graded index transition of cross-sectional SEM image. **b** Simulation and measurement of ARC with graded index. **c** TE and TM reflectance in different angle of incident. Reprinted with permission from [8]

23.2 Nanostructured ARC

Due to their superior AR performances, antireflective nanostructures have recently drawn extensive research efforts. Depending on the geometric features, nanostructured ARC are commonly grouped into two important types:

23.2.1 *Homogeneous Nanostructured ARC*

The homogeneous nanostructured ARC displays an unchanged feature from bottom to top of the nanostructure. In this type of ARC, the refractive index of ARC can be seen as a specific value according to the effective medium theory (EMT) [9–11]. The mathematical models of EMT were proposed by Garnett and Bruggeman, who calculate the effective refractive index (n_{eff}) of a homogeneous nanostructured ARC with the following equations: [9–11]
Maxwell-Garnett model:

$$\left(\frac{n_{eff}^2 - n_1^2}{n_{eff}^2 + 2n_1^2}\right) = (1 - f_1)\left(\frac{n_2^2 - n_1^2}{n_2^2 + 2n_1^2}\right) \tag{23.2}$$

Bruggeman approximation:

$$f_1\left(\frac{n_1^2 - n_{eff}^2}{n_1^2 + 2n_{eff}^2}\right) + (1 - f_1)\left(\frac{n_2^2 - n_{eff}^2}{n_2^2 + 2n_{eff}^2}\right) = 0 \tag{23.3}$$

where n_1 and n_2 are the refractive indices of two constituent layers, and f_1 is the corresponding volume ratio. With EMT models, one can see the n_{eff} of the ARC can be tuned via the filling ratio of the nanostructure. For instance, Wang et al. [12] reported the Si nanopillar arrays fabricated by colloidal lithography combined with metal-assisted chemical etching exhibit excellent AR properties in broadband wavelengths and angles of incident (AOIs) regions. The authors demonstrated that the n_{eff} of the nanopillar arrays can be varied by controlling the diameter of nanopillar, causing different AR performance. The excellent AR properties were achieved by the optimized value of n_{eff}, which minimizes the impedance between the air and the Si substrate and thus facilitates light traveling through the interface. Figure 23.3a–d show the SEM image of the Si nanorods with controlled diameters and the corresponding reflectance spectra over the wavelength range of 200–1,900 nm.

23.2.2 *Inhomogeneous Nanostructured ARC*

In contrast to the homogeneous counterpart, theinhomogeneous nanostructured ARC shows a gradualchange in geometry from the bottom to the top. The changed

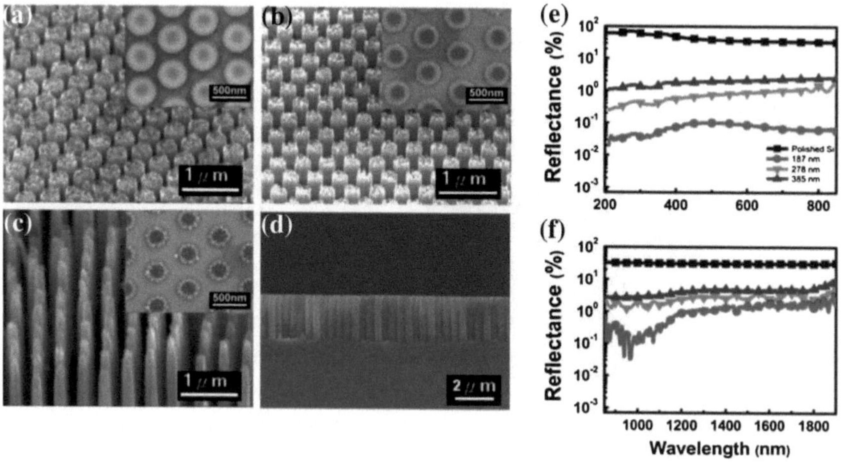

Fig. 23.3 **a–d** SEM of Si nanorods with controlled diameters. **e–f** Reflectance of nanorods. Reprinted with permission from [12]

geometry usually leads to a grading in the n_{eff}, which can further suppress the undesired reflectance in comparison with the performances of the homogeneous ARCs. The other advantage of inhomogeneous ARCs is that their graded index is attained by a single material, not by the deposition of different dielectrics with varied refractive indices, which tends to encounter adhesion problems due to the difference in thermal expansion coefficients. Seeing the merits of inhomogeneous nanostructured ARCs, many research labs have developed the techniques to obtain various inhomogeneous nanostructures. Fan et al. reported that the dual-diameter nanostructures can obtain broadband absorption, as shown in Fig. 23.4 [13]. They systematically analyzed reflectance, transmittance, and absorption of nanorod arrays with varied diameters. The small diameter rod on the top is for minimizing the reflectance and increasing the amount of light entering the structures, and then the large diameter in the base is for maximal optical absorption once the photon enters the nanorod array. As a result, the optimized Ge dual-diameter nanorod array exhibits 95–100 % absorption for wavelength from 300 to 900 nm, which is a drastic improvement over single-diameter nanorod arrays. Yeh et al. also proposed a syringe-like nanorod array with an ultrasharp tip, providing a gradual broadening in diameter, as presented in Fig. 23.5 [14]. The syringe-like nanorods were found to exhibit lower reflectance than that caused by the flat-top nanorods. The results were also attributed to the improved impedance matching at the air/nanorods interface. Chao et al. also reported a flower-like nanorod array, whose nanorod filling ratio gradually increased from the top area to the bottom area, as shown in Fig. 23.6 [15]. Similarly, the flower-like nanorods display improved AR properties, including broadband working range, omnidirectionality, and polarization-insensitivity, as compared with the well-aligned nanorods.

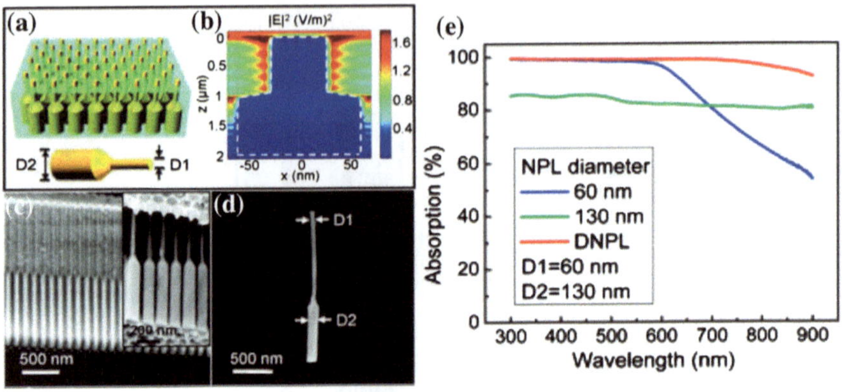

Fig. 23.4 **a** Schematic of Ge nanorods with dual diameters. **b** Simulation of Ge nanorod. **c–d** TEM of Ge nanorods with controlled diameters. **e** Absorption of different Ge nanorods. Reprinted with permission from [13]

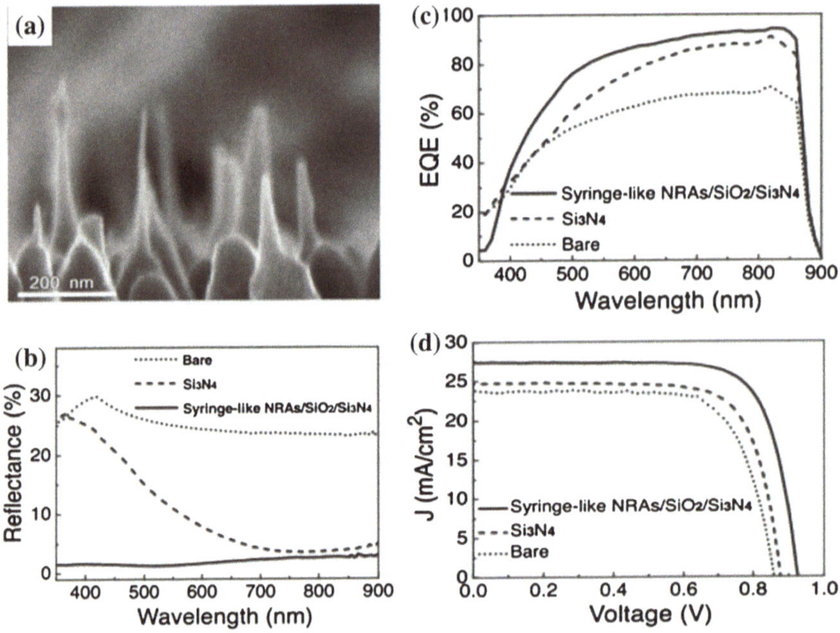

Fig. 23.5 **a** Schematic of syringe-like ZnO nanorods. **b** Reflectance of GaAs solar cells with Si_3N_4 and syringe-like ARC. **c** EQE and **d** I–V measurement of GaAs solar cells. Reprinted with permission [14]

Chang et al. [16] reported wafer-scaled nanowire arrays by one-step galvanic etching method to create the micro-roughened interface underlying the etched nanowires, which exhibit excellent AR properties (Fig. 23.7). Chang et al. compared the reflectance of three samples: (a) the 5.21 ± 0.09 μm nanowire arrays with abrupt

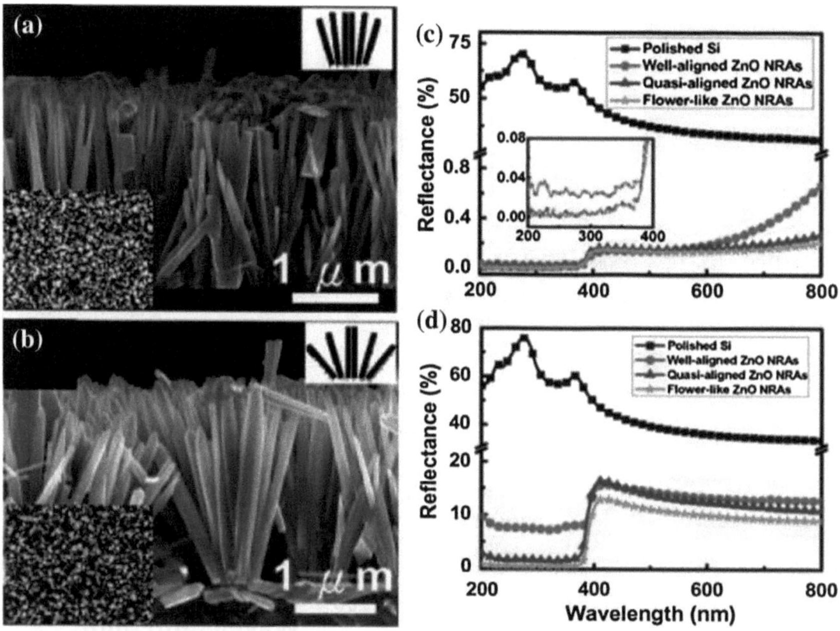

Fig. 23.6 **a** and **b** Cross-sectional SEM of different aligned ZnO nanorods. **c** Specular and **d** total reflectance of ZnO nanorods. Reprinted with permission from [15]

interfaces, (b) the 4.52 ± 0.95 μm nanowire arrays with the rough interfaces, and (c) the 3.68 ± 0.10 μm inverted micropyramid texturing combined with the 100 nm Si_3N_4 layer on the Si solar cell (Fig. 23.7e). The nanowire arrays with the rough interfaces suppressed the specular reflectance to <0.1 % in the wavelengths from 200 to 850 nm, even superior to the surfaces purely consisting of longer nanowires with up to 5.21 ± 0.09 μm in length. The AR performance of nanostructures can be further improved when the slope of the refractive index profile becomes minimum.

23.2.3 Combination of Thin Film ARC and Nanostructures

Although nanostructures hold the advantage of broadband working range, the very low reflectance generally requires lengthened nanostructures (>2 μm), which not only cost much material and fabrication time but also is vulnerable to mechanical damages. To obtain the improved AR performances in a cost-effective way, some research groups proposed that combining quarter-wavelength and nanostructured ARCs is a feasible method [14, 17]. Since a quarter-wavelength layer is typically of the thickness less than 0.5 μm, the composite ARCs can effectively reduce the length of nanostructures while maintain favored AR properties.

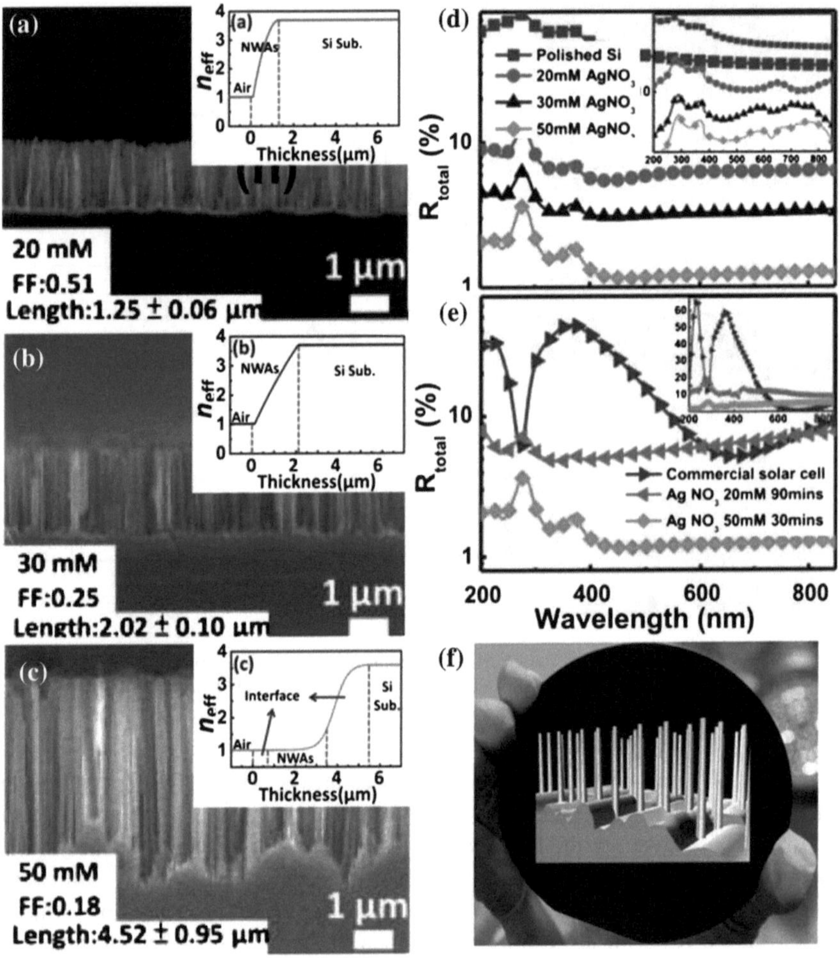

Fig. 23.7 The cross-sectional SEM images of Si nanowire array layers fabricated with **a** 20 mM, **b** 30 mM, and **c** 50 mM $AgNO_3$. The inset shows the calculated refractive index profiles of corresponding Si nanowire arrays. **d** Total reflectance of the nanowire arrays obtained with different $AgNO_3$ concentrations. **e** The total reflectance of the nanowire arrays with abrupt interfaces (height: 5.21 ± 0.09 µm), the nanowire arrays with rough interfaces (height: 4.52 ± 0.95 µm), and the inverted micropyramid texturing in combination with 100 nm-thick Si_3N_4 of a commercial Si solar cell (height: 3.68 ± 0.10 µm). **f** A photographic image of the nanowire arrays with rough interfaces structures on 5-inch wafers. Reprinted with permission from [16]

Lin et al. [17] showed only the 500 nm-long ZnO nanorod arrays grown on Si_3N_4 thin film ARCs reduce the surface reflections over a wide range of wavelengths and incident angles, as show in Fig. 23.8. The AOI limit is pushed to 65°, below which the reflectance is less than 0.3 % for wavelengths of 300–850 nm. The J_{sc} of Si solar cell exhibits strong increase from 32.48 (Si_3N_4/microgrooved structures) to 38.45 mA/cm^2 (ZnO nanorod arrays/Si_3N_4/microgrooved structure).

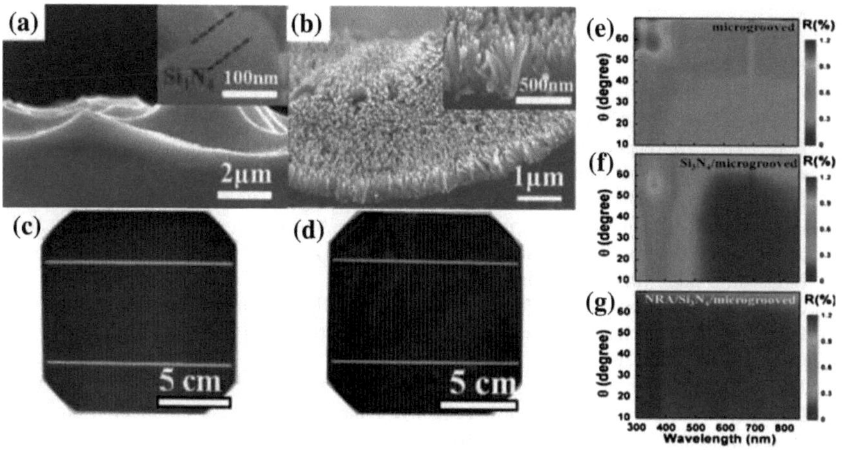

Fig. 23.8 Cross-sectional SEM images of **a** the Si solar cell with 70 nm Si_3N_4-coated microgrooved and **b** the Si solar cell with the ZnO nanorod arrays/Si_3N_4-coated Si microgrooves. A photographic image of **c** the Si solar cell with 70 nm Si_3N_4-coated microgrooved and **d** the Si solar cell with the ZnO nanorod arrays/Si_3N_4-coated Si microgrooves. The reflectance spectra as a function of AOIs and wavelengths on the Si solar cells with the surfaces of **e** bare microgrooves, **f** Si_3N_4/microgrooves and **g** ZnO NRA/Si_3N_4/microgrooves. Reprinted with permission from [17]

In the antireflective structure reported by Yeh et al. [14] a hierarchical ARC made of syringe-like ZnO nanorods and quarter-wavelength Si_3N_4 thin film was fabricated on GaAs solar cells. The Si_3N_4 thin film was designed to suppress the reflectance at 750 nm, which is the wavelength giving peak efficiencies of the solar cell. The 1.5-μm nanorods were then employed to further reduce the reflectance at the wavelengths around 750 nm. Shown in Fig. 23.5, the hierarchical AR structure effectively reduced the reflectance at the wavelengths of 400–900 nm, where external quantum efficiencies of the device were maximized. As a result, the conversion efficiency under the illumination of air mass 1.5 G was enhanced by 32 %.

23.3 Future of ARC

Numerous types of ARCs have been developed for the last few decades. From quarter-wavelength thin films, mirostructures to nano-structures, extremely low reflectance has been achieved in broad wavelengths and in wide range of incident angles. However, challenges still remain. For instance, although many nanostructures are of superior AR properties compared with quarter-wavelength thin films, maintaining the uniformity and the low reflectance over wafer-scale dimensions in a cost-effective way has been shown to be difficult. In future, the advanced ARC structures with improved AR performances should attract continuous research efforts to attain the higher uniformity, the lower cost, and the superior optical properties.

References

1. A. Polman, H.A. Atwater, Photonic design principles for ultrahigh-efficiency photovoltaics. Nat. Mater. **11**, 174–177 (2012)
2. L. Rayleigh, On reflection of vibrations at the confines of two media between which the transition is gradual. Proc. London Math Soc. **s1**, 51–56 (1879)
3. J. Fraunhofer, *Joseph von Fraunhofer Gesannekte Schriften* (Munich, Germany, 1888)
4. H.K. Raut, V.A. Ganesh, A.S. Nair, S. Ramakrishna, Anti-reflective coatings: A critical, in-depth review. Energy Environ. Sci. **4**, 3779–3804 (2011)
5. E. Hecht, *Optics*, 4th edn. (Addison-Wesley, New York, 2001)
6. C.G. Someda, *Electromagnetic Waves* (Chapman & Hall, London, 1998)
7. I.G. Kavakli, K. Kantarli, Single and double-layer antireflection coatings on silicon. Turk. J. Phys. **26**, 349–354 (2002)
8. J.Q. Xi, M.F. Schubert, J.K. Kim, E.F. Schubert, M. Chen, S.-Y. Lin, W. Liu, J.A. Smart, Optical thin-film materials with low refractive index for broadband elimination of Fresnel reflection. Nat. Photon **1**, 176–179 (2007)
9. J.C.M. Garnett, Colours in metal glasses and in metallic films. Phil. Trans. R. Soc. A **203**, 385–420 (1904)
10. J.C.M. Garnett, Colours in metal glasses, in metallic films, and in metallic solutions. II. Phil. Trans. R. Soc. A **205**, 237–288 (1906)
11. D.A.G. Bruggeman, Berechnung verschiedener physikalischer Konstanten von heterogenen Substanzen. I. Dielektrizitätskonstanten und Leitfähigkeiten der Mischkörper aus isotropen Substanzen. Ann. Phys. **416**, 636–664 (1935)
12. H.P. Wang, K.Y. Lai, Y.R. Lin, C.A. Lin, J.H. He, Periodic Si nanopillar arrays fabricated by colloidal lithography and catalytic etching for broadband and omnidirectional elimination of Fresnel reflection. Langmuir **26**, 12855–12858 (2010)
13. Z. Fan, R. Kapadia, P.W. Leu, X. Zhang, Y.-L. Chueh, K. Takei, K. Yu, A. Jamshidi, A.A. Rathore, D.J. Ruebusch, M. Wu, A. Javey, Ordered arrays of dual-diameter nanopillars for maximized optical absorption. Nano Lett. **10**, 3823–3827 (2010)
14. L.K. Yeh, K.Y. Lai, G.J. Lin, P.H. Fu, H.C. Chang, C.A. Lin, J.H. He, Giant efficiency enhancement of GaAs solar cells with graded antireflection layers based on syringelike ZnO nanorod arrays. Adv. Energy Mater. **1**, 505 (2011)
15. Y.C. Chao, C.Y. Chen, C.A. Lin, J.H. He, Light scattering by nanostructured anti-reflection coatings. Energy Environ. Sci. **4**, 3436–3441 (2011)
16. H.C. Chang, K.Y. Lai, Y.A. Dai, H.H. Wang, C.A. Lin, J.H. He, Nanowire arrays with controlled structure profiles for maximizing optical collection efficiency. Energy Environ. Sci. **4**, 2863–2869 (2011)
17. C.A. Lin, K.Y. Lai, W.C. Lien, J.H. He, An efficient broadband and omnidirectional light-harvesting scheme employing a hierarchical structure based on a ZnO nanorod/Si3N4-coated Si microgroove on 5-inch single crystalline Si solar cells. Nanoscale **4**, 6520–6526 (2012)

Chapter 24
Nanorod LED Arrays

Jin-Yi Tan, Liang-Yi Chen and Jian-Jang Huang

Symbol

E_2^H	Phonon mode of material

24.1 Fabrication of Nanorod LED

Wide bandgap GaN-based light emitting diodes (LEDs) have been intensively developed and revolutionized the solid-state lighting market in the past 10 years due to their ability to emit light ranging from ultraviolet to the short-wavelength part over the visible spectrum. However, there are some difficulties that limit the performance of GaN-based LEDs. The low light-extraction efficiency and strain-induced quantum-confined Stark effect (QCSE) are the bottlenecks for high-power LEDs. For a GaN-based LED epi-structure, the lattice mismatch between InGaN and GaN results in strain and induces piezoelectric field. The field leads to the tilt of energy band structure and thus the separation of carriers in the active region, which decreases internal quantum efficiency (IQE) and causes wavelength shift of light emission. Great efforts are needed to overcome these obstacles in order to further improve the performance of LEDs.

Low dimensional structures such as nanorods and nanowires have become prominent in recent years. With a diameter in nanoscale, the diode arrays may have advantages of increased surface to volume ratio due to sidewall etching and strain relaxation due to the vacant spaces among the rod array. Most researches on nanorod structure are restricted at material characterization. The process to

J.-Y. Tan · L.-Y. Chen · J.-J. Huang (✉)
Graduate Institute of Photonics and Optoelectronics,
National Taiwan University, Taipei 106, Taiwan
e-mail: jjhuang@cc.ee.ntu.edu.tw

C.-C. Lee (ed.), *The Current Trends of Optics and Photonics*,
Topics in Applied Physics 129, DOI 10.1007/978-94-017-9392-6_24

fabricate nanorod LEDs is difficult especially on the issue of large leakage current under reversed bias. The synthesis of a nanorods or nanowires structure is generally divided into bottom-up and top-down approaches. In the bottom-up method, the vapor-liquid-solid (VLS) growth has been widely employed. This common approach uses metal nano-particles such as Fe, Au, and Ni as catalysts during the growth to control the critical nucleation and subsequent elongation steps of the nanowires [1–4]. High density GaN nanorods have also been grown by molecular-beam-epitaxy (MBE) or metal organic vapor phase epitaxy (MOVPE) [5–7]. Furthermore, vertically aligned and faceted GaN nanorods have been fabricated on a GaN layer with a patterned SiO_2 template [8]. Despite various proposed and demonstrated growth methods, most bottom-up approaches have difficulty in controlling the exact diameter over the entire sample or achieving excellent spatial alignment of nanostructures. The fabrication of nanorods that contain heterostructure or active layers for light emission is also challenging. In the top-down approach, the process simply uses nanoscale etching masks followed by dry or wet etching to form the nanorod structure. Nickel metal or polystyrene nanoparticles have been widely used as the etching mask [9]. However both of them encountered a lack of high density and uniformity. A practical way to fabricate InGaN/GaN nanorod LED arrays with p-i-n structure is by using nanosphere lithography. By deploying self-assemble SiO nanoparticles into IPA (Isopropanol) solution with particular concentration and followed by spin-coating process, a monolayer of silica nanosphere on the p-GaN surface with high density and uniformity can be achieved [10]. The silica monolayer act as an etching mask for the subsequent reactive ion etching (RIE) process. Figure 24.1 shows the (a) SEM image of the monolayer silica nanoparticle mask and (b) cross-section SEM image of the nanorod arrays [11].

The critical issue in nanorod LED fabrication is to prevent shorting of the p- and n-type semiconductors when depositing the top metal contact. Several solutions have been proposed, including the insertion of spin-on glass (SOG) as the space layer with rod tips exposed by RIE [12], oblique indium tin oxide (ITO)

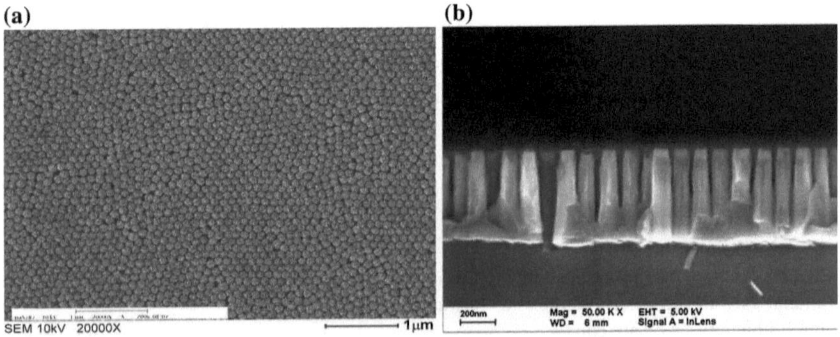

Fig. 24.1 **a** SEM image of the monolayer silica nanoparticle mask and **b** cross-section SEM image of the nanorod arrays

deposition [13], and the photo-enhanced chemical (PEC) wet oxide process. However, these approaches are not practical due to the high leakage current under reversed bias. Some works have reported a reduction of the reverse bias current to the μA range [12, 14, 15]. The optical power of GaN-based nanorod arrays is as high as 3700 mW/cm^2 at an injection current of 20 mA [7]. Despite the report of light emission from nanorod arrays, a reliable manufacturing process with high production yield and excellent performance isn't yet available. The current-voltage curves of such nano-devices reveal that they suffer from large leakage currents, large ideality factors, and low optical output power, as compared with conventional planar GaN-based LEDs. A solution to the high leakage current can be done by employing plasma enhanced chemical vapor deposition (PECVD) grown SiO_2 as the space layer to prevent shorting of the p- and n-type semiconductors and followed by chemical mechanical polishing (CMP) process to remove SiO_2 deposited right on the rod tips. Thus, the contact of current spreading layer can be deposited uniformly. The nano-device achieved a reverse current of 4.77 nA at −5 V, an ideality factor 7.35, and an optical output intensity 6,807 mW/cm^2 at the injection current of 20 mA [16]. Sah-Noyce-Shockley theory [17] suggests ideality factors between 1 and 2 for typical diodes due to the competition between the drift-diffusion and generation-recombination processes. However, for GaN based diodes, the ideality factors usually fall in the range between 5 and 7 [18, 19], which results from the large p-type contact resistance and the polarization-induced triangular band profiles of the quantum barriers [20]. As for nanorod diodes, the formation process using the top-down etching may damage the rod sidewalls and creates a current leakage path; thus larger ideality factors such as 11.2 ± 0.56 in [21] and 18 in [22] were reported. The sidewall passivation and the subsequent CMP process can significantly reduce the sidewall leakage as well as prevent the shorting paths along the nanorods.

24.2 Properties of Nanorod LED

Nanostructure light sources have attracted considerable attention as they may have the potential of better light extraction due to the increased sidewall areas and higher radiation directionality due to the vertical light guiding effect along the rods. In the past couple of years, GaN-based nanorod LED arrays have been fabricated and characterized at room temperature. As mentioned previous, it is well-known that GaN-based LED epi-structure suffer from strain-related QCSE. This phenomenon is related to the piezoelectric field and the strain between the mismatched layers of InGaN and GaN. By analyzing the results of Raman scattering measurement, which is a standard optical characterization technique for studying various aspects of solids such as lattice properties, electronic properties, and magnetic properties, the effect of the nanorod structure on the GaN-based epi-structure can be further verified. It has been found that the InGaN E_2^H phonon mode of nanorods shows a lower wavenumber than that of the planar structure [11]. Since the nanoscale diameter (100–200 nm) nanorods are not small enough for the effect of phonon confinement, the Raman shift toward a lower frequency in the nanorod structure is mainly due to strain relaxation. Strain relaxation in the InGaN/GaN multiple quantum well (MQW) layers will result in a reduced QCSE and thus a nearly constant peak wavelength with various injection currents [12]. The mitigated QCSE can further improve the overlapping ratio between the electron and hole's wave function. The degree of strain relaxation in the MQW layers is related to the etching depth of the nanorod. Generally, a larger nanorod etching depth possesses more relaxed strain in the InGaN/GaN layer and a smaller efficiency droop [23]. However, it should be noticed that during the definition of nanorods, the effect of strain relaxation is accompanied by the formation of sidewall defects picked up from dry etching. The larger etch depth results in the increase of defects states in the sidewall of nanorods, which reduces the effective current for radiative recombination. Both the effects of reduced strain and increased defect states are competing factors that determine light output efficiency of nanorod LED arrays. A shorter nanorod etching depth is preferred for a higher light output. Nevertheless, the longer nanorod structure has a less severe droop effect and a higher operating current, which may eventually lead to higher optical output if the defects can be properly suppressed.

References

1. C.-C. Chen, C.-C. Yeh, Large-scale catalytic synthesis of crystalline gallium nitride nanowires. Adv. Mater. **12**, 738–741 (2000)
2. P. Cheyssac, M. Sacilotti, G. Patriarche, Vapor-liquid-solid mechanisms: challenges for nano-sized quantum cluster/dot/wire materials. J. Appl. Phys. **100**, 044315 (2006)
3. Q. Wu, Z. Hu, X. Wang, Y. Lu, K. Huo, S. Deng, N. Xu, B. Shen, R. Zhang, Y. Chen, Extended vapor–liquid–solid growth and field emission properties of aluminium nitride nanowires. J. Mater. Chem. **8**, 2024–2027 (2003)
4. C.C. Tang, S.S. Fan, M.L. Chapelle, P. Li, Silica-assisted catalytic growth of oxide and nitride nanowires. Chem. Phys. Lett. **333**, 12–15 (2001)
5. Y.H. Kim, J.Y. Lee, S.-H. Lee, J.-E. Oh, H.S. Lee, Synthesis of aligned GaN nanorods on Si (111) by molecular beam epitaxy. Appl. Phys. A **80**, 1635–1639 (2005)
6. L.W. Tu, C.L. Hsiao, T.W. Chi, I. Lo, K.Y. Hsieh, Self-assembled vertical GaN nanorods grown by molecular-beam epitaxy. Appl. Phys. Lett. **82**, 1601–1603 (2003)
7. S.Y. Kuo, C.C. Kei, C.N. Hsiao, C.K. Chao, Growth of gallium nitride nanorods by metalorganic molecular beam epitaxy. J. Vac. Sci. Technol. B **24**, 695–699 (2006)
8. P. Deb, H. Kim, V. Rawat, M. Oliver, S. Kim, M. Marshall, E. Stach, T. Sands, Faceted and vertically aligned GaN nanorod arrays fabricated without catalysts or lithography. Nano Lett. **5**, 1847–1851 (2005)
9. H.W. Huang, C.C. Kao, T.H. Hsueh, C.C. Yu, C.F. Lin, J.T. Chu, H.C. Kuo, S.C. Wang, Fabrication of GaN-based nanorod light emitting diodes using self-assemble nickel nano-mask and inductively coupled plasma reactive ion etching. Mater. Sci. Eng., B **113**, 125–129 (2004)
10. M.Y. Hsieh, C.Y. Wang, L.Y. Chen, M.Y. Ke, J.J. Huang, InGaN–GaN nanorod light emitting arrays fabricated by silica nanomasks. IEEE J. Quantum Electron. **44** (2008)
11. Y.Y. Huang, L.Y. Chen, C.H. Chang, Y.H. Sun, Y.W. Cheng, M.Y. Ke, Y.H. Lu, H.C. Kuo, J.J. Huang, Investigation of low-temperature electroluminescence of InGaN/GaN based nanorod light emitting arrays. Nanotechnology **22**(4), 045202-1–045202-6 (2011)

12. C.Y. Wang, L.Y. Chen, C.P. Chen, Y.W. Cheng, M.Y. Ke, M.Y. Hsieh, H.M. Wu, L.H. Peng, J.J. Huang, GaN nanorod light emitting diode arrays with a nearly constant electroluminescent peak wavelength. Opt. Express **14**, 10556 (2008)
13. Y.J. Lee, S.Y. Lin, C.H. Chiu, T.C. Lu, H.C. Kuo, S.C. Wang, S. Chhajed, J.K. Kim, E.F. Schubert, High output power density from GaN-based two-dimensional nanorod light-emitting diode arrays. Appl. Phys. Lett. **94**, 141111 (2009)
14. C.H. Chiu, T.C. Lu, H.W. Huang, C.F. Lai, C.C. Kao, J.T. Chu, C.C. Yu, H.C. Kuo, S.C. Wang, C.F. Lin, T.H. Hsueh, Fabrication of InGaN/GaN nanorod light-emitting diodes with self-assembled Ni metal islands. Nanotechnology **18**, 445201 (2007)
15. A. Kikuchi, M. Tada, K. Miwa, K. Kishino, Growth and characterization of InGaN/GaN nanocolumn LED. Proc. SPIE **6129**, 612905 (2006)
16. L.Y. Chen, Y.Y. Huang, C.H. Chang, Y.H. Sun, Y.W. Cheng, M.Y. Ke, C.P. Chen, J.J. Huang, High performance InGaN/GaN nanorod light emitting diode arrays fabricated by nanosphere lithography and chemical mechanical polishing processes. Opt. Express **18**, 7664–7669 (2010)
17. C.T. Sah, R.N. Noyce, W. Shockley, Carrier generation and recombination in p-n junctions and p-n junction characteristics. Proc. IRE **45**, 1228 (1957)
18. K. Mayes, A. Yasan, R. McClintock, D. Shiell, S.R. Darvish, P. Kung, M. Razeghi, High-power 280 nm AlGaN light-emitting diodes based on an asymmetric single-quantum well. Appl. Phys. Lett. **84**, 1046 (2004)
19. J.M. Shah, Y.-L. Li, Th Gessmann, E.F. Schubert, Experimental analysis and theoretical model for anomalously high ideality factors ($n \gg 2.0$) in AlGaN/GaN p-n junction diodes. J. Appl. Phys. **94**, 2627 (2003)
20. D. Zhu, J. Xu, A.N. Noemaun, J.K. Kim, E.F. Schubert, M.H. Crawford, D.D. Koleske, The origin of the high diode-ideality factors in GaInN/GaN multiple quantum well light-emitting diodes. Appl. Phys. Lett. **94**, 081113 (2009)
21. P. Deb, H. Kim, Y. Qin, R. Lahiji, M. Oliver, R. Reifenberger, T. Sands, GaN nanorod Schottky and p-n junction diodes. Nano Lett. **6**, 2893–2898 (2006)
22. A. Motayed, A.V. Davydov, M.D. Vaudin, I. Levin, J. Melngailis, S.N. Mohammad, Fabrication of GaNbased nanoscale device structures utilizing focused ion beam induced Pt deposition. J. Appl. Phys. **100**, 024306 (2006)
23. C.H. Chang, L.Y. Chen, L.C. Huang, Y.T. Wang, T.C. Lu, J.J. Huang, Effects of strains and defects on the internal quantum efficiency of InGaN/GaN nanorod light emitting diodes. J. Quantum Electron. **48**(4), 551–555 (2012)

Chapter 25
Plasmonic Nanoslit Arrays for Sensitive Biosensors

Kuang-Li Lee, Shu-Han Wu and Pei-Kuen Wei

Symbols

ε	Dielectric constants of outside environment
ε_m	Dielectric constants of metal
n	The refractive index of outside environment
L_{spp}	Propagating length
k_{spp}	Propagation constant of surface plasmon
k_g	Grating wavevector
E_x	x-component of the electric field
E_z	z-component of the electric field
θ	Angle of incidence in a medium
λ	Wavelength
λ_0	Reference wavelength
RIU	Refractive index unit
P	Period
i, j	Orders of Bragg condition
S_λ	Wavelength sensitivity
S_I	Intensity sensitivity
n_{eff}	Equivalent refractive index of surface plasmon
w	Width of the gold nanoslit
h	Thickness of the gold film
ϕ	Phase shift
R	Integrated response
n_0	Reference refractive index

K.-L. Lee · P.-K. Wei (✉)
Research Center for Applied Sciences, Academia Sinica, Taipei 11529, Taiwan
e-mail: pkwei@sinica.edu.tw

S.-H. Wu · P.-K. Wei
Institute of Biophotonics, National Yang-Ming University, Taipei 11221, Taiwan

C.-C. Lee (ed.), *The Current Trends of Optics and Photonics*,
Topics in Applied Physics 129, DOI 10.1007/978-94-017-9392-6_25

λ_1,λ_2 Integrated wavelength range
T_i ith order of the resonant transmission
$\delta\lambda$ Wavelength resolution of the spectrometer
$\Delta\lambda$ Full width at half maximum (FWHM) bandwidth

25.1 Introduction

There are increasing demands in improving technologies that allow noninvasive, label-free, and high sensitivity with temporal and spatial resolution in high throughput screening (HTS) measurement. Surface plasmon resonance (SPR) biosensor technology has been established for almost three decades due to the advantages of label-free and real-time detection [1]. SPR biosensor provides high sensitivity to the refractive index change near the sensor surface resulting from molecular interaction. The most popular method in SPR detection employs a prism to couple the electromagnetic wave associated with an incident light to surface plasmon polaritons (SPP) at metal/medium interface [2]. Based on this technique, there are some commercial instruments, such as BIAcore, developed as a standard tool for molecular interaction analysis. However, the prism-based SPR method is expensive, bulky and difficult for HTS detections. On the other hand, SPR sensors based on metallic nanostructures permit a scalable sensing area and easy to integrate with other system [3, 4]. In addition to scalability, localized SPP resonance (LSPR) excited around the nanostructures is non- or short propagating, which yields a lateral resolution of a few microns and potentially facilitates SPR imaging [5–7]. Many types of LSPR-based nanoplasmonic sensors have been developed, such as gold nanoparticles [8], non-concentric ring/disk cavity [9], plasmonic nanoparticle clusters [10], planar metamaterial [11] and nanocross structures [12], etc. In this article, we introduce highly sensitive SPR biosensors based on periodic gold nanostructures and their applications for label-free and HTS bio-detections.

25.2 Surface Plasmon Resonance, Excitation and Sensitivity

In 1902, Wood first described the unexplained narrow dark bands in the diffracted spectrum of metallic grating, named Wood's anomaly [13]. This anomalous diffraction is a phenomena resulting from the interaction between light and free electrons on the metal surface. Some years later, Fano [14] re-examined the theory and indicated that the intensity of waves propagating along the metallic surface were influenced by the elements of the wavelength of the light, the angle of incidence, and the geometry of grating. This phenomenon was then

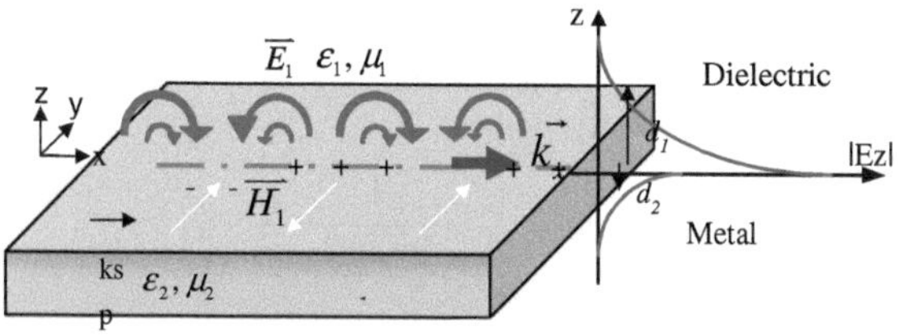

Fig. 25.1 The schematic illustration of charge density oscillations at a metal–dielectric interface in x direction. The electromagnetic fields exponential decay from the interface into dielectric (d_1) and metal (d_2) shown on the *right*

explained in terms of surface plasmon resonance. The works by Wood and Fano undoubtedly made the foundation to the field of surface plasmon and opened the growing interests of this field. Since then, surface plasmon has been intensively studied and the major properties have been assessed. Surface plasmon, also known as surface plasma wave (SPW) or surface plasmon polariton (SPP), is the charge-density oscillation of electromagnetic excitation at the interface between two media with opposite dielectric constants, for instance, a metal and a dielectric. Figure 25.1 illustrates the electromagnetic wave decays evanescently into both media in the perpendicular direction and the field vectors reach their maxima at the interface. These evanescent fields are asymmetric distributed (larger decay in metal; small in the dielectric) and majorly concentrated in the dielectric part.

The electromagnetic wave of surface plasmon can be excited by electrons and photons. In the electrons excitation, the energy from the firing electrons transfers into bulk plasmon. When the component of scattering vector is parallel to the metal surface, it leads to SPPs. On the other hand, the transfer of optical energy into SPPs (photons excitation) is achieved by the coupling mechanism, such as the prism or grating coupling method, in order to match the wave vectors between photon and surface plasmon. It is known that the SPP is a transverse-magnetic (TM polarization or p-polarized) wave and the magnetic vector is perpendicular to the direction of SPP propagation and in parallel with the plane of interface. The physics of SPP can be further explained by the Maxwell's equations with appropriate continuity boundary condition. In the consideration of dispersion relation, the wave vector (k_{spp}) of SPP can be described in following equations.

$$k_{spp} = \frac{\omega}{c}\left[\frac{\varepsilon\varepsilon_m}{(\varepsilon+\varepsilon_m)}\right]^{1/2} = \frac{\omega}{c}\left[\frac{\varepsilon_m n^2}{\varepsilon_m + n^2}\right]^{1/2} \tag{25.1}$$

where ε and ε_m are the dielectric constants of outside environment (the refractive index is n) and the metal, respectively. The SPP loses its energy in the metal due to the ohmic absorption. It also can be understand from the complex dielectric constant of metal ($\varepsilon_m = \varepsilon_{2r} + i\varepsilon_{2i}$) of which ε_{2r} is related to the effective index and

ε_{2i} is associated with the attenuation of surface plasmon in the direction of propagation. The propagating length (L_{spp}) is defined as the length when SPP energy decays to e^{-1} of original intensity.

$$L_{spp} = \frac{1}{2 \times \mathrm{imag}(\mathrm{k_{spp}})} \tag{25.2}$$

The SPP can only be excited under the condition of $\varepsilon_{2r} < -n^2$. Several metals can be used, such as gold (Au), silver (Ag), copper (Cu) and aluminum (Al). Au and Ag are chosen at most of time due to its low ε_{2i}. The major characteristics of the SPP and their comparisons between Au and Ag for SPR excitation are shown in Table 25.1.

SPR is excited by the light wave as the phase matching condition is fulfilled, i.e. The k_x, the wave vector of light in the projection of the x-axis, needs to match the wave vector of SPP. In general, the k_{spp} is larger than k_x. It means that the SPP cannot be excited by the incident light directly. To increase the k_x of incident light reaching the k_{spp}, there are some coupling methods, such as prism, grating or optical waveguide coupling methods. There are two types of prism coupling configuration demonstrated by Kretschmann [15] and Otto [16]. In prism coupling configuration, the phase-matching to the SPP can be achieved by a three layer system including a thin metal layer between two insulators. When the light passes through an optically denser medium, high refractive index (n_p) material, the light will be totally reflected and show minimum intensity of reflection light at a specific angle (θ), where $n_p k_0 \sin\theta = k_{spp}$. The SPP is excited at the interface between metal and dielectric with lower refractive index. It should be noted that the SPP cannot be excited at the interface of metal/prism. In grating coupling configuration [17, 18], the mismatch between the in-plane wave vector k_x and k_{spp} is overcome by the diffraction grating which produces a lateral wave vector, k_g. The phase matching condition is satisfied when $k_{spp} = k_x + k_g$.

The sensor performance is always the priority being considered. It can be understood from the sensitivity and detection limit. Usually, to evaluate the sensitivity, we apply different refractive index solution, which can be prepared by glucose, salt, and glycerol solution at different concentration, into sensor device and monitor the sensor response. The sensitivity can be calculated by the linear fit of sensor response as a function of refractive index to have the relationship between sensor response to environmental refractive index change. The sensitivity is defined by the amount of sensor response changes per refractive index unit (RIU). More precisely, the resolution of instrument should be considered (the resolution of angular, wavelength, intensity). It decides the detection limit which is

Table 25.1 The properties of surface plasma waves at the metal-water interface

Metal layer supporting SPW	Silver (Ag)		Gold (Au)	
Wavelength (nm)	630	850	630	850
Propagation length (μm) (L_{spp})	19	57	3	24
Depth in metal (nm) (d_2)	24	23	29	25
Depth in dielectric (nm) (d_1)	219	443	162	400
Field concentration in dielectric (%)	90	95	85	94

Table 25.2 Theoretical sensitivity and resolution of SPR detection [1]

Detection approach	Angle		Wavelength		Intensity	
Configuration	Sensitivity/resolution					
	λ = 630	λ = 850	λ = 630	λ = 850	λ = 630	λ = 850
Prism	191/5 × 10^{-7}	97/1 × 10^{-6}	970/2 × 10^{-5}	13,800/ 1 × 10^{-6}	3,900/5 × 10^{-5}	15,000/1 × 10^{-5}
Grating	43//2 × 10^{-6}	39//2 × 10^{-6}	390/6 × 10^{-5}	630/ 3 × 10^{-5}	1,100/2 × 10^{-4}	4,400/5 × 10^{-5}

Angle unit Sensitivity/Resolution = degRIU^{-1}/RIU (the resolution: 1 × 10^{-4} deg)
Wavelength unit Sensitivity/Resolution = nm RIU^{-1}/RIU (the resolution: 0.02 nm)
Intensity unit Sensitivity/Resolution = % RIU^{-1}/RIU (the resolution: 0.2 % of optical power)

the minimum detectable refractive index value. Table 25.2 shows the sensitivity of prism and grating based SPR at different interrogations [1]. The prism coupler based SPR has higher sensitivities than grating coupler. The angular detection has a higher resolution than other methods in both prism and grating couplers.

Currently, the development of SPR biosensor are focusing on several objectives: improving sensitivity, minimizing the sensor system, tuning the operation range including wavelength and detection refractive index, high throughput detection, and SPR image [19–21]. The refractive index change on sensor surface associated with the molecular binding on the sensor surface can be detected quantitatively by monitoring the change in intensity, resonance wavelength or angle of the reflected beam. In the sensitivity performance, prism coupler based SPR in the angular interrogation is with the best resolution than others. However, the prism-based SPR method is expensive, bulky and difficult to be scaled up. It is complicated in the optical system, the optical alignment is hard to minimize the device to integrate with other system. In addition to improve the system, high throughput detection and imaging the SPR in 2D are also in the mainstream of SPR development. However, the SPR configurations mentioned before are limited either in the optics, light distortion, or difficult in image reconstruction. Recently developed *Nano-Plasmonics* [22] provides a way to overcome the drawback of conventional SPR sensors such as small-area, chip-based, high-throughput detections, easy to be combined with microfluidic system and simple optical measurement system.

25.3 Surface Plasmon Resonances in Nanostructures

Ebbesen et al. [23] first demonstrated the unique optical property of extraordinary transmission by nanohole arrays coated with silver in 1998. According to Bethe theory [24], the light passes through the aperture with a size smaller than the wavelength, the intensity of light would decay quickly from the aperture exit. However, when the light shines on the nanostructure with a period, there is an

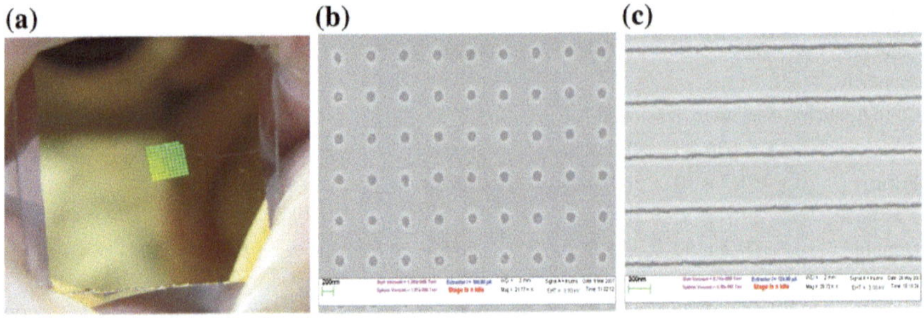

Fig. 25.2 **a** The optical image of a 10 × 10 microarray on a glass slide. Each green area was consisted of a 600 nm-period gold nanoslit array. The area was 150 μm in *square*. **b** The SEM image of a 600 nm-period gold nanohole array. **c** The SEM image of a 600 nm-period gold nanoslit array

extraordinary optical transmission (EOT) at particular wavelengths. This unusual optical property is due to the coupling of plasmons-electronic excitation. The SPPs in these periodic structures are directly excited by subwavelength metallic apertures. There are no needs for prism or gratings. Such chip-based plasmonic nanostructures can be applied for SPR sensors. Currently, there are two major periodic nanostructures for the plasmonic biosensors. They are periodic nanoslits and nanoholes as shown in Fig. 25.2.

In periodic nanostructures, an EOT occurs when the SPP wavelength meets the Bragg condition. It is a Bloch wave surface plasmon polariton (BW-SPP). For normally incident light, the BW-SPP condition for two-dimensional periodic structures is described by

$$\lambda(i,j) = \frac{P}{\sqrt{i^2+j^2}}\left(\frac{\varepsilon_m n^2}{\varepsilon_m + n^2}\right)^{1/2} \tag{25.3}$$

where *i*, *j* are the orders in the x-y directions, *P* is the period of the nanostructure. When the surface molecular density increases, the resonant wavelength is redshifted due to the increase of *n*. Using the peak EOT wavelength as signals, high density sensing arrays have been demonstrated. The theoretical and experimental studies indicate that the wavelength sensitivity (S_λ) is close to the period of the nanostructures, i.e. $S_\lambda \sim P$ nm/*RIU*. On the other hand, the redshift of the resonant wavelength also induces an intensity change at a fixed wavelength near the resonant condition. For a small wavelength shift, the intensity sensitivity (S_I) is proportional to the wavelength sensitivity and the differential of the spectrum.

$$S_I(\lambda) = \frac{\Delta I/I_0}{\Delta n} = \frac{I(\lambda+\Delta\lambda)-I(\lambda)}{I(\lambda)}/\Delta n \cong \frac{I'(\lambda)\Delta\lambda}{I(\lambda)\Delta n} = S_\lambda \frac{I'(\lambda)}{I(\lambda)} \tag{25.4}$$

If we assume a Gaussian profile for the resonant spectrum, $I'(\lambda) = \exp(-(\frac{\lambda-\lambda_0}{d})^2)$, the maximum I' happens at $\lambda = \lambda_0 \pm d/\sqrt{2}$ and the maximum value,

Table 25.3 The SPR sensitivity of gold nanohole arrays

Periodicity (nm)	Diameter (nm)	Depth (nm)	Refractive index sensitivity
600	200	100	**400 nm/RIU**, *Langmuir* 20, 4,813–4,815 (2004)
1,400	200	200	**1,022 nm/RIU**, *Opt. Lett.* 31, 1,528–1,530 (2006)
400	100	50	**200 nm/RIU** *Opt. Express* 15, 18,119–18,129 (2007)
450	150	100	**333 nm/RIU**, *Anal Chem* 79, 4,094–4,100(2007)
600	200	100	**270 and 600 nm/RIU**, *Appl. Phys. Lett.* 90, 243,110 (2007)
1,530	300	150–200	**1,570 nm/RIU (1 x 10^{-5} RIU)**, *Appl. Phys. Lett.* 91, 123,112 (2007)
600	90–250	110	**420 and 561 nm/RIU** *Plasmonics* 3,119–125 (2008)
450	150	220	**1,010%/RIU (Detection limit 2 x 10^{-4} RIU)**, *Nano Lett.* 8(9), 2,718–2,724 (2008)
454	170	100	**350 nm/RIU and 750 %/RIU(3 x 10^{-4} RIU)** Biosensors and Bioelectronics 24, 2,334–2,338 (2009)

$I'(\max) = \pm\sqrt{2}I/d$. It indicates the maximum intensity sensitivity is proportional to the inverse of the bandwidth. The resonant bandwidth is dependent on the size and shape of the metallic structures. For the same period, a small bandwidth will have a large sensitivity in the intensity measurement.

Table 25.3 shows the wavelength and intensity sensitivities for gold nanohole structures by different research groups. The wavelength sensitivities are ~333–1,570 nm/RIU and the intensity sensitivities are ~1,010–4,400 %/RIU. For a 0.1 nm wavelength resolution or 0.2 % intensity resolution, the detection limit is only ~10^{-4}–10^{-5}. As compared to Table 25.2, this limit is comparable with the grating coupling biosensors but still smaller than the prism-based sensors with angular detection. To increase the wavelength sensitivity, the period has to be increased to several micrometer long and the resonant wavelength will be shifted to the infrared region. However, the long-period sensors will suffer from low-transmission light and the absorption of water in the IR range.

Nanohole arrays were first demonstrated and used commonly. The intensity sensitivity of SPR increases by decreasing the linewidth. However, there is a light cut-off effect while the diameter of hole down to the scale of ~100 nm. Alternative, we have demonstrated that nanoslit arrays can overcome the light transmission in small linewidth down to ~20 nm [25]. In the metallic nanoslit, only TM mode can exist. The theory of the TM polarized light propagation and reflection in a nanoslit in metal related with surface plasmon was presented by Gordon [26]. Different from conventional surface plasmon wave, this plasmon is confined in a nano metallic gap, which is the gap plasmon. The x component of the electric field of lowest order mode within the slit can be described:

$$E_x^s = \begin{cases} \cosh\left(\sqrt{k_0^2 - k_{spp}^2}x\right)e^{ik_{spp}z} & if|x| \leq \frac{w}{2} \\ \frac{1}{\varepsilon_2}\cosh\left(\sqrt{k_0^2 - k_{spp}^2}\frac{w}{2}\right)e^{\left[-\sqrt{k_{spp}^2-\varepsilon_2 k_0^2}\left(|x|-\frac{w}{2}\right)\right]}e^{ik_{spp}z} & if|x| > \frac{w}{2} \end{cases} \quad (25.5)$$

where s is the field inside the slit, k_0 is the wave vector in free space, and w is the width of slit. The propagation constant is

$$\tan\left(\sqrt{k_{spp}^2 - k_0^2}\frac{w}{2}\right) = \frac{-\sqrt{k_{spp}^2 - \varepsilon_m k_0^2}}{\varepsilon_m\sqrt{k_{spp}^2 - k_0^2}} \tag{25.6}$$

As we know, the k_{spp} inside the slit is as a function of slit width. The wave vector of gap plasmon increases as the decreasing of slit width and leads increasing the resonance in the metallic slit. This kind of gap plasmon resonance is the cavity resonance [27]. From the Fabry-Perot cavity model, the resonant wavelength can be estimated by

$$\lambda_0 = (4\pi n_{eff} h)/(2\pi - \phi_1 - \phi_2) \tag{25.7}$$

where n_{eff} is the equivalent refractive index of gap plasmon, h is the thickness of the gold film, and ϕ_1 and ϕ_2 are the phase shifts at the top and bottom interfaces, respectively. The increase of refractive index in the slit increases the cavity resonant wavelength.

In nanostructure SPR, there are cavity and BW-SPP modes which are the resonances inside the nanoaperture and on the surface, respectively (Fig. 25.3a). The Fano-resonance can be described by the coupling of cavity and SPR modes [28, 29]. In general, the Fano resonance is understood in terms of coupling of a broadband wave (a continuum state) with the surface-bound state of a periodic array (a discrete state). In the gold nanoslit array, the gap plasmon, i.e. cavity mode in the slits, forms a broadband and localized resonance. On the other hand, the BW-SPP mode in periodic nanostructures has a discrete resonance. The constructive and destructive interferences of these two modes result in the asymmetric Fano-resonance as depicted in the schematic in Fig. 25.3b. In our previous study (Fig. 25.3c), the experimental evidence has shown the Fano spectrum causing by the coupling of cavity plasmon and BW-SPP [30]. If there is only one slit in the structure in terms of only cavity mode existing, a broad band is shown in the spectrum. However, if we introduce some periodic structures, the cavity mode and SPR mode interferes to each other leading to asymmetric Fano resonance.

Different from conventional SPR sensors, the periodic gold nanostructure exhibits complex resonant behaviors due to cavity resonance in the apertures and BW-SPP on the outside surfaces.

The typical transmission spectra of the periodic gold nanoslit arrays (Fig. 25.4) demonstrate two asymmetric resonances at two different interface conditions (medium/gold and substrate/gold). Using P = 500 nm, $\varepsilon_2 = -16 + 1.1i$, $n = 1.3345$ for water and $n = 1.58$ polycarbonate substrate, the predicted BW SPP peaks are 689.8 and 807.5 nm for the water/gold and substrate/gold interfaces, respectively. The observed BW SPP peaks in Fig. 25.4, at 692.2 nm (water/

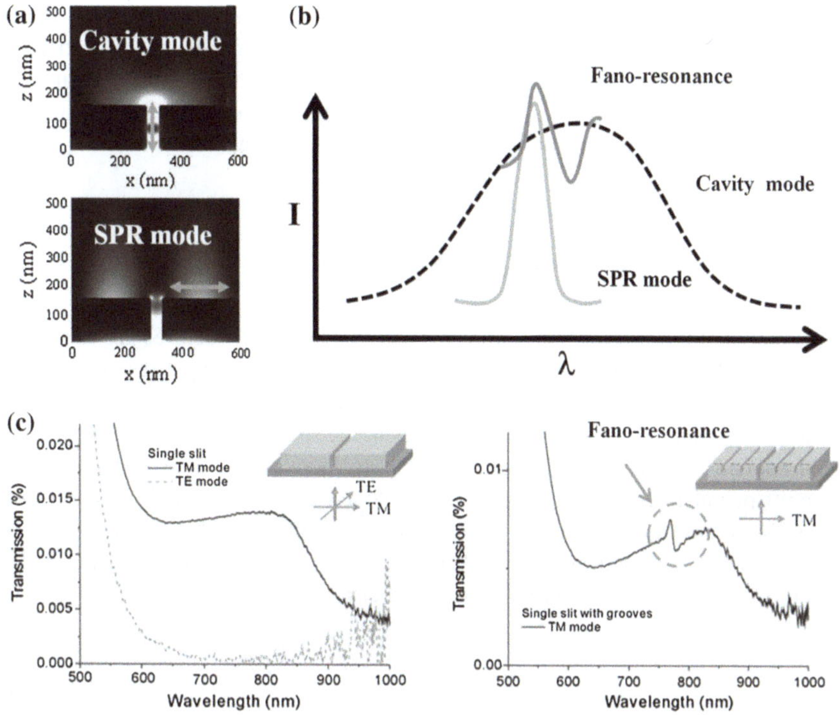

Fig. 25.3 The resonance modes in the nanostructure. **a** There are cavity mode and SPR mode which are inside the aperture and along the surface, respectively. **b** The schematic depiction of Fano-resonance is resulting from the coupling of cavity and SPR modes. **c** The experimental evidence: Fano-like spectrum is shown by inducing the periodic groove structure (SPR mode) around the slit

Fig. 25.4 The measured transmission Fano-like spectra in water and 10 % glycerol/water

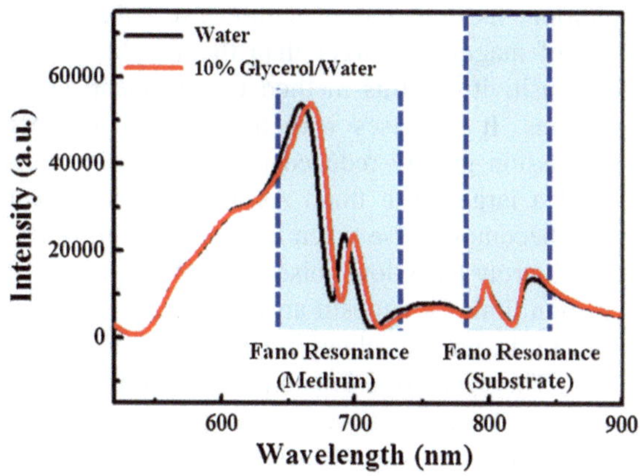

gold interface) and 830.7 nm (substrate/gold interface), respectively, well match the predicted values. When the surrounding solution changes from water to 10 % glycerol in water, the Fano resonance at water/gold interface has an obvious red-shift. Note that the Fano resonance near 790–850 nm is associated with the substrate/gold interface and is not affected by the bulk solution.

25.4 Enhancing Sensitivity of Nanoplasmonic Biosensors

It is noted that the extraordinary transmission in gold nanostructures are caused by different orders of LSPRs, BW-SPPs and their mutual couplings. It is possible to improve the signal-to-noise ratio of the system and enhance the sensing capability by summing up all the transmission changes in the spectrum [31, 32]. The idea has been applied in quasi-3D plasmonic crystals to increase the detection limit [33]. A multispecral integration method was employed for various 2D gold nanosturtcures which consider all the wavelength shifts and intensity changes near the resonances. The integrated response (R) is defined by the root mean square (RMS) difference of the normalized transmission spectra.

$$R(n) = \sqrt{\sum_{\lambda_1}^{\lambda_2} \left| \frac{T_i(n,\lambda) - T_i(n_0,\lambda)}{T_i(n_0,\lambda)} \right|^2 \delta\lambda} \tag{25.8}$$

where $T_i(n,\lambda)$ is the ith order of the resonant transmission under different external refractive index (n), n_0 is the reference refractive index, λ_1 and λ_2 are the integrated wavelength range, and $\delta\lambda$ is the wavelength resolution of the spectrometer. Figure 25.5 shows the experimental results comparing the sensing capabilities of gold nanoslit and nanohole arrays with different aperture sizes using the multispectral integration approach. Under the identical experimental condition, the integration method can enhance detection resolution about one order of magnitude larger than the simple wavelength and intensity methods. The sensitivity of this method is dependent on the aperture size of the nanostructures. It increases with the decrease of apertures. However, the cut-off transmission greatly reduced the transmission intensity of nanoholes and thus caused a large noise floor. As a result, the detection resolution for nanohole arrays becomes worse when hole size is below 100 nm. On the other hand, the nanoslit array has a low noise floor due to its non cut-off transmission. The RIU detection limit for nanoslit array is improved by ~7 times when reducing the slit width to 40 nm. It achieves 4.5×10^{-6} RIU when the intensity resolution is 0.2 %. This resolution is comparable with the bulky prism-based sensors using complicated angular detection method.

The sensitivity of gold nanostructures can be enhanced by using spectral integration method with very narrow nanoslits. Another important factor to enhance

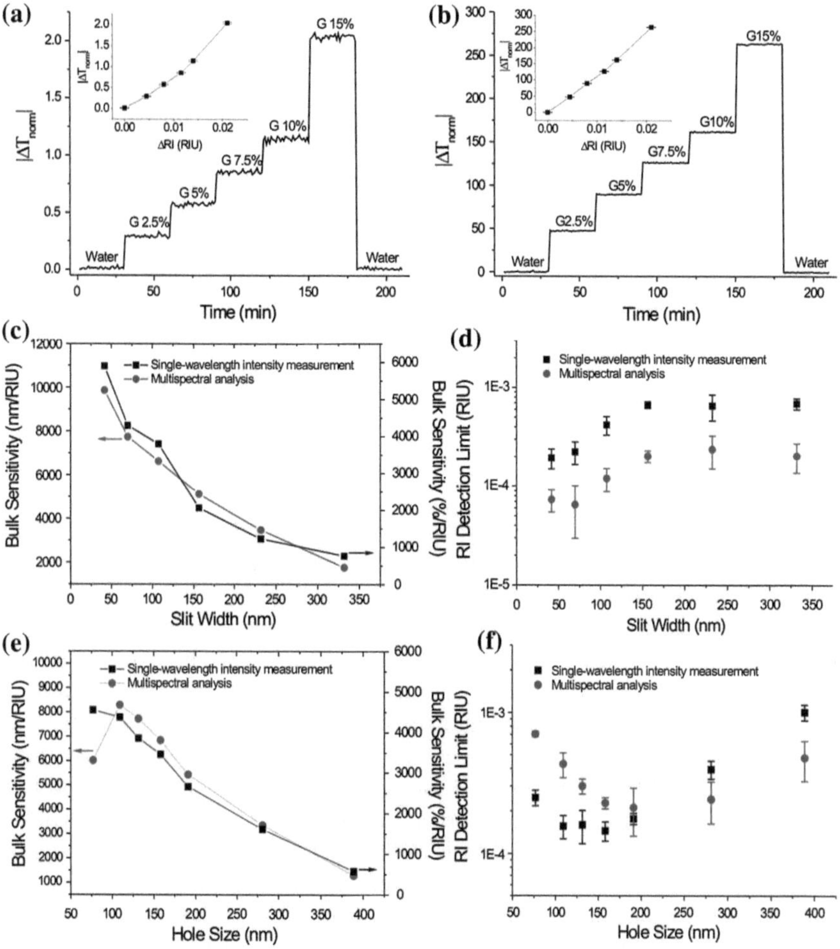

Fig. 25.5 The refractive index sensing using a multispectral integration method. Normal incidence transmission spectra of a 500-nm-period, 65-nm-wide nanoslit array in water and different water/glycerin mixtures for TM polarized wave. **a** Normalized transmission change as a function of different RI mixtures at a 697 nm wavelength. Linear correlation between the normalized transmission change and refractive index changes is shown in the *inset*. **b** Multispectral integration responses as a function of different RI mixtures. The *inset* shows the linear correlation between the responses and refractive index changes. **c** RIU sensitivity as a function of slit width for different analysis approaches, single-wavelength intensity measurement and multispectral analysis. The sensitivity increased with the decrease of the slit width. The sensitivity is highest for 42-nm-wide nanoslit array. **d** Comparisons of the detection limit using intensity measurement and the multispectral analysis for each array. **e** RIU sensitivity as a function of hole size for different analysis approaches, single-wavelength intensity measurement and the multispectral analysis. The sensitivity increased with the decrease of the hole size. The sensitivity is highest for 109-nm-diameter nanohole array. **f** Comparisons of detection limit using intensity measurement and the multispectral analysis for individual nanohole arrays

the sensitivity is to make high-quality gold film, which can reduce the SPP loss and reduce the resonant bandwidth [34, 35]. The figure of merit (FoM) of metallic nanostructures in wavelength unit is often defined as $S_\lambda/\Delta\lambda$, where $\Delta\lambda$ is the full width at half maximum (FWHM) bandwidth. The FWHM is related to the loss of SPR. It is dependent on gold film properties and would be different from different fabrication processes. The majority of the fabrication techniques for gold nanostructures utilize focused ion beam (FIB) milling to directly pattern gold films or electron-beam lithograthy (EBL) combined with a dry etching method or a lift-off process to generate a nanostructure on a substrate. However, these writing techniques are not suitable for massive production. Besides, the gold film made from the vacuum deposition leads to higher SPP loss due to the scattering of electrons at gold grain boundaries. To solve the mass-fabrication problem, many groups have proposed methods, such as interference lithography, nanoimprint lithography, nanosphere lithography and soft lithography techniques, etc. To solve the SPP loss problem, the gold film can be annealed at an elevated temperature to form large gold grains [36, 37].

A simple thermal-annealed template-stripping method [38, 39] can be used to fabricate large-area gold nanoslit arrays on polymer films with low cost and high SPR sensitivities. The template stripping method is a simple and easy fabrication process. Typically, this method make use of the poor adhesion of noble metals on the patterned templates with smooth surfaces. The exposed surface of the metal is then attached to another substrate with an epoxy adhesive. When the template and the substrate are separated, ultra-flat metal surfaces on the substrate can be obtained. It does not require additional photoresist, exposure, etching or lift-off processes. Combining the thermal annealing and template-stripping with epoxy can fabricate nanostructures with smooth gold or silver surfaces and larger grain sizes. Optical epoxy was applied as an adhesion layer to attach the silver film to a clean glass slide and was cured with a UV lamp. After peeling off from the template, the template-stripped grating with sharp bandwidths (~10 nm) in the reflection spectrum were made. The method can also be used for making gold nanostructures on plastic films, such as polycarbonate (PC) films. The gold film is coated on the patterned silicon template and then directly imprinted on polycarbonate films at a temperature higher than the glass transition temperature of PC films (~150 °C). The gold nanostructures are embedded on the PC film after peeling off from the silicon template. Figure 25.6 show an example of gold nanoslit arrays with a 500-nm period and various slit widths, from 30 to 165 nm, made by the thermal-annealed template-stripping technique. The transverse-magnetic (TM) polarized wave in these gold nanostructures generated sharp and asymmetric Fano resonances in transmission spectra. The full FWHMs of Fano resonances decreased with the decrease of the slit width and the narrowest bandwidth was smaller than 10 nm.

To verify the high sensitivity of sensors fabricated by the thermal-annealed template-stripping method, nanoslit-based sensors on a glass substrate was compared by electron-beam lithography combined with a dry etching

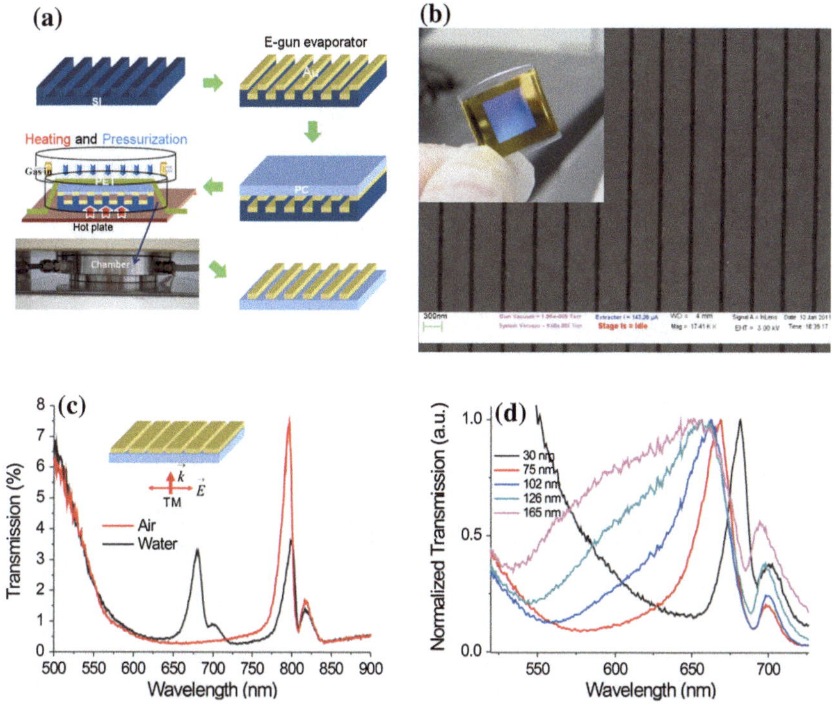

Fig. 25.6 **a** A process flowchart for the fabrication of a silicon template. **b** The SEM and optical images (*inset*) of the fabricated silicon template. The slit width is 105 nm and the area of the slit array is 1 × 1 cm. **c** The measured transmission spectra of the 500-nm-period, 30-nm-wide nanoslit array in air and water for normally-incident TM-polarized light. The *inset* depicts the TM-polarized light with E and k vectors versus the structures. **d** The normalized transmission spectra of 500-nm-period nanoslit arrays with various slit widths. The slit widths are from 30 to 165 nm

technique. Figure 25.7 show the comparisons. Both samples had the same 500-nm-period, 60-nm-wide, and 50-nm-deep nanoslit arrays. The bulk sensitivities of these sensors were measured by injecting purified water mixed with various ratios of glycerin into the microfluidic devices. There are Fano resonant profiles in the spectra. Obviously, the template stripping sample has a higher transmission and shaper Fano resonances. The slopes of the fitting curves show that the intensity sensitivities (wavelength sensitivities) are 5,226 %/RIU (451 nm/RIU) and 7,615 %/RIU (431 nm/RIU) for the dry etching and template stripping methods, respectively. Obviously, the template stripping sample has a higher intensity sensitivity compared with nanostructures made by EBL. It is noted that the intensity sensitivity for 60-nm-wide nanoslits can be further improved by narrowing the slit width. For 30-nm-wide nanoslit array, the intensity sensitivity is increased to 10,367 %/RIU and a FoM up to 55. This value is much higher than those of previous nanoslit and

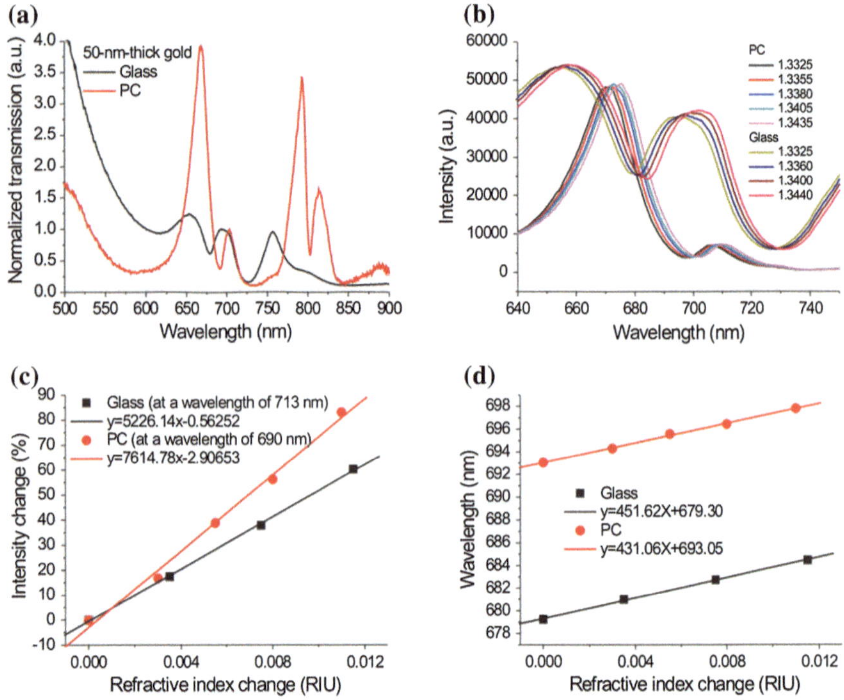

Fig. 25.7 Refractive index sensing capabilities of the 500-nm-period, 60-nm-wide nanoslit nanoslits made by the dry etching and template stripping methods. **a** The measured transmission spectra of the 500-nm-period nanoslit arrays made by both methods in water for normally-incident TM-polarized light. **b** The intensity spectra of the nanoslits made on a glass substrate and a PC substrate with various water/glycerin mixtures for a normally-incident TM-polarized wave. **c** The intensity change against the refractive index of the outside medium. **d** The resonant wavelength against the refractive index of the outside medium

nanohole arrays fabricated by the EBL and FIB methods. From atomic force microscopy images, it is confirmed that such enhanced intensity sensitivity for the template-stripped nanoslit array is attributed to a smoother gold surface and larger grain sizes during the thermal annealing process. It results in a decrease of surface plasmon propagation loss.

25.5 Applications of Nanoplasmonic Biosensors

In the optical measurement system of SPR in the gold nanostructures, there are two ways: The transmission spectrum measurement and the intensity measurement at a fixed wavelength. Figure 25.8a shows the setup for the spectroscopic measurement. A 12 W halogen light was spatially filtered by using an iris diaphragm and

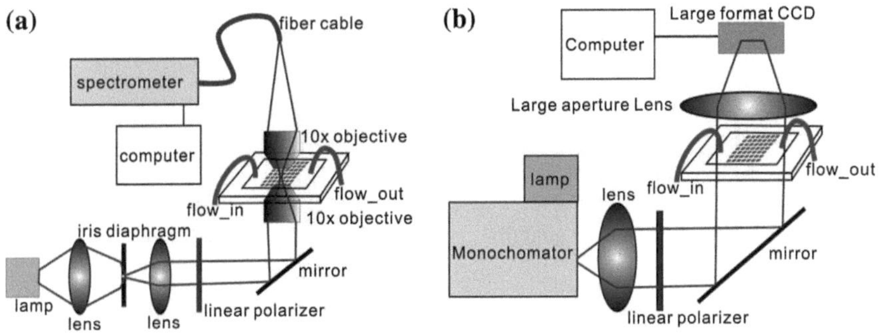

Fig. 25.8 **a** The optical setup for measuring the transmission spectrum of single periodic gold nanostructure in aqueous condition. **b** The optical setup for measuring the transmission intensities of multiple periodic gold nanostructures at a fixed wavelength

a collimation lens. Its incident polarization was controlled by a linear polarizer. The polarized light was then focused on a single array by using a 10× objective lens. The transmission light was then collected by another 10× objective lens and focused on a fiber cable. The transmission spectrum was taken by using a fiber-coupled linear CCD array spectrometer. Figure 25.8b shows the intensity measurement, the array images were taken by using a low-amplification objective and a large-area TE-cooled CCD to reduce the noise. The incident light wavelength was selected near the resonant wavelength.

The nanoplamonic biosensor has advanced tremendously in the past few years [34, 35, 40, 41]. Compared with the conventional SPR sensors, metallic nanostructures permit a scalable sensing area and easier miniaturization. Currently, developments in metallic nanostructures based sensors are low cost, easy fabrication, multiplex sensing with small sample volume, integration with microfluidic system, fast detection, and portable system toward the point-of-care clinical evaluation.

An application for high-throughput and label-free detections of antigen-antibody interactions were shown in Fig. 25.9. In this antigen-antibody study, the interactions between BSA (bovine serum albumin) (Sigma-Aldrich) and anti-BSA (Sigma-Aldrich) were measured. Figure 25.9a shows the setup for the intensity detection of the microarray. In the experiment, the buffer solution, 10 mM phosphate-buffered saline (PBS) (UniRegion Bio-Tech), was first injected to the microarray chip. Then 50 μM BSA was injected on the nanoslit surface. Due to the physical absorption of BSA on gold surface, the BSA will be coated on the slit arrays. The BSA solution flew for 1 h in order to make sufficient BSA fixed on the gold surface. The PBS buffer then washed the chip to remove the unbound proteins. Finally, 5 nM anti-BSA was injected into the microarray surface. After 3 h protein-protein interactions, the unbound anti-BSA was washed away by the PBS buffer. The intensity images of the microarray at different interaction time were taken by using the measurement system shown in Fig. 25.8b. Figure 25.9b shows the intensity images of the 10 × 10 nanoslit arrays captured at three-hours

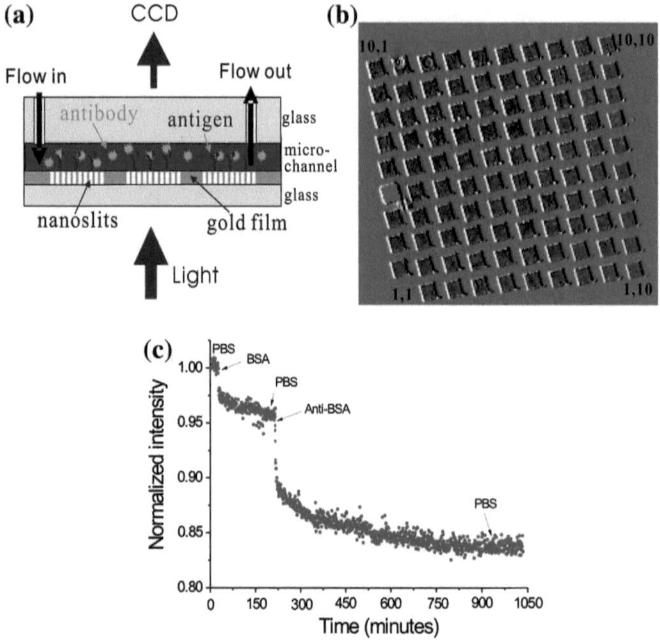

Fig. 25.9 **a** The setup for measuring the antibody-antigen interactions on the microarray. The interactions were monitored by the intensity changes as detected by the CCD camera. **b** The intensity images of the 10 × 10 nanoslit arrays. **c** The normalized intensity as a function of the interaction time for one of the nanoslit array (5,5) shown in Fig. 25.8b. The BSA results in an intensity change of 5 %. The 5 nM Anti-BSA caused an intensity change of 12 %

interaction time. The images were obtained by subtracting the measured time-lapsed images I(x,y;t) with the initial CCD image I(x,y;t = 0). The intensities indicate the amounts of the surface refractive index changes. It can be found that the slit arrays decreased the transmission intensity due to surface binding of BSA and Anti-BSA molecules. There were two defects (1,1) and (4,1) in the microarray, which had no nanoslit structures inside. These two areas did not show any intensity changes during the interactions. This is an evidence that the measured intensity changes are related to the SPR in gold nanostructures. Figure 25.9c shows the normalized transmission intensity as a function of the interaction time for one of the nanoslit array (5,5). The intensity signal is stable with time when the PBS buffer is injected into the microfluidic device. The BSA coated on the gold surface resulted in an intensity change of 5 %. For the Anti-BSA, the 5 nM concentration caused an intensity change up to 12 %. The Anti-BAS has a larger intensity change because the molecular weight of anti-BSA (150 kDa) is much larger than that of BSA (66 kDa). These measurement results verify the multiple dynamic detections of the gold nanoslit arrays. Using the intensity measurement, the detection sensitivity for Anti-BSA molecules is up to 83.3 pM when the intensity stability is 0.2 %.

While SPR in gold films is widely used for sensitive and label-free detection of biomolecular interactions, the application for living cell function analysis is also an advantage of the gold nanostructures for its easy integration with the microfluidic system [42–44]. Figure 25.10a shows a cell culture microfluidic device combined with the nanoslit arrays based SPR detection system to detect the cell secretion in real time. In this work, the gold nanoslits measure the dynamic secretion behavior of a small number of human monocytic leukemia cells (THP-1) without the need of labeling. Upon continuous lipopolysaccharide (LPS) stimulation, matrix metallopeptidase 9 (MMP-9) secretion is detected within 2 h. The polycarbonate substrate is bonded with a PDMS microfluidic channel and the gold nanosilt surface is further coated with an antibody for detecting specific antigen binding. The microfluidic channel immobilizes cells from a suspension to pre-defined locations facing the nanoslit array. A schematic of the PDMS microfluidic channel is shown in Fig. 25.10b which contains an array of U-shaped cell traps. The opening of the each trap is ~18 μm to favorably accommodate single cell [45]. Figure 25.10c shows an optical image of the trap array and individual cells in the cell traps.

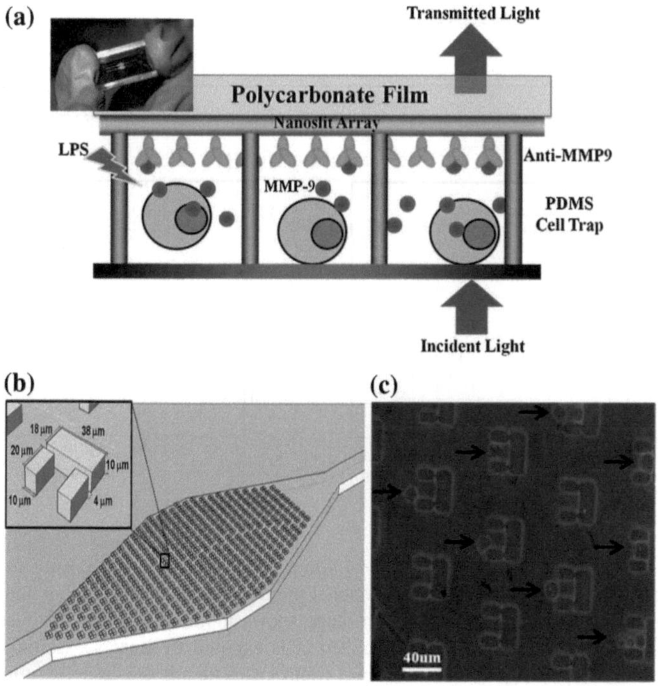

Fig. 25.10 **a** The sensor chip used in this study contains two major components: a nanoslit based SPR sensor and a microfluidic cell culture channel. **b** A schematic drawing of the microfluidic chamber. The height of the chamber is 10 μm. The chamber contains an array of cell traps. The traps are staggered to maximize cell capture. **c** A *bright field* image showing THP-1 cells captured in the traps. Each arrow points to one cell

Figures 25.11a and b show the real-time cell secretion responses measured from different numbers of cells deposited on the chip. The cell responses are grouped according to the total number of cells initially loaded into the devices, 2, ~30, ~500, ~800, and more than 1,500 cells respectively. The number of cells

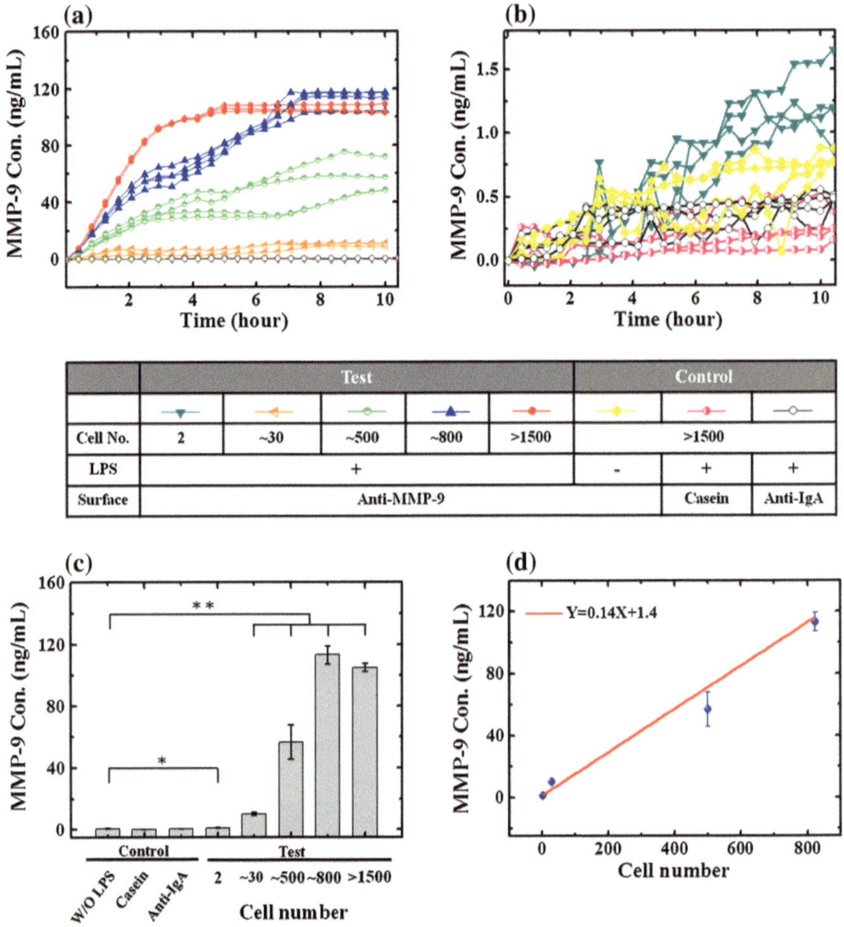

Fig. 25.11 **a** Dynamic secretion of MMP-9 from THP-1 cells after exposure to 10 μg/mL LPS. The cell numbers are determined by optical microscopy after cell loading and flushing. **b** A zoom-in image of the low concentration range shows minimal intensity changes in control groups. Two of the controls are LPS stimulated cells exposed to sensors passivated by casein or coated with non-specific anti-IgA. The third control uses un-stimulated cells cultured with an anti-MMP-9 sensor. The cell numbers in all three control experiments are over 1,500/device. **c** End point MMP-9 concentrations in control and test groups. MMP-9 binding is only observed with the LPS treated cells exposed to the specific antibody. **d** MMP-9 concentration versus cell number up to 800 cells. The linear fitting indicates that each cell secrete 0.14 ng/(mL cell) MMP-9 over the culture period of 13 h

inside the devices is determined by optical microscopy after cell loading and flushing off un-captured cells. The signals are recorded for 13 h, which is shorter than the average doubling time (~26 h) of THP-1 cells, thus cell proliferation is neglected. As observed from the kinetic curves, cells start to secrete MMP-9 almost immediately after stimulation, suggesting an intracellular reservoir. An intermediate shoulder is observed on most kinetic curves involving stimulated cells. It might be related to cell responses, such as a secretion slow down or pause due to reservoir depletion before MMP-9 production. Figure 25.11c summarizes the MMP-9 concentration as a function of cell number in the device. The average secretion from 2, ~30, ~500, ~800 and >1,500 cells after 13 h are 1.20 $\pm$ 0.32, 10.30 $\pm$ 1.02, 56.75 $\pm$ 11.07, 113.28 $\pm$ 5.94, and 105.08 $\pm$ 2.69 ng/mL MMP-9 respectively. When the end point data are compared by student's t-test, we observe significant differences between LPS treated versus untreated cells in terms of MMP-9 secretion, even for a low cell number of 2 cells per device. The early plateau from 1,500 cells after 4 h of measurements is likely a result of saturation of the surface antibody. For the other cell numbers, the end-point secretion increases with the number of cells in the device. Linear fitting of average secretion versus cell number (Fig. 25.11d) shows that the amount of secretion is 0.14 ng/mL per cell. As the volume of microfluidic chamber is 5 μL, each cell secrets around 0.7 pg MMP-9 in the culture period of ~13 h. This value is on the same order of magnitude as that obtained from ELISA (~0.35 pg/cell under 1 ng/mL LPS stimulation in 48 h) [46]. This result supports the utilization of a nanoslit SPR sensor for quantitative cell secretion studies.

25.6 Conclusions

Surface plasmon resonance biosensors are widely used in sensitive chemical, biological and environmental sensing. The common SPR biosensors utilize the prism or grating coupling method to meet the momentum matching conditions. The typical refractive index detection limit is about 10^{-5} refractive index unit for prism-based sensors and 5×10^{-5} RIU for grating-base sensors using intensity interrogation. In addition to the prism or grating methods, the SPR can be directly excited by using metallic nanostructures. Biosensors based on EOT of periodic nanohole arrays and nanoslit arrays in gold have been proposed. These sensors utilize normally incident transmitted light to excited SPR, which provide an easier optical setup than prism-based sensors. Compared to prism-based sensors, the gold nanostructure, combined with a charge-coupled device (CCD), is better for chip-based and high-throughput label-free detections. In periodic metallic nanostructures, the extraordinary transmission is usually accompanied by an asymmetrical resonant spectrum. The observed Fano resonance profile is understood in terms of the coupling of the surface-bound state of a periodic array (a discrete state) with the incoming wave (a continuum state). A higher wavelength sensitivity and a small peak bandwidth will have a higher intensity sensitivity. The resonant bandwidth is related to the structure and dielectric function of metal. There are several ways to increase the detection resolution and bandwidth of the SPR in gold nanostructures:

1. Increasing the period: the wavelength sensitivity is increased proportionally.
2. Decreasing dimension of the nano-apertures: The nanoslit is better than nanohole due to its non cut-off condition for TM wave.
3. Using Fano resonance to reduce the resonance bandwidth and make a sharp slope.
4. Using the spectral integration analysis method to reduce the noise and enhance the S/N ratio.
5. Using thermally annealing nanoimprint method to fabricate smooth gold film: The SPP propagation loss is reduced which results in a sharp SPP resonance.

By using above methods and nanostructures, the detection sensitivity and resolution will be comparable or higher than conventional prism-based SPR system.

In the applications, the gold nanostructure sensors take advantages of small detection area, chip-based/high-throughput detection and ease of integration with microfluidic devices. It can be used for label-free microarray biochips, such as DNA microarray, protein microarray and cell microarray. Different from conventional labeling methods, the nano-plasmonic SPR sensors provide real-time, dynamic responses of the bio-interactions without any labeling. When combining with the microfluidic devices, the new platform can be applied to various kinds of bio-interaction studies in different physical and chemical environments as well as fast diagnosis of various diseases.

References

1. J. Homola, S.S. Yee, G. Gauglitz, Surface plasmon resonance sensors: review. Sens. Actuators B-Chem. **54**, 3–15 (1999)
2. J. Homola, Present and future of surface plasmon resonance biosensors. Anal. Bioanal. Chem. **377**, 528–539 (2003)
3. M. Schena, D. Shalon, R.W. Davis, P.O. Brown, Quantitative monitoring of gene-expression patterns with a complementary-DNA microarray. Science **270**, 467–470 (1995)
4. J.N. Anker, W.P. Hall, O. Lyandres, N.C. Shah, J. Zhao, R.P. Van Duyne, Biosensing with plasmonic nanosensors. Nat. Mater. **7**, 442–453 (2008)
5. J. Homola, Surface plasmon resonance sensors for detection of chemical and biological species. Chem. Rev. **108**, 462–493 (2008)
6. K.A. Tetz, L. Pang, Y. Fainman, High-resolution surface plasmon resonance sensor based on linewidth-optimized nanohole array transmittance. Opt. Lett. **31**, 1528–1530 (2006)
7. K.L. Lee, S.H. Wu, P.K. Wei, Intensity sensitivity of gold nanostructures and its application for high-throughput biosensing. Opt. Express **17**, 23104–23113 (2009)
8. H. Chen, X. Kou, Z. Yang, W. Ni, J. Wang, Shape- and size-dependent refractive index sensitivity of gold nanoparticles. Langmuir **24**, 5233–5237 (2008)
9. F. Hao, N. Sonnefraud, P. Van Dorpe, S.A. Maier, N.J. Halas, P.J. Nordlander, Symmetry breaking in plasmonic nanocavities: subradiant LSPR sensing and a tunable fano resonance. Nano Lett. **8**, 3983–3988 (2008)

10. N.A. Mirin, K. Bao, P. Nordlander, Fano resonances in plasmonic nanoparticle aggregates. J. Phys. Chem. A **113**, 4028–4034 (2009)
11. N. Liu, T. Weiss, M. Mesch, L. Langguth, U. Eigenthaler, M. Hirscher, C. Sonnichsen, H. Giessen, Planar metamaterial analogue of electromagnetically induced transparency for plasmonic sensing. Nano Lett. **10**, 1103–1107 (2010)
12. N. Verellen, P. Van Dorpe, C.J. Huang, K. Lodewijks, G.A.E. Vandenbosch, L. Lagae, V.V. Moshchalkov, Plasmon line shaping using nanocrosses for high sensitivity localized surface plasmon resonance sensing. Nano Lett. **11**, 391–397 (2011)
13. R.W. Wood, On a remarkable case of uneven distribution of light in a diffraction grating spectrum. Phil. Mag. **4**, 396–402 (1902)
14. U. Fano, Theory of anomalous diffraction gratings and of quasi-stationary waves on metallic surface (Sommerfield's waves). J. Opt. Soc. Am. **31**, 213–222 (1941)
15. E. Kretschmann, H. Raether, Radiative decay of non-radiative surface plasmons excited by light. Z. Naturforsch. **23**, 2135–2136 (1968)
16. A. Otto, Excitation of surface plasma waves in silver by the method of frustrated total reflection. Z. Physik **216**, 398–410 (1968)
17. H. Reather, *Surface Plasmons on Smooth and Rough Surfaces and on Gratings*, vol. 111, Springer Tracks Modern Physics (Springer, Berlin, 1983)
18. F.C. Chien, C.Y. Lin, J.N. Yih, K.L. Lee, C.W. Chang, P.K. Wei, C.C. Sun, S.J. Chen, Coupled waveguide-surface plasmon resonance biosensor with subwavelength grating. Biosens. Bioelectron. **22**, 2737–2742 (2007)
19. J.S. Shumaker-Parry, C.T. Campbell, Quantitative methods for spatially resolved adsorption/desorption measurements in real time by surface plasmon resonance microscopy. Anal. Chem. **76**, 907–917 (2004)
20. T.M. Chinowsky, M.S. Grow, K.S. Johnston, K. Nelson, T. Edwards, E. Fu, P. Yager, Compact, high performance surface plasmon resonance imaging system. Biosens. Bioelectron. **22**, 2208–2215 (2007)
21. P.Y. Li, B. Lin, J. Gerstenmaier, B.T. Cunningham, A new method for label-free imaging of biomolecular interactions. Sens. Actuators B: Chem. **99**, 6–13 (2004)
22. J.N. Anker, W.P. Hall, O. Lyandres, N.C. Shah, J. Zhao, R.P. Van Duyne, Biosensing with plasmonic nanosensors. Nat. Mater. **7**, 442–453 (2008)
23. T.W. Ebbesen, H.J. Lezec, H.F. Ghaemi, T. Thio, P.A. Wolff, Extraordinary optical transmission through sub-wavelength hole arrays. Nature **391**, 667–669 (1998)
24. H.A. Bethe, Theory of Diffraction by Small Holes. Phys. Rev. **66**, 163–182 (1944)
25. K.L. Lee, W.S. Wang, P.K. Wei, Comparisons of surface plasmon sensitivities in periodic gold nanostructures. Plasmonics **3**, 119–125 (2008)
26. R. Gordon, Light in a subwavelength slit in a metal: propagation and reflection. Phys. Rev. B **73**, 153405 (2006)
27. K.L. Lee, W.S. Wang, P.K. Wei, Sensitive label-free biosensors by using gap plasmons in gold nanoslits. Biosens. Bioelectron. **24**, 210–215 (2008)
28. A.E. Miroshnichenko, S. Flach, Y.S. Kivshar, Fano resonances in nanoscale structures. Rev. Mod. Phys. **82**, 2257–2298 (2010)
29. C. Wu, A.B. Khanikaev, R. Adato, N. Arju, A.A. Yanik, H. Altug, G. Shvets, Fano-resonant asymmetric metamaterials for ultrasensitive spectroscopy and identification of molecular monolayers. Nat. Mater. **11**, 69–75 (2012)
30. K.-L. Lee, S.-H. Wu, C.-W. Lee, P.-K. Wei*, Sensitive biosensors using Fano resonance in single gold nanoslit with periodic grooves. Opt. Express **19**(24), 24530–24539 (2011)
31. J. Feng, V.S. Siu, A. Roelke, V. Mehta, S.Y. Rhieu, G.T.R. Palmore, D. Pacifici, Nanoscale plasmonic interferometers for multispectral, high-throughput biochemical sensing. Nano Lett. **12**, 602–609 (2012)
32. K.-L. Lee, P.-K. Wei*, Enhancing surface plasmon detection using ultrasmall nanoslits and multispectral integration method. Small **6**, 1900–1907 (2010)

33. M.E. Stewart, N.H. Mack, V. Malyarchuk, J. Soares, T.W. Lee, S.K. Gray, R.G. Nuzzo, J.A. Rogers, Quantitative multispectral biosensing and 1D imaging using quasi-3D plasmonic crystals. Proc. Natl. Acad. Sci. USA **103**, 17143–17148 (2006)
34. A.G. Brolo, R. Gordon, B. Leathem, K.L. Kavanagh, Surface plasmon sensor based on the enhanced light transmission through arrays of nanoholes in gold films. Langmuir **20**, 4813–4815 (2004)
35. J.C. Yang, J. Ji, J.M. Hogle, D.N. Larson, Metallic nanohole arrays on fluoropolymer substrates as small label-free real-time bioprobes. Nano Lett. **8**, 2718–2724 (2008)
36. P. Nagpal, N.C. Lindquist, S.H. Oh, D.J. Norris, Ultrasmooth patterned metals for plasmonics and metamaterials. Science **325**, 594–597 (2009)
37. K.P. Chen, V.P. Drachev, J.D. Borneman, A.V. Kildishev, V.M. Shalaev, Drude relaxation rate in grained gold nanoantennas. Nano Lett. **10**, 916–922 (2010)
38. K.L. Lee, P.W. Chen, S.H. Wu, J.B. Huang, S.Y. Yang, P.K. Wei, Enhancing surface plasmon detection using template-stripped gold nanoslit arrays on plastic films. ACS Nano **6**, 2931–2939 (2012)
39. H. Im, S.H. Lee, J.N. Wittenberg, T.W. Johnson, C.N. Lindquist, P. Nagpal, D.J. Norris, S.H. Oh, Template-stripped smooth ag nanohole arrays with silica shells for surface plasmon resonance biosensing. ACS Nano **5**, 6244–6253 (2011)
40. J.N. Anker, W.P. Hall, O. Lyandres, N.C. Shah, J. Zhao, R.P. Van Duyne, Biosensing with plasmonic nanosensors. Nat. Mater. **7**, 442–453 (2008)
41. R. Gordon, D. Sinton, K.L. Kavanagh, A.G. Brolo, A new generation of sensors based on extraordinary optical transmission. Acc. Chem. Res. **41**, 1049–1057 (2008)
42. A.A. Yanik, M. Huang, O. Kamohara, A. Artar, T.W. Geisbert, J.H. Connor, H. Altug, An optofluidic nanoplasmonic biosensor for direct detection of live viruses from biological media. Nano Lett. **10**, 4962–4969 (2010)
43. S.H. Wu, K.L. Lee, A. Chiou, X.H. Cheng, P.K. Wei, Optofluidic platform for real-time monitoring of live cell secretory activities using fano resonance in gold nanoslits. Small **9**, 3532–3540 (2013)
44. F.B. Myers, L.P. Lee, Innovations in optical microfluidic technologies for point-of-care diagnostics. Lab Chip **8**, 2015–2031 (2008)
45. D. Di Carlo, L.Y. Wu, L.P. Lee, Dynamic single cell culture array. Lab Chip **6**, 1445–1449 (2006)
46. B.M. Wong, W.C. Lumma, A.M. Smith, J.T. Sisko, S.D. Wright, T.Q. Cai, Cell migration and secretion of matrix metalloproteinase 9 by inhibiting geranylgeranylation. J. Leukoc. Biol. **69**, 959–962 (2001)

Part VIII
Biophotonics

Arthur Chiou and Chi-Kuang Sun

In the most generic sense, "Biophotonics," can probably be interpreted as a scientific/technical discipline involving the study of the principles and applications of the interactions of photons with biological samples. Under this generic umbrella, the scope of "Biophotonics" has expanded in many ways in recent years. In the "photonics" arm, the range of energy (or the frequency) spectrum may extend from microwave to x-ray; and in terms of biological/biomedical applications, the samples (or objects) of interest extendfrom single molecule study for the fundamental understanding of the basic molecular interactions at one end to subcellular organelles, live cells, tissues, organ, animals, and humans for the ultimate goals of disease prevention, diagnosis, and therapeutics. In addition, many novel photonics technologies and molecular probes have been developed such that multi-modality studies with unprecedented spatial and temporal resolutions have been realized to provide not only structural, but also functional information. Apparently, it is unrealistic and impractical to give a comprehensive overview of this field in a short chapter like this. Hence, we have deliberately chosen to narrow down our scope, and focus on four specific biophotonics techniques based on laser microscopy, namely FLIM/FRET (Fluorescence Life-time Imaging/Foster Resonance Energy Transfer) Microscopy; OCT (Optical Coherence Tomography); Superresolution Microscopy, and Harmonic Generation Microscopy; hence, contextually, this chapter represents a small, but important subset of biophotonics, which has made tremendous progress in the last decade not only in technological development, but also in niche biological and biomedical applications.

The chapter is subdivided into four relatively independent sections (Chap. 26 through 29), each covers one selected topic listed above. Each section begins with a brief historical/chronological overview to highlight the key technological developments, followed by the underlying principle and the basic experimental configurations, procedures, and analysis. Microscopic images of a wide variety of biological samples obtained by these different approaches show that, to a large extent, these different approaches complement one another in terms of spatial and temporal resolution, field of view, and penetration depth. A fairly extensive list of references is included so that readers interested in a specific technique or application can consult the appropriate references to get detailed information.

Chapter 26
Cellular Autofluorescence Detection Through FLIM/FRET Microscopy

Fu-Jen Kao, Gitanjal Deka and Nirmal Mazumder

26.1 Introduction

The number of publications using FLIM/FRET (Fluorescence lifetime imaging and Forster resonance energy transfer) technique is increasing over the past 10 years [1]. Fluorescence lifetime measurements, a excited state phenomenon, acquired with both spectroscopy and microscopy is a powerful tool in investigating the effects of various biological and chemical factors (DNA sequence, pH, ions, and molecularity) in biological systems [2, 3]. The empire of fluorescence microscopy spread over a wide range of applications in biotechnology, flow cytometry, medical diagnosis, DNA sequencing, forensics, and medical diagnosis [2]. The spatial organization and quantification of protein-protein interaction with FRET using multiphoton FLIM microscopy has emerged a powerful technique both in vivo and in vitro analysis [4, 5]. The assay based on FLIM by using time correlated single photon counting [6] to measure the FRET between two proteins. Noninvasive and nondestructive capability makes auto-fluorescence techniques a very useful candidate in the characterization of live cell and tissue imaging [3, 7, 8]. A fluorophore excited to the higher energy state and return to the ground state with a certain probability based on the decay rates through a number of different radiative and/or nonradiative decay pathways [7, 8]. For a homogenous ensemble, the fluorescence emitted will decay with a time according to a single exponential decay function [3]. Physically, the lifetime is the average time that a molecule spend in the excited state prior to return to the ground state [2]. The fluorescence

F.-J. Kao (✉) · G. Deka · N. Mazumder
Institute of Biophotonics, National Yang-Ming University, Taipei, Taiwan
e-mail: fjkao@ym.edu.tw

C.-C. Lee (ed.), *The Current Trends of Optics and Photonics*,
Topics in Applied Physics 129, DOI 10.1007/978-94-017-9392-6_26

lifetime is a marker for specific fluorophore, but for multi exponential decays the direct correlation between decay components and structural conformations are not simple [9]. Therefore, time-resolved fluorescence anisotropy imaging, essentially polarization resolved FLIM, is a complementary method which has been applied to monitor DNA digestion, mobility of fluorophores in living cells [10, 11].

26.2 Autofluorescence in Biological Specimen

Two photon fluorescence lifetime imaging microscopy (TP-FLIM), is a noninvasive and nondestructive technique, which reveals the sensitivity of endogenous fluorescent elements such as Reduced nicotinamide adenine dinucleotide (NADH) and its phosphate forms (NAD(P)H) along with flavins and flavoproteins in cells as well as collagen, elastin in tissues. TP-FLIM is applied to visualize the alteration of metabolic state in various cancerous tissues [12–17] such as skin, cervix, bladder, breast, stomach, and Alzheimer disease [18–24]. NADPH is an essential cofactor for oxidation-reduction reaction and energy metabolism in living cells [16, 17, 25, 26], while NADH is a major electron donor within the cells for both oxidative phosphorylation and glycolytic pathways. These molecules have their particular lifetime signature and specific functioning in maintaining cellular physiology. NADH and FAD have been extensively exploited to monitor relative cellular metabolic activity [27, 28, 29]. Both NADH and FAD are co-enzymes associated with the electron transport system (ETS) that generates ATP, the most important molecule which holds the energy for a cell in the form of a phosphate bond. While NADH serves as a primary electron donor, FAD is among the electron acceptors. These electron transporters exists in two physiological forms, free and bound to the dehydrogenage of complex-I; one of the mitochondrial inner membrane proteins while they are associated with energy generation [27, 29]. Thus, the knowledge about relative quantities of free and bound species of these two co-enzymes can give an insight to the metabolic state of a cell.

The discovery of fluorescence properties of pyridine nucleotides in the early 1950s [30] led to a new field of study: that uses fluorescence spectroscopy to observe the processes in live mitochondria. The discovery leads to sudden increase in mitochondrial bioenergetics studies mainly with the fluorescence detection. Previous reports suggest that NADH has a single-photon excitation peak at ~340 nm, with the fluorescence emission peak at ~460 nm [2]. Unlike NADH, FAD absorbs blue light, at ~470 nm and emits green-yellow, peak at ~540 nm [2]. However, the autofluorescence spectroscopy technique does not provide enough contrast between free and enzyme-bound forms of each of these molecules. The solution of this problem lies in the use of FLIM technique, which is based on the measurement of the average time a fluorophore spent in the excited state. Lakowicks et al., pioneered the work on NADH FLIM imaging of free and bound form in solution [31]. Fluorescence lifetime measurement is the only technique that gives a noninvasive way to assess the free/bound ration of NADH and

FAD, where the short lifetime ~400 ps is usually attributed to the free and the long lifetime ~2,500 ps to the bound form of NADH [31]. It is reported that the fluorescence lifetime of free FAD is to be around 2.3 ns [32] and fluorescence lifetime of bound FAD is 0.3 ns, which is lower than that of free FAD [29].

Tryptophan is an amino acid essential for human life, which can't be synthesized by our body, and therefore must be part of our diet. Tryptophan has special importance in our body physiology and protein metabolism that can be observed inside the cell. The tryptophan residues has single photon excitation peak at ~270 nm and emission maximum at ~350 nm [2]. Recently, Jyothikumar et al. has reported the use of FLIM/FRET technique between tryptophan and NADH to investigate the metabolic activity of cell [33].

26.3 Multiphoton Excitation

Multiphoton microscopy provides some advantages in autofluorescence imaging than a single photon confocal microcopy, such as photo-bleaching [34, 35], photo-damage [36], less scattering for live cell and tissue imaging. Generally, multiphoton microscopy uses photons at near infra red (NIR) light. It is noted that the fluorescence lifetime measurement from endogenous fluorophores doesn't interfere with the intrinsic metabolism of the cells in tissue. Multiphoton microscopy has the capability of optical biopsy nondestructively under in vivo physiology conditions with a similar resolution as the microscope of the pathologists without any slicing and staining [37, 38]. Multiphoton excitation was proposed and theoretically investigated by [39]. However, the developments in the field of nonlinear optical detection halted due to lack of a suitable laser source. With the invention of femtosecond lasers in the 1980s multiphoton absorption has been materialized. In 1990, the decisive study of Denk, Strickler, and Webb on two-photon laser scanning fluorescence microscopy clearly demonstrated the capability of two-photon excitation microscopy for biology [40]. A multiphoton excitation absorbs more than one, two photon at 2λ (two photon excitation) or three photon at 3λ (three photon excitation), instead of one photon at a given wavelength λ at an instance. Multiphoton excitation requires a higher photon density (100 MW/cm^2 to 100 GW/cm^2), from a picoseconds or femtosecond pulsed laser source. Use of high numerical aperture objective lens can focus the pulse laser beam to gather the power density required for multiphoton excitation without photo damaging effects due to the short time of the laser excitation. The most distinguishing advantage of multiphoton microscopy over one-photon fluorescence imaging is that the one-photon process excite the molecules in a full double-cone through the sample, while the two-photon process excite only a few micron volume around the laser focal point [40]. Unlike one photon microscopy, multiphoton microscopy doesn't require the pinhole to suppress out of focus fluorescence photons for 3D imaging. Thus, it avoids using of special filters, and non-descanned collection possibilities increases number of photons detection. These advantages of multiphoton microscopy make cellular and tissue autofluorescence imaging possible.

26.4 Measurements and Data Analysis

Fluorescence lifetime can be measured in two ways: the time domain and frequency domain. In time domain method, the fluorophores are excited with a train of pulsed light [3]. The width of the pulse is made as small as it become much smaller than the decay time τ of the fluorophore. The time dependant intensity is measured following the excitation pulse, and the decay time τ is calculated from the slope of a plot of log $I(t)$ against t called decay curve [2]. For time domain measurements most common implementation of fluorescence lifetime imaging are based on laser scanning microscopes both confocal and multiphoton mode with single photon lifetime detection methods. These microscopes usually amalgamated with picoseconds or femtosecond pulsed lasers, detectors with same time resolutions (PMT, MCP), and special single photon detection electronics. Time correlated single photon counting (TCSPC) uses a high frequency pulse laser to scan the sample. Data recording is done by detecting single photons of the fluorescence light and determining the arrival times of the photons with respect to the laser pulses as well as the position of the laser beam in the moment of photon detection [3, 41]. A typical TCSPC instrumentation uses a mode-locked Ti:sapphire Mira F-900 laser (Coherent, United States), capable of producing 76 MHz femtosecond laser pulses in the spectral range from 700 to 1,000 nm, pumped by a solid-state frequency-doubled 532 nm Verdi laser (Coherent, United States) for the two-photon excitation of cellular autofluorescence [27, 28] .

Lifetime calculation from the multi-exponential decay was done by mathematical convolution of a model function and the instrument response function (IRF) by fitting with the experimental data. Lifetimes from the composite decays of biomolecules can be derived by convolution of an IRF, I_{instr}, with a double-exponential model function, defined in (26.4.1), with offset correction for the ambient light and/or dark noise I_0 to obtain calculated lifetime decay function $I_c(t)$ in (26.4.2)

$$F(t) = a_1 e^{-t/\tau_1} + a_2 e^{-t/\tau_2} \tag{26.4.1}$$

$$I_c(t) = \int_{-\infty}^{\infty} I_{instru}(t)\{I_o + F(t)\}dt \tag{26.4.2}$$

where $a_1 e^{-t/\tau_1}$ and $a_2 e^{-t/\tau_2}$ denote the contributed fluorescence decays from short and long lifetime components of NADH, respectively; τ_1 and τ_2 represent their corresponding lifetime constants; and a_1 and a_2 refer to the corresponding relative amplitudes. The I_{instr} can be measured experimentally with the periodically poled lithium niobate (PPLN) crystal at 370 nm (i.e. the second harmonic of 740 nm from the Ti:Sapphire laser). The average lifetime was calculated as an amplitude-weighted parameter of the two lifetime components:

$$\tau_m = \frac{a_1\tau_1 + a_2\tau_2}{a_1 + a_2} \tag{26.4.3}$$

The model parameters (i.e. a_i and τ_i) were derived by fitting the decay $I_c(t)$, from (26.4.2), to the actual data $I_a(t)$ by minimizing the goodness-of-fit function defined in (26.4.3), using the Levenberg-Marquardt search algorithm,

$$\chi_R^2 = [\sum\nolimits_{k=0}^{n} [I_a(t) - I_c(t)]^2 / I_a(t)]/(n-p) \qquad (26.4.4)$$

where n denotes the number of data (time) points (equal to 256), and p represents the number of model parameters. A good fit is characterized by a χ^2 close to 1 and residuals showing no noticeable systematic variations. Figure 26.1 schematically represents a typical bi-exponential model fitting of NAD(P)H fluorescence decay standard fitted curve along with representative color-coded image of NAD(P)H lifetimes in Huh7 cells and NAD(P)H fluorescence intensity. Both lifetimes (t_1 and t_2) and amplitudes (a_1 and a_2—population sizes of molecules with the different decay rate) were obtained from fitting optimization software. In case of NADH energetic measurements the ratio of a_1 and a_2 is the best indicator for free and protein-bound states of NADH, which can be used to depict the status and changes in cellular metabolism [2, 42]. Note that higher value of a_2 represents higher fraction

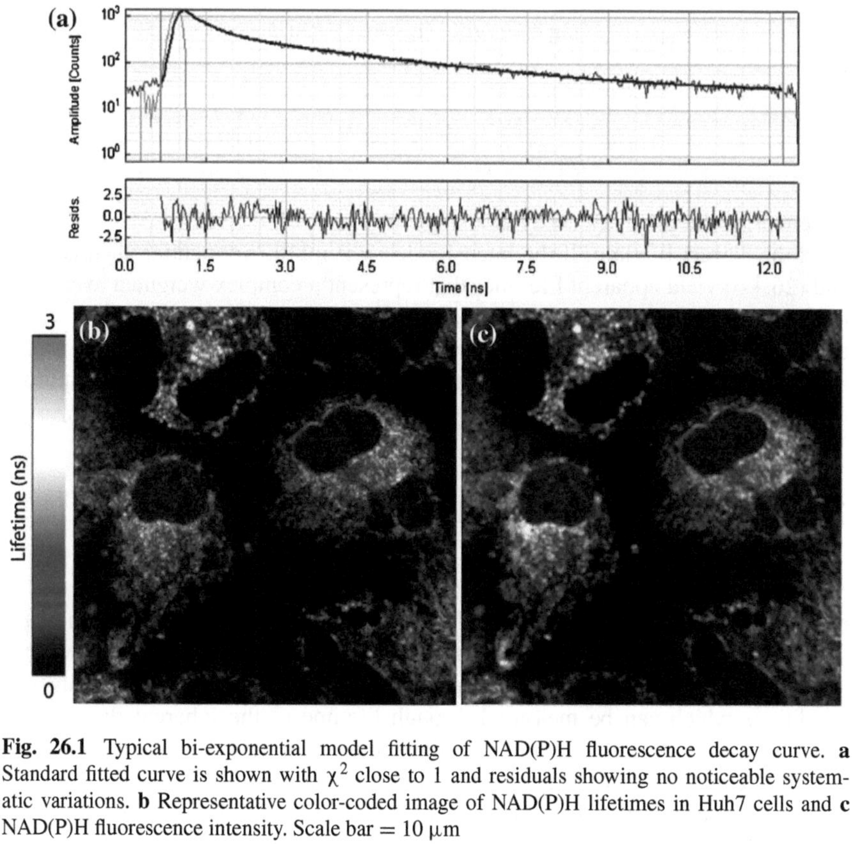

Fig. 26.1 Typical bi-exponential model fitting of NAD(P)H fluorescence decay curve. **a** Standard fitted curve is shown with χ^2 close to 1 and residuals showing no noticeable systematic variations. **b** Representative color-coded image of NAD(P)H lifetimes in Huh7 cells and **c** NAD(P)H fluorescence intensity. Scale bar = 10 μm

of protein bound NADH. Thus, the value of a_1/a_2 ratio is inversely related to the metabolic activity.

In the case of frequency-domain or phase-modulated light, the fluorophores are excited with intensity-modulated light. The intensity of the excited light is modulated at a high frequency typically around 100 MHz, so its reciprocal frequency is comparable to the reciprocal of decay time τ [3]. The fluorescence emits at the same modulation frequency but phase shifted due to the delay caused by the lifetime of the fluorophore relative to the excitation time. This delay is measured as a phase shift (ϕ_ω), which can be used to measure the decay time, where ω is the modulation frequency in radian/sec. The lifetime of the fluorophore also cause demodulation to the modulated excitation beam by a factor m_ω, by introducing changes in the peak-to-peak power as a function of the decay time and light modulation frequency. Suppose the modulation of the excitation is given by b/a, where 'a' is the average intensity and 'b' is the peak-to-peak height of the incident light. And the modulation of the emission is also defined as B/A. The modulation of the emission relative to the excitation is measured as, $m = (B/A)/(b/a)$.The phase angle (ϕ_ω) and the modulation can be employed to calculate the lifetime using (26.4.5) and (26.4.6).

$$\tan\phi = \omega\tau_\phi \quad \tau_\phi = \omega^{-1}\tan\phi \tag{26.4.5}$$

$$m = \frac{1}{\sqrt{1+\omega^2\tau m^2}} \quad \tau_m = \frac{1}{\omega}\left[\frac{1}{m^2} - 1\right]^{1/2} \tag{26.4.6}$$

Note that, if the intensity decay is a single exponential, then (26.4.5) and (26.4.6) yield the correct lifetime. If the intensity decay is multi-exponential, then (26.4.5) and (26.4.6) yield apparent lifetimes that represent a complex weighted average of the decay components [42]. The time-domain and frequency-domain are related to each other through Fourier transform. Time-domain data have intensity values in subsequent time channels, whereas, frequency-domain translate into amplitude and phase values at multiples of the signal repetition rate [3].

26.5 Applications of FLIM/FRET and Significance

Spectrally resolved fluorescence microscopy technique easily discriminate different fluorophores. FLIM has the novelty in discriminating different fraction of the same fluorophore in different state of interaction with environment or in various structural conformations. FLIM technique is very sensitive to the environmental conditions which can be measured through lifetime of the inherent fluorophore. Thus, FLIM is a desirable tool to investigate molecular effects inside cells and tissues by detecting the lifetime of autofluorescent molecules present in unknown at various concentration-in body physiology [2, 42, 43]. The parameters effect the fluorescence lifetime at a specific cellular condition. Some of them are listed below:

1. *Binding of a fluorophore to protein and protein configuration*: The fluorescence lifetime is sensitive to the conformational changes and surroundings of a fluorophore [9]. The fluorophore may bind with biomolecules; membranes, enzymes or proteins in the cellular system which in turn change the internal non-radiative decay profile. The fluorescence lifetimes of endogenous fluorophores NADH and FAD are different upon binding to the protein during the electron transport chain [44, 45]. Both the relative components of protein bound and free as well as their lifetime are sensitive to the metabolic state of cells and tissues [31–33]. Vladimir et al. have used FLIM to monitor cellular metabolism in a culture of HeLa cells by detecting the change in NADH lifetime and relative components of free and bound forms. It have been illustrated the capacity of NADH-FLIM to properly characterize the metabolism of normal and pathological states to reveal the dynamics of NADH lifetime at different cell densities, cell death pathways, and inhibition of mitochondrial electron transport chain [27]. The real time event of bacterial infection on cell in situ has been studied non-invasively by Buryakina et al. They have observed decrease in metabolic activity in terms of NADH free to protein bound ration with time passes after infection, which they compare with STS induced cell death process [28]. A similar kind of study reported by Nirmal Mazumder et al., in alteration of cellular metabolism by domain 2 of the Hepatitis C virus core protein. Their findings suggest that HCV causes a shift of metabolic control away from the use of the co-enzyme in mitochondrial electron transport and towards glycolysis, lipid biosynthesis [46]. FLIM has also contributed in stem cell research. Koning et al. have imaged 3D-stem cell spheroids during differentiation by NADH-FLIM. Their findings elucidate that stem cells increased their autofluorescence lifetimes and decreases their total fluorescence intensity during the adipogenic-differentiation process [47]. Not only in cellular levels, NADH-FLIM has also been used to study different layers of the skin tissue from epidermis to dermis layer; components in skin autofluorescence in various skin strata; imaging and quantification of coexisting drugs and their metabolites; skin pH; nanoparticle zinc oxide skin penetration; liposome delivery of drugs to deeper tissues; and observations in skin ageing and in various skin diseases [37]. Work has also been done on ex vivo human skin deterioration by using NAD(P)H auto-fluorescence intensity and lifetime imaging to assess ischemic necrosis at different temperature [48]. Recently Deka et al., have studied wound healing process in vivo by monitoring the relative metabolic activity of cells involved in the process. They have used rats to create some fresh wounds and studied the time lapsed dynamics of the metabolic activity in terms of relative ratio of the component of free and protein bound NADH. They conclude that, if wound healing can be divided into three main phases, such as inflammation, proliferation and remodeling, then metabolic activity

is higher during first phase and gradually decreases in the successive phases [38]. Multiphoton FLIM is also a powerful tool for the noninvasive characterization and detection of cancers and pre-cancers tissues [49]. Low-grade pre-cancers (mild to moderate dysplasia) and high-grade pre-cancers (severe dysplasia and carcinoma) can be discriminated from normal tissues by their decreased protein-bound NADH lifetime. Inhibition of cellular glycolysis and oxidative phosphorylation in cell mono-layers produces an increase and decrease, respectively, in the protein-bound NADH lifetime, which indicates the decrease in protein-bound NADH lifetime with dysplasia is due to a shift from oxidative phosphorylation to glycolysis, consistent with the predictions of neoplastic metabolism [49]. Recently, it is reported that FLIM-FRET technique can be used to investigate interaction between two endogenous auto-fluorescent molecules; tryptophan and NADH. Higher quenching of tryptophan lifetime is observed in response to the increase in protein-bound NADH as the cells were treated with glucose. This suggests the increase in interaction between tryptophan and NADH [33].

2. *pH indicator*: Many auto-fluorescing molecules in biological system are a protonated and a deprotonated form. The equilibrium depends on the pH, if the protonated and deprotonated forms have different lifetimes; the apparent lifetime is an indicator of the pH. Intracellular pH is one of the most important factor for understanding physiological states of cells, and significant changes in intracellular pH occur in close relationship with a variety of cellular functions such as ion transport and cell cycle [50, 51]. Ogikubo et al., observed that fluorescence lifetime of NADH become shorter monotonically with increasing pH, which indicates that the intracellular pH of a single cell can be evaluated using fluorescence lifetime imaging of NADH [52].
3. *Local viscosity*: Fluorescence lifetime is sensitive to molecular rotation. There are auto-fluorescing molecules that have higher degree of internal flexibility. Change in rotational behavior in a fluorophore can introduce a difference in internal radiation less decay profile. Usually the rotation of a molecule is sensitive to the viscosity of the surrounding solvent medium [53]. Diffusion rate of species in condensed media is a function of the viscosity of the medium. In bio-systems, changes in viscosity have been linked to disease and malfunction at the cellular level [54, 55]. The perturbations in viscosity are caused by changes in mobility of chemicals within the cell, influencing fundamental processes such as signaling and transport and the efficiency of bimolecular processes governed by diffusion of short-lived intermediates, such as the diffusion of reactive oxygen species during an oxidative stress attack. It has been reported on imaging local micro viscosity using the fluorescence lifetime of a molecular rotor [53]. In principle, theoretically changes in autofluorescence lifetime of cellular bio-molecules at deferent biological situation can be a marker of a specific viscosity of the cytosol.
4. *Presence of other molecule, atom or ions in proximity or change in their concentration*: Quenching of autofluorescence lifetime of some bio-molecules can be related to the interaction with intermediate reactive species or ions. In

neuronal system, signaling is mainly mediated through Ca^{2+} and Cl^{+} ion discharge [54, 55]. Influx of these ions can trigger metabolic changes in mitochondria, endoplasmic reticulum or other intra cellular compartments, which in-turn can change their lifetime of intracellular fluorophores. Oxygen is also an efficient fluorescence quencher for a number of fluorophores [2, 42]. Oxygen has strong effect on the fluorescence lifetime of the endogenous fluorophores NADH and FAD. Its effect on the redox state can be measured through the reduced NADH, whilst oxidized NAD+ is not. Interestingly, the oxidized form of the cell redox marker FAD/FADH shows the opposite effect, i.e. FAD is fluorescent, whereas FADH is not. Chance et al. defined 'redox ratio' is a direct indicator of the amount of oxygen used in the mitochondria of the cells [56, 57]. Effects of oxygen probably exist for other endogenous fluorophores as well [58]. Oxygen-induced lifetime changes can be expected to become important research area with the introduction of FLIM into clinical applications.

26.6 Conclusion

FLIM is a unique technique with huge potential and application in biomedical field. However, autofluorescence lifetime comes with few inconveniences, since high photon count is required for best FLIM statistical analysis. Autofluorescence occurs with less photon counts and it causes problematic for double exponential data fitting. Additionally, in case of biological samples there is a possibility for more than one fluorophores in a single pixel, which brings a huge difficulty in interpreting the data. Those problems can be addressed by using an improving the microcopy technique and analyzing algorithm. The photo statistics can be improved with combination of high sensitive detector and cooling system. Introduction of hybrid detectors with GaAsP cathodes can solve the problem [59]. The phasor plot, a new algorithm for the FLIM data analysis, helps to detect different fluorophores at each pixel of the image frame. Application of phasor approach helps to analyze the lifetime decay in great detail even if the number of fluorophore is more than two, where number of photon count is not a serious issue [60]. With the new invention of techniques and algorithm FLIM can be a potential tool in clinical diagnosis, such as skin wound healing study and cancer detections [28, 38]. It will be a great future perspective if FLIM technique become multimodal with coherent anti-Stokes Raman scattering microscopy (CARS) [61], high resolution microscopy systems like stimulated emission depletion microscopy (STED) [62], stochastic optical reconstruction microscopy (STORM), photo-activated localization microscopy (PALM), near-field scanning optical microscopy [63]. TCSPC can investigate the fast dynamic changes in metabolic activity by recording the data in time series by stimulating the sample repeatedly which triggered accumulation of a time series [64].

References

1. S.S. Vogel, C. Thaler, S.V. Koushik, Fanciful FRET. Sci STKE **18**(331), re2 (2006)
2. J.R. Lakowicz, *Principles of Fluorescence Spectroscopy, Chap. 1* (Plenum Press, New York, 1999)
3. W. Becker, Fluorescence lifetime imaging—techniques and applications. J. Microsc. **247**(2), 119–136 (2012)
4. R.V. Krishnan, A. Masuda, V.E. Centonze, B. Herman, Quantitative imaging of protein–protein interactions by multiphoton fluorescence lifetime imaging microscopy using a streak camera. J. Biomed. Opt. **8**(3), 362–367 (2003)
5. Y. Sun, N.M. Hays, A. Periasamy, M.W. Davidson, R.N. Day, Monitoring protein interactions in living cells with fluorescence lifetime imaging microscopy. Methods Enzymol. **504**, 371–391 (2012)
6. Y. Sun, R.N. Day, A. Periasamy, Investigating protein–protein interactions in living cells using fluorescence lifetime imaging microscopy. Nat. Protoc. **6**, 1324–1340 (2011)
7. P. Elumalai, P. Atkins, J. de Paula, *Atkins' Physical Chemistry* (Oxford University Press, Oxford, 2002)
8. A. Jabłoński, Efficiency of Anti-Stokes Fluorescence in Dyes. Nature **131**, 839–840 (1933)
9. Q. Yu, A.A. Heikal, Two-photon autofluorescence dynamics imaging reveals sensitivity of intracellular NADH concentration and conformation to cell physiology at the single-cell level. J. Photochem. Photobiol. B **95**, 46–57 (2009)
10. A.H.A. Clayton, Q.S. Hanley, D.J. Arndt-Jovin, V. Subramaniam, T.M. Jovin, Dynamic fluorescence anisotropy imaging microscopy in the frequency domain (rFLIM). Biophys. J. **83**, 1631–1649 (2002)
11. Z. Cao, C.C. Huang, W. Tan, Nuclease resistance of telomere-like oligonucleotides monitored in live cells by fluorescence anisotropy imaging. Anal. Chem. **78**, 1478–1484 (2006)
12. D.H. Williamson, P. Lund, H.A. Krebs, The redox state of free nicotinamide-adenine dinucleotide in the cytoplasm and mitochondria of rat liver. Biochem. J. **103**(2), 514–527 (1967)
13. A. Mayevsky et al., Brain metabolic responses to ischemia in the mongolian gerbil. In vivo and freeze trapped redox state scanning. Brain Res. **276**, 95–107 (1983)
14. Y. Wu et al., Depth-resolved fluorescence spectroscopy reveals layered structure of tissue. Opt. Exp. **12**(14), 3218–3223 (2004)
15. W.S. Kunz, F.N. Gellerich, Quantification of the content of fluorescent flavoproteins in mitochondria from liver, kidney cortex, skeletal muscle, and brain. Biochem. Med. Metab. Biol. **50**(1), 103–110 (1993)
16. W.S. Kunz et al., Measurement of fluorescence changes of NAD(P)H and of fluorescent flavoproteins in saponin-skinned human skeletal muscle fibers. Anal. Biochem. **216**(2), 322–327 (1994)
17. Y. Wu, J.Y. Qu, Combined depth- and time-resolved autofluorescence spectroscopy of epithelial tissue. Opt. Lett. **31**(12), 1833–1835 (2006)
18. C.J. Gulledge, M.W. Dewhirst, Tumor oxygenation: a matter of supply and demand. Anticancer Res. **16**(2), 741–749 (1996)
19. O. Warburg et al., The metabolism of tumors in the body. J. Gen. Physiol. **8**, 519–530 (1927)
20. J. Benavides et al., Multispectral digital colposcopy for in vivo detection of cervical cancer. Opt. Exp. **11**(10), 1223–1236 (2003)
21. D. Zaakab et al., Ultraviolet-excited (308 nm) autofluorescence for bladder cancer detection. Urology **60**(6), 1029–1033 (2002)
22. P.J. Tadrous et al., Fluorescence lifetime imaging of unstained tissues: early results in human breast cancer. J. Pathol. **199**, 309–317 (2003)
23. B. Mayinger, M. Jordan, T. Horbach, P. Horner, C. Gerlach, S. Mueller, W. Hohenberger, E.G. Hahn, Evaluation of in vivo endoscopic autofluorescence spectroscopy in gastric cancer. Gastrointest. Endosc. **59**, 191–198 (2004)
24. E.B. Hanlon, I. Itzkan, R.R. Dasari, M.S. Feld, R.J. Ferrante, A.C. McKee, D. Lathi, N.W. Kowall, Near-infrared fluorescence spectroscopy detects Alzheimer's disease in vitro. J. Photochem. Photobiol. **70**(2), 236–242 (1999)

25. K. Sato, Y. Nishina, K. Shiga, F. Tanaka, Hydrogen-bonding dynamics of free flavins in benzene and FAD in electron-transferring flavoprotein upon excitation. J. Photochem. Photobiol. B **70**, 67–73 (2003)
26. F. Berger, M.H. Ramirez-Hernandez, M. Ziegler, The new life of a centenarian: signalling functions of NAD(P). Trends Biochem. Sci. **29**, 111–118 (2004)
27. V.V. Ghukasyan, F.J. Kao, Monitoring cellular metabolism with fluorescence lifetime of reduced nicotinamide adenine dinucleotide. J. Phys. Chem. C **113**, 11532–11540 (2009)
28. T.Y. Buryakina et al., Metabolism of HeLa cells revealed through autofluorescence lifetime upon infection with enterohemorrhagic Escherichia coli. J. Biomed. Opt. **17**(10), 101503 (2012)
29. A. Gafni, L. Brand, Fluorescence decay studies of reduced nicotinamide adenine dinucleotide in solution and bound to liver alcohol dehydrogenase. Biochemistry **15**(15), 3165–3171 (1976)
30. M.R. Duchen, A. Surin, J. Jacobson, Imaging mitochondrial function in intact cells. Methods Enzymol. **361**, 353–389 (2002)
31. C.B. Baltscheffsky, Light-scattering and absorption effects caused by addition of adenosine diphosphate to rat-heart-muscle sarcosomes. Biochem. J. **68**(2), 283–295 (1958)
32. N. Nakashima et al., Picosecond fluorescence lifetime of the coenzyme of D-amino acid oxidase. J. Biol. Chem. **255**(11), 5261–5263 (1980)
33. V. Jyothikumar et al., Investigation of tryptophan–NADH interactions in live human cells using three-photon fluorescence lifetime imaging and Förster resonance energy transfer microscopy. J. Biomed. Opt. **18**(6), 060501 (2013)
34. G.H. Patterson, D.W. Piston, Photobleaching in two-photon excitation microscopy. Biophys. J. **78**(4), 2159–2162 (2000)
35. L.M. Tiede, M.G. Nichols, Photobleaching of reduced nicotinamide adenine dinucleotide and the development of highly fluorescent lesions in rat basophilic leukemia cells during multiphoton microscopy. Photochem. Photobiol. **82**, 656–664 (2006)
36. K. König, T.W. Becker, P. Fischer, I. Riemann, K.J. Halbhuber, Pulse-length dependence of cellular response to intense near-infrared laser pulses in multiphoton microscopes. Opt. Lett. **24**(2), 113–115 (1999)
37. A. Zoumi, A. Yeh, B.J. Tromberg, Imaging cells and extracellular matrix in vivo by using second-harmonic generation and two-photon excited fluorescence. Proc. Natl. Acad. Sci. USA **99**(17), 11014–11019 (2002)
38. G. Deka et al., In vivo wound healing diagnosis with second harmonic and fluorescence lifetime imaging. J. Biomed. Opt. **18**(6), 061222 (2013)
39. M. Goppert-Mayer, Uber elementaraktemit wei quatensprungen. Ann. Phys. **9**(3), 273–95 (1931)
40. W. Denk, J.H. Strickler, W.W. Webb, Two-photon laser scanning fluorescence microscopy. Science **248**, 73–76 (1990)
41. W. Becker, *Advanced Time-Correlated Single Photon Counting Techniques* (Springer, Berlin, 2005)
42. J.R. Lakowicz, *Principles of Fluorescence Spectroscopy, Chaps. 4 and 5* (Plenum Press, New York, 1999)
43. M.S. Roberts et al., Non-invasive imaging of skin physiology and precancutaneous penetration using 5D (space, time and anisotropy) fluorescence spectral and lifetime imaging with multiphoton and confocal microscopy. Eur. J. Pharm. Biopharm. **77**, 469–488 (2011)
44. M.C. Skala, K.M. Riching, D.K.Bird, A.G. Fitzpatrick, J. Eickhoff, K.W. Eliceiri, P.J. Keely, N. Ramanujam, Invivo multiphoton fluorescence lifetime imaging of protein-bound and free NADH in normal and pre-cancerous epithelia. J. Biomed. opt. **12**(2), 024014 (2007)
45. R.J. Paul, H. Schneckenburger, Oxygen concentration and the oxidation–reduction state of yeast: determination of free/bound NADH and flavins by time-resolved spectroscopy. Naturwissenschaften **83**, 32–35 (1996)
46. N. Mazumder, R.K. Lyn, A. Ridsdale, D.J. Moffatt, R. Singaravelu, J.W. Hu, H.R. Tsai, J. McLauchlan, A. Stolow, F.J. Kao, J.P. Pezacki, Fluorescence lifetime imaging of alterations to cellular metabolism by domain 2 of the hepatitis C virus core protein. PLoS One **8**(6), e66738 (2013)

47. K. Koning, Multiphoton fluorescence lifetime imaging of 3D stem cell spheroids during differentiation. Microsc. Res. Tech. **74**, 9–17 (2011)
48. W.Y. Sanchez et al., Analysis of the metabolic deterioration of ex vivo skin from ischemic necrosis through the imaging of intracellular NAD(P)H by multiphoton tomography and fluorescence lifetime imaging microscopy. J. Biomed. Opt. **15**(4), 046008 (2010)
49. D.K. Bird, L. Yan, K.M. Vrotsos, K.W. Eliceiri, E.M. Vaughan, P.J. Keely, J.G. White, N. Ramanujam, Metabolic mapping of MCF10A human breast cells via multiphoton fluorescence lifetime imaging of the coenzyme NADH. Cancer Res. **65**(19), 8766–8773 (2005)
50. A.M. Paradiso, R.Y. Tsien, T.E. Machen, Fluorometric calcium measurement. Nature **325**, 447–450 (1987)
51. H.J. Lin, P. Herman, J.R. Lakowicz, Fluorescence lifetime-resolved pH imaging of living cells. Cytometry Part A, **52**(2), 77–89 (2003)
52. S. Ogikubo, Intracellular pH sensing using autofluorescence lifetime microscopy. J. Phys. Chem. B **115**, 10385–10390 (2011)
53. M.K. Kuimova, Molecular rotor measures viscosity of live cells via fluorescence lifetime imaging. J. Am. Chem. Soc. **130**, 6672–6673 (2008)
54. H. Kaneko et al., Chloride accumulation in mammalian olfactory sensory nurons. J. Neurosci. **24**(36), 7931–7938 (2004)
55. K.V. Kuchibhotla et al., Synchronous hyperactivity and intercellular calcium waves in astrocytes in Alzheimer mice. Science **323**(5918), 1211–1215 (2009)
56. B. Chance, Pyridine–nucleotide as an indicator of oxygen requirements for energy-linked functions of mitochondria. Circ. Res. **38**, 31–38 (1976)
57. B. Chance, B. Schoener, R. Oshino, F. Itshak, Y. Nakase, Oxidation–reduction ratio studies of mitochondria in freeze-trapped samples—NADH and flavoprotein fluorescence signals. J. Biol. Chem. **254**, 4764–4771 (1979)
58. D. Schweitzer, S. Schenke, M. Hammer, F. Schweitzer, S. Jentsch, E. Birckner, W. Becker, A. Bergmann, Towards metabolic mapping of the human retina. Microsc. Res. Tech. **70**, 410–419 (2007)
59. W. Becker et al., FLIM and FCS detection in laser-scanning microscopes: increased efficiency by GaAsP hybrid detectors. Micros. Res. Tech. **74**, 804–811 (2010)
60. M.A. Digman et al., The phasor approach to fluorescence lifetime imaging analysis. Biophys. J. **94**, L14–L16 (2008)
61. A.D. Slepkov, A. Ridsdale, H.N. Wan, M.H. Wang, A.F. Pegoraro, D.J. Moffatt, J.P. Pezacki, F.J. Kao, A. Stolow, Forward-collected simultaneous fluorescence lifetime imaging and coherent anti-Stokes Raman scattering microscopy. J. Biomed. Opt. **16**(2), 021103 (2011)
62. P.Y. Lin, Y.C. Lin, C.S. Chang, F.J. Kao, Fluorescence lifetime imaging microscopy with sub-diffraction-limited resolution. Jpn. J. Appl. Phys. **52**(2), 028004 (2013)
63. A. Cadby et al., Mapping the fluorescence decay lifetime of a conjugated polymer in a phase-separated blend using a scanning near-field optical microscope. Nano Lett. **5**, 2232–2237 (2005)
64. W. Becker et al., Spatially resolved recording of transient fluorescence-lifetime effects by line-scanning TCSPC. Proc. SPIE **8226**, 82260C-1–82260C-6 (2012)

Chapter 27
Optical Coherence Tomography for Quantitative Diagnosis in Cardiovascular Disease

Wen-Chuan Kuo

27.1 Introduction

In recent years, biomedical imaging technology has made rapid advances that enable the visualization, quantification, and monitoring of morphology and function. There are several tomography modalities which are currently used in clinics, such as computed tomography (CT), positron emission tomography (PET), magnetic resonance imaging (MRI), ultrasound (US), etc. These modalities have been developed for in vivo structural and functional imaging in humans, but frequently require large, expensive and complex systems. The penetration depth of these tomographic techniques is long, but the spatial resolution is typically on the order of several millimeters or hundred of micrometers.

Currently, the use of optical tomographic techniques for biomedical imaging is gaining considerable attention. Optical coherence tomography (OCT) is one of a class of optical tomographic techniques. It was developed in 1991 and is analogous to ultrasound B-mode imaging except reflections of near-infrared light are detected rather than sound. The attractive features of OCT include high cellular-level resolution, real-time acquisition rates, no exogenous fluorophores are necessary in a compact noninvasive instrument. Repeated OCT imaging also reduces the need to sacrifice specimens to obtain histological images. In particular, the development of new broad bandwidth light sources has led directly to ultrahigh resolution OCT (UHR OCT). For example, the combination of superluminescent diodes with different central wavelengths into one broadband sources [1] and the use of mode-locked solid state lasers [2–4] or supercontinuum light source [5–8] have all been proposed in the literature. The subcellular axial resolution (1–5 μm) now available from these advances has made it possible to obtain in vivo OCT tomograms close to the level of histology, which holds enormous promise for early

W.-C. Kuo (✉)
Institute of Biophotonics, National Yang-Ming University, Taipei, Taiwan
e-mail: wckuo@ym.edu.tw

C.-C. Lee (ed.), *The Current Trends of Optics and Photonics*,
Topics in Applied Physics 129, DOI 10.1007/978-94-017-9392-6_27

cancer detection and the assessment of tissue pathologies [9, 10]. Therefore, OCT can function as a type of "optical biopsy", enabling visualization of tissue microstructure with a resolution approaching that of histology, but without the need for tissue excision and processing.

OCT's development was initially started by researchers who were trying to find a way to see the minute details of the layers of the retina. Initial OCT technology was licensed by MIT to Carl Zeiss, which commercialized the technology for ophthalmic applications, released its first unit in 1996, gained FDA clearance in 2002, and has sold more than 6,000 of its Stratus OCT™ systems to date. Now, OCT has become a key diagnostic technology in the areas of ocular diseases and the microscopic changes associated with treatment. Although OCT was originally developed for the field of optometry, it has now been applied to a wide range of biological, medical, and materials investigations.

27.2 Overview of Optical Coherence Tomography

OCT is based upon low coherence interferometry. In conventional Michelson interferometer with coherence light source, interference of light occurs over a distance of meters. In OCT which using low coherence sources, this interference is shortened to a distance of micrometers. This is because any light that is outside the short coherence length will not interfere. Thus, by scanning the mirror in the reference arm of a Michelson interferometer, a reflectivity profile of the sample can be obtained (this is time domain OCT). This reflectivity profile, called an A-scan, contains information about the spatial dimensions and location of structures within the item of interest. Areas of the sample that reflect back a lot of light will create greater interference than areas that don't. A cross-sectional tomograph (B-scan) may be achieved by laterally combining a series of these axial depth scans (A-scan). En face imaging (C-scan) at an acquired depth is possible depending on the imaging engine used.

Further developments in the refinement of OCT is the transformation of technology from the "time-domain" OCT to the "Fourier domain" OCT (FD-OCT). In FD-OCT, the reference path length is fixed and the detection system is replaced with a spectrometer. FD-OCT of this type has been called Spectro-domain OCT (SD-OCT). The detected intensity spectrum is then Fourier Transformed into the time domain to reconstruct the depth resolved sample structure. FD-OCT can also be performed using a single detector by sweeping the source spectrum and detecting the intensity due to component frequencies. FD-OCT of this type has been called swept source OCT (SS-OCT), and has been demonstrated using a tunable laser. Scientific studies have shown that the change from TD to FD detection enables one to increase the acquisition rate over 100 times. Since that no moving parts are required, the full axial image can be retrieved through a Fourier transform of the spectral interferogram, with the speeds limited by the electronic capture rate of the imaging sensor, currently greater than 300 kHz for both FDOCT approaches can be achieved.

The advances of high-speed and high resolution OCT instruments during the past 5 years have opened new fields of applications and perspectives for further development of OCT imaging. Moreover, the use of fiber-optics and the ability to combine with endoscope and catheters allows OCT to access the small parts of the body. Therefore, doctors can actually use OCT not only on the body but within the body. For example, from the mid-1990s, the ability of intravascular optical coherence tomography (OCT) to provide high-resolution (10–20 μm) cross-sectional images of both in vitro human aorta and coronary arteries was demonstrated [11, 12]. Moreover, OCT has also been shown to quantify plaque macrophage content [13] in lipid-rich plaques and to assess the success of intracoronary stent implantation in patients with coronary artery disease during percutaneous intra-arterial procedures [14]. Also, functional imaging based on Doppler flow, spectroscopy, and polarization is possible. Following sections review our previous works by constructing SS-OCT and PS-OCT systems for quantitative diagnosis in cardiology research.

27.3 Applications of Quantitative OCT

27.3.1 Rapid Quantification of Heartbeat Parameters in Drosophila Using Swept Source Optical Coherence Tomography (SS-OCT)

Drosophila melanogaster has been a useful myocardial model for investigating human heart disease. In previous studies using time-domain (TD) OCT, given the limitation of imaging speed, a number of heartbeat parameters have been obtained by manual counting. By using the recently developed frequency-swept lasers, making real-time B-mode and Doppler OCT imaging possible [15–17], high-throughput studies in *Drosophila* are now feasible. Figure 27.1 demonstrates a two-dimensional (2D) OCT images, which were obtained in the longitudinal direction. The arrows in Fig. 27.1a, b identify the conical chamber (CC) of the heart. After acquiring several 2D images, a 3D reconstructed image can be obtained, as shown in Fig. 27.1c. Moreover, by continuous in-depth scans in the midline of a fly's cardiac chamber over time, an M-mode OCT images were obtained, as demonstrated in Fig. 27.2.

Currently, we proposed a rapid OCT-based method for obtaining quantitative cardiac parameters from adult flies. Briefly, we adopted the random walker algorithm for semiautomatic cardiac chamber segmentation in B-mode OCT images. As shown in Fig. 27.3, Random walker is a semiautomatic segmentation method based on graph theory by Leo Grady [19]; it requires that the user give some labels as input seed points for the cardiac chamber area and background. First, we created an initial set of seed points for the random walker algorithm on the first frame. In the next frame, the cross-sections of the heart tube are automatically segmented in a total of 2,000 frames (i.e. Fig. 27.3e), and the size of the inner

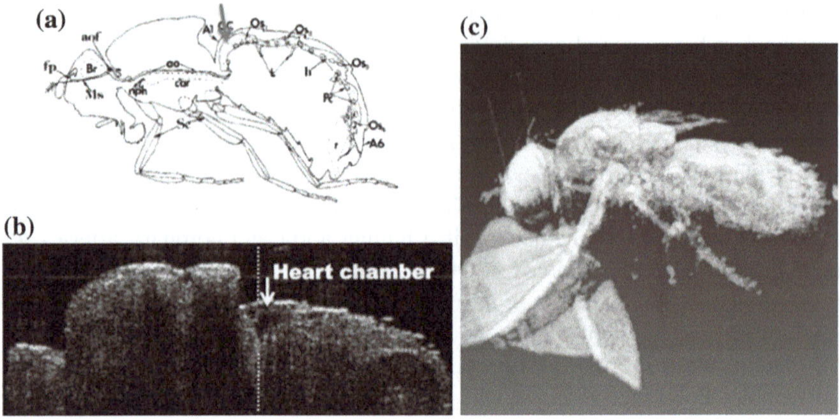

Fig. 27.1 **a** A representative B-mode OCT image. The *arrow* shows the coni-cal heart chamber (CC), **b** in longitudinal orientation. **c** 3D reconstructed OCT image

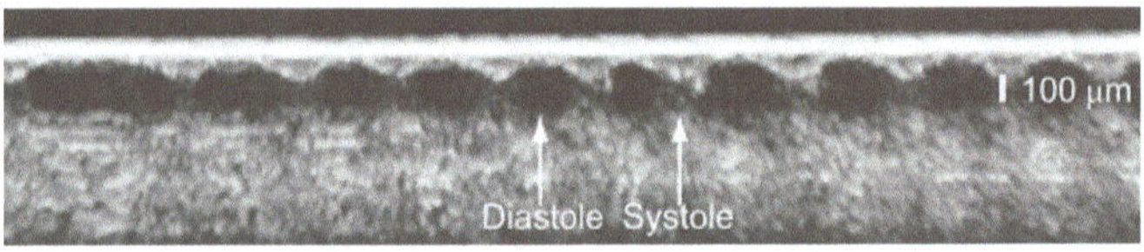

Fig. 27.2 A representative M-mode OCT image in 1-week-old fly [18]

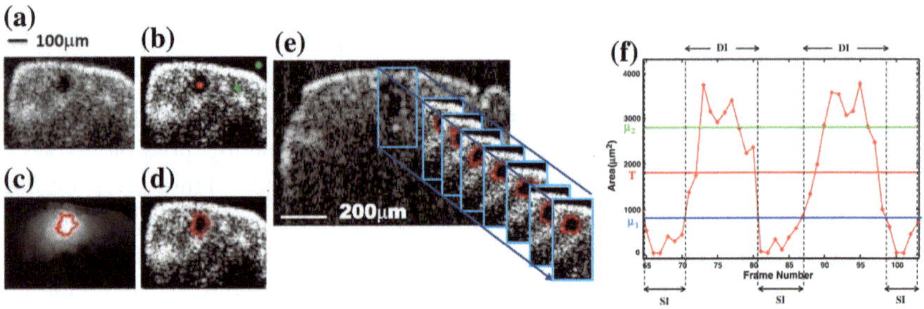

Fig. 27.3 The process of implementation of the autosegmentation algorithm. **a** Original image. **b** Enhanced image. **c** Related probability distribution. **d, e** *Red circle* indicates the heart segmentation result. **f** Change in area inside the segmented boundary of the heart tube in two heart periods [18]

margin is represented by area for each *Drosophila*. Then, depending on the histogram distribution of the changing area during each heartbeat cycle, an optimum threshold can be chosen automatically by an iterative process [20], as demonstrated in Fig. 27.3f. Thus, each heart period (HP) is subdivided into two meaningful sections, which are used to represent diastolic and systolic intervals (DI and SI), respectively. These two parameters are important for detailed heartbeat analysis. The proposed method combines high-speed swept-source OCT (SS-OCT), for optical recording of beating hearts, with robust, semiautomatic analysis to rapidly detect and quantify, on a beat-to-beat basis, not only heart rate (HR) and HP but also DI and SI, and end-diastolic (EDA) and end-systolic area (ESA). These parameters can be used to determine dynamic parameters of heart function, including the arrhythmia index (AI) and percent fractional shortening (%FS) [18].

Figures 27.4 summarize eight heartbeat parameters of male Drosophila w^{1118} derived and quantified by our automated algorithm, in their first, third, fifth, and ninth week, respectively. Data points represent the mean [±standard error of the mean (SEM)] for 10 files per datum point. The quantitative heartbeat parameters measured using our proposed method show that as flies age, HP lengthens (Fig. 27.4a), resulting in an age-related decline in HR (which is the inverse of HP; Fig. 27.4b). We also found that the age-dependent decrease in HR may be due to a disproportionate increase in DI, compared to SI, with age (Fig. 27.4c). Moreover, the current algorithm not only measures HR but also quantitatively expresses the age-dependent increase of heartbeat arrhythmicity (i.e., AI), which may be due to large variations in DI associated with age, particularly when comparing flies in their fifth and ninth weeks of age (Fig. 27.4d).

The Renin-angiotensin system (RAS) is an important regulator of blood pressure homeostasis, in which the protease renin cleaves the angiotensinogen into the angiotensin I (Ang I), followed by further cleaving of the Ang I by the angiotensin-converting enzyme (ACE) into angiotensin II (Ang II), thereby resulting in blood vessel contraction and hypertension [21]. Prior studies have established the necessity of Angiotensin-converting enzyme-related (ACER) gene for the heart morphogenesis of *Drosophila*. Nevertheless, the physiology of ACER has yet to be comprehensively understood. Herein, we employed RNA interference

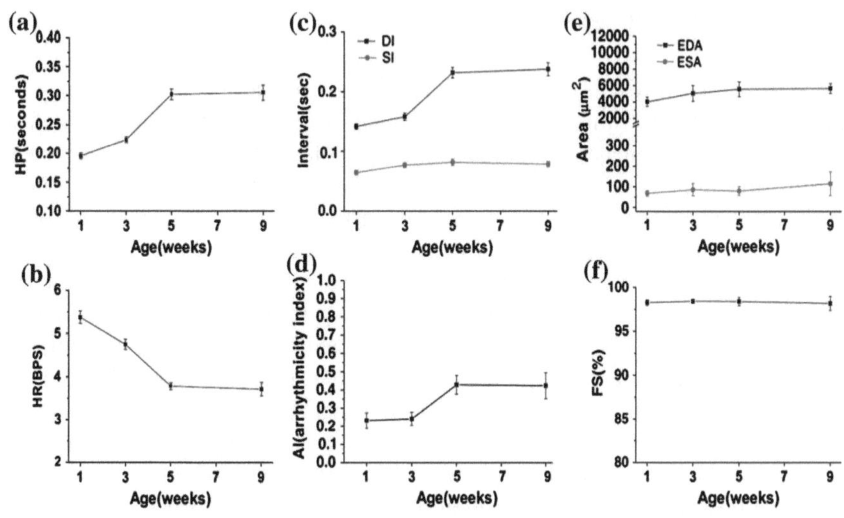

Fig. 27.4 Cardiac parameters in male *Drosophila* w^{1118} at 1, 3, 5, and 9 weeks of age showing **a** HP, **b** HR, **c** DI and SI, **d** AI, **e** EDA and DSA, and **f** FS [18]

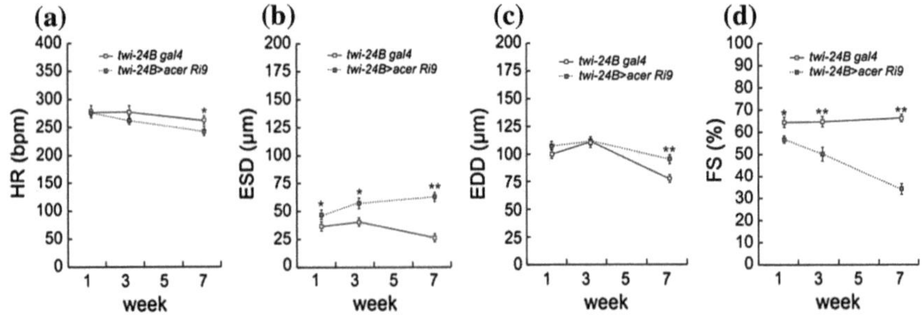

Fig. 27.5 Cardiac parameters in male wild-type control (twi-24B gal4/+) and ACER silencing lines (twi-24B > ACER-Ri9) at 1, 3, and 7 weeks of age showing **a** HR, **b** ESD, **c** EDD, and **d** %FS. *Data points* represent the mean (±SEM) for 20 flies per data point [22]

to down-regulate the expression of ACER in *Drosophila* heart and the abobe mentioned quantitative analysis to assess whether ACER is required for cardiac functions in living adult flies. We found that several contractile parameters of *Drosophila* heart, including the HR, EDD, ESD, FS, and stress-induced cardiac performance, are age dependent. These age dependent cardiac functions declined significantly when ACER was down-regulated. The age-dependent physiological functions of the heart were significantly reduced in ACER knock-down flies, as shown in Fig. 27.5 [22].

27.3.2 PS-OCT Imaging and Quantitative Characterization of Human Atherosclerosis

Polarization sensitive OCT (PS-OCT), a functional mode of OCT, combines the advantages of OCT with additional image contrast obtained by using the birefringence of the specimen as a contrast agent. Many biological tissues have microscopic fibrous structure and as a result of intrinsic birefringence. For, example, Fig. 27.6 is the in vivo swept source PS-OCT images demonstrating intrinsic birefringence properties in a human nail. The reflectivity image (Fig. 27.6a) reveals several structures including epidermis, dermis, cuticle, nail plate, and nail bed. The retardation image (Fig. 27.6b) displays the PR values in color-coding from 0° (blue) to 90° (red). If the sample was non-birefringent, no phase retardation was observed (0°, blue). A value of 90° (red) corresponded to a phase lag of a quarter wavelength between two orthogonal polarization directions due to birefringence. White arrows denote the birefringent layers. Since changes in birefringence may indicate changes in functionality, structure, or viability of tissues in the early stages of disease [23]. From 2004, we presented the first application of PS-OCT in human atherosclerosis, and proposed an approach to characterize a plaque lesion based on its birefringence property [24–26]. For demonstration, two fibrocalcific plaques are shown in

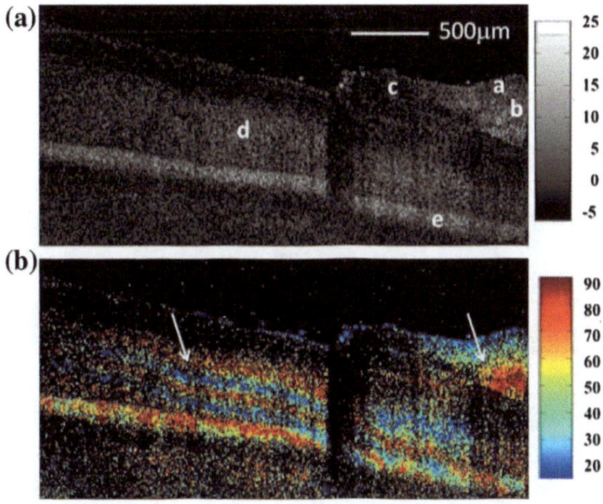

Fig. 27.6 In vivo PS-OCT images of a human nail fold. **a** Reflectivity image: *a* epidermis; *b* dermis; *c* cuticle; *d* nail plate; *e* nail bed. **b** Retardation image. *White arrows* denote the birefringent layers

Fig. 27.7. The PS-OCT image showed a large sharply delineated, signal-rich area of heterogeneous backscattering (Fig. 27.7b, f), and strong birefringence (Fig. 27.7c, g). Different structural orientations were also indicated by the PS-OCT image (i.e., different orientations of fast-axis angle signal in three parts of the tomogram; see Fig. 27.7h) but not by the H and E stained specimen (Fig. 27.7e). Because the calcified lesion was damaged during the sectioning process, only a large empty hole with a few calcified fragments appeared within the calcified plaque. We found that the normal vascular intima has a low intrinsic birefringence property, while change in birefringence characteristics was apparent in fibrous and calcified plaques and differed from that in normal vessels and lipid loaded lesion.

Moreover, we also assessed arterial characteristics in human atherosclerosis by quantitatively determining several optical properties of vessel tissue from PS-OCT images [26–28]. Based on our findings, a quantitative PS-OCT image criterion for plaque characterization was constructed. Briefly, following PS-OCT imaging, an algorithm was used to determine both scattering and birefringence properties of vessel tissue from the PS-OCT images. First, the user selected regions corresponding to those evaluated by histopathology from different area of the images. After laterally (i.e. along the x-axis) delineating and averaging the R, Φ, and β signals within each ROI, the μ_s and root-mean-square scattering angle (θ_{rms}), which can be used to calculate the effective anisotropy factor ($g_{eff} = \cos(\theta_{rms})$), were extracted by fitting the reflectivity signals as a function of depth to an extended Huygens-Fresnel model [27–30]. Furthermore, Δn could be calculated by linear least-squares fitting through the averaged Φ data over the depth of the ROI and then determined its slope. In addition, the mean fast-axis angle calculated by averaging across the width of the ROI at each depth could be determined. The extracted data, μ_s, g_{eff}, Δn, and β are summarized in Fig. 27.8, where each box shows the median, 25th and 75th percentiles, and extreme values within a category. Open circles and stars indicate outlier data.

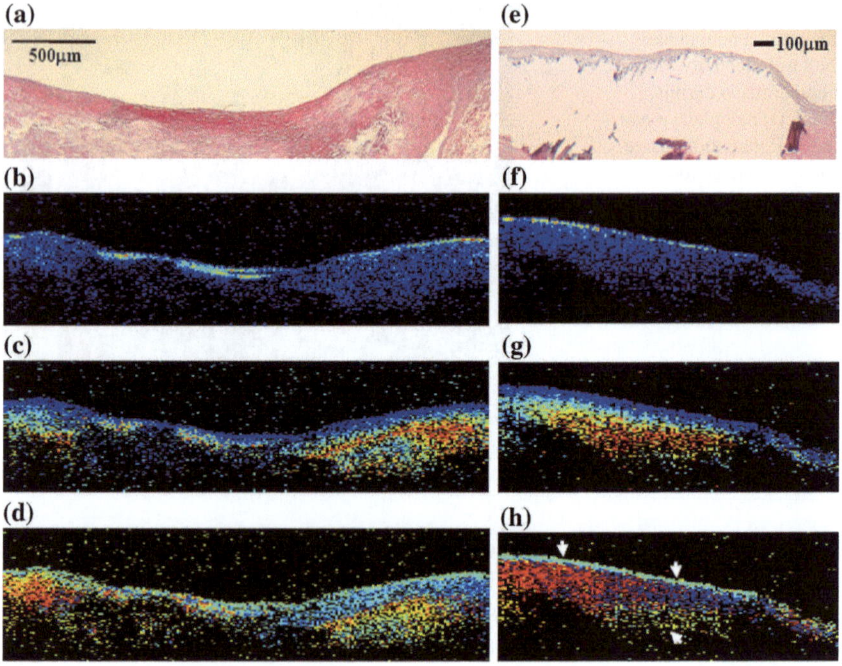

Fig. 27.7 Histological and PS-OCT images of fibroucalcific plaques: **a, e** Histology (H and E); **b, f** Back-scattered intensity image; **c, g** Phase retardation image (linear color scale degrees); **d, h** Fast-axis angle image (linear color scale degrees) [26]

Our preliminary data indicated that more than 80 % normal arterial samples had μ_s between 10 and 39 mm^{-1} and have significant differences from other different types of plaques ($p < 0.05$), which is consistent with results obtained by Levitz [29]. From the multiple comparison tests, we also noticed that the significant difference in scattering property exists between fibrous and fibrocalcific plaques. These findings are consistent with results obtained with qualitative image-based plaque characterization methods where fibrous and fibrocalcific plaques can be distinguished by the signal-rich versus signal poor region [31, 32]. However, the effective anisotropy factor demonstrates no significant difference between normal and other atherosclerotic lesions ($p = 0.104$), perhaps because the g_{eff} of the fibrocalcific and fibrous lesions were correlated with the μ_s and Δn properties, respectively. In relation to the birefringence property of the vessel that has not yet been quantitatively analyzed before. β values in the most atherosclerotic lesion were over 70°, whereas smaller β values were presented in most fit areas of normal vascular intima. The Δn values were small and more concentrated in normal intima, but demonstrated larger variations in the entire atherosclerotic lesion. Both β and Δn values have significant differences between normal arterial vessel and other different types of plaques ($p < 0.05$).

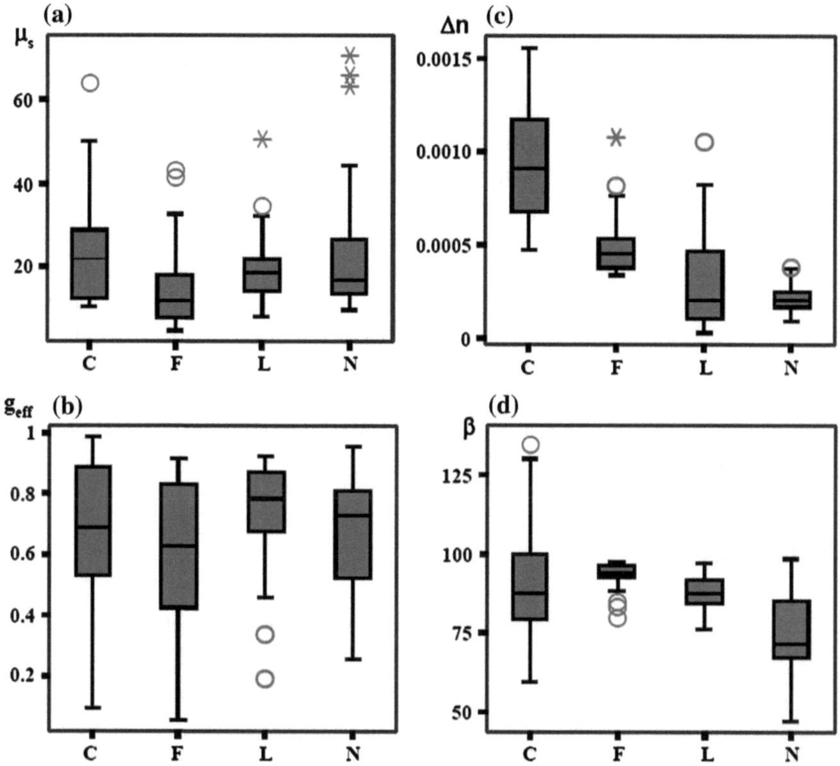

Fig. 27.8 Distributions of μ_s, g_{eff}, Δn, and β in normal vascular intima (N), lipid laden (L), fibrous (F), and fibrocalcific (C) plaques [27]

27.4 Summary

It is well known that optical properties can be used to indicate whether a tissue is in a normal or pathological state. Accurate knowledge of optical properties is essential for optimum use of light in diagnosis and treatment of diseases. OCT can provide not only subsurface morphology but also functionality. Combining with quantitative parameters extraction, OCT can be a quantitative diagnosis tool for monitoring the morphological and functional changes. For example, in our study of quantification of heartbeat parameters in *Drosophila*, our findings are consistent with results obtained with a movement detection algorithm by Fink et al. [33]. Ocorr et al. [34] also suggested that age-related cardiac dysfunction in the fly heart is due to a decrease in the efficacy of cardiac relaxation. Since age-dependent decreases in intrinsic HR, and increases in the incidence of AI, have also been documented in humans [35–38], our observation of the age-related changes in heart function of flies has correlates in humans. Regarding to our study of quantitative characterization of human atherosclerosis, collagen fiber constitutes up to 60 % of the total atherosclerotic plaque protein, uncontrolled collagen accumulation leads to vascular stenosis, whereas excessive collagen breakdown weakens plaques making them prone to rupture. Thus, by quantifying the collagen content in atherosclerotic lesions may provides significant pathophysiological information, and influence clinical decision-making in patients with risk factors. The quantitative information on both arterial scattering and birefringence properties can also be integrated with the qualitative visual information provided by PS-OCT images that will be supportive for facilitating image-based plaque characterization method. Our preliminary results present an important step in validating this new imaging modality and will provide a basis for the interpretation of PS-OCT images obtained from human specimens. However, an analysis from a much larger set of specimens and the analysis considering of the effect of cluster data (i.e. specimens come from the same person) will be required in the future for developing a more suitable prediction model.

References

1. T.H. Ko, D.C. Adler, J.G. Fujimoto et al., Ultrahigh resolution optical coherence tomography imaging with a broadband superluminescent diode light source. Opt. Express **12**, 2112–2119 (2004)
2. B.E. Bouma, G.J. Tearney, I.P. Bilinsky et al., Self-phase-modulated Kerr-lens mode-locked Cr: forsterite laser source for optical coherence tomography. Opt. Lett. **21**, 1839–1841 (1996)
3. W. Drexler, U. Morgner, F.X. Kärtner et al., In vivo ultrahigh-resolution optical coherence tomography. Opt. Lett. **24**, 1221–1223 (1999)
4. R.A. Leitgeb, W. Drexler, A. Unterhuber et al., Ultrahigh resolution Fourier domain optical coherence tomography. Opt. Express **12**, 2156–2165 (2004)
5. B. Povazay, K. Bizheva, A. Unterhuber et al., Submicrometer axial resolution optical coherence tomography. Opt. Lett. **27**, 1800–1802 (2002)
6. A. Aguirre, N. Nishizawa, J. Fujimoto et al., Continnum generation in a novel photonic crystal fiber for ultrahigh resolution optical coherence tomography at 800 nm and 1300 nm. Opt. Express **14**, 1145–1160 (2006)
7. F. Spoler, S. Kray, P. Grychtol et al., Simultaneous dual-band ultra-high resolution optical coherence tomography. Opt. Express **15**, 10832–10841 (2007)
8. P. Cimalla, J. Walther, M. Mehner et al., Simultaneous dual-band optical coherence tomography in the spectral domain for high resolution in vivo imaging. Opt. Express **17**, 19486–19500 (2009)
9. W. Drexler, Ultrahigh-resolution optical coherence tomography. J. Biomed. Opt. **9**, 47–74 (2004)
10. L. Liu, J.A. Gardecki, S.K. Nadkarni et al., Imaging the subcellular structure of human coronary atherosclerosis using micro–optical coherence tomography. Nat. Med. **17**, 1010–1015 (2011)
11. M.E. Brezinski, G.J. Tearney, B.E. Bouma et al., Optical coherence tomography for optical biopsy: properties and demonstration of vascular pathology. Circulation **93**, 1206–1213 (1996)
12. J.G. Fujimoto, M.E. Bresinski, G.J. Tearney et al., Optical biopsy and imaging using optical coherence tomography. Nat. Med. **1**, 970–972 (1995)
13. G.J. Tearney, H. Yabushita, S.L. Houser et al., Quantification of macrophage content in atherosclerotic plaques by optical coherence tomography. Circulation **107**, 113–119 (2003)

14. B.E. Bouma, G.J. Tearney, H. Yabushita et al., Evaluation of intracoronary stenting by intravascular optical coherence tomography. Heart **89**, 317–321 (2003)
15. A. Bradu, L. Ma, J.W. Bloor et al., Dual optical coherence tomography/fluorescence microscopy for monitoring of Drosophila melanogaster larval heart. J. Biophotonic **2**, 380–388 (2009)
16. M.A. Choma, M.J. Suter, B.J. Vakoc et al., Heart wall velocimetry and exogenous contrast-based cardiac flow imaging in Drosophila melanogaster using Doppler optical coherence tomography. J. Biomed. Opt. **15**, 056020-1–056020-6 (2010)
17. M.T. Tsai, F.Y. Chang, C.K. Lee et al., Observations of cardiac beating behaviors of wild-type and mutant Drosophilae with optical coherence tomography. J. Biophotonics **4**, 610–618 (2011)
18. S.Y. Guo, F.T. Liao, M.T. Su et al., Semiautomatic and rapid quantification of heartbeat parameters in Drosophila using optical coherence tomography imaging. J. Biomed. Opt. **18**(2), 26004 (2013)
19. L. Grady, Random walks for image segmentation. IEEE Trans. Pattern Anal. Mach. Intell. **28**(11), 1768–1783 (2006)
20. T.W. Ridler, S. Calvard, Picture thresholding using an iterative selection method. IEEE Trans. Syst. Man Cybern **8**(8), 630–632 (1978)
21. P.C. Santos, J.E. Krieger, A.C. Pereira, Renin-angiotensin system, hypertension, and chronic kidney disease: pharmacogenetic implications. J. Pharmacol. Sci. **120**(2), 77–88 (2012)
22. F.T. Liao, C.Y. Chuang, M.T. Su, W.C. Kuo, Necessity of angiotensin-converting enzyme-related gene for cardiac functions and longevity of Drosophila melanogaster assessed by optical coherence tomography. J. Biomed. Opt. **19**, 011014–011014-6 (2014)
23. J.F. de Boer, T.E. Milner, M.J.C. van Gemert, J.S. Nelson, Two-dimensional birefringence imaging in biological tissue by polarization-sensitive optical coherence tomography. Opt. Lett. **22**, 934–936 (1997)
24. W.C. Kuo, J.J. Shyu, N.K. Chou, et al., Imaging of human aortic atherosclerotic plaques by polarization-sensitive optical coherence tomography, in *Proceedings of IEEE Conference on Engineering in Medicine and Biology*, pp. 1111–1114, San Francisco, 2004
25. W.C. Kuo, J.J. Shyu, N.K. Chou et al., Correlation of collagen synthesis with polarization-sensitive optical coherence tomography imaging of in vitro human atherosclerosis. Proc. SPIE **5690**, 563–571 (2005)
26. W.C. Kuo, N.K. Chou, C. Chou et al., Polarization-sensitive optical coherence tomography for imaging human atherosclerosis. Appl. Opt. **46**, 2520–2527 (2007)
27. W.C. Kuo, M.W. Hsiung, J.J. Shyu et al., Assessment of arterial characteristics in human atherosclerosis by extracting optical properties from polarization-sensitive optical coherence tomography. Opt. Express **16**, 8117–8125 (2008)
28. W.C. Kuo, M.W. Hsiung, J.J. Shyu, et al., Quantitative analysis on optical properties of human atherosclerosis by using polarization-sensitive optical coherence tomography. Proc. SPIE **6842**, 684223-1–684223-9 (2008)
29. D. Levitz, L. Thrane, M.H. Frosz et al., Determination of optical properties of highly-scattering media in optical coherence tomography. Opt. Express **12**, 249–259 (2004)
30. L. Thrane, H.T. Yura, P.E. Andersen, Analysis of optical coherence tomography systems based on the extended Huygens-Fresenel principle. J. Opt. Soc. Am. A **17**, 484–490 (2000)
31. D. Stamper, N.J. Weissman, M. Brezinski, Plaque characterization with optical coherence tomography. J. Am. Coll. Cardiol. **47**, C69–C79 (2006)
32. H. Yabushita, B.E. Bouma, S.L. Houser et al., Characterization of human atherosclerosis by optical coherence tomography. Circulation **106**, 1640–1645 (2002)
33. M. Fink, C. Callol-Massot, A. Chu et al., A new method for detection and quantification of heart beat parameters in Drosophila, zebrafish, and embryonic mouse hearts. Biotechniques **46**, 101–113 (2009)
34. K. Ocorr, N.L. Reeves, R.J. Wessells et al., KCNQ potassium channel mutations cause cardiac arrhythmias in Drosophila that mimic the effects of aging. PNAS **104**(10), 3943–3948 (2007)

35. A.D. Jose, D. Collison, The normal range and determinants of the intrinsic heart rate in man. Cardiovasc. Res. **4**, 160–167 (1970)
36. J.S. Strobel, A.E. Epstein, R.C. Bourge et al., Nonpharmacologic validation of the intrinsic heart rate in cardiac transplant recipients. J. Intervent. Card Electrophysiol. **3**, 15–18 (1999)
37. G. Paternostro, C. Vignola, D.U. Bartsch et al., Age-associated cardiac dysfunction in Drosophila melanogaster. Circ. Res. **88**, 1053–1058 (2001)
38. R.J. Wessells, E. Fitzgerald, J.R. Cypser et al., Insulin regulation of heart function in aging fruit flies. Nat. Genet. **36**, 1275–1281 (2004)

Chapter 28
Introduction to Superresolution Microscopy

Shi-Wei Chu

28.1 Introduction

Since the time of Janssen and Hooke, optical microscopy has made great impact to human civilization. There are several important parameters in optical microscopy, including contrast, resolution, penetration depth, noninvasiveness, etc. Among them, contrast might be the most important factor. During the last century, the most significant progresses in the field of optical microscopy are the advancement in new contrast mechanism, such as phase contrast, and differential interference contrast techniques. On the other hand, spatial resolution, i.e. how small a feature can be distinguished, is another equally important parameter as contrast. However, in late nineteenth century, a German scientist Ernst Abbe proved that the resolution in optical microscopy is restricted by the diffraction limit, which states that the best optical resolution is about half of wavelength.

Rigid mathematical proof of diffraction limit can be found in many textbooks, see for example [1, 2]. Alternatively, here I would like to provide an intuitive explanation toward the concept of diffraction limit. According to Arnold Sommerfeld, diffraction can be broadly defined as "any deviation of light rays from a rectilinear path which cannot be interpreted as reflection or refraction." Experimentally, diffraction refers to the phenomenon of light bending at the edge of an obstacle. When a beam of light passes through a tiny hole in space, a concentric diffraction pattern is formed. It is well known that the smaller the pinhole, the larger is the divergence angle, and thus the larger is the central lobe of the diffraction pattern on screen.

When a parallel beam is focused by a converging lens, it forms a focus at one focal distance from the lens. In the viewpoint of ray optics, the effect of the

S.-W. Chu (✉)
Department of Physics, National Taiwan University, Taipei, Taiwan
e-mail: swchu@phys.ntu.edu.tw

C.-C. Lee (ed.), *The Current Trends of Optics and Photonics*,
Topics in Applied Physics 129, DOI 10.1007/978-94-017-9392-6_28

converging lens is to bend every ray in the beam to form a tight focus. If there is no effect of diffraction, the focus should be infinitesimal. However, with diffraction as a natural property of light, when a light beam is focus into a small region, diffraction results in divergence of the beam, similar to the case of pushing the beam through a pinhole. During the progress of focusing a light beam, the size of the beam becomes smaller and smaller when moving toward the focus, and thus the diverging power from diffraction becomes larger and larger. In the vicinity of focal point, the converging power of the lens reaches equilibrium with the diverging power of diffraction, resulting in a spot size with finite diameter. Therefore, the focus spot is never infinitesimal with the presence of diffraction, and that's why this effect is named diffraction limit.

Under plane wave incidence, the focus exhibits the Airy pattern, whose size can be derived through Fraunhofer diffraction theory. In modern microscopy, the image of a point light source, i.e. point spread function (PSF), also exhibits the Airy pattern. It is because that the emission of the point light source is first transformed into a plane wave by an objective, and then projected to form an image via another camera lens. Considering two adjacent point sources, the resolution of optical microscopy can be defined when the center of the Airy pattern generated by one point source falls exactly on the first zero of the Airy pattern generated by the other, known as Rayleigh criterion [2]. The minimal resolvable separation is approximately $d = \frac{\lambda}{2NA}$, where λ is wavelength, and NA is numerical aperture of the lens system.

During the past century, there have been numerous theoretical and experimental attempts to increase resolution in optical microscopy. In the last decade, several clever techniques are invented to push optical resolution over diffraction barrier. There have been many in-depth reviews of the recent progress of superresolution microscopy, just to list a few [3–8]. The purpose of this review is to illustrate the fundamental principle of superresolution microscopies, and describes the latest progresses in the improvement of contrast mechanism in this active field.

28.2 Fundamentals of Superresolution Microscopy

28.2.1 Spectral Separation with Localization

Currently, most superresolution techniques rely on nonlinear properties of fluorescence, including saturation and switch on/off effects. Let's start from the simplest case to illustrate the basic principles. Imagine a blue laser forms a microscopic focus, and that there are two fluorescent molecules located within the focus, as shown in Fig. 28.1a. Due to diffraction, the image of individual fluorescent molecule, which can be viewed as a point source, exhibits the Airy pattern. When the distance between the two molecules is less than the Rayleigh resolution criterion, the Airy patterns are overlapped and the two molecules cannot be resolved. However, there is one special case that the two molecules can be distinguished,

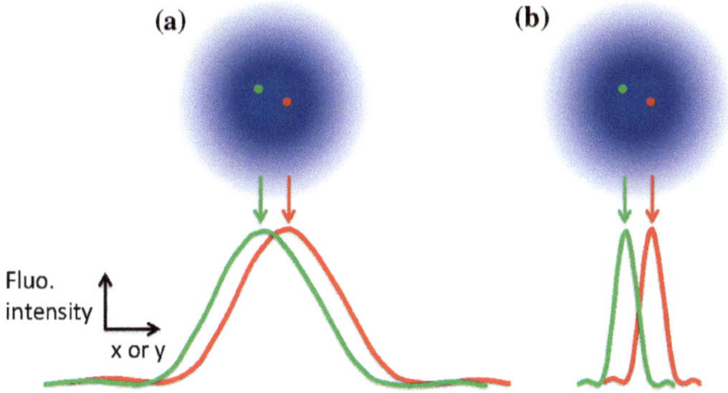

Fig. 28.1 **a** Schematics of two point fluorescent sources (*green* and *red dots*) within a single excitation focus spot (*blue*) and their resulting emission photon distribution before localization (*green* and *red lines*). **b** When the emission wavelengths are different, it is possible to enhance resolution by localization, which can effectively reflect the position of molecule with higher precision

i.e. when their emission wavelengths are different. In this case, the emission of the two molecules can be easily separated via filters, so that we know there are two molecules within the focus. However, this spectral separation along does not increase spatial resolution. Here we shall introduce an important concept called localization, which is the key step toward superresolution.

The fundamental concept behind localization is that the center of the emission profile (Airy pattern) should correspond exactly to the position of the molecule. Therefore, by collecting enough photons from an individual fluorescent molecule, the center of the emission profile can be found by fitting with the two-dimensional Airy function or the Gaussian function. The more photons that are collected, the more precise is this localization process. To the first order of approximation, the precision is proportional to the inverse square root of the number of collected photons. Therefore, the resolution of localization microscopy can be estimated as $d \approx \frac{\lambda}{2NA\sqrt{N_{ph}}}$, where N_{ph} is the number of collected photons from individual fluorescence molecule.

In 1996–1997, Prof Christoph Cremer and his colleagues in Germany demonstrated localization of fluorescent samples, with accuracy as high as 1 nm, laying down the foundation for fluorescence localization microscopy [9]. Later on, they proposed to use fluorescent molecules with different spectral emission wavelength to enhance microscopy resolution, termed as spectral precision distance microscopy (SPDM) [10, 11]. The basic principle is outlined in Fig. 28.2. When there are a lot of fluorescent molecules in the sample, their fluorescence can be distinguished via spectral fingerprint, i.e. absorption spectrum, emission spectrum, and/or fluorescence lifetime, etc. In Fig. 28.2, we show the case that emission spectra are different among fluorophores. In the diffraction-limited imaging in Fig. 28.2a, all different colors are mixed together. With the aid of barrier filters in

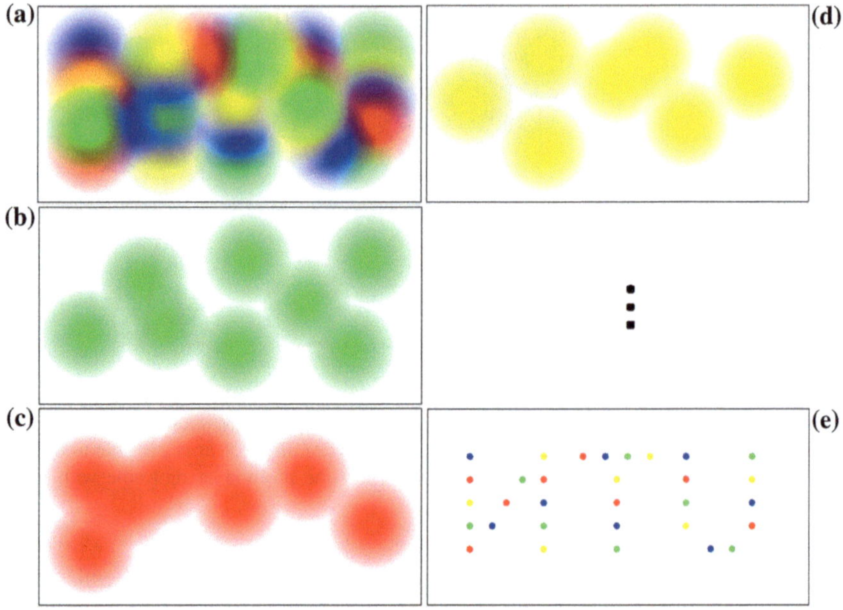

Fig. 28.2 **a** Isolated molecules arranged in an "NTU" pattern, which were labeled with several different colored fluorophores. When observed with a diffraction-limited microscope, the image is significantly blurred. **b–d** With the aid of filters, it is possible to split different wavelength channels. With appropriate labeling strategy, the density of same-color fluorophores is not too high, enabling the localization process for each type of fluorophores. **e** By combining the localization results of individual fluorophores, the detailed distribution of molecules is found

a fluorescent microscope, it is possible to select out specific emission wavelength, as shown in Fig. 28.2b–d. It is important to control the labeling density of same-color fluorophores to be low enough so that the distance between any two of the same-color fluorophores is larger than diffraction limit. In this case, high-precision localization of individual fluorophore is allowed, no matter how many different labelings are used, and the resultant image looks like Fig. 28.2e.

28.2.2 *Switch On/Off (Temporal Separation) of Fluorescence with Localization*

Other than spectral separation, in the case that the two adjacent molecules exhibit same fluorescent wavelengths, we need different methods to distinguish them. A possible way is to separate the fluorescence emission temporally, as shown in Fig. 28.3. In Fig. 28.3a, when both fluorophores emit fluorescence at the same time, it is not possible to resolve them with conventional optics. If there is a

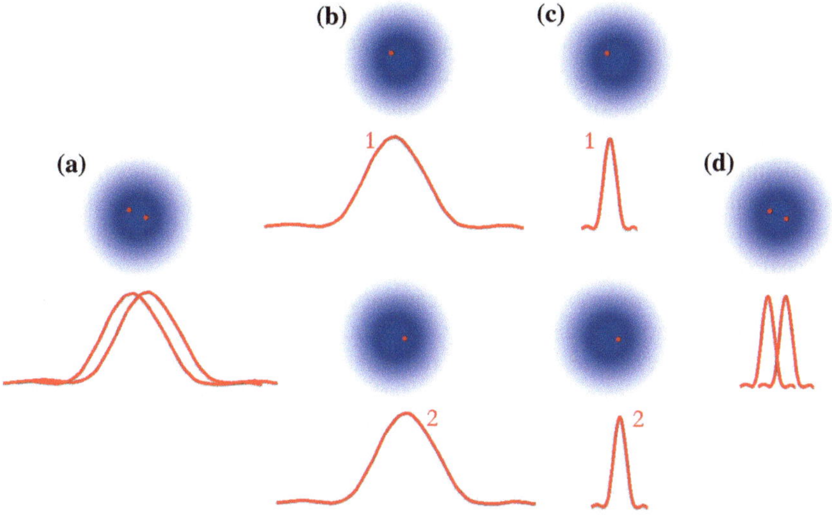

Fig. 28.3 **a** Schematics of two point fluorescent sources within a single focus spot and their resulting emission photon distribution before localization. In this case, the emission wavelengths of the fluorophores are the same. **b** If the fluorescent sources emit light sequentially, the photon distribution profiles can be collected individually. **c** With individually collected PSF, localization helps to reduce the full width at half maximum, and thus to enhance spatial resolution. **d** The resultant profiles are added together, to show a high-resolution distribution of fluorophores

method that allows the two fluorophores to emit at different timing, it is possible to distinguish the fluorescence from them (Fig. 28.3b). Then localization calculation can be performed on individual PSF (Fig. 28.3c), and as the result, the effective resolution is greatly enhanced (Fig. 28.3d).

That is, if the fluorescence emission from individual fluorophores can be distinguished in time, superresolution microscopy can be achieved with the aid of localization. So the key question is how to achieve temporal separation of fluorescence. In principle, under wide-field illumination in conventional microscopy, all fluorescent molecules within the excitation field emit fluorescence simultaneously. The most important factor to achieve such temporal separation is the photoactivable or photoswitchable fluorophores. Photoactivable means that the fluorophore is initially found in a non-fluorescent state ("off" state), and can be activated to become fluorescent ("on" state) via light illumination. On the other hand, photoswitchable in general refers to reversible conversion between "off" state and "on" state, and the conversion is also controlled by light. Thanks to the quick development of biochemical technologies, there has been a large set of photoactivable and photoswitchable fluorescent probes, organic or inorganic, that are suitable for superresolutioin applications [7, 12–14].

To achieve superresolution, in the initial state, all fluorophores have to stay in the "off" state. Then with the aid of an activation beam, a small portion of the fluorophores is turned into the "on" state. It is important that the activation (switch) probability has to be low enough so that within a diffraction-limited region, only one fluorophore is turned "on". The localization calculation is then applied to each the "on" molecule to find its central position. Since the precision of localization is strongly dependent on the number of photons collected from individual fluorophore, the photon emission from the "on" molecules needs to be maximized by driving the molecule until its bleaching. Once these specific fluorophores have been bleached, they can no longer be turned on, so they will not affect the next activation/switch process.

The schematics of resolution enhancement in microscopy by temporal separation and localization are shown in Fig. 28.4. In Fig. 28.4a, which is a diffraction-limited situation, a blurred image with red photoswitchable fluorophores is shown. In Fig. 28.4b–d, the sample is send into dark state initially, and a small group of fluorophores is switched "on" each time. Similar to the illustration in Fig. 28.2, as long as the density of "on" fluorophores is not too high, localization calculation can be applied to each PSF to determine the position of the fluorophore with high precision, effectively reduce the size of PSF. After repeating the switch on and

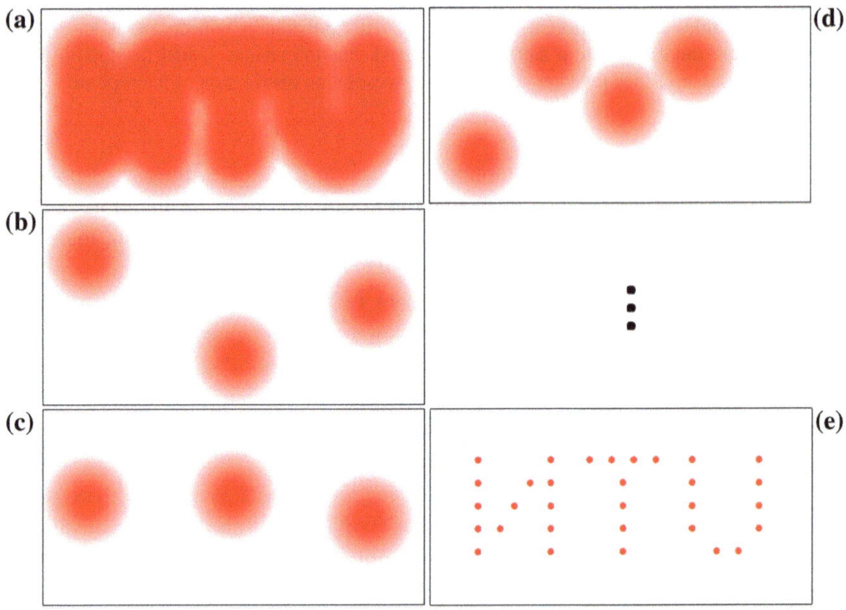

Fig. 28.4 **a** Isolated fluorophores arranged in an "NTU" pattern, where the fluorescence exhibits the same emission wavelength. When observed with a diffraction-limited microscope, the image is significantly blurred. **b–d** With the capability to turn on only a small subset of fluorophores at one time, it is possible to separate the fluorescence among adjacent molecules. Localization calculation is allowed to enhance resolution within this subset of "on" molecules. By repeating the switch process for many times, all fluorophores can be localized, resulting in a high-resolution image in (**e**)

localization process for many times, a high-resolution distribution map of fluorophores is reconstructed, as shown in Fig. 28.4e.

The representative methods in this category are photo-activated localization microscopy (PALM) [15, 16] and stochastic optical reconstruction microscopy (STORM) [17]. Their principles are basically the same: both rely on random activation of fluorophores within the specimen. In PALM, photoactivable fluorescence proteins or dye pairs are used, while in STORM, photoswitchable proteins are used. On the other hand, in STORM, photoblinking is adopted to separate the fluorophores in time, while in PALM, photoactivation and subsequent photobleaching is the key factor to distinguish fluorophores. From the concept of localization, resolution of both techniques can be estimated as $d = \frac{\lambda}{2NA\sqrt{N_{ph}}}$. With efficient fluorophores, typically the resolution of PALM and STORM is around 20–30 nm in lateral direction. Axial localization has also been achieved with 50–80 nm resolution recently by introducing astigmatism [18], by bi-plane acquisition [19], or by adopting a double helix-shaped PSF [20].

The main limitation of such temporally separated fluorescence techniques is the speed of acquisition. Since only a very small subset of fluorophores are activated in a frame, it routinely requires accumulation of hundreds to a few thousands of images to form a complete image of the specimen. Even with a fast real-time camera that can capture 30 frames/sec, typically a few minutes are necessary to finish the superresolution imaging. There are currently various efforts to enhance the acquisition speed of localization microscopy, enabling the potential to monitor fast cell dynamics [21, 22].

28.2.3 Switch On/Off Fluorescence with Spatial Engineering of Beam Focus

In case we have the capability to turn on and turn off fluorescence signals as we wish, it is also possible to achieve superresolution without localization. In a general context, assuming that the fluorescence can be turned on and off optically, and the turn-on and turn-off wavelengths are different, by spatially engineer the geometrical distribution of the two beams, it is possible to eliminate the fluorescence in the peripheral of the turn-on (excitation) focal spot, and thus enhance resolution. This concept can be realized under a laser-scanning scheme. As shown in Fig. 28.5a, when a diffraction-limited focal spot with correct excitation wavelength (shown in blue) scans across a fluorescent molecule, the resulting spatial distribution of fluorescence exhibits the Airy pattern. Imagine that there is a "magic" light with specific wavelength (shown in red) that can turn off fluorescence upon illumination. In Fig. 28.5b, by engineering the beam focus to be a hollow shape, when the turn-off beam scans across the fluorescent molecule, the molecule is only allowed to emit fluorescence in the hollow area, and thus the spatial resolution can be effectively enhanced. However, in Fig. 28.5b, there is no excitation light, so no fluorescence would emit. To realize such superresolution

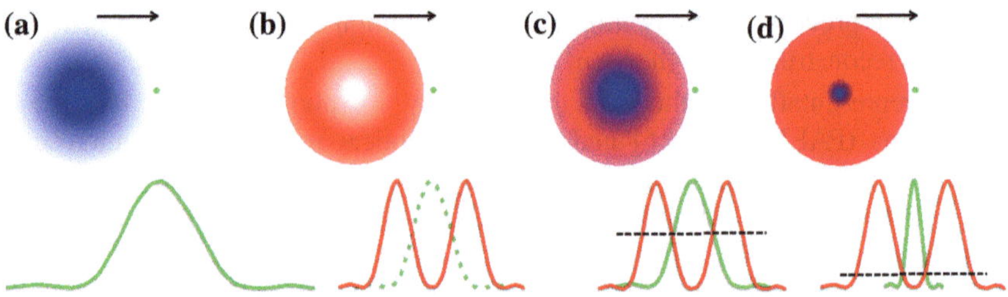

Fig. 28.5 **a** When a focused excitation laser spot (blue color) scan across a tiny fluorescent source (*green dot*), a broad distribution of fluorescence, equivalent to the PSF, is formed in the detection channel (*green line*). **b** For the turn-off light (*in red*) that can suppress fluorescence, at focal point, it is shaped into a donut, where the intensity approaches zero in the center. When this pattern is scanned across the tiny fluorescent source, the fluorescence is only allowed in the center of the donut beam. **c** Combining both excitation and turn-off beams together, and scanning across the fluorescent source, the FWHM of the emission can be effectively reduced; i.e. resolution enhanced. The *black dashed line* represents the level of saturation intensity. **d** To achieve higher resolution, either the intensity of turn-off light is increased, or equivalently, the saturation intensity is reduced

microscopy, the turn-off beam (in donut shape) has to be aligned concentrically with the excitation beam, as shown in Fig. 28.5c. One important concept here is that the higher the intensity of turn-off beam, the better the resulting resolution, since the central hollow region that allows fluorescence emission becomes smaller, as shown in Fig. 28.5d.

Apparently, the key question is how to turn off fluorescence in a well-controlled manner? One of the most successful methods is proposed by Wichmann and Hell in 1994 [23]. Their simple but elegant method is to suppress spontaneous fluorescence emission by stimulated emission. When a fluorescence molecule is at the excited state, it can be driven back to the ground state via stimulated emission with an additional beam whose wavelength located at the red side of the spontaneous fluorescence band. Since the wavelength of stimulated emission is exactly the same as the incident photon that triggers the stimulated emission, it can be easily separated from spontaneous emission by a spectral filter. This technique is termed as stimulated emission depletion (STED) microscopy [24].

It is important to notice that the suppression effect is "saturable". Since the number of excited fluorophores within a PSF is finite, with high intensity of turn-off light, the fluorescence can be completely suppressed. For a specific labeling molecule with a certain density, saturation intensity can be defined as the required turn-off beam intensity that results in 50 % suppression of fluorescence emission. As a result, the resolution of STED microscopy can be estimated as $d = \frac{\lambda}{2NA\sqrt{I/I_S}}$, where I is turn-off beam intensity, I_S is saturation intensity [25, 26].

In Fig. 28.6, the concept of STED microscopy implementation is shown. Figure 28.6a presents the situation of conventional laser scanning microscopy, where a diffraction-limited excitation spot (blue color) scans across the sample, resulting in blurred fluorescent image (green color) in Fig. 28.6b. To implement

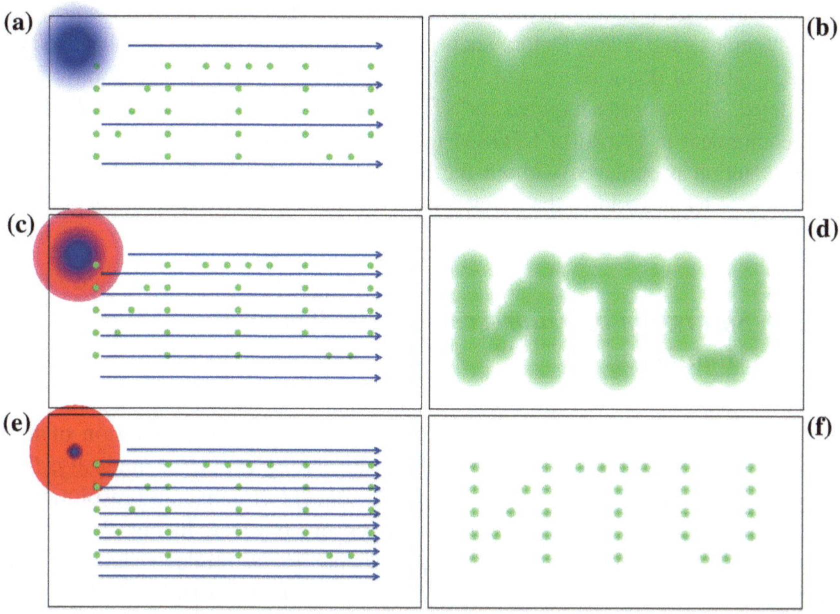

Fig. 28.6 **a** Isolated fluorophores arranged in an "NTU" pattern, where the fluorescence exhibits the same emission wavelength (shown as *green*). In a conventional laser scanning microscope, a diffraction-limited spot is raster scanned across the sample, and the resultant *blurred* image is shown in (**b**). **c** To enhance resolution, the turn-off beam in donut shape is added together with the excitation beam, and scans across the sample. Note that since the effective PSF becomes smaller, the density of sampling should increase. As the result, **d** shows the resultant image with better resolution. **e** To further increase spatial resolution, the intensity of turn-off beam is increased, and the resultant image **f** shows clear distribution of individual fluorophores

STED microscopy, an additional STED beam (red color) is converted into a donut beam via a phase plate or a birefringent plate [27], and is overlapped with the excitation spot. Both beams are scanned together across the sample, as shown in Fig. 28.6c. Since the effective resolution is enhanced with the aid of the STED beam, the scanning density has to be increased to cover the whole sample. Figure 28.6d shows the result of resolution enhancement with STED beam intensity comparable to the saturation intensity. To further enhance spatial resolution, the STED beam intensity has to be much larger than the saturation intensity, and the scanning density has to be high enough, as shown in Fig. 28.6e. As a result, the detailed distribution of fluorescent sources can be revealed.

In principle, the resolution of STED microscopy is unlimited, before sample damage occurs. In most of the applications, STED provides resolution among 30–80 nm. After its invention, this technique has been quickly adopted to study biological samples [28, 29], and has been extended to live-cell experiments [30–32]. Currently, the best resolution that STED achieved is about 6 nm, relying on non-bleaching fluorescence from diamond color centers [33].

Compare to localization microscopy, the setup of STED microscopy is significantly more complex, and many factors such as timing/duration of excitation and STED pulses, quality of donut beam, and wavelength selection of STED beam, have to be considered. The intensity required for STED beam is typically higher than the intensity requirement in PALM or STORM microscopy. However, in STED microscopy, superresolution is achieved without the need of any reconstruction and calculation processes, so it can avoid all the artifacts from the mathematical manipulations.

28.2.4 Saturation of Fluorescence with Temporal Modulation

Another possibility to enhance spatial resolution relies on the saturation of fluorescence. It is known for a long time that with strong excitation intensity, fluorescence can be saturated [34, 35]. The basic concept of enhancing resolution by saturation is not difficult to perceive, as shown in Fig. 28.7. In Fig. 28.7a, once again, a point fluorescent source at microscopic focus produces an Airy pattern, whose intensity is higher in the center than in the peripheral. Therefore, as the excitation intensity grows up, saturation starts from the center, as shown in Fig. 28.7b. When the central part of the PSF is saturated, the peripheral part still maintains linear. So if we can extract the saturated part, the resolution can be effectively enhanced, as shown in Fig. 28.7c.

One way to extract the saturated part to enhance spatial resolution of optical microscopy is invented by Fujita et al. [36], with the name of saturated excitation (SAX) microscopy. The idea is shown in Fig. 28.8. The key concept here is to add a temporal sinusoidal modulation (f_m) to the excitation. In the peripheral part of the PSF, where the intensity is low and no saturation is occurred, the fluorescence intensity follows exactly the sinusoidal modulation. Thus, in the Fourier

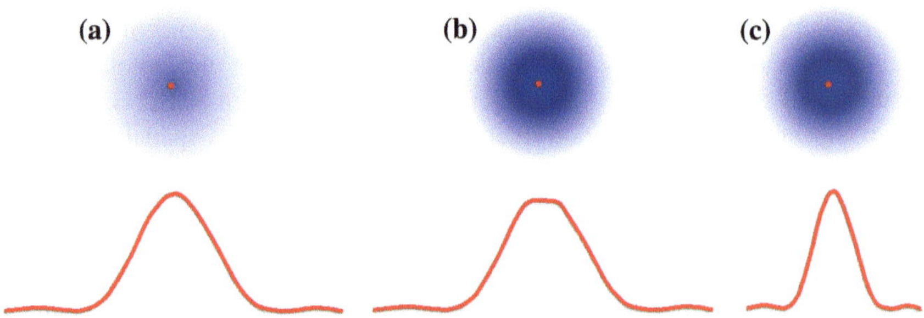

Fig. 28.7 **a** One point fluorescent source is inside an excitation focus spot, and its resultant PSF at detection channel. No saturation is found at low intensity of excitation. **b** When the excitation intensity grows up, saturation of fluorescence starts from the center. **c** By extracting the saturated part, the resolution can be effectively enhanced

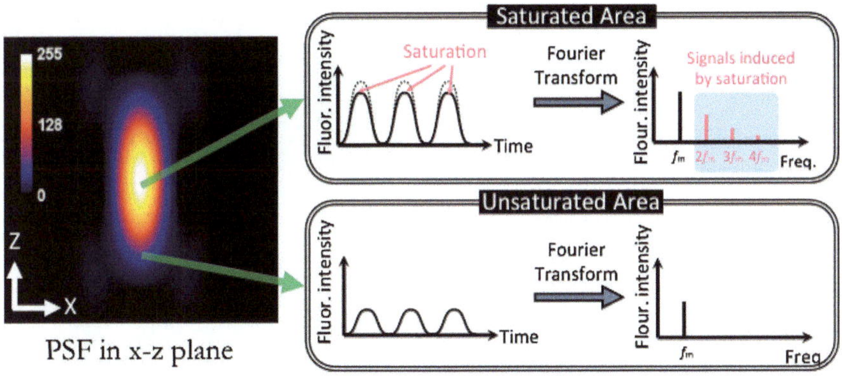

Fig. 28.8 The concept of extracting saturated fluorescence signal with the aid of temporal modulation

domain, there is only one frequency component (f_m) for the fluorescence intensity. On the other hand, in the central part of PSF, where the excitation intensity is higher and saturation can start to occur here, the fluorescence intensity no longer follows perfectly the sinusoidal modulation. As a result, in the Fourier domain, components with harmonics of modulation frequency appear due to the nonlinear saturation. By using the harmonic terms ($2f_m$, $3f_m$, ...), it is equivalently to extract the saturated part, and thus the resolution can be enhanced. Note that the resolution enhancement is achieved in all three dimensions simultaneously.

Similar to STED microscopy, the implementation of SAX microscopy is combined with a laser-scanning scheme. The diffraction-limited situation in laser scanning microscopy is depicted again in Fig. 28.9a, b. With the capability to extract saturated part of fluorescence, the PSF of SAX microscopy is effectively reduced. By raster scanning the smaller PSF across a sample, as shown in Fig. 28.9c, fluorescent image with higher resolution is obtained.

In principle, there is no resolution limit as long as extremely deep saturation can be reached. However, in practice, the saturation behavior, as well as resolution, is limited by the damage threshold of samples. Currently, the best resolution that has been achieved is about 100 nm [37]. Compared to STED and localization microscopy, the resolution enhancement of SAX microscopy is lower. SAX microscopy has also been applied to live-cell observations recently [38].

28.2.5 Saturation of Fluorescence with Spatial Modulation

The last concept we shall introduce is based on saturation of fluorescence signals plus spatial modulation of excitation to enhance spatial resolution. We will start from resolution enhancement by spatial modulation of excitation, and then explain how incorporating saturation of fluorescence significantly improves the resolution.

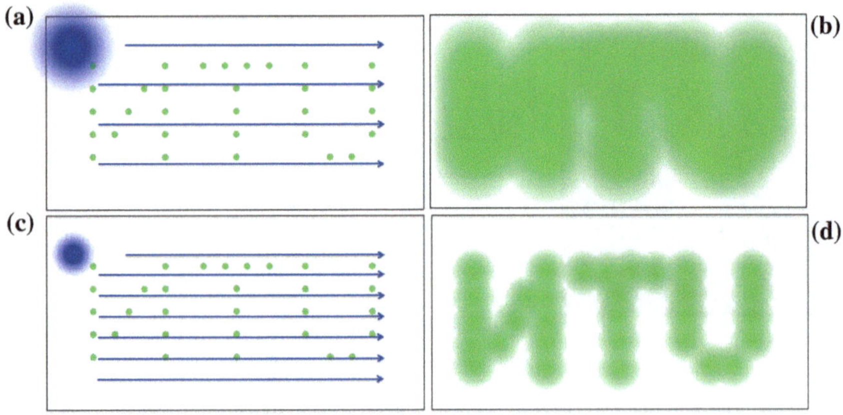

Fig. 28.9 (**a**, **b**) Conventional laser scanning microscopy with a diffraction limited spot. **c** With the aid of SAX, the PSF becomes smaller when fluorescence is driven into saturation, resulting in better resolution in (**d**)

Structured illumination microscopy (SIM) [39] is introduced in year 2,000, and allows resolution enhancement by a factor of two. The basic principle is described in Fig. 28.10. In brief, the sample is illuminated with sinusoidal striped patterns, which is typically generated by a grating placed in a conjugated plane of image. The fine structures of the sample create the so-called Moiré fringe under such patterned illumination. By translating and rotating the striped patterns to acquire multiple Moiré fringes with different phases and angles, it is then possible to find out the high-resolution distribution of the underlying structure via reconstruction algorithms.

The resolution enhancement of SIM is two-fold compared to conventional wide-field microscopy. Competed with other superresolution microscopy, the resolution enhancement is not significant. But one major advantage of SIM is that most standard staining or fluorescent proteins can be used directly, no prerequisite of saturation or switching capability. In addition, it is wide-field observation, and only ten to twenty images are required to reconstruct high-resolution image, so the imaging speed can be relatively fast, especially when compared with wide-field localization microscopy. Axial resolution enhancement can be achieved by generating a special interference pattern, enabling three-dimensional SIM [40, 41].

Apparently, the resolution of SIM can be further enhanced if the period of sinusoidal modulation pattern is smaller. However, the minimal period of the pattern is limited by the diffraction limit, so it is not realistic to reduce the period. Alternatively, the duty cycle of the modulation pattern can be changed by saturation. The concept is shown in Fig. 28.11, where the width of the non-excited region becomes smaller due to saturation of fluorescence signals in the excited region [42, 43]. The technique is named as saturated SIM (SSIM). Similar to the principle of SIM, by phase shift and rotation of the grating sinusoidal patterns,

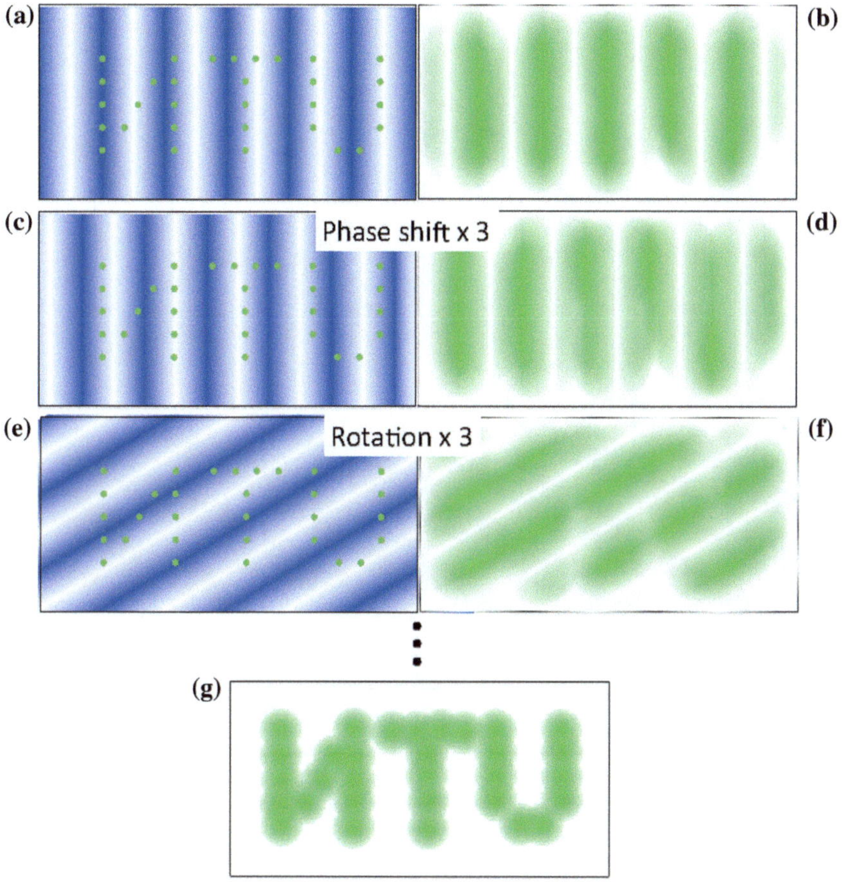

Fig. 28.10 Schematics of SIM. **a** A grating pattern with sinusoidal modulated excitation intensity is formed at the image plane. Excitation light is shown in blue while fluorophores are shown in *green*. **b** The resulting diffraction-limited fluorescence image from a recording camera, showing the blurred image and the modulation pattern. **c, d** To achieve resolution enhancement, the phase of the grating modulation pattern is shifted laterally. It requires at least three different phase shifts to enhance resolution in lateral direction. **e**, **f** The grating modulation pattern has to be rotated in at least three different directions, with additional three phase shift movement in each direction, to accomplish resolution enhancement in vertical direction. **g** In total (shift + rotation), it requires ten to twenty images to reconstruct an image with resolution enhanced by a factor of two

better resolution can be obtained in SSIM via reconstruction. Since the width of non-excited stripes are smaller, more images with finer phase shift and rotation are required.

Similar to SAX, the resolution of SSIM is theoretically unlimited as long as deep saturation can be achieved before photodamage occurs. In practice, SSIM can provide 50-nm spatial resolution [43, 44].

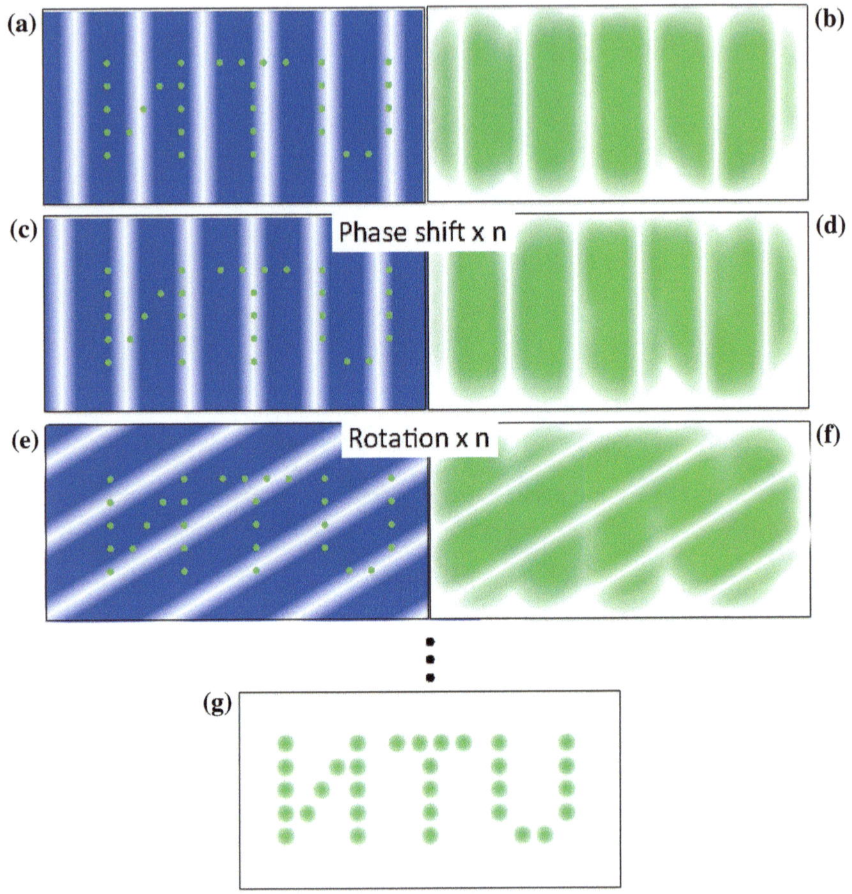

Fig. 28.11 Schematics of SSIM. **a**, **b** The intensity of the grating pattern is increased to reach the saturation threshold, resulting in the narrowing of non-excited region. **c**, **d** Similar to SIM, phase shift is required to enhance lateral resolution. But in SSIM, more shifts are required to retrieve better resolution. **e**, **f** Rotation of the pattern is necessary to enhance resolution in vertical direction. Again, because the width of stripe is smaller, more rotations are required. **g** After repetitive phase shifting in each rotation, a superresolution image is reconstructed

28.3 Realization of Superresolution Microscopy Based on Non-fluorescence Contrast

For all the techniques mentioned above, the key concept to achieve superresolution relies on switching and saturation properties of fluorescence. However, fluorescence exhibits issues of photo-bleaching and switching reversibility after repetitive excitation. Recently, there have been several attempts to achieve superresolution without fluorescence. One possibility is to use saturation of absorption as contrast agents [45]. Our approach is to adopt saturation of scattering [46, 47]. In the following, we shall give a short introduction to our recent results.

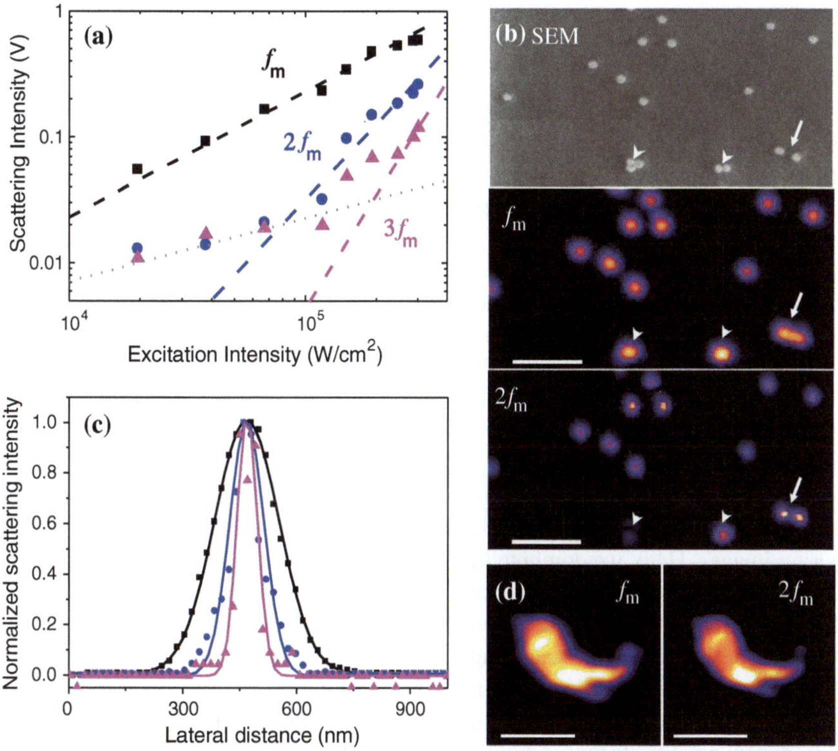

Fig. 28.12 **a** The attenuation spectrum of 100-nm gold nanoparticles in oil, showing a SPR peak around 580 nm. **b** The dependency of scattering on excitation intensity for a single gold nanoparticle with three different excitation wavelengths. With stronger intensities, clear saturation is observed for *green* and *red excitations*, which correspond to plasmonic bands in (**a**)

It is well known that scattering from plasmonic particles is particularly strong with resonant excitation wavelength due to surface plasmon resonance (SPR), so we will examine the nonlinearity in scattering from metallic nanoparticles. The study of SPR has recently attracted extensive interests because it provides light manipulation capability for photonic integrated circuits, nano laser, biosensing, and near-field superresolution imaging applications [48–51]. The spatial resolution of plasmonic nano-imaging can be improved by incorporating nonlinear optical phenomena, such as coherent anti-Stokes Raman scattering [52, 53] and saturation. To our knowledge, neither saturation, nor switching of scattering from SPR structures has been reported, but there are plenty of reports on saturable absorption of plasmonic nanoparticles embedded in dielectric matrix [54–56]. Since in nanoparticles, scattering and absorption are closely linked via Mie theory [57], we set out to examine saturable scattering in a single gold nanoparticle.

Figure 28.12a shows the attenuation spectrum of 100-nm nanoparticles, manifesting a clear SPR peak. Three different wavelengths, namely 405, 532, and 671 nm, are selected for scattering measurement, and the corresponding saturation behaviors are given in Fig. 28.12b. Among the excitation sources, 532 and 671 nm are located inside the plasmonic band, while 405 nm is not. The attenuation at 405 nm is mainly the consequence of the bulk absorption of gold.

For each excitation wavelength, a dashed line with slope equal to unity is shown to highlight linear dependency of scattering versus excitation intensity. Deviation from this line manifests saturation behavior, and the 50 % saturation intensity can be defined as the input intensity at which half of the maximal saturated output signal is obtained. From Fig. 28.12b, the 50 % saturation intensity is lowest for the excitation closest to the peak of SPR, i.e. the 532-nm line. The 50 % saturation intensity with green laser is ~10^5 W/cm^2, which is easily achievable with <1-mW laser power at the focus of a high-NA objective. As the excitation wavelength moves away from SPR, the required intensity for saturation quickly increases. For blue excitation, which is not in the plasmonic band, saturation of scattering is not observed even with 10^8 W/cm^2 intensity level, revealing that the saturation is dominated by SPR.

Now we know that scattering from plasmonic particles can be saturated, so it is straightforward to adopt a superresolution technique to enhance resolution of scattering images. Here, we combine the saturation of scattering and SAX microscopy. To demonstrate resolution enhancement, 100-nm particles were used with 532-nm excitation. Figure 28.13a shows the experimental verification of emergence of modulation harmonics in scattering signal when the excitation intensity increases over the threshold intensity. The gray dotted line represents the shot noise level of the detection system. It should be noted that for higher order harmonics, the intensity dependence is steeper. As a consequence, the peripheral of PSF is suppressed

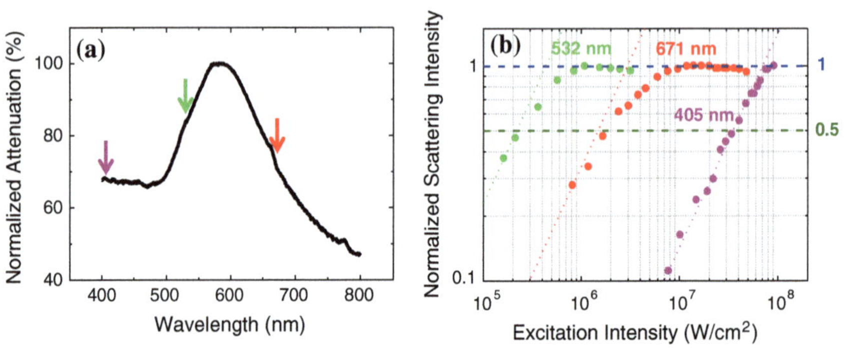

Fig. 28.13 Scattering saturation and resolution enhancement for a 100-nm gold nanoparticle excited with a 532-nm laser. **a** As the excitation intensity grows, the nonlinear part of scattering emerges. **b** By extracting scattering with modulation harmonics, the resolution is significantly enhanced. The topmost panel is an scanning electron microscope (*SEM*) image for reference. Other panels correspond to signals acquired with harmonics of modulation frequency. **c** The line profiles showing resolution enhancement, where *black*, *blue*, and *purple lines* represent fundamental, second-, and third-harmonic modulation frequencies, respectively. **d** Images of irregularly shaped gold nanostructures, showing significantly improved spatial resolution for plasmonic patterns (copyright permission by APS)

by this nonlinear dependence, so the optical resolution is enhanced, as shown in Fig. 28.13b, c. In Fig. 28.13b, an image from scanning electron microscope in the corresponding area is provided for comparison. It is evident that resolution enhancement is universal for isolated particles. The white arrow shows two particles that can only be resolved when we took signals from modulation harmonics. In Fig. 28.13c, the FWHM with fundamental modulation frequency is 230 nm, and is reduced to 110 and 65 nm when extracting second and third harmonic frequencies, respectively. Since only scattering is involved, no photobleaching is expected/observed for prolonged observation here.

One more note from Fig. 28.13b is that when two or more particles are closely packed to each other, our technique cannot distinguish these aggregated particles because plasmonic coupling occurs [58]. The arrowheads mark particle aggregations, showing that the $2f_m$ and $3f_m$ intensities drop very fast due to SPR wavelength shift. The sub-70-nm FWHM we are claiming is for plasmonic modes, not for particle itself. As can be seen from the arrowheads in Fig. 28.13b, the intensities of 2nd and 3rd harmonics are significantly weaker than those of isolated particles, since the plasmonic coupling results in resonant band shift.

One expected applications of our technique is to provide high-resolution inspection of plasmonic patterns in nanostructures. In Fig. 28.13d, we show an example with an irregularly shaped gold nanostructure, which is found within the same solution of the gold nanoparticles for our experiment. It can be the aggregate of nanoparticles or the result of fabrication imperfection to form a large particle. As can be seen in the leftmost image in Fig. 28.13d (with fundamental modulation frequency f_{m}), there are some regions with strong scattering on the gold structure, reflecting the position of SPR. This is similar to the structure of a random gold island film, where multiple plasmonic hot spots can be formed on such randomly-shaped gold nanostructures [59]. Comparing the images with fundamental modulation frequency f_m, second harmonic $2\,f_m$, and third harmonic $3\,f_m$, it is obvious that the resolution of plasmonic imaging is improved when taking signals with higher harmonics.

28.4 Future Perspective

In the field of optical microscopy, there are several vital parameters, including contrast, resolution, imaging depth, imaging speed, noninvasiveness, etc. In the past decade, great advances in respect to spatial resolution have been achieved, overcoming the long-lasting diffraction barrier. The key concepts of superresolution microscopy are saturation and switching of imaging contrast agents. Less than 30-nm resolution has been demonstrated with several different techniques. All these superresolution techniques have been quickly adopted in biology researches, allowing researchers to visualize biochemical/biophysical processes with unprecedented clarity. Nevertheless, there are still many challenges and opportunities in this active field. For example, in terms of contrast, although two-color imaging has been demonstrated in most superresolution techniques [60–62], multicolor imaging capability is still a grand challenge, especially toward the optically complex system such as STED. In addition, the photostability is one of the major issues when selecting fluorescence probes [63], and thus it is vital to find slow-bleaching fluorescent species, or alternatively, to explore other non-fluorescent contrast agents, as we mentioned in the last section.

In terms of resolution, although the resolution in cellular environment has been improved to more than one order less than diffraction limit, it is still less than electron microscopy or scanning probe microscopy. There have been efforts in combining superresolution imaging with correlative electron microscopy [64]. For no doubt, better resolution with optical modality itself is still highly desirable. One way to achieve this goal is to find fluorescent probes with less switching threshold and higher quantum yield.

In terms of imaging depth, most superresolution techniques are limited to thin samples. It is possible to achieve deep-tissue imaging by means of post processing or adaptive optics [65]. Two-photon excitation has been incorporated into different superresolution modalities to enhance penetration depth [66, 67]. Recently, more than one millimeter depth has been demonstrated with three-photon fluorescence microscopy [45]. It will be very interesting to combine the deep penetration of three-photon imaging and superresolution imaging.

In terms of imaging speed, for scanning-based superresolution techniques, such as STED and SAX, it is possible to reach video-rate recording [68]. For wide-field based superresolution techniques, such as PALM, STORM, and SSIM, much longer acquisition time is required since they need many frames to reconstruct the details. There have been many efforts in speeding up localization microscopy to allow image acquisition on the sub-second scale [22, 62], but not yet video rate.

Finally, in terms of noninvasiveness, live cell imaging has been achieved in most superresolution techniques, with compromises in resolution and versatility. Since high-intensity laser is typically required to achieve superresolution, special take has to be taken in sample preparation. To reduce phototoxicity loading of biological cells, more sensitive detection with brighter labeling will be very helpful.

Within a decade of development, superresolution techniques have now made strong impacts in not only cell biology, but also material science and semiconductor lithography fields. The overall trend is toward development of high-contrast, nanometer-resolution, deep-imaging, fast-acquisition, and noninvasive observation tools. In the near future, more novel microscope architecture along with more exciting new insights in biological structure/function correlations are to be expected.

References

1. M. Born, E. Wolf, *Principles of Optics: Electromagnetic Theory of Propagation, Interference and Diffraction of Light* (Cambridge University Press, Cambridge, 1999)
2. J. Goodman, *Introduction to Fourier Optics* (Roberts & Company Publishers, Greenwood Village, 2004)
3. S.W. Hell, Far-field optical nanoscopy. Science **316**, 1153–1158 (2007)

4. B. Huang, M. Bates, X. Zhuang, Super-resolution fluorescence microscopy. Ann. Rev. Biochem. **78**, 993–1016 (2009)
5. M. Heilemann, Fluorescence microscopy beyond the diffraction limit. J. Biotechnol. **149**, 243–251 (2010)
6. L. Schermelleh, R. Heintzmann, H. Leonhardt, A guide to super-resolution fluorescence microscopy. J. Cell Biol. **190**, 165–175 (2010)
7. S. van de Linde, M. Heilemann, M. Sauer, Live-cell super-resolution imaging with synthetic fluorophores. Ann. Rev. Phys. Chem. **63**, 519–540 (2012)
8. U. Agrawal, D.T. Reilly, C.M. Schroeder, Zooming in on biological processes with fluorescence nanoscopy. Curr. Opin. Biotech. **24**, 646–653 (2013)
9. R. Heintzmann, H. Münch, C. Cremer, High-precision measurements in epifluorescent microscopy—simulation and experiment. Cell Vision **4**, 252–253 (1997)
10. H. Bornfleth, K. Satzler, R. Eils, C. Cremer, High-precision distance measurements and volume-conserving segmentation of objects near and below the resolution limit in three-dimensional confocal fluorescence microscopy. J. Microsc. Oxf. **189**, 118–136 (1998)
11. C. Cremer, R. Kaufmann, M. Gunkel, S. Pres, Y. Weiland, P. Muller, T. Ruckelshausen, P. Lemmer, F. Geiger, S. Degenhard, C. Wege, N.A.W. Lemmermann, R. Holtappels, H. Strickfaden, M. Hausmann, Superresolution imaging of biological nanostructures by spectral precision distance microscopy. Biotech. J. **6**, 1037–1051 (2011)
12. X.X. Zhou, M.Z. Lin, Photoswitchable fluorescent proteins: ten years of colorful chemistry and exciting applications. Curr. Opin. Chem. Biol. **17**, 682–690 (2013)
13. T. Ha, P. Tinnefeld, Photophysics of fluorescent probes for single-molecule biophysics and super-resolution imaging. Ann. Rev. Phys. Chem. **63**, 595–617 (2012)
14. Z.Y. Tian, W.W. Wu, A.D.Q. Li, Photoswitchable fluorescent nanoparticles: preparation, properties and applications. Chem. Phys. Chem. **10**, 2577–2591 (2009)
15. E. Betzig, G.H. Patterson, R. Sougrat, O.W. Lindwasser, S. Olenych, J.S. Bonifacino, M.W. Davidson, J. Lippincott-Schwartz, H.F. Hess, Imaging intracellular fluorescent proteins at nanometer resolution. Science **313**, 1642–1645 (2006)
16. S.T. Hess, T.P.K. Girirajan, M.D. Mason, Ultra-high resolution imaging by fluorescence photoactivation localization microscopy. Biophys. J. **91**, 4258–4272 (2006)
17. M.J. Rust, M. Bates, X.W. Zhuang, Sub-diffraction-limit imaging by stochastic optical reconstruction microscopy (STORM). Nat. Methods **3**, 793–795 (2006)
18. B. Huang, S.A. Jones, B. Brandenburg, X.W. Zhuang, Whole-cell 3D STORM reveals interactions between cellular structures with nanometer-scale resolution. Nat. Methods **5**, 1047–1052 (2008)
19. M.F. Juette, T.J. Gould, M.D. Lessard, M.J. Mlodzianoski, B.S. Nagpure, B.T. Bennett, S.T. Hess, J. Bewersdorf, Three-dimensional sub-100 nm resolution fluorescence microscopy of thick samples. Nat. Methods **5**, 527–529 (2008)
20. S.R.P. Pavani, M.A. Thompson, J.S. Biteen, S.J. Lord, N. Liu, R.J. Twieg, R. Piestun, W.E. Moerner, Three-dimensional, single-molecule fluorescence imaging beyond the diffraction limit by using a double-helix point spread function. Proc. Natl. Acad. Sci. USA **106**, 2995–2999 (2009)
21. C. Flors, J. Hotta, H. Uji-I, P. Dedecker, R. Ando, H. Mizuno, A. Miyawaki, J. Hofkens, A stroboscopic approach for fast photoactivation-localization microscopy with Dronpa mutants. J. Am. Chem. Soc. **129**, 13970–13977 (2007)
22. L. Zhu, W. Zhang, D. Elnatan, B. Huang, Faster STORM using compressed sensing. Nat. Methods **9**, 721–U286 (2012)
23. S.W. Hell, J. Wichmann, Breaking the diffraction resolution limit by stimulated-emission-depletion fluoresecence microscopy. Opt. Lett. **19**, 780–782 (1994)
24. T.A. Klar, S. Jakobs, M. Dyba, A. Egner, S.W. Hell, Fluorescence microscopy with diffraction resolution barrier broken by stimulated emission. Proc. Natl. Acad. Sci. USA **97**, 8206–8210 (2000)
25. S.W. Hell, M. Dyba, S. Jakobs, Concepts for nanoscale resolution in fluorescence microscopy. Cur. Opin. Neurobiol. **14**, 599–609 (2004)

26. B. Harke, J. Keller, C.K. Ullal, V. Westphal, A. Schoenle, S.W. Hell, Resolution scaling in STED microscopy. Opt. Exp. **16**, 4154–4162 (2008)
27. M. Reuss, J. Engelhardt, S.W. Hell, Birefringent device converts a standard scanning microscope into a STED microscope that also maps molecular orientation. Opt. Exp. **18**, 1049–1058 (2010)
28. R.J. Kittel, C. Wichmann, T.M. Rasse, W. Fouquet, M. Schmidt, A. Schmid, D.A. Wagh, C. Pawlu, R.R. Kellner, K.I. Willig, S.W. Hell, E. Buchner, M. Heckmann, S.J. Sigrist, Bruchpilot promotes active zone assembly, Ca^{2+} channel clustering, and vesicle release. Science **312**, 1051–1054 (2006)
29. K.I. Willig, S.O. Rizzoli, V. Westphal, R. Jahn, S.W. Hell, STED microscopy reveals that synaptotagmin remains clustered after synaptic vesicle exocytosis. Nature **440**, 935–939 (2006)
30. B. Hein, K.I. Willig, S.W. Hell, Stimulated emission depletion (STED) nanoscopy of a fluorescent protein-labeled organelle inside a living cell. Proc. Natl. Acad. Sci. USA **105**, 14271–14276 (2008)
31. C. Eggeling, C. Ringemann, R. Medda, G. Schwarzmann, K. Sandhoff, S. Polyakova, V.N. Belov, B. Hein, C. von Middendorff, A. Schonle, S.W. Hell, Direct observation of the nanoscale dynamics of membrane lipids in a living cell. Nature **457**, 1159–U1121 (2009)
32. C. Eggeling, K.I. Willig, F.J. Barrantes, STED microscopy of living cells - new frontiers in membrane and neurobiology. J. Neurochem. **126**, 203–212 (2013)
33. E. Rittweger, K.Y. Han, S.E. Irvine, C. Eggeling, S.W. Hell, STED microscopy reveals crystal colour centres with nanometric resolution. Nature Photon. **3**, 144–147 (2009)
34. A. Bril, On the saturation of fluorescence with cathode-ray excitation. Physica **15**, 361–379 (1949)
35. R. Altkorn, R.N. Zare, Effects of saturation on laser-induced fluorescence measurements of population and polarization. Ann. Rev. Phys. Chem. **35**, 265–289 (1984)
36. K. Fujita, M. Kobayashi, S. Kawano, M. Yamanaka, S. Kawata, High-resolution confocal microscopy by saturated excitation of fluorescence. Phys. Rev. Lett. **99**, 228105 (2007)
37. M. Yamanaka, Y.K. Tzeng, S. Kawano, N.I. Smith, S. Kawata, H.C. Chang, K. Fujita, SAX microscopy with fluorescent nanodiamond probes for high-resolution fluorescence imaging. Biomed. Opt. Exp. **2**, 1946–1954 (2011)
38. M. Yamanaka, K. Saito, N.I. Smith, S. Kawata, T. Nagai, K. Fujita, Saturated excitation of fluorescent proteins for subdiffraction-limited imaging of living cells in three dimensions. Interface Focus **3**(6), 20130007 (2013)
39. M.G.L. Gustafsson, Surpassing the lateral resolution limit by a factor of two using structured illumination microscopy. J. Microsc. Oxf. **198**, 82–87 (2000)
40. M.G.L. Gustafsson, L. Shao, P.M. Carlton, C.J.R. Wang, I.N. Golubovskaya, W.Z. Cande, D.A. Agard, J.W. Sedat, Three-dimensional resolution doubling in wide-field fluorescence microscopy by structured illumination. Biophys. J. **94**, 4957–4970 (2008)
41. L. Schermelleh, P.M. Carlton, S. Haase, L. Shao, L. Winoto, P. Kner, B. Burke, M.C. Cardoso, D.A. Agard, M.G.L. Gustafsson, H. Leonhardt, J.W. Sedat, Subdiffraction multicolor imaging of the nuclear periphery with 3D structured illumination microscopy. Science **320**, 1332–1336 (2008)
42. R. Heintzmann, T.M. Jovin, C. Cremer, Saturated patterned excitation microscopy—a concept for optical resolution improvement. J. Opt. Soc. Am. A Opt. Image Sci. Vis. **19**, 1599–1609 (2002)
43. M.G.L. Gustafsson, Nonlinear structured-illumination microscopy: wide-field fluorescence imaging with theoretically unlimited resolution. Proc. Natl. Acad. Sci. USA **102**, 13081–13086 (2005)
44. E.H. Rego, L. Shao, J.J. Macklin, L. Winoto, G.A. Johansson, N. Kamps-Hughes, M.W. Davidson, M.G.L. Gustafsson, Nonlinear structured-illumination microscopy with a photoswitchable protein reveals cellular structures at 50-nm resolution. Proc. Natl. Acad. Sci. USA **109**, E135–E143 (2012)
45. P. Wang, M.N. Slipchenko, J. Mitchell, C. Yang, E.O. Potma, X.F. Xu, J.X. Cheng, Far-field imaging of non-fluorescent species with subdiffraction resolution. Nat. Photon. **7**, 450–454 (2013)

46. T.Y. Su, R. Oketani, Y.T. Huang, H.Y. Wu, Y. Yonemaru, M. Yamanaka, H. Lee, G.Y. Zhuo, M.Y. Lee, S. Kawata, K. Fujita, Measurement of a saturated emission of optical radiation from gold nanoparticles: application to an ultrahigh resolution microscope. Phys. Rev. Lett. **112**, 017402 (2014)
47. S.W. Chu, H.Y. Wu, Y.T. Huang, T.Y. Su, H. Lee, Y. Yonemaru, M. Yamanaka, R. Oketani, S. Kawata, S. Shoji, K. Fujita, Saturation and reverse saturation of scattering in a single plasmonic nanoparticle. ACS Photon. **1**, 32–37 (2014)
48. B. Lee, S. Kim, H. Kim, Y. Lim, The use of plasmonics in light beaming and focusing. Prog. Quant. Electron. **34**, 47–87 (2010)
49. S. Lal, S. Link, N.J. Halas, Nano-optics from sensing to waveguiding. Nat. Photon. **1**, 641–648 (2007)
50. M.P. Nezhad, A. Simic, O. Bondarenko, B. Slutsky, A. Mizrahi, L.A. Feng, V. Lomakin, Y. Fainman, Room-temperature subwavelength metallo-dielectric lasers. Nat. Photon. **4**, 395–399 (2010)
51. N. Fang, H. Lee, C. Sun, X. Zhang, Sub-diffraction-limited optical imaging with a silver superlens. Science **308**, 534–537 (2005)
52. T. Ichimura, N. Hayazawa, M. Hashimoto, Y. Inouye, S. Kawata, Tip-enhanced coherent anti-Stokes Raman scattering for vibrational nanoimaging. Phys. Rev. Lett. **92**, 220801 (2004)
53. K.M. Hajek, B. Littleton, D. Turk, T.J. McIntyre, H. Rubinsztein-Dunlop, A method for achieving super-resolved widefield CARS microscopy. Opt. Exp. **18**, 19263–19272 (2010)
54. T.S. Ahmadi, S.L. Logunov, M.A. El-Sayed, Picosecond dynamics of colloidal gold nanoparticles. J. Phys. Chem. **100**, 8053–8056 (1996)
55. I. Ros, P. Schiavuta, V. Bello, G. Mattei, R. Bozio, Femtosecond nonlinear absorption of gold nanoshells at surface plasmon resonance. Phys. Chem. Chem. Phys. **12**, 13692–13698 (2010)
56. S.S. Chong, W. Min, X.S. Xie, Ground-state depletion microscopy: detection sensitivity of single-molecule optical absorption at room temperature. J. Phys. Chem. Lett. **1**, 3316–3322 (2010)
57. C.F. Bohren, D.R. Huffman, *Absorption and scattering of light by small particles* (Wiley, Weinheim, 1983)
58. K.H. Su, Q.H. Wei, X. Zhang, J.J. Mock, D.R. Smith, S. Schultz, Interparticle coupling effects on plasmon resonances of nanogold particles. Nano Lett. **3**, 1087–1090 (2003)
59. A.M. Schwartzberg, J.Z. Zhang, Novel optical properties and emerging applications of metal nanostructures. J. Phys. Chem. C **112**, 10323–10337 (2008)
60. H. Shroff, C.G. Galbraith, J.A. Galbraith, H. White, J. Gillette, S. Olenych, M.W. Davidson, E. Betzig, Dual-color superresolution imaging of genetically expressed probes within individual adhesion complexes. Proc. Natl. Acad. Sci. USA **104**, 20308–20313 (2007)
61. J. Tonnesen, F. Nadrigny, K.I. Willig, R. Wedlich-Soldner, U.V. Nagerl, Two-color STED microscopy of living synapses using a single laser-beam pair. Biophys. J. **101**, 2545–2552 (2011)
62. S.A. Jones, S.H. Shim, J. He, X.W. Zhuang, Fast, three-dimensional super-resolution imaging of live cells. Nat. Methods **8**, 499–U496 (2011)
63. T. Grotjohann, I. Testa, M. Leutenegger, H. Bock, N.T. Urban, F. Lavoie-Cardinal, K.I. Willig, C. Eggeling, S. Jakobs, S.W. Hell, Diffraction-unlimited all-optical imaging and writing with a photochromic GFP. Nature **478**, 204–208 (2011)
64. B.G. Kopek, G. Shtengel, J.B. Grimm, D.A. Clayton, H.F. Hess, Correlative photoactivated localization and scanning electron microscopy. PLoS One **8**, e77209 (2013)
65. N.T. Urban, K.I. Willig, S.W. Hell, U.V. Nagerl, STED nanoscopy of actin dynamics in synapses deep inside living brain slices. Biophys. J. **101**, 1277–1284 (2011)
66. P. Bethge, R. Chereau, E. Avignone, G. Marsicano, U.V. Nagerl, Two-photon excitation STED microscopy in two colors in acute brain slices. Biophys. J. **104**, 778–785 (2013)
67. A. Vaziri, J.Y. Tang, H. Shroff, C.V. Shank, Multilayer three-dimensional super resolution imaging of thick biological samples. Proc. Natl. Acad. Sci. USA **105**, 20221–20226 (2008)
68. V. Westphal, S.O. Rizzoli, M.A. Lauterbach, D. Kamin, R. Jahn, S.W. Hell, Video-rate far-field optical nanoscopy dissects synaptic vesicle movement. Science **320**, 246–249 (2008)

Chapter 29
Harmonic Generation Microscopy

Szu-Yu Chen and Chi-Kuang Sun

29.1 Introduction

Since the second-harmonic generation (SHG) and two-photon fluorescence (2PF) microscopies were demonstrated in 1961 [1], two-photon microscopy has emerged as one of the most common technique used in biomedical applications. Based on the various endogenous fluorophores or extrinsic fluorescent staining, 2PF microscopy can provide cellular, morphological, and molecular information in bio-tissues [2, 3]. On the other hand, SHG is known to arise from the non-centro-symmetric materials which have non-zero second-order susceptibility, $\chi^{(2)}$. In bio-tissues, SHG can be applied to the studies of microtubule polarity, nerve fibers, muscles, collagen fibrils, and so on [4–7]. With a quadratic dependence of excitation probability, the two-photon excitation is restricted to the tiny focal volume resulting in rejection of off-focus excitation and photodamages, and thus optical sectioning can be achieved. However, since 2PF microscopy is a fluorescence-based technique, issues of photodamage and photobleaching are always of concern in the imaging process. To avoid such fluorescence-induced concerns, third harmonic generation microscopy (THG), a third-order nonlinear optical microscopy is used instead of 2PF microscopy. THG microscopy is indicated to be a general-purpose microscopy which has interface sensitivity and can be used to provide morphological information of the tissues [8–10]. Recently, THG microscopy has been applied to various biological investigation like microscopic

S.-Y. Chen (✉)
Department of Optics and Photonics, National Central University, Zhongli, Taiwan
e-mail: sychen@dop.ncu.edu.tw

C.-K. Sun
Department of Electrical Engineering, National Taiwan University, Taipei, Taiwan
e-mail: sun@cc.ee.ntu.edu.tw

C.-C. Lee (ed.), *The Current Trends of Optics and Photonics*,
Topics in Applied Physics 129, DOI 10.1007/978-94-017-9392-6_29

imaging of zebrafish embryos, drosophila embryo, mouse skin, mouse eye, human skin, human oral cavity, and so on [9–13]. SHG and THG have the characteristics of virtual-level transition and the generation process of SHG and THG follows the energy conservation rules. Since there is no energy deposition during the process, the problems of fluorescence-induced photodamage and photobleaching could be avoided so as to reduce the optical invasiveness. Due to its higher-order nonlinearity of SHG and THG microscopy, both of them can provide the optical section power for investigations of thick bio-tissues. In this chapter, the harmonic generation microscopy (HGM), combining SHG and THG microscopy will be introduced, including the principles, system setup, and the biomedical applications accomplished.

29.1.1 Principle

SHG and THG are both higher-order nonlinear processes which imply the polarization intensity of a molecule produced has a nonlinear dependence on the electric field of the excitation light. Induced by the electrical field, the polarization intensity of a molecule can be described by [14]

$$\begin{aligned}\tilde{P}(t) &= \varepsilon_0 \chi^{(1)} \tilde{E}(t) + \varepsilon_0 \chi^{(2)} \tilde{E}(t)^2 + \varepsilon_0 \chi^{(3)} \tilde{E}(t)^3 + \cdots \\ &\equiv \tilde{P}^{(1)}(t) + \tilde{P}^{(2)}(t) + \tilde{P}^{(3)}(t) + \cdots ,\end{aligned} \tag{29.1.1}$$

where $\chi^{(2)}$ and $\chi^{(3)}$ are the second-order and third-order nonlinear susceptibilities, respectively. According to (29.1.1), the intensity of second-order nonlinear polarization, $\tilde{P}^{(2)}(t)$, has a quadratic dependence on the excitation electric field $\tilde{E}(t)$, while the intensity of third-order nonlinear polarization, $\tilde{P}^{(3)}(t)$, has a cubic dependence. This dependence indicates that the excitation volume is confined right near the focal spot and high axial resolution can be achieved. Compared with the second-order process, the third-order process is less efficient but can provide higher spatial resolution than the second-order process.

SHG is a second-order nonlinear optical process described by $\chi^{(2)}$ (2ω: ω, ω). Because each element of tensor $\chi^{(2)}$ vanishes in the materials with inversion symmetry, SHG is well-known to occur only in a material that is non-centrosymmetric (without inversion symmetry). On the other hand, THG is a third-order nonlinear optical process described by terms $\chi^{(3)}$ (3ω: ω, ω, ω) involving three electric fields with the same frequency. Third-order susceptibility $\chi^{(3)}$ is non-vanishing regardless of the symmetry of materials. However, due to the Gouy phase shift under strong focusing conditions, THG is found to vanish in isotropic materials with a negative phase mismatch (normal dispersion). Efficient THG can only be produced at interfaces, where constructive interference possibly occurs due to changes of dispersion or to the nonlinear susceptibility of materials. As illustrated in Fig. 29.1a, b, two photons and three photons with a frequency ω are up-converted to a photon with a frequency 2ω and 3ω in SHG and THG processes,

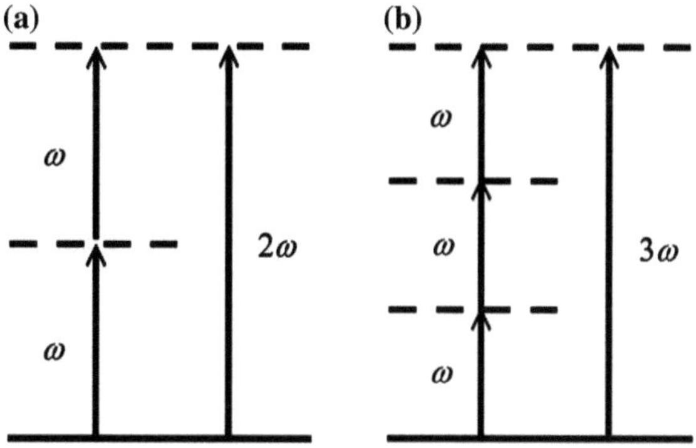

Fig. 29.1 **a** Energy-level diagram of second-harmonic generation. **b** Energy-level diagram of third-harmonic generation

respectively and only virtual-level transition is involved. Since the up-conversion processes obey the energy-conservation rule, no energy is deposited in the interacting material.

29.1.2 System Setup

Because of the higher-order nonlinearity, extremely high excitation intensity is needed to generate SHG and THG. Tightly focusing of the excitation laser and point-scanning geometry are basically required. The HGM is composed of four main parts—excitation laser, scanning units, microscope, and detectors. For pixel-by-pixel imaging, a scanning unit is used to achieve real-time 2D beam scanning. Before guided into a scanning system, the excitation beam has to be shaped and collimated by a pair of telescopes to fill the scan mirror size. Depending on the applications, an upright microscope or an inverted microscope can be connected to the scanning system with an aperture fitting tube lens. After passing through the tube lens and the optics in the microscope, the scanned laser beam is focused onto the specimen by a high NA objective and the laser beam has to fill the objective back aperture. With the forward-collection geometry, the generated SHG and THG signals are collected by a condenser and guided into photomultipliers (PMTs) for detection (Fig. 29.2a). A dichroic beamsplitter with a suitable cut-off wavelength should be used to separate SHG and THG signals and two individual PMTs are used to obtain SHG and THG images. In front of the PMTs, a color filter is inserted to filter out the laser radiation to avoid noise and damages of PMT, while band-pass filters are inserted to block the background noise to increase the signal-to-noise ratio (SNR). For the backward-collection geometry, the excited epi-SHG and epi-THG signals are epi-collected by the same objective (Fig. 29.2b). The collected signals are then reflected by a dichroic beamsplitter, which can transmit the laser beams and reflect the HG signals, which are then directed into PMTs. The signal detection geometry is the same as that used in the forward-collection

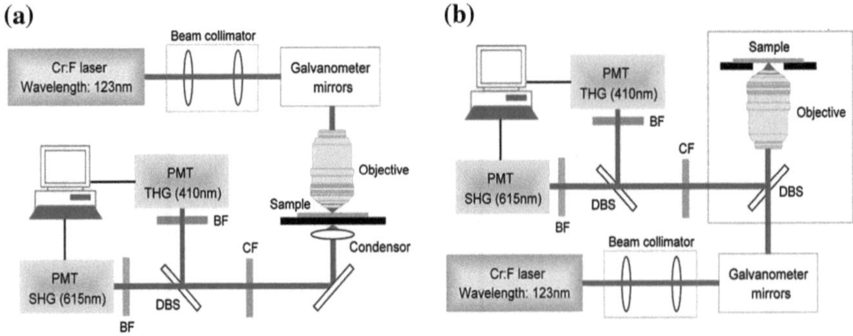

Fig. 29.2 Schematic diagrams of the HGM with **a** forward-collection and **b** backward-collection geometry. *DBS* dichroic beamsplitter; *CF* color filter; *BF* bandpass filter; *PMT* photomultiplier

imaging system. To obtain simultaneous SHG and THG images, two PMTs should be perfectly synchronized with the scanning system for intensity mapping. As the laser beam is 2D scanned over the specimen, the signals are record point-by-point simultaneously to form 2D SHG and THG images.

In this chapter, a home-made Cr:forsterite (Cr:F) 1,230 nm laser is used for excitation. In contrast to the Ti:sapphire (Ti:S) laser, which is commonly used for multiphoton microscopy [7, 15, 16], the Cr:F laser provides three advantages in HGM. Firstly, with the 1,230 nm excitation, the THG wavelength is shifted into the visible region, while the THG wavelength falls within the deep UV region under a 800 nm (Ti:S) excitation [17]. The serious absorption and scattering of THG signals in the bio-tissue can thus be avoided to increase the signal collection efficiency [9, 18]. Secondly, the attenuation (combination of scattering and absorption) is found to reach a minimum value in the bio-tissues around 1,200–1,300 nm [19]. The Cr:F 1,230 nm laser, well within this penetration window, is an optimal laser source for reducing the attenuation and increasing the imaging penetrability [9, 20, 21]. As shown in Fig. 29.3, Cr:F-based HGM (forward-collection geometry) was applied for in vivo imaging of zebrafish embryo with a diameter of 1.5-mm thick [9]. From the upper chorion structures (Fig. 29.3a), top cellular layer (Fig. 29.3b), yolk cells (Fig. 29.3c–e), bottom cellular layer (Fig. 29.3f), to the bottom chorion structures (Fig. 29.3g), the cellular structures of the whole embryo (Fig. 29.3h) can be clearly observed with a penetration depth of ~ 1.5 mm. Finally, the previous studies show that mouse embryos under 10 min continuous observation with >120-mW average power and >21.6-J total exposure per embryo have survival rates same as the non-imaged embryos [22, 23]. In contrast to the exposure energy limit of 0.2 and 2 J under 730–800-nm and 1,047-nm excitation, respectively [24, 25], photodamages are found to be obviously reduced under Cr:F laser excitation.

The imaging system is adapted from a commercial confocal scanning system (FV300, Olympus) combined with an upright microscope (BX51, Olympus) or an inverted microscope (IX71, Olympus). All optical components in the system are

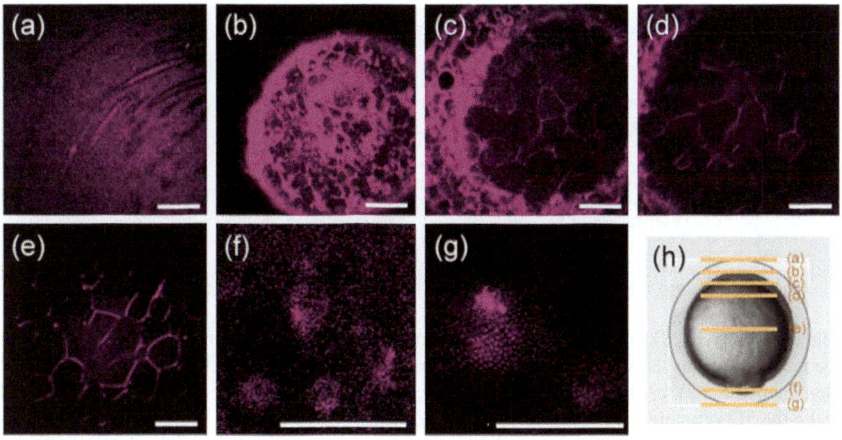

Fig. 29.3 In vivo combined SHG/THG images of a zebrafish embryo obtained at different depths beneath the top chorion surface. The structures of the **a** top chorion, **b** top cellular layer, **c–e** yolk cells, **f** bottom cellular layer, and **g** bottom chorion can be observed within a depth of ~1.5 mm. The corresponding imaging planes are indicated in **h**. THG is represented by purple colors. *Scale bar* 50 μm

modified for the excitation of near-infrared light (~1,230 nm). Objectives with NA 1.2 (UPLSAPO 60XW, Olympus) or NA 0.9 (LUMPLFL 60 W/IR2, Olympus) are used for focusing the laser beam, depending on the applications. Sing the wavelengths of SHG and THG are 615 and 410 nm, dichroic beamsplitters with cut-off wavelength of 665 nm (FF665-Di02, Semrock) and 490 nm (490DRXR, Chroma Technology) are used to reflect the HG signals and separate SHG and THG signals, respectively.

29.1.3 3D Spatial Resolution

In contrast to the confocal pinhole in a confocal microscope, the restriction of SHG and THG excitation can be considered to be a virtual pinhole. Not only the off-focus information can be eliminated to increase the spatial resolution but also the off-focus photodamages can be reduced. In HGM, image can be considered as a convolution result of the excitation PSF and the features of the specimens. Since the SHG and THG intensity has a quadratic and cubic dependence on the excitation light intensity, the size of the effective SHG and THG PSF is reduced by a factor of $\sqrt{2}$ and $\sqrt{3}$, respectively [26]. The lateral resolution is given by

$$r_{\mathrm{SHG}}=\frac{1}{\sqrt{2}}\cdot\frac{0.51\lambda}{NA}\approx\frac{0.36\lambda}{NA},\quad r_{\mathrm{THG}}=\frac{1}{\sqrt{3}}\cdot\frac{0.51\lambda}{NA}\approx\frac{0.3\lambda}{NA},\qquad (29.1.2)$$

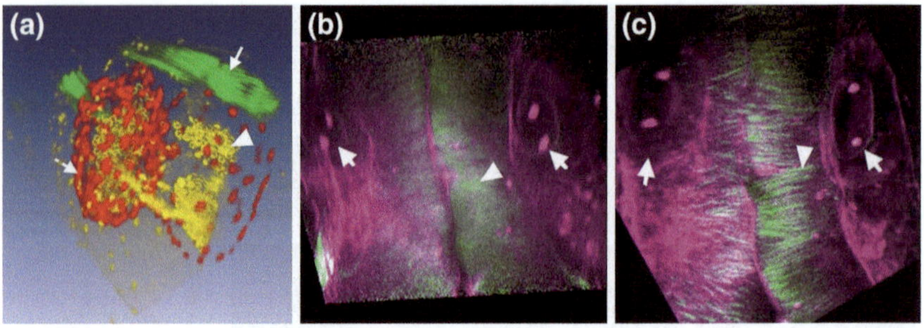

Fig. 29.4 In vivo 3D images of **a** a zebrafish heart and **b**–**c** a zebrafish brain. The 3D images are reconstructed from stacks of HGM images versus depth. **a** The 3D structures of the cardiac muscles (*arrow*; SHG), cardiac cells (*dashed arrow*; 2PF), and red blood cells (*arrowhead*; THG); **b** and **c** the 3D structures of the neural tube (THG), otic vesicles (*arrows*; THG), and the nerve fibers (*arrowhead*; SHG) can be observed with a submicron resolution. *Image size* 240 × 240 μm^2

and the axial resolution is given by

$$d_{z;\mathrm{SHG}}=\frac{1}{\sqrt{2}}\cdot\frac{n\lambda}{NA^2}\approx\frac{0.7n\lambda}{NA^2},\quad d_{z;\mathrm{SHG}}=\frac{1}{\sqrt{3}}\cdot\frac{n\lambda}{NA^2}\approx\frac{0.57n\lambda}{NA^2}. \tag{29.1.3}$$

Based on the high axial resolution, virtual optical sectioning of bulky tissues can be easily achieved and 3D images can be reconstructed from the obtained stack of HGM images versus depth. Figure 29.4a shows the 3D image of the zebrafish heart [27], including the 3D structures of the cardiac muscle, cardiac cells, and red blood cells; while Fig. 29.4b, c show the 3D structures of the zebrafish brain, including the 3D structures of the nerve fibers, neural tubes, and otic vesicles (arrows) [21].

29.1.4 Imaging Contrasts of SHG and THG in Bio-tissues

In the previous ex vivo and in vivo studies, various imaging contrasts of SHG and THG were revealed in different animal models and human tissues. In the studies of zebrafish embryos [9, 21, 28], SHG was found to arise from non-centrosymmetric structures like spindle fibers (arrow in Fig. 29.5a), skeleton muscles (arrow in Fig. 29.5b), cardiac muscles (arrow in Fig. 29.5c), and nerve fibers (arrow in Fig. 29.5d), while THG was found to arise from cell membranes (arrowhead in Fig. 29.5a) and tissue inhomogeneity. In the in vivo investigations, the structural changes during the embryo development can be revealed by THG microscopy [21, 28]. In the studies of mouse embryos [22],

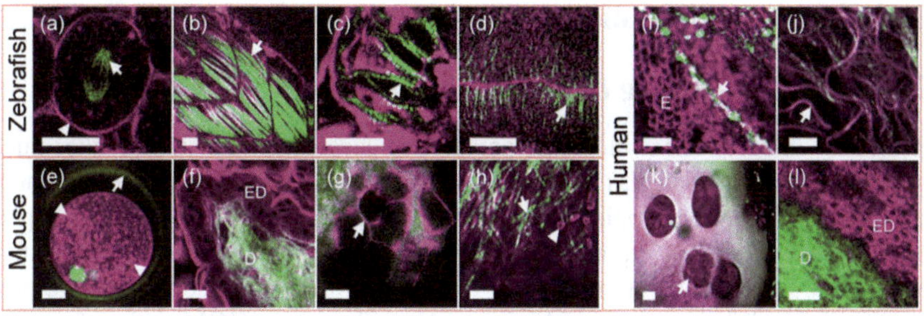

Fig. 29.5 **a–d** In vivo HGM images of zebrafish embryos. THG contrasts are provided by cell membranes (*arrowhead* in **a**) and tissue inhomogeneity; SHG contrasts are provided by spindle fibers (*arrow* in **a**), skeleton muscles (*arrow* in **b**), cardiac muscles (*arrow* in **c**), and nerve fibers (*arrow* in **d**). **e** In vivo HGM image of a mouse embryo. THG contrasts are contributed by cytoplasmic organelles (*arrowheads* in **e**); SHG contrasts are contributed by zona pellucida (*arrow* in **e**). **f–g** Ex vivo and **h** in vivo HGM image of mouse skin. THG contrasts are provided by cytoplasm of keratinocytes in epidermis (*ED*), and adipocytes (*arrow* in **g**) and red blood cells (*arrowhead* in **h**) in dermis; SHG contrasts are provided by collagen fibers (*arrow* in **h**) in dermis (*D*). **i–l** Ex vivo HGM images of excised human **i** teeth, **j** lung, **k** cartilage, and **l** skin. THG microscopy reveals rod structures of the tooth enamel (**i**), elastic fibers in the lung tissues (*arrow* in **j**), chondrocytes in the cartilage (*arrow* in **k**), and cellular morphology of epidermis (*ED*) of the skin (**l**); strain status of the abnormal enamel (*arrow* in **i**), type II collagen in the cartilage (**k**) and collagenous structures of dermis (*D*) of the skin (**l**) can be revealed by SHG microscopy. *Scale bar* 20 μm

SHG can reveal zona pellucida (arrow in Fig. 29.5e) and mitotic spindles of the mouse embryos, while THG contrasts were contributed by cytoplasmic organelles (arrowheads in Fig. 29.5e). In mouse skin [29], strong SHG contrast were observed from collagen fibers (Fig. 29.5f–h), while THG contrasts from cytoplasm of keratinocytes (Fig. 29.5f), adipocytes (Fig. 29.5g), and red blood cells (Fig. 29.5h) revealed the cellular morphology of epidermis (ED) and dermis (D). HGM can also be applied to human tissues like teeth [30], lung [11], cartilage [31], eye [10, 21], and skin [12]; [32]. The strain status of the abnormal enamel (arrow in Fig. 29.5i), type II collagen in the cartilage (Fig. 29.5k) and collagenous structures of the skin (Fig. 29.5l) can be revealed by SHG microscopy, while THG microscopy was shown to be able to reveal rod structures of the tooth enamel (Fig. 29.5i), elastic fibers in the lung tissues (arrow in Fig. 29.5j), chondrocytes in the cartilage (arrow in Fig. 29.5k), and cellular morphology of the skin (Fig. 29.5l). Based on the rich imaging contrasts of SHG and THG in bio-tissues, HGM shows the capability for various biological and medical applications. Additionally, since the contrast sources of SHG are different from those of THG, the SHG-sensitive structures can be easily identified in bio-tissues through HGM.

29.2 Biomedical Applications of HGM

29.2.1 HGM Imaging of Zebrafish Embryo

Zebrafish embryos have many advantages in developmental studies like much genetic material the same as humans', similar but simpler embryonic developmental programs, rapid developing rate, precisely-defined developing stages, transparency and small size of embryos, and external development. Therefore, zebrafish embryos are used as an animal model for studies of complex developmental processes of vertebrate embryos. Observing the embryos by HGM with a forward-collection geometry, SHG modality provides valuable information, including mitosis spindle fibers [9, 21, 28], nerve fibers [21], muscle fibers [9, 28], and stacked membranes [21], to study the embryonic development of different systems at different stages. THG modality, which can reflect the structural information of the embryos, can help to localize the SHG signals and identify and contrast sources of the SHG signals.

In the in vivo investigation of cell mitosis in zebrafish embryos, strong SHG can be observed from centrosomes and mitotic spindles which are known to be made up of spatially organized dynamic microtubules and of which the optical centro-symmetry is broken [7]. On the other hand, based on the sensitivity to optical inhomogeneity, THG contrast can reflect various interfaces inside a single cell [33–35] such as nuclear membranes, cell membranes, and the cytoplasmic organelles [22]. Under the HGM, the dynamic changes of spindle and membrane between two daughter cells can be imaged in vivo without any exogenous markers (Fig. 29.6). At the initial prophase stage, two centrosomes are revealed by SHG signals, while THG signals show a circular cell nuclear membrane (Fig. 29.6a). At the prometaphase, the microtubules, revealed by SHG signals, begin to elongate from the centrioles to form the spindle (Fig. 29.6b). During the anaphase, the separation of spindle fibers and the alternation of the cell contour are picked up by the SHG and THG modalities, respectively (Fig. 29.6c). At the telophase, the end of the mitosis, since the spindle microtubules disperse into the cells and exhibit no more crystalline characteristic, SHG signals vanish at this stage (Fig. 29.6d).

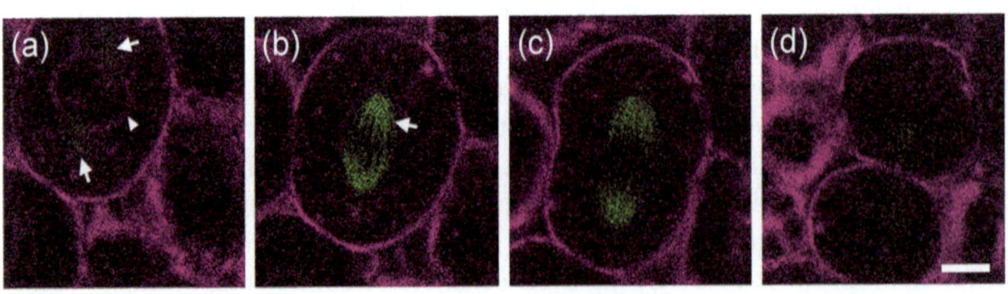

Fig. 29.6 SHG/THG time series of the mitosis process in the 1-k-cell-stage zebrafish embryo. During **a** the prophase stage, **b** the metaphase, **c** the anaphase, and **d** the telophase, the cell nuclear membrane (*arrowhead*), centrosomes (*arrows* in **a**), and mitotic spindles (*arrow* in **b**) can be visualized through THG and SHG respectively. *Scale bar* 10 μm

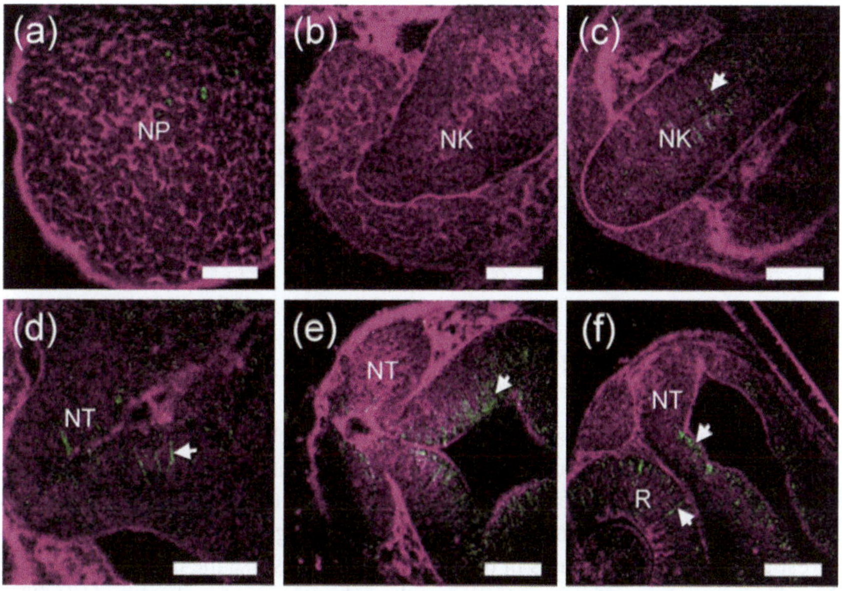

Fig. 29.7 Continuous 20-h in vivo HGM imaging of the brain development in the same zebrafish embryo from the bud-stage to the prim-15-stage. The structural changes of brain from **a** neural plate (*NP*) at the bud stage; **b** neural keel (*NK*) at the 3-somite stage; **c** NK at the 5-somite stage; **d** neural tube (*NT*) at the 14 somite-stage; **e** midbrain at the 22-somite stage; to **f** NT and eye at the prim-15 stage can be revealed by THG microscopy. The nerve fibers (*arrows*) in both the NT and the retina (*R*) can be revealed by SHG. Scale bar 50 μm

Utilizing the SHG contrast from nerve fibers and spindle microtubules, HGM is a good tool for investigation of the zebrafish embryonic brain development. Without any fluorescence markers, >20-h long-term observation of the brain development in the same live zebrafish embryo hours was performed by using HGM [21]. At the bud stage, the right beginning of the brain development, the plate-like neural plate is observed to be formed by a cluster of cells which are still under differentiation and division with no neural fibers. Through strong SHG signals from spindles, the cells under mitosis for differentiation or division can be clearly recognized, while THG images help to localize the position of those cells (Fig. 29.7a). After 1 h, at the 3-somite stage, the thickened neural plate with a middle line, named neural keel, can be revealed through the THG modality (Fig. 29.7b). At this stage, the rate of cell mitosis is retarded and no nerve fibers have been developed. Almost no SHG signals can be observed. As the embryo developed into the 5-somite stage, the first neuron inside the neural keel begins to differentiate and grow the nerve fibers [36]. SHG signals which arise from the nerve fibers begin to appear in the middle of the neural keel and gradually extend to be linear-like (Fig. 29.7c). From the 5-somite to prim-15 stage (Fig. 29.7d–f), SHG modality dynamically records the growth of the nerve fibers and the SHG signals become much stronger and more linear-like. On the other hand, the morphological changes from neural keel, rod-like neural rod, to neural tube with a hollow structure can all be revealed by THG modality. During the whole process of the brain development, interface-sensitive THG provides the 3D sketch of different structures from a neural plate to a hollow neural tube; nerve-fiber- and mitotic-spindle-sensitive SHG can tell us much more stories about how cells behave during the development process. In addition, the axons elongated from the ganglia in the retina were observed through the SHG, while THG also showed the structure of the eyes, including the retina and the lens (Fig. 29.7f).

29.2.2 HGM Imaging of Mouse Tissues

Mouse is popularly-used as the standard animal model for many biomedical studies. Applying HGM to mouse tissues can help to identify various image contrasts of SHG and THG and examine the imaging capability of HGM in mammal tissues. The strong SHG signals arising from the collagen fibers can be utilized to investigate the fibril structures in the connective tissues like the dermis of skin [12, 32], the stroma layer of cornea [10], and the sub-mucosa of oral cavity [13]. THG modality can generally provide morphological information to reveal the cellular structures of the skin, cornea, lens, oral mucosa, cartilage, and so on. Moreover, THG is found to be enhanced by specific kinds of molecule like hemoglobin [37–39], lipid [40], and elastin [11] through the mechanisms of real-level absorption resonance enhancement. Therefore, THG modality is suitable for the investigation of red blood cells (RBC), adipocytes, and elastic fibers. In the studies of the mouse tissues, HGM with a backward-collection geometry was used.

By obtaining the optical sections of mouse abdomen skin at different depths, the cellular structures of the epidermis and the collagenous structures of the dermis can be revealed by THG and SHG modality, respectively (Fig. 29.8). Since THG is well-known to be sensitive to local optical inhomogeneity [34, 41, 42] and lipid [40], the outermost stratum corneum, composed of multi-layers of lipid and corneocytes, can be revealed by THG modality with ultra-strong THG contrast (Fig. 29.8a, b). Within 40 μm beneath the skin surface, THG modality shows the cellular structures of the epidermis based on the contrast from cytoplasmic organelles (Fig. 29.8a–d) [22]. The cytoplasm of the keratinocytes appears THG-bright, while the nuclei of the keratinocytes appear THG-dark. The strong SHG contrasts from the type I collagen fibers help to show the collagenous structures of the dermis (Fig. 29.8c–h). At the depths of 60 and 100 μm, SHG-revealed collagenous structures sketch the hair follicle in the dermis, while the hair root in the hair follicle is shown by THG modality (Fig. 29.8e, f). In addition, since THG signals can be enhanced by lipid and the sebaceous gland is made of lipid-filled cells, the sebaceous gland shows the strong THG contrasts. In the skin, it is believed that there are discrete populations of epidermal stem cells in the stratum basale, hair follicles, sebaceous glands, apocrine glands, and eccrine glands. To study the activation of these stem cells, an imaging tool with the ability for in vivo observation is desired.

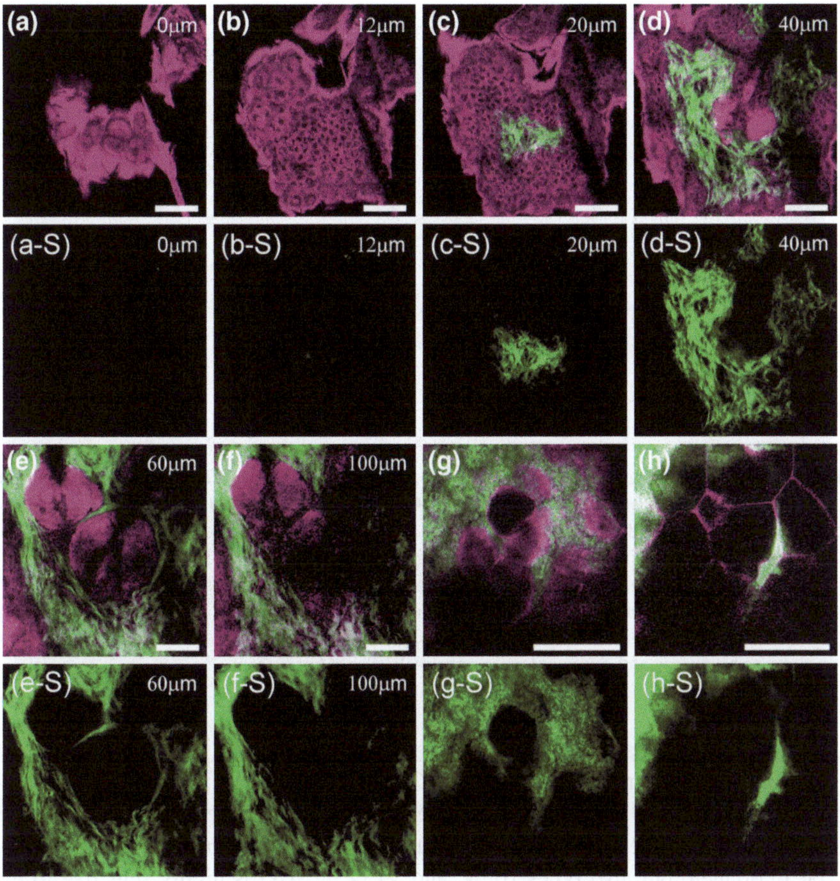

Fig. 29.8 **a–f** An exampled series of SHG/THG images of the mouse ear skin obtained at **a** 0 μm, **b** 12 μm, **c** 20 μm, **d** 40 μm, **e** 60 μm, and **f** 100 μm beneath the skin surface. The sebaceous glands surrounding the hair follicles in the ear skin were observed at a more superficial layer of the dermis and were shown with stronger THG intensity and higher density. **g–h** The fat cells in the hypodermis were observed at **g** 70 μm and **h** 120 μm in the abdomen skin of a fat mouse. Even at 120 μm, the submicron of THG microscopy was still preserved. **a-S–h-S** The separated SHG images corresponding to **a-h**. *Scale bar* 50 μm

Our results of the mouse skin show that THG microscopy has strong imaging contrast in the hair follicles and the sebaceous glands. This also indicates the capability of THG microscopy for studies of the epidermal stem cells. On the other hand, since more abundant fat cells are in the mouse abdomen skin, the polygonal fat cells are easily revealed by THG modality (Fig. 29.8g), while the loose collagen fibers surrounding the fat cells are revealed by SHG modality. Even at 120 μm (Fig. 29.8h), the fat cells can still be observed with a submicron spatial resolution.

Mouse eye has similar structures to human eye, including cornea, aqueous, lens, vitreous, and retina. The outermost layer of eye, cornea, can be roughly

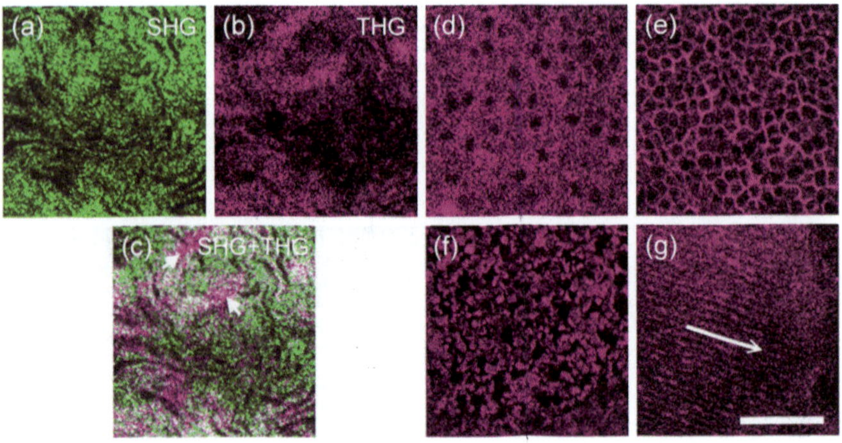

Fig. 29.9 SHG/THG images of an excised mouse eye obtained at different depths beneath the corneal surface. **a** and **b** show the SHG and THG images of the corneal stroma and **c** shows the corresponding combined SHG/THG image. The collagenous structures can be revealed by SHG, while the keratocytes lying within the collagen fibers can be recognized from the non-SHG-overlapped THG signals (*arrows*). Through THG modality, the cellular structures of the **d** upper corneal epithelium, **e** deeper corneal epithelium, and **f** corneal endothelium can be revealed, while **g** the lens fibers in the lens were shown clearly along the direction of the *arrow*. THG and SHG are represented by *purple* and *green* colors, respectively. *Scale bar* 50 μm

divided into three main components—the corneal epithelium (EP), corneal stroma (CS), and corneal endothelium (ED) [43, 44]. The EP and ED are layers composed of epithelium and endothelium cells, while the CS between them consists of hundreds of layers of regularly-organized collagen fibers, including mainly type I collagen, but also types III, V, and VI [45]. SHG is highly sensitive to the type I collagen fibers [46–48] and is a good tool for revealing the collagen fiber arrangements in the stroma. On the other hand, more structural and cellular information in the CS can be given by THG modality, as shown in Fig. 29.9a, b. In the combined SHG/THG image (Fig. 29.9c), the SHG-overlapped THG signals are recognized to arise from the collagen fibers, while the non-SHG-overlapped THG signals reflect the keratocytes lying within the collagen fiber meshes (arrows in Fig. 29.9c). By combining SHG with THG modality, not only the collagenous structures can be observed but also the cellular information in the CS can be distinguished and obtained more easily. In addition to the CS, the EP and ED are responsible for protection and governing the fluid transportation of the eye, respectively. ED covers the anterior surface of the cornea and has 5–6 cell layers thick, while the ED is a mono-layer of flattened and polygonal ED cells. Based on the THG contrast arising from the cytoplasmic organelles [22], the cytoplasm of the EP cells appeared bright in contrast to the dark nuclei (Fig. 29.9d). The morphology of the EP cells (Fig. 29.9d–e) is shown to be consistent with the histology results [44]. Figure 29.9f shows the THG image of the ED at 120 μm where the uniformly-sized polygonal cells in this monolayer are revealed. Beneath the cornea is the aqueous humor (AH), which is mainly composed of water and has no THG contrast due to its optical homogeneity. Passing through the AH is the lens (L), consisting of three main parts: the membrane-like lens capsule (LC), cellular lens epithelium (LE), and lens fibers (LF). At a depth of 430 μm beneath the anterior corneal surface, the lens fibers (Fig. 29.9g) with a width of ~5 μm (along the arrow in Fig. 29.9g) can be highly-resolved through THG signals. No significant SHG signals can be observed from the lens fibers. Even at a depth of >700 μm, the structure of the lens fibers can still be revealed through THG. Since the total thickness of the human cornea is about 535 μm [49], a penetration depth greater than 700 μm indicates the ability to investigate even the deepest part of the human cornea. With different imaging contrasts of SHG and THG, not only the collagenous structures of the CS but also the cellular morphology of the EP, ED, and the structures of the lens fibers can be resolved with high resolution. Based on the imaging capability, HGM is shown to provide a suitable tool for cornea diagnoses.

29.2.3 HGM Imaging of Human Skin

From the HGM results of animal models, the imaging capabilities of SHG and THG modalities are demonstrated. With the SHG contrasts in the collagen fibers and the ability of THG to reveal the cellular structures, HGM is shown to be a potential clinical tool for skin diseases diagnosis [23, 32, 50]. In following preliminary studies of human skin, backward-collection geometry is used to construct the HGM for future clinical applications.

Using an excised normal human skin as sample, Fig. 29.10 shows a series of HGM images obtained at the epidermis (Fig. 29.10a–c), the demo-epidermal junction (Fig. 29.10d), and the dermis (Fig. 29.10e–h). Epidermis can be divided into four cellular layers—the stratum corneum, stratum granulosum, stratum spinosum, and the stratum basale. The cellular morphology of the epidermis is clearly revealed by the THG modality. At the skin surface, the dead and cornified stratum corneum (arrow in Fig. 29.10a) shows strong THG contrasts which are caused by the interfaces of the multi-layered structure [8, 51] and the lipids within the corneocytes [40]. Due the THG contrasts from cytoplasmic organelles [22], the nuclei of the epidermal cells, including the stratum corneum (arrowheads in Fig. 29.10a), granular cells (arrows in Fig. 29.10b), spinous cells (arrowheads in Fig. 29.10b), and basal cells (dashed arrows in Fig. 29.10b) appear dark in contrast to the bright cytoplasm. The honeycomb cellular structures are found throughout all levels of the epidermis. In addition, the progressively changes of nuclear diameter, cell diameter, and cell density revealed in different layers can clearly show the keratinization process in the skin. At the bottom layer of the skin, the stratum basale (Fig. 29.10c), clusters of basale cells (arrowheads) show very strong THG contrasts relative to the surrounding cells. These strong THG contrasts are found to strongly correlate with the distribution of melanin and are identified to be contributed by the absorption resonance enhancement of the melanin [23]. The dermis of

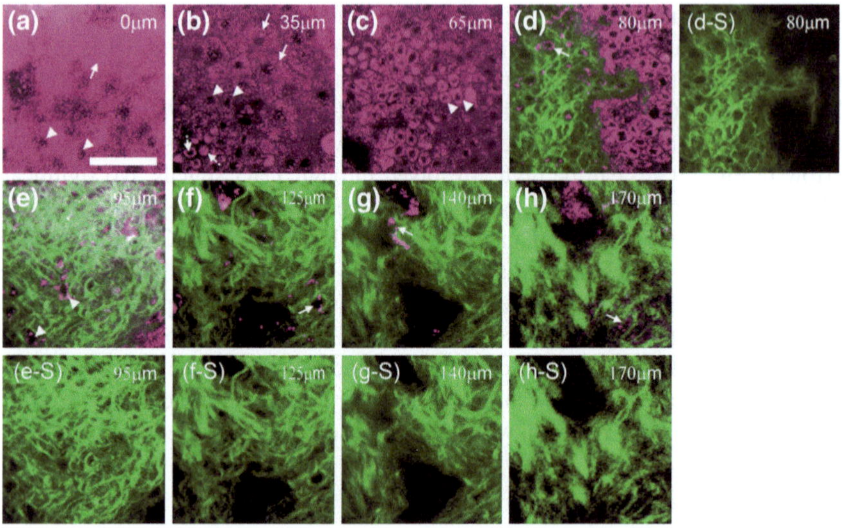

Fig. 29.10 A series of SHG/THG images of the normal human skin obtained at **a** 0 μm, **b** 35 μm, **c** 65 μm, **d** 80 μm, **e** 95 μm, **f** 120 μm, **g** 140 μm, and **h** 170 μm beneath the skin surface. In **a**, the stratum corneum (*arrow*) with ultra-strong THG contrast and THG-dark nuclei of the granular cells (arrowheads) are shown; **b** shows the granular cells (*arrows*), spinous cells (*arrowheads*), and basal cells (*dashed arrows*). **c** The THG-brighter basal cells (*arrowheads*) covering the peak of the dermal papilla. **d** At the dermo-epidermal junction, the collagen fibers in the peak of the dermal papilla surrounded by the basal cells are shown by SHG microscopy and the inactivated melanocyte (*arrow*) can be found in the dermis. **e** Fibroblasts (arrowheads) can be observed within the collagen fibers in the papillary dermis. **f–h** The collagenous structures of the reticular dermis and the fibroblasts (*arrows*) within the fibers. **d-S–h-S** The separated SHG images corresponding to **d-h**. *Scale bar* 50 μm

the human skin can be divided into two layers—the papillary dermis and reticular dermis. In the papillary dermis (Fig. 29.10d, e), the collagen fibers are shown to be loose and areolar; in the thick reticular dermis, the fibers are found to be dense and irregular (Fig. 29.10f–h). In addition to the collageneous structures, the intra-dermal cellular information, including the inactivate melanocytes (arrow in Fig. 29.10d) and the fibroblasts (arrowheads in Fig. 29.10e) can be revealed by THG. Depending on the different ages, sexes, positions, colors, and so on, the penetrability in the skin samples can vary within 270–300 μm. The penetrability is simply defined as the maximum depth at which the fine collagenous structures or the sub-cellular structures can still be distinguished.

Based on the sensitivity to the collageneous and cellular structures, HGM was also demonstrated on the diseased human skin, including three different pigmented skin lesions—superficial spreading melanoma (SSM), compound nevus, and pigmented basal cell carcinoma (BCC). These three pigmented skin lesions have different pathological changes and major diagnostic characteristics but are usually misdiagnosed due to their similar clinical and dermoscopical presentation [52].

Using HGM to reveal the typical characteristics of each skin lesion can help to increase the diagnosis accuracy and accurately distinguish between benign and malignant skin lesions.

Superficial spreading melanoma (SSM) is the most common form of melanoma [53, 54]. In the lesional region, large epithelioid melanocytes, called melanoma cells, can be found to distribute singularly or in nests within all levels of epidermis. The melanoma cells disrupt the normal architectures of epidermis and replace the normal epidermal cells. The epidermis accumulation of the melanoma cells causes the thickening of the epidermis and the increasing of the melanin concentration. Figure 29.11a–d show the HGM images of the freshly-excised SSM specimen. Due to the stronger THG attenuation resulting from absorption of melanin, the imaging depth in the SSM specimen is limited to around 130 μm. THG signals reveal the larger nuclear diameter and varied cell shapes of the superficially-spreading melanoma cells which can be used as diagnostic characteristics. Because of the thickened epidermis, dermis is far beyond the penetration depth and no SHG signals from collagen fibers can be observed. Therefore, the THG-revealed information becomes much more significant for diagnosis.

Figure 29.11e–h show the HGM images obtained from an excised compound nevus specimen at different depth. Compound nevus is a common melanocytic lesion which is often misdiagnosed as pigmented basal cell carcinoma. According to the pathological evidences [55, 56], compound nevus demonstrates nevomelanocytes (i.e., nevus cells) at both the dermo-epidermal junction and the superficial dermis. At a depth of 180 μm in the specimen, SHG modality helps to reveal the thickening of fiber bundles while THG modality shows the increasing of fibroblasts (Fig. 29.11e). These two morphological characteristics both indicate the hosts response with fibrosis in the nevus. In the deeper dermis (Fig. 29.11f–h), intra-dermal nevomelanocytes which are the most significant characteristic for compound nevus, are revealed by THG modality. The nevomelanocytes are shown to group in round clusters (arrows) and be surrounded by the SHG-revealed collagen fibers. Even deep inside the reticular dermis at 300 μm, the grouping nevomelanocytes can still be observed by THG modality because of the much stronger THG contrast enhanced by melanin.

BCC is the most common type of skin cancer and pigmented BCC is a variant of nodular BCC. Due to the abundant pigment in this lesion, it is sometimes misdiagnosed as malignant melanoma. From the histology studies [57], pigmented BCC is characterized by tightly-packed tumor cells bud from epidermis to papillary dermis. The tumor cells form a nodular pattern which is enclosed by collagen fibers and the nuclei of the tumor cells appear as a parallel arrangement (palisading). Throughout the tumor nodules, melanin pigment is non-uniformly distributed. Figure 29.11i–l shows the HGM images obtained from an excised pigmented BCC specimen at different depth. In the superficial epidermis (Fig. 29.11i), the honeycomb pattern of the granular cells is shown by THG modality to remain the same as that in the normal skin. At the dermo-epidermal junction, the cells are shown with elongated and polarized shapes (arrowheads in Fig. 29.11j, k), while the collagen fibers are shown with a much wavier pattern (arrows in Fig. 29.11j, k) and

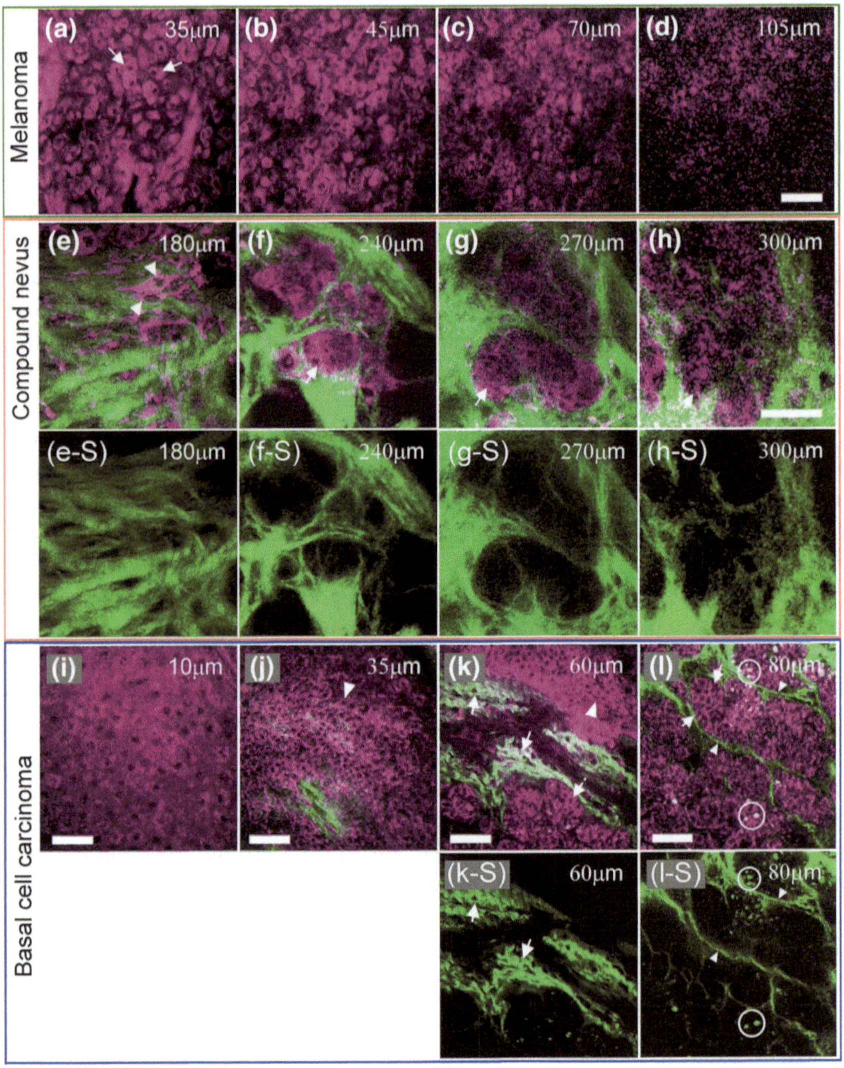

failed to form the dermal papilla architectures. In the reticular dermis, the tumor nodules with parallel cell arrangement (palisading) which are the most significant diagnostic characteristic of BCC are clearly revealed by THG modality (dotted arrow in Fig. 29.11k; arrows in Fig. 29.11l). SHG modality shows the loose collagen fibers enclosing the tumor nodules (arrowheads in Fig. 29.11k, l). Moreover, within the tumor nodules, bright THG spots (circled in Fig. 29.11l) are frequently found to be randomly distributed and these spots with isolated THG signal enhancements indicate the high melanin contents and also the melanophages in the nodules.

Fig. 29.11 **a–d** Ex vivo SHG/THG images of the freshly-excised SSM specimen. Throughout the first 100-μm layer in the darker region, THG-bright melanoma cells (*arrows* in **a**) with varied shapes and larger nuclei were irregularly distributed to replace the normal keratinocytes and **d** no collagenous structure can be observed due to increased thickness of epidermis and limited penetrability. **e–h** Ex vivo SHG/THG images of a freshly-excised compound nevus specimens obtained at **e** 180 μm; **f** 240 μm; **g** 270 μm; and **h** 300 μm beneath the skin surface. In contrast to the normal skin, the increased number of the fibroblasts (*arrowheads* in **e**), clustered nevomelanocytes in the dermis (*arrows* in **f–g**), and dermal fibrosis with thickened collagen fiber bundles and can be observed in the compound nevus specimen. **i–l** Ex vivo SHG/THG images of freshly-excised pigmented BCC specimen. The spindle-like keratinocytes (*arrowheads* in **b** and **c**) were usually found in epidermis and collagen fibers in a much wavier pattern (*arrows* in **c** and **d**) can be found at the dermo-epidermal junction. Without the normal pattern of rete ridges, tumor nodules (*dashed arrow* in **c**) enclosed by collagen fibers (*arrowheads* in **d**) were found to occupy the normal dermis. The parallel arrangement of the tumor cells (palisading; *arrows* in **d**) can be identified at the edges of the nodules and bright HG spots (*circled* in **d**) can suggest the high melanin contents. **e-S–h-S** and **k-S–l-S** The separated SHG images corresponding to each **e-h** and **k-l** respectively. *Scale bar* 50 μm

29.3 Conclusion

In this section, the HGM, combining SHG and THG modalities, has been introduced, including the principle, system setup, and biomedical applications. In this system, a Cr:F laser with a wavelength located within 1,200–1,300 nm is used for excitation. Based on the Cr:F excitation, the wavelength of THG locats within the visible range to facilitate the efficient detection of THG signals. The Cr:F excitation (1,230 nm) can also reduce both the scattering and absorption in the bio-tissues, which can help to increase the imaging penetrability and reduce the photodamages. Since the SHG and THG processes obey the energy conservation and no energy is deposited in the interacted tissues, the HGM is said to be noninvasive. Utilizing the different characteristics of SHG and THG microscopy, various image contrasts can be provided by SHG and THG microscopy respectively to reveal valuable information in the bio-tissues. Based on its noninvasiveness nature, HGM is very suitable for in vivo and clinical investigation. Compared with fluorescence-based microscopy, no issues about photodamges and photobleaching have to be taken into consideration in HGM even when the illumination power is around 100mW.

Making use of the strong SHG contrasts provided by collagen fibers in the dermis and the THG contrasts from cellular structures, the HGM can help to reveal both the connective tissue distribution in the dermis and the cellular morphology in the epidermis of human skin. In addition, since THG contrasts can be enhanced by melanin through real-level resonance, THG microscopy can provide additional information on the pigmentation distribution. All previous studies thus show that SHG and THG microscopy are with the capability to reveal the diagnostic characteristics of various skin diseases such as nevus, melanoma, and BCC. With the diagnostic significance, its noninvasiveness nature, >300-μm imaging penetrability, and sub-micron spatial resolution, the HGM is an undoubted desirable tool for clinical diagnosis. For further practical usage in clinical diagnosis, a reliable data base for disease diagnosis has to be first established and robust systems for both imaging and data analysis are also required.

References

1. P.A. Franken, A.E. Hill, C.W. Peters, G. Weinreich, Generation of optical harmonics. Phys. Rev. Lett. **7**, 118–119 (1961)
2. B.R. Masters, P.T.C. So, E. Gratton, Multiphoton excitation fluorescence microscopy and spectroscopy of in vivo human skin. Biophys. J. **72**, 2405–2412 (1997)
3. K. Konig, Clinical multiphoton tomography. J. Biophoton. **1**(1), 13–23 (2008)
4. Y. Guo, P.P. Ho, H. Savage, D. Harris, P. Sack, S. Schantz, F. Liu, N. Zhadin, R.R. Alfano, Second-harmonic tomography of tissues. Opt. Lett. **22**, 133–135 (1997)
5. D.A. Dombeck, K.A. Kasischke, H.D. Vishwasrao, B.T. Hyman, W.W. Webb, Uniform polarity microtubule assemblies imaged in native brain tissue by second-harmonic generation microscopy. Proc. Natl. Acad. Sci. USA **100**(2), 7081–7086 (2003)
6. S.W. Chu, S.Y. Chen, G.W. Chern, T.H. Tsai, Y.C. Chen, B.L. Lin, C.K. Sun, Studies of $\chi^{(2)}/\chi^{(3)}$ tensors in submicron-scaled bio-tissues by polarization harmonics optical microscopy. Biophys. J. **86**(6), 3914–3922 (2004)
7. P.J. Campagnola, A.C. Millard, M. Terasake, P.E. Hoppe, C.J. Malone, W.A. Mohler, Three-dimensional high-resolution second-harmonic generation imaging of endogenous structural proteins in biological tissues. Biophys. J. **81**, 493–508 (2002)
8. T.Y.F. Tsang, Optical third-harmonic generation at interfaces. Phy. Rev. A **52**(5), 4116–4125 (1995)
9. S.W. Chu, S.Y. Chen, T.H. Tsai, T.M. Liu, C.Y. Lin, H.J. Tsai, C.K. Sun, In vivo developmental biology study using noninvasive multi-harmonic generation microscopy. Opt. Express **11**(23), 3093–3099 (2003)
10. S.Y. Chen, H.C. Yu, I.J. Wang, C.K. Sun, Infrared-based third and second harmonic generation imaging of cornea. J. Biomed. Opt. **14**(4), 044012 (2009)
11. C.H. Yu, S.P. Tai, C.T. Kung, I.J. Wang, H.C. Yu, H.J. Huang, W.J. Lee, Y.F. Chan, C.K. Sun, In vivo and ex vivo imaging of intra-tissue elastic fibers using third-harmonic-generation microscopy. Opt. Express **15**(18), 11167–11177 (2007)
12. S.P. Tai, T.H. Tsai, W.J. Lee, D.B. Shieh, Y.H. Liao, H.Y. Huang, K. Zhang, H.L. Liu, C.K. Sun, Optical biopsy of fixed human skin with backward-collected optical harmonics signals. Opt. Express **13**(20), 8231–8242 (2005)
13. S.P. Tai, W.J. Lee, D.B. Shieh, P.C. Wu, H.Y. Huang, C.H. Yu, C.K. Sun, In vivo optical biopsy of hamster oral cavity with epi-third-harmonic-generation microscopy. Opt. Express **14**(13), 6178–6187 (2006)
14. H.A. Haus, *Waves and fields in optoelectronics* (Prentice-Hall Inc., Englewood Cliffs, 1984)
15. E. Brown, T. McKee, E. diTomaso, A. Pluen, B. Seed, Y. Boucher, R.K. Jain, Dynamic imaging of collagen and its modulation in tumors in vivo using second harmonic generation. Nat. Med. **9**, 796–800 (2003)
16. A. Zoumi, A. Yeh, B.J. Tromberg, Imaging cells and extracellular matrix in vivo by using second-harmonic generation and two-photon excited fluorescence. Proc. Natl. Acad. Sci. USA **99**(17), 11014–11019 (2002)
17. D. Yelin, D. Oron, E. Korkotian, M. Segal, Y. Silberberg, Third-harmonic microscopy with a titanium-sapphire laser. Appl. Phys. B **74**(9), 97–101 (2002)
18. C.K. Sun, C.C. Chen, S.W. Chu, T.H. Tsai, Y.C. Chen, B.L. Lin, Multi-harmonic generation biopsy of skin. Opt. Lett. **28**(24), 2488–2490 (2003)
19. R.R. Anderson, J.A. Parish, The optics of human skin. J. Invest. Dermatol. **77**(1), 13–19 (1981)

20. T. Yasui, Y. Takahashi, M. Ito, S. Fukushima, T. Araki, Ex vivo and in vivo second-harmonic-generation imaging of dermal collagen fiber in skin: comparison of imaging characteristics between mode-locked Cr:forsterite and Ti: sapphire lasers. Appl. Opt. **48**(10), D88–D95 (2009)
21. S.Y. Chen, C.S. Hsieh, S.W. Chu, Noninvasive harmonics optical microscopy for long-term observation of embryonic nervous system development in vivo. J. Biomed. Opt. **11**(5), 054022 (2006)
22. C.S. Hsieh, S.U. Chen, Y.W. Lee, Y.S. Yang, C.K. Sun, Higher harmonic generation microscopy of in vitro cultured mammal oocytes and embryos. Opt. Express **16**(15), 11574–11588 (2008)
23. S.Y. Chen, S.U. Chen, H.Y. Wu, W.J. Lee, Y.H. Liao, C.K. Sun, In vivo virtual biopsy of human skin by using noninvasive higher harmonic generation microscopy. IEEE J. Select. Topic Quantum Electron. **16**(3), 478–492 (2010)
24. K. Konig, P.T.C. So, W.W. Mantulin, E. Gratton, Cellular response to near-infrared femtosecond laser pulses in two-photon microscopes. Opt. Lett. **22**(2), 135–136 (1997)
25. J.M. Squirrell, D.L. Wokosin, J.G. White, B.D. Bavister, Long-term two-photon fluorescence imaging of mammalian embryos without compromising viability. Nat. Biotechnol. **17**, 763–767 (1999)
26. J. Squier, M. Müller, High resolution nonlinear microscopy: a review of sources and methods for achieving optimal imaging. Rev. Sci. Instrum. **72**(7), 2855–2867 (2001)
27. C.T. Kung, C.C. Chuang, Y.K. Huang, H.J. Tsai, C.K. Sun, In vivo continuous observation of vertebrate cardiac valve for congenital heart disease study and drug screening using third harmonic generation microscopy. Conference on Lasers and Electro-Optics/Quantum Electronics and Laser Science Conference/Conference on Photonic Applications, Baltimore, 2007, paper CTuP4
28. C.K. Sun, S.W. Chu, S.Y. Chen, T.H. Tsai, T.M. Liu, C.Y. Lin, H.J. Tsai, Higher harmonic generation microscopy for developmental biology. J. Struct. Biol. **147**(1), 19–30 (2004)
29. J.H. Lee, S.Y. Chen, S.W. Chu, L.F. Wang, C.K. Sun, B.L. Chiang, Noninvasive in vitro and in vivo assessment of epidermal hyperkeratosis and dermal fibrosis in atopic dermatitis. J. Biomed. Opt. **14**(1), 014008 (2009)
30. S.Y. Chen, C.Y.S. Hsu, C.K. Sun, Epi-third and second harmonic generation microscopic imaging of abnormal enamel. Opt. Express **16**(15), 11670–11679 (2008)
31. M.R. Tsai, C.H. Chen, C.K. Sun, in Third and second harmonic generation imaging of human articular cartilage. *Proceeding of SPIE*, San Jose, 2009, p. 71831V-1–71831V
32. S.Y. Chen, H.Y. Wu, C.K. Sun, In vivo harmonic generation biopsy of human skin. J. Biomed. Opt. **14**(6), 060505 (2009)
33. S.W. Chu, I.H. Chen, T.M. Liu, P.C. Cheng, C.K. Sun, B.L. Lin, Multimodal nonlinear spectral microscopy based on a femtosecond Cr:forsterite laser. Opt. Lett. **26**(23), 1909–1911 (2001)
34. D. Yelin, Y. Silberberg, Laser scanning third-harmonic-generation microscopy in biology. Opt. Express **3**(8), 169–175 (1999)
35. J.M. Schins, T. Schrama, J. Squier, G.J. Brakenhoff, M. Müller, Determination of material properties by use of third-harmonic generation microscopy. J. Opt. Soc. Am. B **19**(7), 1627–1634 (2002)
36. C.B. Kimmel, W.W. Ballard, S.R. Kimmel, B. Ullmann, T.F. Schilling, Stages of embryonic development of the zebrafish. Dev. Dyn. **203**(3), 253–310 (1995)
37. C.O. Clay, A.C. Millard, C.B. Schaffer, J. Aus-der-Au, P.S. Tsai, J.A. Squier, D. Kleinfeld, Spectroscopy of third-harmonic generation: evidence for resonances in model compounds and ligated hemoglobin. J. Opt. Soc. Am. B **23**(5), 932–950 (2006)
38. S.P. Tai, C.H. Yu, T.M. Liu, Y.C. Wen, C.K. Sun, in *In Vivo Molecular-Resonant Third Harmonic Generation Microscopy of Hemoglobin*. Conference on Lasers and Electro-Optics/ Quantum Electronics and Laser Science Conference, Baltimore, 2007, paper CTuF4
39. C.F. Chang, C.H. Yu, C.K. Sun, Multi-photon resonance enhancement of third harmonic generation in human oxyhemoglobin and deoxyhemoglobin. J. Biophoton. **3**, 678–658 (2010)

40. D. Debarre, W. Supatto, A.M. Pena, A. Fabre, T. Tordjmann, L. Combettes, M.C. Schanne-Klein, E. Beaurepaire, Imaging lipid bodies in cells and tissues using third-harmonic generation microscopy. Nat. Methods **3**, 47–53 (2006)
41. J.A. Squier, M. Muller, G.J. Brakenhoff, K.R. Wilson, Third harmonic generation microscopy. Opt. Express **3**(9), 315–324 (1998)
42. G. Peleg, A. Lewis, M. Linial, L.M. Loew, Non-linear optical measurement of membrane potential around single molecules at selected cellular sites. Proc. Natl. Acad. Sci. USA **96**(12), 6700–6704 (1999)
43. D.M. Maurice, *The Eye*, 3rd edn. (Academic Press, London, 1984)
44. K. Ramaesh, T. Ramaesh, J.D. West, B. Dhillon, Immunolocalisation of leukaemia inhibitory factor in the cornea. Eye **18**(10), 1006–1009 (2004)
45. R.L. McCally, R.A. Farrell, *Noninvasive Diagnostic Techniques in Ophthalmology* (Springer, New York, 1990)
46. B.G. Wang, I. Riemann, H. Schubert, D. Schweitzer, K. Konig, K.J. Halbhuber, Multiphoton microscopy for monitoring intratissue femtosecond laser surgery effects. Lasers Surg. Med. **39**(6), 527–533 (2007)
47. C.K. Sun, Higher harmonic generation microscopy. Adv Biochem Engin Biotechnol **95**:17–56 (2005)
48. P.J. Campagnola, L.M. Loew, Second-harmonic imaging microscopy for visualizing biomolecular arrays in cells, tissues and organisms. Nat. Biotechnol. **21**(11), 1356–1360 (2003)
49. I. Jalbert, F. Stapleton, E. Papas, D.F. Sweeney, M. Coroneo, In vivo confocal microscopy of the human cornea. Br. J. Ophthalmol. **87**(2), 225–236 (2003)
50. R. Cicchi, D. Kapsokalyvas, V. de Giorgi, V. Maio, A. van Wiechen, D. Massi, T. Lotti, F.S. Pavone, Scoring of collagen organization in healthy and diseased human dermis by multiphoton microscopy. J. Biophoton. **3**, 34–43 (2010)
51. Y. Barad, H. Eisenberg, M. Horowitz, Y. Silberberg, Nonlinear scanning laser microscopy by third harmonic generation. Appl. Phys. Lett. **70**(8), 922–924 (1997)
52. E. Makino, T. Uchida, Y. Matsushita, M. Inaoki, W. Fujimoto, Melanocytic nevi clinically simulating melanoma. J. Dermatol. **34**(1), 52–55 (2007)
53. R.G.B. Langley, T.B. Fitzpatric, A. Sober, *Clinical Characteristics. Cutaneous Melanoma* (Quality Medical Publishing, St Louis, 1998)
54. S.B. Forman, T.C. Ferringer, S.J. Peckham, S.R. Dalton, G.T. Sasaki, L.F. Libow, D.M. Elston, Is superficial spreading melanoma still the most common form of malignant melanoma? J. Am. Acad. Dermatol. **58**(6), 1013–1020 (2008)
55. M.R. Okun, Silhouette symmetry - an unsupportable histologic criterion for distinguishing Spitz nevi and compound nevi from malignant melanoma. Arc. Pathol. Lab. Med. **121**(1), 48–53 (1997)
56. S. Gonzalez, K. Swindells, M. Rajadhyaksha, A. Torres, Changing paradigms in dermatology: Confocal microscopy in clinical and surgical dermatology. Clin. Dermatol. **21**(5), 359–369 (2003)
57. A.L.C. Agero, K.J. Busam, C. Benvenuto-Andrade, A. Scope, M. Gill, A.A. Marghoob, S. Gonzalez, A.C. Halpern, Reflectance confocal microscopy of pigmented basal cell carcinoma. J. Am. Acad. Dermatol. **54**(4), 638–643 (2006)

Index

C.-C. Lee (ed.), *The Current Trends of Optics and Photonics*,
Topics in Applied Physics 129, DOI 10.1007/978-94-017-9392-6

MIX
Papier aus verantwortungsvollen Quellen
Paper from responsible sources
FSC® C105338

If you have any concerns about our products,
you can contact us on
ProductSafety@springernature.com

In case Publisher is established outside the EU,
the EU authorized representative is:
Springer Nature Customer Service Center GmbH
Europaplatz 3, 69115 Heidelberg, Germany

Printed by Libri Plureos GmbH
in Hamburg, Germany